Matrix Completions, Moments, and Sums of Hermitian Squares

Matrix Completions, Moments, and Sums of Hermitian Squares

Mihály Bakonyi and Hugo J. Woerdeman

PRINCETON UNIVERSITY PRESS
PRINCETON AND OXFORD

Published by Princeton University Press,
41 William Street, Princeton, New Jersey 08540

In the United Kingdom: Princeton University Press,
6 Oxford Street, Woodstock, Oxfordshire OX20 1TW

press.princeton.edu

ISBN 978-0-691-12889-4
Library of Congress Control Number: 2011922035

British Library Cataloging-in-Publication Data is available

The publisher would like to acknowledge the authors of this volume
for providing the camera-ready copy from which this book was
printed

To our beloved wives and supportive partners,
Gina and Dara

Front cover: It is easy to find a minimal rank completion of the partial matrix on the front cover, but how about a completion with minimal normal defect?

Contents

Preface

Over the past twenty years the related areas of matrix completions, moment problems, and sums of squares have undergone a spectacular development. In the area of matrix completion problems, various questions regarding positive definite, contractive, minimal rank completion problems have been solved, and applied in several areas. In the area of moment problems, several multivariable questions that for a long time have been open have either been completely resolved or seen some substantial gained understanding. In the area of sums of squares (in this book we focus on the sums of Hermitian squares, which are expressions of the form $\sum_{i=1}^{k} A_i(z)^* A_i(z)$, with $A_i(z)$ a polynomial in $z_1, \ldots, z_d$; when $k = 1$ and $A_1(z)$ is square we refer to it as a factorization), the gained understanding has led to significantly improved optimization algorithms over the class of positive (trigonometric) polynomials. Although active research continues, the time seemed ripe for a comprehensive account of the subject, interlacing old and new results together. This was our aim, and we hope to have at least partially succeeded.

We tried to cover a lot in this book, but hopefully we did not go overboard. For instance, in the class of matrix completion problems we restricted ourselves to notions (positive semidefiniteness, contractivity, rank, etc.) that are invariant under a change of (an orthonormal) basis, and that typically have an interpretation for operator matrices as well. The only exception is the distance matrix completion problem, which we have included as the techniques and results are very close to those in the positive semidefinite completion problem. Completion problems that are not included in the book are those that involve notions like M-matrices, P-matrices, and so on. In addition, we tried to be as self contained as possible, relying only on generally known results in Linear Algebra, Operator Theory, Measure Theory, and Complex Analysis. Though we did not completely succeed (we quote a few nonstandard results without proof, in which case we provide a reference), we hope that the reader will be able to master the material without having to consult many other sources.

We included some Matlab code in the book. We do not claim any proficiency here, and in fact we expect many to be able to improve on our code. We have found, though, that even the most basic use of Matlab can help tremendously in this area of research, and we hope that by including some of our code we lower the threshold to make use of this wonderful resource.

Many parts of the book can be read independently from other parts. In addition, not all parts require the same background material. For instance, with only a Linear Algebra background one will be able to understand almost everything in Chapters 1 and 5. If in addition, one knows what an atomic measure is and one has some basic understanding of graph theory, one will be able to understand all of Chapters 1 and 5. In Chapter 2 the problems that appear are typically operator-valued problems, however, if one pretends that all the underlying spaces are finite dimensional one will be able to appreciate this chapter as well. In addition, the techniques in Chapter 2 do not rely on those in Chapter 1, so that in principle one can start with Chapter 2 if so desired. Chapters 3 and 4 do rely on results in Chapter 2. Indeed, Chapter 3 consists in large part of multivariable generalizations of results in Chapter 2, and Chapter 4 uses Wielandt's observation that

$$\begin{pmatrix} I & B^* \\ B & I \end{pmatrix} \geq 0 \Leftrightarrow \|B\| \leq 1$$

to solve contractive completion problems based on positive semidefinite ones.

We have designed the book so that it can be used as a topics course for graduate students or for (quite) advanced undergraduate students. Clearly, such a course can not cover the full length of the book. Based on the authors' experience, one would rather cover 8-12 sections in one term. With this possibility in mind, we included a large number of exercises. The exercises range in difficulty, so some care is required in assigning them. The addition of exercises has also allowed us to include (indications of) results in the literature that are related to the material in the text, but whose inclusion in the text would have made the book too long. We should emphasize that by converting a result into an exercise we do not mean to indicate that this result is easy. In fact, some exercises may be quite challenging without consulting the paper on which the exercise is based. We chose to keep the text free from references to the literature (except for a handful of instances where we have decided not to include a full proof in the text, but rather give a reference), and instead to discuss the literature at the end of each chapter in a section called "Notes." In this way we felt more free to include a wider variety of references than we otherwise would have. We hope that our notes and references can serve as a guide of recommended reading material to the various aspects (from theoretical to applied) of the research in the area covered by this book.

There are many people to whom we would like to express our gratitude. First of all, to all of you who we have interacted with us during our education and professional career: Thank you very much! This includes anybody we discussed mathematics with, all our teachers, all our anonymous referees, all our reviewers, anyone whose lecture we heard, all those whose papers we read, anyone who has hosted one of us at some point, all our wonderful co-authors, and, of course, our inspiring Ph.D. advisors: Rien Kaashoek, Israel Gohberg (for Hugo), and Charles Johnson (for Mihály). Without all of you, we would not have come so far!

There are also many to thank who have been involved one way or another with the writing of this book. First of all we thank Vickie Kearn from

Princeton University Press for planting the seed of considering a book. We would also like to thank the two anonymous referees who were positive about our book proposal. Your reviews gave us the impetus to indeed go for it! We would like to thank our students who took the special topics course based on some of the material of the book. The interaction helped shape the format of the book. In particular, we would like to thank Ph.D. student David Kimsey, who contributed the first draft of Section 3.8 and some of the related exercises. Special thanks are also due to all of you who have commented on our manuscript: Leiba Rodman, James Rovnyak, Dan Timotin, David Kimsey, Selcuk Koyuncu, Ilya Spitkovsky, Geir Naevdal, Jean-Pierre Gabardo, Ron Smith, and the anonymous reviewers. In addition, we would like to thank Nick Higham for sharing useful LaTeX tricks with us. Finally, we would like to thank all of you who have helped with the technical aspects of producing this book, which include Byron Greene, Robert Henry, and Michael Newby.

We also appreciate very much the institutional support we received. First of all our home institutions, Georgia State University (Atlanta) and Drexel University (Philadelphia), where we performed most of our work on this book. In addition, the first named author, while preparing the book, was on leave at Centro des Estruturas Lineares e Combinatórias, University of Lisbon, Portugal, and is very thankful for their support. The second author would like to express his appreciation to the National Science Foundation, which has supported his research over the years, most recently through grants DMS 0500678 and DMS 0901628.

We also wish to thank both our families for all their love and support.

Finally, we would like to thank you, our reader, for picking up this book. We appreciate greatly your sharing our interest in this area of mathematics, and we hope very much to communicate with you. Whether it is a typo, a serious mistake, an omission in the references (please, forgive us!), thoughts on a course using our book, solutions to exercises, useful Matlab code, new results, or whatever: we hope to hear about it![1] Thank you very much!

Mihály Bakonyi,
May 2010, Atlanta

Hugo J. Woerdeman,
May 2010, Philadelphia

Postscript. On August 7, 2010, my co-author and friend, Mihály Bakonyi, passed away. Writing this monograph with him was an exciting and very rewarding experience. After Mihály's death only minor changes were made, mostly in response to reviewers' comments. He is missed by all those who knew him as a mathematician, a teacher, and a friend.

Hugo J. Woerdeman,
September 2010, Philadelphia

---

[1] We will set up whatever online outlets are most appropriate to communicate anything related to this area of research. Please just use your favorite search engine to find us.

Standard sets and definitions

Here we list some often used notation and definitions:

- $\mathbb{N} = \{1, 2, 3, \ldots\}$
- $\mathbb{N}_0 = \{0, 1, 2, \ldots\}$
- $\mathbb{Z}$ = the set of all integers
- $\mathbb{Q}$ = the field of rational numbers
- $\mathbb{R}$ = the field of real numbers
- $\mathbb{C}$ = the field of complex numbers
- $\mathbb{T} = \{z \in \mathbb{C} : |z| = 1\}$
- $\mathbb{D} = \{z \in \mathbb{C} : |z| < 1\}$
- $\overline{\mathbb{D}} = \{z \in \mathbb{C} : |z| \leq 1\}$
- $\mathcal{D}(A)$ = the domain of the (unbounded) operator A
- A^T = the transpose of the matrix A
- A^* = the adjoint of the operator A; for matrices it corresponds to the conjugate transpose of A (as an underlying orthonormal basis is assumed).
- $\text{Ran}A$ denotes the range of an operator (or matrix) A.
- $\text{Ker}A$ denotes the kernel (or nullspace) of an operator (or matrix) A.
- $A \geq 0$ denotes that A is positive semidefinite
- $A > 0$ denotes that A is positive definite
- $A \geq B$ and $B \leq A$ denote that $A - B$ is positive semidefinite
- $A > B$ and $B < A$ denote that $A - B$ is positive definite
- $\sigma(A) = \{z \in \mathbb{C} : zI - A \text{ is not invertible}\}$ = the spectrum of the operator (or matrix) A
- $\lambda_j(A)$ = the jth eigenvalue of $A = A^*$, where $\lambda_1(A)$ is the largest eigenvalue
- $s_j(A)$ = the jth singular value of the operator A, where $s_1(A) = \|A\|$ is the largest singular value
- An operator-valued polynomial $P(z)$ is called *stable* if $P(z)$ is invertible for all $z \in \overline{\mathbb{D}}^d$.
- $\text{row}(A_i)_{i=1}^n = \begin{pmatrix} A_1 & \cdots & A_n \end{pmatrix}$ is the row matrix with entries $A_1, \ldots, A_n$. Similarly, $\text{col}(A_i)_{i=1}^n$ is the column matrix with entries $A_1, \ldots, A_n$.

Chapter One

Cones of Hermitian matrices and trigonometric polynomials

In this chapter we study cones in the real Hilbert spaces of Hermitian matrices and real valued trigonometric polynomials. Based on an approach using such cones and their duals, we establish various extension results for positive semidefinite matrices and nonnegative trigonometric polynomials. In addition, we show the connection with semidefinite programming and include some numerical experiments.

1.1 CONES AND THEIR BASIC PROPERTIES

Let $\mathcal{H}$ be a Hilbert space over $\mathbb{R}$ with inner product $\langle \cdot, \cdot \rangle$. A nonempty subset $\mathcal{C}$ of $\mathcal{H}$ is called a *cone* if

(i) $\mathcal{C} + \mathcal{C} \subseteq \mathcal{C}$,

(ii) $\alpha\mathcal{C} \subseteq \mathcal{C}$ for all $\alpha > 0$.

Note that a cone is *convex* (i.e., if $C_1, C_2 \in \mathcal{C}$ then $sC_1 + (1-s)C_2 \in \mathcal{C}$ for $s \in [0,1]$). The cone $\mathcal{C}$ is *closed* if $\mathcal{C} = \overline{\mathcal{C}}$, where $\overline{\mathcal{C}}$ denotes the closure of $\mathcal{C}$ (in the topology induced by the norm $\|\cdot\|$ on $\mathcal{H}$, where, as usual, $\|\cdot\| = \langle \cdot, \cdot \rangle^{1/2}$). For example, the set $\{(x,y) : 0 \le x \le y\}$ is a closed cone in $\mathbb{R}^2$. This cone is the intersection of the cones $\{(x,y) : |x| \le y\}$ and $\{(x,y) : x, y \ge 0\}$. In general, intersections and sums of cones are again cones:

Proposition 1.1.1 *Let $\mathcal{H}$ be a Hilbert space and let $\mathcal{C}_1$ and $\mathcal{C}_2$ be cones in $\mathcal{H}$. Then the following are also cones:*

(i) $\mathcal{C}_1 \cap \mathcal{C}_2$, and

(ii) $\mathcal{C}_1 + \mathcal{C}_2 = \{C_1 + C_2 \;:\; C_1 \in \mathcal{C}_1, C_2 \in \mathcal{C}_2\}$.

Proof. The proof is left to the reader (see Exercise 1.6.1). □

Given a cone $\mathcal{C}$ the *dual* $\mathcal{C}^*$ of $\mathcal{C}$ is defined via

$$\mathcal{C}^* = \{L \in \mathcal{H} \;:\; \langle L, K \rangle \ge 0 \text{ for all } K \in \mathcal{C}\}.$$

The notion of a dual cone is of special importance in optimization problems, as optimality conditions are naturally formulated using the dual cone. As a simple example, it is not hard to see that the dual of $\{(x, y) : 0 \leq x \leq y\}$ is the cone $\{(x, y) : x \geq 0 \text{ and } y \geq -x\}$. We call a cone $\mathcal{C}$ *selfdual* if $\mathcal{C}^* = \mathcal{C}$. For example, $\{(x, y) : x, y \geq 0\}$ is selfdual.

Lemma 1.1.2 *The dual of a cone is again a cone, which happens to be closed. Any subspace* W *of* $\mathcal{H}$ *is a cone, and its dual is equal to its orthogonal complement (i.e., if* W *is a subspace then* $W^* = W^\perp := \{h \in \mathcal{H} : \langle h, w \rangle = 0 \text{ for all } w \in W\}$*). Moreover, for cones* $\mathcal{C}$, $\mathcal{C}_1$, *and* $\mathcal{C}_2$ *we have that*

(i) if $\mathcal{C}_1 \subseteq \mathcal{C}_2$ *then* $\mathcal{C}_2^* \subseteq \mathcal{C}_1^*$;

(ii) $(\mathcal{C}^*)^* = \overline{\mathcal{C}}$;

(iii) $\mathcal{C}_1^* + \mathcal{C}_2^* \subseteq (\mathcal{C}_1 \cap \mathcal{C}_2)^*$;

(iv) $(\mathcal{C}_1 + \mathcal{C}_2)^* = \mathcal{C}_1^* \cap \mathcal{C}_2^*$.

Proof. The proof of this proposition is left as an exercise (see Exercise 1.6.3). □

An *extreme ray of a cone* $\mathcal{C}$ is a subset of $\mathcal{C} \cup \{0\}$ of the form $\{\alpha K : \alpha \geq 0\}$, where $0 \neq K \in \mathcal{C}$ is such that

$$K = A + B,\ A, B \in \mathcal{C} \Rightarrow A = \alpha K \text{ for some } \alpha \in [0, \infty).$$

The cone $\{(x, y) : 0 \leq x \leq y\}$ has the extreme rays $\{(x, 0) : x \geq 0\}$ and $\{(x, x) : x \geq 0\}$.

There are two cones that we are particularly interested in: the cone PSD of positive semidefinite matrices and the cone $\mathbb{T}\text{Pol}^+$ of trigonometric polynomials that take on nonnegative values on the unit circle. We will study some of the basics of these cones in this section, and in the remainder of this chapter we will study related cones. The cone $\mathbb{R}\text{Pol}^+$ of polynomials that are nonnegative on the real line is of interest as well. Some of its properties will be developed in the exercises.

1.1.1 The cone PSD_n

Let $\mathbb{C}^{n \times n}$ be the Hilbert space over $\mathbb{C}$ of $n \times n$ matrices endowed with the inner product

$$\langle A, B \rangle = \text{tr}(AB^*),$$

where $\text{tr}(M)$ denotes the trace of the square matrix M and B^* denotes the complex conjugate transpose of the matrix B. Consider the subset of Hermitian matrices $\mathcal{H}_n = \{A \in \mathbb{C}^{n \times n} : A = A^*\}$, which itself is a Hilbert space over $\mathbb{R}$ under the same inner product $\langle A, B \rangle = \text{tr}(AB^*) = \text{tr}(AB)$. When we restrict ourselves to real matrices we will write $\mathcal{H}_{n,\mathbb{R}} = \mathcal{H}_n \cap \mathbb{R}^{n \times n}$. The following results on the cone PSD_n of positive semidefinite matrices are well known.

Lemma 1.1.3 *The set* PSD_n *of all* $n \times n$ *positive semidefinite matrices is a selfdual cone in the Hilbert space* $\mathcal{H}_n$.

Proof. Let A and B be $n \times n$ positive semidefinite matrices. We denote by $A^{\frac{1}{2}}$ the unique positive semidefinite matrix the square of which is A. Then, by a well-known property of the trace, $\mathrm{tr}(AB) = \mathrm{tr}(A^{\frac{1}{2}}BA^{\frac{1}{2}}) \geq 0$. Thus $\mathrm{PSD}_n \subseteq (\mathrm{PSD}_n)^*$.

For the converse, let $A \in \mathcal{H}_n$ be such that $\mathrm{tr}(AB) \geq 0$ for every positive semidefinite matrix B. For all $v \in \mathbb{C}^n$, the rank 1 matrix $B = vv^*$ is positive semidefinite, thus $0 \leq \mathrm{tr}(Avv^*) = \langle Av, v\rangle$, so A is positive semidefinite. □

Proposition 1.1.4 *The extreme rays of* PSD_n *are given by* $\{\alpha vv^* : \alpha \geq 0\}$, *where* v *is a nonzero vector in* $\mathbb{C}^n$. *In other words, all extreme rays of* PSD_n *are generated by rank 1 matrices.*

Proof. Let $v \in \mathbb{C}^n$ and suppose that $vv^* = A + B$ with A and B positive semidefinite. If $w \in \mathbb{C}^n$ is orthogonal to v then we get that $0 = w^*vv^*w = w^*Aw + w^*Bw$, and as both A and B are positive semidefinite we get that $w^*Aw = 0 = w^*Bw$. Thus $\|A^{\frac{1}{2}}w\| = 0 = \|B^{\frac{1}{2}}w\|$. This yields that $A^{\frac{1}{2}}w = 0 = B^{\frac{1}{2}}w$, and thus $Aw = 0 = Bw$. As this holds for all w orthogonal to v, we get that A and B have at most rank 1. If they have rank 1, the eigenvector corresponding to the single nonzero eigenvalue must be a multiple of v, and this implies that both A and B are a multiple of vv^*.

Conversely, if $K \in \mathrm{PSD}_n$ write $K = v_1v_1^* + \cdots + v_kv_k^*$, where $v_1, \ldots, v_k$ are nonzero vectors in $\mathbb{C}^n$ and k is the rank of K (use, for instance, a *Cholesky factorization* $K = LL^*$, and let $v_1, \ldots, v_k$ be the nonzero columns of the Cholesky factor L). But then it follows immediately that if $k \geq 2$, the matrix K does not generate an extreme ray of PSD_n. □

1.1.2 The cone $\mathbb{T}\mathrm{Pol}_n^+$

We consider *trigonometric polynomials* $p(z) = \sum_{k=-n}^{n} p_k z^k$ with the inner product

$$\langle p, g\rangle = \frac{1}{2\pi}\int_0^{2\pi} p(e^{i\theta})\overline{g(e^{i\theta})}d\theta = \sum_{k=-n}^{n} p_k\overline{g_k},$$

where $g(z) = \sum_{k=-n}^{n} g_k z^k$. We are particularly interested in the Hilbert space $\mathbb{T}\mathrm{Pol}_n$ over $\mathbb{R}$ consisting of those trigonometric polynomials $p(z)$ that are real valued on the unit circle, that is, $p \in \mathbb{T}\mathrm{Pol}_n$ if and only if p is of degree $\leq n$ and $p(z) \in \mathbb{R}$ for $z \in \mathbb{T}$. It is not hard to see that the latter condition is equivalent to $p_k = \overline{p_{-k}}$, $k = 0, \ldots, n$. For such real-valued trigonometric polynomials we have that $\langle p, g\rangle = \sum_{k=-n}^{n} p_k g_{-k}$. The cone that we are interested in is formed by those trigonometric polynomials that are nonnegative on the unit circle:

$$\mathbb{T}\mathrm{Pol}_n^+ = \{p \in \mathbb{T}\mathrm{Pol}_n : p(z) \geq 0 \text{ for all } z \in \mathbb{T}\}.$$

A crucial property of these nonnegative trigonometric polynomials is that they factorize as the modulus squared of a single polynomial. This property stated in the next theorem is known as the *Fejér-Riesz factorization* property; it is in fact the key to determining the dual of $\mathbb{T}\mathrm{Pol}_n^+$ and its extreme rays. We call a polynomial q *outer* if all its roots are outside the open unit disk; that is, q is outer when $q(z) \neq 0$ for $z \in \mathbb{D}$. The degree n polynomial $q(z) = \sum_{k=0}^{n} q_n z^n$ is called *co-outer* if all its roots are inside the closed unit disk and $q_n \neq 0$ (in some contexts it is convenient to interpret $q_n \neq 0$ to mean that ∞ is not a root of $q(z)$). One easily sees that $q(z)$ is outer if and only if $z^n \overline{q(1/\overline{z})}$ is co-outer.

Theorem 1.1.5 *The trigonometric polynomial $p(z) = \sum_{k=-n}^{n} p_k z^k$, $p_n \neq 0$, is nonnegative on the unit circle $\mathbb{T}$ if and only if there exists a polynomial $q(z) = \sum_{k=0}^{n} q_k z^k$ such that*

$$p(z) = q(z)\overline{q(1/\overline{z})}, \ z \in \mathbb{C}, \ z \neq 0.$$

In particular, $p(z) = |q(z)|^2$ for all $z \in \mathbb{T}$. The polynomial q may be chosen to be outer (co-outer) and in that case q is unique up to a scalar factor of modulus 1.

When the factor q is outer (co-outer), we refer to the equality $p = |q|^2$ as an *outer (co-outer) factorization.* To make the outer factorization unique one can insist that $q_0 \geq 0$, in which case one can refer to q as *the* outer factor of p. Similarly, to make the co-outer factorization unique one can insist that $q_n \geq 0$, in which case one can refer to q as *the* co-outer factor of p.

Proof. The "if" statement is trivial, so we focus on the "only if" statement. So let $p \in \mathbb{T}\mathrm{Pol}_n^+$. Using that $p_{-k} = \overline{p}_k$ for $k = 0, \ldots, n$, we see that the polynomial

$$g(z) = z^n p(z) = \overline{p}_n + \cdots + p_0 z^n + \cdots + p_n z^{2n}$$

satisfies the relation

$$g(z) = z^{2n} \overline{g(1/\overline{z})}, \ z \neq 0. \tag{1.1.1}$$

Let $z_1, \ldots, z_{2n}$ be the set of all zeros of g counted according to their multiplicities. Since all z_j are different from zero, it follows from (1.1.1) that $1/\overline{z}_j$ is also a zero of g of the same multiplicity as z_j. We prove that if $z_j \in \mathbb{T}$, in which case $z_j = 1/\overline{z}_j$, then $z_j = e^{it_j}$ has even multiplicity. The function $\phi : \mathbb{R} \to \mathbb{C}$ defined by $\phi(t) = p(e^{it})$ is differentiable and nonnegative on $\mathbb{R}$. This implies that the multiplicity of t_j as a zero of ϕ is even since ϕ has a local minimum at t_j, and consequently z_j is an even multiplicity zero of g. So we can renumber the zeros of g so that $z_1, \ldots, z_n, 1/\overline{z}_1, \ldots, 1/\overline{z}_n$ represent the $2n$ zeros of g. Note that we can choose $z_1, \ldots, z_n$ so that $|z_k| \geq 1$, $k = 1, \ldots, n$ (or, if we prefer, so that $|z_k| \leq 1$, $k = 1, \ldots, n$). Then we have

$$g(z) = p_n \prod_{j=1}^{n} (z - z_j) \prod_{j=1}^{n} \left(z - \frac{1}{\overline{z}_j} \right),$$

so

$$p(z) = z^{-n} g(z) = p_n \prod_{j=1}^{n} (z - z_j) \prod_{j=1}^{n} \left(1 - \frac{1}{z\overline{z}_j} \right) \tag{1.1.2}$$

$$= c\prod_{j=1}^{n}(z-z_j)\prod_{j=1}^{n}\left(\frac{1}{z}-\overline{z}_j\right),$$

where $c = p_n(\prod_{j=1}^{n}(-\overline{z}_j))^{-1}$. Since p is nonnegative on $\mathbb{T}$ and positive somewhere, at $\alpha \in \mathbb{T}$, say, we get that $c > 0$ (as $c = \frac{\prod_j |\alpha - z_j|^2}{p(\alpha)}$). With $q(z) = \sqrt{c}\prod_{j=1}^{n}(z - z_j)$, (1.1.2) implies that $p(z) = q(z)\overline{q(1/\overline{z})}$. Note that choosing $|z_k| \geq 1$, $k = 1, \dots, n$, leads to an outer factorization, and that choosing $|z_k| \leq 1$, $k = 1, \dots, n$, leads to a co-outer factorization. The uniqueness of the (co-)outer factor up to a scalar factor of modulus 1 is left as an exercise (see Exercise 1.6.5). □

To obtain a description of the dual cone of $\mathbb{T}\mathrm{Pol}_n^+$, it will be of use to rewrite the inner product $\langle p, g\rangle$ in terms of the factorization $p = |q|^2$. For this purpose, introduce the *Toeplitz matrix*

$$T_g := (g_{i-j})_{i,j=0}^{n} = \begin{pmatrix} g_0 & g_{-1} & \cdots & g_{-n} \\ g_1 & g_0 & \cdots & g_{-n+1} \\ \vdots & \ddots & \ddots & \vdots \\ g_n & g_{n-1} & \cdots & g_0 \end{pmatrix},$$

where $g(z) = \sum_{k=-n}^{n} g_k z^k$. It is now a straightforward calculation to see that $p(z) = |q_0 + \cdots + q_n z^n|^2$, $z \in \mathbb{T}$, yields

$$\langle p, g\rangle = \begin{pmatrix}\overline{q_0} & \cdots & \overline{q_n}\end{pmatrix} \begin{pmatrix} g_0 & g_1 & \cdots & g_{-n} \\ g_1 & g_0 & \cdots & g_{-n+1} \\ \vdots & \ddots & \ddots & \vdots \\ g_n & g_{n-1} & \cdots & g_0 \end{pmatrix} \begin{pmatrix} q_0 \\ \vdots \\ q_n \end{pmatrix} = x^* T_g x, \tag{1.1.3}$$

where $x = \begin{pmatrix} q_0 & \cdots & q_n \end{pmatrix}^T$ and the superscript T denotes taking the transpose. From this it is straightforward to see that the dual cone of $\mathbb{T}\mathrm{Pol}_n^+$ consists exactly of those trigonometric polynomials g for which the associated Toeplitz matrix T_g is positive semidefinite (notation: $T_g \geq 0$).

Lemma 1.1.6 *The dual cone of* $\mathbb{T}\mathrm{Pol}_n^+$ *is given by*

$$(\mathbb{T}\mathrm{Pol}_n^+)^* = \{g \in \mathbb{T}\mathrm{Pol}_n : T_g \geq 0\}. \tag{1.1.4}$$

Proof. Clearly, if $p = |q|^2 \in \mathbb{T}\mathrm{Pol}_n^+$ and $g \in \mathbb{T}\mathrm{Pol}_n$ is such that $T_g \geq 0$, we obtain by (1.1.3) that $\langle p, g\rangle \geq 0$. This gives $\supseteq$ in (1.1.4). For the converse, suppose that $g \in \mathbb{T}\mathrm{Pol}_n$ is such that T_g is not positive semidefinite. Then choose $x = \begin{pmatrix} q_0 & \cdots & q_n \end{pmatrix}^T$ such that $x^* T_g x < 0$, and let $p \in \mathbb{T}\mathrm{Pol}_n$ be given via $p(z) = |q_0 + \cdots + q_n z^n|^2$, $z \in \mathbb{T}$. Then (1.1.3) yields that $\langle p, g\rangle < 0$, and thus $g \notin (\mathbb{T}\mathrm{Pol}_n^+)^*$. This yields $\subseteq$ in (1.1.4). □

Next, we show that the extreme rays of $\mathbb{T}\mathrm{Pol}_n^+$ are generated exactly by those nonnegative trigonometric polynomials that have $2n$ roots (counting multiplicity) on the unit circle.

Proposition 1.1.7 *The trigonometric polynomial $q(z) = \sum_{k=-n}^{n} q_k z^k$ in $\mathbb{T}\mathrm{Pol}_n^+$ generates an extreme ray of $\mathbb{T}\mathrm{Pol}_n^+$ if and only if $q_n \neq 0$ and all the roots of q are on the unit circle; equivalently, $q(z)$ is of the form*

$$q(z) = c \prod_{k=1}^{n} (z - e^{it_k}) \left(\frac{1}{z} - e^{-it_k} \right), \tag{1.1.5}$$

for some $c > 0$ and $t_k \in \mathbb{R}$, $k = 1, \ldots, n$.

Proof. First suppose that $q(z)$ is of the form (1.1.5) and let $q = g + h$ with $g, h \in \mathbb{T}\mathrm{Pol}_n^+$. As $q(e^{it_k}) = 0$, $k = 1, \ldots, n$, and g and h are nonnegative on the unit circle, we get that $g(e^{it_k}) = h(e^{it_k}) = 0$, $k = 1, \ldots, n$. Notice that $q \geq g$ on $\mathbb{T}$ implies that the multiplicity of a root on $\mathbb{T}$ of g must be at least as large as the multiplicity of the same root of q (use the construction of the function ϕ in the proof of Theorem 1.1.5). Using this one now sees that g and h must be multiples of q.

Next, suppose that q has fewer that $2n$ roots on the unit circle (counting multiplicity). Using the construction in the proof of Theorem 1.1.5 we can factor q as $q = gh$, where $g \in \mathbb{T}\mathrm{Pol}_m^+$, $m < n$, has all its roots on the unit circle, $h \in \mathbb{T}\mathrm{Pol}_k^+$ has none of its roots on the unit circle, and $k + m \leq n$. If h is not a constant, we can subtract a positive constant ϵ, say, from h while remaining in $\mathbb{T}\mathrm{Pol}_k^+$ (e.g., take $\epsilon = \min_{z \in \mathbb{T}} h(z) > 0$). Then $q = \epsilon g + (h - \epsilon) g$ gives a way of writing q as a sum of elements in $\mathbb{T}\mathrm{Pol}_n^+$ that are not scalar multiples of q (as h was not a constant function). In case $h \equiv c$ is a constant function, we can write

$$q(z) = g(z)h(z) = g(z) \left(\frac{1}{2}c - \delta z - \frac{\delta}{z} \right) + g(z) \left(\frac{1}{2}c + \delta z + \frac{\delta}{z} \right),$$

which for $0 < \delta < \frac{c}{4}$ gives a way of writing q as a sum of elements in $\mathbb{T}\mathrm{Pol}_{m+1}^+ \subseteq \mathbb{T}\mathrm{Pol}_n^+$ that are not scalar multiples of q. This proves that q does not generate an extreme ray of $\mathbb{T}\mathrm{Pol}_n^+$. □

Notice that if we consider the cone of nonnegative trigonometric polynomials without any restriction on the degree, the argument in the proof of Proposition 1.1.7 shows that none of its elements generate an extreme ray. In other words, the cone $\mathbb{T}\mathrm{Pol}^+ = \bigcup_{n=0}^{\infty} \mathbb{T}\mathrm{Pol}_n^+$ does not have any extreme rays.

1.2 CONES OF HERMITIAN MATRICES

A natural cone to study is the cone of positive semidefinite matrices that have zeros in certain fixed locations, that is, positive semidefinite matrices with a given sparsity pattern. As we shall see, the dual of this cone consists of those Hermitian matrices which can be made into a positive semidefinite matrix by changing the entries in the given locations. As such, this dual cone is related to the positive semidefinite completion problem. While in this section we analyze this problem from the viewpoint of cones and their properties, we shall directly investigate the completion problem in detail in the next chapter, then in the context of operator matrices.

For presenting the results of this section we first need some graph theoretical preliminaries. An *undirected graph* is a pair $G = (V, E)$ in which V, the *vertex set*, is a finite set and the *edge set* E is a symmetric binary relation on V such that $(v, v) \notin E$ for all $v \in V$. The *adjacency set* of a vertex v is denoted by $\mathrm{Adj}(v)$, that is, $w \in \mathrm{Adj}(v)$ if $(v, w) \in E$. Given a subset $A \subseteq V$, define the *subgraph induced* by A by $G|A = (A, E|A)$, in which $E|A = \{(x, y) \in E \ : \ x \in A \text{ and } y \in A\}$. The *complete graph* is a graph with the property that every pair of distinct vertices is adjacent. A subset $K \subseteq V$ is a *clique* if the induced graph $G|K$ on K is complete. A clique is called a *maximal clique* if it is not a proper subset of another clique.

A subset $P \subseteq \{1, \dots, n\} \times \{1, \dots, n\}$ with the properties

(i) $(i, i) \in P$ for $i = 1, \dots, n$,

(ii) $(i, j) \in P$ if and only if $(j, i) \in P$

is called an $n \times n$ *symmetric pattern*. Such a pattern is said to be a *sparsity pattern* (location of required zeros) for a matrix $A \in \mathcal{H}_n$, if every $1 \leq i, j \leq n$ such that $(i, j) \notin P$ implies that the (i, j) and (j, i) entries of A are 0.

With an $n \times n$ symmetric pattern P we associate the undirected graph $G = (V, E)$, with $V = \{1, \dots, n\}$ and $(i, j) \in E$ if and only if $(i, j) \in P$ and $i \neq j$. For a pattern with associated graph $G = (V, E)$, we introduce the following subspace of $\mathcal{H}_n$:

$$\mathcal{H}_G = \{A \in \mathcal{H}_n : A_{ij} = 0 \text{ for all } (i, j) \notin P\},$$

that is, the set of all $n \times n$ Hermitian matrices with sparsity pattern P. It is easy to see that

$$\mathcal{H}_G^{\perp} = \{A \in \mathcal{H}_n : A_{ij} = 0 \text{ for all } (i, j) \in P\}.$$

When A is an $n \times n$ matrix $A = (A_{ij})_{i,j=1}^{n}$ and $K \subseteq \{1, \dots, n\}$, then $A|K$ denotes the $\mathrm{card}K \times \mathrm{card}K$ principal submatrix

$$A|K = (A_{ij})_{i,j \in K} = (A_{ij})_{(i,j) \in K \times K}.$$

Here $\mathrm{card}K$ denotes the cardinality of the set K.

We now define the following cones in $\mathcal{H}_G$:

$$\begin{aligned}
\mathrm{PPSD}_G &= \{ A \in \mathcal{H}_G : A|K \geq 0 \text{ for all cliques } K \text{ of } G \}, \\
\mathrm{PSD}_G &= \{ X \in \mathcal{H}_G : X \geq 0 \}, \\
\mathcal{A}_G &= \{ Y \in \mathcal{H}_G : \text{ there is a } W \in \mathcal{H}_G^{\perp} \text{ such that } Y + W \geq 0 \}, \\
\mathcal{B}_G &= \{ B \in \mathcal{H}_G : B = \sum\nolimits_{i=1}^{n_B} B_i \text{ where} \\
&\qquad\qquad B_1, \dots, B_{n_B} \in \mathrm{PSD}_G \text{ are of rank } 1 \}.
\end{aligned}$$

The abbreviation PPSD stands for partially positive semidefinite.

We have the following descriptions of the duals of PPSD_G and PSD_G. For Hermitian matrices we introduce the Loewner order $\leq$, defined by $A \leq B$ if and only if $B - A \geq 0$ (i.e., $B - A$ is positive semidefinite).

Proposition 1.2.1 *Let $G = (V, E)$ be a graph. The cones PPSD_G, PSD_G, $\mathcal{A}_G$, and $\mathcal{B}_G$ are closed, and their duals (in $\mathcal{H}_G$) are identified as*

$$(\mathrm{PSD}_G)^* = \mathcal{A}_G, \text{ and } (\mathrm{PPSD}_G)^* = \mathcal{B}_G.$$

Moreover,

$$(\mathrm{PPSD}_G)^* \subseteq \mathrm{PSD}_G \subseteq (\mathrm{PSD}_G)^* \subseteq \mathrm{PPSD}_G. \tag{1.2.1}$$

Proof. The closedness of PPSD_G, $\mathcal{A}_G$, and PSD_G is trivial. For the closedness of $\mathcal{B}_G$, first observe that if $J_1, \ldots, J_p$ are all the maximal cliques of G, then any $B \in \mathcal{B}_G$ can be written as $B = B_1 + \cdots + B_p$, where $B_k \geq 0$ and the nonzero entries of B_k lie in $J_k \times J_k$, $k = 1, \ldots, p$. Note that $B_k \leq B$, $k = 1, \ldots, p$. Let now $A^{(n)} = \sum_{k=1}^{p} A_k^{(n)} \in \mathcal{B}_G$ be so that the nonzero entries of $A_k^{(n)}$ lie in $J_k \times J_k$, $k = 1, \ldots, p$, and assume that $A^{(n)}$ converges to A as $n \to \infty$. We need to show that $A \in \mathcal{B}_G$. Then $A_1^{(n)}$ is a bounded sequence of matrices, and thus there is a subsequence $\{A_1^{(n_k)}\}_{k\in\mathbb{N}}$, that converges to A_1, say. Note that A_1 is positive semidefinite and has only nonzero entries in $J_1 \times J_1$. Next take a subsequence $\{A_2^{(n_{k_l})}\}_{l\in\mathbb{N}}$, of $\{A_2^{(n_k)}\}_{k\in\mathbb{N}}$ that converges to A_2, say, which is automatically positive semidefinite and has only nonzero entries in $J_2 \times J_2$. Repeating this argument, we ultimately obtain $m_1 < m_2 < \cdots$ so that $\lim_{j\to\infty} A_k^{(m_j)} = A_k \geq 0$ with A_k having only nonzero entries in $J_k \times J_k$. But then

$$A = \lim_{n\to\infty} A^{(n)} = \sum_{k=1}^{p} \lim_{j\to\infty} A_k^{(m_j)} = A_1 + \cdots + A_p \in \mathcal{B}_G,$$

proving the closedness of $\mathcal{B}_G$.

By Lemma 1.1.2 it is true that if $\mathcal{C}$ is a cone and W is a subspace, then $(\mathcal{C} \cap W)^* = \mathcal{C}^* + W^\perp$. This implies the duality between PSD_G and $\mathcal{A}_G$ since $\mathrm{PSD}_G = \mathrm{PSD}_n \cap \mathcal{H}_G$ and PSD_n is selfdual.

For the duality of PPSD_G and $\mathcal{B}_G$ note that if $X \in \mathrm{PSD}_G$ and rank $X = 1$, then $X = ww^*$, where w is a vector with support in a clique of G. This shows that any element in PPSD_G lies in the dual of $\mathcal{B}_G$. Next if $A \notin \mathrm{PPSD}_G$, then there is a clique K such that $A|K$ is not positive semidefinite. Thus there is a vector v so that $v^*(A|K)v < 0$. But then one can pad the matrix vv^* with zeros, and obtain a positive semidefinite rank 1 matrix B with all its nonzero entries in the clique K, so that $\langle A, B\rangle < 0$. As $B \in \mathcal{B}_G$, this shows that A is not in the dual of $\mathcal{B}_G$. □

It is not hard to see that equality between PSD_G and $(\mathrm{PSD}_G)^*$ holds only in case the maximal cliques of G are disjoint, and in that case all four cones are equal.

In order to explore when the cones PSD_G^* and PPSD_G are equal, it is convenient to introduce the following graph theoretic notions. A *path* $[v_1, \ldots, v_k]$ in a graph $G = (V, E)$ is a sequence of vertices such that $(v_j, v_{j+1}) \in E$ for $j = 1, \ldots, k-1$. The path $[v_1, \ldots, v_k]$ is referred to as a path between v_1 and v_k. A graph is called *connected* if there exists a path between any two different vertices in the graph. A *cycle* of length $k > 2$ is a path $[v_1, \ldots, v_k, v_1]$

in which $v_1, \ldots, v_k$ are distinct. A graph G is called *chordal* if every cycle of length greater than 3 possesses a *chord*, that is, an edge joining two nonconsecutive vertices of the cycle.

Below are two graphs and the corresponding symmetric patterns (pictured as partial matrices). The graph on the left is chordal, while the one on the right (the four cycle) is not.

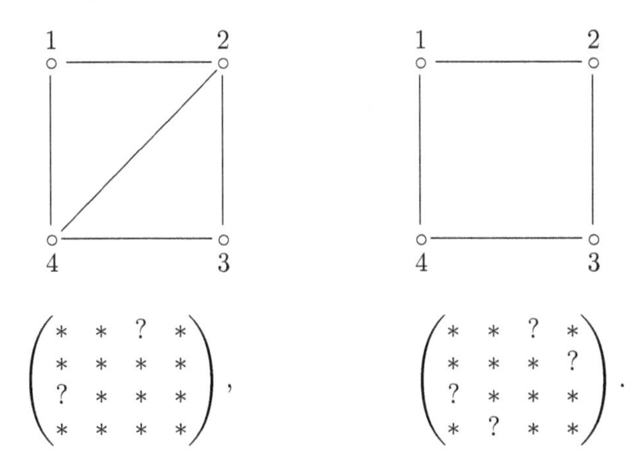

$$\begin{pmatrix} * & * & ? & * \\ * & * & * & * \\ ? & * & * & * \\ * & * & * & * \end{pmatrix}, \qquad \begin{pmatrix} * & * & ? & * \\ * & * & * & ? \\ ? & * & * & * \\ * & ? & * & * \end{pmatrix}.$$

An ordering $\sigma = [v_1, \ldots, v_n]$ of the vertices of a graph is called a *perfect vertex elimination scheme (or perfect scheme)* if each set

$$S_i = \{v_j \in \operatorname{Adj}(v_i) \ : \ j > i\} \tag{1.2.2}$$

is a clique. A vertex v of G is said to be *simplicial* when $\operatorname{Adj}(v)$ is a clique. Thus $\sigma = [v_1, \ldots, v_n]$ is a perfect scheme if each v_i is simplicial in the induced graph $G|\{v_i, \ldots, v_n\}$.

A subset $S \subset V$ is an *$a - b$ vertex separator* for nonadjacent vertices a and b if the removal of S from the graph separates a and b into distinct connected components. If no proper subset of S is an $a - b$ separator, then S is called a *minimal $a - b$ vertex separator.*

Proposition 1.2.2 *Every minimal vertex separator of a chordal graph is complete.*

Proof. Let S be a minimal $a - b$ separator in a chordal graph $G = (V, E)$, and let A and B be the connected components in $G|V - S$ containing a and b, respectively. Let $x, y \in S$. Each vertex in S must be connected to at least one vertex in A and at least one in B. We can choose minimal length paths $[x, a_1, \ldots, a_r, y]$ and $[y, b_1, \ldots, b_s, x]$, such that $a_i \in A$ for $i = 1, \ldots, r$ and $b_j \in B$ for $j = 1, \ldots, s$. Then $[x, a_1, \ldots, a_r, y, b_1, \ldots, b_s, x]$ is a cycle in G of length at least 4. Chordality of G implies it must have a chord. Since $(a_i, b_j) \notin E$ by the definition of a vertex separator, $(a_i, a_k) \notin E$ and $(b_k, b_l) \notin E$ by the minimality of r and s, the only possible chord is (x, y). Thus S is a clique. $\square$

The next result is known as Dirac's lemma.

Lemma 1.2.3 *Every chordal graph G has a simplicial vertex, and if G is not a clique, then it has two nonadjacent simplicial vertices.*

Proof. We proceed by induction on n, the number of vertices of G. When $n = 1$ or $n = 2$, the result is trivial. Assume $n \geq 3$ and the result is true for graphs with fewer than n vertices. Assume $G = (V, E)$ has n vertices. If G is complete, the result holds. Assume G is not complete, and let S be a minimal $a - b$ separator for two nonadjacent vertices a and b. Let A and B be the connected components of a respectively b in $G|V - S$.

By our assumption, either the subgraph $G|A \cup S$ has two nonadjacent simplicial vertices one of which must be in A (since by Proposition 1.2.2, $G|S$ is complete), or $G|(A \cup S)$ is complete and any vertex in A is simplicial in $G|A \cup S$. Since $\mathrm{Adj}(A) \subseteq A \cup S$, a simplicial vertex of $G|A \cup S$ in A is simplicial in G. Similarly, B contains a simplicial vertex of G, and this proves the lemma. □

The following result is an algorithmic characterization of chordal graphs.

Theorem 1.2.4 *An undirected graph is chordal if and only if it has a perfect scheme. Moreover, any simplicial vertex can start a perfect scheme.*

Proof. Assume $G = (V, E)$ be a chordal graph with n vertices. Assume every chordal graph with fewer than n vertices has a perfect scheme. (For $n = 1$ the result is trivial.) By Lemma 1.2.3, G has a simplicial vertex x. Let $[v_1, \ldots, v_{n-1}]$ be a perfect scheme for $G|V - \{x\}$. Then $[x, v_1, \ldots, v_{n-1}]$ is a perfect scheme for G.

Let $G = (V, E)$ be a graph with a perfect scheme $\sigma = [v_1, \ldots, v_n]$ and assume C is a cycle of length at least 4 in G. Let x be the vertex of C with the smallest index in σ. By the definition of a perfect scheme, the two vertices in C adjacent to x must be connected by an edge, so C has a chord. □

The following result is an important consequence of Theorem 1.2.4. We will use the following lemma.

Lemma 1.2.5 *Let $A > 0$. Then $\begin{pmatrix} A & B \\ B^* & C \end{pmatrix}$ is positive (semi)definite if and only if $C - B^*A^{-1}B$ is.*

Proof. This follows immediately from the following factorization:

$$\begin{pmatrix} A & B \\ B^* & C \end{pmatrix} = \begin{pmatrix} I & 0 \\ B^*A^{-1} & I \end{pmatrix} \begin{pmatrix} A & 0 \\ 0 & C - B^*A^{-1}B \end{pmatrix} \begin{pmatrix} I & A^{-1}B \\ 0 & I \end{pmatrix}. \qquad \square$$

In Section 2.4 we will see that $C - B^*A^{-1}B$ is a so-called "Schur complement," and that more general versions of this lemma hold.

Proposition 1.2.6 *Let $A \in \mathrm{PSD}_G$, where G is a chordal graph with n vertices. Then A can be written as $A = \sum_{i=1}^{r} w_i w_i^*$, where $r = \mathrm{rank} A$ and $w_i \in \mathbb{C}^n$ are nonzero vectors such that $w_i w_i^* \in \mathrm{PSD}_G$ for $i = 1, \ldots, r$. In particular, we have $\mathrm{PSD}_G = (\mathrm{PPSD}_G)^*$.*

Proof. We use induction on n. For $n = 1$ the result is trivial and we assume it holds for $n-1$. Let $A \in \text{PSD}_G$, where $G = (V, E)$ is a chordal graph with n vertices, and let $r = \text{rank}A$. We can assume without loss of generality that the vertex 1 is simplicial in G (otherwise we reorder the rows and columns of A in an order that starts with a simplicial vertex). If $a_{11} = 0$, then the first row and column of A are zero, so the result follows from the assumption for $n - 1$. Otherwise, let $w_1 \in \mathbb{C}^n$ be the first column of A multiplied by $\frac{1}{\sqrt{a_{11}}}$. An entry (k, l), $2 \leq k, l \leq n$ of $w_1 w_1^*$ is nonzero if $(1, k)$ and $(1, l)$ are in E. Since the vertex 1 is simplicial, we have $(k.l) \in E$. Then, by Lemma 1.2.5, $A - w_1 w_1^*$ is a positive semidefinite matrix of rank $r - 1$, and has its first row and column equal to zero, and $(A - w_1 w_1^*)|\{2, \ldots, n\} \in \text{PSD}_{G|\{2,\ldots,n\}}$. Since $G|\{2, \ldots, n\}$ is also chordal, by our assumption for $n-1$, we have that $A - w_1 w_1^* = \sum_{i=2}^{r} w_i w_i^*$, where each $w_i \in \mathbb{C}^n$ is a nonzero vector with its first component equal to 0. This completes our proof. □

Proposition 1.2.6 has two immediate corollaries.

Corollary 1.2.7 *Let P be a symmetric pattern with associated graph $G = (V, E)$. Then Gaussian elimination can be carried out on every $A \in \text{PSD}_G$ such that in the process no entry corresponding to $(i, j) \notin P$ is changed even temporarily to a nonzero, if and only if $\sigma = [1, 2, \ldots, n]$ is a perfect scheme for G.*

Corollary 1.2.8 *Let $G = (V, E)$ be a graph. Then the lower-upper Cholesky factorization $A = LL^*$ of every $A \in \text{PSD}_G$ satisfies $L_{ij} = 0$ for $1 \leq j < i \leq n$ such that $(i, j) \notin E$, if and only if $\sigma = [1, 2, \ldots, n]$ is a perfect scheme for G.*

Proof. Follows immediately from the fact that $L = \begin{pmatrix} w_1 & w_2 & \cdots w_n \end{pmatrix}$, where $w_1, \ldots, w_n$ are obtained recursively as in the proof of Proposition 1.2.6. □

Proposition 1.2.9 *Let $G = (V, E)$ be a nonchordal graph. Then $(\text{PSD}_G)^*$ is a proper subset of PPSD_G.*

Proof. For $m \geq 4$ define the $m \times m$ Toeplitz matrix

$$A_m = \begin{pmatrix} 1 & 1 & 0 & 0 & \cdots & -1 \\ 1 & 1 & 1 & 0 & \cdots & 0 \\ 0 & 1 & 1 & 1 & \cdots & 0 \\ \vdots & \ddots & \ddots & \ddots & \ddots & \vdots \\ -1 & 0 & 0 & 0 & \cdots & 1 \end{pmatrix}, \tag{1.2.3}$$

the graph of which is the chordless cycle $[1, \ldots, m, 1]$. Each A_m is partially positive semidefinite since both matrices $\left(\begin{smallmatrix} 1 & 1 \\ 1 & 1 \end{smallmatrix}\right)$ and $\left(\begin{smallmatrix} 1 & -1 \\ -1 & 1 \end{smallmatrix}\right)$ are positive semidefinite. We cannot modify the zeros of A_m in any way to obtain a positive semidefinite matrix, since the only positive semidefinite matrix with the three middle diagonals equal to 1 is the matrix of all 1's.

Let $G = (V, E)$, $V = \{1, \ldots, n\}$, be a nonchordal graph and assume without loss of generality that it contains the chordless cycle $[1, \ldots, m, 1]$, $4 \leq m \leq n$. Let A be the matrix with graph G having 1 on its main diagonal, the matrix A_m in (1.2.3) as its $m \times m$ upper-left corner, and 0 on any other position. Then $A \in \text{PPSD}_G$, but $A \notin (PSD_G)^*$, since the zeros in its upper left $m \times m$ corner cannot be modified such that this corner becomes a positive semidefinite matrix. □

The following result summarizes the above results and shows that equality between $(\text{PPSD}_G)^*$ and PSD_G (or equivalently, between $(\text{PSD}_G)^*$ and PPSD_G) occurs exactly when G is chordal.

Theorem 1.2.10 *Let $G = (V, E)$ be a graph. Then the following are equivalent.*

(i) *G is chordal.*

(ii) $\text{PPSD}_G = (\text{PSD}_G)^*$.

(iii) $(\text{PPSD}_G)^* = \text{PSD}_G$.

(iv) *There exists a permutation σ of $[1, \ldots, n]$ such that after reordering the rows and columns of every $A \in \text{PSD}_G$ by the order σ, A has the lower-upper Cholesky factorization $A = LL^*$ with $L_{ij} = 0$ for every $1 \leq i < j \leq n$ such that $(i, j) \notin E$.*

(v) *The only matrices that generate extreme rays of* PSD_G *are rank* 1 *matrices.*

Proof. (i) ⇒ (iv) follows from Corollary 1.2.8, (iv) ⇒ (iii) from Corollary 1.2.8 and Proposition 1.2.6, (iii) ⇒ (ii) from Lemma 1.1.2 and the closedness of the cones, while (ii) ⇒ (i) follows from Proposition 1.2.9. The implications (i) ⇒ (iii) ⇒ (v) follow from Proposition 1.2.6. □

Let us give an example of a higher rank matrix that generates an extreme ray in the nonchordal case.

Example 1.2.11 Let G be given by the graph representing the chordless cycle $[1, 2, 3, 4, 1]$, and let

$$K = \begin{pmatrix} 1 & 1 & 0 & 1 \\ 1 & 2 & 1 & 0 \\ 0 & 1 & 1 & -1 \\ 1 & 0 & -1 & 2 \end{pmatrix}.$$

Then K generates an extreme ray for PSD_G. Indeed, suppose that $K = A + B$ with $A, B \in \text{PSD}_G$. Notice that if we let

$$V = \begin{pmatrix} 1 & 1 & 0 & 1 \\ 0 & 1 & 1 & -1 \end{pmatrix} = \begin{pmatrix} v_1 & v_2 & v_3 & v_4 \end{pmatrix},$$

then $K = V^*V$. As $K = A + B$ and $A, B \geq 0$, it follows that the nullspace of K lies in the nullspaces of both A and B. Using this, it is not hard to see that A and B must be of the form

$$A = V^*\widetilde{A}V, \ B = V^*\widetilde{B}V,$$

where $\widetilde{A}$ and $\widetilde{B}$ are 2×2 positive semidefinite matrices. As $A, B \in \mathrm{PSD}_G$, we have that

$$v_1^*\widetilde{A}v_3 = 0 = v_1^*\widetilde{B}v_3 \ \text{ and } \ v_2^*\widetilde{A}v_4 = 0 = v_2^*\widetilde{B}v_4.$$

But then both $\widetilde{A}$ and $\widetilde{B}$ are orthogonal to the matrices

$$v_3v_1^* = \begin{pmatrix} 0 & 0 \\ 1 & 0 \end{pmatrix}, \ v_1v_3^* = \begin{pmatrix} 0 & 1 \\ 0 & 0 \end{pmatrix},$$

$$v_2v_4^* = \begin{pmatrix} 1 & -1 \\ 1 & -1 \end{pmatrix}, \ v_4v_2^* = \begin{pmatrix} 1 & 1 \\ -1 & -1 \end{pmatrix}.$$

But then $\widetilde{A}$ and $\widetilde{B}$ must be multiples of the identity, and thus A and B are multiples of K.

1.3 CONES OF TRIGONOMETRIC POLYNOMIALS

Similarly to Section 1.2, we introduce in this section four cones, which now contain real-valued trigonometric polynomials. We establish the relationship between these cones, identify their duals, and discuss the situation when they are pairwise equal.

Let $d \geq 1$ and let S be a finite subset of $\mathbb{Z}^d$. We consider the vector space Pol_S of trigonometric polynomials p with Fourier support in S, that is, $p(z) = \sum_{\lambda \in S} p_\lambda z^\lambda$, where $z = (e^{it_1}, \ldots, e^{it_d})$, $\lambda = (\lambda_1, \ldots, \lambda_d)$, and $z^\lambda = (e^{i\lambda_1 t_1}, \ldots, e^{i\lambda_d t_d})$. Note that in the notation Pol_S we do not highlight the number of variables; it is implicit in the set S. In subsequent notation we have also chosen not to attach the number of variables, as it would make the notation more cumbersome. We hope that this does not lead to confusion. The complex numbers $\{p_\lambda\}_{\lambda \in S}$ are referred to as the *(Fourier) coefficients* of p. The *Fourier support* of p is given by $\{\lambda : p_\lambda \neq 0\}$. We endow Pol_S with the inner product

$$\langle p, q \rangle = \frac{1}{(2\pi)^d} \int_{[0,2\pi]^d} p(e^{it_1}, \ldots, e^{it_d})\overline{q(e^{it_1}, \ldots, e^{it_d})}dt_1 \ldots dt_d = \sum_{\lambda \in S} p_\lambda \overline{q_\lambda},$$

where $\{p_\lambda\}_{\lambda \in S}$ and $\{q_\lambda\}_{\lambda \in S}$ are the coefficients of p and q, respectively. For a set $S \subset \mathbb{Z}^d$ satisfying $S = -S$ we let $\mathbb{T}\mathrm{Pol}_S$ be the vector space over $\mathbb{R}$ consisting of all trigonometric polynomials p with Fourier support in S such that $p(z) \in \mathbb{R}$ for $z \in \mathbb{T}^d$. Equivalently, $p \in \mathbb{T}\mathrm{Pol}_S$ if $p(z) = \sum_{s \in S} p_s z^s$ satisfies $p_s = \overline{p_{-s}}$ for $s \in S$. With the inner product inherited from Pol_S the space $\mathbb{T}\mathrm{Pol}_S$ is a Hilbert space over $\mathbb{R}$. We will be especially interested in the case when S is of the form $S = \Lambda - \Lambda$, where $\Lambda \subset \mathbb{N}_0^d$.

Let us introduce the following cones:

$$\mathbb{T}\mathrm{Pol}_S^+ = \{p \in \mathbb{T}\mathrm{Pol}_S : p(z) \geq 0 \text{ for all } z \in \mathbb{T}^d\},$$

the cone consisting of all nonnegative valued trigonometric polynomials with support in S, and

$$\mathbb{T}\mathrm{Pol}_\Lambda^2 = \{p \in \mathbb{T}\mathrm{Pol}_S : p = \sum_{j=1}^{r} |q_j|^2, \text{ with } q_j \in \mathrm{Pol}_\Lambda\},$$

consisting of all trigonometric polynomials that are sums of squares of absolute values of polynomials with support in Λ. Obviously, $\mathbb{T}\mathrm{Pol}_\Lambda^2 \subseteq \mathbb{T}\mathrm{Pol}_{\Lambda-\Lambda}^+$.

Next we introduce two cones that are defined via Toeplitz matrices built from Fourier coefficients. In general, when we have a finite subset Λ of $\mathbb{Z}^d$ and a trigonometric polynomial $p(z) = \sum_{s\in S} p_s z^s$, where $S = \Lambda - \Lambda$, we can define the (multivariable) Toeplitz matrix

$$T_{p,\Lambda} = (p_{\lambda-\nu})_{\lambda,\nu\in\Lambda}.$$

It should be noticed that in this definition of the multivariable Toeplitz matrix there is some ambiguity in the order of rows and columns. We will only be interested in properties (primarily positive semidefiniteness) of the matrix that are not dependent on the order of the rows and columns, as long as the same order is used for both the rows and the columns. For example, consider $\Lambda = \{(0,0),(0,1),(1,0)\} \subseteq \mathbb{Z}^2$. If we order the elements of Λ as indicated, we get that

$$(p_{\lambda-\nu})_{\lambda,\nu\in\Lambda} = \begin{pmatrix} p_{0,0} & p_{0,-1} & p_{-1,0} \\ p_{0,1} & p_{0,0} & p_{-1,1} \\ p_{1,0} & p_{1,-1} & p_{0,0} \end{pmatrix}.$$

If we order Λ as $\{(0,1),(0,0),(1,0)\}$ we get the matrix

$$(p_{\lambda-\nu})_{\lambda,\nu\in\Lambda} = \begin{pmatrix} p_{0,0} & p_{0,1} & p_{-1,1} \\ p_{0,-1} & p_{0,0} & p_{-1,0} \\ p_{1,-1} & p_{1,0} & p_{0,0} \end{pmatrix},$$

which is of course the previous matrix with rows and columns 1 and 2 permuted. Later on, we may refer, for instance, to the $(0,0)$th row and the $(0,1)$th column, and even the $((0,0),(0,1))$th element of the matrix. In that terminology, we have that the (λ,ν)th element of the matrix $T_{p,\Lambda}$ is the number $p_{\lambda-\nu}$, explaining the convenience of this terminology. Of course, when $\Lambda = \{1,\ldots,n\}$ and the numbers 1 through n are ordered as usual, we get a classical Toeplitz matrix and the associated terminology is the usual one. In general, though, these "multivariable" Toeplitz matrices and the associated terminology take some getting used to. Even in the case when $\Lambda \subset \mathbb{Z}$ some care is needed in these definitions. For instance, for both subsets $\Lambda = \{0,1,3\}$ and $\Lambda' = \{0,1,2,3\}$ of $\mathbb{Z}$, we have that $S = \Lambda - \Lambda = \Lambda' - \Lambda' = \{-3,-2,-1,0,1,2,3\}$. For these sets we have that

$$T_{p,\Lambda} = \begin{pmatrix} p_0 & p_{-1} & p_{-3} \\ p_1 & p_0 & p_{-2} \\ p_3 & p_2 & p_0 \end{pmatrix}, \tag{1.3.1}$$

while

$$T_{p,\Lambda'} = \begin{pmatrix} p_0 & p_{-1} & p_{-2} & p_{-3} \\ p_1 & p_0 & p_{-1} & p_{-2} \\ p_2 & p_1 & p_0 & p_{-1} \\ p_3 & p_2 & p_1 & p_0 \end{pmatrix}. \tag{1.3.2}$$

The positive semidefiniteness of (1.3.1) does not guarantee positive semidefiniteness of (1.3.2). For example, when we take $p(z) = \frac{7}{10z^3} + \frac{7}{10z} + 1 + \frac{7z}{10} + \frac{7z^3}{10}$ it is easy to check that $T_{p,\Lambda}$ is positive semidefinite but $T_{p,\Lambda'}$ is not.

It should be noted that the condition $T_{p,\Lambda} \geq 0$ is equivalent to the statement that

$$\sum_{\lambda,\mu \in \Lambda} p_{\lambda-\mu} c_\mu \overline{c}_\lambda \geq 0 \tag{1.3.3}$$

for every complex sequence $\{c_\lambda\}_{\lambda\in\Lambda}$. Relation (1.3.3) transforms $T_{p,\Lambda} \geq 0$ into a condition that is independent of the ordering of Λ, and in the literature one may see the material addressed in this section presented in this way. We chose to use the presentation with the positive semidefinite Toeplitz matrices because it parallels the setting of positive semidefinite matrices as discussed in the previous section.

Next, we define the notion of extendability. Given $p(z) = \sum_{s\in S} p_s z^s$, we say that p is *extendable* if we can find p_m, $m \in \mathbb{Z}^d \setminus S$, such that for every finite set $J \subset \mathbb{Z}^d$ the Toeplitz matrix $(p_{\lambda-\nu})_{\lambda,\nu\in J}$ is positive semidefinite. Obviously, if $S = \Lambda - \Lambda$, then p being extendable implies that $T_{p,\Lambda} \geq 0$.

We are now ready to introduce the next two cones:

$$\mathcal{A}_\Lambda = \{p \in \mathbb{T}\mathrm{Pol}_{\Lambda-\Lambda} : T_{p,\Lambda} \geq 0\}$$

and

$$\mathcal{B}_S = \{p \in \mathbb{T}\mathrm{Pol}_S : p \text{ is extendable}\}.$$

Clearly, $\mathcal{B}_{\Lambda-\Lambda} \subseteq \mathcal{A}_\Lambda$. In general, though, we do not have equality. For instance,

$$p(z) = \frac{7}{10z^3} + \frac{7}{10z} + 1 + \frac{7z}{10} + \frac{7z^3}{10} \in \mathcal{A}_{\{0,1,3\}} \setminus \mathcal{B}_{\{-3,\ldots,3\}}. \tag{1.3.4}$$

The relationship between the four convex cones introduced above is the subject of the following result.

Proposition 1.3.1 *Let $\Lambda \subset \mathbb{Z}^d$ be finite and put $S = \Lambda - \Lambda$. Then the cones $\mathbb{T}\mathrm{Pol}^2_\Lambda$, $\mathbb{T}\mathrm{Pol}^+_S$, $\mathcal{B}_S$, and $\mathcal{A}_\Lambda$ are closed in $\mathbb{T}\mathrm{Pol}_S$, and we have*

$$\mathbb{T}\mathrm{Pol}^2_\Lambda \subseteq \mathbb{T}\mathrm{Pol}^+_S \subseteq \mathcal{B}_S \subseteq \mathcal{A}_\Lambda. \tag{1.3.5}$$

Proof. The closedness of $\mathbb{T}\mathrm{Pol}^+_S$, $\mathcal{B}_S$, and $\mathcal{A}_\Lambda$ is trivial. For the closedness of $\mathbb{T}\mathrm{Pol}^2_\Lambda$, observe first that $\dim(\mathbb{T}\mathrm{Pol}^+_S) = \mathrm{card} S =: m$. First we show that every element of $\mathbb{T}\mathrm{Pol}^2_\Lambda$ is a sum of at most m squares. Let $p \in \mathbb{T}\mathrm{Pol}^2_\Lambda$, and write

$$p = \sum_{j=1}^{r} |q_j|^2, \tag{1.3.6}$$

with each $q_j \in \mathbb{T}\mathrm{Pol}_\Lambda$. If $r \leq m$ we are done, so assume that $r > m$. Since each $|q_j|^2$ is in $\mathbb{T}\mathrm{Pol}_S^+$, there exist real numbers r_j, not all 0, such that

$$\sum_{j=1}^{r} r_j |q_j|^2 = 0. \tag{1.3.7}$$

Without loss of generality we may assume that $|r_1| \geq |r_2| \geq \cdots \geq |r_m|$ and $|r_1| > 0$. Solving (1.3.7) for $|q_1|^2$ and substituting that into (1.3.6) we obtain

$$p = \sum_{j=2}^{r} \left(1 - \frac{r_j}{r_1}\right) |q_j|^2,$$

which is the sum of $r-1$ squares. Repeating these arguments we see that every $p \in \mathbb{T}\mathrm{Pol}_\Lambda^2$ is the sum of at most m squares.

If $p_n \in \mathbb{T}\mathrm{Pol}_\Lambda^2$ for $n \geq 1$ and p_n converges uniformly on $\mathbb{T}^d$ to p, there exist $q_{j,n} \in \mathrm{Pol}_\Lambda$ such that

$$p_n = \sum_{j=1}^{m} |q_{j,n}|^2, \quad n \geq 1. \tag{1.3.8}$$

The sequence p_n is uniformly bounded on Γ; hence so are $q_{j,n}$ by (1.3.8). Fix $k \in \Lambda$. It follows that there is a sequence $n_i \to \infty$ such that the kth Fourier coefficients of q_{j,n_i} converge to $g_{j,k}$, say, for $j = 1, \ldots, m$. Defining

$$\widetilde{q}_j(z) = \sum_{k \in \Lambda} g_{j,k} z^k$$

for $j = 1, \ldots, m$, we obviously have $p = \sum_{j=1}^{m} |\widetilde{q}_j|^2$. Thus $p \in \mathbb{T}\mathrm{Pol}_\Lambda^2$, and this proves $\mathbb{T}\mathrm{Pol}_\Lambda^2$ is closed.

The first and third inclusions in (1.3.5) are trivial. For the second inclusion, let $p \in \mathbb{T}\mathrm{Pol}_S^+$ and put $p_\nu = 0$ for $\nu \notin S$. With this choice we claim that $T_{p,J} \geq 0$ for all finite $J \subset \mathbb{Z}^d$. Indeed, for a complex sequence $v = (v_j)_{j \in J}$ define the polynomial $g(z) = \sum_{j \in J} v_j z^j$. Then

$$v^* T_{p,J} v = \frac{1}{(2\pi)^d} \int_{[0,2\pi]^d} p(e^{ij_1 t_1}, \ldots, e^{ij_d t_d}) |g(e^{ij_1 t_1}, \ldots, e^{ij_d t_d})|^2 dt \geq 0,$$

as $p(z) \geq 0$ for all $z \in \mathbb{T}^d$. This yields that p is extendable; thus $p \in \mathcal{B}_S$. □

The next result identifies duality relations between the four cones.

Theorem 1.3.2 *With the earlier notation, the following duality relations hold in* $\mathbb{T}\mathrm{Pol}_S$*:*

(i) $(\mathbb{T}\mathrm{Pol}_\Lambda^2)^* = \mathcal{A}_\Lambda$*;*

(ii) $(\mathbb{T}\mathrm{Pol}_S^+)^* = \mathcal{B}_S$.

In order to prove this theorem we need a particular case of a result that is known as Bochner's theorem, which will be proved in its full generality as

Theorem 3.9.2. The following result is its particular version for $\mathbb{Z}^d$. For a finite Borel measure μ on $\mathbb{T}^d$, its *moments* are defined as

$$\widehat{\mu}(n) = \frac{1}{(2\pi)^d} \int_{\mathbb{T}^d} z^{-n} d\mu(z),\ n \in \mathbb{Z}^d.$$

All measures appearing in this book are assumed to be regular.

Theorem 1.3.3 *Let μ be a finite positive Borel measure μ on $\mathbb{T}^d$. Then the Toeplitz matrices*

$$(\widehat{\mu}(j-k))_{j,k\in J}$$

are positive semidefinite for all finite sets $J \subset \mathbb{Z}^d$. Conversely, if $(c_j)_{j\in\mathbb{Z}^d}$ is a sequence of complex numbers such that for all finite $J \subset \mathbb{Z}^d$, the Toeplitz matrices

$$(c_{j-k})_{j,k\in J}$$

are positive semidefinite, then there exists a finite positive Borel measure μ on $\mathbb{T}^d$ such that $\widehat{\mu}(j) = c_j$, $j \in \mathbb{Z}^d$.

Proof. This result is a particular case of Theorem 3.9.2. A proof for this version may be found in [314]. □

We also need measures with finite support. For $\alpha \in \mathbb{T}^d$, let δ_α denote the *Dirac mass at α* (or *evaluation measure at α*). Thus for a Borel set $E \subseteq \mathbb{T}^d$ we have

$$\delta_\alpha(E) = \begin{cases} 1 & \text{if } \alpha \in E, \\ 0 & \text{otherwise.} \end{cases}$$

We will frequently use the fact that for a Borel measurable function $f : \mathbb{T}^d \to \mathbb{C}$, $\int_{\mathbb{T}^d} f d\delta_\alpha = f(\alpha)$. Applying this to the monomials $f(z) = z^{-n}$ we get that $\widehat{\delta}_\alpha(n) = \alpha^{-n}$ for every $\alpha \in \mathbb{T}^d$. A positive Borel measure on $\mathbb{T}^d$ is said to be supported on the finite subset $\{\alpha_1, \ldots, \alpha_n\} \subset \mathbb{T}^d$ if there exist positive constants $\rho_1, \ldots, \rho_n$ such that $\mu = \sum_{k=1}^n \rho_k \delta_{\alpha_k}$.

Proof of Theorem 1.3.2. It is clear that for $p \in \mathbb{T}\mathrm{Pol}_S$, we have that $p \in (\mathbb{T}\mathrm{Pol}^2_\Lambda)^*$ if and only if $\langle p, |q|^2\rangle \geq 0$ for all $q \in \mathrm{Pol}_\Lambda$. Let $q(z) = \sum_{\lambda\in\Lambda} c_\lambda z^\lambda$; then $\langle p, |q|^2\rangle = \sum_{\lambda,\mu\in\Lambda} p_{\lambda-\mu} c_\mu \bar{c}_\lambda = v^* T_{p,\Lambda} v$, where $v = (c_\lambda)_{\lambda\in\Lambda}$. Thus $\langle p, |q|^2\rangle \geq 0$ for all $q \in \mathrm{Pol}_\Lambda$ is equivalent to $T_{p,\Lambda} \geq 0$. But then it follows that $p \in (\mathbb{T}\mathrm{Pol}^2_\Lambda)^*$ if and only if $p \in \mathcal{A}_\Lambda$.

For proving (ii), let $p \in \mathcal{B}_S$, where $p(z) = \sum_{s\in S} p_s z^s$. As p is extendible, we can find p_ν, $\nu \in \mathbb{Z}^d \setminus S$ so that $(p_{j-k})_{j,k\in J} \geq 0$ for all finite $J \subset \mathbb{T}^d$. Apply Theorem 1.3.3 to obtain a finite positive Borel measure μ on $\mathbb{T}^d$ so that $\widehat{\mu}(j) = c_j$, $j \in \mathbb{Z}^d$. For every $q \in \mathbb{T}\mathrm{Pol}^+_S$, we have that

$$\langle q, p\rangle = \sum_{s\in S} q_s \widehat{\mu}(-s) = \int_{\mathbb{T}^d} q(z) d\mu(z) \geq 0.$$

This yields $\mathbb{T}\mathrm{Pol}^+_S \subseteq (\mathcal{B}_S)^*$. For the reverse inclusion, let $q \in (\mathcal{B}_S)^*$ and let $\alpha \in \mathbb{T}^d$. Put $p_{\alpha,S}(z) = \sum_{s\in S} \alpha^{-s} z^s$. Since $p_{\alpha,S} \in \mathcal{B}_S$, we have that

$0 \leq \langle q, p_{\alpha,S} \rangle = q(\alpha)$. This yields that $q(\alpha) \geq 0$ for all $\alpha \in \mathbb{T}^d$ and thus $q \in \mathbb{T}\mathrm{Pol}_S^+$. Hence we obtain $(\mathcal{B}_S)^* \subseteq \mathbb{T}\mathrm{Pol}_S^+$, which proves relation (ii). □

For one of the four cones it is easy to identify its extreme rays.

Theorem 1.3.4 *Let $\Lambda \subset \mathbb{Z}^d$ be a finite set and put $S = \Lambda - \Lambda$. The extreme rays of $\mathcal{B}_S$ are generated by the trigonometric polynomials $p_{\alpha,S}(z) = \sum_{s \in S} \alpha^{-s} z^s$, where $\alpha \in \mathbb{T}^d$.*

Proof. Let $\mathcal{C}_S$ be the closed cone generated by the trigonometric polynomials $p_{\alpha,S}$, $\alpha \in \mathbb{T}^d$. Clearly, $\mathcal{C}_S \subseteq \mathcal{B}_S$. Next, note that $q \in (\mathcal{C}_S)^*$ if only if $\langle q, p_{\alpha,S} \rangle = q(\alpha) \geq 0$ for all $\alpha \in \mathbb{T}^d$. But the latter is equivalent to the statement that $q \in \mathbb{T}\mathrm{Pol}_S^+$, and thus $(\mathcal{C}_S)^* = \mathbb{T}\mathrm{Pol}_S^+$. Taking duals we obtain from Theorem 1.3.2 that $\mathcal{C}_S = \mathcal{B}_S$. It remains to observe that each $p_{\alpha,S}$ generates an extreme ray of $\mathcal{B}_S$. This follows from the observation that the rank 1 positive semidefinite Toeplitz matrix $T_{p_{\alpha,S},\Lambda}$ cannot be written as a sum of positive semidefinite rank 1 matrices other than by using multiples of $T_{p_{\alpha,S},\Lambda}$ (see the proof of Proposition 1.1.4). □

As a consequence we obtain the following useful decomposition result.

Theorem 1.3.5 *Let $\Lambda \subset \mathbb{Z}^d$ be a finite set and put $S = \Lambda - \Lambda$. The trigonometric polynomial p belongs to $\mathcal{B}_S$ if and only if $T_{p,\Lambda}$ can be written as*

$$T_{p,\Lambda} = \sum_{j=1}^{r} \rho_j v_j v_j^*,$$

where $r \leq \mathrm{card} S$, $\rho_j > 0$ and $v_j = (\alpha_j^\lambda)_{\lambda \in \Lambda}$ with $\alpha_j \in \mathbb{T}^d$ for $j = 1, \ldots, r$.

Proof. Assume that $m > \mathrm{card} S$ and let $v = \sum_{j=1}^{m} \rho_j v_j v_j^*$, where $\rho_j > 0$ and $v_j = (\alpha_j^\lambda)_{\lambda \in \Lambda}$ with $\alpha_j \in \mathbb{T}^d$ for $j = 1, \ldots, m$. Since $m > \mathrm{card} S$, the matrices $v_j v_j^*$ are linearly dependent over $\mathbb{R}$. Without loss of generality, we may assume there exist $1 \leq l \leq m-1$ and nonnegative numbers $a_1, \ldots, a_l, b_{l+1}, \ldots, b_m$ such that

$$\sum_{i=1}^{l} a_i v_i v_i^* - \sum_{j=l+1}^{m} b_j v_j v_j^* = 0. \tag{1.3.9}$$

We order decreasingly $\frac{\rho_j}{b_j}$ for $b_j \neq 0$, and without loss of generality we assume that among them $\frac{\rho_m}{b_m}$ is the smallest. From (1.3.9) we obtain that

$$v_m v_m^* = \left(\sum_{i=1}^{l} a_i v_i v_i^* - \sum_{j=l+1}^{m-1} b_j v_j v_j^* \right) \frac{1}{b_m},$$

which implies that

$$\sum_{i=1}^{m} \rho_j v_j v_j^* = \sum_{i=1}^{l} \left(\rho_i + \frac{\rho_m a_i}{b_m} \right) v_i v_i^* + \sum_{j=l+1}^{m-1} \left(\rho_j - \frac{\rho_m b_j}{b_m} \right) v_j v_j^*.$$

Clearly all coefficients for $1 \leq i \leq l$ are positive. The minimality of $\frac{\rho_m}{b_m}$ implies that the coefficients for $l+1 \leq j \leq m-1$ are also positive, proving that every positive linear combination of matrices of the form $v_j v_j^*$ is a positive linear combination of at most card$S$ such matrices. Let $\mathcal{D}_S$ denote the cone of all matrices v of the form $\sum_{j=1}^r \rho_j v_j v_j^*$ with $\rho_j > 0$ and $r \leq \text{card}S$. Using a similar argument as for proving the closedness of $\mathcal{B}_G$ in Proposition 1.2.1, one can show that $\mathcal{D}_S$ is closed. The details of this are left as Exercise 1.6.17. Using Theorem 1.3.4, we have that $p \in \mathcal{B}_S$ if and only if $T_{p,\Lambda} \in \mathcal{D}_S$, which is equivalent to the statement of theorem. □

1.3.1 The extension property

An important question is when equality occurs between the cones we have introduced. More specifically, for what $\Lambda \subseteq \mathbb{Z}^d$ do we have that $\mathbb{T}\text{Pol}_\Lambda^2 = \mathbb{T}\text{Pol}_{\Lambda-\Lambda}^+$ (or equivalently, $\mathcal{B}_{\Lambda-\Lambda} = \mathcal{A}_\Lambda$)? When these equalities occur, we say Λ has the *extension property.* It is immediate that this property is translation invariant, namely, that if Λ has the extension property then any set of the form $a + \Lambda$, $a \in \mathbb{Z}^d$, has it. Also, if $h : \mathbb{Z}^d \to \mathbb{Z}^d$ is a group isomorphism, then Λ has the extension property if and only if $h(\Lambda)$ does; we leave this as an exercise (see Exercise 1.6.30).

In case $\Lambda = \{0, 1, \dots, n\} \subset \mathbb{Z}$, by the Fejér-Riesz factorization theorem (Theorem 1.1.5) we have that $\mathbb{T}\text{Pol}_\Lambda^2 = \mathbb{T}\text{Pol}_{\Lambda-\Lambda}^+$. This implies the following classical Carathéodory-Fejér theorem, which solves the so-called *truncated trigonometric moment problem.*

Theorem 1.3.6 *Let $c_j \in \mathbb{C}$, $j = -n, \dots, n$, with $c_j = \overline{c_{-j}}$, $j = 0, \dots, n$. Let T_n be the Toeplitz matrix $T_n = (c_{i-j})_{i,j=1}^n$. Then the following are equivalent.*

(i) There exists a finite positive Borel measure μ on $\mathbb{T}$ such that $c_k = \widehat{\mu}(k), k = -n, \dots, n,$.

(ii) The Toeplitz matrix T_n is positive semidefinite.

(iii) The Toeplitz matrix T_n can be factored as $T_n = RDR^*$, where R is a Vandermonde matrix,

$$R = \begin{pmatrix} 1 & 1 & \dots & 1 \\ \alpha_1 & \alpha_2 & \dots & \alpha_r \\ \vdots & \vdots & & \vdots \\ \alpha_1^n & \alpha_2^n & \dots & \alpha_r^n \end{pmatrix},$$

with $\alpha_j \in \mathbb{T}$ for $j = 1, \dots, r$, and $\alpha_j \neq \alpha_p$ for $j \neq p$, and D is a $r \times r$ positive definite diagonal matrix.

(iv) there exist $\alpha_j \in \mathbb{T}$, $j = 1, \dots, r$, with $\alpha_j \neq \alpha_p$, and $\rho_j > 0$, $j = 1, \dots, r$, such that

$$c_l = \sum_{j=1}^r \rho_j \alpha_j^l, \quad |l| \leq n. \tag{1.3.10}$$

In case T_n is singular, $r = \text{rank}T_n$, and the α_j are uniquely determined as the roots of the polynomial

$$p(z) = \det \begin{pmatrix} c_0 & \overline{c}_1 & \dots & \overline{c}_r \\ c_1 & c_0 & \dots & \overline{c}_{r-1} \\ \vdots & \vdots & \vdots & \vdots \\ c_{r-1} & c_{r-2} & \dots & \overline{c}_1 \\ 1 & z & \dots & z^r \end{pmatrix}.$$

In case T_n is nonsingular, one may choose $c_{n+1} = \overline{c}_{-n-1}$ such that $T_{n+1} = (c_{i-j})_{i,j=0}^{n+1}$ is positive semidefinite and singular, and apply the above to c_j, $|j| \leq n+1$. The measure μ can be chosen to equal $\mu = \sum_{j=1}^{r} \rho_j \delta_{\alpha_j}$, where ρ_j and α_j are as above.

Proof. Let $\Lambda = \{0, \dots, n\}$. By Theorem 1.1.5 we have that $\mathbb{T}\text{Pol}_\Lambda^2 = \mathbb{T}\text{Pol}_{\Lambda-\Lambda}^+$, and thus by duality $\mathcal{A}_\Lambda = \mathcal{B}_{\Lambda-\Lambda}$. Now (i) $\Rightarrow$ (ii) follows from Theorem 1.3.3. (ii) $\Rightarrow$ (iii) follows as $p \in \mathcal{A}_\Lambda$ implies $p \in \mathcal{B}_{\Lambda-\Lambda}$, which in turn implies (iii) by Theorem 1.3.5. The equivalence of (iii) and (iv) is immediate, which leaves the implication (iv) $\Rightarrow$ (i). The latter follows from Theorem 1.3.3. □

In Exercise 1.6.20 we give an algorithm on how to find a solution to the truncated trigonometric moment problem that is a finite positive combination of Dirac masses.

We can also draw the following corollary, which will be useful in Section 5.7.

Corollary 1.3.7 *Let $z_1, \dots, z_n \in \mathbb{C} \setminus \{0\}$. Define $\sigma_{-n}, \dots, \sigma_n$ via*

$$\sigma_k = \sum_{j=1}^{n} z_j^k, \ k = -n, \dots, n.$$

Then $z_1, \dots, z_n$ are all distinct and lie on the unit circle $\mathbb{T}$ if and only if $\sigma = (\sigma_{i-j})_{i,j=0}^n \geq 0$ and $\text{rank}\, \sigma = n$.

Proof. First suppose that $z_1, \dots, z_n$ are all distinct and lie on the unit circle. Put

$$V = \begin{pmatrix} 1 & 1 & \dots & 1 \\ z_1 & z_2 & \dots & z_n \\ \vdots & \vdots & & \vdots \\ z_1^n & z_2^n & \dots & z_n^n \end{pmatrix}.$$

Then $\sigma = VV^* \geq 0$, and $\text{rank}\, \sigma = \text{rank}\, V = n$.

Next suppose that $\sigma \geq 0$ and $\text{rank}\, \sigma = n$. By Theorem 1.3.6 we can write $\sigma = RDR^*$, with R and D as in Theorem 1.3.6 (where $r = n$). Put V as above, and

$$W = \begin{pmatrix} 1 & 1 & \dots & 1 \\ z_1^{-1} & z_2^{-1} & \dots & z_n^{-1} \\ \vdots & \vdots & & \vdots \\ z_1^{-n} & z_2^{-n} & \dots & z_n^{-n} \end{pmatrix}.$$

Then $\sigma = VW^T$. As $\operatorname{rank} \sigma = n$, we get that V and W must be of full rank. This yields that $z_1, \ldots, z_n$ are different. Next, as $\sigma = VW = RDR^*$ has a one-dimensional kernel, there is a nonzero row vector $y = \begin{pmatrix} p_0 & \cdots & p_n \end{pmatrix}$ such that $y\sigma = 0$, and y is unique up to a multiplying with a nonzero scalar. But then we get that $yV = 0 = yR$, yielding that the n different numbers $z_1, \ldots, z_n$ and the n different numbers $\alpha_1, \ldots, \alpha_n$ are all roots of the nonzero polynomial $p(z) = \sum_{i=0}^n p_i z^i$. But then we must have that $\{z_1, \ldots, z_n\} = \{\alpha_1, \ldots, \alpha_n\} \subset \mathbb{T}$, finishing the proof. □

In the one variable case ($d = 1$), the finite subsets of $\mathbb{Z}$ which have the extension property are characterized by the following result.

Theorem 1.3.8 *Let Λ be a nonempty finite subset of $\mathbb{Z}$. Then Λ has the extension property if and only if $\Lambda = \{a, a+b, a+2b, \ldots, a+kb\}$ for some $a, b, k \in \mathbb{Z}$ (namely, Λ is an arithmetic progression).*

Before we prove this result we will develop a useful technique to identify subsets of $\mathbb{Z}^d$ that do not have the extension property. For $\Lambda \subseteq \mathbb{Z}^d$ and $\emptyset \neq A \subset \mathbb{T}^d$, define

$$\Omega_\Lambda(A) = \{b \in \mathbb{T}^d \ : \ p(b) = 0 \text{ for all } p \in \mathrm{Pol}_\Lambda \text{ with } p|A \equiv 0\}, \qquad (1.3.11)$$

where $p|A \equiv 0$ is short for $p(a) = 0$ for all $a \in A$. For $a \in \mathbb{T}^d$ and $\Lambda \subseteq \mathbb{Z}^d$, denote $L_a = \mathrm{row}(a^\lambda)_{\lambda \in \Lambda}$. Note that if $p(z) = \sum_{\lambda \in \Lambda} p_\lambda z^\lambda$, then $p(a) = L_a P$, where $P = \mathrm{col}(p_\lambda)_{\lambda \in \Lambda}$. Thus one may think of L_a as the linear functional on Pol_Λ that evaluates a polynomial at the point $a \in \mathbb{T}^d$. With this notation, we have

$$\Omega_\Lambda(A) = \{b \in \mathbb{T}^d \ : \ L_b P = 0 \text{ for all } P \in \mathrm{Pol}_\Lambda \text{ with } L_a P = 0 \text{ for all } a \in A\}.$$

The following is easy to verify.

Lemma 1.3.9 *For $\Lambda \subseteq \mathbb{Z}^d$ and $\emptyset \neq A, B \subset \mathbb{T}^d$, we have*

(i) $A \subseteq \Omega_\Lambda(A)$;

(ii) $A \subseteq B$ *implies* $\Omega_\Lambda(A) \subseteq \Omega_\Lambda(B)$;

(iii) $\Omega_\Lambda(A) = \Omega_{\Lambda+a}(A)$ *for all* $a \in \mathbb{Z}^d$.

For a nonempty set $S \subseteq \mathbb{Z}^d$ we let $G(S)$ denote the smallest (by inclusion) subgroup of $\mathbb{Z}^d$ containing S.

In case A is a singleton and $G(\Lambda - \Lambda) = \mathbb{Z}^d$, it is not hard to determine $\Omega_\Lambda(A)$.

Proposition 1.3.10 *Let $\Lambda \subseteq \mathbb{Z}^d$ be such that $G(\Lambda - \Lambda) = \mathbb{Z}^d$. Then for all $a \in \mathbb{Z}^d$ we have that $\Omega_\Lambda(\{a\}) = \{a\}$.*

Proof. Without loss of generality $0 \in \Lambda$ (use Lemma 1.3.9 (iii)). Let $b \in \Omega_\Lambda(\{a\})$, and let $0 \neq \lambda \in \Lambda$ (which must exist as $G(\Lambda - \Lambda) = \mathbb{Z}^d$). Introduce the polynomial $p(z) = z^\lambda - a^\lambda$. As $b \in \Omega_\Lambda(\{a\})$, we must have that $p(b) = 0$.

Thus we obtain that $a^\lambda = b^\lambda$ for all $\lambda \in \Lambda$. By taking products and inverses, and using that $G(\Lambda - \Lambda) = \mathbb{Z}^d$, we obtain that $a^\lambda = b^\lambda$ for all $\lambda \in \mathbb{Z}^d$. In particular, this holds for $\lambda = e_i$, $i = 1, \ldots, d$, where $e_i \in \mathbb{Z}^d$ has all zeros except for a 1 in the ith position. It now follows that $a_i = b_i$, $i = 1, \ldots, d$, and thus $a = b$. □

Exercise 1.6.31 shows that when $G(\Lambda - \Lambda) \neq \mathbb{Z}^d$ then $\Omega_\Lambda(\{a\})$ contains more than one element.

Theorem 1.3.11 *Let $\Lambda \subseteq \mathbb{Z}^d$ be a finite set such that $G(\Lambda - \Lambda) = \mathbb{Z}^d$. If Λ has the extension property, then for all finite subsets A of $\mathbb{T}^d$ either $\Omega_\Lambda(A) = A$ or $\mathrm{card}\Omega_\Lambda(A) = \infty$.*

Before starting the proof of Theorem 1.3.11, we need several auxiliary results. The range and kernel of a matrix T are denoted by $\mathrm{Ran}\, T$ and $\mathrm{Ker}\, T$, respectively.

Lemma 1.3.12 *Let $T \geq 0$, $0 \neq v \in \mathrm{Ran}\, T$. Let x be so that $Tx = v$. Then $\mu(v) := x^*Tx > 0$ is independent of the choice of x. Moreover, $T - \frac{1}{\mu(v)} vv^* \geq 0$ and $x \in \mathrm{Ker}\,(T - \frac{1}{\mu(v)} vv^*)$.*

Proof. Clearly, $x^*Tx \geq 0$. If $x^*Tx = 0$, we get that $\|T^{\frac{1}{2}}x\|^2 = x^*Tx = 0$ and thus $Tx = T^{\frac{1}{2}}T^{\frac{1}{2}}x = 0$, which is a contradiction. Thus $x^*Tx > 0$. Next, if $Tx = v = T\widetilde{x}$, then $x - \widetilde{x} \in \mathrm{Ker}\, T$, and thus

$$x^*Tx - \widetilde{x}^*T\widetilde{x} = (x - \widetilde{x})Tx + \widetilde{x}^*T(x - \widetilde{x}) = 0.$$

Next,

$$\begin{pmatrix} T & v \\ v^* & \mu(v) \end{pmatrix} = \begin{pmatrix} I \\ x^* \end{pmatrix} T \begin{pmatrix} I & x \end{pmatrix} \geq 0,$$

and thus, by Lemma 1.2.5, $T - \frac{vv^*}{\mu(v)} \geq 0$.

Finally,

$$x^* \left(T - \frac{vv^*}{\mu(v)} \right) x = x^*Tx - \frac{(x^*Tx)^2}{\mu(v)} = 0,$$

and since $T - \frac{vv^*}{\mu(v)} \geq 0$ it follows that $x \in \mathrm{Ker}\,(T - \frac{vv^*}{\mu(v)})$. □

Lemma 1.3.13 *Let*

$$T = (c_{\lambda - \mu})_{\lambda, \mu \in \Lambda} \geq 0,$$

and

$$\Sigma_T = \{a \in \mathbb{T}^d \ : L_a P = 0 \text{ for all } P \in \mathrm{Ker}\, T\}.$$

Then $a \in \Sigma_T$ if and only if $L_a^* \in \mathrm{Ran}\, T$.

Proof. As $T = T^*$, we have $\mathrm{Ran}\, T = (\mathrm{Ker}\, T)^\perp$, and the lemma follows. □

The extension property implies that $\Sigma_T \neq \emptyset$ and has a strong consequence regarding the possible forms of the decompositions as in Theorem 1.3.5.

Proposition 1.3.14 *Let $\Lambda \subseteq \mathbb{Z}^d$ be a finite set that has the extension property, and let*

$$0 \neq T = (c_{\lambda-\mu})_{\lambda,\mu\in\Lambda} \geq 0.$$

Put $m = \text{rank}T$, and let Σ_T as in Lemma 1.3.13. Then $\Sigma_T \neq \emptyset$, and for any $b_1 \in \Sigma_T$ there exist $b_2, \dots, b_m \in \Sigma_T$ and $\mu_1, \dots, \mu_m > 0$ such that

$$T = \sum_{k=1}^{m} \frac{1}{\mu_k} L_{b_k}^* L_{b_k}.$$

Note that necessarily $L_{b_1}, \dots L_{b_m}$ are linearly independent.

Proof. We first show that $\Sigma_T \neq \emptyset$. As Λ has the extension property, we have that $\mathcal{B}_{\Lambda-\Lambda} = \mathcal{A}_\Lambda$. As $T \in \mathcal{A}_\Lambda = \mathcal{B}_{\Lambda-\Lambda}$, we obtain from Theorem 1.3.5 that we may represent T as

$$T = \sum_{k=1}^{r} \rho_k L_{a_k}^* L_{a_k},$$

where $a_1, \dots, a_r \in \mathbb{T}^d$ and $\rho_1, \dots, \rho_r > 0$. Now, if $P \in \text{Ker}\, T$, we get that $P^*TP = \sum_{k=1}^{r} \rho_k |L_{a_k} P|^2 = 0$ and thus $L_{a_k} P = 0$, $k = 1, \dots, r$. This yields that $a_1, \dots, a_r \in \Sigma_T$.

Let now $b_1 \in \Sigma_T$. By Lemma 1.3.13, $0 \neq L_{b_1}^* = Tx$ for some $x \neq 0$. If we put $\mu_1 = x^*Tx$, we have by Lemma 1.3.12 that $\mu_1 > 0$, $\widetilde{T} := T - \frac{1}{\mu_1} L_{b_1}^* L_{b_1} \geq 0$, and $x \in \text{Ker}\, \widetilde{T}$. In particular, since $Tx = L_{b_1} \neq 0$ and $\text{Ker}\, T \subseteq \text{Ker}\, \widetilde{T}$ (as $0 \leq \widetilde{T} \leq T$), we get that $\dim\, \text{Ker}\, \widetilde{T} = \dim\, \text{Ker}\, T + 1$, and thus $\text{rank}\, \widetilde{T} = \text{rank}\, T - 1$.

If $\widetilde{T} = 0$, we are done. If not, note that $\Sigma_{\widetilde{T}} \subseteq \Sigma_T$, and repeat the above with $\widetilde{T}$ instead of T. As in each step the rank decreases by 1, we are done in at most m steps. □

Lemma 1.3.15 *Let $x = (x_1, \dots, x_m)^T \in \mathbb{C}^m$, and suppose x has all nonzero components. Then there is no nonempty finite subset $F \subseteq \mathbb{C}^m$ with the properties that $0 \notin F$ and that for every $d_1, \dots, d_m > 0$ there exists $y = (y_1, \dots, y_m)^T \in F$ satisfying $\sum_{i=1}^{m} d_i x_i \overline{y}_i = 0$.*

Proof. We prove the claim by induction. When $m = 1$ the statement is obviously true. Suppose now that the result holds for m, and we prove it for $m+1$. Let $x = (x_1, \dots, x_{m+1})^T \in \mathbb{C}^{m+1}$ be given with $x_i \neq 0$, $i = 1, \dots, m+1$, and suppose $F \subset \mathbb{C}^{m+1}$ is finite. We can assume $(y_1, \dots, y_m) \neq (0, \dots, 0)$ for any $(y_1, \dots, y_m, y_{m+1})^T \in F$. Indeed, otherwise we would have that $d_{m+1} x_{m+1} \overline{y}_{m+1} \neq 0$ for all $d_{m+1} > 0$, which is impossible since $x_{m+1} \neq 0$ and $y_{m+1} \neq 0$ (as $0 \notin F$). But then we can apply our induction hypothesis, and conclude that there exist $d_1^0, \dots, d_m^0 > 0$ such that

$$\sum_{i=1}^{m} d_i^0 x_i \overline{y}_i \neq 0$$

for all $(y_1, \ldots, y_m, y_{m+1})^T \in F$. If the equation

$$\sum_{i=1}^{m} d_i^0 x_i \overline{y}_i + d_{m+1} x_{m+1} \overline{y}_{m+1} = 0$$

holds, then necessarily $y_{m+1} \neq 0$ and

$$d_{m+1} = -\frac{\sum_{i=1}^{m} d_i^0 x_i \overline{y}_i}{x_{m+1} \overline{y}_{m+1}}.$$

Since the right-hand side can take on only a finite number of values, we get that there must exists a $d_{m+1}^0 > 0$ so that

$$\sum_{i=1}^{m+1} d_i^0 x_i \overline{y}_i \neq 0$$

for all $(y_1, \ldots, y_m, y_{m+1})^T \in F$, proving our claim. □

We are now ready to prove Theorem 1.3.11.

Proof of Theorem 1.3.11. We do this by induction on $m = \text{card} A$. When $m = 1$ it follows from Proposition 1.3.10. Next suppose that the result has been proven for sets up to cardinality $m - 1$.

Let $A = \{a_1, \ldots, a_m\}$. If $L_{a_1}, \ldots, L_{a_m}$ are linearly dependent, then, without loss of generality, L_{a_m} belongs to the span of $L_{a_1}, \ldots, L_{a_{m-1}}$. But then, for $p(z) = \sum_{\lambda \in \Lambda} p_\lambda z^\lambda \in \text{Pol}_\Lambda$ and $P = \text{col}(p_\lambda)_{\lambda \in \Lambda}$, we have that $p(a_k) = L_{a_k} P = 0$, $k = 1, \ldots, m-1$, implies that $p(a_m) = L_{a_m} P = 0$. This gives that $\Omega_\Lambda(\{a_1, \ldots, a_m\}) = \Omega_\Lambda(\{a_1, \ldots, a_{m-1}\})$, and one can finish this case by using the induction assumption.

Next assume that $L_{a_1}, \ldots, L_{a_m}$ are linearly independent, and that the set $\Omega_\Lambda(\{a_1, \ldots, a_m\})$ is finite (otherwise we are done). By Lemma 1.3.9(iii) this implies that $\Omega_\Lambda(\{a_1, \ldots, a_{m-1}\})$ is a finite set as well. As $L_{a_1}, \ldots, L_{a_m}$ are linearly independent, we can find $U_1, \ldots, U_m$ so that

$$\begin{pmatrix} L_{a_1} \\ \vdots \\ L_{a_m} \end{pmatrix} \begin{pmatrix} U_1 & \cdots & U_m \end{pmatrix} = I_m,$$

or, equivalently, $L_{a_i} U_j = \delta_{ij}$ with δ_{ij} being the Kronecker delta. Let $b_1 \in \Omega_\Lambda(\{a_1, \ldots, a_m\})$. We claim that $L_{b_1} \in \text{Span}\{L_{a_1}, \ldots, L_{a_m}\}$. It suffices to show that $L_{b_1} P = 0$ for all P with $L_{a_1} P = \cdots = L_{a_m} P = 0$. But this follows directly from the definition of $\Omega_\Lambda(\{a_1, \ldots, a_m\})$.

Next, suppose that $L_{b_1} U_i = 0$ for some $i = 1, \ldots, m$. Without loss of generality, $L_{b_1} U_m = 0$. As L_{b_1} is in the span of $L_{a_1}, \ldots, L_{a_m}$, we may write $L_{b_1} = \sum_{i=1}^m z_i L_{a_i}$ for some complex numbers z_i. But then $0 = L_{b_1} U_m = \sum_{i=1}^m z_i L_{a_i} U_m = z_m$. Thus it follows that L_{b_1} lies in the span of $L_{a_1}, \ldots, L_{a_{m-1}}$. This, as before, gives that $b_1 \in \Omega_\Lambda(\{a_1, \ldots, a_{m-1}\})$. As $\text{card}\Omega_\Lambda(\{a_1, \ldots, a_{m-1}\}) < \infty$ we get by the induction assumption that $b_1 \in \Omega_\Lambda(\{a_1, \ldots, a_{m-1}\}) = \{a_1, \ldots, a_{m-1}\}$, and we are done.

Finally, we are in the case when $L_{b_1}U_i \neq 0$ for all $i = 1, \dots, m$. We will show that this case cannot occur as we will reach a contradiction. Let $D = \text{diag}(d_i)_{i=1}^m > 0$, and consider

$$T(D) = \sum_{i=1}^m d_i L_{a_i}^* L_{a_i}.$$

Note that $\Sigma_{T(D)} = \Omega_\Lambda(\{a_1, \dots, a_m\})$, where Σ_T is defined in Proposition 1.3.14. By Proposition 1.3.14 there exist $\mu_k = \mu_k(D) > 0$, $k = 1, \dots, m$, and $b_2(D), \dots, b_m(D) \in \Omega_\Lambda(\{a_1, \dots, a_m\})$ such that

$$T(D) = \frac{1}{\mu_1(D)} L_{b_1}^* L_{b_1} + \sum_{k=2}^m \frac{1}{\mu_k(D)} L_{b_k(D)}^* L_{b_k(D)}.$$

Thus

$$\sum_{i=1}^m d_i L_{a_i}^* L_{a_i} = \frac{1}{\mu_1(D)} L_{b_1}^* L_{b_1} + \sum_{k=2}^m \frac{1}{\mu_k(D)} L_{b_k(D)}^* L_{b_k(D)}.$$

Multiplying this equation with $U = \begin{pmatrix} U_1 & \cdots & U_m \end{pmatrix}$ on the right and with U^* on the left, we get that

$$D = Q^* \text{diag}(\mu_i(D))_{i=1}^m Q,$$

where

$$Q = \begin{pmatrix} L_{b_1} \\ L_{b_2(D)} \\ \vdots \\ L_{b_m(D)} \end{pmatrix} U.$$

Thus

$$I_m = (Q^* \text{diag}(\mu_i(D))_{i=1}^m)(QD^{-1}),$$

and as all matrices in this equation are $m \times m$ we also get

$$I_m = (QD^{-1})(Q^* \text{diag}(\mu_i(D))_{i=1}^m).$$

Looking at the $(1,2)$ entry of this equality (remember $m \geq 2$) and dividing by $\mu_2(D) > 0$, we get that for all for all $D = \text{diag}(d_i)_{i=1}^m > 0$ there exists $b_2(D) \in \Omega_\Lambda(\{a_1, \dots, a_m\})$ such that

$$\begin{pmatrix} \frac{1}{d_1} L_{b_1} U_1 & \cdots & \frac{1}{d_m} L_{b_1} U_m \end{pmatrix} U^* L_{b_2(D)}^* = 0.$$

As $L_{b_1}U_i \neq 0$, $i = 1, \dots, m$, we get by Lemma 1.3.15 that there must be infinitely many different vectors $U^* L_{b_2(D)}$. But as each $b_2(D)$ lies in the finite set $\Omega_\Lambda(\{a_1, \dots, a_m\})$, we have a contradiction. □

Corollary 1.3.16 *Assume $\Lambda \subset \mathbb{Z}^d$ is a finite subset such that $G(\Lambda - \Lambda) = \mathbb{Z}^d$. If there exist k linearly independent polynomials $p_1, \dots, p_k \in \text{Pol}_\Lambda$ such that the set $A = \cap_{i=1}^k \{a \in \mathbb{T}^d : p_i(a) = 0\}$ is a finite set with cardinality greater than $\text{card}\Lambda - k$, then Λ fails to have the extension property.*

Proof. The card$A$ × cardΛ matrix $\text{col}(L_a)_{a\in A}$ has a kernel of dimension at least k, and thus rank $\text{col}(L_a)_{a\in A} \leq \text{card}\Lambda - k < \text{card}A$. Thus the set of linear functionals $\{L_a\}_{a\in A}$ are linearly dependent. Let $\{a_1, \ldots, a_m\}$ be a maximal subset of A for which the set $\{L_{a_i}\}_{i=1}^m$ is linearly independent. Then $A = \Omega_\Lambda(\{a_1, \ldots, a_m\})$ is a finite set of cardinality greater than card$A$. Thus by Theorem 1.3.11, Λ does not have the extension property. □

We now have the techniques to easily prove Theorem 1.3.8.

Proof of Theorem 1.3.8. Since having the extension property is a translation invariant property, we may assume that $a = 0$. When $b = 1$, $\Lambda = \{0, 1, \ldots, k\}$ has the extension property by Theorem 1.3.6. Next replacing z by z^b, it easily follows that $\Lambda = \{0, b, 2b, \ldots, kb\}$ has the extension property.

Assume now that $\Lambda \subset \mathbb{Z}$ has the extension property, $G(\Lambda - \Lambda) = \mathbb{Z}$, 0 is the smallest element of Λ, and N is the largest. Then $z^N - 1 \in \text{Pol}_\Lambda$ has N roots on $\mathbb{T}$, and consequently Corollary 1.3.16 implies that Λ has at least $N + 1$ elements. Thus $\Lambda = \{0, 1, \ldots, N\}$. If $G(\Lambda - \Lambda) \neq \mathbb{Z}$, then, letting b be the greatest common divisor of the nonzero elements of Λ, it is easy to see that that $G(\Lambda - \Lambda) = b\mathbb{Z}$. By considering polynomials in z^b as opposed to polynomials in z, one easily reduces it to the above case and sees that Λ must equal $\{0, b, 2b, \ldots, kb\}$, where $k = \frac{N}{b}$. □

In the multivariable case we consider finite sets Λ of the form

$$R(N_1, \ldots, N_d) := \{0, \ldots, N_1\} \times \cdots \times \{0, \ldots, N_d\}, \tag{1.3.12}$$

where R stands for rectangular.

Now that the one-variable case has been settled, let us first consider a three-variable case.

Theorem 1.3.17 *The extension property fails for* $\Lambda = R(N_1, N_2, N_3)$, *for every* $N_1, N_2, N_3 \geq 1$.

Proof. The set of the common zeros of the polynomials $(z_1 - 1)(z_2 z_3 + 1)$, $(z_2 - 1)(z_1 z_2 + 1)$, and $(z_3 - 1)(z_1 z_2 + 1)$ consist of the points in $\mathbb{T}^3$:

$$\alpha_1 = (1, 1, 1),\ \alpha_2 = (1, -1, -1),\ \alpha_3 = (-1, 1, -1),$$

$$\alpha_4 = (-1, -1, 1),\ \alpha_5 = (i, i, i),\ \alpha_6 = (-i, -i, -i).$$

This implies that the set of common zeros in $\mathbb{T}^3$ of the polynomials $(z_1^{N_1} - 1)(z_2^{N_2} z_3^{N_3} + 1)$, $(z_2^{N_2} - 1)(z_1^{N_1} z_3^{N_3} + 1)$, and $(z_3^{N_3} - 1)(z_1^{N_1} z_2^{N_2} - 1)$ belonging to Pol_Λ has cardinality $6N_1N_2N_3$. Since card$\Lambda = (N_1 + 1)(N_2 + 1)(N_3 + 1)$, Corollary 1.3.16 with $k = 3$ that the extension property fails for $R(N_1, N_2, N_3)$ whenever

$$6N_1N_2N_3 > (N_1 + 1)(N_2 + 1)(N_3 + 1) - 3,$$

an inequality that is true when $N_i \geq 1$, $i = 1, 2, 3$. □

The remainder of this section is devoted to proving the following characterization of all the finite subsets of $\mathbb{Z}^2$ that have the extension property. We say that two finite subsets Λ_1 and Λ_2 of $\mathbb{Z}^2$ are *isomorphic* if there exists a group isomorphism Φ on $\mathbb{Z}^2$ so that $\Phi(\Lambda_1) = \Lambda_2$.

Theorem 1.3.18 *A finite set $\Lambda \subseteq \mathbb{Z}^2$ with $G(\Lambda - \Lambda) = \mathbb{Z}^2$ has the extension property if and only if for some $a \in \mathbb{Z}^2$ the set $\Lambda - a$ is isomorphic to $R(0,n)$, $R(1,n)$, or $R(1,n) \setminus \{(1,n)\}$, for some $n \in \mathbb{N}$.*

We prove this result in several steps. We use the following terminology. Let $R \subset \mathbb{Z}^d$. We say that $(c_k)_{k\in R-R}$ is a *positive semidefinite sequence* with respect to R if $(c_{k-l})_{k,l\in L} \geq 0$ for all finite subsets L of R. Clearly, if R is finite it suffices to check whether $(c_{k-l})_{k,l\in R} \geq 0$. Often the choice of the set R is clear, in which case we will just say that $(c_k)_{k\in R-R}$ is a positive semidefinite sequence.

Theorem 1.3.19 *For every $n \geq 1$, $R(1,n)$ has the extension property.*

Proof. Let $S = R(1,n) - R(1,n)$ and let $(c_{kl})_{(k,l)\in S}$ be a positive semidefinite sequence. Thus

$$\widetilde{C}_n = \begin{pmatrix} C_0 & C_1^* & \cdots & C_{n-1}^* & C_n^* \\ C_1 & C_0 & \ddots & & C_{n-1}^* \\ \vdots & \ddots & \ddots & \ddots & \vdots \\ C_{n-1} & & \ddots & \ddots & C_1^* \\ C_n & C_{n-1} & \cdots & C_1 & C_0 \end{pmatrix} \geq 0, \tag{1.3.13}$$

where

$$C_j = \begin{pmatrix} c_{0j} & c_{1j} \\ c_{-1,j} & c_{0j} \end{pmatrix}, \quad 0 \leq j \leq n.$$

Consider now the partial matrix

$$\begin{pmatrix} c_{00} & P^*C_1^* & \cdots & P^*C_n^* & X^* \\ C_1P & C_0 & \cdots & C_{n-1}^* & C_n^* \\ \vdots & \vdots & \ddots & \vdots & \vdots \\ C_nP & C_{n-1} & \cdots & C_0 & C_1^* \\ X & C_n & \cdots & C_1 & C_0 \end{pmatrix}, \tag{1.3.14}$$

where $P = \binom{0}{1}$ and X is the unknown. Since $\widetilde{C}_n$ is positive semidefinite, the partial matrix (1.3.14) is partially positive semidefinite. By Theorem 1.2.10 (i) $\Rightarrow$ (ii) we can find an $X = \binom{\beta}{\alpha}$ so that (1.3.14) is positive semidefinite. Using the Toeplitz structure of C_j, $0 \leq j \leq n$, it is now not hard to see that the following matrix is positive semidefinite as well:

$$\begin{pmatrix} C_0 & C_1^* & \cdots & C_n^* & Y^* \\ C_1 & C_0 & \cdots & C_{n-1}^* & C_n^*Q \\ \vdots & \vdots & \ddots & \vdots & \vdots \\ C_n & C_{n-1} & \cdots & C_0 & C_1^*Q \\ Y & Q^*C_n & \cdots & Q^*C_1 & c_{00} \end{pmatrix}, \tag{1.3.15}$$

where $Q = \binom{1}{0}$ and $Y = \begin{pmatrix} \alpha & \beta \end{pmatrix}$. Indeed, (1.3.15) may be obtained from (1.3.14) by reversing the rows and columns and taking complex conjugates entrywise. Next consider

$$\widetilde{C}_{n+1} = \begin{pmatrix} C_0 & C_1^* & \cdots & C_n^* & C_{n+1}^* \\ C_1 & C_0 & \cdots & C_{n-1}^* & C_n^* \\ \vdots & \vdots & \ddots & \vdots & \vdots \\ C_n & C_{n-1} & \cdots & C_0 & C_1^* \\ C_{n+1} & C_n & \cdots & C_1 & C_0 \end{pmatrix},$$

where

$$C_{n+1} = \begin{pmatrix} \alpha & \beta \\ \gamma & \alpha \end{pmatrix}$$

with γ as an unknown. As (1.3.14) and (1.3.15) are positive semidefinite, we have that $\widetilde{C}_{n+1}$ is partially positive semidefinite. Applying again Theorem 1.2.10 (i) $\Rightarrow$ (ii) we can find a γ so that $\widetilde{C}_{n+1}$ is positive semidefinite. Define now $c_{0,n+1} = \alpha$, $c_{1,n+1} = \beta$, and $c_{-1,n+1} = \gamma$. This way we have extended the given positive sequence to a positive sequence supported in the set $R(1, n+1) - R(1, n+1)$. We repeat successively the previous construction for $n+1, n+2, \ldots$. This process produces a positive semidefinite extension of the given sequence supported in the infinite band $I_1 = R_1 - R_1$, where $R_1 = \{0, 1\} \times \mathbb{N}$. Thus, by Exercise 1.6.34, $R(1, n)$ has the extension property. This completes the proof. $\square$

In Exercise 2.9.33 we will point out an alternative proof for Theorem 1.3.19 that will make use of the Fejér-Riesz factorization theorem (Theorem 2.4.16).

Our next result introduces a new class of sets with the extension property.

Theorem 1.3.20 *For every $n \geq 1$, $R(1, n) \backslash \{(1, n)\}$ has the extension property.*

Proof. Let $(c_{kl})_{(k,l) \in R(1,n) \backslash \{(1,n\} - R(1,n) \backslash \{(1,n)\}}$ be an arbitrary positive semidefinite sequence. By Theorem 1.3.19, it is sufficient to prove that the sequence can be extended to a positive semidefinite sequence supported in $S = R(1, n) - R(1, n)$. The Toeplitz form associated with this sequence is

$$\widetilde{D}_n = \begin{pmatrix} C_0 & C_1^* & \cdots & C_{n-1}^* & D_n^* \\ C_1 & C_0 & \cdots & C_{n-2}^* & D_{n-1}^* \\ \vdots & \vdots & \ddots & \vdots & \vdots \\ C_{n-1} & C_{n-2} & \cdots & C_0 & D_1^* \\ D_n & D_{n-1} & \cdots & D_1 & c_{00} \end{pmatrix},$$

where

$$C_j = \begin{pmatrix} c_{0j} & c_{1j} \\ c_{-1,j} & c_{0j} \end{pmatrix}$$

and $D_j = \begin{pmatrix} c_{0j} & c_{1j} \end{pmatrix}$.

It is easy to observe that deleting the first row and first column (resp., the last row and last column) of the matrix (1.3.13) of a sequence supported in $R(1,n) - R(1,n)$, one obtains (a permuted version of) the matrix $\widetilde{D}_n$. Thus every positive semidefinite matrix $\widetilde{D}_n$ can be extended to a positive semidefinite matrix of the form (1.3.13), and this completes the proof. $\square$

We would like to make a remark at this point.

Remark 1.3.21 Theorems 1.3.19 and 1.3.20 are true for scalar matrices only. They are not true when the entries are 2×2 or larger. Take for instance in Theorem 1.3.20, $\Lambda = \{(0,0),(0,1),(1,0)\}$, $C_{00} = \left(\begin{smallmatrix} 1 & 0 \\ 0 & 1 \end{smallmatrix}\right)$, and C_{01} and C_{10} to be two noncommuting unitary matrices. Then, in order to be positive semidefinite with respect to Λ, by Lemma 1.2.5, a sequence must satisfy $C_{-1,1} = C_{01}^* C_{10}$. We will show that no C_{11} exists for which the matrix of the sequence restricted to $\Lambda' = \{(0,0),(0,1),(1,0),(1,1)\}$ is positive semidefinite. Indeed, by a Schur complement type argument, we must have that C_{11} is unitary and $C_{11} = C_{01}C_{10} = C_{10}C_{01}$. The last equality does not hold, because C_{01} and C_{10} do not commute by assumption.

Theorem 1.3.22 *The set $R(N_1, N_2)$ has the extension property if and only if* $\min(N_1, N_2) = 1$.

We first need an auxiliary result.

Lemma 1.3.23 *The real polynomial*

$$F(s,t) = s^2t^2(s^2 + t^2 - 1) + 1$$

is strictly positive but it is not the sum of squares of polynomials with real coefficients.

Proof. It is an easy exercise (see Exercise 1.6.25) to show that F takes on the minimum value $\frac{26}{27}$ at $s = \pm\frac{1}{\sqrt{3}}$ and $t = \pm\frac{1}{\sqrt{3}}$.

Assume that

$$F = F_1^2 + \cdots + F_n^2 \tag{1.3.16}$$

for some real polynomials $F_1,\ldots,F_n$. From $F(s,0) = F(0,t) = 1$ it follows that $F_i(s,0)$ and $F_i(0,t)$ are constant for $i = 1,\ldots,n$, so

$$F_i(s,t) = a_i + stH_i(s,t) \tag{1.3.17}$$

for some constant a_i and some polynomial H_i of degree at most one. Substituting (1.3.17) into (1.3.16) and comparing the coefficients, we obtain

$$s^2t^2(s^2+t^2-1) = s^2t^2 \sum_{k=1}^{n} H_i^2(s,t).$$

This is a contradiction as the right side is always nonnegative while the left is negative if $st \neq 0$ and $s^2 + t^2 < 1$. $\square$

Proof of Theorem 1.3.22. We prove the theorem by showing that for $\Lambda = R(N_1, N_2)$ and $S = \Lambda - \Lambda$, $\mathbb{T}\mathrm{Pol}_S^+ \neq \mathbb{T}\mathrm{Pol}_\Lambda$.

Let X be the linear space of all real polynomials of the form $F(s,t) = \sum_{m=0}^{2N_1}\sum_{n=0}^{2N_2} a_{mn}s^m t^n$. Every $p \in \mathbb{T}\mathrm{Pol}_S$ is of the form

$$p(z,w) = \sum_{m=-N_1}^{N_1}\sum_{n=-N_2}^{N_2} c_{mn}z^m w^n$$

with $c_{-m,-n} = \overline{c}_{mn}$. Define the linear map $\Psi : \mathbb{T}\mathrm{Pol}_S \to X$ by

$$(\Psi(f))(s,t) = (1+s^2)^{N_1}(1+t^2)^{N_2} f\left(\frac{s+i}{s-i}, \frac{t+i}{t-i}\right).$$

It is clear Ψ is one-to-one, and since $\dim_{\mathbb{R}}(\mathbb{T}\mathrm{Pol}_S) = \dim_{\mathbb{R}} X = (2N_1 + 1)(2N_2+1)$, it follows that Ψ is a linear isomorphism of $\mathbb{T}\mathrm{Pol}_S$ onto X.

Assume that $f \in \mathbb{T}\mathrm{Pol}^2_\Lambda$. Then $f = \sum_{j=1}^r |g_j|^2$, where each $g_j \in \mathrm{Pol}_\Lambda$. Define for $j = 1, \ldots, r$,

$$G_j(s,t) = (s-i)^{N_1}(t-i)^{N_2} g_j\left(\frac{s+i}{s-i}, \frac{t+i}{t-i}\right).$$

Each G_j is a complex polynomial of degree at most N_1 in s and at most N_2 in t, and

$$(\Psi(f))(s,t) = \sum_{j=1}^r |G_j(s,t)|^2.$$

Define $F = \Psi(f)$ and let $G_j = u_j + iv_j$, where u_j and v_j are polynomials with real coefficients. We have then $F \in X$, and $F = \sum_{j=1}^r (u_j^2 + v_j^2)$, so F is a sum of squares. This cannot happen when F is the polynomial in Lemma 1.3.23, and this implies $\mathbb{T}\mathrm{Pol}^+_S \neq \mathbb{T}\mathrm{Pol}^2_\Lambda$. □

Theorem 1.3.24 *Let $\Lambda \subset \mathbb{Z}^2$ be a finite subset containing the points $(0,0)$, $(m,0)$, $(0,n)$, and (m,n), where $m,n \geq 2$. If $\mathrm{card}\Lambda \leq (m+1)(n+1)$ and $G(S) = \mathbb{Z}^2$, then the extension property fails for Λ. In particular, the extension property fails for $R(m,n)$ for $m,n \geq 2$.*

Proof. The polynomial $1 + z_1 + z_2$ has two roots on $\mathbb{T}^2$, $(\alpha, \overline{\alpha})$ and $(\overline{\alpha}, \alpha)$, where $\alpha = e^{\frac{2\pi i}{3}}$. Therefore, the polynomial $1 + z_1^m + z_2^n \in \mathrm{Pol}_\Lambda$ has $2mn$ roots on $\mathbb{T}^2$. If $m,n \geq 2$ and either $m \geq 3$ or $n \geq 3$, we have that $\mathrm{card}(\Lambda) \leq (m+1)(n+1) \leq 2mn$, and so Λ fails the extension property by Corollary 1.3.16. When $m = n = 2$, the condition in Corollary 1.3.16 with $k = 2$ is verified by the polynomials $1 - z_1^2 z_2^2$ and $z_1^2 - z_2^2$, which have 8 common zeros on $\mathbb{T}^2$, proving $R(2,2)$ also fails the extension property. □

A *partial Toeplitz matrix* is a partial matrix that is Toeplitz at the extent to which it is specified; that is, all specified entries lie along certain *specified diagonals*, and all the entries along a specified diagonal have the same value. Positive semidefinite Toeplitz completions are of special interest. The trigonometric moment problem is equivalent to the fact that every partially positive Toeplitz matrix with entries specified in a band $|i-j| \leq k$, for some specified k, admits a positive Toeplitz completion.

By the *pattern* of a partial Toeplitz matrix we mean the set of its specified diagonals. The main diagonal is always assumed to be specified. Thus the pattern of a partially positive semidefinite Toeplitz matrix can be considered to be a subset of $\{1, 2, \ldots, n\}$. A pattern P is said to be *(positive semidefinite) completable* if every partial positive Toeplitz matrix with pattern P admits a positive semidefinite completion. We will use the following result.

Theorem 1.3.25 *A pattern $P \subseteq \{1, 2, \ldots, n\}$ is completable if and only if it is an arithmetic progression.*

Proof. The proof may be found in [336]. □

Although the statement above and Theorem 1.3.8 seem to be quite close, they are rather different. For instance, if we let $\Lambda = P = \{1, 2, 4\}$, then Theorem 1.3.8 yields that there exist c_k, $k \in \Lambda - \Lambda = \{-3, \ldots, 3\}$, such that

$$\begin{pmatrix} c_0 & c_{-1} & c_{-3} \\ c_1 & c_0 & c_{-2} \\ c_3 & c_2 & c_0 \end{pmatrix} \geq 0$$

and

$$\begin{pmatrix} c_0 & c_{-1} & c_{-2} & c_{-3} \\ c_1 & c_0 & c_{-1} & c_{-2} \\ c_2 & c_1 & c_0 & c_{-1} \\ c_3 & c_2 & c_1 & c_0 \end{pmatrix} \not\geq 0.$$

On the other hand, Theorem 1.3.25 yields in this case that there exist $c_k = \overline{c_k}$, $k = 0, 1, 2, 4$, so that the partial matrix

$$\begin{pmatrix} c_0 & c_{-1} & c_{-2} & ? & c_{-4} \\ c_1 & c_0 & c_{-1} & c_{-2} & ? \\ c_2 & c_1 & c_0 & c_{-1} & c_{-2} \\ ? & c_2 & c_1 & c_0 & c_{-1} \\ c_4 & ? & c_2 & c_1 & c_0 \end{pmatrix}$$

is partially positive semidefinite but no positive semidefinite Toeplitz completion exists.

Next, we will introduce a canonical form for a finite set $\Lambda \subset \mathbb{Z}^2$. We will be interested in those satisfying $G(\Lambda - \Lambda) = \mathbb{Z}^2$. Exercise 1.6.35 gives an easy method to determine whether a finite set $\Lambda \subset \mathbb{Z}^2$ satisfies $G(\Lambda - \Lambda) = \mathbb{Z}^2$ or not. By Exercise 1.6.30, a finite subset $\Lambda \subset \mathbb{Z}^2$ has the extension property if and only if every translation of Λ by a vector in $a \in \mathbb{Z}^2$ and every set isomorphic to Λ has the same property.

Let $d_1 = \max\{p_1 - p_2 : (p_1, q), (p_2, q) \in \Lambda\}$, $d_2 = \max\{q_1 - q_2 : (p, q_1), (p, q_2) \in \Lambda\}$, and $d = \max\{d_1, d_2\}$. By using a translation and maybe also interchanging the order of coordinates in $\mathbb{Z}^2$, we may assume that $(0, 0), (d, 0) \in \Lambda$, and that for each $k < 0$, $\max\{p_1 - p_2 : (p_1, k), (p_2, k) \in \Lambda\} < d$. If Λ has the above properties, then we say it is in *canonical form*. This notion plays a crucial role in our considerations. For Λ in canonical form, let $m = -\min\{l : (k, l) \in \Lambda\}$, and $M = \max\{l : (k, l) \in \Lambda\}$. Without loss of generality, we assume that $M \geq m$ for every Λ in canonical form.

If Λ is in canonical form and $M = m = 0$, then Λ is isomorphic to a subset of $\mathbb{Z}$ for which we may apply Theorem 1.3.8. Therefore we may assume that $M \geq 1$.

Lemma 1.3.26 *Suppose that Λ is in canonical form and has the extension property. Let $\Sigma = \{k : (k,0) \in \Lambda - \Lambda\}$ and $\Lambda(l) = \{(k,l) \in \Lambda\}$, $l \in \mathbb{Z}$. Then Σ forms an arithmetic progression. Moreover, there exists an l such that $\Sigma = \{k : (k,0) \in \Lambda(l) - \Lambda(l)\}$.*

Proof. Let $N = \text{card}\Sigma$. Define the sequence $\{c_{k,l}\}_{(k,l)\in\Lambda-\Lambda}$ by $c_{k,l} = c_k$ whenever $l = 0$ and $c_{k,l} = 0$ whenever $l \neq 0$. The sequence c_k will be specified later in the proof. The matrix of the sequence $\{c_{k,l}\}_{(k,l)\in\Lambda-\Lambda}$ can be written in a block diagonal form with $m + M + 1$ diagonal blocks (one for each row in Λ). Let Θ_j, $j = -m, \ldots, M$ denote these diagonal blocks. All the entries of the matrices Θ_j are terms of the sequence $\{c_k\}$. The sequence $\{c_{k,l}\}_{(k,l)\in\Lambda-\Lambda}$ is positive semidefinite if and only if all matrices Θ_j are positive semidefinite. The matrices Θ_j can be viewed also as fully specified principal submatrices of the partial Toeplitz matrix $(c_{s-t})_{s,t=0}^{d}$. If $\{k_j\}_{j=1}^{N} = \Sigma$ is not an arithmetic progression, then by Theorem 1.3.25 we can choose c_k such that all the matrices Θ_j are positive semidefinite but the sequence c_k does not have a positive semidefinite Toeplitz extension. This implies that the sequence $\{c_{k,l}\}$ constructed earlier in the proof is positive semidefinite and does not have a positive semidefinite extension. This proves that Σ must be an arithmetic progression.

Each Θ_l is an $n(l) \times n(l)$ matrix, where $n(l) = \text{card } \Lambda(l)$ and $c_{k,0}$ is an entry of Θ_l if and only if $(k,0) \in \Lambda(l) - \Lambda(l)$. Let $n = \max_l n(l)$. Then we can redefine the sequence $\{c_k\}$ used above with $c_0 = 1$ and $c_k = -1/(n-1)$ if $k \neq 0$. Then each matrix Θ_j is positive semidefinite by diagonal dominance. It can easily be verified that if $N > n$ the sequence $\{c_k\}$ is not positive semidefinite. Therefore we must have $n = N$. But then the theorem follows since $\Lambda(l) - \Lambda(l) \subseteq \Sigma$ for all l. □

Lemma 1.3.27 *Suppose that Λ is in canonical form, $G(\Lambda - \Lambda) = \mathbb{Z}^2$, and Λ has the extension property. Then there exist p and q such that*

$$(p,q), (p+1,q), \ldots, (p+d,q) \in \Lambda.$$

In the remainder of the section $\{(p,q), (p+1,q), \ldots, (p+d,q)\}$ will be referred to as the *full row*.

Proof. The lemma is immediate for $d \leq 1$, so we assume that $d > 1$.

Let k be such that $(k, M) \in \Lambda$. Since the polynomial

$$1 + w^d + z^M w^k = 0 \tag{1.3.18}$$

has $2Md$ zeros on $\mathbb{T}^2$, then by Theorem 1.3.16, Λ does not have the extension property unless it contains more than $2Md$ points.

Assume that no row Λ contains $d+1$ elements. Then the maximum number of elements in Λ is $(M + m + 1)d$. The inequality $(M + m + 1)d \leq 2Md$ is true if $m \leq M - 1$, therefore the lemma is certainly true unless $M = m$.

Let q be a row of Λ containing the largest number of elements among all rows. If $\{(l_0,q),(l_1,q),\ldots,(l_m,q)\}$ is the set of elements on this row, then by Lemma 1.3.26, $l_0,l_1,\ldots,l_m$ must be an arithmetic progression. Therefore, if Λ does not have a full row, each row of Λ contains at most $\frac{d}{2}+1$ elements. All rows $q<0$ contain then at most $\frac{d}{2}$ elements. Assuming $M=m$, we get that the maximal numbers of elements in Λ is $(M+1)(\frac{d}{2}+1)+M\frac{d}{2}$. Using again the polynomial in (1.3.18), we have that Λ does not have the extension property unless $(M+1)(\frac{d}{2}+1)+M\frac{d}{2}\le 2Md$, which holds if and only if

$$\frac{M+1}{M-\frac{1}{2}}\le d.$$

The function $f(M)=\frac{M+1}{M-\frac{1}{2}}$ is strictly decreasing. Since $f(2)=2$, the inequality holds for any $d>1$ if $M\ge 2$. It remains to consider the case when $M=1$. Since $f(1)=4$, we can restrict the consideration to the case when $d=2$ or $d=3$. If $d=3$, there will be at most 2 points in each row by Lemma 1.3.26 unless there is a row with four elements. There are at most 5 elements in Λ in this case. By counting the number of zeros on $\mathbb{T}^2$ of the polynomial (1.3.18), this possibility is ruled out. If $d=2$, the same reasoning holds if there are fewer than 5 elements in Λ. On the other hand, if there are 5 elements in Λ, three of them must be $(k,1)$, $(k+2,1)$, and $(w,-1)$. Since the polynomial $z^kw+z^{k+2}w+z^lw^{-l}=0$ has 8 zeros on $\mathbb{T}^2$, the proof is complete. □

Lemma 1.3.28 *Suppose* $\{(0,0),(1,0)\}\subset\Lambda$ *and that* Λ *is in canonical form. Then* Λ *does not have the extension property unless it contains elements of the form* (p_j,j) *for every* j *such that* $-m\le j\le M$.

Proof. Suppose there exists $-m\le q\le M$ such that there are no elements of the form (p,q) in Λ.

Since $G(\Lambda-\Lambda)=\mathbb{Z}^2$ it is clear that the set $Q=\{q:(p,q)\in\Lambda \text{ for some } p\}$ is not an arithmetic progression. Then it follows from Theorem 1.3.8 that there exists a data set $\{c_j, j\in Q-Q\}$, where $c_0=1$ without loss of generality, such that

$$A=\left(c_{q_1-q_2}\right)_{q_1,q_2\in Q}$$

is positive semidefinite, but its elements are not extendable to a positive semidefinite sequence on the set $[-m,M]$.

Let $J_{k\times k}$ be the $k\times k$ matrix with all entries entries equal to one, where

$$k=\max_{j,j_1,j_2}\{\,|p_{j_1}-p_{j_2}|:(p_{j_1},j),(p_{j_2},j)\in\Lambda\,\}.$$

The *Kronecker product* $B=A\otimes J_{k\times k}$ is a positive semidefinite matrix. Moreover, there exists a principal submatrix of B which gives a positive semidefinite sequence $(c_k)_{k\in\Lambda-\Lambda}$.

Let $c_{1,0}=c_{0,0}=1$. Since

$$\begin{pmatrix} c_{0,0} & c_{1,0} & c_{p_j,j}\\ \overline{c}_{1,0} & c_{0,0} & c_{p_j-1,j}\\ \overline{c}_{p_j,j} & \overline{c}_{p_j-1,j} & c_{0,0}\end{pmatrix}$$

must be positive semidefinite for any positive semidefinite sequence on $\mathbb{Z}^2$, it is clear that for all extensions of $(c_{j,l})_{(j,l)\in\Lambda-\Lambda}$ the $c_{j,l}$ must be constant along each row. Since the elements of A are not extendable to a positive semidefinite sequence on $\mathbb{Z}$, it follows that we have constructed a positive sequence on $\Lambda - \Lambda$ which is not extendable to a positive sequence on $\mathbb{Z}^2$. □

Lemma 1.3.29 *Suppose Λ is a canonical form with $d = 1$. Then Λ is homomorphic to $R(1,1)$, to $R(1,1) \setminus (1,1)$, to a set Λ' in canonical form with $d > 1$, or to a set $\Lambda = \{(0,0),(1,0),(0,1),(p,q)\}$ which does not have the extension property.*

Proof. By Lemma 1.3.28 there exists an element $(r,1) \in \Lambda$; it is possible that $(r+1,1) \in \Lambda$. The isomorphism Φ defined by the matrix

$$\begin{pmatrix} 1 & -r \\ 0 & 1 \end{pmatrix} \tag{1.3.19}$$

maps Λ onto a set Δ which contains the elements $(0,0),(1,0),(0,1)$, and possibly also $(1,1)$. If Λ does not contain any additional element the proof is complete. Therefore, we may assume that Λ contains another element which is mapped to $(p,q) \in \Delta$ by Φ. If $(1,1) \in \Delta$ and $\text{card}\Delta > 4$, then Δ, or a translation of it, is isomorphic to a set containing an element $(d,0)$, $d \geq 2$. The latter follows by Exercise 1.6.36 since $\gcd((p,q)-(i,j)) \geq 2$ for at least one choice of (i,j) among the elements $(0,0),(1,0),(0,1),(1,1)$. This proves the lemma if $(1,1) \in \Delta$.

If $S = \{(0,0),(1,0),(0,1),(p_1,q_1),(p_2,q_2)\} \subseteq \Delta$ it is clear that there exist $s_1, s_2 \in S$ such that $\gcd\{p,q\} \geq 2$ where $(p,q) = s_1 - s_2$. Then by Exercise 1.6.36, Δ is isomorphic to a set Λ' with $d > 1$.

It remains to discuss the case when $\Delta = \{(0,0),(1,0),(0,1),(p,q)\}$. If $(p,q) \in \{(1,1),(1,-1),(-1,1),(-1,-1)\}$, then Δ is isomorphic to $R(1,1)$. If $(p,q) \notin \{(1,1),(1,-1),(-1,1),(-1,-1)\}$, then Δ does not have the extension property. This follows from the fact that $c_{0,0} = c_{1,0} = c_{0,1} = c_{-1,1} = 1$, and $c_{p,q} = c_{p-1,q} = c_{p,q-1} = 0$ is a positive semidefinite sequence without a positive semidefinite extension. □

Remark 1.3.30 If the set Λ is in canonical form, there is a maximal j such that (p_0, j) and (p_d, j) are elements in Λ with $p_d - p_0 = d$. We may assume that $j \leq M - m$. Otherwise translate (p_0, j) into $(0,0)$, and change the sign of the first coordinate of the indices. Then we obtain a new set in canonical form. For this new set we have $M_1 = m + j$, and the maximal j has the same value as in the original set. We want to maximize M, and since we are free to work with any of the two sets introduced, we may assume $M \geq M_1$, i.e. $j \leq M - m$. Therefore, for Λ in canonical form we may assume that $\text{card}\Lambda \leq (j+1)(d+1) + (M-j+m)d$.

The above remark leads to the following lemma.

Lemma 1.3.31 *Suppose Λ is in canonical form with $d > 1$. Then Λ does not have the extension property if $M - m > 1$ and $\max\{M-m, d\} \geq 3$.*

Proof. The polynomial (1.3.18) has $2Md$ zeros on $\mathbb{T}^2$. By Theorem 1.3.16, Λ does not have the extension property unless card$\Lambda \leq 2Md$. Λ has at most $(j+1)(d+1) + (M-j+m)d$ elements. The latter number is maximized by $j = M - m$, and we get the inequality $(M-m+1)(d+1) + (M-(M-m)+m)d \leq 2Md$, which can be rewritten as

$$\frac{M-m+1}{M-m-1} \leq d.$$

Let $k = M - m$, and consider the function $f(k) = (k+1)/(k-1)$. Since this function is strictly decreasing and $f(2) = 3$ and $f(3) = 2$, the result follows. □

Lemma 1.3.32 *Suppose Λ is a set in canonical form with $m = 0$, $M = 1$, and $d > 1$. Then Λ has the extension property if and only if it is isomorphic to one of the sets $R(1,d)$ or $R(1,d) \setminus \{(1,d)\}$.*

Proof. By Lemma 1.3.27 at least one of the rows must be full. Let us assume, without loss of generality, that it is the zeroth row. Furthermore, using the isomorphism (1.3.19) we may assume that $(0,1) \in \Lambda$. Using the polynomial (1.3.18) with $M = 1$ and $p_M = 0$ and Theorem 1.3.16 shows that Λ does not have the extension property unless its cardinality is at least $2d+1$. Then $\Lambda = R(1,d) \setminus \{(1,d)\}$, if $(d,1) \notin \Lambda$. If $(d,1) \in \Lambda$, the two polynomials $1 - z^d w = 0$ and $z^d - w = 0$ have $2d$ common zeros on $\mathbb{T}^2$. Then Theorem 1.3.16 and the definition of the canonical form imply that Λ does not have the the extension property unless $\Lambda = R(1,d)$. The proof of the lemma is concluded by the fact that $R(1,d)$ and $R(1,d) \setminus (1,d)$ have the extension property. □

Lemma 1.3.33 *Let Λ be in canonical form with $M + m \geq 2$ and $d > 1$. If there exist p and q, $p \neq q$, such that $(p, M), (q, M) \in \Lambda$ or $(p, -m), (q, -m) \in \Lambda$, then Λ does not have the extension property.*

Proof. If $m = 0$, then Lemma 1.3.31 implies that Λ does not have the extension property except maybe for $M = d = 2$. If $M = d = 2$ the polynomial (1.3.18) has 8 zeros on $\mathbb{T}^2$. This implies that if Λ has the extension property, then it contains at least 9 elements. If Λ contains 9 or more elements, there exists p_2 such that $\{(0,0), (2,0), (p_2, 2), (p_2+2, 2)\} \subset \Lambda$. The polynomials $1 - z^{p_2+2}w^2 = 0$ and $z^2 - w^2 z^{p_2} = 0$ have 8 common zeros on $\mathbb{T}^2$, and then Theorem 1.3.16 implies that Λ does not have the extension property. Therefore we may assume $m \geq 1$ for the rest of the proof.

We claim that if there are at most two elements in the Mth and $-m$th row of Λ, and for either $i = -m$ or M the elements in the ith row are (p_i, i) and $(p_i + k, i)$ with $k > \lfloor \frac{d}{2} \rfloor$, then Λ does not have the extension property. Using these two elements together with an element from the other extreme row we can construct a polynomial of a similar form as (1.3.18) with $2k(M+m)$ zeros on $\mathbb{T}^2$. Let j be as in 1.3.30. Then there are at most $4 + (j+1)(d+1) + (M-j+m-2)d$ elements in Λ. By Theorem 1.3.16, Λ

does not have the extension property if

$$4 + (j + 1)(d + 1) + (M - j + m - 2)d \leq 2k(M + m).$$

The left-hand side of the above inequality is maximized by $j = M - m$ which implies

$$(M + m - 1)d + M - m + 5 \leq 2k(M + m). \tag{1.3.20}$$

If d is even then $k \geq \frac{d}{2} + 1$. Using this lower bound for k, inequality (1.3.20) simplifies to

$$5 \leq d + M + 3m,$$

which holds since $d \geq 2$ and $m \geq 1$. If d is odd, $k \geq \frac{d}{2} + \frac{1}{2}$. Using this lower bound for k, inequality (1.3.20) simplifies to

$$5 \leq d + 2m,$$

which holds since $d \geq 3$ and $m \geq 1$. This proves the claim.

To complete the proof of the lemma we will produce a nonextendable positive semidefinite data set in the case when Λ contains a pair of elements of the form $(p_i, i), (p_i + k, i)$ with $i = -m$ or $i = M$ and $0 < k \leq \lfloor \frac{d}{2} \rfloor$. If there are more than two elements in row $-m$ or M, such a pair must exist. If there are at most two elements in these rows we may assume that such a pair exists by the previous claim. To produce a nonextendable positive data set we define $c_{0,0} = 1$, and $c_{p,q} = 0$ for any $(p, q) \neq (0, 0)$ such that $q \neq M - (-m)$. Let us represent the matrix

$$(c_{p_1 - p_2, q_1 - q_2})_{(p_1, q_1), (p_2, q_2) \in \Lambda}$$

in the block form

$$\begin{pmatrix} T_{M,M} & T_{M,M-1} & \cdots & \cdots & T_{M,-m} \\ & T_{M-1,M-1} & \ddots & & \vdots \\ & & \ddots & \ddots & \vdots \\ & & & T_{-m+1,-m+1} & T_{-m+1,-m} \\ & & & & T_{-m,-m} \end{pmatrix}, \tag{1.3.21}$$

where the block $T_{i,n}$ contains all the elements of the form $c_{p-q,i-n}$ for a fixed pair of second coordinates i and n. By our choice of values for $c_{p,q}$, all block diagonal elements are identity matrices, which sizes depend on the number of elements $(p, q) \in \Lambda$ for a fixed q. By Lemma 1.3.27 at least one of these block diagonal matrices is an $(d + 1) \times (d + 1)$ identity matrix. The off-diagonal blocks are, except for the $T_{M,-m}$ block, zero matrices of appropriate dimension. The dimension of both square block matrices $T_{M,M}$ and $T_{-m,-m}$ is less than $d + 1$. The lemma now follows by choosing entries in $T_{M,-m}$ such that the matrix (1.3.21) is positive semidefinite, but when extending the matrices $T_{M,M}$, $T_{-m,-m}$, and $T_{M,-m}$ to $(d + 1) \times (d + 1)$ matrices we get a 2×2 block matrix which is not positive semidefinite. The matrices $T_{M,M}$ and $T_{-m,-m}$ are both submatrices of $(c_{i-n,0})_{i,n \in \{0,1,\ldots,d+1\}}$, which equals the $(d+1) \times (d+1)$ identity matrix. The matrix $T_{M,-m}$ will be a submatrix of a Toeplitz matrix of the form

$$A = \begin{pmatrix} t_0 & t_1 & \dots & \dots & t_{\lfloor \frac{d}{2} \rfloor} \\ t_{-1} & t_0 & t_1 & \dots & t_{\lfloor \frac{d}{2} \rfloor - 1} \\ \vdots & \ddots & \ddots & \ddots & \vdots \\ t_{-\lfloor \frac{d}{2} \rfloor + 1} & \dots & t_{-1} & t_0 & t_1 \\ t_{-\lfloor \frac{d}{2} \rfloor} & \dots & \dots & t_{-1} & t_0 \end{pmatrix}.$$

If Λ has the extension property, it must be possible to embed $T_{M,-m}$ into the bigger Toeplitz contraction of the form

$$T = \begin{pmatrix} t_0 & t_1 & \dots & \dots & t_d \\ t_{-1} & t_0 & t_1 & \dots & t_{d-1} \\ \vdots & \ddots & \ddots & \ddots & \vdots \\ t_{-d+1} & \dots & t_{-1} & t_0 & t_1 \\ t_{-d} & \dots & \dots & t_{-1} & t_0 \end{pmatrix}.$$

The block matrix (1.3.21) is positive semidefinite if and only if $T_{M,-m}$ is a contraction. If $T_{M,-m}$ has only one column or one row, define $c_{p_i - n_1, i-n_2} = c_{p_{i+k} - n_1, i - n_2} = \frac{1}{\sqrt{2}}$ for $i = M$ and $n_2 = -m$, and all other entries equal to zero. If $i = -m$, make a similar construction. Then the information we are given about T prevents it from being a contraction. If $T_{M,-m}$ has more than one row and column, let $t_i = t_n = 1$, where n is the index of the lower left corner of $T_{M,-m}$ and i is the smallest index such that t_i is not an element of the first column or last row of $T_{M,-m}$. Let the other t_l of $T_{M,-m}$ be zero. Since $|i-n| \leq d$ it is clear that this choice prevents T from being a contraction. □

Using the definition of canonical form, the following is an immediate consequence of Lemma 1.3.33.

Corollary 1.3.34 *Let Λ be such that $G(\Lambda - \Lambda) = \mathbb{Z}^2$, and suppose that Λ is in canonical form with $m = 0$, $M \geq 2$, and $d \geq 2$. Then Λ does not have the extension property.*

Lemma 1.3.35 *There is no set Λ in canonical form with $d = 2$ and $M = m = 1$ which has the extension property.*

Proof. By Lemma 1.3.27 and Lemma 1.3.33 it is sufficient to consider the case when Λ is of the form $\Lambda = \{(p_{-1}, -1), (0,0), (1,0), (2,0), (p_1, 1)\}$. Let $\{c_k\}_{k \in \Lambda - \Lambda}$ be a positive semidefinite sequence. The matrix of the sequence is of the form

$$\begin{pmatrix} c_{0,0} & c_{-p_1,-1} & c_{1-p_1,-1} & c_{2-p_1,-1} & c_{p_{-1}-p_1,-2} \\ & c_{0,0} & c_{1,0} & c_{2,0} & c_{p_{-1},-1} \\ & & c_{0,0} & c_{1,0} & c_{p_{-1}-1,-1} \\ & & & c_{0,0} & c_{p_{-1}-2,-1} \\ & & & & c_{0,0} \end{pmatrix}.$$

Let $c_{0,0} = 1$, $c_{1,0} = c_{2,0} = 0$, $c_{1-p_1,-1} = c_{2-p_1,-1} = \frac{1}{\sqrt{2}}$, $c_{k,-1} = 0$ for $k \neq 1 - p_1, 2 - p_1$, and

$$c_{p_{-1}-p_1,-2} = c_{-p_1,-1}c_{p_{-1},-1} + c_{1-p_1}c_{p_{-1}-1,-1} + c_{2-p_1,-1}c_{p_{-1}-2,-1}.$$

Then the matrix above is positive semidefinite. Still this data set does not have a positive extension as the Hermitian matrix

$$\begin{pmatrix} c_{0,0} & c_{1,0} & c_{1-p_1,-1} & c_{2-p_1,-1} \\ & c_{0,0} & c_{-p_1,-1} & c_{1-p_1,-1} \\ & & c_{0,0} & c_{1,0} \\ & & & c_{0,0} \end{pmatrix}$$

is not positive semidefinite. □

Lemma 1.3.36 *Suppose Λ is in canonical form with $d > 1$. In addition we assume that there are only one p such that $(p, M) \in \Lambda$ and only one q such that $(q, -m) \in \Lambda$. Then Λ does not have the extension property if $d + M - m \geq 3$.*

Proof. The proof follows the same lines as the proof of Lemma 1.3.31, but now there are at most $(j + 1)(d + 1) + (M - j + m - 2)d + 2$ elements in Λ. This number is maximized if $j = M - m$ and must be less than $2Md$. As a consequence we have the inequality

$$\frac{M - m + 3}{M - m + 1} \leq d.$$

The function $f(k) = (k + 3)/(k + 1)$ is strictly decreasing, and the lemma follows since $f(0) = 3$, $f(1) = 2$ and $f(2) < 2$. □

Corollary 1.3.37 *Every Λ in canonical form with $d \geq 2$, $m \geq 1$, and $d + M - m \geq 3$ does not have the extension property.*

Proof. The result follows from the definition of the canonical form, Lemma 1.3.33, and Lemma 1.3.36. □

Lemma 1.3.38 *Suppose Λ is in canonical form with $m \geq 2$ and Λ has a unique element on each of the rows $-m$, $-m + 1$, $M - 1$, and M. Then Λ does not have the extension property if $d > 1$.*

Proof. The proof is similar to the proof of Lemma 1.3.36. We have to consider in this case the inequality

$$(j + 1)(d + 1) + (M - j + m - 4)d + 4 \leq 2Md.$$

The left-hand side of the inequality is maximized when $j = M - m$. This leads to the inequality

$$\frac{M - m + 5}{M - m + 3} \leq d.$$

The function $f(k) = (k+5)/(k+3)$ is strictly decreasing. The lemma follows since $f(0) = 5/3 < 2$. □

Lemma 1.3.39 *Let Λ be in canonical form with $M + m \geq 4$ and $d > 1$. If Λ does not have a unique element on each of the rows $-m$, $-m+1$, $M-1$, and M, then Λ does not have the extension property.*

Proof. By a similar ordering of the elements as in the proof of Lemma 1.3.33 the matrix

$$(c_{p_1-p_2,q_1-q_2})_{(p_1,q_1),(p_2,q_2)\in\Lambda}$$

can be written in the block form

$$\begin{pmatrix} T_{M,M} & T_{M,M-1} & \cdots & \cdots & T_{M,-m} \\ & T_{M-1,M-1} & \ddots & & \vdots \\ & & \ddots & \ddots & \vdots \\ & & & T_{-m+1,-m+1} & T_{-m+1,-m} \\ & & & & T_{-m,-m} \end{pmatrix}, \quad (1.3.22)$$

where the block $T_{i,j}$ contains all the elements of the form $c_{p-q,i-j}$ for fixed second coordinates i and j.

Define $c_{0,0} = 1$ and all other entries of the matrix which are not in the $T_{M,-m+1}$ or $T_{M-1,-m}$ blocks to be zero. By our choice of values for $c_{i,j}$, all block diagonal elements are identity matrices, whose sizes depend on the number of elements $(p,q) \in \Lambda$ for a fixed q. By Lemma 1.3.27, at least one of these block diagonal matrices is a $(d+1) \times (d+1)$ identity matrix. By Lemma 1.3.33, Λ does not have the extension property unless it contains exactly one element of both the forms (p,k), for $k = -m, M$. We can therefore assume this. This implies that both block matrices $T_{M,M}$ and $T_{-m,-m}$ are 1×1 matrices, and the $T_{M,-m+1}$ and $T_{M-1,-m}$ blocks are a single row and a single column, respectively. Both blocks $T_{M,-m+1}$ and $T_{M-1,-m}$ are submatrices of Toeplitz matrices (not necessarily the same one) of the form

$$T = \begin{pmatrix} t_0 & t_1 & \cdots & \cdots & t_d \\ t_{-1} & t_0 & t_1 & \cdots & t_{d-1} \\ \vdots & \ddots & \ddots & \ddots & \vdots \\ t_{-d+1} & \cdots & t_{-1} & t_0 & t_1 \\ t_{-d} & \cdots & \cdots & t_{-1} & t_0 \end{pmatrix}.$$

This means that some entries in a row or column of T are specified. A positive semidefinite extension of the matrix (1.3.22) exists if and only if there exists a contractive extension of T.

Now, if $t_j = t_i = 1/\sqrt{2}$ with $|j-i| \leq \frac{d}{2}$ and $t_k = 0$ for $k \neq i,j$ the matrix (1.3.22) is a positive semidefinite matrix, but T cannot be extended to a contraction. If there are more than three elements of the form $(i_{M-1}, M-1) \in \Lambda$ (or of the form $(i_{-m+1}, -m+1) \in \Lambda$) or two elements with distance less than $d/2$ in row $M-1$ or $-m+1$, this construction is possible.

It remains to prove the lemma in the case when there are at most two elements in both rows $M-1$ and $-m+1$, and that the distance between them is more than $d/2$. By this assumption there are at most

$$2 + 4 + (M-1)(d+1) + (m-2)d = 5 + M + (M+m-3)d$$

elements in Λ. Moreover, there exists a polynomial of similar type as (1.3.18) with $2k(M+m-1)$ zeros on $\mathbb{T}^2$, where k is the distance between the elements in one of the rows $M-1$ or $-m+1$. We may assume that $k \geq \frac{d}{2}+\frac{1}{2}$ if d is odd, and $k \geq \frac{d}{2}+1$ if d is even. The proof is concluded by Theorem 1.3.16 since

$$M+5+(M+m-3)d \leq 2\left(\frac{d}{2}+\frac{1}{2}\right)(M+m-1)$$

holds if d is odd, $d \geq 3$, and $M \geq m \geq 2$, and

$$M+5+(M+m-3)d \leq 2\left(\frac{d}{2}+1\right)(M+m-1)$$

holds if d is even and $M \geq m \geq 2$. □

Corollary 1.3.40 *Suppose Λ is a canonical form with $d > 1$ and $m \geq 2$. Then Λ does not have the extension property.*

Proof. The result follows from Lemma 1.3.38 and Lemma 1.3.39 . □

d	m	$M-m$	
1	≥ 0	≥ 0	Lemma 1.3.29
2	0	$0,1$	Theorem 1.3.8 and Lemma 1.3.32
2	0	≥ 2	Corollary 1.3.34
2	1	0	Lemma 1.3.35
2	1	≥ 1	Corollary 1.3.37
2	≥ 2	≥ 0	Corollary 1.3.40
3	0	0,1	Theorem 1.3.8 and Lemma 1.3.32
3	0	≥ 2	Corollary 1.3.34
3	≥ 1	≥ 0	Corollary 1.3.37
≥ 4	0	0,1	Theorem 1.3.8 and Lemma 1.3.32
≥ 4	0	≥ 2	Corollary 1.3.34
≥ 4	≥ 1	≥ 0	Corollary 1.3.37

Table 1.1 Cases considered in the proof of Theorem 1.3.18.

Proof of Theorem 1.3.18. The sets Λ written in canonical form can be divided into the cases in Table 1.1, depending on the size d, m, and $M-m$. The table indicates which of the results cover the particular case under consideration. □

1.4 DETERMINANT AND ENTROPY MAXIMIZATION

In this section we study the following functions, defined on the interior of the cones PSD_n and $\mathbb{T}\mathrm{Pol}_n^+$, respectively:

$$\log\det : \mathrm{int}(\mathrm{PSD}_n) \to \mathbb{R}, \tag{1.4.1}$$

$$\mathcal{E} : \mathrm{int}(\mathbb{T}\mathrm{Pol}_n^+) \to \mathbb{R},\ \mathcal{E}(p) = \frac{1}{2\pi}\int_0^{2\pi} \log p(e^{it})dt. \tag{1.4.2}$$

These functions will be useful in the study of these cones. We will show that both functions are strictly concave on their domain and we will obtain optimality conditions for their unique maximizers. These functions are examples of *barrier functions* as they tend to minus infinity when the argument approaches the boundary of the domain.

1.4.1 The log det function

Let PD_n denote the open cone of $n \times n$ positive definite matrices, or equivalently, PD_n is the interior of PSD_n. In other words,

$$\mathrm{PD}_n = \{A \in \mathbb{C}^{n\times n} \ :\ A > 0\}.$$

Let $F_0, \dots, F_m$ be $n\times n$ Hermitian matrices, and put $F(x) = F_0 + \sum_{i=1}^m x_i F_i$, where $x = (x_1, \dots, x_m)^T \in \mathbb{R}^m$. We assume that $F_1, \dots, F_m$ are linearly independent. Let $\Delta = \{x \in \mathbb{R}^m \ :\ F(x) \in \mathrm{PD}_n\}$, which we will assume to be nonempty. By a translation of the vector $(x_1, \dots, x_n)$, we may assume without loss of generality that $0 \in \Delta$. Note that Δ is convex, as $F(x) > 0$ and $F(y) > 0$ imply that $F(sx + (1-s)y) = sF(x) + (1-s)F(y) > 0$ for $0 \le s \le 1$. Define now the function

$$\phi(x) = \log\det F(x), \quad x \in \Delta.$$

We start by computing its gradient and Hessian.

Lemma 1.4.1 *We have that*

$$\frac{\partial \phi}{\partial x_i}(x) = \mathrm{tr}(F(x)^{-1}F_i), \quad \frac{\partial^2 \phi}{\partial x_i \partial x_j}(x) = -\,\mathrm{tr}(F(x)^{-1}F_iF(x)^{-1}F_j),$$

and consequently,

$$\nabla\phi(x) = \begin{pmatrix}\mathrm{tr}(F(x)^{-1}F_1) & \cdots & \mathrm{tr}(F(x)^{-1}F_m)\end{pmatrix}, \tag{1.4.3}$$

$$\nabla^2\phi(x) = \begin{pmatrix} -\,\mathrm{tr}\,F(x)^{-1}F_1F(x)^{-1}F_1 & \cdots & -\,\mathrm{tr}\,F(x)^{-1}F_1F(x)^{-1}F_m \\ \vdots & & \vdots \\ -\,\mathrm{tr}\,F(x)^{-1}F_mF(x)^{-1}F_1 & \cdots & -\,\mathrm{tr}\,F(x)^{-1}F_mF(x)^{-1}F_m \end{pmatrix}. \tag{1.4.4}$$

Proof. Without loss of generality we take $x = 0$ (one can always do a translation so that 0 is the point of interest). Let us take $m = 1$ and let us compute $\phi'(0)$. Note that

$$\phi(x) = \log\det(F_0(I + xF_0^{-1}F_1)) \tag{1.4.5}$$
$$= \log\det F_0 + \log\det(I + xF_0^{-1}F_1) \tag{1.4.6}$$
$$= \log\det F_0 + \log(1 + x\,\mathrm{tr}(F_0^{-1}F_1) + O(x^2)) \tag{1.4.7}$$
$$= \log\det F_0 + x\,\mathrm{tr}(F_0^{-1}F_1) + O(x^2) \quad (x \to 0). \tag{1.4.8}$$

But then $\phi'(0) = \lim_{x\to 0} \frac{\phi(x)-\phi(0)}{x} = \mathrm{tr}(F_0^{-1}F_1)$. The general expression for $\frac{\partial\phi}{\partial x_i}(x)$ follows now easily. The second derivatives are obtained similarly. □

Corollary 1.4.2 *The function* log det *is strictly concave on* PD_n.

Proof. Note that the entries in the Hessian of ϕ can be rewritten as

$$-\,\mathrm{tr}(F(x)^{-1/2}F_iF(x)^{-1/2}F(x)^{-1/2}F_jF(x)^{-1/2}).$$

Thus for $y \in \mathbb{R}^m$ we have that

$$y^*\nabla^2\phi(x)y = -\sum_{i,j=1}^{m} y_i\,\mathrm{tr}(F(x)^{-1/2}F_iF(x)^{-1/2}F(x)^{-1/2}F_jF(x)^{-1/2})y_j \tag{1.4.9}$$

$$= -\,\mathrm{tr}(F(x)^{-1/2}\left(\sum_{i=1}^{m} y_iF_i\right)F(x)^{-1/2})^2 \tag{1.4.10}$$

$$= -\|F(x)^{-1/2}\left(\sum_{i=1}^{m} y_iF_i\right)F(x)^{-1/2}\|_2 \leq 0, \tag{1.4.11}$$

and equality holds if and only if $\sum_{i=1}^{m} y_iF_i = 0$. Here $\|\cdot\|_2$ stands for the *Frobenius norm of a matrix*, i.e., $\|A\|_2 = \sqrt{\mathrm{tr}(A^*A)}$. As $F_1, \ldots, F_m$ are linearly independent, this happens only if $y_1 = \cdots = y_m = 0$. □

The following is the main result of this subsection.

Theorem 1.4.3 *Let $\mathcal{W}$ be a linear subspace of $\mathcal{H}_n$, with* 0 *the only positive semidefinite matrix contained in it. For every $A > 0$ and $B \in \mathcal{H}_n$, there is a unique $F \in (A + \mathcal{W}) \cap \mathrm{PD}_n$ such that $F^{-1} - B \perp \mathcal{W}$. Moreover, F is the unique maximizer of the function*

$$f(X) = \log\det X - \mathrm{tr}(BX)\ ,\ X \in (A + \mathcal{W}) \cap \mathrm{PD}_n. \tag{1.4.12}$$

If A and B are real matrices then so is F.

Using the Hahn-Banach separation theorem one can easily show that when 0 is the only positive semidefinite matrix in $\mathcal{W}$, for every Hermitian B we have that $(B + \mathcal{W}^\perp) \cap \mathrm{PSD}_n \neq \emptyset$. Indeed, if the set is empty then the separation theorem yields the existence of a Hermitian Φ and a real number α such that $\mathrm{tr}(P\Phi) > \alpha$ for all $P > 0$ and $\mathrm{tr}(X\Phi) \leq \alpha$ for all $X \in B + \mathcal{W}^\perp$. Since the positive definite matrices form a cone, it is easy to see that we must have $\alpha \leq 0$, and since $\mathcal{W}^\perp$ is a subspace it is easy to see that we must have $\Phi \in \mathcal{W}$. But then it follows that $\Phi \geq 0$ and $\Phi \neq 0$. Thus the content of the theorem is not weakened when one restricts oneself to $B > 0$.

Proof. First note that $(A + \mathcal{W}) \cap \mathrm{PD}_n$ is convex. Next, since 0 is the only positive semidefinite matrix in $\mathcal{W}$, $(A+\mathcal{W}) \cap \mathrm{PD}_n$ is a bounded set. Indeed, as each nonzero $W \in \mathcal{W}$ has a negative eigenvalue, we have that for each $0 \neq W \in \mathcal{W}$ the set $(A+\mathbb{R}W) \cap \mathrm{PSD}_n$ is bounded. Let $g(W) = \max\{a : A + aW \in \mathrm{PSD}_n\}$. As $(A+\mathcal{W}) \cap \mathrm{PSD}_n$ is convex, we obtain that g is a continuous function on the unit ball $\mathcal{B}$ in $\mathcal{W}$. Using now the finite dimensionality and thus the compactness of $\mathcal{B}$, we get that g attains a maximum on $\mathcal{B}$, M say. But then we obtain that we must have that $(A+\mathcal{W}) \cap \mathrm{PD}_n \subset (A+\mathcal{W}) \cap \mathrm{PSD}_n$ lies in $A + M\mathcal{B}$, which proves the boundedness. Next, by Corollary 1.4.2, log det is strictly concave on PD_n, and since $\mathrm{tr}(BX)$ is linear in X, $f(X)$ is strictly concave. Moreover, $f(X)$ tends to $-\infty$ when X approaches the boundary of $(A + \mathcal{W}) \cap \mathrm{PD}_n$ (as $\det X$ tends to 0 as X approaches the boundary). Thus f takes on a unique maximum on $(A + \mathcal{W}) \cap \mathrm{PD}_n$ for a matrix denoted by F, say.

Fix an arbitrary $W \in \mathcal{W}$. Consider the function $f_{F,W}(x) = \log\det(F + xW) - \mathrm{tr}(B(F+xW))$ defined in a neighborhood of 0 in $\mathbb{R}$. Then $f'_{F,W}(0) = 0$ (since f has its maximum at F). Similarly as in the proof of Lemma 1.4.1, we obtain that

$$f'_{F,W}(0) = \left.\frac{(\det(I + xF^{-1}W))'}{\det(I + xF^{-1}W)}\right|_{x=0} - (\mathrm{tr}(B(F + xW)))'|_{x=0}$$

$$= \mathrm{tr}(F^{-1}W) - \mathrm{tr}(BW) = \mathrm{tr}((F^{-1} - B)W) = 0.$$

Since W is an arbitrary element of $\mathcal{W}$ we have that $F^{-1} - B \perp \mathcal{W}$. Assume that $G \in (A + \mathcal{W}) \cap \mathrm{PD}_n$ and $G^{-1} - A \perp \mathcal{W}$. Then $f'_{G,W}(0) = 0$ for any $W \in \mathcal{W}$ and since f is strictly convex it follows that $G = F$. This proves the uniqueness of F.

In case A and B are real matrices, one can restrict attention to real matrices, and repeat the above argument. The resulting matrix F will also be real. □

Corollary 1.4.4 *Given is a positive definite matrix $A = (A_{ij})_{i,j=1}^n$, a Hermitian matrix $B = (B_{ij})_{i,j=1}^n$, and an $n \times n$ symmetric pattern P with associated graph G. Then there exists a unique positive definite matrix $F \in A + \mathcal{H}_G^\perp$ such that $(F^{-1})_{ij} = B_{ij}$ for every $(i,j) \notin P$. Moreover, F maximizes the function $f(X) = \mathrm{logdet} X - \mathrm{tr}(BX)$ over the set of all positive definite matrices $X \in A + \mathcal{H}_G^\perp$. In case A and B are real, F is also real.*

Proof. Consider in Theorem 1.4.3, $\mathcal{W} = \mathcal{H}_G^\perp$. Then evidently the only positive definite matrix in $\mathcal{W}$ is 0. By Theorem 1.4.3 there exists a unique $F \in (A + \mathcal{W}) \cap \mathrm{PD}_n$ such that $F^{-1} - B \perp \mathcal{W}$. For any $(k,j) \notin P$, consider the matrix $W_R^{(k,j)} \in \mathcal{W}$ having all its entries 0 except those in positions (k,j) and (j,k), which equal 1, and the matrix $W_I^{(k,j)}$ having i in position (k,j), $-i$ in position (j,k), and 0 elsewhere. The conditions

$\mathrm{tr}((F^{-1}-B)W_R^{(k,j)}) = \mathrm{tr}((F^{-1}-B)W_I^{(k,j)}) = 0$ imply that $(F^{-1})_{kj} = B_{kj}$ for any $(k,j) \in P$. □

Corollary 1.4.5 *Let P be a symmetric pattern with associated graph G and let A be a Hermitian matrix such that there exists a positive definite matrix $B \in A + \mathcal{H}_G^\perp$. Then among all such positive definite matrices B there exists a unique one, denoted $\widetilde{A}$, which maximizes the determinant. Also, $\widetilde{A}$ is the only positive definite matrix in $A + \mathcal{H}_G^\perp$ the inverse of which has 0 in all positions corresponding to entries $(i,j) \notin P$.*

Proof. Follows immediately from Corollary 1.4.4 when $B = 0$. □

Let us end this subsection with an example for Corollary 1.4.5.

Example 1.4.6 Find

$$\max_{x_1,x_2} \det \begin{pmatrix} 3 & 2 & x_1 & -1 \\ 2 & 3 & 1 & x_2 \\ \overline{x_1} & 1 & 3 & 1 \\ -1 & \overline{x_2} & 1 & 3 \end{pmatrix},$$

the maximum being taken over those x_1 and x_2 for which the matrix is positive definite. Notice that for $x_1 = x_2 = 0$ the matrix above is positive semidefinite by diagonal dominance (or by using the Geršgorin disk theorem). As one can easily check, for $x_1 = x_2 = 0$ the matrix is invertible (and thus positive definite). Consequently, the above problem will have a solution. By Corollary 1.4.4 the maximum is obtained at real x_1 and x_2, and thus we can suffice with optimizing over the reals. The following Matlab script solves the problem. The script is a *damped Newton's method* with α as damping factor. Notice that y and H below correspond to the gradient and the Hessian, respectively.

```
F0=[3 2 0 -1; 2 3 1 0; 0 1 3 1 ; -1 0 1 3];
F1=[0 0 1 0 ; 0 0 0 0; 1 0 0 0; 0 0 0 0];
F2=[0 0 0 0 ; 0 0 0 1; 0 0 0 0; 0 1 0 0];
x=[0 ; 0];
G=inv(F0);
y = [G(1,3); G(2,4)];
while norm(y) > 1e-7,
  H=[G(1,3)*G(3,1)+G(1,1)*G(3,3) G(1,4)*G(2,3)+G(2,1)*G(3,4) ; ...
  G(1,4)*G(2,3)+G(2,1)*G(3,4) G(2,4)*G(4,2)+G(2,2)*G(4,4)];
  v=H\y;
  delta = sqrt(v'*y);
  if delta < 1/4 alpha=1; else alpha=1/(1+delta); end;
  x=x+alpha*v;
  Fx=F0+x(1)*F1+x(2)*F2;
  G=inv(Fx);
  y = [G(1,3); G(2,4)];
```

end;

The solution we find is

$$\begin{pmatrix} 3 & 2 & 0.3820 & -1 \\ 2 & 3 & 1 & -0.3820 \\ 0.3820 & 1 & 3 & 1 \\ -1 & -0.3820 & 1 & 3 \end{pmatrix}$$

with a determinant equal to 29.2705.

1.4.2 The entropy function

Let

$$\mathbb{T}\mathrm{Pol}_n^{++} = \{p \in \mathbb{T}\mathrm{Pol}_n \ : \ p(z) > 0, z \in \mathbb{T}\}.$$

Equivalently, $\mathbb{T}\mathrm{Pol}_n^{++}$ is the interior of $\mathbb{T}\mathrm{Pol}_n^{+}$. Let $p^{(0)}, \ldots, p^{(m)}$ be elements of $\mathbb{T}\mathrm{Pol}_n$, and put $p_x = p^{(0)} + \sum_{i=1}^m x_i p^{(i)}$, where $x = (x_1, \ldots, x_m)^T \in \mathbb{R}^m$. We will assume that $p^{(1)}, \ldots, p^{(m)}$ are linearly independent. Let $\Delta = \{x \in \mathbb{R}^m \ : \ p_x \in \mathbb{T}\mathrm{Pol}_n^{++}\}$, which we assume to be nonempty. It is easy to see that Δ is convex. Define now the function

$$\psi(x) = \mathcal{E}(p_x) := \frac{1}{2\pi} \int_0^{2\pi} \log p_x(e^{it}) dt. \tag{1.4.13}$$

Lemma 1.4.7 *We have that*

$$\frac{\partial \psi}{\partial x_j}(x) = \frac{1}{2\pi} \int_0^{2\pi} \frac{p^{(j)}(e^{it})}{p_x(e^{it})} dt,$$

$$\frac{\partial^2 \psi}{\partial x_j \partial x_k}(x) = -\frac{1}{2\pi} \int_0^{2\pi} \frac{p^{(j)}(e^{it}) p^{(k)}(e^{it})}{p_x(e^{it})^2} dt.$$

Proof. Let us take $m = 1$ and let us compute $\psi'(0)$ (assuming that $0 \in \Delta$). Note that

$$\psi(x) = \frac{1}{2\pi} \int_0^{2\pi} \left[\log p^{(0)}(e^{it}) - \log \left(1 + x \frac{p^{(1)}(e^{it})}{p^{(0)}(e^{it})} \right) \right] dt.$$

Writing $\log(1 + z) = z + O(z^2)$ $(z \to 0)$, one readily finds that

$$\psi'(0) = \lim_{x \to 0} \frac{\psi(x) - \psi(0)}{x} = \frac{1}{2\pi} \int_0^{2\pi} \frac{p^{(1)}(e^{it})}{p^{(0)}(e^{it})} dt.$$

The general expression for $\frac{\partial \psi}{\partial x_i}(x)$ follows now easily. The second derivatives are obtained similarly. □

Corollary 1.4.8 *The function $\mathcal{E}$ is strictly concave on* $\mathbb{T}\mathrm{Pol}_n^{++}$.

Proof. For $y \in \mathbb{R}^m$ we have that

$$y^T \nabla^2 \psi(x) y = -\frac{1}{2\pi} \int_0^{2\pi} \frac{(\sum_{j=1}^m y_j p^{(j)}(e^{it}))^2}{p_x(e^{it})^2} dt \leq 0,$$

and equality holds if and only if $\sum_{j=1}^m y_j p^{(j)} \equiv 0$. As $p^{(1)}, \ldots, p^{(m)}$ are linearly independent, this happens only if $y_1 = \cdots = y_m = 0$. □

The following is the main result of this subsection.

Theorem 1.4.9 *Let $\mathcal{W}$ be a linear subspace of $\mathbb{T}\mathrm{Pol}_n$, with $\mathcal{W} \cap \mathbb{T}\mathrm{Pol}_n^+ = \{0\}$. For every $p \in \mathbb{T}\mathrm{Pol}_n^{++}$, there is a unique $g \in (p + \mathcal{W}) \cap \mathbb{T}\mathrm{Pol}_n^{++}$ such that $\frac{1}{g} - h \perp \mathcal{W}$. Moreover, g is the unique maximizer of the function*

$$\alpha(q) = \mathcal{E}(q) + \langle q, h \rangle \ , \ q \in (p + \mathcal{W}) \cap \mathbb{T}\mathrm{Pol}_n^{++}. \tag{1.4.14}$$

If p and h have real coefficients then so does g.

The proof is analogous to that of Theorem 1.4.3. We leave this as an exercise for the reader (see Exercise 1.6.48).

Example 1.4.10 Let $p_x(z) = \frac{1}{2z^3} + \frac{2}{z^2} + \frac{\overline{x}}{z} + 6 + xz + 2z^2 + \frac{z^3}{2}$. Determine

$$\max_x \mathcal{E}(p_x).$$

As the given data are real it suffices to consider real x. In order to perform the maximization, we need a way to determine the Fourier coefficients of $\frac{1}{p_x}$. We do this by using Matlab's command `fft`. Indeed, by using the fast Fourier transform we obtain a vector containing the values of p_x on a grid of the unit circle. Taking this vector and entry wise computing the reciprocal yields a vector containing the values of $\frac{1}{p_x}$ on a grid on $\mathbb{T}$. Performing an inverse fast Fourier transform now yields an approximation of the Fourier coefficients of $\frac{1}{p_x}$. We can now execute the maximization:

```
p=[.5;2;0;6;0;2;.5];
p1=[0;0;1;0;1;0;0];
x=0;
hh=fft([zeros(61,1);p;zeros(60,1)]);
hh=ones(128,1)./hh;
hh2=ifft(hh);
y=hh2(64);
while abs(y)>1e-5,
  ll=fft([zeros(62,1);[1;0;2;0;1];zeros(61,1)]);
  ll=ll.*abs(hh).*abs(hh);
  ll=ifft(ll);
  H=ll(65);
  v=H\y;
  delta=sqrt(v'*y);
  if delta < 1/4 alpha=1; else alpha = 1/(1+delta); end;
  x=x+alpha*v;
  p=p+x*p1;
  hh=fft([zeros(61,1);p;zeros(60,1)]);
  hh=ones(128,1)./hh;
  hh2=ifft(hh);
  y=hh2(64);
end;
```

The solution we find is $x = 0.4546$.

1.5 SEMIDEFINITE PROGRAMMING

Many of the problems we have discussed in this chapter can be solved numerically very effectively using semidefinite programming (SDP). We will briefly outline some of the main ideas behind semidefinite programming and present some examples. There is a general theory on "conic programming" that goes well beyond the cone of positive semidefinite matrices, but to discuss this in that generality is beyond the scope of this book.

Let $F_0, \ldots, F_m$ be given Hermitian matrices, and let $F(x) = F_0 + \sum_{i=1}^m x_i F_i$. Let also $c \in \mathbb{R}^m$ be given, and write $x = (x_1, \ldots, x_m)^T \in \mathbb{R}^m$. We consider the problem

$$\begin{array}{c} \text{minimize } \; c^T x \\ {\scriptstyle \text{subject to } F(x) \geq 0} \end{array} . \tag{1.5.1}$$

We will call this the *primal* problem. We will denote

$$p_* = \inf\{c^T x \; : F(x) \geq 0\}.$$

The primal problem has a so-called *dual problem*, which is the following:

$$\begin{array}{c} \text{maximize } \; -\operatorname{tr}(F_0 Z) \\ {\scriptstyle \text{subject to } Z \geq 0} \\ {\scriptstyle \text{and } \operatorname{tr}(F_i Z) = c_i, i=1,\ldots,m} \end{array} . \tag{1.5.2}$$

We will denote

$$d_* = \sup\{-\operatorname{tr}(F_0 Z) \; : Z \geq 0 \text{ and } \operatorname{tr}(F_i Z) = c_i, i = 1, \ldots, m\}.$$

We will call the set $\{x \; : \; F(x) \geq 0\}$ the *feasible set for the primal problem*, and the set $\{Z \geq 0 \; : \; \operatorname{tr}(F_i Z) = c_i, i = 1, \ldots, m\}$ the *feasible set for the dual problem.* We say that a (primal) feasible x is *strictly feasible* if $F(x) > 0$. Similarly, a (dual) feasible Z is *strictly feasible* if $Z > 0$. Observe that when x is primal feasible and Z is dual feasible, we have that

$$c^T x + \operatorname{tr}(F_0 Z) = \sum_{i=1}^m \operatorname{tr}(F_i Z) x_i + \operatorname{tr}(F_0 Z) = \operatorname{tr}(F(x) Z) \geq 0$$

as $Z, F(x) \geq 0$. Thus

$$-\operatorname{tr}(F_0 Z) \leq c^T x.$$

Consequently, when one can find feasible x and Z, one immediately sees that

$$-\operatorname{tr}(F_0 Z) \leq d_* \leq p_* \leq c^T x.$$

Consequently, when one can find feasible x and Z such that

$$\text{duality gap} := c^T x + \operatorname{tr}(F_0 Z)$$

is small, then one knows that one is close to the optimum. We state here without proof, that if a strict feasible x or a strict feasible Z exists, one has in fact that $d_* = p_*$, so in that case one will be able to make the duality gap arbitrary small.

The log det function, which we discussed in the previous section, has three very useful properties: (i) it is strictly concave, and thus any maximizer will be unique; (ii) it tends to minus infinity as the argument approaches the boundary, and thus it will be easy to stay within the set of positive definite matrices; and (iii) its gradient and Hessian are easily computable. These three properties do not change when we add a linear function to log det. This now leads to modified primal and dual problems. Let $\mu > 0$, and introduce

$$\operatorname*{minimize}_{\text{subject to } F(x)>0} \ \mu \log\det F(x) + c^T x \tag{1.5.3}$$

$$\operatorname*{maximize}_{\substack{\text{subject to } Z>0 \\ \text{and } \operatorname{tr}(F_i Z)=c_i, i=1,\ldots,m}} \ -\operatorname{tr}(F_0 Z) + \mu \log\det Z \ . \tag{1.5.4}$$

When strictly feasible x and Z exist, the above problems will have unique optimal solutions which may for instance be found via *Newton's algorithm.* If the duality gap for these solutions is small, we will have found an approximate solution to our original problem. Of course, finding the right parameter μ (or rather the right updates for $\mu \to 0$) and finding efficient algorithms to implement these ideas are still an art, especially as closer to the optimal value the matrix becomes close to a singular one. It is beyond the scope of this book to go into further detail on semidefinite programming. In the notes we will refer the reader to references on the subject.

We end this section with two examples. We will make use of Matlab's LMI (Linear Matrix Inequalities) Toolbox to solve these problems numerically. The LMI Toolbox uses semidefinite programming.

Example 1.5.1 Find

$$\max_{x_1,x_2} \lambda_{\min} \begin{pmatrix} 1 & 2 & x_1 & -1 \\ 2 & 1 & 1 & x_2 \\ \overline{x_1} & 1 & 1 & 1 \\ -1 & \overline{x_2} & 1 & 1 \end{pmatrix}. \tag{1.5.5}$$

We first observe that as the given data are all real, it suffices to consider real x_1 and x_2. Using Matlab's command **gevp** (generalized eigenvalue problem) command, we get the following.

```
F0=[1 2 0 -1; 2 1 1 0; 0 1 1 1; -1 0 1 1]
F1=[0 0 1 0 ; 0 0 0 0; 1 0 0 0 ; 0 0 0 0]
F2=[0 0 0 0 ; 0 0 0 1; 0 0 0 0; 0 1 0 0]
setlmis([])
X1 = lmivar(1,[1 1]);
X2 = lmivar(1,[1 1]);
lmiterm([1 1 1 0],-F0)
lmiterm([1 1 1 X1],-F1,1)
lmiterm([1 1 1 X2],-F2,1)
```

```
lmiterm([-1 1 1 0],eye(4))
lmis = getlmis
[alpha,popt]=gevp(lmis,1,[1e-10 0 0 0 0 ])
x1 = dec2mat(lmis,popt,X1)
x2 = dec2mat(lmis,popt,X2)
```

Running this script leads to $\alpha = 1.0000, x_1 = 1.0000, x_2 = -1.0000$. This leads to

$$\alpha I + \begin{pmatrix} 1 & 2 & x_1 & -1 \\ 2 & 1 & 1 & x_2 \\ \overline{x}_1 & 1 & 1 & 1 \\ -1 & \overline{x}_2 & 1 & 1 \end{pmatrix} = \begin{pmatrix} 2 & 2 & 1 & -1 \\ 2 & 2 & 1 & -1 \\ 1 & 1 & 2 & 1 \\ -1 & -1 & 1 & 2 \end{pmatrix} \geq 0,$$

with α optimal. So the answer to (1.5.5) is -1.

Example 1.5.2 Find the outer factorization of $p(z) = 2z^{-2} - z^{-1} + 6 - z + z^2$. In principle one can determine the roots of $p(z)$ and use the approach presented in the proof of Theorem 1.1.5 to compute the outer factor. Here we will present a different approach based on semidefinite programming that we will be able to use in the matrix values case as well (see Section 2.4). We are looking for $g(z) = g_0 + g_1 z + g_2 z^2$ so that

$$p(z) = |g(z)|^2, \ z \in \mathbb{T}. \tag{1.5.6}$$

If we introduce the matrix

$$F := (F_{ij})_{i,j=0}^2 = \begin{pmatrix} g_0 \\ g_1 \\ g_2 \end{pmatrix} \begin{pmatrix} \overline{g_0} & \overline{g_1} & \overline{g_2} \end{pmatrix},$$

we have that $F \geq 0$, $\operatorname{tr}(F) = |g_0|^2 + |g_1|^2 + |g_2|^2 = 6$, $F_{10} + F_{21} = g_1\overline{g_0} + g_2\overline{g_1} = -1$, and $F_{20} = g_2\overline{g_0} = 2$. Here we used (1.5.6). Thus we need to find a positive semidefinite matrix F with trace equal to 6, the sum of the entries in diagonal 1 equal to -1, and the (2,0) entry equal to 2. In addition, we must see to it that g does not have any roots in the open unit disk. One way to do this is to maximize the value $|g(0)| = |g_0|$ which is the product of the absolute values of the roots of g. As we have seen in the proof of Theorem 1.1.5, the roots of any factor g of $p = |g|^2$ are the results of choosing one out each pair $(\alpha, 1/\overline{\alpha})$ of roots of p. When we choose the roots that do not lie in the open unit disk, we obtain an outer g. This is exactly achieved by maximizing $|g_0|^2$. In conclusion, we want to determine

$$\max\ F_{00} \ \text{ subject to } \begin{cases} F = (F_{ij})_{i,j=0}^2 \geq 0, \\ \operatorname{tr} F = 6, \ F_{01} + F_{21} = -1, \ F_{20} = 2. \end{cases}$$

When we write a Matlab script to find the appropriate $F \geq 0$ we get the following:

```
F0=[6 -1 2; -1 0 0 ; 2 0 0]   % Notice that any
F1=[1 0 0; 0 -1 0 ; 0 0 0]    % affine combination
```

```
F2=[0 0 0; 0 1 0; 0 0 -1]      % F =F0 + x1*F1 + x2*F2 + x3*F3
F3=[0 1 0; 1 0 -1 ; 0 -1 0]    % satisfies the linear constraints.
setlmis([])
x1 = lmivar(1,[1 1]); % x1, x2, x3 are a real scalars
x2 = lmivar(1,[1 1]);
x3 = lmivar(1,[1 1]);
lmiterm([-1 1 1 0],F0)
lmiterm([-1 1 1 x1],F1,1) % these lmiterm commands
lmiterm([-1 1 1 x2],F2,1) % create the LMI
lmiterm([-1 1 1 x3],F3,1) % 0 ≤ F0 + x1*F1 + x2*F2 + x3*F3
lmis = getlmis
[copt,xopt] = mincx(lmis,[-1;0;0],[1e-6 0 0 0 0 ]) % minimize -x1
F=F0+xopt(1)*F1+xopt(2)*F2 + xopt(3)*F3
chol(F)
```

Executing this script leads to

$$F = \begin{pmatrix} 5.1173 & -0.7190 & 2.0000 \\ -0.7190 & 0.1010 & -0.2810 \\ 2.0000 & -0.2810 & 0.7817 \end{pmatrix},$$

which factors as vv^* with $v^* = \begin{pmatrix} 2.2621 & -0.3178 & 0.8841 \end{pmatrix}$. Thus $p(z) = |2.2621 - 0.3178z + 0.8841z^2|^2$, $z \in \mathbb{T}$. It is no coincidence that the optimal matrix $F(x)$ is of rank 1 (see Exercise 1.6.49).

1.6 EXERCISES

1 Prove Proposition 1.1.1.

2 For the following subsets of $\mathbb{R}^2$ (endowed with the usual inner product), check whether the following sets are (1) cones; (2) closed. In the case of a cone determine their dual and their extreme rays.

(a) $\{x, y) \in \mathbb{R}^2 \ : \ x, y \geq 0\}$.

(b) $\{x, y) \in \mathbb{R}^2 \ : \ xy > 0\}$.

(c) $\{x, y) \in \mathbb{R}^2 \ : \ 0 \leq x \leq y\}$.

(d) $\{x, y) \in \mathbb{R}^2 \ : \ x + y \geq 0\}$.

3 Prove Lemma 1.1.2. (For the proof of part (ii) one needs the following corollary of the Hahn-Banach separation theorem: given a closed convex set $A \subset \mathcal{H}$ and a point $h_0 \notin A$, then there exist a $y \in \mathcal{H}$ and an $\alpha \in \mathbb{R}$ such that $\langle h_0, y \rangle < \alpha$ and $\langle a, y \rangle \geq \alpha$ for all $a \in A$.)

4 Show that $p(z) = \sum_{k=-n}^{n} p_k z^k$ is real for all $z \in \mathbb{T}$ if and only if $p_k = \overline{p_{-k}}$, $k = 0, \ldots, n$.

5 Prove the uniqueness up to a constant factor of modulus 1 of the outer (co-outer) factor of a polynomial $p \in \mathbb{T}\mathrm{Pol}_n^+$.

6 Show that $\mathrm{PSD}_G = (\mathrm{PSD}_G)^*$ if and only if all the maximal cliques in the graph G are disjoint.

7 For the following matrices A and patterns P determine whether A belongs to each of $\mathrm{PPSD}_G, \mathrm{PSD}_G, (\mathrm{PSD}_G)^*$, and $(\mathrm{PPSD}_G)^*$. Here G is the graph associated with P.

(a) $P = \{(i,j) \ ; \ 1 \leq i,j \leq 3 \text{ and } |i-j| \leq 1\}$, and $A = \begin{pmatrix} 5 & 1 & 0 \\ 1 & 2 & \frac{1}{2} \\ 0 & \frac{1}{2} & 1 \end{pmatrix}$.

(b) $P = \{(i,j) \ ; \ 1 \leq i,j \leq 3 \text{ and } |i-j| \leq 1\}$, and $A = \begin{pmatrix} 1 & 1 & 0 \\ 1 & 1 & \frac{1}{2} \\ 0 & \frac{1}{2} & \frac{1}{4} \end{pmatrix}$.

(c) $P = (\{1,\ldots,4\} \times \{1,\ldots,4\}) \setminus \{(1,3),(2,4),(3,1),(4,2)\}$ and

$$A = \begin{pmatrix} 1 & 1 & 0 & 1 \\ 1 & 2 & 1 & 0 \\ 0 & 1 & 1 & -1 \\ 1 & 0 & -1 & 2 \end{pmatrix}.$$

(d) $P = \{(i,j)\ ;\ 1 \le i,j \le 5 \text{ and } |i-j| \ne 2\}$ and

$$A = \begin{pmatrix} 1 & 2 & 0 & 4 & 5 \\ 2 & 4 & 6 & 0 & 10 \\ 0 & 6 & 9 & 12 & 0 \\ 4 & 0 & 12 & 16 & 20 \\ 5 & 10 & 0 & 20 & 25 \end{pmatrix}.$$

(e) $P = \{(i,j)\ ;\ 1 \le i,j \le 3 \text{ and } |i-j| \le 1\}$, and $A = \begin{pmatrix} 1 & 1 & 0 \\ 1 & 2 & 1 \\ 0 & 1 & 1 \end{pmatrix}$.

(f) $P = \{(i,j)\ ;\ 1 \le i,j \le 3 \text{ and } |i-j| \le 1\}$, and $A = \begin{pmatrix} 1 & 1 & 0 \\ 1 & 1 & 1 \\ 0 & 1 & 1 \end{pmatrix}$.

(g) $P = (\{1,\ldots,4\} \times \{1,\ldots,4\}) \setminus \{(1,3),(2,4),(3,1),(4,2)\}$ and

$$A = \begin{pmatrix} 1 & 1 & 0 & 1 \\ 1 & 1 & 1 & 0 \\ 0 & 1 & 1 & 1 \\ 1 & 0 & 1 & 1 \end{pmatrix}.$$

8 Can a chordal graph have 0,1,2,3 simplicial vertices? If yes, provide an example of such a graph. If no, explain why not.

9 Let

$$C_n = \{A = (A_{ij})_{i,j=1}^n \in \mathrm{PSD}_n\ :\ A_{ii} = 1, i = 1, \ldots, n\}$$

be the set of correlation matrices.

(a) Show that C_n is convex.

(b) Recall that $E \in C_n$ is an *extreme point* of C_n if $E \pm \Delta \in C_n$ implies that $\Delta = 0$. Show that if $E \in C_3$ is an extreme point, then E has rank 1.

(c) Show that the convex set of real correlation matrices $C_3 \cap \mathbb{R}^{3\times 3}$ has extreme points of rank 2. (In fact, they are exactly the rank 2 real correlation matrices whose off-diagonal entries have absolute value less than one.)

(d) Show that if $E \in C_n$ is an extreme point of rank k then $k^2 \le n$.

(e) Show that if $E \in C_n \cap \mathbb{R}^{n\times n}$ is an extreme point of rank k then $k(k+1) \le 2n$.

10 Let G be an undirected graph on n vertices. Let $\beta(G)$ denote the minimal number of edges to be added to G to obtain a chordal graph. For instance, if G is the four cycle then $\beta(G) = 1$. Prove the following.

(a) If A generates an extreme ray of $\mathrm{PSD}_G \cap \mathbb{R}^{n\times n}$, then $\mathrm{rank} A \le \beta(G)+1$. (Hint: Let $\widehat{G}$ be a chordal graph obtained from G by adding $\beta(G)$ edges. Let $A \in \mathrm{PSD}_G \cap \mathbb{R}^{n\times n}$ be of rank $r > \beta(G)+1$. Write $A = \sum_{k=1}^{r} A_k$ with $\mathrm{rank} A_k = 1$ and $A_k \in \mathrm{PSD}_{\widehat{G}} \cap \mathbb{R}^{n\times n}$. Show now that there exist positive $c_1, \ldots, c_r$, not all equal, such that $\sum_{k=1}^{r} c_k A_k \in \mathrm{PSD}_G \cap \mathbb{R}^{n\times n}$. Conclude that A does not generate an extreme ray.)

(b) If A generates an extreme ray of PSD_G, then $\mathrm{rank} A \le 2\beta(G)+1$.

(c) Show that the bound in (a) is not sharp. (Hint: Consider the ladder graph G on $2m$ vertices $\{1, \ldots, 2m\}$ where the edges are $(2i-1, 2i), i = 1, \ldots, m$, and $(2i-1, 2i+1), (2i, 2i+2), i = 1, \ldots, m-1$, and show that $\beta(G) = m-1$ while all extreme rays in $\mathrm{PSD}_G \cap \mathbb{R}^{n\times n}$ are generated by matrices of rank ≤ 2.)

11 Let G be the loop of n vertices. Show that $\mathrm{PSD}_G \cap \mathbb{R}^{n\times n}$ has extreme rays generated by elements of rank $n-2$, and that elements of higher rank do not generate an extreme ray.

12 Let G be the $(3,2)$ full bipartite graph. Show that PSD_G has extreme rays generated by rank 3 matrices, while all extreme rays in $\mathrm{PSD}_G \cap \mathbb{R}^{5\times 5}$ are generated by matrices of rank 2 or less.

13 This exercise concerns the dependance on Λ of $\mathbb{T}\mathrm{Pol}_\Lambda^2$.

(a) Find a sequence $\{p_n\}_{n=-3}^{3}$ for which is the matrix (1.3.1) is positive semidefinite but the matrix (1.3.2) is not.

(b) Find a trigonometric polynomial $q(z) = q_{-3}z^{-3} + \cdots + q_3 z^3 \ge 0$ such that $q(z) = |h(z)|^2$, where $h(z)$ is some polynomial $h(z) = h_0 + h_1 z + h_2 z^2 + h_3 z^3$, but $q(z)$ cannot be written as $|\widetilde{h}(z)|^2$ with $\widetilde{h}(z) = \widetilde{h}_0 + \widetilde{h}_1 z + \widetilde{h}_3 z^3$.

(c) Explain why the existence of the sequence $\{p_n\}_{n=-3}^{3}$ in (a) implies the existence of the trigonometric polynomial $q(z)$ in (b), and vice versa.

14 Let $T_2 = (c_{i-j})_{i,j=0}^{2}$ be positive semidefinite with $\dim \mathrm{Ker} T_2 = 1$. Show that $c_{-2} = \frac{1}{c_0}(c_{-1}^2 + (c_0^2 - |c_1|^2)e^{it})$ for some $t \in \mathbb{R}$.

15 Let $T_n = (c_{i-j})_{i,j=0}^{n}$ be positive definite, and define $c_{n+1} = \overline{c}_{-n-1}$ via

$$c_{n+1} := \begin{pmatrix} c_n & \cdots & c_1 \end{pmatrix} \begin{pmatrix} c_0 & \cdots & c_{-n-1} \\ \vdots & \ddots & \vdots \\ c_{n-1} & \cdots & c_0 \end{pmatrix}^{-1} \begin{pmatrix} c_1 \\ \vdots \\ c_n \end{pmatrix} + \frac{\det T_n}{\det T_{n-1}} e^{it_0}, \tag{1.6.1}$$

where $t_0 \in \mathbb{R}$ may be chosen arbitrarily. Show that $T_{n+1} = (c_{i-j})_{i,j=0}^{n+1}$ is positive semidefinite and singular.

16 Let μ be the measure given by $\mu(\theta) = \frac{1}{|e^{2i\theta}-2|^2} d\theta$. Find its moments c_k for $k = 0, 1, 2, 3, 4$, and show that $T_4 = (c_{i-j})_{i,j=0}^4$ is strictly positive definite.

17 Prove that the cone $\mathcal{D}_S$ in the proof of Theorem 1.3.5 is closed.

18 Consider the cone of positive maps $\Phi : \mathbb{C}^{n\times n} \to \mathbb{C}^{m\times m}$ (i.e., Φ is linear and $\Phi(\mathrm{PSD}_n) \subseteq \mathrm{PSD}_m$). Show that the maps $X \mapsto AXA^*$ and $X \mapsto AX^T A^*$ generate extreme rays of this cone.

19 Write the following positive semidefinite Toeplitz matrices as the sum of rank 1 positive semidefinite Toeplitz matrices. (Hint: any $n \times n$ rank 1 positive semidefinite Toeplitz matrix has the form

$$c \begin{pmatrix} 1 \\ \overline{\alpha} \\ \vdots \\ \overline{\alpha}^{n-1} \end{pmatrix} \begin{pmatrix} 1 & \alpha & \cdots & \alpha^{n-1} \end{pmatrix},$$

where $c > 0$ and $\alpha \in \mathbb{T}$.)

(a) $T = \begin{pmatrix} 2 & 1-i & 0 \\ 1+i & 2 & 1-i \\ 0 & 1+i & 2 \end{pmatrix}$.

(b) $T = \begin{pmatrix} 3 & -1+\sqrt{2}(1-i) & 1-2i \\ -1+\sqrt{2}(1+i) & 3 & -1+\sqrt{2}(1-i) \\ 1+2i & -1+\sqrt{2}(1+i) & 3 \end{pmatrix}$.

20 Let $c_j = \overline{c}_{-j}$ be complex numbers such that

$$T_n = (c_{j-k})_{j,k=0}^n$$

is positive semidefinite.

(a) Assume that T_n has a one-dimensional kernel, and let $(p_j)_{j=0}^n$ be a nonzero vector in this kernel. Show that $p(z) := \sum_{j=0}^n p_j z^j$ has all its roots on the unit circle $\mathbb{T}$.

(b) Show that

$$\det \begin{pmatrix} c_0 & \cdots & c_{-n+1} & 1 \\ \vdots & & \vdots & \vdots \\ c_n & \cdots & c_1 & z^n \end{pmatrix}$$

has the same roots as p.

(c) Show that the roots of p are all different (start with the cases $n = 1$ and $n = 2$).

(d) Denote the roots of p by $\alpha_1, \ldots, \alpha_n$. Let $g^{(j)}(z) = g_0^{(j)} + \cdots + g_n^{(j)} z^n$ be the unique polynomial of degree n such that

$$g^{(j)}(z) = \begin{cases} 1 & \text{when } z = \alpha_j, \\ 0 & \text{when } z = \alpha_k, \ k \neq j. \end{cases}$$

Put $\rho_j = v_j^* T_n v_j$, where $v_j = (g_k^{(j)})_{k=0}^n$. Show that $c_j = \sum_{k=1}^n \rho_k \alpha_k^j$ for $j = -n, \ldots, n$.

(e) Show how the above procedure gives a way to find a measure of finite support for the truncated trigonometric moment problem (see Theorem 1.3.6) in the case that T_n has a one-dimensional kernel.

(f) Adjust the above procedure to the case when T_n has a kernel of dimension $k > 1$. (Hint: apply the above procedure to T_{n-k+1}.)

(g) Adjust the above procedure to the case when T_n positive definite. (Hint: choose a $c_{n+1} = \overline{c}_{-n-1}$ so that T_{n+1} is positive semidefinite with a one-dimensional kernel; see Exercise 1.6.15.)

21 Given a finite positive Borel measure μ on $\mathbb{T}^d$ we consider integrals of the form

$$I(f) = \int_{\mathbb{T}^d} f d\mu.$$

A *cubature* (or *quadrature*) formula for $I(f)$ is an expression of the form

$$C(f) = \sum_{j=1}^{n} \rho_j f(\alpha_j),$$

with fixed nodes $\alpha_1, \ldots, \alpha_n \in \mathbb{T}^n$ and fixed weights $\rho_1, \ldots, \rho_n > 0$, such that $C(f)$ is a "good" approximation for $I(f)$ on a class of functions of interest. Show that if we express the positive semidefinite Toeplitz matrix

$$(\widehat{\mu}(\lambda - \nu))_{\lambda, \nu \in \Lambda}$$

as

$$\left(\sum_{j=1}^{n} \rho_j \alpha_j^{\lambda - \nu} \right)_{\lambda, \nu \in \Lambda}$$

(which can be done due to Theorem 1.3.5), then we have that

$$I(p) = C(p) \text{ for all } p \in \mathrm{Pol}_{\Lambda - \Lambda}.$$

22 Use the previous two exercises to find a cubature formula for the measure $\mu(z) = \frac{1}{2\pi i} \frac{dz}{z|2-z|^2}$ on $\mathbb{T}$ that is exact for trigonometric polynomials of degree 2.

23 Consider

$$A = \begin{pmatrix} 1 & \bar{t}_1 & \bar{t}_2 \\ t_1 & 1 & \bar{t}_1 \\ t_2 & t_1 & 1 \end{pmatrix}, \ B = \begin{pmatrix} 1 & \bar{t}_1 & \bar{t}_3 \\ t_1 & 1 & \bar{t}_2 \\ t_3 & t_2 & 1 \end{pmatrix}.$$

Assume that $|t_1| = 1$. Show that $A \geq 0$ if and only if $t_2 = t_1^2$ and that $B \geq 0$ if and only if $|t_2| \leq 1$ and $t_3 = t_1 t_2$.

24 For $\gamma \in [0, \pi]$, consider the cone $\mathcal{C}_\gamma$ of real valued trigonometric polynomials $q(z)$ on $\mathbb{T}$ of degree n so that $q(e^{i\theta}) \geq 0$ for $-\gamma \leq \theta \leq \gamma$.

(a) Show that $q \in \mathcal{C}_\gamma$ if and only if there exist polynomials p_1 and p_2 of degree at most n and $n-1$, respectively, so that

$$q(z) = |p_1(z)|^2 + \left(z + \frac{1}{z} - 2\cos\gamma\right)|p_2(z)|^2, \ z \in \mathbb{T}.$$

(Hint: Look at all the roots of $q(z)$ and conclude that the roots on the unit circle where $q(z)$ switches sign lie in the set $\{e^{i\theta} : \theta \in [\gamma, 2\pi - \gamma]\}$ and that there exist an even number of them. Notice that the other roots appear in pairs $\alpha, \frac{1}{\bar{\alpha}}$, so that $q(e^{i\theta}) = q_1(e^{i\theta})\prod_{k=1}^{2r}\sin\frac{t-t_k}{2}$, where $t_1, \ldots, t_{2r} \in [\gamma, 2\pi - \gamma]$ and q_1 is nonnegative on $\mathbb{T}$.)

(b) Determine the extreme rays of the cone $\mathcal{C}_\gamma$.

(c) Show that the dual cone of $\mathcal{C}_\gamma$ is given by those real valued trigonometric polynomials $g(z) = \sum_{k=-n}^{n} g_k z^k$ so that the following Toeplitz matrices are positive semidefinite:

$$(g_{i-j})_{i,j=0}^{n} \geq 0 \ \text{ and } \ (g_{i-j+1} + g_{i-j-1} - 2g_{ij}\cos\gamma)_{i,j=0}^{n-1} \geq 0.$$

25 Prove the first statement of the proof of Lemma 1.3.23.

26 Using positive semidefinite matrices, provide an alternative proof for the fact that

$$F(s,t) = s^2t^2(s^2 + t^2 - 1) + 1$$

is not a sum of squares. (Hint: suppose that $F(s,t) = \sum_{i=1}^{k} g_i(s,t)^2$. Show that $g_i(s,t) = a_i + b_i st + c_i s^2 t + d_i st^2$ for some real numbers a_i, b_i, c_i and d_i. Now rewrite $F(s,t)$ as

$$F(s,t) = \begin{pmatrix} 1 & st & s^2t & st^2 \end{pmatrix} \left(\sum_{i=1}^{k} \begin{pmatrix} a_i \\ b_i \\ c_i \\ d_i \end{pmatrix} \begin{pmatrix} a_i & b_i & c_i & d_i \end{pmatrix} \right) \begin{pmatrix} 1 \\ st \\ s^2t \\ st^2 \end{pmatrix}.$$

Thus we have written

$$F(s,t) = \begin{pmatrix} 1 & st & s^2t & st^2 \end{pmatrix} A \begin{pmatrix} 1 \\ st \\ s^2t \\ st^2 \end{pmatrix},$$

where A is a positive semidefinite real matrix. Now argue that this is impossible.)

27 For the following polynomials f determine polynomials g_i with real coefficients so that $f = \sum_{i=1}^{k} g_i^2$, or argue that such a representation is impossible. For those for which the representation is impossible, try to find a representation as above with g_i rational. (Hint: try $1 + x^2$, $1 + x^2 + y^2$, etc., as the numerator of g_i.)

(a) $f(x) = x^4 - 2x^3 + x^2 + 16$.

(b) $f(x, y) = 1 + x^2y^4 + x^4y^2 - 3x^2y^2$.

(c) $f(x, y) = 2x^4 + 2xy^3 - x^2y^2 + 5x^4$.

(d) $f(x, y) = x^2y^4 + x^4y^2 - 2x^3y^3 - 2xy^2 + 2x^2y + 1$.

(e) $f(x, y, z) = x^4 - (2yz + 1)x^2 + y^2z^2 + 2yz + 2$.

(f) $f(x, y) = \frac{1}{2}x^2y^4 - (2x^3 - \beta x^2)y^3 + [(\beta^2 + 2)x^4 + \beta x^3 + (\beta^2 + 9)x^2 + 2\beta x]y^2 - (8\beta x^2 - 4x)y + 4(\beta^2 + 1)$. (Hint: consider the cases $0 < |\beta| < 2$, $\beta = 0$, and $|\beta| \geq 2$, separately.)

28 Recall that $p(x, y, z)$ is a homogeneous polynomial of degree n if $p(ax, ay, az) = a^n p(x, y, z)$ for all x, y, z, and a.

(a) Show that if $p(x, y, z)$ is a homogeneous polynomial in three variables of degree 3, then

$$4p(0, 0, 1) + p(1, 1, 1) + p(-1, 1, 1) + p(1, -1, 1) + p(-1, -1, 1)$$

$$-2(p(1, 0, 1) + p(-1, 0, 1) + p(0, 1, 1) + p(0, -1, 1)) = 0.$$

(b) Show that if $f(x, y, z)$ is a sum of squares of homogeneous polynomials of degree 3, then

$$f(1, 1, 1) + f(-1, 1, 1) + f(1, -1, 1) + f(-1, -1, 1) + f(1, 0, 1)$$

$$+f(-1, 0, 1) + f(0, 1, 1) + f(0, -1, 1) - \frac{4}{5}f(0, 0, 1) \geq 0.$$

(c) Use (b) to show that the so-called Robinson polynomial

$$x^6 + y^6 + z^6 - (x^4(y^2 + z^2) + y^4(x^2 + z^2) + z^4(x^2 + y^2)) + 3x^2y^2z^2$$

is not a sum of squares of polynomials.

(d) Show that the Robinson polynomial only takes on nonnegative values on $\mathbb{R}^3$.

29 Theorem 1.3.8 may be interpreted in the following way. Given a nonempty finite $S \subseteq \mathbb{Z}$ of the form $S = \Lambda - \Lambda$, then $\mathcal{B}_S = \{p : T_{p,\Lambda} \geq 0\}$ if and only if $S = \{jb : j = -k, \dots, k\}$ for some $k, b \in \mathbb{N}$. For general finite S satisfying $S = -S$ one may ask whether the equality

$$\mathcal{B}_S = \{p : T_{p,\Lambda} \geq 0 \text{ for all } \Lambda \text{ with } \Lambda - \Lambda \subseteq S\}$$

holds. Show that for the set $S = \{-4, -3, -1, 0, 1, 3, 4\}$ equality does not hold.

30 Let $h : \mathbb{Z}^d \to \mathbb{Z}^d$ be a group isomorphism, and let Λ be a finite subset of $\mathbb{Z}^d$. Show that Λ has the extension property if and only if $h(\Lambda)$ has the extension property.

31 Prove that when $a \in \mathbb{T}^d$ and $G(\Lambda - \Lambda) \neq \mathbb{Z}^d$, then $\Omega_\Lambda(\{a\})$ contains more than one element.

32 The entropy of a continuous function $f \in C(\mathbb{T}^d)$, $f > 0$ on $\mathbb{T}^d$ is defined as

$$E(f) = \frac{1}{(2\pi)^d} \int_{[0,2\pi]^d} \log f(e^{i\langle k,t\rangle})dt.$$

Let $n \geq 1$, and let $\Lambda = R(0, n)$, $R(1, n)$, or $R(1, n) \setminus \{(1, n)\}$, and $\{c_k\}_{k\in\Lambda-\Lambda}$ be a positive sequence of complex numbers. We call a function

$$f(e^{i\langle k,t\rangle}) = \sum_{k\in\mathbb{Z}^d} c_k(f)e^{i\langle k,t\rangle}, \quad f \in C(\mathbb{T}^d),$$

a *positive definite extension of the sequence* $\{c_k\}_{k\in\Lambda-\Lambda}$ if $f > 0$ on $\mathbb{T}^d$ and $c_k(f) = c_k$ for $k \in \Lambda - \Lambda$. Assume that $\{c_k\}_{k\in\Lambda-\Lambda}$ admits an extension $f \in C(\mathbb{T}^d)$, $f > 0$ on $\mathbb{T}^d$. Prove that among all positive continuous extensions of $\{c_k\}_{k\in\Lambda-\Lambda}$, there is a unique one denoted f_0 which maximizes the entropy. Moreover, f_0 is the unique positive extension of the form $\frac{1}{P}$, with $P \in \Pi_{\Lambda-\Lambda}$.

33 Let $\{c_{kl}\}_{(k,l)\in R(1,n)-R(1,n)}$ be a positive semidefinite sequence. Show that there exists a unique positive definite extension $\{d_{kl}\}$, $k \in \{-1, 0, 1\}$, $l \in \mathbb{Z}$ of $\{c_{kl}\}_{(k,l)\in R(1,n)-R(1,n)}$ which maximizes the entropy $E(\{d_{kl}\})$, and that this unique extension is given via the formula

$$F_0(e^{it}, e^{iw}) = \frac{f_0(e^{iw})[f_0^2(e^{iw}) - |f_1(e^{iw})|^2]}{|f_0(e^{iw}) - e^{it}f_1(e^{iw})|^2},$$

where $f_0(e^{iw}) = \sum_{j=-\infty}^{\infty} d_{0j}e^{ijw}$ and $f_1(e^{iw}) = \sum_{j=-\infty}^{\infty} d_{1j}e^{ijw}$.

34 Prove that every positive semidefinite sequence on $S = \{-1, 0, 1\} \times \mathbb{Z}$ with respect to $R_1 = \{0, 1\} \times \mathbb{N}$ admits a positive semidefinite extension to $\mathbb{Z}^2$.

35 Suppose $\Lambda = \{(p_i, q_i)_{i=1}^n\} \subset \mathbb{Z}^2$ such that

$$A = \begin{pmatrix} p_1 & p_2 & \cdots & p_n \\ q_1 & q_2 & \cdots & q_n \end{pmatrix}$$

has rank 2. Prove the following.

(a) There exist integer matrices P and Q with $|\det P| = |\det Q| = 1$ such that

$$PAQ = \begin{pmatrix} f_1 & 0 & 0 & \dots & 0 \\ 0 & f_2 & 0 & \dots & 0 \end{pmatrix} = F,$$

where $f_1 = \gcd\{p_1, \dots, p_n, q_1, \dots, q_n\}$ and

$$f_2 = \frac{1}{f_1} \gcd \left\{ \det \begin{pmatrix} p_i & p_j \\ q_i & q_j \end{pmatrix}, 1 \le i < j \le n \right\}.$$

(b) $G(\Lambda - \Lambda) = \mathbb{Z}^2$ if and only if $f_1 = f_2 = 1$.

36 Let $\Lambda = \{(p_i, q_i) : i = 1, \dots, n\} \subset \mathbb{Z}^2$ and define

$$d = \max\{\gcd(p_i, q_i) : i = 1, \dots, n\}. \tag{1.6.2}$$

Show that there exists an isomorphism $\Phi : \mathbb{Z}^2 \to \mathbb{Z}^2$ such that $(d, 0) \in \Phi(\Lambda)$. In addition, show that d is the largest number with this property.

(Hint: if the maximum is obtained at $i = 1$, there exist integers k, l such that $p_1 k - q_1 l = d$. Define Φ via the matrix

$$\begin{pmatrix} k & -l \\ \frac{q_1}{d} & -\frac{p_1}{d} \end{pmatrix}.)$$

37 Let $f(x) = \sum_{k=0}^{m} f_k x^k$ be a polynomial such that $f(x) \ge 0$, $x \in \mathbb{R}$, and $f_m \ne 0$.

(a) Show that $m = 2n$ for some $n \in \mathbb{N}_0$.

(b) Show that $f_k \in \mathbb{R}$, $k = 0, \dots, 2n$.

(c) Show that α is a root of f if and only if $\overline{\alpha}$ is a root of f, and show that if α is a real root, then it has even multiplicity.

(d) Show that $f(x) = |g(x)|^2$ for some polynomial g of degree n with possibly complex coefficients.

(e) Show that g above can be chosen so that all its roots have nonnegative (nonpositive) imaginary part, and that with that choice g is unique up to a constant of modulus 1.

(f) Show that $f = h_1^2 + h_2^2$ for some polynomials h_1 and h_2 of degree $\le n$ with real coefficients. (Hint: take g as above and let h_1 and h_2 be real valued on the real line so that $g = h_1 + ih_2$ on the real line.)

(g) Use ideas similar to those above to show that a polynomial $\psi(x)$ with $\psi(x) \ge 0$, $x \ge 0$, can be written as

$$\psi(x) = h_1(x)^2 + h_2(x)^2 + x(h_3(x)^2 + h_4(x)^2),$$

where h_1, h_2, h_3, h_4 are polynomials with real coefficients.

38 Define the $(n+1)\times(n+1)$ matrices Φ_n and Ψ_n such that

$$\Phi_n \begin{pmatrix} 1 \\ u \\ \vdots \\ u^n \end{pmatrix} = \begin{pmatrix} (1+iu)^0(1-iu)^n \\ (1+iu)^1(1-iu)^{n-1} \\ \vdots \\ (1+iu)^n(1-iu)^0 \end{pmatrix},$$

$$\Psi_n \begin{pmatrix} 1 \\ z \\ \vdots \\ z^n \end{pmatrix} = \begin{pmatrix} (i-iz)^0(1+z)^n \\ (i-iz)^1(1+z)^{n-1} \\ \vdots \\ (i-iz)^n(1+z)^0 \end{pmatrix}.$$

For instance, when $n=2$ we obtain the matrix

$$\Phi_2 = \begin{pmatrix} 1 & -2i & -1 \\ 1 & 0 & 1 \\ 1 & 2i & -1 \end{pmatrix}.$$

(i) Show that $\Psi_n\Phi_n = 2^n I_{n+1} = \Phi_n\Psi_n$.

(ii) Prove that if $H = (h_{i+j})_{i,j=0}^n$ is a Hankel matrix then $T = \Phi_n H \Phi_n^*$ is a Toeplitz matrix, and vice versa. (A *Hankel matrix* $H = (h_{ij})$ is a matrix that satisfies $h_{ij} = h_{i+1,j-1}$ for all applicable i and j.)

39 Let $\mathbb{R}\mathrm{Pol}_n = \{g : g(x) = \sum_{k=0}^n g_k x^k, g_k \in \mathbb{R}\}$ be the space of real valued degree $\leq n$ polynomials, and let $\mathbb{R}\mathrm{Pol}_n^+ = \{g \in \mathbb{R}\mathrm{Pol}_n : g(x) \geq 0 \text{ for all } x \in \mathbb{R}\}$ be the cone of nonnegative valued degree $\leq n$ polynomials. On $\mathbb{R}\mathrm{Pol}_n$ define the inner product

$$\langle g, h\rangle = \sum_{i=0}^n g_i h_i, \quad \text{for } g(x) = \sum_{k=0}^n g_k x^k \text{ and } h(x) = \sum_{k=0}^n h_k x^k.$$

For $h(x) = \sum_{i=0}^{2n} h_i x^i$ define the Hankel matrix

$$H_h = (h_{i+j})_{i,j=0}^n.$$

(a) Show that $\mathbb{R}\mathrm{Pol}_{2n+1}^+ = \mathbb{R}\mathrm{Pol}_{2n}^+$ (see also part (a) of Exercise 1.6.37).

(b) Show that for $g(x) = \sum_{k=0}^n g_k x^k$ and $h(x) = \sum_{k=0}^{2n} h_k x^k$ we have that

$$\langle g^2, h\rangle = \begin{pmatrix} \overline{g_0} & \cdots & \overline{g_n} \end{pmatrix} H_h \begin{pmatrix} g_0 \\ \vdots \\ g_n \end{pmatrix}.$$

(c) Show that the dual cone of $\mathbb{R}\mathrm{Pol}_{2n}^+$ is the cone

$$\mathcal{C} = \{h \in \mathbb{R}\mathrm{Pol}_{2n} : H_h \geq 0\}.$$

(d) Show that the extreme rays of $\mathbb{R}\mathrm{Pol}_{2n}^+$ are generated by the polynomials in $\mathbb{R}\mathrm{Pol}_{2n}^+$ that have only real roots.

(e) Show that the extreme rays of $\mathcal{C}$ are generated by those polynomials h for which H_h has rank 1. (This part may be easier to do after reading the proof of Theorem 2.7.9.)

40 Let $p(x) = p_0 + \cdots + p_{2n}x^{2n}$ with $p_j \in \mathbb{R}$.

(a) Show that there exist polynomials $g_j(x) = g_0^{(j)} + \cdots + g_n^{(j)}x^n$, $j = 1, \ldots, m$, with real coefficients such that $p = \sum_{j=1}^m g_j^2$, if and only if there exists a positive semidefinite matrix $F = (F_{jk})_{j,k=0}^n$ satisfying

$$\sum_{l=0}^{k} F_{k-l,l} = p_k, \quad k = 0, \ldots, 2n. \tag{1.6.3}$$

(b) Let

$$F_0 = \begin{pmatrix} p_0 & \frac{1}{2}p_1 & & \\ \frac{1}{2}p_1 & p_2 & \ddots & \\ & \ddots & \ddots & \frac{1}{2}p_{2n-1} \\ & & \frac{1}{2}p_{2n-1} & p_{2n} \end{pmatrix} \in \mathbb{C}^{(n+1)\times(n+1)}.$$

Show that $F = F^*$ satisfies (1.6.3) if and only if there is an $n \times n$ Hermitian matrix $Y = (Y_{jk})_{j,k=1}^n$ so that

$$F = F_0 + \begin{pmatrix} 0_{n\times 1} & iY \\ 0_{1\times 1} & 0_{n\times 1} \end{pmatrix} + \begin{pmatrix} 0_{1\times n} & 0_{1\times 1} \\ -iY & 0_{n\times 1} \end{pmatrix}.$$

Here $0_{j\times k}$ is the $j \times k$ zero matrix.

(c) Let

$$A = -i \begin{pmatrix} 0 & & & \\ 1 & 0 & & \\ & \ddots & \ddots & \\ & & 1 & 0 \end{pmatrix} \in \mathbb{C}^{n\times n}, B = -i \begin{pmatrix} 1 \\ 0 \\ \vdots \\ 0 \end{pmatrix} \in \mathbb{C}^{n\times 1}.$$

Show that

$$\begin{pmatrix} 0_{1\times n} & 0_{1\times 1} \\ -iY & 0_{n\times 1} \end{pmatrix} = \begin{pmatrix} 0_{1\times 1} & 0_{1\times n} \\ YB & YA \end{pmatrix}.$$

(d) Assume that $p_0 > 0$ and write F_0 as

$$F_0 = \begin{pmatrix} G_{11} & G_{12} \\ G_{21} & G_{22} \end{pmatrix},$$

where $G_{11} = p_0$ and $G_{22} \in \mathbb{C}^{n\times n}$. Show that F in (b) is positive semidefinite if and only if

$$A^*Y + YA - (YB + G_{21})G_{11}^{-1}(B^*Y + G_{12}) + G_{22} \geq 0. \tag{1.6.4}$$

In addition, show that F is of rank 1 if and only equality holds in (1.6.4).

(e) The equation

$$P^*Y + YP - YQY + R = 0 \tag{1.6.5}$$

for known matrices P, $Q = Q^*$, $R = R^*$, and unknown $Y = Y^*$ is referred to as the *Algebraic Riccati Equation.* Show that this Riccati equation has a solution Y if and only if the range of the matrix $\begin{pmatrix} I_n \\ Y \end{pmatrix}$ is an invariant subspace of the so-called *Hamiltonian matrix*

$$H = \begin{pmatrix} -P^* & Q \\ R & P \end{pmatrix}.$$

(Hint: rewrite the Riccati equation as $\begin{pmatrix} -Y & I_n \end{pmatrix} H \begin{pmatrix} I_n \\ Y \end{pmatrix} = 0$.)

(f) For $p(x) = 1 + 4x + 10x^2 + 12x^3 + 9x^4$, determine the appropriate Riccati equation and its Hamiltonian. Use the eigenvectors of the Hamiltonian to determine an invariant subspace of H of the appropriate form, and obtain a solution to the Riccati equation. Use this solution to determine a rank 1 positive semidefinite F satisfying (1.6.3), and obtain a factorization $p(x) = |g(x)|^2$. Notice that if one chooses the invariant subspace of the Hamiltonian by using eigenvectors whose eigenvalues lie in the right half-plane, the corresponding g has all its roots in the closed left half-plane

Remark: The Riccati equation is an important tool in control theory to determine factorizations of matrix-valued functions with desired properties (see also Section 2.7).

41 Consider

$$\mathcal{C} = \{A = (a_{ij})_{i,j=1}^3 \in \mathbb{R}^{3\times 3} \;:\; A = A^T \geq 0 \text{ and } a_{12} = a_{33}\}.$$

(a) Show that $\mathcal{C}$ is a closed cone.

(b) Show that for $x, y \in \mathbb{R}$, not both 0, the matrix

$$\begin{pmatrix} x^2 \\ y^2 \\ xy \end{pmatrix} \begin{pmatrix} x^2 & y^2 & xy \end{pmatrix} \tag{1.6.6}$$

generates an extreme ray of $\mathcal{C}$.

(c) Show that all extreme rays are generated by a matrix of the type (1.6.6). (Hint: the most involved part is to show that any rank 2 element in $\mathcal{C}$ does not generate an extreme ray. To prove this, first show that if

$$\begin{pmatrix} a_1 \\ a_2 \end{pmatrix}, \begin{pmatrix} b_1 \\ b_2 \end{pmatrix}, \begin{pmatrix} c_1 \\ c_2 \end{pmatrix} \in \mathbb{R}^2$$

are such that $a_1 b_1 + a_2 b_2 = c_1^2 + c_2^2$, then there exist α and x_1, x_2, y_1, y_2 in $\mathbb{R}$ so that

$$\begin{pmatrix} \cos\alpha & -\sin\alpha \\ \sin\alpha & \cos\alpha \end{pmatrix} \begin{pmatrix} a_1 & b_1 & c_1 \\ a_2 & b_2 & c_2 \end{pmatrix} = \begin{pmatrix} x_1^2 & y_1^2 & x_1 y_1 \\ x_2^2 & y_2^2 & x_2 y_2 \end{pmatrix}.$$

The correct α will satisfy $\tan(2\alpha) = \frac{a_1 b_1 - a_2 b_2 - c_1^2 + c_2^2}{a_1 b_2 + a_2 b_1 - 2c_1 c_2}$.)

(d) Determine the dual cone of $\mathcal{C}$. What are its extreme rays?

42 Let $A = (A_{ij})_{i,j=1}^n$ be positive definite. Prove that there exists a unique positive definite matrix $F = (F_{ij})_{i,j=1}^n$ such that $F_{ij} = A_{ij}$, $i \neq j$, $\text{tr}F = \text{tr}A$ and the diagonal entries of the inverse of F are all the same. In case A is real, F is also real. (Hint: consider in Theorem 1.4.3 $B = 0$ and let $\mathcal{W}$ be the set of all real diagonal $n \times n$ matrices with their trace equal to 0.)

43 Consider the partial matrix

$$\mathcal{A} = \begin{pmatrix} 2 & ? & 1 & 1 \\ ? & 2 & ? & 1 \\ 1 & ? & 2 & ? \\ 1 & 1 & ? & 2 \end{pmatrix}.$$

Find the positive definite completion for which the determinant is maximal. Notice that this completion is not Toeplitz. Next, find the positive definite Toeplitz completion for which the determinant is maximal.

44 Find the representation (1.3.10) for the matrix

$$\begin{pmatrix} 2 & -1 & & & -1 \\ -1 & 2 & -1 & & \\ & \ddots & \ddots & \ddots & \\ & & -1 & 2 & -1 \\ -1 & & & -1 & 2 \end{pmatrix}. \tag{1.6.7}$$

45 Let A be a positive semidefinite Toeplitz matrix. Find a closest (in the Frobenius norm) rank 1 positive semidefinite Toeplitz matrix A_0. When is the solution unique? (Hint: use (1.3.10).) Is the solution unique in the case of the matrix (1.6.7)?

46 Let $A = (A_{j-i})_{i,j=1}^n$ be a positive definite Toeplitz matrix, $0 < p_1 < p_2 < \cdots < p_r < n$ and $\alpha_1, \ldots, \alpha_r \in \mathbb{C}$. Then there exists a unique Toeplitz matrix $F = (F_{j-i})_{i,j=1}^n$ with $F_q = A_q$, $|q| \neq p_1, \ldots, p_r$, and

$$\sum_{j-i=p_k} (F^{-1})_{ij} = \alpha_k \ , \ k = 1, \ldots, r.$$

In case A and $\alpha_1, \ldots, \alpha_r$ are real, the matrix F is also real. (Hint: Consider in Theorem 1.4.3 $B = (B_{ij})_{i,j=1}^n$ with $B_{1p_k} = B_{p_k 1}^* = \alpha_k$, $k = 1, \ldots, r$ and $B_{ij} = 0$ otherwise, and let $\mathcal{W}$ be the set of all Hermitian Toeplitz matrices which have entries equal to 0 on all diagonals except $\pm p_1, \ldots, \pm p_r$.)

47 Recall that $s_1(X) \geq s_2(X) \geq \cdots$ denote the singular values of the matrix X.

(a) Show that for positive definite $n \times n$ matrices A and B it holds that

$$\operatorname{tr}(A-B)(B^{-1}-A^{-1}) \geq \frac{\sum_{j=1}^n s_j(A-B)^2}{s_1(A)s_1(B)}.$$

(b) Use (a) to give an alternative proof of Corollary 1.4.4 that a positive definite matrix is uniquely determined by some of its entries and the complementary entries of its inverse (in fact, the statement is slightly stronger than Corollary 1.4.4, as we do not require the given entries in the matrix to include the main diagonal).

48 Prove Theorem 1.4.9.

49 Let $p_j = \overline{p}_{-j}$ be complex numbers and introduce the convex set

$$\mathcal{F} = \left\{ F = (F_{j,k})_{j,k=0}^n \geq 0 \ : \ \sum_{l=0}^{n-j} F_{l+j,l} = p_j, j = 0, \ldots, n \right\},$$

and assume that $\mathcal{F} \neq \emptyset$. Let $F_{\text{opt}} \in \mathcal{F}$ be so that $(F_{\text{opt}})_{00} \geq F_{00}$ for all $F \in \mathcal{F}$. Show that F_{opt} is of rank 1. (Hint: Let

$$F_{\text{opt}} = \begin{pmatrix} a & 0_{1\times n} \\ b & F_1 \end{pmatrix} \begin{pmatrix} \overline{a} & b^* \\ 0_{n\times 1} & F_1^* \end{pmatrix}$$

be a Cholesky factorization of F_{opt}. If F_1 has a nonzero upper left entry, argue that

$$F_{\text{opt}} - \begin{pmatrix} 0_{1\times 1} & 0_{1\times n} \\ 0_{n\times 1} & F_1F_1^* \end{pmatrix} + \begin{pmatrix} F_1F_1^* & 0_{n\times 1} \\ 0_{1\times n} & 0_{1\times 1} \end{pmatrix}$$

lies in $\mathcal{F}$ but has a larger upper left entry than F_{opt}. Adjust the argument to also include the case when $F_1 \neq 0$ but happens to have a zero upper left entry.)

Remark. Notice that the above argument provides a simple iterative method to find $F \in \mathcal{F}$ whose upper left entry is maximal. In principle one could employ this naive procedure to find a solution to the problem in Example 1.5.2.

1.7 NOTES

Section 1.1

The general results about cones and their duals, as well as those about positive (semi)definite matrices in Section 1.1 can be found in works like [91], [485], [108], [21], [319], and [538]. The proof of the Fejér-Riesz theorem dates back to [213] and can also be found in many contemporary texts (see, e.g., [53] and [510]). In Section 2.4 we present its generalization for operator-valued polynomials.

Section 1.2

Matrices which are inverses of tridiagonal ones were first characterized in [80] and matrices which are the inverses of banded ones in [81]. The theory of positive definite completions started in [114] with the basic 3×3 block matrix case. Next, banded partially positive definite block matrices were considered in [192]. Theorems 2.1.1 and 2.1.4 were first proven there in the finite dimensional case.

In [285] undirected graphs were associated with the patterns of partially positive semidefinite matrices and the importance of chordal graphs in this context was discovered. Proposition 1.2.9 is also taken from [285]. The approach based on cones and duality to study positive semidefinite extensions originates in [452] (see also [315] for the block tridiagonal case). Theorem 1.2.10 was first obtained there. The equality $(\mathrm{PSD}_G)^* = \mathrm{PPSD}_G$ for a chordal graph G was also established there, in a way similar to that in Section 1.2. The extreme rays of PSD_G for nonchordal graphs G were studied in [9], [309], [311], and [518]. Example 1.2.11 is taken from [9]. All notation, definitions, and results in Section 1.2 concerning graph theory follow the book [276].

Section 1.3

The approach in Section 1.3 for characterizing finite subsets of $\mathbb{Z}^d$ which have the extension property was developed in [115] and [503] (see also [510]). The characterization in [503] reduces the problem to the equivalence between $\mathbb{T}\mathrm{Pol}_\Lambda^2 = \mathbb{T}\mathrm{Pol}_S^+$ and $\mathcal{B}_S = \mathcal{A}_\Lambda$. Theorem 1.3.3 was taken from [314], which contains the simplest version of Bochner's theorem, namely the one for $G = \mathbb{Z}$.

Theorem 1.3.6 was originally proved by Carathéodory and Fejér in [120] by using a result in [119] by which the extremal rays of the cone of $n \times n$ positive semidefinite Toeplitz matrices are generated by rank 1 such matrices; see also [281, Section 4.1] for an excellent account. Corollary 1.3.7 may be found in [133].

Theorem 1.3.5 was first obtained in [240] by a different method. The theorem replicates a result by Naimark [17] characterizing the extremal solutions of the Hamburger moment problem. Theorems 1.3.8 through Corollary 1.3.16 were also first proven in [240]. However, we found an inaccuracy in the original statement and proof of Lemma 1.3.15. Still, Theorem 1.3.11

holds, which is the key for the proof of the main result, Corollary 1.3.16. We thank the author of that article for his kind help.

Theorem 1.3.20 comes from [58]. The first proofs of Corollary 1.3.34, Corollary 1.3.40, and Theorem 1.3.18 were obtained in [59]. A reference on the Smith normal forms for matrices with integer entries is [415]. Theorem 1.3.25 is due to [336].

Theorem 1.3.22 was proved for $R(N, N)$, $N \geq 3$, in [115] and [503]. That $R(2, 2)$ and $R(1, 1, 1)$ (see Theorem 1.3.17) do not have the extension property appears in [506] (see also [507]). A theorem by Hilbert [317] states that for every $N \geq 3$, there exist real polynomials $F(s, t)$ of degree $2N$ which are positive for every (s, t), but can not be represented as sums of squares of polynomials. The first explicit example of such polynomial was found in [426] (see Exercise 1.6.27(b)). Other parts from Exercise 1.6.27 came from [447], [398], and [103]. Many other examples appear in [483]. The polynomial in Lemma 1.3.23 and its proof can be found in [90] (see also [510]). Its use makes Theorem 1.3.22 be true for $R(2, 2)$ as well ([510]).

For application of positive trigonometric polynomials to signal processing, see, for instance, [188].

Section 1.4

Theorem 1.4.3 and Corollary 1.4.4 are taken from [69]. Corollary 1.4.4 appears earlier in [525, Theorem 1] as a statement on Gaussian measures with prescribed margins. The modified Newton algorithm used in Example 1.5.1 can be found in [433] (see also [106]). For a semidefinite programming approach, see [26]. Corollaries 1.4.4 and 1.4.5 were first proved in [219] (see also [285]). Exercise 1.6.47, which is based on [219], shows the reasoning there. In [70] an operator-valued version of Corollary 1.4.5 appears. The results in Subsection 1.4.2 are presented for polynomials of a single variable, but they similarly hold for polynomials of several variables as well. For a matrix-valued correspondent of Theorem 1.4.9 see Theorem 2.3.9. The fact that diagonally strictly dominant Hermitian matrices are positive definite follows directly from the Geršgorin disk theorem; see for example [323].

Section 1.5

Semidefinite programming (and the more general area of conic optimization) is an area that has developed developed during the past two decades. It is not our aim to give here a detailed account on this subject. The articles and books [106], [433], [107], [554], [108], [258], and [170] are good sources for information on this topic.

Exercises

Exercise 1.6.9 is based on [286] (parts (c), (e)) and [131] (part (b), (d)); see also [409] and [404] for more results on extreme points of the set of correlation matrices.

Exercise 1.6.10 (a) and (b) are based on [518, Theorem B], which solved a conjecture from [309]. Part (c) of this exercise is based on [309]. As a separate note, let us remark that the computation of $\beta(G)$ is an NP-complete problem; see [582]. In addition, all graphs for which the extreme rays are generated by matrices of rank at most 2 are characterized in [395].

The answer to Exercise 1.6.11 can be found in [9].

Exercise 1.6.12 is based on a result in [309].

Exercise 1.6.18 is based on [583]. In [316] the problem of finding a closest correlation matrix is considered, motivated by a problem in finance.

Exercise 1.6.23 is based on the proof of Lemma 2.2 in [336].

Exercise 1.6.24 is a result proved by N. Akhiezer and M.G. Krein (see [376]; in [25] the matrix-valued case appears), while Exercises 1.6.42 and 1.6.46 are results from [69].

Exercise 1.6.28 is based on [95]. The Robinson polynomial appears first in the paper [484]; see also [483] for a further discussion.

Exercises 1.6.32 and 1.6.33 come from [58].

The operations Φ and Ψ from Exercise 1.6.38 are discussed in [330].

Exercise 1.6.43 is based on an example from [412], and the answer for Exercise 1.6.45 may be found in [533]. The matrix (1.6.7), as well as minor variations, is discussed in [531].

Additional notes

The standard approach to the *frequency-domain design* of one-dimensional *recursive digital filters* begins by finding the *squared magnitude response* of a filter that best approximates the desired response. *Spectral factorization* is then performed to find the (unique) minimum-phase filter that has that magnitude response, which is guaranteed to be stable. The design problem, embodied by the determination of the squared magnitude function, and the implementation problem, embodied by the spectral factorization, are decoupled.

Consider a class of systems whose input u and output y satisfy a *linear*

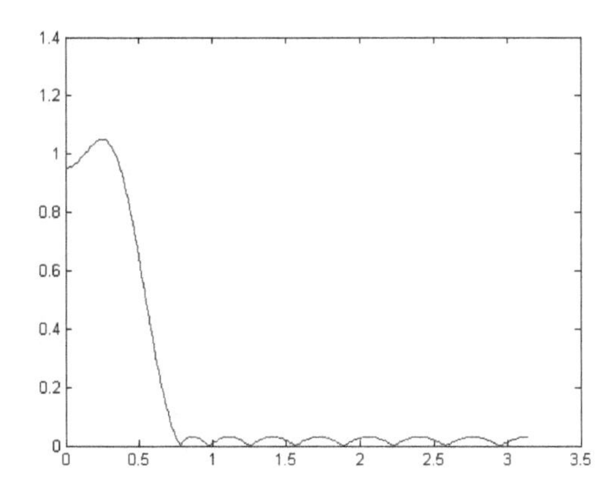

constant coefficient difference equation of the form

$$\sum_{k=0}^{n} a_k y_{t-k} = \sum_{k=0}^{n} b_k u_k. \tag{1.7.1}$$

The frequency response, that is, the system transfer function evaluated on the unit circle, has the form

$$f(z) = \frac{\sum_{k=0}^{n} a_k z^k}{\sum_{k=0}^{n} b_k z^k}.$$

In order to design a *low pass filter*, the constraints we have to satisfy are best written using the squared magnitude of the filter frequency response $|f(z)|^2$. In designing a lowpass filter one would like $|f(z)|^2 \approx 1$ when $z \in \{e^{it} : t \in [0, s]\}$ for some $0 < s < \pi$, and $|f(z)|^2 \approx 0$ when $z \in \{e^{it} : t \in [s, \pi]\}$. A typical squared magnitude response of a low pass filter looks like the graph above (see for example [581] or [244]).

Spectral factorization is now performed to find the coefficients a_k, b_k of the minimum-phase filter that has the desired magnitude response, which is guaranteed to be stable. The coefficients may then be used to implement the filter. General references on the subject include [347], [584], [27], [561], [469], [446], [589], [333], [107], and [432].

Chapter Two

Completions of positive semidefinite operator matrices

In this chapter we are concerned with positive definite and semidefinite completions of partial operator matrices, and we consider the banded case in Section 2.1, the chordal case in Section 2.2, the Toeplitz case in Section 2.3, and the generalized banded case and the operator-valued positive semidefinite chordal case in Section 2.6. In Section 2.4 we introduce the Schur complement and use it to derive an operator-valued Fejér-Riesz factorization. Section 2.5 is devoted to describing the structure of positive semidefinite operator matrices. In Section 2.7 we study the Hamburger problem based on positive semidefinite completions of Hankel matrices. Finally, in Section 2.8 we indicate the connection with linear prediction.

2.1 POSITIVE DEFINITE COMPLETIONS: THE BANDED CASE

We prove in this section that every partially positive definite banded operator matrix admits a positive definite completion. Among these completions, we show that there exists a distinguished one, the inverse of which admits a certain structured factorization. This particular completion is also characterized by a "maximum entropy" property, which in the finite dimensional case implies that it is the unique positive definite completion which maximizes the determinant. The section also contains a linear fractional parametrization for the set of all positive definite completions of a banded partially positive definite matrix, the parameter being a structured "upper-banded" contraction.

We denote by $\mathcal{L}(\mathcal{H}, \mathcal{H}')$ the set of all bounded linear operators acting from the Hilbert space $\mathcal{H}$ into the Hilbert space $\mathcal{H}'$; we abbreviate this to $\mathcal{L}(\mathcal{H})$ when $\mathcal{H} = \mathcal{H}'$. Recall that $A \in \mathcal{L}(\mathcal{H})$ is *positive definite* ($A > 0$) if $\langle Ax, x\rangle > 0$ for $0 \neq x \in \mathcal{H}$, and A has a bounded inverse. The operator is called *positive semidefinite* ($A \geq 0$) if $\langle Ax, x\rangle \geq 0$ for all $x \in \mathcal{H}$. A *partial matrix* is an $m \times n$ array $A = (A_{ij})_{i=1,j=1}^{m,n}$ of bounded linear operators, $A_{ij} \in \mathcal{L}(\mathcal{H}_j, \mathcal{H}_i)$, some of these operators being specified (known), and others being unspecified (unknown). In some particular cases we consider the entries to be matrices of appropriate size, or even scalars. We denote the unspecified entries of a partial matrix typically by ?, X, x, Y, y, and so on.

A *completion* or *extension* of a partial matrix is a specification of its unspecified entries, resulting in a conventional (operator) matrix. In this chapter we are interested in positive semidefinite completions of partial matrices. Since every principal submatrix of a positive semidefinite matrix inherits this property, we introduce the following notion. A partial operator matrix $A = (A_{ij})_{i,j=1}^n$ is called *partially positive semidefinite* when the following three conditions hold:

(i) All diagonal entries are specified.

(ii) A_{ij} is specified if and only if A_{ji} is specified and $A_{ji} = A_{ij}^*$.

(iii) All fully specified principal minors of A are positive semidefinite.

Note that in the scalar case this notion and the one of partial matrices which admit positive semidefinite completions are similar to that of the cones PPSD_G and $\mathcal{A}_G$ defined for the graph G of the pattern of the partial matrix in Section 1.2. Conditions (i) and (iii) imply that $A_{ii} \geq 0$, $i = 1, \ldots, n$. Similarly as above, we say A is *partially positive definite* when all fully specified minors of A are positive definite. Although one does not need to require that all diagonal entries be specified, it turns out that the principal submatrix where all entries are specified plays the determining role in this problem. This is indicated in Exercise 2.9.1

In this section we consider partial operator matrices of the form

$$K = \begin{pmatrix} A_{11} & \cdots & A_{1,1+p} & & ? \\ \vdots & & & \ddots & \\ A_{p+1,1} & & & & A_{n-p,n} \\ & \ddots & & & \vdots \\ ? & & A_{n,n-p} & \cdots & A_{nn} \end{pmatrix}, \tag{2.1.1}$$

which we call *p-banded.* A partial matrix K as in (2.1.1) is partially positive definite if and only if

$$\begin{pmatrix} A_{ii} & \cdots & A_{i,i+p} \\ \vdots & & \vdots \\ A_{i+p,i} & \cdots & A_{i+p,i+p} \end{pmatrix} > 0, \; i = 1, \ldots, n-p. \tag{2.1.2}$$

An operator matrix $H = (H_{ij})_{i,j=1}^n$ is called *p-banded* if $H_{ij} = 0$ for all $1 \leq i, j \leq n$ such that $|i - j| > p$. We first construct a positive definite completion for a partially positive definite p-banded operator matrix. This particular completion is the unique one which has a p-banded inverse. In the scalar case the existence of such a completion follows from Corollary 1.4.5.

Theorem 2.1.1 *Let K be a partially positive definite p-banded operator matrix as in* (2.1.1). *For $q = 1, \ldots, n$, let*

$$\begin{pmatrix} Y_{qq} \\ \vdots \\ Y_{\beta(q),q} \end{pmatrix} = \begin{pmatrix} A_{qq} & \cdots & A_{q,\beta(q)} \\ \vdots & & \vdots \\ A_{\beta(q),q} & \cdots & A_{\beta(q),\beta(q)} \end{pmatrix}^{-1} \begin{pmatrix} I_q \\ 0 \\ \vdots \\ 0 \end{pmatrix}, \tag{2.1.3}$$

and

$$\begin{pmatrix} X_{\gamma(q),q} \\ \vdots \\ X_{q,q} \end{pmatrix} = \begin{pmatrix} A_{\gamma(q),\gamma(q)} & \cdots & A_{\gamma(q),q} \\ \vdots & & \vdots \\ A_{q,\gamma(q)} & \cdots & A_{qq} \end{pmatrix}^{-1} \begin{pmatrix} 0 \\ \vdots \\ 0 \\ I_q \end{pmatrix}, \tag{2.1.4}$$

where $\beta(q) = \min\{n, p+q\}$ *and* $\gamma(q) = \max\{1, q-p\}$. *Let the* $n \times n$ *triangular operator matrices* U *and* V *be defined by*

$$V_{ij} = \begin{cases} Y_{ij} Y_{jj}^{-\frac{1}{2}} & \textit{when } j \leq i \leq \beta(j), \\ 0 & \textit{elsewhere}, \end{cases} \tag{2.1.5}$$

$$U_{ij} = \begin{cases} X_{ij} X_{jj}^{-\frac{1}{2}} & \textit{when } \gamma(j) \leq i \leq j, \\ 0 & \textit{elsewhere}. \end{cases} \tag{2.1.6}$$

Then the $n \times n$ *operator matrix* F *defined by the factorizations*

$$F := U^{*-1} U^{-1} = V^{*-1} V^{-1} \tag{2.1.7}$$

is the unique positive definite completion of K *such that* F^{-1} *is* p*-banded.*

Proof. Let $X = (X_{ij})_{i,j=1}^{n}$ be the banded upper triangular matrix with X_{ij} within the band given by (2.1.4). For any $n \times n$ matrix, we refer to the main diagonal as the 0 diagonal, those below it as diagonals $1, 2, \ldots, n-1$, and those above it as $-1, \ldots, -n+1$; thus, entry a_{ij} appears in the $(i-j)$th diagonal. Notice that KX has identities on the main diagonal and zeros in diagonals $1, \ldots, p$. Let P denote the strict upper triangular part of KX, and put $H = -PX^{-1}$. We now let $F = K + H + H^*$. Notice that $F = F^*$ and that the diagonals $-p, \ldots, 0 \ldots, p$ of F coincide with those of K. Then

$$FX = KX + HX + H^*X =: I + L,$$

where L is strictly lower triangular. Then $X^*FX = \operatorname{diag}(X_{jj})_{j=1}^{n} + \widetilde{L}$, with $\widetilde{L}$ strictly lower triangular. As X^*FX is selfadjoint, we must have that $\widetilde{L} = 0$. That gives $F = X^{*-1} \operatorname{diag}(X_{jj})_{j=1}^{n} X^{-1} = U^{*-1} U^{-1}$. Notice that $F^{-1} = UU^*$ is p-banded. In a similar way one can show that $V^{*-1}V^{-1}$ has the entries A_{ij} within the band $|i-j| \leq p$, and has an inverse VV^* that is p-banded.

Suppose now that $\widetilde{F} > 0$ has the entries A_{ij} within the band $|i-j| \leq p$, and that $\widetilde{F}^{-1}$ is p-banded. We will show that $\widetilde{F} = F$, yielding the uniqueness claim. Writing $F - \widetilde{F} = L + L^*$, with L lower triangular, gives that L has only nonzero diagonals $-n, \ldots, -p-1$. Factor $\widetilde{F}^{-1} = \widetilde{V}\widetilde{V}^*$ with $\widetilde{V}$ lower triangular. Then V is p-banded (as $\widetilde{F}^{-1}$ is). Write $M = \widetilde{F}^{-1} - F^{-1}$, which is p-banded. Then $L + L^* = F - \widetilde{F} = F(\widetilde{F}^{-1} - F^{-1})\widetilde{F} = U^{*-1}U^{-1}M\widetilde{V}^{*-1}\widetilde{V}^{-1}$. Thus $U^*(L + L^*)\widetilde{V} = U^{-1}M\widetilde{V}^{*-1}$, which is a matrix that has zeros in diagonals $-n, \ldots, -p-1$, as M is p-banded and U^{-1} and V^{*-1} are upper triangular. But then we must have that $U^*L\widetilde{V} = 0$, and thus $L = 0$. This gives that $F = \widetilde{F}$ is the unique positive definite completion of K which has a p-banded inverse. Consequently, we also find that $U^{*-1}U^{-1} = V^{*-1}V^{-1}$. □

A positive definite extension F of K is called a *band* or *central extension* of K if F^{-1} is p-banded. Note that the conditions (2.1.2) are clearly necessary for the existence of a positive extension of K, and by Theorem 2.1.1, these conditions are also sufficient for the existence of a unique band extension. We now obtain the following description for the set of all positive definite extensions of a given partially positive definite p-banded operator matrix.

Theorem 2.1.2 *Let K be a p-banded partial operator matrix as in (2.1.1). Then K admits a positive definite completion if and only if*

$$\begin{pmatrix} A_{ii} & \cdots & A_{i,i+p} \\ \vdots & & \vdots \\ A_{i+p,i} & \cdots & A_{i+p,i+p} \end{pmatrix} > 0, \; i = 1, \ldots, n-p.$$

Assume that the latter conditions hold. Let U and V be the the operator matrices defined in (2.1.3) - (2.1.6). *Then each positive definite extension F of K is of the form*

$$T(G) = (G^*V^* + U^*)^{-1}(I - G^*G)(VG + U)^{-1}, \tag{2.1.8}$$

where G is a strictly contractive (i.e., $\|G\| < 1$) $n \times n$ operator matrix with $G_{ij} = 0$, $j - i \leq p$. Furthermore, formula (2.1.8) *gives a one-to-one correspondence between all such G and all positive definite extensions F of K.*

Proof. Let $F_0 = U^{*-1}U^{-1} = V^{*-1}V^{-1}$ be as in Theorem 2.1.1 and write $F_0 = C + C^*$ with $C = (C_{ij})_{i,j=1}^n$ upper triangular and $C_{ii} = \frac{1}{2}A_{ii}$, $i = 1, \ldots, n$. Define

$$S(G) := (-C^*VG + CU)(VG + U)^{-1}$$

for all G for which $VG + U$ is invertible. Notice that when G is as above, VG is strictly upper triangular, and thus $VG + U$ is invertible. In fact, for such G we have that $S(G)$ is upper triangular. It is straightforward to check that $S(G) + S(G)^*$ equals

$$\begin{aligned} &(-C^*VG + CU)(VG + U)^{-1} + (G^*V^* + U^*)^{-1}(-G^*V^*C + U^*C^*) \\ &= (VG + U)^{*-1}(U^*CU + U^*C^*U - G^*V^*CVG - G^*V^*C^*VG)(VG + U)^{-1} \\ &= (G^*V^* + U^*)^{-1}(I - G^*G)(VG + U)^{-1} =: T(G). \end{aligned}$$

Thus $\|G\| < 1$ if and only if $T(G)$ is positive definite. Also, observe that

$$\begin{aligned} S(G) &= [(C - F)VG + CU](VG + U)^{-1} \\ &= C - FVG(VG + U)^{-1} \\ &= C - V^{*-1}G(VG + U)^{-1} \\ &= C - V^{*-1}(I + GU^{-1}V)^{-1}GU^{-1}. \end{aligned}$$

Notice that when G is a strictly contractive (in operator norm) $n \times n$ operator matrix with $G_{ij} = 0$, $j - i \leq p$, then $S(G) - C$ is upper triangular with diagonals $0, \ldots, p$ equal to zero. Thus in that case $T(G)$ is a positive definite extension of K.

Conversely, if F is a positive definite extension of K, write $F = Z + Z^*$, where $Z = (Z_{ij})_{i,j=1}^n$ is upper triangular with $Z_{ii} = \frac{1}{2} A_{ii}$. Put $W = Z - C$ and notice that W is upper triangular with zeros in diagonals $0, \ldots, p$. Next, notice that V^*WV is strictly upper triangular, and thus $I + V^*WV$ is invertible. Put $G = -(I + V^*WV)^{-1} V^*WU$, which is upper triangular with zeros in diagonals $0, \ldots, p$. It is straightforward to check that $S(G) = Z$, and thus $T(G) = F$. Thus $T(G)$ is positive definite and therefore $\|G\| < 1$. This proves the result. □

Remark 2.1.3 We remark that if K in the above theorem is Toeplitz, then only part of the operators U and V show a Toeplitz structure. However, it turns out (see Exercise 2.9.4) that $T(G)$ is Toeplitz if and only if G is Toeplitz.

Similarly to the Cholesky factorization in the scalar case, a positive definite operator matrix $A = (A_{ij})_{i,j=1}$ can be factored as $A = L^*L$, with $L = (L_{ij})_{i,j=1}^n$ an invertible lower triangular operator matrix (i.e., $L_{ij} = 0$ for $i > j$), and as $A = U^*U$, with $U = (U_{ij})_{i,j=1}^n$ an invertible upper triangular operator matrix (i.e., $U_{ij} = 0$ for $j > i$). The factors L and U are unique up to multiplying on the left with a diagonal unitary operator matrix. Indeed, if $U^*U = \widetilde{U}^*\widetilde{U}$ with $\widetilde{U}$ also upper triangular and invertible, then $U\widetilde{U}^{-1} = \widetilde{U}^{*-1}U^*$ is unitary and simultaneously upper and lower triangular. Thus $U\widetilde{U}^{-1} = \widetilde{U}^{*-1}U^*$ must be a diagonal unitary operator matrix. This shows that U is unique up to multiplying on the left by a diagonal unitary factor. Therefore, we can associate with A the following uniquely determined diagonal factors. When $A = L^*L$ with $L = (L_{ij})_{i,j=1}^n$ a lower triangular operator matrix, we define

$$\Delta_l(A) := \operatorname{diag}(L_{ii}^* L_{ii})_{i=1}^n \tag{2.1.9}$$

to be the *left diagonal factor* of A. Similarly, when $A = U^*U$ with $U = (U_{ij})_{i,j=1}^n$ an invertible upper triangular operator matrix, we define

$$\Delta_r(A) := \operatorname{diag}(U_{ii}^* U_{ii})_{i=1}^n \tag{2.1.10}$$

to be the *right diagonal factor* of A. It is not hard to see that the left and right diagonal factors are uniquely defined. In addition, when A is positive definite, the diagonal entries $(\Delta_l(A))_{ii}$ of $\Delta_l(A)$ are given by

$$A_{ii} - \begin{pmatrix} A_{i,i+1} & \cdots & A_{in} \end{pmatrix} \begin{pmatrix} A_{i+1,i+1} & \cdots & A_{i+1,n} \\ \vdots & & \vdots \\ A_{n,i+1} & \cdots & A_{nn} \end{pmatrix}^{-1} \begin{pmatrix} A_{i+1,i} \\ \vdots \\ A_{ni} \end{pmatrix}$$

for $i = 1, \ldots, n-1$, and

$$(\Delta_l(A))_{nn} = A_{nn}.$$

Similarly, $(\Delta_r(A))_{ii}$ is given by

$$A_{ii} - \begin{pmatrix} A_{i1} & \cdots & A_{i,i-1} \end{pmatrix} \begin{pmatrix} A_{11} & \cdots & A_{1,i-1} \\ \vdots & & \vdots \\ A_{1,i-1} & \cdots & A_{i-1,i-1} \end{pmatrix}^{-1} \begin{pmatrix} A_{1i} \\ \vdots \\ A_{i-1,i} \end{pmatrix}$$

for $i = 2, \ldots, n$, and

$$(\Delta_r(A))_{11} = A_{11}.$$

In fact, as we shall see in Section 2.4, the diagonal entries of $\Delta_l(A)$ and $\Delta_r(A)$ are Schur complements.

Recall that the *Loewner ordering* is a partial ordering on selfadjoint operators given by $A \geq B$ if and only if $A - B \geq 0$ (i.e., $A - B$ is positive semidefinite). We also write $B \leq A$ instead of $A \geq B$. In addition, the notation $A > B$ (or $B < A$) means that $A - B$ is positive definite. In the following theorem we show that among all positive definite extensions of a p-banded partially positive definite matrix, the unique band extension has the maximal right and left diagonal factors (in the Loewner ordering). We refer to this as the *maximum entropy property* of the band extension.

Theorem 2.1.4 *Let K be a partially positive definite p-banded operator matrix as in (2.1.1). For $i = 1, \ldots, n$ define*

$$(M_i)^{-1} := \begin{pmatrix} 0 & \cdots & 0 & I \end{pmatrix} \begin{pmatrix} A_{\gamma(i),\gamma(i)} & \cdots & A_{\gamma(i),i} \\ \vdots & & \vdots \\ A_{i,\gamma(i)} & \cdots & A_{ii} \end{pmatrix}^{-1} \begin{pmatrix} 0 \\ \vdots \\ 0 \\ I \end{pmatrix}, \tag{2.1.11}$$

where $\gamma(i) = \max\{1, i-p\}$. Then for the right multiplicative diagonal $\Delta_r(A)$ of a positive definite extension A of K we have that

$$\Delta_r(A) \leq \operatorname{diag}(M_i)_{i=1}^n. \tag{2.1.12}$$

Moreover, equality holds if and only if A is the unique band extension of K.

Also, we let

$$(\widehat{M}_i)^{-1} := \begin{pmatrix} I & 0 & \cdots & 0 \end{pmatrix} \begin{pmatrix} A_{ii} & \cdots & A_{i,\beta(i)} \\ \vdots & & \vdots \\ A_{\beta(i),i} & \cdots & A_{\beta(i),\beta(i)} \end{pmatrix}^{-1} \begin{pmatrix} I \\ 0 \\ \vdots \\ 0 \end{pmatrix}, \tag{2.1.13}$$

where $\beta(i) = \min\{n, p+i\}$. Then for the left multiplicative diagonal $\Delta_l(A)$ of a positive definite extension A of K we have that

$$\Delta_l(A) \leq \operatorname{diag}(\widehat{M}_i)_{i=1}^n. \tag{2.1.14}$$

Moreover, equality holds if and only if A is the unique band extension of K.

Proof. We only prove the statement for $\Delta_r(A)$ as the statement for $\Delta_l(A)$ follows in a similar manner. For an operator matrix $B = (B_{ij})_{i,j=1}^n$ we let $\text{diag}(B)$ denote the diagonal operator matrix $\text{diag}(B_{ii})_{i=1}^n$. Note that if $B \geq 0$, then $\text{diag}(B) = 0$ if and only if $B = 0$.

Let A be a positive definite extension of K and let F be the unique band extension of K. Factor

$$A = (I + W)^*\Delta_r(A)(I + W), \quad F = (I + X)^*\Delta_r(F)(I + X),$$

with W and X strictly upper triangular. Notice that as F^{-1} is p-banded, $(I+X)^{-1}$ is upper triangular p-banded. Write $A = F+(A-F)$ and observe that

$$(I + X)^{*-1}(I + W)^*\Delta_r(A)(I + W)(I + X)^{-1}$$

$$= \Delta_r(F) + (I + X)^{*-1}(A - F)(I - X)^{-1}. \tag{2.1.15}$$

As A and F are both positive extensions of K, we get that $A - F = Z + Z^*$, where Z is some upper triangular operator matrix with zeros in diagonals $0, \ldots, p$. Then $(I + X)^{*-1}Z$ is strictly upper triangular (as $(I + X)^{*-1}$ has zeros in diagonals $-n, \ldots, -p - 1$), and thus also $(I + X)^{*-1}Z(I + X)^{-1}$. But then

$$(I + X)^{*-1}(A - F)(I + X)^{-1} = (I + X)^{*-1}(Z + Z^*)(I + X)^{-1}$$

has zero diagonal entries. Write $(I + W)(I + X)^{-1} = I + Y$ with Y strictly upper triangular. Looking at the main diagonal of (2.1.15) we get

$$\text{diag}((I + Y)^*\Delta_r(A)(I + Y)) = \Delta_r(F),$$

and thus

$$\Delta_r(A) + \text{diag}(Y^*\Delta_r(A)Y) = \Delta_r(F).$$

But then $\Delta_r(A) \leq \Delta_r(F)$ as $Y^*\Delta_r(A)Y \geq 0$. Moreover, equality holds if and only if $\text{diag}(Y^*\Delta_r(A)Y) = 0$, which holds if only if $Y^*\Delta_r(A)Y = 0$. But the latter is equivalent to $(\Delta_r(A))^{\frac{1}{2}}Y = 0$, which by the invertibility of $\Delta_r(A)$ is equivalent to $Y = 0$. Thus $\Delta_r(A) = \Delta_r(F)$ if and only if $A = F$. It remains to observe that $\Delta_r(F) = \text{diag}(M_i)_{i=1}^n$, which follows directly from (2.1.7) and (2.1.4). □

In case K is a partial block matrix (all Hilbert spaces $\mathcal{H}_1, \ldots, \mathcal{H}_n$ are finite dimensional), Theorem 2.1.4 implies the following.

Corollary 2.1.5 *Let K be a partially positive definite block-matrix as in (2.1.1). Then the band extension F of K is the only extension that maximizes the determinant over the set of all positive definite extensions of K.*

Proof. This follows immediately from Theorem 2.1.4, since for every positive definite extension A of K, $\det A = \prod_{k=1}^n (\Delta_l(A))_{kk}$, and if X and Y are matrices such that $0 < X \leq Y$, then $\det X = \det Y$ implies $X = Y$ (see Exercise 2.9.3). □

We note that Corollary 2.1.5 also follows from Corollary 1.4.5.

2.2 POSITIVE DEFINITE COMPLETIONS: THE CHORDAL CASE

In this section we show that several of the results of the previous section go through for partially positive definite operator matrices whose pattern has an underlying graph that is chordal. Not all results go through; for example, Theorem 2.1.2 does not have a direct analog in the chordal pattern case.

We start the section by analyzing in more detail the basic positive definite completion problem, namely that of a 3×3 partial matrix with only its two corners unspecified. We then associate a graph to each partially positive definite operator matrix. The importance of chordal graphs in this context is already known from Section 1.2. Using a certain graph construction, we prove that every partially positive definite operator matrix with a chordal graph admits a positive definite completion. The proof is done by a one entry at a time procedure, using at each step the basic 3×3 construction. Among all positive definite completions of a partially positive definite operator matrix with a chordal graph, we show that there exists a distinguished one, the inverse of which admits a certain structured factorization. This particular completion is also characterized by a "maximum entropy" property, which in the finite dimensional case is equivalent to the fact that it is the unique positive definite completion which maximizes the determinant.

The following lemma is a useful characterization of positive definite 2×2 operator matrices.

Lemma 2.2.1 *The following are equivalent:*

(i) The operator matrix $\begin{pmatrix} A & B \\ B^* & C \end{pmatrix}$ *is positive definite.*

(ii) $A > 0$ *and* $C - B^*A^{-1}B > 0$.

(iii) $C > 0$ *and* $A - BC^{-1}B^* > 0$.

Proof. The result follows immediately from the *Frobenius-Schur factorizations* (see Exercise 2.9.2), which are

$$\begin{pmatrix} A & B \\ B^* & C \end{pmatrix} = \begin{pmatrix} I & 0 \\ B^*A^{-1} & I \end{pmatrix} \begin{pmatrix} A & 0 \\ 0 & C - B^*A^{-1}B \end{pmatrix} \begin{pmatrix} I & A^{-1}B \\ 0 & I \end{pmatrix} \tag{2.2.1}$$

and

$$\begin{pmatrix} A & B \\ B^* & C \end{pmatrix} = \begin{pmatrix} I & BC^{-1} \\ 0 & I \end{pmatrix} \begin{pmatrix} A - BC^{-1}B^* & 0 \\ 0 & C \end{pmatrix} \begin{pmatrix} I & 0 \\ C^{-1}B^* & I \end{pmatrix}. \tag{2.2.2}$$

□

Corollary 2.2.2 *Let* $H = \begin{pmatrix} A & B \\ B^* & C \end{pmatrix} \in \mathcal{L}(\mathcal{H} \oplus \mathcal{H}')$. *Then* $H > 0$ *if and only if* $A > 0$, $C > 0$, *and there exists* $G \in \mathcal{L}(\mathcal{H}', \mathcal{H})$, $\|G\| < 1$, *such that* $B = A^{\frac{1}{2}}GC^{\frac{1}{2}}$.

Proof. Assume $A > 0$, $C > 0$, and $B = A^{\frac{1}{2}}GC^{\frac{1}{2}}$, where $G \in \mathcal{L}(\mathcal{H}', \mathcal{H})$, $\|G\| < 1$. Then $C - B^*A^{-1}B = C^{\frac{1}{2}}(I - G^*G)C^{\frac{1}{2}} > 0$, since $\|G\| < 1$ implies $I - G^*G > 0$. Then Lemma 2.2.1 implies $H > 0$.

If $H > 0$, then we certainly must have $A > 0$ and $C > 0$. Define $G = A^{-\frac{1}{2}}BC^{-\frac{1}{2}}$. By Lemma 2.2.1 we have that

$$0 < C - B^*A^{-1}B = C^{\frac{1}{2}}(I - G^*G)C^{\frac{1}{2}},$$

which implies $I - G^*G > 0$, so $\|G\| < 1$. □

The basic step in the completion problem of partially positive definite operator matrices is the analysis of a 3×3 partial matrix with only one entry unspecified. This situation is described by the next theorem, which deals with a particular case of Theorem 2.1.2. The proof provided here is more straightforward and outlines the insight of this basic problem.

Theorem 2.2.3 *Every partially positive definite operator matrix of the form*

$$K = \begin{pmatrix} A & B & ? \\ B^* & C & D \\ ? & D^* & E \end{pmatrix} \tag{2.2.3}$$

admits a positive definite completion. In particular, one of them is obtained by defining the $(1,3)$ *entry of* K *to equal* $BC^{-1}D$*. In fact, this choice for the* $(1,3)$ *entry is the unique choice for which the completion* F *satisfies* $(F^{-1})_{13} = 0$*.*

Proof. The partial matrix K is partially positive definite when $\left(\begin{smallmatrix} A & B \\ B^* & C \end{smallmatrix}\right) > 0$ and $\left(\begin{smallmatrix} C & D \\ D^* & E \end{smallmatrix}\right) > 0$. By Corollary 2.2.2 there exist strict contractions G_1 and G_2 such that $B = A^{\frac{1}{2}}G_1C^{\frac{1}{2}}$ and $D = C^{\frac{1}{2}}G_2E^{\frac{1}{2}}$. Let F be the Hermitian operator matrix obtained by defining the $(1,3)$ entry of K to equal $A^{\frac{1}{2}}G_1G_2E^{\frac{1}{2}} = BC^{-1}D$. The factorization

$$F = \begin{pmatrix} A & A^{\frac{1}{2}}G_1C^{\frac{1}{2}} & A^{\frac{1}{2}}G_1G_2E^{\frac{1}{2}} \\ C^{\frac{1}{2}}G_1^*A^{\frac{1}{2}} & C & C^{\frac{1}{2}}G_2E^{\frac{1}{2}} \\ E^{\frac{1}{2}}G_2^*G_1^*A^{\frac{1}{2}} & E^{\frac{1}{2}}G_2^*C^{\frac{1}{2}} & E \end{pmatrix} \tag{2.2.4}$$

$$= \begin{pmatrix} A^{\frac{1}{2}} & 0 & 0 \\ 0 & C^{\frac{1}{2}} & 0 \\ 0 & 0 & E^{\frac{1}{2}} \end{pmatrix} \begin{pmatrix} I & G_1 & G_1G_2 \\ G_1^* & I & G_2 \\ G_2^*G_1^* & G_2^* & I \end{pmatrix} \begin{pmatrix} A^{\frac{1}{2}} & 0 & 0 \\ 0 & C^{\frac{1}{2}} & 0 \\ 0 & 0 & E^{\frac{1}{2}} \end{pmatrix}$$

$$= \begin{pmatrix} A^{\frac{1}{2}} & 0 & 0 \\ 0 & C^{\frac{1}{2}} & 0 \\ 0 & 0 & E^{\frac{1}{2}} \end{pmatrix} \begin{pmatrix} I & 0 & 0 \\ G_1^* & I & 0 \\ G_2^*G_1^* & G_2^* & I \end{pmatrix} \begin{pmatrix} I & 0 & 0 \\ 0 & I - G_1^*G_1 & 0 \\ 0 & 0 & I - G_2^*G_2 \end{pmatrix}$$

$$\times \begin{pmatrix} I & G_1 & G_1G_2 \\ 0 & I & G_2 \\ 0 & 0 & I \end{pmatrix} \begin{pmatrix} A^{\frac{1}{2}} & 0 & 0 \\ 0 & C^{\frac{1}{2}} & 0 \\ 0 & 0 & E^{\frac{1}{2}} \end{pmatrix}$$

implies that F is a positive definite completion of (2.2.3).

Next, denote $F^{-1} = (\Phi_{ij})_{i,j=1}^{3}$. Using the above factorization of F one sees immediately that $\Phi_{13} = 0$. In addition, if F is a completion of K with $\Phi_{13} = 0$ then

$$B^*\Phi_{11} + C\Phi_{12} = 0, \quad F_{31}\Phi_{11} + D^*\Phi_{12} = 0,$$

which implies that $F_{31} = -D^*\Phi_{12}\Phi_{11}^{-1} = D^*C^{-1}B^*$, yielding the uniqueness claim. □

If A is an $n \times n$ partially positive semidefinite matrix, then *the graph* $G = (V, E)$ *of* A is defined via

$$V = \{1, \dots, n\}, \; E = \{(i, j) : i \neq j, \; A_{ij} \text{ is specified}\}.$$

The following is the main result of this section.

Theorem 2.2.4 *Every partially positive definite operator matrix the graph of which is chordal admits a positive definite completion.*

For the proof of Theorem 2.2.4 we need the following additional graph theoretical result. Given a graph $G = (V, E)$, its *complement* is defined as $G = (V, E^c)$, where $E^c = \{(u, v) : u \neq v, (u, v) \notin E\}$.

Lemma 2.2.5 *Let $G = (V, E)$ be a noncomplete chordal graph. Then there exist $u, v \in V$ such that $(u, v) \notin E$ and the graph $G' = (V, E \cup \{(u, v)\})$ is also chordal. Moreover, G' contains a unique maximal clique which is not a clique in G.*

Proof. Let $\sigma = [v_1, \dots, v_n]$ be a perfect scheme for the chordal graph G and let k be the largest number such that $G|\{v_k, \dots, v_n\}$ is not complete. Let $l \in \{k+1, \dots, n\}$ be such that $(v_k, v_l) \notin E$. Define $E' = E \cup \{(v_k, v_l)\}$ and let $G' = (V, E')$. Then by definition, $\sigma = [v_1, \dots, v_n]$ is a perfect scheme for G' as well, so G' is chordal by Theorem 1.2.4.

Assume G' contains two distinct maximal cliques, K and K', which are not cliques in G. Then $\{v_k, v_l\} \subset K \cap K'$, and there exist $v_s \in K - K'$ and $v_t \in K' - K$. This implies $[v_k, v_s, v_l, v_t]$ is a chordless cycle of length 4 in G, contradicting its chordality. □

Given a chordal graph $G = (V, E)$, by Lemma 2.2.5 there exists a sequence $G_j = (V, E_j)$, $j = 0, \dots, t$, of chordal graphs, such that $G_0 = G$, G_t is the complete graph on the set V, and each E_j is obtained by adding a single edge (u_j, v_j) to E_{j-1}, for $j = 1, \dots, t$. We call such a sequence an *admissible sequence* for G.

Proof of Theorem 2.2.4. Let $A = (A_{ij})_{i,j=1}^{n}$ be a partially positive definite operator matrix, the graph $G = (V, E)$ of which is chordal. Let $G_j = (V, E_j)$, $j = 0, \dots, t$ be an admissible sequence for G, where $E_j = E_{j-1} \cup \{(u_j, v_j)\}$ for $j = 1, \dots, t$. Let K_j be the unique maximal clique of G_j which is not a clique in G_{j-1}. We construct a positive definite completion of A by specifying at each step j the entry A_{u_j,v_j} of A. Assume we have already extended A to a partially positive definite matrix $A^{(j-1)}$ with graph G_{j-1}. To K_j there corresponds a partially positive definite matrix $A^{(j)}|K_j$ in which only the entries corresponding to (u_j, v_j) and (v_j, u_j) are unspecified. We can reorder the rows and columns of this partial matrix such that it becomes of the form

$$\begin{pmatrix} M & N & X_{u_j,v_j} \\ N^* & P & R \\ X^*_{u_j,v_j} & R^* & Q \end{pmatrix}, \tag{2.2.5}$$

which by Theorem 2.2.3 admits a positive definite completion. This completion defines the (u_j, v_j) and (v_j, u_j) entries of a partially positive definite extension $A^{(j)}$ of A, the graph of which is G_j. At the end of this process, at step t, we obtain a positive definite completion of A. □

Let $\mathcal{H}_1, \dots, \mathcal{H}_n$ be Hilbert spaces and $G = (V, E)$ a graph with $V = \{1, \dots, n\}$. In line with the notation in Chapter 1, we denote by PD_G the set of all positive definite operator matrices $A = (A_{ij})_{i,j=1}^n$ acting on $\mathcal{H}_1 \oplus \cdots \oplus \mathcal{H}_n$, such that $A_{ij} = 0$ whenever $(i, j) \notin E$.

The next proposition is a slight modification of Proposition 1.2.9. It shows that Theorem 2.2.4 holds for chordal graphs only.

Proposition 2.2.6 *Let $G = (V, E)$ be a nonchordal graph. Then there exists a partially positive definite matrix K with graph G which does not admit positive definite completions.*

Proof. The proof is left to the reader (see Exercise 2.9.7). □

The following result is an operator version of Corollary 1.2.8.

Theorem 2.2.7 *Let $G = (V, E)$ be a chordal graph such that $[1, \dots, n]$ is a perfect scheme for G. Then $H \in PD_G$ if and only if H admits the factorization*

$$H = LL^*, \tag{2.2.6}$$

where L is lower triangular with $L_{ij} = 0$ whenever $i > j$ and $(i, j) \notin E$.

Proof. Assume H admits the factorization (2.2.6), and let $1 \le i, j \le n$ be such that $(i, j) \notin E$. Then

$$H_{ij} = \sum_{k=1}^{n} L_{ik}(L^*)_{kj}.$$

If L_{ik} and $(L^*)_{kj}$ are nonzero, we have $i \ge k$, $j \ge k$, and $(i, k), (j, k) \in E$. Since the vertex k is simplicial in the graph $G|\{k, \dots, n\}$, we obtain that $(i, j) \in E$, which is false. Thus for every $k = 1, \dots, n$, $L_{ik} = 0$ or $(L^*)_{kj} = 0$, so $H \in PD_G$.

Conversely, let $H \in PD_G$. Write H in the form

$$H = \begin{pmatrix} M & N \\ N^* & P \end{pmatrix} = \begin{pmatrix} I & 0 \\ N^*M^{-1} & I \end{pmatrix} \begin{pmatrix} M & 0 \\ 0 & P - N^*M^{-1}N \end{pmatrix} \begin{pmatrix} I & M^{-1}N \\ 0 & I \end{pmatrix} \tag{2.2.7}$$

with

$$M = H_{11},\ N = \begin{pmatrix} H_{12} & \cdots & H_{1n} \end{pmatrix},\ \text{and } P = H|\{2, \ldots, n\}.$$

Let $2 \leq i, j \leq n$ be such that $(i, j) \notin E$. Then

$$(P - N^*M^{-1}N)_{ij} = -H_{i1}H_{11}^{-1}H_{1j},$$

and since the vertex 1 is simplicial in G, we have that $(1, i) \notin E$, or $(1, j) \notin E$, so $(P - N^*M^{-1}N)_{ij} = 0$. This implies that $P - N^*M^{-1}N \in PD_{G|\{2,\ldots,n\}}$.

Consider $\binom{M}{N^*} M^{-\frac{1}{2}}$ to be the first column of L. The factorization (2.2.7) of $P - N^*M^{-1}N$ similarly defines the second column of L. We continue this way until we eliminate all vertices of G and finally obtain the factorization (2.2.6) of H. □

We call the factorization (2.2.6) of an operator matrix $H \in PD_G$ its *right triangular G-factorization.*

Theorem 2.2.8 *Let $G = (V, E)$ be a chordal graph such that $[1, \ldots, n]$ is a perfect scheme for G. Let $K = (A_{ij})_{i,j=1}^n$ be a partially positive definite operator matrix with graph G. For each $j = 1, \ldots, n$, let*

$$S_j = \{k \in \text{Adj}(j) | k > j\} = \{j_1, \ldots, j_s\}.$$

Define

$$\begin{pmatrix} X_{jj} \\ X_{jj_1} \\ \vdots \\ X_{jj_s} \end{pmatrix} = (K|\{j\} \cup S_j)^{-1} \begin{pmatrix} I \\ 0 \\ \vdots \\ 0 \end{pmatrix}. \tag{2.2.8}$$

Let $L = (L_{ij})_{i,j=1}^n$ be the $n \times n$ lower triangular operator matrix be defined by

$$L_{ij} = \begin{cases} X_{jj}^{\frac{1}{2}} & \textit{when } i = j, \\ X_{ij}X_{jj}^{-\frac{1}{2}} & \textit{when } j < i \textit{ and } (i,j) \in E, \\ 0 & \textit{elsewhere.} \end{cases} \tag{2.2.9}$$

Then the operator matrix F defined by

$$F := L^{*-1}L^{-1} \tag{2.2.10}$$

is the unique positive definite completion of K the inverse of which admits a right triangular G-factorization.

Proof. We first prove that F is a positive definite completion of K. From the construction of F it immediately follows that $F_{nn} = A_{nn}$. Let $1 \leq k < n$ and assume that for $k+1 \leq i, j \leq n$, such that $(i, j) \in E$ or $i = j$, we have that $F_{ij} = A_{ij}$. Let

$$K|\{k\} \cup S_k = \begin{pmatrix} M_k & N_k \\ N_k^* & P_k \end{pmatrix},$$

with

$$M_k = A_{kk},\ N_k = [A_{kk_1}, \ldots, A_{kk_s}],\ \text{and}\ P_k = K|S_k,$$

where $S_k = \{k_1, \ldots, k_s\}$. Denoting

$$X^{(k)} = \begin{pmatrix} X_{kk_1} \\ \vdots \\ X_{kk_s} \end{pmatrix},$$

(2.2.8) implies that

$$\begin{pmatrix} M_k & N_k \\ N_k^* & P_k \end{pmatrix} \begin{pmatrix} X_{kk} \\ X^{(k)} \end{pmatrix} = \begin{pmatrix} I \\ 0 \end{pmatrix}. \tag{2.2.11}$$

Let

$$F|\{k\} \cup S_k = \begin{pmatrix} M_k' & N_k' \\ N_k'^* & P_k \end{pmatrix},$$

with

$$M_k' = F_{kk},\ N_k' = [F_{kk_1}, \ldots, F_{kk_s}],\ \text{and}\ P_k = F|S_k = K|S_k.$$

Since $FL = L^{*-1}$ and L^{*-1} is upper triangular, from the definition of L we have that

$$\begin{pmatrix} M_k' & N_k' \\ N_k'^* & P_k \end{pmatrix} \begin{pmatrix} X_{kk} \\ X^{(k)} \end{pmatrix} = \begin{pmatrix} I \\ 0 \end{pmatrix}. \tag{2.2.12}$$

As X_{kk} is invertible, a straightforward computation shows that

$$N_k' = N_k = -X_{kk}^{-1} X^{(k)*} P_k,$$

and

$$M_k' = M_k = X_{kk}^{-1} + X_{kk}^{-1} X^{(k)*} P_k X^{(k)} X_{kk}^{-1},$$

implying that $F_{ij} = A_{ij}$ whenever $k \leq i, j \leq n$ and $(i, j) \in E$ or $i = j$. Thus, by induction, F is a positive definite completion of K, and $F^{-1} = LL^*$ is a right triangular G-factorization.

It remains to prove the uniqueness part. Assume F' is another positive definite completion of K such that F'^{-1} admits the right triangular G-factorization $F'^{-1} = L'L'^*$. For every $j = 1, \ldots, n$ we have $F'|\{j\} \cup S_j = K|\{j\} \cup S_j$, and consequently $F'L' = L'^{*-1}$ implies

$$(K|\{j\} \cup S_j) \begin{pmatrix} L_{jj}' \\ L_{jj_1}' \\ \vdots \\ L_{jj_s}' \end{pmatrix} = \begin{pmatrix} L_{jj}'^{-1} \\ 0 \\ \vdots \\ 0 \end{pmatrix}. \tag{2.2.13}$$

Using relations (2.2.8), (2.2.9), and (2.2.13), an elementary computation shows the equality

$$\begin{pmatrix} L_{jj} \\ L_{jj_1} \\ \vdots \\ L_{jj_s} \end{pmatrix} = \begin{pmatrix} L_{jj}' \\ L_{jj_1}' \\ \vdots \\ L_{jj_s}' \end{pmatrix}$$

for all $j = 1, \ldots, n$, so $L' = L$, which implies the uniqueness of F. □

Corollary 2.2.9 *Every partially positive definite operator matrix K the graph G of which is chordal admits a unique positive definite completion F such that $F^{-1} \in PD_G$.*

Proof. Reorder the rows and columns of K by the same permutation so that $[1, \ldots, n]$ becomes a perfect vertex elimination scheme for its graph and use Theorem 2.2.8 to construct the positive definite completion $\widetilde{F}$ of it. Let F be the matrix obtained by changing the rows and columns of $\widetilde{F}$ back to their initial positions. Then F is a positive definite completion of K which by Theorem 2.2.7 has the property that $K^{-1} \in PD_G$. □

We next prove that the distinguished completion F given by Theorem 2.2.8 verifies a maximum entropy property similar to the one for the band case described by Theorem 2.1.4.

Theorem 2.2.10 *Let $G = (V, E)$ be a chordal graph such that $[1, \ldots, n]$ is a perfect scheme for G. Let K be a partially positive definite operator matrix with graph G, and let F be the unique positive definite completion of K such that F^{-1} admits a right triangular G-factorization. Then for every positive definite completion B of K we have that*

$$\Delta_l(B) \leq \Delta_l(F), \tag{2.2.14}$$

with equality if and only if $B = F$.

Proof. The proof is very similar to that of Theorem 2.1.4. Let

$$F = (I + X)^* \Delta_l(F)(I + X), \text{ and } B = (I + W)^* \Delta_l(B)(I + W),$$

where X and W are lower triangular with a zero diagonal. Then the result follows as in Theorem 2.1.4 if we prove that the operator matrix $(I + X)^{*-1}(B - F)(I + X)^{-1}$ has a zero diagonal. Indeed, we have that

$$[(I + X)^{*-1}(B - F)(I + X)^{-1}]_{jj} = \sum_{i,k=1}^{n} (I + X)^{*-1}_{ji}(B - F)_{ik}(I + X)^{-1}_{kj},$$

so if $(I + X)^{*-1}_{ji} \neq 0$, and $(I + X)^{-1}_{kj} \neq 0$, then $i \geq j$, $(i, j) \in E$, $k \geq j$, and $(k, j) \in E$. Since j is simplicial in $G|\{j, j+1, \ldots, n\}$, we have that $(i, k) \in E$, so $F_{ik} = B_{ik} = A_{ik}$; thus $(B - F)_{ik} = 0$, and the result follows. □

The following is a consequence of Theorem 2.2.10. It generalizes Corollary 2.1.5 to the chordal case and it also follows from Corollary 1.4.5.

Corollary 2.2.11 *Let K be a partially positive definite block matrix the graph G of which is chordal. Then the unique positive definite completion F of K such that $F^{-1} \in PD_G$ is also the unique completion of K which maximizes the determinant over the set of all positive definite completions of F.*

Proof. Reorder the rows and columns of K by the same permutation so that $[1, \ldots, n]$ becomes a perfect vertex elimination scheme for its graph.

The result follows then immediately from Theorem 2.2.10, since for every positive definite extension B of K, we have that $\det B = \prod_{k=1}^{n}(\Delta_l(B))_{kk}$, and if X and Y are matrices such that $0 < X \leq Y$, then $\det X = \det Y$ implies $X = Y$ (see Exercise 2.9.3). $\square$

Similarly, one can define its *left triangular G-factorization* of a matrix $H \in PD_G$ as $H = UU^*$, where U is upper triangular with U_{ij} whenever $i > j$ and $(i,j) \notin E$.

We can also define for every positive definite completion B of a partially positive definite matrix K with an associated chordal graph the quantity $\Delta_r(B)$, which verifies $\Delta_r(B) \leq \Delta_(F)$ with equality if and only if $B = F$.

2.3 POSITIVE DEFINITE COMPLETIONS: THE TOEPLITZ CASE

In this section we will consider the Carathéodory-Toeplitz trigonometric moment problem but now for operator-valued coefficients. We will consider the strictly positive definite case as it will allow us to construct a particularly elegant operator-valued measure. Szegő's result on roots of orthogonal polynomials will be connected to the theory.

The Hilbert spaces $\mathcal{H}, \mathcal{H}'$ in this section are separable Hilbert spaces. Let $C_k \in \mathcal{L}(\mathcal{H})$, $k = -n, \ldots, 0, \ldots, n$, be given such that $C_{-j} = C_j^*$ for $j = 0, \ldots, n$. We are interested in finding an operator-valued positive measure $\mu(\theta)$ on the unit circle so that its moments $\widehat{\mu}(k)$ coincide with C_k, $k = -n, \ldots, n$. As in the previous section one can easily argue that the positive semidefiniteness of the Toeplitz operator matrix

$$T_n = \begin{pmatrix} C_0 & C_{-1} & \cdots & C_{-n} \\ C_1 & C_0 & \cdots & C_{-n+1} \\ \vdots & \vdots & \ddots & \vdots \\ C_n & C_{n-1} & \cdots & C_0 \end{pmatrix}$$

is necessary for the existence of a solution. For the remainder of this section we will assume that $T_n > 0$. In the next sections, where we will deal with general positive semidefinite completions, we will develop techniques to deal with the singular case as well. As we will see, the positive definiteness will yield the existence of a measure that is absolutely continuous with respect to the Lebesgue measure with a weight function that is the inverse of a trigonometric polynomial. This implies that given an operator-valued polynomial $C(z) = \frac{1}{2}C_0 + \sum_{k=1}^{n} C_k z^k$ such that $T_n > 0$, there exist so-called positive definite real part extensions of $C(z)$, that is, bounded analytic functions $F(z) = \frac{1}{2}F_0 + \sum_{k=0}^{\infty} F_k z^k$ in $\mathbb{D}$ such that $F_k = C_k$ for $k = 0, \ldots, n$, and $(F + F^*)(e^{it}) \geq \epsilon I$, almost everywhere on $\mathbb{T}$, for a certain $\epsilon > 0$. We notice that the problem of finding such a function F corresponds to the completion of T_n to an infinite positive definite kernel $(F_{i-j})_{i,j=0}^{\infty}$. The meaning of this is that for any solution F of the above problem and any $q \geq n$, the Toeplitz matrix $T_q = (F_{i-j})_{i,j=0}^q$, where $F_{-j} = F_j^*$, verifies $T_q \geq \epsilon I$. We include a linear fractional parametrization for the set of all such extensions $F(z)$ of $C(z)$, and show that in the matrix-valued case, the distinguished extension the inverse of which is a trigonometric polynomial of degree n, is also the unique extension which maximizes a certain entropy function.

Let $L^\infty(\mathcal{L}(\mathcal{H}, \mathcal{H}'))$ denote the Banach algebra of all *weakly measurable* functions $F : \mathbb{T} \to \mathcal{L}(\mathcal{H}, \mathcal{H}')$ such that $\|F\|_\infty = \text{ess sup}_{z\in\mathbb{T}}\|F(z)\| < \infty$. On $L^\infty(\mathcal{L}(\mathcal{H}))$ we have an involution $*$ defined via $F^*(z) = F(z)^*$. As the norm satisfies $\|F^*F\|_\infty = \|F\|_\infty^2$ for all $F \in L^\infty(\mathcal{L}(\mathcal{H}))$, we have that $L^\infty(\mathcal{L}(\mathcal{H}))$ is a C^*-algebra. The elements F of $L^\infty(\mathcal{L}(\mathcal{H}, \mathcal{H}'))$ can be expanded as $F(e^{it}) = \sum_{k=-\infty}^{\infty} F_k e^{ikt}$ with the series converging weakly for almost all $e^{it} \in \mathbb{T}$. As in the scalar case, the elements F_k are called the *Fourier coefficients of* F. A function $P : \mathbb{T} \to \mathcal{L}(\mathcal{H}, \mathcal{H}')$ of the form $P(e^{it}) = \sum_{k=-n}^{n} P_k e^{ikt}$ is referred to as an $\mathcal{L}(\mathcal{H}, \mathcal{H}')$*-valued trigonometric polynomial of degree* n, while $Q : \mathbb{D} \to \mathcal{L}(\mathcal{H}, \mathcal{H}')$, of the form $Q(z) = \sum_{k=0}^{n} Q_k z^k$ as an $\mathcal{L}(\mathcal{H}, \mathcal{H}')$*-valued analytic polynomial of degree* n. A function $F : \mathbb{D} \to \mathcal{L}(\mathcal{H}, \mathcal{H}')$ is called *analytic* if for every $h \in \mathcal{H}$ and $k \in \mathcal{H}'$, $\langle F(z)h, k\rangle$ is an analytic function in the unit disk. Such a function can be represented as $F(z) = \sum_{n=0}^{\infty} F_n z^n$, where $F_n \in \mathcal{L}(\mathcal{H}, \mathcal{H}')$ for $n \geq 0$, the series converging weakly in $\mathbb{D}$. For an analytic function $F : \mathbb{D} \to \mathcal{L}(\mathcal{H}, \mathcal{H}')$, define $\|F\|_\infty := \sup_{z\in\mathbb{D}} \|F(z)\|$. The space of all analytic F such that $\|F\|_\infty < \infty$ is denoted by $H^\infty(\mathcal{L}(\mathcal{H}, \mathcal{H}'))$ (the $\mathcal{L}(\mathcal{H}, \mathcal{H}')$-valued Hardy space). $H^\infty(\mathcal{L}(\mathcal{H}, \mathcal{H}'))$ is identified in a natural way with a closed subspace of $L^\infty(\mathcal{L}(\mathcal{H}, \mathcal{H}'))$.

Let $\mathcal{B}(\mathbb{T})$ denote the class of all Borel measurable subsets of $\mathbb{T}$. A function $\mu : \mathcal{B}(\mathbb{T}) \to \mathcal{L}(\mathcal{H})$ is called an *operator-valued positive measure* (or *semi-spectral measure*) on $\mathbb{T}$, if for each $h \in \mathcal{H}$, $\mu_h(\sigma) = \langle \mu(\sigma)h, h\rangle$ is a positive Borel measure on $\mathbb{T}$. It can easily be shown (see Exercise 2.9.14) that for each $j \in \mathbb{Z}$, there exists $\widehat{\mu}(j) \in \mathcal{L}(\mathcal{H})$, called the j*th moment* of μ, such that for each $h \in \mathcal{H}$, $\langle \widehat{\mu}(j)h, h\rangle$ is the jth moment of the measure μ_h. We say μ is *absolutely continuous* when for each $h \in \mathcal{H}$, μ_h is absolutely continuous with respect to the Lebesgue measure on $\mathbb{T}$. In this case there exists a weakly measurable function $F : \mathbb{T} \to \mathcal{L}(\mathcal{H})$ with $F(e^{it}) \geq 0$ a.e. on $\mathbb{T}$, such that for each $h \in \mathcal{H}$, $d\langle \mu(t)h, h\rangle = \langle F(e^{it})h, h\rangle dt$. In this case the jth moment of μ equals $\widehat{F}(j)$, the jth Fourier coefficient of F. Moreover, $F(e^{it}) = \sum_{j=-\infty}^{\infty} \widehat{F}(j)e^{ijt}$, where the series converges weakly in the L^1 norm; see Exercise 2.9.15.

The main result of this section is the following theorem. It can be viewed as a Toeplitz version of Theorem 2.1.1. It also represents an operator-valued generalization of the Carathéodory-Fejér truncated trigonometric moment problem (Theorem 1.3.6). We denote $\mathbb{C}_\infty = \mathbb{C} \cup \{\infty\}$. The expression that $Y(z)$ is invertible for $z = \infty$ means that $\lim_{z\to\infty} Y(z)$ exists and is invertible.

Theorem 2.3.1 *Let $C_j \in \mathcal{L}(\mathcal{H})$, $j = -n, \ldots, n$, be such that the Toeplitz* *operator matrix*

$$T_n := \begin{pmatrix} C_0 & C_{-1} & \cdots & C_{-n} \\ C_1 & C_0 & \cdots & C_{-n+1} \\ \vdots & \vdots & \ddots & \vdots \\ C_n & C_{n-1} & \cdots & C_0 \end{pmatrix} \quad (2.3.1)$$

is positive definite. Let

$$\begin{pmatrix} X_0 \\ X_1 \\ \vdots \\ X_n \end{pmatrix} = T_n^{-1} \begin{pmatrix} I \\ 0 \\ \vdots \\ 0 \end{pmatrix}, \quad \begin{pmatrix} Y_{-n} \\ \vdots \\ Y_{-1} \\ Y_0 \end{pmatrix} = T_n^{-1} \begin{pmatrix} 0 \\ \vdots \\ 0 \\ I \end{pmatrix}, \quad (2.3.2)$$

and put $X(z) = \sum_{k=0}^{n} X_k z^k$ *and* $Y(z) = \sum_{k=-n}^{0} Y_k z^k$. *Then* $X(z)$ *is invertible for* $z \in \overline{\mathbb{D}}$, *and* $Y(z)$ *is invertible for* $z \in \mathbb{C}_\infty \setminus \mathbb{D}$. *Moreover, the operator-valued function*

$$F(z) := X(z)^{*-1} X_0 X(z)^{-1} = Y(z)^{*-1} Y_0 Y(z)^{-1}, \quad z \in \mathbb{T}, \quad (2.3.3)$$

with Fourier coefficients $\widehat{F}(k)$, $k \in \mathbb{Z}$, *has the property that*

$$\widehat{F}(k) = C_k, \ k = -n, \dots, n. \quad (2.3.4)$$

Moreover, $\widehat{F}(k) = \widehat{F}(-k)^* =: C_k$, $|k| > n$, *is given inductively by*

$$C_k = \begin{pmatrix} C_{k-1} & \cdots & C_{k-n} \end{pmatrix} T_{n-1}^{-1} \begin{pmatrix} C_1 \\ \vdots \\ C_n \end{pmatrix}. \quad (2.3.5)$$

The equations (2.3.2) are sometimes referred to as Yule-Walker equations. Let us illustrate the result by considering the following simplest case.

Example 2.3.2 Let $T_1 = \begin{pmatrix} c_0 & c_{-1} \\ c_1 & c_0 \end{pmatrix} : \mathbb{C}^2 \to \mathbb{C}^2$ be positive definite. Then $x(z) = \frac{c_0 - c_1 z}{c_0^2 - |c_1|^2}$, so either $x(z)$ has no roots (when $c_1 = 0$), or the single root $\frac{c_0}{c_1}$ which is of modulus greater than one (due to the positive definiteness of T_1). Similarly, $y(z) = \frac{c_0 - c_{-1} z^{-1}}{c_0^2 - |c_1|^2}$ has no roots in $\mathbb{C}_\infty \setminus \mathbb{D}$. Furthermore, for $z \in \mathbb{T}$,

$$\begin{aligned} f(z) = \frac{c_0(c_0^2 - |c_1|^2)}{|c_0 - c_1 z|^2} &= \frac{c_1 z}{1 - \frac{c_1}{c_0} z} + \frac{c_0}{1 - \frac{c_{-1}}{c_0 z}} \\ &= \cdots + \frac{c_{-1}^2}{c_0 z^2} + \frac{c_{-1}}{z} + c_0 + c_1 z + \frac{c_1^2 z^2}{c_0} + \cdots, \end{aligned}$$

which has the desired Fourier coefficients.

We first establish the following useful lemma.

Lemma 2.3.3 *For* $G_0, \dots, G_{n-1} \in \mathcal{L}(\mathcal{H})$, *introduce the monic polynomial*

$$G(z) = z^n I + z^{n-1} G_{n-1} + \cdots + z G_1 + G_0$$

and its companion matrix

$$C_G := \begin{pmatrix} 0 & I & \cdots & 0 \\ \vdots & \vdots & \ddots & \vdots \\ 0 & 0 & \cdots & I \\ -G_0 & -G_1 & \cdots & -G_{n-1} \end{pmatrix} : \mathcal{H}^n \to \mathcal{H}^n. \qquad (2.3.6)$$

Then $G(z)$ is invertible if and only if $zI - C_G$ is. Similarly, if $U_1, \dots, U_n \in \mathcal{L}(\mathcal{H})$, put

$$H_U = \begin{pmatrix} -U_1 & I & \cdots & 0 \\ \vdots & \vdots & \ddots & \vdots \\ -U_{n-1} & 0 & \cdots & I \\ -U_n & 0 & \cdots & 0 \end{pmatrix} : \mathcal{H}^n \to \mathcal{H}^n.$$

Then $I - zH_U$ is invertible if and only if $I + zU_1 + \cdots + z^n U_n$ is.

Proof. Multiplying $I - zH_U$ on the left by the upper triangular Toeplitz matrix

$$\begin{pmatrix} I & zI & \cdots & z^{n-1}I \\ \vdots & \ddots & \ddots & \vdots \\ 0 & \cdots & I & zI \\ 0 & \cdots & 0 & I \end{pmatrix}$$

yields

$$\begin{pmatrix} I + zU_1 + \cdots + z^n U_n & 0 & \cdots & 0 \\ * & I & \cdots & 0 \\ \vdots & \vdots & \ddots & \vdots \\ * & 0 & \cdots & I \end{pmatrix}.$$

The second statement of the lemma follows immediately. The first statement is proven similarly. □

Notice that $zI - C_G$ is a linear operator polynomial, which is sometimes referred to as a *linearization* of the operator polynomial $G(z)$. The operator matrix C_G is called a *companion matrix* of the polynomial $G(z)$. In Exercise 2.9.16 we will describe the eigenvectors of C_G.

We are now ready to prove Theorem 2.3.1.

Proof of Theorem 2.3.1. We start the proof by showing that $X(z)$ is invertible for $|z| \leq 1$, which will be a major part of the proof.

Since T_n is positive definite, so is T_n^{-1}. But then X_0, which is the top left entry of T_n^{-1}, is also positive definite. Put $U_j = X_j X_0^{-1}$, $j = 0, \dots, n$. We start by proving that

$$\begin{pmatrix} C_{l-1} & \cdots & C_{l-n} \\ \vdots & & \vdots \\ C_{n+l-2} & \cdots & C_{l-1} \end{pmatrix} \begin{pmatrix} -U_1 \\ \vdots \\ -U_n \end{pmatrix} = \begin{pmatrix} C_l \\ \vdots \\ C_{n+l-1} \end{pmatrix}, \ l \geq 1, \qquad (2.3.7)$$

where C_k, $k > n$, is given by (2.3.5). For the case when $l = 1$ observe that multiplying the first equation in (2.3.2) on the left by T_n and on the right by X_0^{-1} gives that

$$\begin{pmatrix} C_0 \\ C_1 \\ \vdots \\ C_n \end{pmatrix} + \begin{pmatrix} C_{-1} & \cdots & C_{-n} \\ C_0 & \cdots & C_{-n+1} \\ \vdots & \ddots & \vdots \\ C_{n-1} & \cdots & C_0 \end{pmatrix} \begin{pmatrix} U_1 \\ \vdots \\ U_n \end{pmatrix} = \begin{pmatrix} X_0^{-1} \\ 0 \\ \vdots \\ 0 \end{pmatrix}.$$

The bottom n rows of this equation yield

$$T_{n-1} \begin{pmatrix} -U_1 \\ \vdots \\ -U_n \end{pmatrix} = \begin{pmatrix} C_1 \\ \vdots \\ C_n \end{pmatrix}, \tag{2.3.8}$$

which is exactly equation (2.3.7) for the case when $l = 1$. Let now $l \geq 1$, and suppose that (2.3.7) holds up to l. Looking at the bottom $n-1$ rows of (2.3.7) we get

$$\begin{pmatrix} C_l & \cdots & C_{l-n+1} \\ \vdots & & \vdots \\ C_{n+l-2} & \cdots & C_{l-1} \end{pmatrix} \begin{pmatrix} -U_1 \\ \vdots \\ -U_n \end{pmatrix} = \begin{pmatrix} C_{l+1} \\ \vdots \\ C_{n+l-1} \end{pmatrix}. \tag{2.3.9}$$

When we combine (2.3.5) for $k = n + l$ with (2.3.8) we get

$$\begin{pmatrix} C_{n+l-1} & \cdots & C_l \end{pmatrix} \begin{pmatrix} -U_1 \\ \vdots \\ -U_n \end{pmatrix} = \begin{pmatrix} C_{n+l-1} & \cdots & C_l \end{pmatrix} T_{n-1}^{-1} \begin{pmatrix} C_1 \\ \vdots \\ C_n \end{pmatrix} = C_{n+l}. \tag{2.3.10}$$

Putting (2.3.9) and (2.3.10) together yields (2.3.7), where l is replaced by $l+1$. This proves (2.3.7) for all $l \geq 1$. From (2.3.7) we now easily get that for $l \geq 1$,

$$\begin{pmatrix} C_0 & \cdots & C_{-n-l+2} \\ \vdots & & \vdots \\ C_{n+l-2} & \cdots & C_0 \end{pmatrix} \begin{pmatrix} -U_1 \\ \vdots \\ -U_n \\ 0 \\ \vdots \\ 0 \end{pmatrix} = \begin{pmatrix} C_1 \\ \vdots \\ C_{n+l-1} \end{pmatrix}. \tag{2.3.11}$$

Combining this with (2.3.8) gives that

$$\begin{aligned} &\begin{pmatrix} C_{n+l-1} & \cdots & C_1 \end{pmatrix} T_{n+l-2}^{-1} \begin{pmatrix} C_1 \\ \vdots \\ C_{n+l-1} \end{pmatrix} \\ &= \begin{pmatrix} C_{n+l-1} & \cdots & C_l \end{pmatrix} \begin{pmatrix} -U_1 \\ \vdots \\ -U_n \end{pmatrix} = C_{n+l}. \end{aligned} \tag{2.3.12}$$

Note that by repeated use of Theorem 2.2.3 in combination with (2.3.12) gives that $T_k = (C_{i-j})_{i,j=0}^k$ is positive definite for all $k \in \mathbb{N}$.

We now let K be the companion type matrix

$$K = \begin{pmatrix} -U_1 & I & \cdots & 0 \\ \vdots & \vdots & \ddots & \vdots \\ -U_{n-1} & 0 & \cdots & I \\ -U_n & 0 & \cdots & 0 \end{pmatrix}.$$

Then, by (2.3.7) with $l = 1$, we get

$$T_{n-1}K = \begin{pmatrix} C_1 & \cdots & C_{-n+1} \\ \vdots & \ddots & \vdots \\ C_n & \cdots & C_1 \end{pmatrix}.$$

Using (2.3.7) in a similar way for $l = 2, 3, \ldots, n$, we get

$$T_{n-1}K^n = \begin{pmatrix} C_n & \cdots & C_1 \\ \vdots & \ddots & \vdots \\ C_{2n-1} & \cdots & C_n \end{pmatrix}. \tag{2.3.13}$$

But then

$$\begin{pmatrix} T_{n-1} & K^{*n}T_{n-1} \\ T_{n-1}K^n & T_{n-1} \end{pmatrix} = T_{2n-1}, \tag{2.3.14}$$

which is positive definite. By Lemma 2.2.1, we obtain

$$T_{n-1} - K^{*n}T_{n-1}T_{n-1}^{-1}T_{n-1}K^n = T_{n-1} - K^{*n}T_{n-1}K^n > 0. \tag{2.3.15}$$

Multiplying on the left and right with $T_{n-1}^{-\frac{1}{2}}$, gives that

$$I - (T_{n-1}^{-\frac{1}{2}}K^{n*}T_{n-1}^{\frac{1}{2}})(T_{n-1}^{\frac{1}{2}}K^nT_{n-1}^{-\frac{1}{2}}) > 0,$$

and thus

$$\rho := \|T_{n-1}^{\frac{1}{2}}K^nT_{n-1}^{-\frac{1}{2}}\| < 1.$$

But then the spectrum of $T_{n-1}^{\frac{1}{2}}K^nT_{n-1}^{-\frac{1}{2}}$ lies in the open disk $\{z : |z| < \rho\}$, and by similarity so does the spectrum of K^n. The spectral mapping theorem now implies that the spectrum of K lies in the open disk $\{z : |z| < r\}$, where $r = \sqrt[n]{\rho} \in (0, 1)$. Thus $I - zK$ is invertible for $|z| \leq \frac{1}{r}$. By Lemma 2.3.3 we now obtain that $X(z) = (I + zU_1 + \cdots + z^nU_n)X_0$ is invertible for $|z| \leq \frac{1}{r}$, and so $X(z)^{-1}$ is analytic in an open disk which contains the closed unit disc.

The proof of the invertibility of $Y(z)$ for $z \in \mathbb{C}_\infty \setminus \mathbb{D}$ is similar, and is left to the reader (see Exercise 2.9.17).

We now prove (2.3.4). For this introduce $C(z) = \sum_{k=-n}^{n} C_kz^k$ and let

$$C(z)X(z) = \sum_{k=-n}^{2n} G_kz^k.$$

Notice that from (2.3.2) we have that $G_0 = I$ and $G_i = 0$, $i = 1, \dots, n$. Define

$$H(z) = \sum_{k=n+1}^{\infty} H_k z^k := -(\sum_{k=n+1}^{2n} G_k z^k) X(z)^{-1},$$

where we used that $X(z)^{-1}$ is analytic on an open disk containing the closed unit disc. Then the function $\widetilde{F}(z) = H(z)^* + C(z) + H(z)$ on $\mathbb{T}$ admits the expansion

$$\widetilde{F}(z) = \sum_{k=-\infty}^{-n-1} H_{-k}^* z^k + \sum_{k=-n}^{n} C_k z^k + \sum_{k=n+1}^{\infty} H_k z^k.$$

Next,

$$\widetilde{F}(z)X(z) = H(z)^* X(z) + C(z)X(z) - \sum_{k=n+1}^{2n} G_k z^k$$

$$= \left(\sum_{k=-\infty}^{-n-1} H_{-k}^* z^k \right) X(z) + \sum_{k=-n}^{-1} G_k z^k + I.$$

But then

$$X(z)^* \widetilde{F}(z) X(z) = \sum_{k=-\infty}^{-1} Z_k z^k + X_0^*$$

for some $Z_{-1}, Z_{-2}, \dots$. As $\widetilde{F}(z)$ is selfadjoint for $z \in \mathbb{T}$, so is $X(z)^* \widetilde{F}(z) X(z)$. But then we must have that $Z_k = 0$, $k = -1, -2, \dots$. Therefore we obtain $X(z)^* \widetilde{F}(z) X(z) = X_0^* = X_0$, and thus $\widetilde{F}(z) = X(z)^{*-1} X_0 X(z)^{-1} = F(z)$ satisfies (2.3.4). The proof that $Y(z)^{*-1} Y_0 Y(z)^{-1}$ has its kth Fourier coefficient equal to C_k, $k = -n, \dots, n$, is similar.

It remains to prove the second equality in (2.3.3). For this consider

$$L(z) := X(z)^{*-1} X_0 X(z)^{-1} - Y(z)^{*-1} Y_0 Y(z)^{-1},$$

which has the form

$$L(z) = \sum_{k=-\infty}^{-n-1} L_{-k}^* z^k + \sum_{k=n+1}^{\infty} L_k z^k \tag{2.3.16}$$

for some L_k, $k \geq n+1$. Write

$$Y(z) Y_0^{-1} Y(z)^* - X(z) X_0^{-1} X(z)^* = \sum_{k=-n}^{n} M_k z^k.$$

As $L(z) = X(z)^{*-1} X_0 X(z)^{-1} (\sum_{k=-n}^{n} M_k z^k) Y(z)^{*-1} Y_0 Y(z)^{-1}$, we get that

$$X(z)^* L(z) Y(z) = X_0 X(z)^{-1} \left(\sum_{k=-n}^{n} M_k z^k \right) Y(z)^{*-1} Y_0$$

has its Fourier coefficients equal to 0 whenever $k < -n$. Thus (2.3.16) yields that $X(z)^*(\sum_{k=-\infty}^{-n-1} L_{-k}^* z^k)Y(z) \equiv 0$. This gives that $L_i = 0$, $i \geq n+1$, and thus $L(z) \equiv 0$. This yields the second equality in (2.3.3). □

An operator-valued polynomial $X(z)$ such that $X(z)$ is invertible for all $|z| \leq 1$ is called *stable*. An alternative way to write (2.3.3) is

$$F(z) = P(z)^{*-1}P(z)^{-1} = Q(z)^{-1}Q(z)^{*-1}, \tag{2.3.17}$$

where

$$P(z) = X(z)X_0^{-\frac{1}{2}}, Q(z) = Y_0^{-\frac{1}{2}}Y\left(\frac{1}{\bar{z}}\right)^* = Y_0^{-\frac{1}{2}}(Y_0^* + Y_{-1}^*z + \cdots + Y_{-n}^*z^n), \tag{2.3.18}$$

with X and Y given via (2.3.2). Note that $P(z)$ and $Q(z)$ are both stable polynomials and that $P(z)P(z)^* = Q(z)^*Q(z)$ holds for $z \in \mathbb{T}$. In the scalar-valued case the polynomials P and Q are equal.

With P and Q so closely related, several relations exists. A very important one is the following relation, which is known as the *Christoffel-Darboux formula.*

Proposition 2.3.4 *Let the positive definite Toeplitz operator matrix be given as in* (2.3.1)*, and define the stable polynomials P and Q as in* (2.3.18)*. Then there exist operator-valued polynomials $G_j(z) = \sum_{k=0}^{j} G_k^{(j)} z^k$, $j = 0, \dots, n-1$, of degree j such that*

$$P(z)P(w)^* - \overleftarrow{Q}(z)\overleftarrow{Q}(w)^* = (1 - z\overline{w}) \sum_{j=0}^{n-1} G_j(z)G_j(w)^*, \quad z, w \in \mathbb{C}, \tag{2.3.19}$$

where $\overleftarrow{Q}(z) = z^n Q(\frac{1}{\bar{z}})^*$. In fact, the polynomials $G_j(z)$, $j = 0, \dots, n-1$, can be found via a Cholesky factorization of T_{n-1}^{-1} as follows:

$$T_{n-1}^{-1} = \begin{pmatrix} G_0^{(0)} & \cdots & G_0^{(n-1)} \\ & \ddots & \vdots \\ 0 & & G_{n-1}^{(n-1)} \end{pmatrix} \begin{pmatrix} G_0^{(0)*} & & 0 \\ \vdots & \ddots & \\ G_0^{(n-1)*} & \cdots & G_{n-1}^{(n-1)*} \end{pmatrix}. \tag{2.3.20}$$

Proof. Equation (2.2.1) implies that

$$\begin{pmatrix} A & B \\ B^* & C \end{pmatrix}^{-1} = \begin{pmatrix} (A - BC^{-1}B^*)^{-1} & * \\ * & * \end{pmatrix} = \begin{pmatrix} * & * \\ * & (C - B^*A^{-1}B)^{-1} \end{pmatrix},$$

provided all the appropriate inverses exist. Applying this general observation to $T_n = \begin{pmatrix} A & B \\ B^* & C \end{pmatrix}^{-1}$, we get two different expressions for T_{n-1}^{-1}. We write

(2.3.20) as $T_{n-1}^{-1} = GG^*$. This implies that

$$T_n^{-1} = \begin{pmatrix} P_0 & 0 \\ \begin{pmatrix} P_1 \\ \vdots \\ P_n \end{pmatrix} & G \end{pmatrix} \begin{pmatrix} P_0^* & (P_1^* & \cdots & P_n^*) \\ 0 & & G^* & \end{pmatrix}$$

$$= \begin{pmatrix} G & \begin{pmatrix} Q_n^* \\ \vdots \\ Q_1^* \end{pmatrix} \\ 0 & Q_0^* \end{pmatrix} \begin{pmatrix} & G^* & & 0 \\ (Q_n & \cdots & Q_1) & Q_0 \end{pmatrix}, \tag{2.3.21}$$

where

$$P(z) = \sum_{k=0}^{n} P_k z^k, \quad Q(z) = \sum_{k=0}^{n} Q_k z^k.$$

Let $\ell_n(z) = \begin{pmatrix} 1 & z & \cdots & z^n \end{pmatrix}$. Multiplying (2.3.21) on the left by $\ell_n(z)$ and on the right by $\ell_n(w)^*$ yields

$$P(z)P(w)^* + \sum_{j=0}^{n-1} zG_j(z)\overline{w}G_j(w)^* = \sum_{j=0}^{n-1} G_j(z)G_j(w)^* + \overleftarrow{Q}(z)\overleftarrow{Q}(w)^*.$$

Equation (2.3.19) now follows. □

Remark 2.3.5 It should be noticed that $G_j(z)$ is the same as the polynomial $Y(z)$ defined in Theorem 2.3.1 for T_j.

With the coefficients of P and Q from the Christoffel-Darboux formula one can obtain simple expressions for the inverse of T_{n-1}.

Theorem 2.3.6 *Let T_n be as in Theorem 2.3.1 and let X_k and Y_{-k}, $k = 0, \dots, n$, be given via (2.3.2). Put $P_k = X_k X_0^{-\frac{1}{2}}$, $Q_k = Y_0^{-\frac{1}{2}} Y_{-k}^*$, $k = 0, \dots, n$. Then*

$$T_{n-1}^{-1} = \begin{pmatrix} P_0 & & 0 \\ \vdots & \ddots & \\ P_{n-1} & \cdots & P_0 \end{pmatrix} \begin{pmatrix} P_0^* & \cdots & P_{n-1}^* \\ & \ddots & \vdots \\ 0 & & P_0^* \end{pmatrix}$$

$$- \begin{pmatrix} Q_n^* & & 0 \\ \vdots & \ddots & \\ Q_1^* & \cdots & Q_n^* \end{pmatrix} \begin{pmatrix} Q_n & \cdots & Q_1 \\ & \ddots & \vdots \\ 0 & & Q_n \end{pmatrix} \tag{2.3.22}$$

$$= \begin{pmatrix} Q_0^* & \cdots & Q_{n-1}^* \\ & \ddots & \vdots \\ 0 & & Q_0^* \end{pmatrix} \begin{pmatrix} Q_0 & & 0 \\ \vdots & \ddots & \\ Q_{n-1} & \cdots & Q_0 \end{pmatrix}$$

$$- \begin{pmatrix} P_n & \cdots & P_1 \\ & \ddots & \vdots \\ 0 & & P_n \end{pmatrix} \begin{pmatrix} P_n^* & & 0 \\ \vdots & \ddots & \\ P_1^* & \cdots & P_n^* \end{pmatrix}. \tag{2.3.23}$$

Proof. Using that $T_{n-1}^{-1} = GG^*$ as in (2.3.20), equation (2.3.21) yields that

$$\begin{pmatrix} T_{n-1}^{-1} & \begin{pmatrix} 0 \\ \vdots \\ 0 \end{pmatrix} \\ \begin{pmatrix} 0 & \cdots & 0 \end{pmatrix} & 0 \end{pmatrix} - \begin{pmatrix} 0 & \begin{pmatrix} 0 & \cdots & 0 \end{pmatrix} \\ \begin{pmatrix} 0 \\ \vdots \\ 0 \end{pmatrix} & T_{n-1}^{-1} \end{pmatrix}$$

$$= \begin{pmatrix} P_0 \\ \vdots \\ P_n \end{pmatrix} \begin{pmatrix} P_0 & \cdots & P_n^* \end{pmatrix} - \begin{pmatrix} Q_n^* \\ \vdots \\ Q_0^* \end{pmatrix} \begin{pmatrix} Q_n & \cdots & Q_0 \end{pmatrix}.$$

Letting

$$Z = \begin{pmatrix} 0 & 0 & \cdots & 0 \\ I & 0 & \cdots & 0 \\ \vdots & \ddots & \ddots & \vdots \\ 0 & \cdots & I & 0 \end{pmatrix}$$

be the forward shift operator acting on $\bigoplus_{k=1}^{2n} \mathcal{H}$, we get

$$\begin{pmatrix} T_{n-1}^{-1} & 0 \\ 0 & -T_{n-1}^{-1} \end{pmatrix} = \sum_{k=0}^{n-1} Z^k \left(\begin{pmatrix} T_{n-1}^{-1} & 0 \\ 0 & 0 \end{pmatrix} - Z \begin{pmatrix} T_{n-1}^{-1} & 0 \\ 0 & 0 \end{pmatrix} Z^* \right) Z^k$$
$$= R(P)R(P)^* - R(\overleftarrow{Q})R(\overleftarrow{Q})^*,$$

where for a polynomial $A(z) = \sum_{k=0}^{n} A_k z^k$ we define

$$R(A) = \begin{pmatrix} A_0 & \cdots & 0 \\ \vdots & \ddots & \\ A_{n-1} & \cdots & A_0 \\ A_n & \cdots & A_1 \\ \vdots & \ddots & \vdots \\ 0 & \ddots & A_n \end{pmatrix}.$$

Now (2.3.22) and (2.3.23) directly follow. □

We say that $F \in H^\infty(\mathcal{L}(\mathcal{H}))$ has *positive definite real part* if $F + F^* \geq \epsilon I$ for some $\epsilon > 0$, where $(F + F^*)(z) := F(z) + F(z)^*$ is viewed as a member of $L^\infty(\mathcal{L}(\mathcal{H}))$.

Before stating our next results, we need to define one more function space. For a weakly measurable function $f : \mathbb{T} \to \mathcal{H}$, define

$$\|f\|_2 = \left(\frac{1}{2\pi} \int_0^{2\pi} \|f(e^{it})\|^2 dt \right)^{\frac{1}{2}}.$$

Let $L^2(\mathcal{H})$ be the set of such functions with $\|f\|_2 < \infty$. The space $L^2(\mathcal{H})$ can be identified with $L^2(\mathbb{T}) \otimes \mathcal{H}$ and also with $\bigoplus_{n \in \mathbb{Z}} \mathcal{H}$. Every $f \in L^2(\mathcal{H})$ can thus be written as $f(e^{it}) = \sum_{k=-\infty}^{\infty} f_k e^{ikt}$, with $\|f\|^2 = \sum_{n=-\infty}^{\infty} \|f_k\|^2$. Let $F(e^{it}) = \sum_{k=-\infty}^{\infty} F_k e^{ikt} \in L^\infty(\mathcal{L}(\mathcal{H},\mathcal{H}'))$. The *multiplication operator with symbol* F is defined as $M_F : L^2(\mathcal{H}) \to L^2(\mathcal{H}')$, $(M_F f)(e^{it}) = F(e^{it})f(e^{it})$. The following proposition shows that the norm of a multiplication operator equals the essential norm of its symbol.

Proposition 2.3.7 *Let M_F be defined as above. Then $\|M_F\| = \|F\|_\infty$.*

Proof. First note that for $g \in L^2(\mathcal{H})$ we have

$$\|M_f g\|^2 = \frac{1}{2\pi}\int_0^{2\pi} \|F(e^{it})g(e^{it})\|^2 dt \leq \|F\|_\infty^2 \left(\frac{1}{2\pi}\int_0^{2\pi} \|g(e^{it})\|^2 dt\right),$$

and thus $\|M_F\| \leq \|F\|_\infty$ follows.

For the converse inequality, note that when $\|F\|_\infty = 0$, we have that $M_F = 0$, and thus the result holds. Next, assume that $A := \|F\|_\infty > 0$. Let $\epsilon > 0$. Then the measure of the set $B = \{t \in [0, 2\pi] \ : \ \|F(e^{it})\| > A - \epsilon\}$ is greater than zero. Now let $g(e^{it}) = I$ for $t \in B$ and zero elsewhere. Then $\|g\|$ equals the Lebesgue measure $m(B)$ of B, while $\|Fg\| \geq (A-\epsilon)m(B)$. Thus $\|M_F\| \geq A - \epsilon$. As ϵ was arbitrary, $\|M_F\| \geq A = \|F\|_\infty$ follows. Thus $\|M_F\| = \|F\|_\infty$. □

Next, observe that with respect to the direct sum decompositions $L^2(\mathcal{H}) = \bigoplus_{n\in\mathbb{Z}} \mathcal{H}$ and $L^2(\mathcal{H}') = \bigoplus_{n\in\mathbb{Z}} \mathcal{H}'$, M_F is represented by the doubly infinite Toeplitz matrix $(F_{i-j})_{i,j=-\infty}^{\infty}$.

We can use the polynomials $P(z)$ and $Q(z)$ constructed above to solve the following interpolation problem: "Given the polynomial $C(z) = \frac{1}{2}C_0 + \sum_{k=1}^{n} C_k z^k$, find all functions $F(z) = \frac{1}{2}F_0 + \sum_{k=1}^{\infty} F_k z^k$ in $H^\infty(\mathcal{L}(\mathcal{H}))$ with $F_k = C_k$, $k = 0, \ldots, n$, so that F has positive definite real part." We shall call such an F a *positive definite real part extension* of C.

Theorem 2.3.8 *Let $C_j \in \mathcal{L}(\mathcal{H})$, $j = 0, \ldots, n$, and let $C(z) = \frac{1}{2}C_0 + \sum_{k=1}^{n} C_k z^k$. There exists a positive definite real part extension of C if and only if the Toeplitz operator matrix T_n defined in (2.3.1) is positive definite. In that case let X_i and Y_{-i}, $i = 0, \ldots, n$, be given via (2.3.2) and put*

$$P(z) = \sum_{k=0}^{n} z^k X_k X_0^{-\frac{1}{2}}, \ Q(z) = \sum_{k=1}^{n} z^k Y_0^{-\frac{1}{2}} Y_{-k}^*.$$

Next, let

$$K(z) = P(z)^{*-1}P(z)^{-1} =: \sum_{k=-\infty}^{\infty} K_k z^k, \ K_+(z) = \frac{1}{2}K_0 + \sum_{k=1}^{\infty} K_k z^k.$$

Then F is a positive definite real part extension of C if and only if

$$F(z) = (-z^{n+1}K_+(z)^*Q^*(z)G(z) + K_+(z)P(z))(z^{n+1}Q^*(z)G(z) + P(z))^{-1}, \tag{2.3.24}$$

for some $G \in H^\infty(\mathcal{L}(\mathcal{H}))$ with $\|G\|_\infty < 1$. Moreover, this correspondence is one-to-one.

In addition, C has a unique positive definite real part extension F with the property that $(F+F^*)^{-1}$ is an $\mathcal{L}(\mathcal{H})$-valued trigonometric polynomial of degree $\leq n$, and this unique F is obtained by choosing $G=0$ in (2.3.24).

Proof. Assume first that F is a positive definite real part extension of C. Then the multiplication operator with symbol $F+F^*$ on $L^2(\mathcal{H})$ is positive definite, and thus $(F_{i-j})_{i,j=-\infty}^{\infty}$ is positive definite. Consequently, the finite Toeplitz operator T_n is positive definite. Now suppose that $T_n > 0$ and make P, Q, and K_+ as indicated. Since $P(z)$ is a polynomial, it follows that $K(z)$ is in the *Wiener class* (i.e., $\sum_{k=-\infty}^{\infty} \|K_k\| < \infty$), so by the Bochner-Phillips generalization of Wiener's theorem K_+ is in this class as well, implying $K_+ \in H^\infty(\mathcal{L}(\mathcal{H}))$. Let $S(G)$ denote the right-hand side of (2.3.24). Notice that

$$\begin{aligned} S(G) &= (-z^{n+1}K_+(z)^*Q^*(z)G(z) + K_+(z)P(z) + z^{n+1}K_+(z)Q^*(z)G(z) \\ &\quad -z^{n+1}K_+(z)Q^*(z)G(z))\left(z^{n+1}Q^*(z)G(z) + P(z)\right)^{-1} \\ &= K_+ - z^{n+1}Q(z)^{-1}\left(I + z^{n+1}G(z)P(z)^{-1}Q(z)^*\right)^{-1}G(z)P(z)^{-1}, \end{aligned} \tag{2.3.25}$$

where we used that $K_+(z) + K_+(z)^* = Q(z)^{-1}Q(z)^{*-1}$. Also, it is straightforward to compute that

$$\begin{aligned} S(G) + S(G)^* = \left(z^{-(n+1)}G(z)^*Q(z) + P(z)^*\right)^{-1}(I - G(z)^*G(z)) \\ \times \left(z^{n+1}Q^*(z)G(z) + P(z)\right)^{-1}. \end{aligned} \tag{2.3.26}$$

Suppose now that $G \in H^\infty(\mathcal{L}(\mathcal{H}))$ with $\|G\|_\infty < 1$. We first observe that $P(z)^{-1}Q(z)^* = Q(z)^{-1}P(z)^*$ and thus $P(z)^{-1}Q(z)^*$ is unitary. As $\|G\|_\infty < 1$, we also get that $\|P^{-1}(z)Q^*(z)G(z)\| < 1$. Thus $I + z^{n+1}G(z)P(z)^{-1}Q(z)^*$ is invertible, and also $z^{n+1}Q^*(z)G(z) + P(z)$. Now by (2.3.26) we have that $S(G) + S(G)^* > 0$. Moreover, as $z^{n+1}Q(z)^*$ is a polynomial, we get that $Q(z)^{-1}(I + z^{n+1}G(z)P(z)^{-1}Q(z)^*)^{-1}G(z)P(z)^{-1} \in H^\infty(\mathcal{L}(\mathcal{H}))$, and thus by (2.3.25) we get that $S(G)$ is a positive definite real part extension of F.

Conversely, let F be a positive definite real part extension of C. Put $W = F - K_+$. Then $QWQ^* \in zH^\infty(\mathcal{L}(\mathcal{H}))$ as $W \in z^{n+1}H^\infty(\mathcal{L}(\mathcal{H}))$ and $z^nQ(z)^*$ is a polynomial. We will first show that $I + QWQ^*$ is an invertible element of $H^\infty(\mathcal{L}(\mathcal{H}))$. First observe, using (2.3.17), that

$$I + QWQ^* + I + QW^*Q^* = Q(K + F + F^*)Q^*$$

is a positive definite member of the C^*-algebra $L^\infty(\mathcal{L}(\mathcal{H}))$. Thus we can write $2I + QWQ^* + QW^*Q^* = AA^*$, where A is an invertible element of $L^\infty(\mathcal{L}(\mathcal{H}))$. Choose $\epsilon > 0$ so that $\|\epsilon^{\frac{1}{2}}(I + QWQ^*)A^{*-1}\|_\infty < 1$. Then there exists an invertible $B \in L^\infty(\mathcal{L}(\mathcal{H}))$ such that

$$I - \epsilon A^{-1}(I + QW^*Q^*)(I + QWQ^*)A^{*-1} = BB^*.$$

Now

$$I - (I - \epsilon(I + QW^*Q^*))(I - \epsilon(I + QWQ^*)) = \epsilon ABB^*A^* > 0.$$

Thus $\|I - \epsilon(I + QWQ^*)\|_\infty < 1$, and so

$$I + QWQ^* = \frac{1}{\epsilon}(I - (I - \epsilon(I + QWQ^*)))$$

is invertible, and its inverse lies in $H^\infty(\mathcal{L}(\mathcal{H}))$ (use the von Neumann series $(e - z)^{-1} = \sum_{k=0}^{\infty} z^k$ which converges in a Banach algebra with unit e for $\|z\| < 1$).

Now let

$$G = -z^{-n-1}(I + QWQ^*)^{-1}QWP.$$

Then $G \in H^\infty(\mathcal{L}(\mathcal{H}))$. Next,

$$\begin{aligned} z^{n+1}Q^*G + P &= -Q^*(I + QWQ^*)^{-1}QWP + P \\ &= Q^*(I + QWQ^*)^{-1}(-QW + (I + QWQ^*)Q^{*-1})P \\ &= Q^*(I + QWQ^*)^{-1}Q^{*-1}P = (I + Q^*QW)^{-1}P. \end{aligned}$$

Consequently, $z^{n+1}Q^*G+P$ is invertible, and thus $S(G)$ is well defined. Next we check that $S(G) = F$. Indeed, using the general observation $(I-AB)^{-1} = I + A(I - BA)^{-1}B$, which can be checked by direct multiplication, we get

$$\begin{aligned} S(G) &= [-K_+^*Q^*(I + QWQ^*)^{-1}QWP + K_+P]P^{-1}(I + Q^*QW) \\ &= [K_+^*Q^*(I - (I - QWQ^*)^{-1})Q^{*-1}P + K_+P]P^{-1}(I + Q^*QW) \\ &= [-K_+^*(I + Q^*QW)^{-1}P + K_+^*P + K_+P]P^{-1}(I + Q^*QW) \\ &= -K_+^* + K(I + Q^*QW) = -K_+^* + K + W = F. \end{aligned}$$

Next, the fact that $\|G\|_\infty < 1$ follows from the observation that

$$\begin{aligned} 0 < F + F^* &= S(G) + S(G)^* \\ &= (z^{-n-1}G^*Q + P^*)^{-1}(I - G^*G)(z^{n+1}Q^*G + P)^{-1}. \end{aligned}$$

Finally, the one-to-one correspondence follows from showing that $S(G) = F$ if and only if $G = -z^{-n-1}(I + QWQ^*)^{-1}QWP$, which can be checked in a straightforward manner.

For the last part, first observe that $(S(0) + S(0)^*)^{-1} = PP^* = Q^*Q$ is a trigonometric polynomial of degree $\leq n$. Write $S(0)+S(0)^* = \sum_{k=-\infty}^{\infty} C_k z^k$. Suppose that $\widetilde{F} + \widetilde{F}^* = \sum_{k=-\infty}^{\infty} \widetilde{C}_k z^k$, with $\widetilde{C}_k = C_k$, $k = -n, \ldots, n$, also has the property that $(\widetilde{F} + \widetilde{F}^*)^{-1}$ is a trigonometric polynomial of degree $\leq n$. We need to show that $\widetilde{C}_k = C_k$ for all $k \in \mathbb{Z}$. For this observe that

$$(C_{i-j})_{i,j=-\infty}^{\infty} \text{ and } (\widetilde{C}_{i-j})_{i,j=-\infty}^{\infty} \tag{2.3.27}$$

both have banded inverses of bandwidth n, and that they coincide on the band of bandwidth n. If we decompose the operator matrices in (2.3.27) with respect to the decomposition

$$\left(\bigoplus_{j=-\infty}^{-1} \mathcal{H}\right) \oplus \mathcal{H} \oplus \left(\bigoplus_{j=1}^{n} \mathcal{H}\right) \oplus \mathcal{H} \oplus \left(\bigoplus_{j=n+2}^{\infty} \mathcal{H}\right),$$

we get 5×5 matrices $(A_{ij})_{i,j=1}^5$ and $(\widetilde{A}_{ij})_{i,j=1}^5$, with inverses $(B_{ij})_{i,j=1}^5$ and $(\widetilde{B}_{ij})_{i,j=1}^5$, respectively, that satisfy

(i) $A_{22} = \widetilde{A}_{22}$, $A_{23} = \widetilde{A}_{23}$, $A_{32} = \widetilde{A}_{32}$, $A_{33} = \widetilde{A}_{33}$, $A_{34} = \widetilde{A}_{34}$, $A_{43} = \widetilde{A}_{43}$, $A_{44} = \widetilde{A}_{44}$.

(ii) $B_{14}, \widetilde{B}_{14}$, $B_{15}, \widetilde{B}_{15}$, $B_{24}, \widetilde{B}_{24}$, $B_{25}, \widetilde{B}_{25}$, $B_{41}, \widetilde{B}_{41}$, $B_{42}, \widetilde{B}_{42}$, $B_{51}, \widetilde{B}_{51}$, $B_{52}, \widetilde{B}_{52}$, are all equal to zero.

It follows from (ii) that $A_{24} = A_{23}A_{33}^{-1}A_{34}$, $A_{42} = A_{43}A_{33}^{-1}A_{34}$, $\widetilde{A}_{24} = \widetilde{A}_{23}\widetilde{A}_{33}^{-1}\widetilde{A}_{34}$ and $\widetilde{A}_{42} = \widetilde{A}_{43}\widetilde{A}_{33}^{-1}\widetilde{A}_{34}$. Combining this with (i) yields that $A_{24} = \widetilde{A}_{24}$, and $A_{42} = \widetilde{A}_{42}$. But then it follows that $C_{n+1} = \widetilde{C}_{n+1}$ and $C_{-n-1} = \widetilde{C}_{-n-1}$. Repeating this argument gives that $C_k = \widetilde{C}_k$, $k \in \mathbb{Z}$. This proves the uniqueness. □

Given $C(z)$ as in Theorem 2.3.8, the unique positive definite real part extension F of C with the property that $(F + F^*)^{-1}$ is an $\mathcal{L}(\mathcal{H})$-valued trigonometric polynomial of degree $\leq n$ is called the *band extension of* C. This band extension is characterized by the maximum entropy property described by the following result, which is a matrix-valued generalization of Theorem 1.4.9.

Theorem 2.3.9 *Let $C(z) = \frac{1}{2}C_0 + \sum_{k=1}^{n} C_k z^k$ be a matrix-valued polynomial as in Theorem 2.3.8 such that the Toeplitz matrix (2.3.1) is positive definite. Then among all positive definite real part extensions of C, the band extension is the unique one which maximizes the entropy function*

$$\mathcal{E}(F) = \frac{1}{2\pi}\int_0^{2\pi} \log\, \det(F + F^*)(e^{it})dt.$$

Proof. By Corollary 1.4.2, $\log\det$ is a strictly concave function on PD_n, and consequently so is $\mathcal{E}(F)$ over the set of all positive definite real part extensions of C. Thus $\mathcal{E}(F)$ is maximized by a unique such extension F. Similarly to the proof of Theorem 1.4.3, consider for a fixed $k > n$ and matrix X of corresponding size the functions

$$g_{F,X,k}(x,t) = (F + F^*)(e^{it}) + x(e^{ikt}X^* + e^{-ikt}X)$$

and

$$f_{F,X,k}(x) = \frac{1}{2\pi}\int_0^{2\pi} \log\det g_{F,X,k}(x,t)dt,$$

both defined for x in a neighborhood of 0 in $\mathbb{R}$. Since F maximizes $\mathcal{E}(F)$, we have that $f'_{F,X,k}(0) = 0$, which just as the proof of Theorem 1.4.3 implies

$$\mathrm{tr}(G_k X + G_k^* X^*) = 0, \tag{2.3.28}$$

where $(F+F^*)^{-1}(e^{it}) = \sum_{j=-\infty}^{\infty} G_j e^{ijt}$. Repeating the above argument for the matrix iX instead of X, we obtain together with (2.3.28) that $\mathrm{tr}(G_k X) = 0$. Since X was arbitrary, it follows that $G_k = 0$, and since $k > n$ was arbitrary, it follows that F is the unique band extension of C. □

As a corollary to Theorem 2.3.8 we get an operator-valued Fejér-Riesz factorization result for the positive definite case.

Corollary 2.3.10 *Let $F(z) = \sum_{k=-n}^{n} F_k z^k$ be a $\mathcal{L}(\mathcal{H})$-valued trigonometric polynomial that takes on positive definite values on the unit circle, that is, $F(z) > 0$, $z \in \mathbb{T}$. Then there exist stable $\mathcal{L}(\mathcal{H})$-valued polynomials $P(z) = \sum_{k=0}^{n} P_k z^k$ and $Q(z) = \sum_{k=0}^{n} Q_k z^k$ so that $F(z) = P(z)P(z)^* = Q(z)^*Q(z)$, $z \in \mathbb{T}$. The polynomials P and Q are unique up to multiplication by a constant unitary operator in $\mathcal{L}(\mathcal{H})$ on the right and on the left, respectively.*

Proof. Write $F(z)^{-1} = \sum_{k=-\infty}^{\infty} C_k z_k$, and let T_n be given via (2.3.1). Since F is continuous on $\mathbb{T}$ and $F(z) > 0$, $z \in \mathbb{T}$, we have that $T_n > 0$. Let X_k and Y_k be given via (2.3.2) and put $P(z) = \sum_{k=0}^{n} z^k X_k X_0^{-\frac{1}{2}}$, $Q(z) = \sum_{k=0}^{n} z^k Y_0^{-\frac{1}{2}} Y_{-k}^*$. By the uniqueness statement in Theorem 2.3.8, we get that $F(z)^{-1} = P(z)^{-1}P(z)^{*-1} = Q(z)^{*-1}Q(z)^{-1}$. Thus $F(z) = P(z)P(z)^* = Q(z)^*Q(z)$. In addition, using Theorem 2.3.1 we get that P and Q are stable. Finally, for the uniqueness suppose that $P(z)P(z)^* = \widetilde{P}(z)\widetilde{P}(z)^*$, $z \in \mathbb{T}$, with P and $\widetilde{P}$ stable polynomials of degree n. Then $\widetilde{P}(z)^{-1}P(z) = \widetilde{P}(\frac{1}{\bar{z}})^* P(\frac{1}{\bar{z}})^{*-1}$ defines an entire function that is unitary on the unit circle; thus it must equal a constant unitary U. This gives that $P(z) = \widetilde{P}(z)U$. The uniqueness statement for $Q(z)$ is proved in a similar way. □

We end this section with Szegő's result on the roots of orthogonal polynomials. The classical starting point is a positive measure on the unit circle. Our starting point will be a finite positive definite Toeplitz matrix (which may be built from the moments of a positive measure on the unit circle).

Let $T_n = (c_{i-j})_{i,j=0}^{n}$ be a positive definite $(n+1)\times(n+1)$ complex Toeplitz matrix. The Toeplitz matrix induces an inner product on the vector space $\mathcal{P}_n$ of polynomials of degree $\leq n$, via the equalities

$$\langle z^k, z^l \rangle = c_{k-l}, \quad 0 \leq k, l \leq n.$$

Let now $p_i(z)$, $i = 0, \ldots, n$, be polynomials of degree i such that

$$\langle p_i(z), z^j \rangle = 0, j = 0, \ldots, i-1.$$

The polynomials $p_0, \ldots, p_n$ are referred to as *orthogonal polynomials* with respect to $\langle \cdot, \cdot \rangle$, and they are unique up to multiplication with a nonzero scalar. The orthogonal polynomials can be obtained by performing the *Gram-Schmidt orthonormalization process* on the basis $\{1, z, \ldots, z^n\}$ of $\mathcal{P}_n$. We now have the following result, known as *Szegő's theorem.*

Theorem 2.3.11 *Let T_n be a positive definite Toeplitz matrix. Then the orthogonal polynomials with respect to the inner product induced by T_n have all their roots in $\mathbb{D}$.*

Proof. Notice that if $p(z) = p_0 + \cdots + p_n z^n$ and $q(z) = q_0 + \cdots + q_n z^n$, then

$$\langle p, q \rangle = \begin{pmatrix} \overline{q}_0 & \cdots & \overline{q}_n \end{pmatrix} \begin{pmatrix} c_0 & \cdots & c_{-n} \\ \vdots & \ddots & \vdots \\ c_n & \cdots & c_0 \end{pmatrix} \begin{pmatrix} p_0 \\ \vdots \\ p_n \end{pmatrix}.$$

But then it is easy to see that $p_i(z) = p_0^{(i)} + \cdots + p_i^{(i)} z^n$ may be found via

$$\begin{pmatrix} p_i^{(i)} \\ \vdots \\ p_1^{(i)} \\ p_0^{(i)} \end{pmatrix} = T_i^{-1} \begin{pmatrix} 0 \\ \vdots \\ 0 \\ 1 \end{pmatrix}. \tag{2.3.29}$$

Thus, we can find $p_i(z)$ by applying Theorem 2.3.1 to T_i and take $p_i(z) = Y(\frac{1}{z})$. But then it follows from Theorem 2.3.1 that the roots of p_i lie in $\mathbb{D}$. □

In the exercises we will indicate an elementary proof for Szegő's theorem (see Exercise 2.9.22).

2.4 THE SCHUR COMPLEMENT AND FEJÉR-RIESZ FACTORIZATION

The notion of Schur complement is a relatively simple one that is surprisingly strong and has many useful consequences. In this section we will introduce the notion and derive some of its properties; we will also show how it can be used to prove the Fejér-Riesz factorization theorem for operator-valued trigonometric polynomials.

Let $\mathcal{H}$ be a Hilbert space and $T \in \mathcal{L}(\mathcal{H})$, $T \geq 0$. Let $\mathcal{M}$ be a closed subspace of $\mathcal{H}$, and let $P_{\mathcal{M}} : \mathcal{H} \to \mathcal{M}$ be the orthogonal projection onto $\mathcal{M}$. The operator $S : \mathcal{M} \to \mathcal{M}$ is called the *Schur complement of T supported on $\mathcal{M}$* if

(i) $T - P_{\mathcal{M}}^* S P_{\mathcal{M}} \geq 0$;

(ii) (maximality) if $\widetilde{S} : \mathcal{M} \to \mathcal{M}$ is such that $T - P_{\mathcal{M}}^* \widetilde{S} P_{\mathcal{M}}$ is positive semidefinite, then $\widetilde{S} \leq S$.

We shall denote the Schur complement of T supported on $\mathcal{M}$ by $S(T; \mathcal{M})$.

Theorem 2.4.1 *Let $T \in \mathcal{L}(\mathcal{H})$, $T \geq 0$. Let $\mathcal{M}$ be a closed subspace of $\mathcal{H}$. Then the Schur complement $S(T; \mathcal{M})$ of T supported on $\mathcal{M}$ exists, is unique, and is itself a positive semidefinite operator.*

We will need some lemmas to prove this result. We start with a famous one, called the Douglas lemma.

Lemma 2.4.2 *Let $A \in \mathcal{L}(\mathcal{H}, \mathcal{H}_1)$ and $B \in \mathcal{L}(\mathcal{H}, \mathcal{H}_2)$. Then*

$$A^* A \leq B^* B \tag{2.4.1}$$

if and only if there exists a contraction $G : \overline{\text{Ran} B} \to \overline{\text{Ran} A}$ such that $A = GB$. Moreover, G is unique and G is an isometry if and only if equality holds in (2.4.1).

Notice that in writing $A = GB$ we have a slight abuse of notation, as A takes values in $\mathcal{H}_1$ and B takes values in $\mathcal{H}_2$. If, however, we view G as acting from $\mathcal{H}_2$ into $\mathcal{H}_1$, where on the orthogonal complement of $\overline{\text{Ran}B}$ and into the orthogonal complement of $\overline{\text{Ran}A}$ it acts as the zero operator, the equation $A = GB$ is well defined. We will take the liberty to continue to make this slight abuse of notation for the ease of the presentation.

Proof. The "if" statements are trivial in both cases. We focus on the "only if" parts. Let $y \in \text{Ran}B$. Then $y = Bx$ for some $x \in \mathcal{H}$. Define $Gy := Ax$. We need to check that G is well defined on $\text{Ran}B$. For this purpose, suppose that $y = Bx = B\widetilde{x}$. Then $B(x - \widetilde{x}) = 0$. Thus, by (2.4.1),

$$0 \leq \langle A(x - \widetilde{x}), A(x - \widetilde{x}) \rangle \leq \langle B(x - \widetilde{x}), B(x - \widetilde{x}) \rangle = 0. \tag{2.4.2}$$

So $A(x - \widetilde{x}) = 0$ and thus $Gy = Ax = A\widetilde{x}$, so G is well defined. Next, let $y \in \overline{\text{Ran}B}$ and let $y_n \in \text{Ran}B$ such that $\lim_{n\to\infty} y_n = y$, and let $x_n \in \mathcal{H}$ be so that $Bx_n = y_n$. Then

$$\|Gy_n - Gy_m\| = \|Ax_n - Ax_m\| \leq \|Bx_n - Bx_m\| = \|y_n - y_m\|. \tag{2.4.3}$$

As $\{y_n\}_{n=1}^{\infty}$ is a Cauchy sequence, so is $\{Gy_n\}_{n=1}^{\infty}$. By completeness of the space $\mathcal{H}_1$, we have that

$$Gy := \lim_{n\to\infty} Gy_n$$

exists.

The argument in (2.4.3) also shows that $\|Gy\| \leq \|y\|$ for every $y \in \text{Ran}B$, and by continuity also for every $y \in \overline{\text{Ran}B}$. Thus G is a contraction. The uniqueness of G is also clear from this construction, as the equation $A = GB$ demands that when $y = Bx$ we must have that $Gy = Ax$.

Finally, assume that $A^*A = B^*B$. Then $y = Bx$ and $Gy = Ax$ imply that $\|Gy\|^2 = \langle Ax, Ax \rangle = \langle Bx, Bx \rangle = \|y\|^2$. This shows that $\|Gy\| = \|y\|$ on $\text{Ran}B$, but then by continuity this also holds on $\overline{\text{Ran}B}$. Thus $G : \overline{\text{Ran}B} \to \overline{\text{Ran}A}$ is an isometry. $\square$

Lemma 2.4.2 can be formulated equivalently in terms of adjoint operators as follows.

Corollary 2.4.3 *Let $A \in \mathcal{L}(\mathcal{H}_1, \mathcal{H})$ and $B \in \mathcal{L}(\mathcal{H}_2, \mathcal{H})$. Then*

$$AA^* \leq BB^*$$

if and only if there exists a contraction $G : \overline{\text{Ran}A^*} \to \overline{\text{Ran}B^*}$ such that $A = BG$. Moreover, G is unique and G is a co-isometry if and only if equality holds in (2.4.1).

The following lemma is the generalization of Corollary 2.2.2 for positive semidefinite matrices.

Lemma 2.4.4 *Let*

$$T = \begin{pmatrix} A & B \\ B^* & C \end{pmatrix} \in \mathcal{L}(\mathcal{H}_1 \oplus \mathcal{H}_2). \tag{2.4.4}$$

Then $T \geq 0$ if and only if $A \geq 0$ and $C \geq 0$ and $B = A^{\frac{1}{2}} G C^{\frac{1}{2}}$ for some contraction G, which is uniquely determined when chosen to act from $\overline{\text{Ran}C}$ into $\overline{\text{Ran}A}$. Moreover, in case $T \geq 0$, one can choose any factorizations $A = P^*P$ and $C = Q^*Q$ and write $B = P^*GQ$ for some contraction G that is uniquely defined when G is chosen to act from $\overline{\text{Ran}Q}$ into $\overline{\text{Ran}P}$. In that case, T admits the Cholesky factorizations

$$T = \begin{pmatrix} P^* & 0 \\ Q^*G^* & Q^*D_G^* \end{pmatrix} \begin{pmatrix} P & GQ \\ 0 & D_G Q \end{pmatrix} \tag{2.4.5}$$

and

$$T = \begin{pmatrix} P^*D_{G^*}^* & P^*G \\ 0 & Q^* \end{pmatrix} \begin{pmatrix} D_{G^*}P & 0 \\ G^*P & Q \end{pmatrix}, \tag{2.4.6}$$

where for a contraction M, $D_M = (I - M^*M)^{\frac{1}{2}}$.

Proof. For the "if" statement, write (2.4.4) as

$$\begin{pmatrix} P^* & 0 \\ 0 & Q^* \end{pmatrix} \begin{pmatrix} I & G \\ G^* & I \end{pmatrix} \begin{pmatrix} P & 0 \\ 0 & Q \end{pmatrix}. \tag{2.4.7}$$

As $\|G\| \leq 1$, we have that

$$\left\langle \begin{pmatrix} I & G \\ G^* & I \end{pmatrix} \begin{pmatrix} h_1 \\ h_2 \end{pmatrix}, \begin{pmatrix} h_1 \\ h_2 \end{pmatrix} \right\rangle = \|h_1\|^2 + \|h_2\|^2 + \langle G^*h_1, h_2\rangle + \langle Gh_2, h_1\rangle$$

$$\geq \|h_1\|^2 + \|h_2\|^2 - 2\|h_1\|\|h_2\| \geq 0.$$

Thus the middle operator in (2.4.7) is positive semidefinite. But then the product in (2.4.7), which equals (2.4.4), is positive semidefinite.

For the "only if" statement, let $\mathcal{H} = \mathcal{H}_1 \oplus \mathcal{H}_2$ and let

$$\begin{pmatrix} P & Q \end{pmatrix} : \mathcal{H}_1 \oplus \mathcal{H}_2 \to \mathcal{H}$$

be the square root of (2.4.4). Thus

$$\begin{pmatrix} P^* \\ Q^* \end{pmatrix} \begin{pmatrix} P & Q \end{pmatrix} = \begin{pmatrix} A & B \\ B^* & C \end{pmatrix},$$

and consequently $P^*P = A^{\frac{1}{2}} A^{\frac{1}{2}}$ and $Q^*Q = C^{\frac{1}{2}} C^{\frac{1}{2}}$. Applying Lemma 2.4.2 we obtain that there exist isometries $G_1 : \overline{\text{Ran}A} \to \overline{\text{Ran}P}$ and $G_2 : \overline{\text{Ran}C} \to \overline{\text{Ran}Q}$ such that $P = G_1 A^{\frac{1}{2}}$ and $Q = G_2 C^{\frac{1}{2}}$. But then $B = P^*Q = A^{\frac{1}{2}} G_1^* G_2 C^{\frac{1}{2}}$. Letting $G = G_1^* G_2$, we have the desired contraction. □

The above lemma can be seen as yielding the answer to the simplest positive semidefinite matrix completion problem: Let $A \in \mathcal{L}(\mathcal{H}_1)$ and $C \in \mathcal{L}(\mathcal{H}_2)$ be given operators. The existence of $X \in \mathcal{L}(\mathcal{H}_2, \mathcal{H}_1)$ such that

$$\begin{pmatrix} A & X \\ X^* & C \end{pmatrix} \geq 0$$

is equivalent to $A \geq 0$ and $C \geq 0$, and in that case the set of all solutions is parameterized in a one-to-one way by the set of all contractions $G : \overline{\text{Ran}C} \to \overline{\text{Ran}A}$, via $X = A^{\frac{1}{2}} G C^{\frac{1}{2}}$.

The next result is the positive semidefinite version of the 3×3 completion problem in Theorem 2.2.3.

Theorem 2.4.5 *Every partially positive semidefinite operator matrix of the form*

$$K = \begin{pmatrix} A & B & ? \\ B^* & C & D \\ ? & D^* & E \end{pmatrix} \tag{2.4.8}$$

admits a positive semidefinite completion. In particular, if $G_1 : \overline{\text{Ran}C} \to \overline{\text{Ran}A}$ *and* $G_2 : \overline{\text{Ran}E} \to \overline{\text{Ran}C}$ *are contractions such that* $B = A^{\frac{1}{2}}G_1C^{\frac{1}{2}}$ *and* $D = C^{\frac{1}{2}}G_2E^{\frac{1}{2}}$, *then choosing the* $(1,3)$ *entry of* K *to equal* $A^{\frac{1}{2}}G_1G_2E^{\frac{1}{2}}$ *provides a positive semidefinite completion of (2.4.8) (which will be referred to as the central completion).*

Proof. This is the same as the proof of Theorem 2.2.3, by using the factorization (2.2.4). □

We are now ready to prove the existence and uniqueness of the Schur complement.

Proof of Theorem 2.4.1. Write T as

$$T = \begin{pmatrix} A & B \\ B^* & C \end{pmatrix} : \begin{matrix} \mathcal{M} \\ \oplus \\ \mathcal{M}^\perp \end{matrix} \to \begin{matrix} \mathcal{M} \\ \oplus \\ \mathcal{M}^\perp \end{matrix}. \tag{2.4.9}$$

Applying Lemma 2.4.4 we can write $B = A^{\frac{1}{2}}GC^{\frac{1}{2}}$ with G a contraction. We claim that $S := A^{\frac{1}{2}}(I - GG^*)A^{\frac{1}{2}}$ is the Schur complement. For this, first notice that

$$\begin{pmatrix} A - S & B \\ B^* & C \end{pmatrix} = \begin{pmatrix} A^{\frac{1}{2}}G \\ C^{\frac{1}{2}} \end{pmatrix} \begin{pmatrix} G^*A^{\frac{1}{2}} & C^{\frac{1}{2}} \end{pmatrix} \geq 0.$$

Next, suppose that

$$\begin{pmatrix} A - \widetilde{S} & A^{\frac{1}{2}}GC^{\frac{1}{2}} \\ C^{\frac{1}{2}}G^*A^{\frac{1}{2}} & C \end{pmatrix}$$

is positive semidefinite. Let $x \in \mathcal{M}$. Then $G^*A^{\frac{1}{2}}x \in \overline{\text{Ran}C} = \overline{\text{Ran}C^{\frac{1}{2}}}$, so there exist $y_n \in \mathcal{M}^\perp$ satisfying $\lim_{n\to\infty} C^{\frac{1}{2}}y_n = G^*A^{\frac{1}{2}}x$. Now

$$\begin{aligned} 0 &\leq \left\langle \begin{pmatrix} A - \widetilde{S} & A^{\frac{1}{2}}GC^{\frac{1}{2}} \\ C^{\frac{1}{2}}G^*A\frac{1}{2} & C \end{pmatrix} \begin{pmatrix} x \\ -y_n \end{pmatrix}, \begin{pmatrix} x \\ -y_n \end{pmatrix} \right\rangle \\ &= \langle (A - \widetilde{S})x, x\rangle - \langle C^{\frac{1}{2}}y_n, G^*A^{\frac{1}{2}}x\rangle - \langle G^*A^{\frac{1}{2}}x, C^{\frac{1}{2}}y_n\rangle + \langle C^{\frac{1}{2}}y_n, C^{\frac{1}{2}}y_n\rangle \\ &\to \langle (A - \widetilde{S})x, x\rangle - \langle G^*A^{\frac{1}{2}}x, G^*A^{\frac{1}{2}}x\rangle = \langle (A^{\frac{1}{2}}(I - GG^*)A^{\frac{1}{2}} - \widetilde{S})x, x\rangle. \end{aligned}$$

But this yields that $A^{\frac{1}{2}}(I - GG^*)A^{\frac{1}{2}} \geq \widetilde{S}$, and thus we obtain the maximality of $S = A^{\frac{1}{2}}(I - G^*G)A^{\frac{1}{2}}$. This proves our claim that S is the Schur complement. □

Remark 2.4.6 The above proof implies that $S(T; \mathcal{M}) = A^{\frac{1}{2}}(I - GG^*)A^{\frac{1}{2}}$. It is also customary to call this the *Schur complement of* C *in* T. In case $C > 0$, we have that $S(T; \mathcal{M}) = A - BC^{-1}B^*$. Likewise, $S(T; \mathcal{M}^\perp) = C^{\frac{1}{2}}(I - G^*G)C^{\frac{1}{2}}$ (see Exercise 2.9.26), is also called the Schur complement of A in T. In case $A > 0$, we have that $S(T; \mathcal{M}^\perp) = C - B^*A^{-1}B$.

We now develop some properties of Schur complements. The first property may be viewed as an inheritance principle.

Proposition 2.4.7 *Let $T \in \mathcal{L}(\mathcal{H})$, $T \geq 0$, and let $\mathcal{M}_1$ and $\mathcal{M}_2$ be subspaces of $\mathcal{H}$ such that $\mathcal{M}_2 \subseteq \mathcal{M}_1$. Then $S(T;\mathcal{M}_2) = S(S(T;\mathcal{M}_1);\mathcal{M}_2)$.*

Proof. Denote $S_1 = S(T;\mathcal{M}_1)$, $S_2 = S(T;\mathcal{M}_2)$ and

$$S_{12} = S(S(T;\mathcal{M}_1);\mathcal{M}_2) = S(S_1,\mathcal{M}_2).$$

We need to prove that $S_{12} = S_2$. As $T - P^*_{\mathcal{M}_1} S_1 P_{\mathcal{M}_1} \geq 0$ and $S_1 - P^*_{\mathcal{M}_2} S_{12} P_{\mathcal{M}_2} \geq 0$, we get that $T - P^*_{\mathcal{M}_1} P^*_{\mathcal{M}_2} S_{12} P_{\mathcal{M}_2} P_{\mathcal{M}_1} = T - P^*_{\mathcal{M}_2} S_{12} P_{\mathcal{M}_2} \geq 0$. But then it follows from the maximality of the Schur complement that $S_{12} \leq S_2$.

Next, observe that $T - P^*_{\mathcal{M}_2} S_2 P_{\mathcal{M}_2} = T - P^*_{\mathcal{M}_1} P^*_{\mathcal{M}_2} S_2 P_{\mathcal{M}_2} P_{\mathcal{M}_1} \geq 0$, and thus by the maximality of S_1 we have that $S_1 - P^*_{\mathcal{M}_2} S_2 P_{\mathcal{M}_2} \geq 0$. But then, by the maximality of S_{12} we get that $S_2 \leq S_{12}$. □

The following proposition shows how the Schur complement supported on a subspace may be built from Schur complements on smaller subspaces.

Proposition 2.4.8 *Let*

$$M = \begin{pmatrix} A & B & C \\ B^* & D & E \\ C^* & E^* & F \end{pmatrix} \in \mathcal{L}(\mathcal{H}_1 \oplus \mathcal{H}_2 \oplus \mathcal{H}_3),$$

$M \geq 0$. Then

$$S(M;\mathcal{H}_1 \oplus \mathcal{H}_2) = \begin{pmatrix} S\left(\begin{pmatrix} A & C \\ C^* & F \end{pmatrix};\mathcal{H}_1\right) & Q \\ Q^* & S\left(\begin{pmatrix} D & E \\ E^* & F \end{pmatrix};\mathcal{H}_2\right) \end{pmatrix} \tag{2.4.10}$$

for some $Q \in \mathcal{L}(\mathcal{H}_2,\mathcal{H}_1)$.

Proof. Denote

$$S(M;\mathcal{H}_1 \oplus \mathcal{H}_2) = \begin{pmatrix} P & Q \\ Q^* & R \end{pmatrix}, \quad S_1 = S\left(\begin{pmatrix} A & C \\ C^* & F \end{pmatrix};\mathcal{H}_1\right),$$

and

$$S_2 = S\left(\begin{pmatrix} D & E \\ E^* & F \end{pmatrix};\mathcal{H}_2\right).$$

Then

$$\begin{pmatrix} A-P & B-Q & C \\ B^*-Q^* & D-R & E \\ C^* & E^* & F \end{pmatrix} \geq 0,$$

and thus

$$\begin{pmatrix} A-P & C \\ C^* & F \end{pmatrix} \geq 0, \quad \begin{pmatrix} D-R & E \\ E^* & F \end{pmatrix} \geq 0.$$

This gives that $P \leq S_1$ and $R \leq S_2$.

Next, consider the matrix

$$N(X) = \begin{pmatrix} A - S_1 & X & C \\ X^* & D - S_2 & E \\ C^* & E^* & F \end{pmatrix}.$$

As

$$\begin{pmatrix} A - S_1 & C \\ C^* & F \end{pmatrix} \geq 0, \quad \begin{pmatrix} D - S_2 & E \\ E & F \end{pmatrix} \geq 0,$$

it follows (after a simple permutation similarity) from Theorem 2.4.5 that there exists an $X \in \mathcal{L}(\mathcal{H}_2, \mathcal{H}_1)$ such that $N(X) \geq 0$. Put now $\widetilde{Q} = B - X$. Then $N(B - \widetilde{Q}) \geq 0$. This shows that

$$\begin{pmatrix} S_1 & \widetilde{Q} \\ \widetilde{Q}^* & S_2 \end{pmatrix} \leq S(M; \mathcal{H}_1 \oplus \mathcal{H}_2) = \begin{pmatrix} P & Q \\ Q^* & R \end{pmatrix},$$

yielding $S_1 \leq P$ and $S_2 \leq R$. In the first paragraph of the proof we obtained the other inequalities, so we must have $S_1 = P$ and $S_2 = R$. □

Notice that the proof shows that the operator Q can be obtained by solving a 3×3 positive semidefinite completion problem that apparently has a unique solution (as the proof shows that we must have that $Q = \widetilde{Q} = B - X$) .

The following lemma connects the Schur complement to Cholesky factorizations.

Lemma 2.4.9 *Let $T \in \mathcal{L}(\mathcal{H}_1 \oplus \mathcal{H}_2)$ be factored as*

$$\begin{pmatrix} A & B \\ B^* & C \end{pmatrix} = \begin{pmatrix} P^* & Q^* \\ 0 & R^* \end{pmatrix} \begin{pmatrix} P & 0 \\ Q & R \end{pmatrix}. \tag{2.4.11}$$

Then $S(T; \mathcal{H}_1) \geq P^*P$. Moreover, $S(T; \mathcal{H}_1) = P^*P$ if and only if $\text{Ran}Q \subseteq \overline{\text{Ran}R}$. *In addition, for any P such that $P^*P = S(T; \mathcal{H}_1)$ and any R such that $R^*R = C$, there exists a unique Q such that* (2.4.11) *holds. In that case,*

$$\overline{\text{Ran}} \begin{pmatrix} P & 0 \\ Q & R \end{pmatrix} = \overline{\text{Ran}P} \oplus \overline{\text{Ran}R}. \tag{2.4.12}$$

Proof. Clearly,

$$\begin{pmatrix} A - P^*P & B \\ B^* & C \end{pmatrix} = \begin{pmatrix} Q^* \\ R^* \end{pmatrix} \begin{pmatrix} Q & R \end{pmatrix} \geq 0,$$

and thus $S(T; \mathcal{H}_1) \geq P^*P$.

Suppose now that $\text{Ran}Q \subseteq \overline{\text{Ran}R}$. As (2.4.11) is positive semidefinite, we have by Lemma 2.4.4 that $B = A^{\frac{1}{2}}GR$ for some contraction $G : \overline{\text{Ran}R} \to \overline{\text{Ran}P}$. Also $B = Q^*R$. As $\text{Ran}Q \subseteq \overline{\text{Ran}R}$, we obtain from $A^{\frac{1}{2}}GR = B = Q^*R$ that $A^{\frac{1}{2}}G = Q^*$. But then it follows that $P^*P = A - Q^*Q = A^{\frac{1}{2}}(I - GG^*)A^{\frac{1}{2}}$, which by Remark 2.4.6 equals $S(T; \mathcal{H}_1)$.

Next, suppose that $S(T;\mathcal{H}_1) = P^*P$. Let $\mathcal{M} = \overline{\text{Ran}R}$ and decompose Q and R as follows:

$$Q = \begin{pmatrix} Q_1 \\ Q_2 \end{pmatrix} : \mathcal{H}_1 \to \begin{matrix} \mathcal{M}^\perp \\ \oplus \\ \mathcal{M} \end{matrix}, \quad R = \begin{pmatrix} 0 \\ R_1 \end{pmatrix} : \mathcal{H}_2 \to \begin{matrix} \mathcal{M}^\perp \\ \oplus \\ \mathcal{M} \end{matrix}.$$

Then

$$\begin{pmatrix} A - P^*P - Q_1^*Q_1 & B \\ B^* & C \end{pmatrix} = \begin{pmatrix} Q_2^* \\ R_1^* \end{pmatrix} \begin{pmatrix} Q_2 & R_1 \end{pmatrix} \geq 0,$$

and thus $P^*P = S(T;\mathcal{H}_1) \geq P^*P + Q_1^*Q_1$. This gives that $Q_1 = 0$, and thus $\text{Ran}Q \subseteq \mathcal{M} = \overline{\text{Ran}R}$.

Notice that when $\text{Ran}Q \subseteq \mathcal{M} = \overline{\text{Ran}R}$, we immediately have (2.4.12). The uniqueness of Q follows directly from applying Lemma 2.4.4. □

Lemma 2.4.10 *Suppose*

$$M = \begin{pmatrix} A & B & C \\ B^* & D & E \\ C^* & E^* & F \end{pmatrix} = \begin{pmatrix} P^* & Q^* & R^* \\ 0 & S^* & T^* \\ 0 & 0 & U^* \end{pmatrix} \begin{pmatrix} P & 0 & 0 \\ Q & S & 0 \\ R & T & U \end{pmatrix}, \tag{2.4.13}$$

where M is acting on $\mathcal{H}_1 \oplus \mathcal{H}_2 \oplus \mathcal{H}_3$. Then

$$S(M;\mathcal{H}_1 \oplus \mathcal{H}_2) - S(M;\mathcal{H}_1) = \begin{pmatrix} Q^* \\ S^* \end{pmatrix} \begin{pmatrix} Q & S \end{pmatrix} \tag{2.4.14}$$

if and only if

$$\text{Ran}Q \subseteq \overline{\text{Ran}S} \qquad \text{and} \qquad \text{Ran}T \subseteq \overline{\text{Ran}U}. \tag{2.4.15}$$

Furthermore there exists a factorization of M as in (2.4.13) with the factors operators on $\mathcal{H}_1 \oplus \mathcal{H}_2 \oplus \mathcal{H}_3$ such that (2.4.14) and (2.4.15) hold,

$$\begin{pmatrix} P^* & Q^* \\ 0 & S^* \end{pmatrix} \begin{pmatrix} P & 0 \\ Q & S \end{pmatrix} = S(M;\mathcal{H}_1 \oplus \mathcal{H}_2),$$

and

$$P^*P = S(M;\mathcal{H}_1) = S(S(M;\mathcal{H}_1 \oplus \mathcal{H}_2);\mathcal{H}_1).$$

Proof. To begin with, suppose (2.4.14) holds. Then if $\widetilde{P}^*\widetilde{P} = S(M;\mathcal{H}_1)$, we have

$$S(M;\mathcal{H}_1 \oplus \mathcal{H}_2) = \begin{pmatrix} \widetilde{P}^* & Q^* \\ 0 & S^* \end{pmatrix} \begin{pmatrix} \widetilde{P} & 0 \\ Q & S \end{pmatrix}.$$

As $U^*U = F$, by Lemma 2.4.9 there exist $\widetilde{R}$ and $\widetilde{T}$ such that

$$M = \begin{pmatrix} \widetilde{P}^* & Q^* & \widetilde{R}^* \\ 0 & S^* & \widetilde{T}^* \\ 0 & 0 & U^* \end{pmatrix} \begin{pmatrix} \widetilde{P} & 0 & 0 \\ Q & S & 0 \\ \widetilde{R} & \widetilde{T} & U \end{pmatrix}.$$

By Lemma 2.4.9

$$\text{Ran} \begin{pmatrix} Q \\ \widetilde{R} \end{pmatrix} \subseteq \overline{\text{Ran} \begin{pmatrix} S & 0 \\ \widetilde{T} & U \end{pmatrix}}$$

and

$$\operatorname{Ran}\begin{pmatrix}\widetilde{R} & \widetilde{T}\end{pmatrix} \subseteq \overline{\operatorname{Ran} U}.$$

Hence $\operatorname{Ran}\widetilde{T} \subseteq \overline{\operatorname{Ran} U}$ and so

$$\overline{\operatorname{Ran}}\begin{pmatrix} S & 0 \\ \widetilde{T} & U\end{pmatrix} = \overline{\operatorname{Ran} S} \oplus \overline{\operatorname{Ran} U}.$$

Thus $\operatorname{Ran} Q \subseteq \overline{\operatorname{Ran} S}$.

Next observe that $D = S^*S + T^*T = S^*S + \widetilde{T}^*\widetilde{T}$ and so there is an isometry $V_T : \overline{\operatorname{Ran} T} \to \overline{\operatorname{Ran} \widetilde{T}}$ such that $T^* = \widetilde{T}^* V_T^*$. Also $\operatorname{Ran}\widetilde{T} \subseteq \overline{\operatorname{Ran} U}$ implies that $\operatorname{Ran} V_T \subseteq \overline{\operatorname{Ran} U}$. Thus V_T is an isometry from $\overline{\operatorname{Ran} T}$ into $\overline{\operatorname{Ran} U}$. But $U^*T = E^* = U^*\widetilde{T} = U^*V_T T$, so $V_T = 1_{\overline{\operatorname{Ran} T}}$ and $\operatorname{Ran} T \subseteq \overline{\operatorname{Ran} U}$.

Now conversely assume we have a factorization of M as in (2.4.13) where (2.4.15) holds. Set

$$L = \begin{pmatrix} D & E \\ E^* & F\end{pmatrix} = \begin{pmatrix} S^* & T^* \\ 0 & U^*\end{pmatrix}\begin{pmatrix} S & 0 \\ T & U\end{pmatrix}.$$

Using Lemma 2.4.9, suppose $\widetilde{G}$ is any other operator matrix satisfying $\widetilde{G}^*\widetilde{G} = M$ with

$$\widetilde{G} = \begin{pmatrix} \widetilde{P} & 0 & 0 \\ \widetilde{Q} & \widetilde{S} & 0 \\ \widetilde{R} & \widetilde{T} & U \end{pmatrix}, \tag{2.4.16}$$

where

$$S(M; \mathcal{H}_1 \oplus \mathcal{H}_2) = \begin{pmatrix} \widetilde{P}^* & \widetilde{Q}^* \\ 0 & \widetilde{S}^* \end{pmatrix}\begin{pmatrix} \widetilde{P} & 0 \\ \widetilde{Q} & \widetilde{S}\end{pmatrix}$$

and $\widetilde{P}$ is chosen so that $S(S(M; \mathcal{H}_1 \oplus \mathcal{H}_2); \mathcal{H}_1) = \widetilde{P}^*\widetilde{P}$. Note that

$$L = \begin{pmatrix} \widetilde{S}^* & \widetilde{T}^* \\ 0 & U^* \end{pmatrix}\begin{pmatrix} \widetilde{S} & 0 \\ \widetilde{T} & U\end{pmatrix}.$$

Since by assumption $\operatorname{Ran} T \subseteq \overline{\operatorname{Ran} U}$, we have $S^*S = S(L; \mathcal{H}_2) \geq \widetilde{S}^*\widetilde{S}$. On the other hand, since

$$S(M; \mathcal{H}_1 \oplus \mathcal{H}_2) \geq \begin{pmatrix} P^* & Q^* \\ 0 & S^* \end{pmatrix}\begin{pmatrix} P & 0 \\ Q & S\end{pmatrix},$$

we also have $\widetilde{S}^*\widetilde{S} \geq S^*S$. Hence $\widetilde{S}^*\widetilde{S} = S^*S$. Thus $VS = \widetilde{S}$ for some isometry $V : \overline{\operatorname{Ran} S} \to \overline{\operatorname{Ran} \widetilde{S}}$. Since we have chosen $S(S(M; \mathcal{H}_1 \oplus \mathcal{H}_2); \mathcal{H}_1) = \widetilde{P}^*\widetilde{P}$, by Lemma 2.4.9, $\operatorname{Ran}\widetilde{Q} \subseteq \overline{\operatorname{Ran} \widetilde{S}}$. Moreover,

$$0 \leq \begin{pmatrix} \widetilde{P}^*\widetilde{P} + \widetilde{Q}^*\widetilde{Q} & \widetilde{Q}^*S \\ S^*\widetilde{Q} & S^*S \end{pmatrix} - \begin{pmatrix} P^*P + Q^*Q & Q^*S \\ S^*Q & S^*S\end{pmatrix}$$

and $\widetilde{S}^*\widetilde{S} \geq S^*S$ imply that $0 = \widetilde{Q}^*\widetilde{S} - Q^*S = (\widetilde{Q}^*V - Q^*)S$. As $\operatorname{Ran} Q \subseteq \overline{\operatorname{Ran} S}$ it follows that $\widetilde{Q}^*V = Q^*$. Thus, in particular, $\widetilde{Q}^*\widetilde{Q} = Q^*Q$. But then we obtain that

$$\widetilde{P}^*\widetilde{P} \geq P^*P. \tag{2.4.17}$$

Observe that (2.4.17) will be true no matter what the original factorization of M in (2.4.13) is as long as the range conditions in (2.4.15) are satisfied.

Now instead consider the factorization $M = G'^*G'$, where

$$G' = \begin{pmatrix} P' & 0 & 0 \\ Q' & S & 0 \\ R' & T & U \end{pmatrix}$$

with $P'^*P' = S(M; \mathcal{H}_1)$. Such a factorization is possible by Lemma 2.4.9. Since by assumption $\operatorname{Ran} T \subseteq \overline{\operatorname{Ran}} U$, we have

$$\overline{\operatorname{Ran}} \begin{pmatrix} S & 0 \\ T & U \end{pmatrix} \subseteq \overline{\operatorname{Ran}} S \oplus \overline{\operatorname{Ran}} U.$$

Also by Lemma 2.4.9, then

$$\operatorname{Ran} \begin{pmatrix} Q' \\ R' \end{pmatrix} \subseteq \overline{\operatorname{Ran}} S \oplus \overline{\operatorname{Ran}} U,$$

and hence $\operatorname{Ran} Q' \subseteq \overline{\operatorname{Ran}} S$. So the conditions in (2.4.15) are satisfied for this factorization, and hence, as noted above, we must have $\widetilde{P}^*\widetilde{P} \geq P'^*P'$. But by definition of the Schur complement, $P'^*P' \geq \widetilde{P}^*\widetilde{P}$, so we have equality. Consequently, (2.4.14) holds.

Finally, using Lemma 2.4.9, there is a factorization

$$L = \begin{pmatrix} S^* & T^* \\ 0 & U^* \end{pmatrix} \begin{pmatrix} S & 0 \\ T & U \end{pmatrix},$$

where $\operatorname{Ran} T \subseteq \overline{\operatorname{Ran}} U \subseteq \mathcal{H}_3$, so that $\overline{\operatorname{Ran}} \left(\begin{smallmatrix} S & T \\ 0 & U \end{smallmatrix} \right) = \overline{\operatorname{Ran}} S \oplus \overline{\operatorname{Ran}} U \subseteq \mathcal{H}_2 \oplus \mathcal{H}_3$. Again by Lemma 2.4.9, there exists $P : \mathcal{H}_1 \to \mathcal{H}_1$ such that $P^*P = S(M; \mathcal{H}_1)$ and (2.4.13) holds. Consequently $\operatorname{Ran} \left(\begin{smallmatrix} Q \\ R \end{smallmatrix} \right) \subseteq \overline{\operatorname{Ran}} S \oplus \overline{\operatorname{Ran}} U$, giving $\operatorname{Ran} Q \subseteq \overline{\operatorname{Ran}} S$.

It is now clear that the factorization in (2.4.13) with these choices of P, Q, R, S, T, and U satisfies the last statement of the theorem. □

Corollary 2.4.11 *Let $M = (M_{ij})_{i,j=1}^n : \bigoplus_{j=1}^n \mathcal{H}_j \to \bigoplus_{j=1}^n \mathcal{H}_j$ be an $n \times n$ positive semidefinite operator matrix, and $J \subseteq K \subseteq \{1, \dots, n\}$. Then*

$$S(M; \oplus_{j\in J}\mathcal{H}_j) = S(S(M; \oplus_{j\in K}\mathcal{H}_j); \oplus_{j\in J}\mathcal{H}_j). \tag{2.4.18}$$

Proof. Let $I_1 = J$, $I_2 = K \setminus J$ and $I_3 = \{1, \dots, n\} \setminus K$. Writing M as a 3×3 block matrix with respect to the partition

$$\{0, \dots, n-1\} = I_1 \cup I_2 \cup I_3,$$

the corollary follows directly from Lemma 2.4.10. □

Corollary 2.4.12 *Given $M = (M_{ij})_{i,j=1}^n : \bigoplus_{j=1}^{n-1} \mathcal{H}_j \to \bigoplus_{j=1}^{n-1} \mathcal{H}_j$ an $n \times n$ positive semidefinite operator matrix, there exists a factorization $M = P^*P$ where*

$$P = \begin{pmatrix} P_{11} & 0 & \cdots & \cdots & 0 \\ P_{21} & P_{22} & 0 & \cdots & 0 \\ \vdots & \vdots & \ddots & \ddots & \vdots \\ \vdots & \vdots & \ddots & \ddots & 0 \\ P_{n,1} & P_{n,2} & \cdots & \cdots & P_{n,n} \end{pmatrix},$$

$\overline{\mathrm{Ran}}P = \overline{\mathrm{Ran}}P_{11} \oplus \cdots \oplus \overline{\mathrm{Ran}}P_{n,n}$, *and such that if* P_k *is the truncation of* P *to the upper left* $k \times k$ *corner, then* $S(M; \bigoplus_{j=1}^k \mathcal{H}_j) = P_k^* P_k$, $k = 1, \ldots, n$.

Lemma 2.4.13 *Let*

$$\begin{pmatrix} P^* & Q^* \\ 0 & R^* \end{pmatrix} \begin{pmatrix} P & 0 \\ Q & R \end{pmatrix} = \begin{pmatrix} \widetilde{P}^* & \widetilde{Q}^* \\ 0 & \widetilde{R}^* \end{pmatrix} \begin{pmatrix} \widetilde{P} & 0 \\ \widetilde{Q} & \widetilde{R} \end{pmatrix}, \tag{2.4.19}$$

and suppose $\mathrm{Ran}Q \subseteq \overline{\mathrm{Ran}}R$. *Then there is a unique isometry*

$$\begin{pmatrix} V_{11} & 0 \\ V_{21} & V_{22} \end{pmatrix}$$

acting on $\overline{\mathrm{Ran}}P \oplus \overline{\mathrm{Ran}}R$ *such that*

$$\begin{pmatrix} \widetilde{P} & 0 \\ \widetilde{Q} & \widetilde{R} \end{pmatrix} = \begin{pmatrix} V_{11} & 0 \\ V_{21} & V_{22} \end{pmatrix} \begin{pmatrix} P & 0 \\ Q & R \end{pmatrix}. \tag{2.4.20}$$

Proof. It is a standard result that $A^*A = B^*B$ if and only there exists an isometry $V : \overline{\mathrm{Ran}\,B} \to \overline{\mathrm{Ran}\,A}$ such that $VB = A$. The operator V is uniquely determined by setting $V(Bx) = Ax$ for every $Bx \in \mathrm{Ran}\,B$, and extending V to $\overline{\mathrm{Ran}\,B}$ by continuity. Thus (2.4.19) implies the existence of an isometry $V = (V_{ij})_{i,j=1}^2$ satisfying

$$\begin{pmatrix} \widetilde{P} & 0 \\ \widetilde{Q} & \widetilde{R} \end{pmatrix} = \begin{pmatrix} V_{11} & V_{12} \\ V_{21} & V_{22} \end{pmatrix} \begin{pmatrix} P & 0 \\ Q & R \end{pmatrix}. \tag{2.4.21}$$

It remains to show that $V_{12} = 0$. Note that (2.4.21) implies that $V_{22}R = \widetilde{R}$. Combining this with (2.4.19) we get that

$$R^*R = \widetilde{R}^*\widetilde{R} = R^*V_{22}^*V_{22}R,$$

and thus

$$R^*(I_{\overline{\mathrm{Ran}\,R}} - V_{22}^*V_{22})R = 0. \tag{2.4.22}$$

As $\mathrm{Ran}\,Q \subseteq \overline{\mathrm{Ran}\,R}$ we have that

$$\overline{\mathrm{Ran}} \begin{pmatrix} P & 0 \\ Q & R \end{pmatrix} = \overline{\mathrm{Ran}\,P} \oplus \overline{\mathrm{Ran}\,R}.$$

Thus V_{22} and V_{12} act on $\overline{\mathrm{Ran}\,R}$. From (2.4.22) we now obtain that V_{22} is an isometry on $\overline{\mathrm{Ran}\,R}$. But then, since V is an isometry, we must have that $V_{12} = 0$. □

Lemma 2.4.14 *Let* $M = (M_{ij})_{i,j=1}^4 : \bigoplus_{j=1}^4 \mathcal{H}_j \to \bigoplus_{j=1}^4 \mathcal{H}_j$ *be a positive semidefinite operator matrix. Then we have that*

$$\left[S\left(M; \bigoplus_{j=1}^3 \mathcal{H}_j \right) \right]_{21} = 0 \tag{2.4.23}$$

if and only if

$$S\left(M; \bigoplus_{j=1}^3 \mathcal{H}_j \right) = S(M; \mathcal{H}_1 \oplus \mathcal{H}_2) + S(M; \mathcal{H}_1 \oplus \mathcal{H}_3) - S(M; \mathcal{H}_1). \tag{2.4.24}$$

Proof. The direction (2.4.24) $\Rightarrow$ (2.4.23) is trivial.

By Corollary 2.4.12, there is a lower triangular 3×3 operator matrix

$$P=\begin{pmatrix}P_{00}&0&0\\P_{10}&P_{11}&0\\P_{20}&P_{21}&P_{22}\end{pmatrix}$$

such that $S(M;\oplus_{j=1}^3\mathcal{H}_j)=P^*P$,

$$S(M;\oplus_{j=1}^2\mathcal{H}_j)=\begin{pmatrix}P_{00}^*&P_{10}^*\\0&P_{11}^*\end{pmatrix}\begin{pmatrix}P_{00}&0\\P_{10}&P_{11}\end{pmatrix},$$

and $S(M;\mathcal{H}_1)=P_{00}^*P_{00}$. Also, $\operatorname{Ran}P_{10}\subseteq\overline{\operatorname{Ran}}\,P_{11}$ and $\operatorname{Ran}P_{20},\operatorname{Ran}P_{21}\subseteq\overline{\operatorname{Ran}}\,P_{22}$. Thus $[S(M;\bigoplus_{j=1}^3\mathcal{H}_j)]_{21}=0$ is equivalent to $P_{21}=0$. Interchanging the order of rows 1 and 2 and columns 1 and 2, yielding $\widetilde{M}$, say, we have

$$S(\widetilde{M};\mathcal{H}_1\oplus\mathcal{H}_3\oplus\mathcal{H}_2)=\begin{pmatrix}P_{00}^*&P_{20}^*&P_{10}^*\\0&P_{22}^*&0\\0&0&P_{11}^*\end{pmatrix}\begin{pmatrix}P_{00}&0&0\\P_{20}&P_{22}&0\\P_{10}&0&P_{11}\end{pmatrix}. \tag{2.4.25}$$

Since $\operatorname{Ran}\begin{pmatrix}P_{10}&0\end{pmatrix}\subseteq\overline{\operatorname{Ran}}\,P_{11}$, by Lemma 2.4.9,

$$S(\widetilde{M};\mathcal{H}_1\oplus\mathcal{H}_3)=\begin{pmatrix}P_{00}^*&P_{20}^*\\0&P_{22}^*\end{pmatrix}\begin{pmatrix}P_{00}&0\\P_{20}&P_{22}\end{pmatrix}. \tag{2.4.26}$$

A direct calculation verifies the equality in (2.4.24). $\square$

By relabeling and grouping as we did in the proof of Corollary 2.4.11, we obtain the following.

Corollary 2.4.15 *Suppose* $M=(M_{ij})_{i,j=1}^n:\bigoplus_{j=1}^n\mathcal{H}_j\to\bigoplus_{j=1}^n\mathcal{H}_j$ *is an* $n\times n$ *positive semidefinite operator matrix,* $K\cup J=N\subseteq\{1,\dots n\}$. *Then*

$$S(M;\oplus_{j\in N}\mathcal{H}_j)=S(M;\oplus_{j\in K}\mathcal{H}_j)+S(M;\oplus_{j\in J}\mathcal{H}_j)-S(M;\oplus_{j\in K\cap J}\mathcal{H}_j) \tag{2.4.27}$$

if and only if

$$[S(M;\oplus_{j\in N}\mathcal{H}_j)]_{k,j}=0,\quad (k,j)\in(N\times N)\setminus((K\times K)\cup(J\times J)). \tag{2.4.28}$$

We apply the theory of Schur complements to prove the Fejér-Riesz factorization result for operator-valued positive semidefinite trigonometric polynomials. In the scalar case the result was proved in Theorem 1.1.5. The positive definite operator-valued version was proved by different methods in Corollary 2.3.10. We need the following definition. We call the operator-valued polynomial $P(z)=\sum\limits_{k=0}^m P_kz^k$ *outer* if

$$\operatorname{Ran}\begin{pmatrix}P_1\\P_2\\\vdots\end{pmatrix}\subseteq\overline{\operatorname{Ran}}\begin{pmatrix}P_0&0&\cdots&\\P_1&P_0&0&\cdots\\\vdots&\ddots&\ddots&\ddots\end{pmatrix}.$$

We say that $P(z)=\sum_{k=0}^m P_kz^k$, with $P_m\neq 0$, is *co-outer* if $z^mP(\frac{1}{\bar z})^*$ is outer.

Theorem 2.4.16 *Let $Q_j \in \mathcal{L}(\mathcal{H})$, $j = -m, \dots, m$, be such that $Q(z) := \sum_{j=-m}^{m} Q_j z^j \geq 0$ for $z \in \mathbb{T}$. Then there exist operators $P_j \in \mathcal{L}(\mathcal{H})$, $j = 0, \dots, n$ such that $P(z) := \sum_{j=0}^{m} P_j z^j$ satisfies*

$$P(z)^* P(z) = Q(z), \ z \in \mathbb{T}.$$

Moreover, $P(z)$ can be chosen to be outer and to satisfy $P_0 \geq 0$, and in that case $P(z)$ is unique. When $Q(z) > 0$, $z \in \mathbb{T}$, this unique outer factor has the property that $P(z)$ is invertible for $z \in \overline{\mathbb{D}}$. Alternatively, $P(z)$ can be chosen to be co-outer and to satisfy $P_0 \geq 0$, which again leads to a unique choice that is invertible on $\mathbb{C} \setminus \mathbb{D}$ when $Q(z)$ is invertible on $\mathbb{T}$.

Before proving Theorem 2.4.16 we need some definitions and a lemma.

A function $f : \mathbb{D} \to \mathcal{H}$ is called *analytic* if for all $h \in \mathcal{H}$, the function $\langle f(z), h \rangle$ is analytic in $\mathbb{D}$. Such functions can be represented as $f(z) = \sum_{k=0}^{\infty} f_k z^k$, with $f_k \in \mathcal{H}$, the series converging in $\mathcal{H}$ for all $z \in \mathbb{D}$. Define for such a function $\|f\|_2^2 = \sum_{k=0}^{\infty} \|f_k\|^2$, and let $H^2(\mathcal{H})$ be the set of analytic functions $f : \mathbb{D} \to \mathcal{H}$ such that $\|f\|_2 < \infty$. Then $H^2(\mathcal{H})$ can be identified in a natural way with a closed subspace of $L^2(\mathcal{H})$, and also with $H^2 \otimes \mathcal{H}$, or $l^2(\mathcal{H}) = \bigoplus_{n=0}^{\infty} \mathcal{H}$. Given $F(e^{it}) = \sum_{k=-\infty}^{\infty} F_k e^{ikt} \in L^\infty(\mathcal{L}(\mathcal{H}, \mathcal{H}'))$, we define the *Toeplitz operator with symbol F* is defined by

$$T_F : H^2(\mathcal{H}) \to H^2(\mathcal{H}'), \ T_F f = P_+(Ff),$$

where P_+ is the orthogonal projection of $L^2(\mathcal{H}')$ onto $H^2(\mathcal{H}')$. The following proposition shows that the norm of the Toeplitz operator equals the essential norm of its symbol.

Proposition 2.4.17 *Let T_F be defined as above. Then $\|T_F\| = \|F\|_\infty$.*

Proof. Note that when $\|F\|_\infty = 0$, we have that $T_F = 0$, and thus the result holds. Next, assume that $A := \|F\|_\infty > 0$. Let $M_F : L^2(\mathcal{H}) \to L^2(\mathcal{H}')$ be the multiplication operator with symbol F. Clearly, $\|T_F\| \leq \|M_F\| = A$, where for the last equality we use Proposition 2.3.7. Let $\epsilon > 0$ and let $g \in L^2(\mathcal{H})$ be of norm 1 so that $\|M_F g\| \geq A - \frac{\epsilon}{3}$. Choose an $n \in \mathbb{Z}$, so that $\|z^n g(z) - P_+(z^n g(z))\| < \frac{\epsilon}{3A}$. Put $h(z) = P_+(z^n g(z))$. Then $\|F(z)h(z)\| > A - \frac{2\epsilon}{3}$. Next choose $m \in \mathbb{N}_0$, so that $\|P_+(z^m F(z)h(z))\| > A - \epsilon$. Now it follows that $\|T_F(z^m h(z))\| > A - \epsilon$. As $\|P_+(z^m h(z))\| \leq 1$, it now follows that $\|T_F\| > A - \epsilon$. □

With respect to the direct sum decompositions $H^2(\mathcal{H}) = \bigoplus_{n=0}^{\infty} \mathcal{H}$ and $H^2(\mathcal{H}') = \bigoplus_{n=0}^{\infty} \mathcal{H}'$, T_F has the matrix decomposition

$$T_F = (F_{i-j})_{i,j=0}^{\infty} = \begin{pmatrix} F_0 & F_{-1} & F_{-2} & \cdots \\ F_1 & F_0 & F_{-1} & \ddots \\ F_2 & F_1 & F_0 & \ddots \\ \vdots & \ddots & \ddots & \ddots \end{pmatrix}. \tag{2.4.29}$$

For $F \in L^\infty(\mathcal{L}(\mathcal{H}))$, we have that $T_F \geq 0$ if and only if $F(e^{it}) \geq 0$ almost everywhere on $\mathbb{T}$. If $Q(z) = \sum_{k=-m}^{m} Q_k z^k$, then in the decomposition (2.4.29) of T_Q we have $Q_k = 0$ for $|k| > m$, and $T_Q > 0$ if and only if $Q(e^{it}) > 0$ for $e^{it} \in \mathbb{T}$.

Lemma 2.4.18 *Let $Q(z) = \sum_{k=-m}^{m} Q_k z^k$ be such that $Q(e^{it}) \geq 0$ for $e^{it} \in \mathbb{T}$, and denote $S(k) = S(T_Q; \bigoplus_{j=0}^{k} \mathcal{H})$. Then*

$$S(k) = \begin{pmatrix} A_k & B_k \\ B_k^* & S(k-1) \end{pmatrix}, \tag{2.4.30}$$

for some A_k and B_k. If $Q_j = 0$ for $|j| > m$, then

$$A_j = Q_0, \; B_k = \begin{pmatrix} Q_{-1} & \cdots & Q_{-k} \end{pmatrix}, \; k \geq m.$$

Proof. This immediately follows from Proposition 2.4.8 and the Toeplitz structure of T_Q. Note that $A_k = S(\widetilde{T}; \bigoplus_{j=0}^{0} \mathcal{H})$ where $\widetilde{T}$ is obtained from T_Q by keeping rows and columns indexed $0, k+1, k+2, \ldots$ (and thus removing rows and columns indexed $1, \ldots, k$). That means that A_k is the Schur complement of T_Q in $\begin{pmatrix} Q_0 & 0 \\ 0 & T_Q \end{pmatrix}$, which happens to be true. □

Proof of Theorem 2.4.16. We use the notation $S(k)$ as in Lemma 2.4.18. We will construct the coefficients of $P(z)$ inductively via the Schur complements $S(k)$ as follows.

Let $P_0 = S(0)^{\frac{1}{2}}$. Suppose now that $P_0, \ldots, P_k$ have been found so that

$$S(k) = \begin{pmatrix} P_0^* & \cdots & P_k^* \\ & \ddots & \vdots \\ 0 & & P_0^* \end{pmatrix} \begin{pmatrix} P_0 & & 0 \\ \vdots & \ddots & \\ P_k & \cdots & P_0 \end{pmatrix} \tag{2.4.31}$$

and

$$\operatorname{Ran} \begin{pmatrix} P_1 \\ \vdots \\ P_k \end{pmatrix} \subseteq \overline{\operatorname{Ran}} \begin{pmatrix} P_0 & & 0 \\ \vdots & \ddots & \\ P_{k-1} & \cdots & P_0 \end{pmatrix}. \tag{2.4.32}$$

Consider now $S(k+1)$, which by Lemma 2.4.18 has the form (2.4.30). Apply now Lemma 2.4.9 and Proposition 2.4.8 with $P = S(0)^{\frac{1}{2}} = P_0$ and

$$R = \begin{pmatrix} P_0 & & 0 \\ \vdots & \ddots & \\ P_k & \cdots & P_0 \end{pmatrix}$$

to obtain

$$S(k+1) = \begin{pmatrix} P_0^* & \widetilde{Q}^* \\ 0 & R^* \end{pmatrix} \begin{pmatrix} P_0 & 0 \\ \widetilde{Q} & R \end{pmatrix},$$

with

$$\widetilde{Q} = \begin{pmatrix} \widetilde{Q}_1 \\ \vdots \\ \widetilde{Q}_{k+1} \end{pmatrix}, \quad \operatorname{Ran}\widetilde{Q} \subseteq \overline{\operatorname{Ran}} R. \tag{2.4.33}$$

We would like to show that $\widetilde{Q}_i = P_i$, $i = 1, \ldots, k$. By Proposition 2.4.7 we have that

$$S\left(S(k+1), \bigoplus_{j=0}^{k} \mathcal{H}\right) = S(k). \tag{2.4.34}$$

Note that (2.4.33) implies

$$\operatorname{Ran}\begin{pmatrix} \widetilde{Q}_1 \\ \vdots \\ \widetilde{Q}_k \end{pmatrix} \subseteq \overline{\operatorname{Ran}}\begin{pmatrix} P_0 & & 0 \\ \vdots & \ddots & \\ P_{k-1} & \cdots & P_0 \end{pmatrix}. \tag{2.4.35}$$

But then Lemma 2.4.9 implies that $S\left(S(k+1), \oplus_{j=0}^{k}\mathcal{H}\right)$ equals

$$\begin{pmatrix} P_0^* & \widetilde{Q}_1^* & \cdots & \widetilde{Q}_k^* \\ & P_0^* & \cdots & P_{k-1}^* \\ & & \ddots & \vdots \\ 0 & & & P_0^* \end{pmatrix}\begin{pmatrix} P_0 & & & 0 \\ \widetilde{Q}_1 & P_0 & & \\ \vdots & \vdots & \ddots & \\ \widetilde{Q}_k & P_{k-1} & \cdots & P_0 \end{pmatrix}. \tag{2.4.36}$$

This together with (2.4.31) and (2.4.34) implies that

$$\begin{pmatrix} P_0^* & \cdots & P_{k-1}^* \\ & \ddots & \vdots \\ 0 & & P_0^* \end{pmatrix}\begin{pmatrix} \widetilde{Q}_1 \\ \vdots \\ \widetilde{Q}_k \end{pmatrix} = \begin{pmatrix} P_0^* & \cdots & P_{k-1}^* \\ & \ddots & \vdots \\ 0 & & P_0^* \end{pmatrix}\begin{pmatrix} P_1 \\ \vdots \\ P_k \end{pmatrix}.$$

Using (2.4.33) and (2.4.35) we now get that $\widetilde{Q}_i = P_i$, $i = 1, \ldots, k$. This proves the existence of P_k, $k \geq 0$, so that (2.4.31) and (2.4.32) hold for all k. Moreover, the construction shows that P_k, $k \geq 0$, are unique when one requires $P_0 \geq 0$, (2.4.31), and (2.4.32).

Next, observe that Lemma 2.4.18 yields that

$$S(m) = \begin{pmatrix} Q_0 & \begin{pmatrix} Q_{-1} & \cdots & Q_{-m} \end{pmatrix} \\ \begin{pmatrix} Q_1 \\ \vdots \\ Q_m \end{pmatrix} & S(m-1) \end{pmatrix}.$$

Together with (2.4.31), this gives that

$$Q_j = P_0^* P_j + \cdots + P_{m-j}^* P_m, \ \ j = 0 \ldots, m,$$

which implies $Q(z) = P(z)^* P(z)$, $z \in \mathbb{T}$.

Finally, suppose that $Q(z) > 0$, $z \in \mathbb{T}$, and write $F(z) := Q(z)^{-1} = \sum_{k=-\infty}^{\infty} C_k z^k$, $z \in \mathbb{T}$. Apply Theorem 2.3.1 and set $P(z) = Y_0^{-\frac{1}{2}} z^m Y(\frac{1}{\bar{z}})^*$ to obtain a polynomial $P(z)$ that is invertible for $z \in \overline{\mathbb{D}}$ and satisfies

$$P(z)^{-1} P(z)^{*-1} = \sum_{k=-\infty}^{\infty} \widetilde{C}_k z^k,$$

with $\widetilde{C}_k = C_k$, $k = -m, \ldots, m$. By the last statement in Theorem 2.3.8 that $P(z)^{-1}P(z)^{*-1} = F(z)$, and thus $Q(z) = P(z)^* P(z)$. As P is outer with

$P_0 > 0$, we obtain that $P(z)$ is the unique outer factor of $Q(z)$. Thus the unique outer factor of $Q(z)$ is invertible for $z \in \overline{\mathbb{D}}$.

The co-outer statement follows analogously. □

The following result provides a way to find the outer and co-outer factorizations via finding the extreme solutions to a linear matrix inequality (namely, the linear matrix inequality $A(X) \geq 0$ below).

Theorem 2.4.19 *Let $Q(z) = Q_{-n}z^{-n} + \cdots + Q_n z^n \geq 0, z \in \mathbb{T}$, where $Q_j \in \mathcal{L}(\mathcal{H})$. Let*

$$Z = \begin{pmatrix} Q_0 & Q_{-1} & \dots & Q_{-n} \\ Q_1 & 0 & \dots & 0 \\ \vdots & \vdots & & \vdots \\ Q_n & 0 & \dots & 0 \end{pmatrix},$$

and introduce the convex set

$$\mathcal{G} = \left\{ X : \mathcal{H}^n \to \mathcal{H}^n : A(X) := Z - \begin{pmatrix} X & 0 \\ 0 & 0_{\mathcal{H}} \end{pmatrix} + \begin{pmatrix} 0_{\mathcal{H}} & 0 \\ 0 & X \end{pmatrix} \geq 0 \right\}.$$

Then $\mathcal{G}$ has elements $X_{\max}$ and $X_{\min}$ that are maximal and minimal with respect to the Loewner ordering, respectively; that is, $X_{\max}, X_{\min} \in \mathcal{G}$ have the property that $X \in \mathcal{G}$ implies $X_{\min} \leq X \leq X_{\max}$. In fact, $X_{\max}$ is the Schur complement of T_Q supported in the first n rows and columns, and

$$X_{\min} = \begin{pmatrix} Q_0 & \dots & Q_{-n+1} \\ \vdots & \ddots & \vdots \\ Q_{n-1} & \dots & Q_0 \end{pmatrix} - \widetilde{S},$$

where $\widetilde{S}$ is the Schur complement of

$$\widetilde{T}_Q = \begin{pmatrix} \ddots & \ddots & \vdots \\ \ddots & Q_0 & Q_{-1} \\ \cdots & Q_1 & Q_0 \end{pmatrix}$$

supported in the last n rows and columns.

Moreover, consider the set

$$\mathcal{A} = \left\{ A = (A_{ij})_{i,j=0}^n : \mathcal{H}^{n+1} \to \mathcal{H}^{n+1} \ : \ A = A(X) \ for\ some\ X \in \mathcal{G} \right\}.$$

Then $A(X_{\max})$ is the unique element in $\mathcal{A}$ so that A_{nn} is maximal (or equivalently, A_{00} is minimal) in the Loewner order. Also, $A(X_{\min})$ is the unique element in $\mathcal{A}$ so that A_{00} is maximal (or equivalently, A_{nn} is minimal) in the Loewner order. Finally, if we factor $A(X_{\max})$ and $A(X_{\min})$ as

$$A(X_{\max}) = \begin{pmatrix} H_0^* \\ \vdots \\ H_n^* \end{pmatrix} \begin{pmatrix} H_0 & \dots & H_n \end{pmatrix}, \ A(X_{\min}) = \begin{pmatrix} K_0^* \\ \vdots \\ K_n^* \end{pmatrix} \begin{pmatrix} K_0 & \dots & K_n \end{pmatrix},$$

with $H_i, K_i \in \mathcal{B}(\mathcal{K}), i = 0, \dots, n$, and put $H(z) = \sum_{k=0}^n H_k z^k$, $K(z) = \sum_{k=0}^n K_k z^k$, then $H(z)$ is co-outer and $K(z)$ is outer, and

$$Q(z) = H(z)^* H(z) = K(z)^* K(z), \quad z \in \mathbb{T}.$$

Proof. Let $X_{\max}$ be the Schur complement of T_Q supported in the first n rows and columns. Then by Lemma 2.4.18 the operator matrix

$$Z + \begin{pmatrix} 0_{\mathcal{K}} & 0 \\ 0 & X_{\max} \end{pmatrix} = \begin{pmatrix} Q_0 & Q_{-1} & \cdots & Q_{-n} \\ Q_1 & (X_{\max})_{11} & \cdots & (X_{\max})_{1n} \\ \vdots & \vdots & & \vdots \\ Q_n & (X_{\max})_{n1} & \cdots & (X_{\max})_{nn} \end{pmatrix} \tag{2.4.37}$$

is the Schur complement of T_Q supported in the first $n+1$ rows and columns (one can also check this easily from the definition). As

$$T_Q - \begin{pmatrix} X_{\max} & 0 & 0 \\ 0 & 0_{\mathcal{K}} & 0 \\ 0 & 0 & 0 \end{pmatrix} \geq 0$$

it follows that by definition of the Schur complement that

$$\begin{pmatrix} X_{\max} & 0 \\ 0 & 0_{\mathcal{K}} \end{pmatrix} \leq S(T_Q; \mathcal{H}^{n+1}) = Z + \begin{pmatrix} 0_{\mathcal{K}} & 0 \\ 0 & X_{\max} \end{pmatrix},$$

and thus $X_{\max} \in \mathcal{G}$. Take now $X = (X_{ij})_{i,j=1}^{n} \in \mathcal{G}$. Then

$$\begin{pmatrix} X_{11} & \cdots & X_{1n} & 0 & 0 & \cdots \\ \vdots & & \vdots & \vdots & & \\ X_{n1} & \cdots & X_{nn} & 0 & 0 & \cdots \\ 0 & \cdots & 0 & 0 & & \\ 0 & & 0 & & 0 & \\ \vdots & & \vdots & & & \ddots \end{pmatrix} \leq \begin{pmatrix} Q_0 & Q_{-1} & \cdots & Q_{-n} & 0 & \cdots \\ Q_1 & X_{11} & \cdots & X_{1n} & 0 & \cdots \\ \vdots & \vdots & & \vdots & \vdots & \\ Q_n & X_{n1} & \cdots & X_{nn} & 0 & \cdots \\ 0 & 0 & \cdots & 0 & 0 & \\ \vdots & \vdots & & \vdots & & \ddots \end{pmatrix}. \tag{2.4.38}$$

Applying the above inequality for the right-hand side term of it, we obtain

$$\begin{pmatrix} Q_0 & Q_{-1} & \cdots & Q_{-n} & 0 & \cdots \\ Q_1 & X_{11} & \cdots & X_{1n} & 0 & \cdots \\ \vdots & \vdots & & \vdots & \vdots & \\ Q_n & X_{n1} & \cdots & X_{nn} & 0 & \cdots \\ 0 & 0 & \cdots & 0 & 0 & \\ \vdots & \vdots & & \vdots & & \ddots \end{pmatrix}$$

$$\leq \begin{pmatrix} Q_0 & \cdots & Q_{-n} & 0 & 0 & \cdots & \\ \vdots & Q_0 & \cdots & Q_{-n} & 0 & \cdots & \\ & \vdots & X_{11} & \cdots & X_{1n} & 0 & \cdots \\ Q_n & & \vdots & & & \vdots & \\ 0 & Q_n & X_{n1} & \cdots & X_{nn} & 0 & \cdots \\ 0 & 0 & 0 & \cdots & 0 & 0 & \\ \vdots & \vdots & \vdots & & \vdots & & \ddots \end{pmatrix}.$$

Repeating this argument, we get that

$$\begin{pmatrix} X & 0 \\ 0 & 0 \end{pmatrix} = \begin{pmatrix} X_{11} & \dots & X_{1n} & 0 & \dots \\ \vdots & & \vdots & \vdots & \\ X_{n1} & \dots & X_{nn} & 0 & \dots \\ 0 & \dots & 0 & 0 & \dots \\ \vdots & & \vdots & \vdots & \ddots \end{pmatrix} \leq \begin{pmatrix} Q_0 & \dots & Q_{-n} & \dots \\ \vdots & \ddots & & \ddots \\ Q_n & & & \\ \vdots & \ddots & & \end{pmatrix} = T_Q.$$

Thus

$$T_Q - \begin{pmatrix} X & 0 \\ 0 & 0 \end{pmatrix} \geq 0. \tag{2.4.39}$$

By definition of the Schur complement, $X_{\max}$ is the maximal operator to satisfy (2.4.39), and thus

$$X \leq X_{\max}.$$

By reversing rows and columns, the same argument gives that when

$$\begin{pmatrix} 0 & \dots & 0 & Q_{-n} \\ \vdots & \ddots & \vdots & \vdots \\ 0 & \dots & 0 & Q_{-1} \\ Q_n & \dots & Q_1 & Q_0 \end{pmatrix} - \begin{pmatrix} 0 & 0 & \dots & 0 \\ 0 & S_{11} & \dots & S_{1n} \\ \vdots & \vdots & & \vdots \\ 0 & S_{n1} & \dots & S_{nn} \end{pmatrix}$$

$$+ \begin{pmatrix} S_{11} & \dots & S_{1n} & 0 \\ \vdots & & \vdots & \vdots \\ S_{n1} & \dots & S_{nn} & 0 \\ 0 & \dots & 0 & 0 \end{pmatrix} \geq 0 \tag{2.4.40}$$

then

$$S = (S_{ij})_{i,j=1}^n \leq \widetilde{S},$$

where $\widetilde{S}$ is as in the statement of the theorem. As $X \in \mathcal{G}$ implies that

$$S := \begin{pmatrix} Q_0 & \dots & Q_{-n+1} \\ \vdots & \ddots & \vdots \\ Q_{n-1} & \dots & Q_0 \end{pmatrix} - X$$

satisfies (2.4.40), we get that $X \in \mathcal{G}$ implies

$$\begin{pmatrix} Q_0 & \dots & Q_{-n+1} \\ \vdots & \ddots & \vdots \\ Q_{n-1} & \dots & Q_0 \end{pmatrix} - X \leq \widetilde{S}.$$

Thus

$$X \geq \begin{pmatrix} Q_0 & \dots & Q_{-n+1} \\ \vdots & & \vdots \\ Q_{n-1} & \dots & Q_0 \end{pmatrix} - \widetilde{S} =: X_{\min}.$$

So $X_{\min}$ is as desired.

To prove the statements on $A(X_{\max})$ note that $(A(X))_{nn} = X_{nn}$, so that the existence of a maximal element in $\mathcal{G}$ implies the existence of an element $A_{\max}$ in $\mathcal{A}$ that maximizes A_{nn} with respect to the Loewner ordering. As $A_{\max} \in \mathcal{A}$ we have that $A_{\max} = A(X)$ for some $X \in \mathcal{G}$. As $X_{nn} = (A(X))_{nn} = A(X_{\max})_{nn} = (X_{\max})_{nn}$ and $X \leq X_{\max}$, we have that $X_{in} = (X_{\max})_{nn}$ and $X_{ni} = (X_{\max})_{ni}$, $i = 1, \dots, n$. As

$$A(X_{\max}) - \begin{pmatrix} T & 0 \\ 0 & 0_{\mathcal{K}} \end{pmatrix} \geq 0 \text{ and } T \geq 0 \implies T_Q - \begin{pmatrix} X_{\max} + T & 0 \\ 0 & 0 \end{pmatrix} \geq 0,$$

we get that $T = 0$. Thus the Schur complement of $A(X_{\max})$ supported in the first n rows equals 0. One now obtains that

$$X - X_{\max} =: \begin{pmatrix} E & 0 \\ 0 & 0_{\mathcal{K}} \end{pmatrix} = \begin{pmatrix} E_{11} & \cdots & E_{1,n-1} & 0 \\ \vdots & & \vdots & \vdots \\ E_{n-1,1} & \cdots & E_{n-1,n-1} & 0 \\ 0 & \cdots & 0 & 0_{\mathcal{K}} \end{pmatrix}$$

satisfies

$$\begin{pmatrix} 0_{\mathcal{K}} & 0 \\ 0 & E \end{pmatrix} - \begin{pmatrix} E & 0 \\ 0 & 0_{\mathcal{K}} \end{pmatrix} \geq 0. \tag{2.4.41}$$

But then it follows easily that $E_{ij} = 0$ for $i, j = 1, \dots, n-1$, and thus $X = X_{\max}$. □

2.5 SCHUR PARAMETERS

We start this section by a result describing the structure of a row contraction, based on Lemma 2.4.2. The main result is a theorem about the structure of an $n \times n$ positive semidefinite operator matrix in terms of the so-called Schur parameters. For a better understanding, the result is preceded by a simple computation illustrating the $n = 3$ case. As a consequence, one obtains an explicit construction of the Cholesky factorization of an $n \times n$ positive semidefinite operator matrix, which in the block matrix case implies a formula for the determinant of the matrix in terms of the Schur parameters. For a contraction $T \in \mathcal{L}(\mathcal{H}_1, \mathcal{H}_2)$, we denote by $D_T = (I - T^*T)^{\frac{1}{2}}$ and $D_{T^*} = (I - TT^*)^{\frac{1}{2}}$ the so-called *defect operators* of T, and by $\mathcal{D}_T = \overline{\text{Ran} D_T}$ and $\mathcal{D}_{T^*} = \overline{\text{Ran} D_{T^*}}$ the so-called *defect spaces* of T. We claim that $TD_T = D_{T^*}T$. Indeed, let p_n be a sequence of polynomials that converges uniformly to the square root function on $[0, 1]$. As $Tp_n(I - T^*T) = p_n(I - TT^*)T$, we obtain by taking limits that $TD_T = D_{T^*}T$. Similarly, one shows $T^*D_{T^*} = D_T^*T$.

Proposition 2.5.1 *Let $H_i \in \mathcal{L}(\mathcal{H}_i, \mathcal{H})$, $i = 1, \dots, n$. Here $\mathcal{H}_1, \dots \mathcal{H}_n, \mathcal{H}$ are Hilbert spaces. Then $\left\| \begin{pmatrix} H_1 & \cdots & H_n \end{pmatrix} \right\| \leq 1$ if and only if there exist contractions $G_i : \mathcal{H}_i \to \mathcal{D}_{G_{i-1}}$, $i = 1, \dots, n$, with $G_0 = 0 : \mathcal{H} \to \mathcal{H}$, so that*

$$H_i = D_{G_1^*} \cdots D_{G_{i-1}^*} G_i, \quad i = 1, \dots, n. \tag{2.5.1}$$

The correspondence $(H_1, \ldots, H_n) \leftrightarrow (G_1, \ldots, G_n)$ *is one-to-one and onto. In addition*

$$I - \begin{pmatrix} H_1 & \cdots & H_n \end{pmatrix} \begin{pmatrix} H_1^* \\ \vdots \\ H_n^* \end{pmatrix} = D_{G_1^*} \cdots D_{G_n^*}^2 \cdots D_{G_1^*}, \tag{2.5.2}$$

and

$$I - \begin{pmatrix} H_1^* \\ \vdots \\ H_n^* \end{pmatrix} \begin{pmatrix} H_1 & \cdots & H_n \end{pmatrix} = D_L^* D_L, \tag{2.5.3}$$

where

$$D_L = \begin{pmatrix} D_{G_1} & -G_1^* G_2 & -G_1^* D_{G_2} G_3 & \cdots & -G_1^* D_{G_2} \cdots D_{G_{n-1}} G_n \\ & D_{G_2} & -G_2^* G_3 & \cdots & -G_2^* D_{G_3} \cdots D_{G_{n-1}} G_n \\ & & \ddots & & \vdots \\ & & & & D_{G_n} \end{pmatrix}.$$

Proof. We proceed by induction on n. For $n = 1$ the proposition is immediate, and we assume (2.5.1), (2.5.2), and (2.5.3) hold for $k = 1, \ldots, n$. Assume now that $\| \begin{pmatrix} H_1 & \cdots & H_{n+1} \end{pmatrix} \| \leq 1$. The latter is equivalent to

$$0 \leq I - \begin{pmatrix} H_1 & \cdots & H_{n+1} \end{pmatrix} \begin{pmatrix} H_1^* \\ \vdots \\ H_{n+1}^* \end{pmatrix} = D_{G_1^*} \cdots D_{G_n^*}^2 \cdots D_{G_1^*} - H_{n+1} H_{n+1}^*,$$

which by Corollary 2.4.3 is true if and only if $H_{n+1} = D_{G_1^*} \cdots D_{G_n^*} G_{n+1}$ for a certain contraction $G_{n+1} : \mathcal{H} \to \mathcal{D}_{G_n^*}$. The relations (2.5.2) and (2.5.3) for $n+1$ follow by direct computation. □

For a better understanding of the general structure of an $n \times n$ positive semidefinite operator matrix and of the proof leading to this result, we first establish this for $n = 3$ by direct computation.

Consider

$$A = \begin{pmatrix} A_{11} & A_{12} & A_{13} \\ A_{12}^* & A_{22} & A_{23} \\ A_{13}^* & A_{23}^* & A_{33} \end{pmatrix} \in \mathcal{L}(\mathcal{H}_1 \oplus \mathcal{H}_2 \oplus \mathcal{H}_3).$$

If $A \geq 0$, then $A_{ii} \geq 0$, $i = 1, 2, 3$, and by Lemma 2.4.4 there exist contractions $G_1 : \overline{\text{Ran} A_{22}} \to \overline{\text{Ran} A_{11}}$, $G_2 : \overline{\text{Ran} A_{33}} \to \overline{\text{Ran} A_{22}}$, and $G : \overline{\text{Ran} A_{33}} \to \overline{\text{Ran} A_{11}}$, such that $A_{12} = A_{11}^{\frac{1}{2}} G_1 A_{22}^{\frac{1}{2}}$, $A_{23} = A_{22}^{\frac{1}{2}} G_2 A_{33}^{\frac{1}{2}}$, and $A_{13} = A_{11}^{\frac{1}{2}} G A_{33}^{\frac{1}{2}}$. Using the factorization

$$A = \begin{pmatrix} A_{11}^{\frac{1}{2}} & & \\ & A_{22}^{\frac{1}{2}} & \\ & & A_{33}^{\frac{1}{2}} \end{pmatrix} \begin{pmatrix} I & G_1 & G \\ G_1^* & I & G_2 \\ G^* & G_2^* & I \end{pmatrix} \begin{pmatrix} A_{11}^{\frac{1}{2}} & & \\ & A_{22}^{\frac{1}{2}} & \\ & & A_{33}^{\frac{1}{2}} \end{pmatrix},$$

the condition $A \geq 0$ is equivalent to

$$\begin{pmatrix} I & G_1 & G \\ G_1^* & I & G_2 \\ G^* & G_2^* & I \end{pmatrix} \geq 0. \tag{2.5.4}$$

Using again Lemma 2.4.4, (2.5.4) is equivalent to

$$\begin{pmatrix} G_1 & G \end{pmatrix} = K \begin{pmatrix} I & G_2 \\ G_2^* & I \end{pmatrix}^{\frac{1}{2}} \tag{2.5.5}$$

for a certain contraction $K : \overline{\text{Ran}M} \to \mathcal{H}_1$, where

$$M = \begin{pmatrix} I & G_2 \\ G_2^* & I \end{pmatrix}^{\frac{1}{2}}.$$

Since

$$\begin{pmatrix} I & G_2 \\ G_2^* & I \end{pmatrix} = \begin{pmatrix} I & 0 \\ G_2^* & D_{G_2} \end{pmatrix} \begin{pmatrix} I & G_2 \\ 0 & D_{G_2} \end{pmatrix} = M^2,$$

by Lemma 2.4.2, there exists an isometry $U : \mathcal{H}_2 \oplus \mathcal{D}_{G_2} \to \overline{\text{Ran}M}$, such that

$$M = U \begin{pmatrix} I & G_2 \\ 0 & D_{G_2} \end{pmatrix}.$$

Then (2.5.5) implies

$$\begin{pmatrix} G_1 & G \end{pmatrix} = \widetilde{K} \begin{pmatrix} I & G_2 \\ 0 & D_{G_2} \end{pmatrix}, \tag{2.5.6}$$

where $\widetilde{K} : \mathcal{H}_1 \oplus \mathcal{D}_{G_2} \to \mathcal{H}_3$ is a contraction. By Exercise 2.9.25(a), we have that $\widetilde{K} = \begin{pmatrix} G_1 & D_{G_1^*}\Gamma \end{pmatrix}$, where $\Gamma : \mathcal{D}_{G_2} \to \mathcal{D}_{G_1^*}$ is a uniquely determined contraction. Then (2.5.6) implies that

$$G = G_1 G_2 + D_{G_1^*} \Gamma D_{G_2}. \tag{2.5.7}$$

This computation leads to the following conclusion about 3×3 positive semidefinite operator matrices.

Proposition 2.5.2 *A 3×3 operator matrix*

$$A = \begin{pmatrix} A_{11} & A_{12} & A_{13} \\ A_{12}^* & A_{22} & A_{23} \\ A_{13}^* & A_{23}^* & A_{33} \end{pmatrix}$$

is positive semidefinite if and only if $A_{ii} \geq 0$, $i = 1, 2, 3$, and there exist uniquely determined contractions

$$\Gamma_{12} : \overline{\text{Ran}A_{22}} \to \overline{\text{Ran}A_{11}},\ \Gamma_{23} : \overline{\text{Ran}A_{33}} \to \overline{\text{Ran}A_{22}},\ \textit{and}$$

$$\Gamma_{13} : \mathcal{D}_{\Gamma_{23}} \to \mathcal{D}_{\Gamma_{12}^*},$$

such that

$$A_{12} = A_{11}^{\frac{1}{2}} \Gamma_{12} A_{22}^{\frac{1}{2}},\quad A_{23} = A_{22}^{\frac{1}{2}} \Gamma_{23} A_{33}^{\frac{1}{2}},\quad \text{and}$$

$$A_{13} = A_{11}^{\frac{1}{2}} (\Gamma_{12}\Gamma_{23} + D_{\Gamma_{12}^*} \Gamma_{13} D_{\Gamma_{23}}) A_{33}^{\frac{1}{2}}.$$

The methods for establishing the general structure $n \times n$ positive semidefinite operator matrices are similar to the ones used during the computations that led to Proposition 2.5.2. They involve the use of Lemma 2.4.4, the Cholesky factorization on an $n \times n$ operator matrix, and the general structure of a row contraction in Proposition 2.5.1.

Given a contraction $T \in \mathcal{L}(\mathcal{H}, \mathcal{K})$, the unitary operator $J(T) : \mathcal{H} \oplus \mathcal{D}_{T^*} \to \mathcal{K} \oplus \mathcal{D}_T$ defined by

$$J(T) = \begin{pmatrix} T & D_{T^*} \\ D_T & -T^* \end{pmatrix}$$

is called the *Julia operator* (or *elementary rotation*) defined by T. It represents a building block for dilation theory.

For an easier presentation, we assume in the remainder of this section that the operator matrix $(A_{ij})_{i,j=1}^n$ acts on the direct sum of n identical copies of the Hilbert space $\mathcal{H}$. The results hold similarly in case in acts on $\mathcal{H}_1 \oplus \cdots \oplus \mathcal{H}_n$. The parameters used for describing a positive semidefinite operator matrix $(A_{ij})_{i,j=1}^n$ form an upper triangular array $\{\Gamma_{ij}\}_{1 \leq i < j \leq n}$ of contractions, in which $\Gamma_{i,i+1} : \overline{\text{Ran} A_{i+1,i+1}} \to \overline{\text{Ran} A_{ii}}$ for $i = 1, \ldots, n-1$, and for $j \geq i+2$, $\Gamma_{ij} : \mathcal{D}_{\Gamma_{i+1,j}} \to \mathcal{D}_{\Gamma^*_{i,j-1}}$. Such an array of contractions is called an $(A_{ii})_{i=1}^n$*-choice triangle*. When $A_{11} = \cdots = A_{nn} = I_{\mathcal{H}}$, we call it an $\mathcal{H}$*-choice triangle*. In both cases, the individual contractions $\{\Gamma_{ij}\}_{1 \leq i < j \leq n}$ are called the *Schur parameters* of the matrix $(A_{ij})_{i,j=1}^n$.

With a fixed $\mathcal{H}$-choice triangle we associate the following operators for $1 \leq i < j \leq n$:

$$J_{ij}^1 : (\mathcal{H} \oplus \mathcal{D}_{\Gamma^*_{i,i+1}}) \oplus \mathcal{D}_{\Gamma_{i,i+2}} \oplus \cdots \oplus \mathcal{D}_{\Gamma_{ij}} \tag{2.5.8}$$

$$\to (\mathcal{H} \oplus \mathcal{D}_{\Gamma_{i,i+1}}) \oplus \mathcal{D}_{\Gamma_{i,i+2}} \oplus \cdots \oplus \mathcal{D}_{\Gamma_{ij}},$$

$$J_{ij}^1 = J(\Gamma_{i,i+1}) \oplus I,$$

and for $j - 1 \geq k > 1$,

$$J_{ij}^k : \mathcal{H} \oplus \mathcal{D}_{\Gamma_{i+1,i+2}} \oplus \cdots \oplus (\mathcal{D}_{\Gamma_{i+1,i+k}} \oplus \mathcal{D}_{\Gamma^*_{i,i+k}}) \oplus \cdots \oplus \mathcal{D}_{\Gamma_{ij}} \tag{2.5.9}$$

$$\to \mathcal{H} \oplus \mathcal{D}_{\Gamma_{i+1,i+2}} \oplus \cdots \oplus (\mathcal{D}_{\Gamma^*_{i+1,i+k-1}} \oplus \mathcal{D}_{\Gamma_{i,i+k}}) \oplus \cdots \oplus \mathcal{D}_{\Gamma_{ij}},$$

$$J_{ij}^k = I \oplus J(\Gamma_{i,i+k}) \oplus I.$$

We also need the following operators:

$$L_{ij} : \mathcal{H} \oplus \mathcal{D}_{\Gamma_{i+1,i+2}} \oplus \mathcal{D}_{\Gamma_{i+1,i+3}} \oplus \cdots \oplus \mathcal{D}_{\Gamma_{i+1,j}} \oplus \mathcal{D}_{\Gamma^*_{ij}} \tag{2.5.10}$$

$$\to \mathcal{H} \oplus \mathcal{D}_{\Gamma_{i,i+1}} \oplus \cdots \oplus \mathcal{D}_{\Gamma_{i,j-1}} \oplus \mathcal{D}_{\Gamma_{ij}},$$

$$L_{ij} = J_{ij}^1 J_{ij}^2 \cdots J_{ij}^{j-i},$$

and the following row and column contractions (see Proposition 2.5.1 and Exercise 2.9.37):

$$X_{ij} : \mathcal{H} \oplus \mathcal{D}_{\Gamma_{i+1,i+2}} \oplus \cdots \oplus \mathcal{D}_{\Gamma_{i+1,j}} \to \mathcal{H}, \tag{2.5.11}$$

$$X_{ij} = \begin{pmatrix} \Gamma_{i,i+1} & D_{\Gamma i,i+1^*}\Gamma_{i,i+2} & \cdots & D_{\Gamma^*_{i,i+1}} \cdots D_{\Gamma^*_{i,j-1}}\Gamma_{ij} \end{pmatrix},$$

$$\widetilde{X}_{ij} : \mathcal{H} \to \mathcal{H} \oplus \mathcal{D}_{\Gamma^*_{j-2,j-1}} \oplus \cdots \oplus \mathcal{D}_{\Gamma^*_{i,j-1}}, \tag{2.5.12}$$

$$\widetilde{X}_{ij} = \begin{pmatrix} \Gamma_{j-1,j} & \Gamma_{j-2,j}D_{\Gamma_{j-1,j}} & \cdots & \Gamma_{ij}D_{\Gamma_{i+1,j}} \cdots D_{\Gamma_{j-1,j}} \end{pmatrix}^T.$$

The definitions easily imply that

$$L_{ij} = \begin{pmatrix} X_{ij} & D_{\Gamma^*_{i,i+1}}D_{\Gamma^*_{i,i+2}} \cdots D_{\Gamma^*_{ij}} \\ D_{ij} & -Y_{ij} \end{pmatrix}, \tag{2.5.13}$$

where $D_{i,i+1} = D_{\Gamma_{i,i+1}}$, and for $j \geq i+2$,

$$D_{ij} = \begin{pmatrix} D_{i,j-1} & -Y_{ij}\Gamma_{ij} \\ 0 & D_{\Gamma_{ij}} \end{pmatrix} \tag{2.5.14}$$

and

$$Y_{ij} = \begin{pmatrix} \Gamma^*_{i,i+1}D_{\Gamma^*_{i,i+2}} \cdots D_{\Gamma^*_{ij}} & \Gamma^*_{i,i+2}D_{\Gamma^*_{i,i+3}} \cdots D_{\Gamma^*_{ij}} & \cdots & \Gamma^*_{ij} \end{pmatrix}^T. \tag{2.5.15}$$

Also, for $i = 1, \ldots, n$, consider the operators $U_{ii} = I_{\mathcal{H}}$, and for $1 \leq i < j \leq n$

$$\begin{aligned} U_{ij} : \mathcal{H} \oplus \mathcal{D}_{\Gamma^*_{j-1,j}} \oplus \cdots \oplus \mathcal{D}_{\Gamma^*_{i+1,j}} \oplus \mathcal{D}_{\Gamma^*_{ij}} \\ \to \mathcal{H} \oplus \mathcal{D}_{\Gamma_{i,i+1}} \oplus \cdots \oplus \mathcal{D}_{\Gamma_{ij}}, \end{aligned} \tag{2.5.16}$$

$$U_{ij} = L_{ij}(U_{i+1,j} \oplus I_{\mathcal{D}_{\Gamma^*_{ij}}});$$

$F_{ii} = I_{\mathcal{H}}$, for $i = 1, \ldots, n$, and, for $1 \leq i < j \leq n$,

$$F_{ij} : \bigoplus_{k=1}^{j-i+1} \mathcal{H} \to \mathcal{H} \oplus \mathcal{D}_{\Gamma_{i,i+1}} \oplus \cdots \oplus \mathcal{D}_{\Gamma_{ij}}, \tag{2.5.17}$$

$$F_{ij} = \begin{pmatrix} F_{i,j-1} & U_{i,j-1}\widetilde{X}_{ij} \\ 0 & D_{\Gamma_{ij}} \cdots D_{\Gamma_{j-1,j}} \end{pmatrix}.$$

Lemma 2.5.3 *For $1 \leq i < j \leq n$, we have that*

$$F_{ij} = L_{ij} \begin{pmatrix} X^*_{ij} & F_{i+1,j} \\ D_{\Gamma^*_{ij}} \cdots D_{\Gamma^*_{i,i+1}} & 0 \end{pmatrix}.$$

Proof. Using (2.5.13) and (2.5.16), we obtain that $U_{i,j-1}\widetilde{X}_{ij}$ equals

$$\begin{pmatrix} X_{i,j-1}U_{i+1,j-1}\widetilde{X}_{i+1,j} + D_{\Gamma^*_{i,i+1}} \cdots D_{\Gamma^*_{i,j-1}}\Gamma_{ij}D_{\Gamma_{i+1,j}} \cdots D_{\Gamma_{j-1,j}} \\ D_{i,j-1}U_{i+1,j-1}\widetilde{X}_{i+1,j} - Y_{i,j-1}\Gamma_{ij}D_{\Gamma_{i+1,j}} \cdots D_{\Gamma_{j-1,j}} \end{pmatrix}. \tag{2.5.18}$$

Then, for every fixed i, we prove by induction (the general step using (2.5.18)) the following equality:

$$F_{ij} = \begin{pmatrix} I_{\mathcal{H}} & X_{ij}F_{i+1,j} \\ 0 & D_{ij}F_{i+1,j} \end{pmatrix}. \tag{2.5.19}$$

Using (2.5.13) again, we have

$$L_{ij} \begin{pmatrix} X^*_{ij} & F_{i+1,j} \\ D_{\Gamma^*_{ij}} \cdots D_{\Gamma^*_{i,i+1}} & 0 \end{pmatrix} = \begin{pmatrix} I_{\mathcal{H}} & X_{ij}F_{i+1,j} \\ 0 & D_{ij}F_{i+1,j} \end{pmatrix},$$

and then the desired equality follows from (2.5.19). □

The following result describes the structure of $n \times n$ positive semidefinite operator matrices with a unit diagonal.

Theorem 2.5.4 *There exists a one-to-one correspondence between the set of all positive semidefinite operator matrices $(B_{ij})_{i,j=1}^n$ with $B_{ii} = I_{\mathcal{H}}$ and the set of all $\mathcal{H}$-choice triangles $\{\Gamma_{ij}\}_{1 \le i < j \le n}$, given by the formulas*

$$B_{i,i+1} = \Gamma_{i,i+1} \text{ for } i = 1, \dots, n-1,$$

and

$$B_{ij} = X_{i,j-1}U_{i+1,j-1}\widetilde{X}_{i+1,j} + D_{\Gamma^*_{i,i+1}} \cdots D_{\Gamma^*_{i,j-1}} \Gamma_{ij} D_{\Gamma_{i+1,j}} \cdots D_{\Gamma_{j-1,j}}$$

for $j \ge i+2$.

Proof. Let $(B_{ij})_{i,j=1}^n$ be a positive semidefinite matrix with $B_{ii} = I_{\mathcal{H}}$ for $i = 1, \dots, n$. By Lemma 2.4.4, $B|\{i, i+1\}$ is positive semidefinite if and only if $B_{i,i+1}$ is a contraction on $\mathcal{H}$; define $\Gamma_{i,i+1} = B_{i,i+1}$ for $i = 1, \dots, n-1$.

For a fixed $i \ge 1$, we prove each of the following relations by induction on $j > i$:

$$\begin{pmatrix} B_{i,i+1} & B_{i,i+2} & \cdots & B_{ij} \end{pmatrix} = X_{ij}F_{i+1,j}, \tag{2.5.20}$$

$$\begin{pmatrix} B_{ij} & B_{i+1,j} & \cdots & B_{j-1,j} \end{pmatrix}^T = F^*_{i,j-1}U_{i,j-1}\widetilde{X}_{ij}, \tag{2.5.21}$$

$$B|\{i, i+1, \dots, j\} = F^*_{ij}F_{ij}, \tag{2.5.22}$$

and there exists a unique contraction $\Gamma_{i,j+1} : \mathcal{D}_{\Gamma_{i+1,j+1}} \to \mathcal{D}_{\Gamma^*_{ij}}$ such that

$$B_{i,j+1} = X_{ij}U_{i+1,j}\widetilde{X}_{i+1,j+1} + D_{\Gamma^*_{i,i+1}} \cdots D_{\Gamma^*_{i,j}} \Gamma_{i,j+1} D_{\Gamma_{i+1,j+1}} \cdots D_{\Gamma_{j,j+1}}. \tag{2.5.23}$$

When $j = i+1$, relations (2.5.20), (2.5.21), and (2.5.22) follow immediately. Lemma 2.4.4 implies that $B_{i,i+1} = \Gamma_{i,i+1}$, for a contraction $\Gamma_{i,i+1} \in \mathcal{L}(\mathcal{H})$. By Proposition 2.5.2, there exists a uniquely determined contraction $\Gamma_{i,i+2} : \mathcal{D}_{\Gamma_{i+1,i+2}} \to \mathcal{D}_{\Gamma^*_{i,i+1}}$ such that

$$B_{i,i+2} = \Gamma_{i,i+1}\Gamma_{i+1,i+2} + D_{\Gamma^*_{i,i+1}} \Gamma_{i,i+2} D_{\Gamma_{i+1,i+2}}, \quad i = 1, \dots, n-2,$$

which is (2.5.23) for $j = i+2$.

For the general step we have

$$\begin{aligned}&\begin{pmatrix} B_{i,i+1} & \cdots & B_{ij} & B_{i,j+1}\end{pmatrix}\\&= \begin{pmatrix} X_{ij} & D_{\Gamma^*_{i,i+1}} \cdots D_{\Gamma^*_{ij}} \Gamma_{i,j+1}\end{pmatrix} \begin{pmatrix} F_{i+1,j} & U_{i+1,j}\widetilde{X}_{i+1,j+1} \\ 0 & D_{\Gamma_{i+1,j+1}} \cdots D_{\Gamma_{j,i+1}}\end{pmatrix}\\&= X_{i,j+1}F_{i+1,j+1},\end{aligned}$$

where we used (2.5.20) and (2.5.23) for j. Next,

$$\begin{aligned}&\begin{pmatrix} B_{ij} & B_{i+1,j} & \cdots & B_{j-1,j}\end{pmatrix}^T\\&= \begin{pmatrix} X_{ij}U_{i+1,j} & D_{\Gamma^*_{i,i+1}} \cdots D_{\Gamma^*_{ij}} \\ F^*_{i+1,j}U_{i+1,j} & 0\end{pmatrix} \widetilde{X}_{i,j+1},\end{aligned}$$

where we used (2.5.21) and (2.5.23) for j. Now (2.5.21) for $j+1$ follows from Lemma 2.5.3.

For (2.5.22), note that

$$\begin{aligned}B|\{i,\ldots,j+1\} &= \begin{pmatrix} B|\{i,\ldots,j\} & \begin{pmatrix} B_{i,j+1} & \cdots & B_{j,j+1}\end{pmatrix}^T \\ \begin{pmatrix} B^*_{i,j+1} & \cdots & B^*_{j,j+1}\end{pmatrix} & I\end{pmatrix}\\&= \begin{pmatrix} F^*_{ij}F_{ij} & F^*_{ij}U_{ij}\widetilde{X}_{i,j+1} \\ \widetilde{X}^*_{i,j+1}U^*_{ij}F_{ij} & I\end{pmatrix} = F^*_{i,j+1}F_{i,j+1},\end{aligned}$$

where we used (2.5.22) for j, and (2.5.21) for $j+1$. If $B|\{i,\ldots,j+2\} \geq 0$, then

$$B|\{i+1,\ldots,j+2\} \geq \begin{pmatrix} B_{i,i+1} & \cdots & B_{i,j+2}\end{pmatrix}^* \begin{pmatrix} B_{i,i+1} & \cdots & B_{i,j+2}\end{pmatrix},$$

and using (2.5.22) for $j+1$, there exists a contraction $K = \begin{pmatrix} K_1 & \cdots & K_{j-i+1}\end{pmatrix}$ such that

$$\begin{aligned}&\begin{pmatrix} B_{i,i+1} & \cdots & B_{i,j+2}\end{pmatrix}\\&= \begin{pmatrix} K_1 & \cdots & K_{j-i+1}\end{pmatrix} \begin{pmatrix} F_{i+1,j+1} & U_{i+1,j+1}\widetilde{X}_{i+1,j+2} \\ 0 & D_{\Gamma_{i+1,j+2}} \cdots D_{\Gamma_{j+1,j+2}}\end{pmatrix},\end{aligned}$$

so $\begin{pmatrix} K_1 & \cdots & K_{j-i}\end{pmatrix} = X_{i,j+1}$, and by Proposition 2.5.1 there exists a uniquely determined contraction $\Gamma_{i,j+2} : \mathcal{D}_{\Gamma_{i+1,j+2}} \to \mathcal{D}_{\Gamma^*_{i,j+1}}$ such that

$$\begin{aligned}B_{i,j+2} &= X_{i,j+1}U_{i+1,j+1}\widetilde{X}_{i+1,j+2}\\&\quad + D_{\Gamma^*_{i,i+1}} \cdots D_{\Gamma^*_{i,j+1}} \Gamma_{i,j+2} D_{\Gamma_{i+1,j+2}} \cdots D_{\Gamma_{j+1,j+2}}.\end{aligned}$$

□

Let $(A_{ij})_{i,j=1}^n$ be a positive semidefinite operator matrix the diagonals of which are not necessarily equal to I. By Lemma 2.4.4, for each $1 \leq i < j \leq n$ there exists a unique contraction $B_{ij} : \overline{\operatorname{Ran} A_{jj}} \to \overline{\operatorname{Ran} A_{ii}}$ such that

$A_{ij} = A_{ii}^{\frac{1}{2}} B_{ij} A_{jj}^{\frac{1}{2}}$. This implies the factorization

$$\begin{pmatrix} A_{11} & \cdots & A_{1n} \\ \vdots & & \vdots \\ A_{n1} & \cdots & A_{nn} \end{pmatrix} = \begin{pmatrix} A_{11}^{\frac{1}{2}} & & \\ & \ddots & \\ & & A_{nn}^{\frac{1}{2}} \end{pmatrix} \begin{pmatrix} B_{11} & \cdots & B_{1n} \\ \vdots & & \vdots \\ B_{n1} & \cdots & B_{nn} \end{pmatrix} \begin{pmatrix} A_{11}^{\frac{1}{2}} & & \\ & \ddots & \\ & & A_{nn}^{\frac{1}{2}} \end{pmatrix},$$

where the identities on the main diagonal of $(B_{ij})_{i,j=1}^n$ act on $\overline{\text{Ran}}(A_{ii})$, $i = 1, \ldots, n$. This yields the following more general version of Theorem 2.5.4.

Theorem 2.5.5 *There exists a one-to-one correspondence between the set of all positive semidefinite operator matrices $(A_{ij})_{i,j=1}^n$ with fixed diagonal entries and the set of all $(A_{ii})_{i=1}^n$-choice triangles $\{\Gamma_{ij}\}_{1 \le i < j \le n}$, given by the formulas*

$$A_{i,i+1} = A_{ii}^{\frac{1}{2}} \Gamma_{i,i+1} A_{i+1,i+1}^{\frac{1}{2}}, \quad i = 1, \ldots, n-1,$$

and

$$\begin{aligned} A_{ij} =& A_{ii}^{\frac{1}{2}} (X_{i,j-1} U_{i+1,j-1} \widetilde{X}_{i+1,j} \\ &+ D_{\Gamma_{i,i+1}^*} \cdots D_{\Gamma_{i,j-1}^*} \Gamma_{ij} D_{\Gamma_{i+1,j}} \cdots D_{\Gamma_{j-1,j}}) A_{jj}^{\frac{1}{2}} \end{aligned}$$

for $1 \le i < j \le n$.

The proof of Theorem 2.5.4 implies the inheritance of the Schur parameters by principal submatrices, i.e., for every $1 \le k < l \le n$, the Schur parameters of $A|\{k, \ldots, l\}$ are $\{\Gamma_{ij}\}_{k \le i < j \le l}$. In particular, we have the formula

$$A_{i,i+2} = A_{ii}^{\frac{1}{2}} (\Gamma_{i,i+1} \Gamma_{i+1,i+2} + D_{\Gamma_{i,i+1}^*} \Gamma_{i,i+2} D_{\Gamma_{i+1,i+2}}) A_{i+2,i+2}^{\frac{1}{2}} \tag{2.5.24}$$

for $i = 1, \ldots, n-2$.

The operators F_{ij} defined by (2.5.17) being upper triangular, formula (2.5.22) provides the lower-upper Cholesky factorization of each principal submatrix $A|\{i, \ldots, j\}$, $1 \le i < j \le n$, of an $n \times n$ positive semidefinite operator matrix $(A_{ij})_{i,j=1}^n$. Let us illustrate this first for $n = 3$, the Schur parameters of the matrix being Γ_{12}, Γ_{13}, and Γ_{23}. We have then

$$F_{13} = V = \begin{pmatrix} A_{11}^{\frac{1}{2}} & \Gamma_{12} A_{22}^{\frac{1}{2}} & (\Gamma_{12} \Gamma_{23} + D_{\Gamma_{12}^*} \Gamma_{13} D_{\Gamma_{23}}) A_{33}^{\frac{1}{2}} \\ 0 & D_{\Gamma_{12}} A_{22}^{\frac{1}{2}} & (D_{\Gamma_{12}} \Gamma_{23} - \Gamma_{12}^* \Gamma_{13} D_{\Gamma_{23}}) A_{33}^{\frac{1}{2}} \\ 0 & 0 & D_{\Gamma_{13}} D_{\Gamma_{23}} A_{33}^{\frac{1}{2}} \end{pmatrix}, \tag{2.5.25}$$

and similarly for the upper-lower Cholesky factorization $A = W^* W$,

$$W = \begin{pmatrix} D_{\Gamma_{13}^*} D_{\Gamma_{12}^*} A_{11}^{\frac{1}{2}} & 0 & 0 \\ (D_{\Gamma_{23}^*} G_{12}^* - \Gamma_{23} \Gamma_{13}^* D_{\Gamma_{12}^*}) A_{11}^{\frac{1}{2}} & D_{\Gamma_{23}^*} A_{22}^{\frac{1}{2}} & 0 \\ (\Gamma_{23}^* \Gamma_{12}^* + D_{\Gamma_{23}} \Gamma_{13}^* D_{\Gamma_{12}^*}) A_{11}^{\frac{1}{2}} & \Gamma_{23}^* A_{22}^{\frac{1}{2}} & A_{33}^{\frac{1}{2}} \end{pmatrix}. \tag{2.5.26}$$

Using relations like $T^*\mathcal{D}_{T^*} \subseteq \mathcal{D}_T$ for a contraction T, one obtains that $\overline{\text{Ran}V_{ij}} \subseteq \overline{\text{Ran}V_{ii}}$ and $\overline{\text{Ran}W_{ij}} \subseteq \overline{\text{Ran}W_{ii}}$ for all i and j. The triangularity of V and W yields

$$\overline{\text{Ran}V} = \overline{\text{Ran}A_{11}} \oplus \mathcal{D}_{\Gamma_{12}} \oplus \mathcal{D}_{\Gamma_{13}}, \quad \overline{\text{Ran}W} = \mathcal{D}_{\Gamma_{13}^*} \oplus \mathcal{D}_{\Gamma_{23}^*} \oplus \overline{\text{Ran}A_{33}}. \tag{2.5.27}$$

The relation $A = V^*V = W^*W$ implies the existence of a unitary $U : \overline{\text{Ran}W} \to \overline{\text{Ran}V}$ with $UW = V$. A straightforward computation gives us the following explicit expression of U:

$$U = \begin{pmatrix} D_{\Gamma_{12}^*}D_{\Gamma_{13}^*} & \Gamma_{12}D_{\Gamma_{23}^*} - D_{\Gamma_{12}^*}\Gamma_{13}\Gamma_{23}^* & \Gamma_{12}\Gamma_{23} + D_{\Gamma_{12}^*}\Gamma_{13}D_{\Gamma_{23}} \\ -\Gamma_{12}^*D_{\Gamma_{13}^*} & D_{\Gamma_{12}}D_{\Gamma_{23}^*} - \Gamma_{12}\Gamma_{13}\Gamma_{23}^* & D_{\Gamma_{12}}\Gamma_{23} - \Gamma_{12}^*\Gamma_{13}D_{\Gamma_{23}} \\ -\Gamma_{13}^* & -D_{\Gamma_{13}}\Gamma_{23}^* & D_{\Gamma_{13}}D_{\Gamma_{23}^*} \end{pmatrix}. \tag{2.5.28}$$

We note that when Γ_{12} and Γ_{23} are fixed, then for $\Gamma_{13} = 0$ the closures of the ranges of the Cholesky factors V and W given by formulas (2.5.27) are as large as possible and the $(3,1)$ entry of U is 0 if and only if $\Gamma_{13} = 0$.

When $A = (A_{ij})_{i,j=1}^n$ is a positive semidefinite operator matrix with Schur parameters $\{\Gamma_{ij}\}_{1\le i<j\le n}$ and $A = V^*V = W^*W$ are its lower-upper and upper-lower Cholesky factorizations, then using (2.5.22) for $i = 1$ and $j = n$ and (2.5.17), one obtains that

$$V : \bigoplus_{i=1}^n \overline{\text{Ran}A_{ii}} \to \overline{\text{Ran}A_{11}} \oplus \left(\bigoplus_{k=2}^n \mathcal{D}_{\Gamma_{1k}} \right), \tag{2.5.29}$$

V has dense range, and

$$V_{ii} = D_{\Gamma_{1i}} \cdots D_{\Gamma_{i-1,i}} A_{ii}^{\frac{1}{2}}. \tag{2.5.30}$$

Similarly, one can prove that

$$W : \bigoplus_{i=1}^n \overline{\text{Ran}A_{ii}} \to \left(\bigoplus_{k=1}^{n-1} D_{\Gamma_{kn}^*} \right) \oplus \overline{\text{Ran}A_{nn}}, \tag{2.5.31}$$

W has dense range, and

$$W_{ii} = D_{\Gamma_{1n}^*} \cdots D_{\Gamma_{i-1,i}^*} A_{ii}^{\frac{1}{2}}. \tag{2.5.32}$$

Formula (2.5.30) implies the following condition for an operator matrix to be positive definite. By a *strict contraction* we mean an operator of norm strictly less than 1.

Corollary 2.5.6 *Let $A = (A_{ij})_{i,j=1}^n$ be a positive semidefinite operator matrix with $\{\Gamma_{ij}\}_{1\le i<j\le n}$ as its Schur parameters. Then $A = (A_{ij})_{i,j=1}^n$ is positive definite if and only each A_{ii} is positive definite and each Γ_{ij} is a strict contraction.*

Theorem 2.5.7 *Let $A = (A_{ij})_{i,j=1}^n$ be positive definite block matrix in which each A_{ij} is an $r \times r$ matrix. If $\{\Gamma_{ij}\}_{1\le i<j\le n}$ are the Schur parameters of A, then*

$$\det A = \prod_{i=1}^n \det A_{ii} \prod_{1\le i<j\le n} (\det D_{\Gamma_{ij}})^2.$$

Proof. Let $A = V^*V$ be the lower-upper Cholesky factorization of A. The theorem follows immediately from (2.5.30). □

Assume now that the positive semidefinite operator matrix $(A_{ij})_{i,j=0}^n$ is in particular a Toeplitz matrix $(A_{i-j})_{i,j=0}^n$. By the remark after Theorem 2.5.5 about the inheritance of the Schur parameters, it follows that the parameters of $(A_{i-j})_{i,j=0}^n$ also have a Toeplitz-like structure, namely $\Gamma_{ij} = \Gamma_{kl}$ for $1 \leq i < j \leq n$ and $1 \leq k < l \leq n$ such that $j - i = l - k$. Denoting then $\Gamma_{i,i+k}$ by Γ_k, we obtain the following Toeplitz version of Theorem 2.5.5.

Theorem 2.5.8 *There exists a one-to-one correspondence between the set of all positive semidefinite Toeplitz matrices $(A_{i-j})_{i,j=0}^n$ with fixed diagonal and the set of all contractions $\{\Gamma_k\}_{k=1}^n$, such that $\Gamma_1 \in \mathcal{L}(\overline{\text{Ran} A_0})$, and $\Gamma_k \in \mathcal{L}(\mathcal{D}_{\Gamma_{k-1}}, \mathcal{D}_{\Gamma_{k-1}^*})$ for $k = 1, \ldots, n$. In case $\mathcal{H}$ is finite dimensional, we have*

$$\det A = (\det A_0)^{n+1} \prod_{k=1}^{n} (\det D_{\Gamma_k})^{2(n+1-k)}.$$

In particular, the formulas in Theorem 2.5.5 imply that

$$A_1 = A_0^{\frac{1}{2}} \Gamma_1 A_0^{\frac{1}{2}}$$

and

$$A_2 = A_0^{\frac{1}{2}} (\Gamma_1^2 + D_{\Gamma_1^*} \Gamma_2 D_{\Gamma_1}) A_0^{\frac{1}{2}}.$$

2.6 THE CENTRAL COMPLETION, MAXIMUM ENTROPY, AND INHERITANCE PRINCIPLE

In the first part of this section we prove facts about generalized banded partially positive operator-valued matrices, like a new characterization of the central completion, a maximum entropy result, and a linear fractional parametrization for the set of all solutions. In the second part of the section, we prove that for a partially positive semidefinite operator matrix with a chordal graph we still have an inheritance principle, which generalizes the results from the scalar case. There exists a central completion which does not depend on the order we complete the matrix and is characterized by an extremal property.

2.6.1 The generalized banded case

We call $S \subset \{1, \ldots, n\} \times \{1, \ldots, n\}$ *a generalized banded pattern* if

(1) $(i, i) \in S$, $i = 1, \ldots, n$;

(2) if $(i, j) \in S$, then $(j, i) \in S$;

(3) if $(i, j) \in S$ and $i \leq p \leq q \leq j$, then $(p, q) \in S$.

It is easy to show that generalized banded patterns are the ones the associated graphs of which are so-called *proper interval graphs*, a particular case of chordal graphs (see Exercise 2.9.38). Thus every generalized banded partially positive semidefinite operator matrix admits a positive semidefinite completion. In this section we present a parametrization for the set of all such completions, generalizing Theorem 2.1.2 to the generalized banded positive semidefinite case. We give two different characterizations of a distinguished completion, one of them generalizing Theorem 2.1.4, and being a particular case of Theorem 2.2.10. In the partially positive definite block matrix case this completion is also the only one that maximizes the determinant over the set of all positive definite completions.

Let $K = \{A_{ij} : (i, j) \in S\}$ be a given generalized banded partially positive semidefinite operator matrix. Here $A_{ij} : \mathcal{H}_j \to \mathcal{H}_i$ are Hilbert space operators. Based on the recursive nature of the Schur parameters given by Theorem 2.5.5, K uniquely determines the parameters $\{\Gamma_{ij} : 1 \leq i \leq j \leq n, (i, j) \in S\}$. Making a positive semidefinite completion of K corresponds to choosing the parameters $\{\Gamma_{ij} : 1 \leq i \leq j \leq n, (i, j) \notin S\}$. Thus there exists a one-to-one correspondence between the set of all positive semidefinite completions of K and the set of completions of $\{\Gamma_{ij} : i \leq j, (i, j) \in S\}$ to an $(A_{ii})_{i=1}^n$-choice triangle.

The completion of a generalized banded partially positive semidefinite operator matrix $K = \{A_{ij} : (i, j) \in S\}$ corresponding to the choice $\Gamma_{ij} = 0$ whenever $1 \leq i \leq j \leq n$ with $(i, j) \notin S$ is called the *central completion* of K. We denote it by F_c, where the subscript c stands for central. We will show that this generalizes the notion of band (or central) completion defined in Section 2.1 to the case of generalized banded partially positive semidefinite matrices.

An alternative way to obtain the central completion is described below. For a given $n \times n$ partially positive semidefinite generalized banded partial matrix $K = \{A_{ij}, (i, j) \in S\}$ one can proceed as follows. Choose a position $(i_0, j_0) \notin S$, $i_0 \leq j_0$, such that $S \cup \{(i_0, j_0), (j_0, i_0)\}$ is also generalized banded. Choose A_{i_0, j_0} such that $(A_{ij})_{i,j=i_0}^{j_0}$ is the central completion of $\{A_{ij}, (i, j) \in S$ and $i_0 \leq i, j \leq j_0\}$. This is a 3×3 problem and A_{i_0, j_0} can be found via a formula as in the proof of Theorem 2.4.5, which corresponds to the choice of $\Gamma = 0$ in (2.5.7). Proceed in the same way with the thus obtained partial matrix until all positions are filled. We will prove that the resulting positive semidefinite completion is the central completion F_c. Note that for $(i_0, j_0) \notin S$, $i_0 \leq j_0$, the entry A_{i_0, j_0} depends only upon $\{A_{ij}, (i, j) \in S$ and $i_0 \leq i, j \leq j_0\}$. This implies that the submatrix of F_c located in the rows and columns $\{k, k+1, \ldots, l\}$ is precisely the central completion of $\{A_{ij}, (i, j) \in S \cap \{k, k+1, \ldots, l\} \times \{k, k+1, \ldots, l\}\}$. This principle is referred to as the *inheritance principle*.

Our first result gives four equivalent conditions which characterize the central completion. We will use the left and right diagonal factors, Δ_l and

Δ_r, whose definitions may be found in (2.1.9) and (2.1.10), respectively.

Theorem 2.6.1 *Let S be a generalized banded pattern and F a positive semi-*
definite completion of $K = \{A_{ij}, (i,j) \in S\}$. Let $F = V^*V = W^*W$ be lower-upper and upper-lower Cholesky factorizations of F. Then the following are equivalent:

(i) F is the central completion of K.

(ii) $\Delta_l(F) \geq \Delta_l(\widetilde{F})$ for all positive semidefinite completions $\widetilde{F}$ of K.

(iii) $\Delta_r(F) \geq \Delta_r(\widetilde{F})$ for all positive semidefinite completions $\widetilde{F}$ of K.

(iv) The unitary $U : \overline{\text{Ran} W} \to \overline{\text{Ran} V}$ with $UW = V$ verifies $U_{ij} = 0$ for $i \geq j, (i,j) \notin S$.

Note that the uniqueness of the central completion implies that $\Delta_l(F) = \Delta_l(\widetilde{F})$ (or $\Delta_r(F) = \Delta_r(\widetilde{F})$) yields $F = \widetilde{F}$. For a banded partially positive definite matrix, the equivalence of (i), (ii), and (iii) was proven in Theorem 2.1.4.

Proof. The equivalence of (i) and (ii) can be read off immediately from (2.5.30), and similarly the equivalence of (i) and (iii) can be read off immediately from (2.5.32).

We prove the equivalence of (i) and (iv) by induction on the number of missing entries in the pattern S. For the 3×3 problem (2.4.8), formula (2.5.28) proves the equivalence immediately.

Let $S \subseteq \{1, \ldots, n\} \times \{1, \ldots, n\}$ be a generalized banded pattern and let $K = \{A_{ij}, (i,j) \in S\}$ be partially positive semidefinite. Let F_c denote the central completion of K, and let V_c and W_c be upper and lower triangular operator matrices such that

$$F_c = V_c^* V_c = W_c^* W_c.$$

Consider the unitary operator matrix $U : \overline{\text{Ran} W_c} \to \overline{\text{Ran} V_c}$ so that $UW_c = V_c$. Let $\widehat{S} = S \cap (\{1, \ldots, n-1\} \times \{1, \ldots, n-1\})$, and $\widehat{F} = (\widehat{F}_{ij})_{ij=1}^{n-1}$ obtained from F_c by compressing its last two rows and columns. So, $\widehat{F}_{ij} = (F_c)_{ij}$ for $i, j \leq n-1$,

$$\widehat{F}_{i,n-1} = \widehat{F}^*_{n-1,i} = ((F_c)_{i,n-1} \ (F_c)_{in}), \quad i < n-1,$$

and

$$\widehat{F}_{n-1,n-1} = \begin{pmatrix} (F_c)_{n-1,n-1} & (F_c)_{n-1,n} \\ (F_c)_{n,n-1} & (F_c)_{nn} \end{pmatrix}.$$

Consider the data $\{\widehat{F}_{ij}, (i,j) \in \widehat{S}\}$. From the way the central completion is defined one sees that $\widehat{F}(= F_c)$ is the central completion of $\{\widehat{F}_{ij}, (i,j) \in \widehat{S}\}$. Now, in the same way, consider the operator matrices $\widehat{U} = (\widehat{U}_{ij})_{i,j=1}^n$, $\widehat{V} =$

$(\widehat{V}_{ij})_{i,j=1}^{n}$, and $\widehat{W} = (\widehat{W}_{ij})_{i,j=1}^{n}$ obtained by the compression of the last two rows and columns of U, V_c and W_c, respectively. We obtain by the induction hypothesis that $\widehat{U}_{ij} = 0$ for $(i,j) \notin \widehat{S}$ with $i > j$. Thus it remains to show that $U_{nj} = 0$ for j with $(n,j) \notin S$ and $(n-1,j) \in S$. For this purpose let $\gamma = \min\{j, (n,j) \in S\}$ and consider the decomposition

$$U = \widehat{U} = \begin{pmatrix} \Sigma_{11} & \Sigma_{12} & \Sigma_{13} \\ \Sigma_{21} & \Sigma_{22} & \Sigma_{23} \\ \Sigma_{31} & \Sigma_{32} & \Sigma_{33} \end{pmatrix},$$

with $\Sigma_{11} = (U_{ij})_{i,j=1}^{\gamma-1}$, $\Sigma_{22} = (U_{ij})_{i,j=\gamma}^{n-1}$, and $\Sigma_{33} = U_{nn}$. Consider also the corresponding decomposition of $F_c = (\phi_{ij})_{i,j=1}^{3}$. Again we have that F_c is also the central completion of

$$\begin{pmatrix} \phi_{11} & \phi_{12} & ? \\ \phi_{21} & \phi_{22} & \phi_{23} \\ ? & \phi_{32} & \phi_{33} \end{pmatrix}.$$

But then from the 3×3 case we obtain that $\Sigma_{13} = 0$ and, consequently, $U_{nj} = 0$ for $j \leq \gamma - 1$, proving (iv).

Implication (iv) $\Rightarrow$ (i) can be proved by the same type of induction process. One needs to use the observation that if S_1 and S_2 are two generalized banded patterns, F is the central completion of both $\{A_{ij} : (i,j) \in S_1\}$, and $\{A_{ij} : (i,j) \in S_2\}$, then F is the central completion of $\{A_{ij} : (i,j) \in S_1 \cap S_2\}$. We omit the details. □

Proposition 2.6.2 *Let $S \subseteq \{1, \ldots, n\} \times \{1, \ldots, n\}$ be a generalized banded pattern and let $K = \{A_{ij} : (i,j) \in S\}$ be a partially positive definite operator matrix. Then the central completion F_c of K is also positive definite.*

Proof. By Corollary 2.5.6, all specified Schur parameters $\{\Gamma_{ij} : i \leq j, (i,j) \in S\}$ of F_c are strict contractions. All the other parameters $\{\Gamma_{ij} : i \leq j, (i,j) \notin S\}$ of F_c are 0, so again by Corollary 2.5.6, F_c is positive definite. □

Theorem 2.6.3 *Let $S \subseteq \{1, \ldots, n\} \times \{1, \ldots, n\}$ be a generalized banded pattern, and let $K = \{A_{ij} : (i,j) \in S\}$ be a partially positive semidefinite operator matrix. Let F_c denote the central completion of K, and V_c and W_c be upper and lower triangular operator matrices such that*

$$F_c = V_c^* V_c = W_c^* W_c. \tag{2.6.1}$$

Further, let $U : \overline{\text{Ran} W_c} \to \overline{\text{Ran} V_c}$ be the unitary operator matrix such that

$$U W_c = V_c.$$

Then each positive semidefinite completion of K is of the form

$$\begin{aligned} \mathcal{T}(G) &= V_c^*(I + UG)^{*-1}(I - G^*G)(I + UG)^{-1}V_c \\ &= W_c^*(I + GU)^{-1}(I - GG^*)(I + GU)^{*-1}W_c, \end{aligned} \tag{2.6.2}$$

where $G = (G_{ij})_{i,j=1}^{n} : \overline{\text{Ran} V_c} \to \overline{\text{Ran} W_c}$ is a contraction with $G_{ij} = 0$ whenever $i \geq j$ or $(i,j) \in S$. Moreover, the correspondence between the set of all positive semidefinite completions and all such contractions G is one-to-one.

Based on (2.5.29) and (2.5.31), we have the decompositions

$$\overline{\text{Ran}V} = \overline{\text{Ran}A_{11}} \oplus \left(\bigoplus_{k=2}^{n} \mathcal{D}_{\Gamma_{1k}} \right) \tag{2.6.3}$$

and

$$\overline{\text{Ran}W} = \left(\bigoplus_{k=1}^{n-1} \mathcal{D}_{\Gamma_{kn}^*} \right) \oplus \overline{\text{Ran}A_{nn}}. \tag{2.6.4}$$

Before starting the proof, we need additional results.

Proposition 2.6.4 *Let $S \subseteq \{1, \dots, n\} \times \{1, \dots, n\}$ be a generalized banded pattern and let $K = \{A_{ij} : (i,j) \in S\}$ be a partially positive semidefinite operator matrix. Let F_c denote the central completion and F an arbitrary positive semidefinite completion of K. Then*

$$\text{Ran}F^{1/2} \subseteq \text{Ran}F_c^{1/2}, \quad \text{Ker}F_c \subseteq \text{Ker}F. \tag{2.6.5}$$

We remark first that if ϕ is an operator on $\mathcal{H}$ and $A = \phi^*\phi$ then there exists a unitary $U : \overline{\text{Ran}\phi} \to \overline{\text{Ran}A^{1/2}}$ such that $A^{1/2} = U\phi$, and thus $\text{Ran}\phi^* = \text{Ran}A^{1/2}$. Thus (2.6.5) is equivalent to the fact that if $F_c = \Sigma_c^*\Sigma_c$ and $F = \Sigma^*\Sigma$ with Σ_c and Σ upper (lower) triangular, then $\text{Ran}\Sigma^* \subseteq \text{Ran}\Sigma_c^*$.

Proof. We start the proof with the 3×3 problem (2.2.3). Let F be the positive semidefinite completion of (2.2.3) corresponding to the Schur parameter Γ_{13}. Then, as we have already seen, $F = V^*V$, where V is given by (2.5.25). Thus

$$V = \begin{pmatrix} I & 0 & D_{\Gamma_{12}^*}\Gamma_{13} \\ 0 & I & -\Gamma_{12}^*\Gamma_{13} \\ 0 & 0 & D_{\Gamma_{13}} \end{pmatrix} V_c, \tag{2.6.6}$$

which yields $\text{Ran}V^* \subseteq \text{Ran}V_c^*$. The result now follows from the remark preceding the proof.

Let now a generalized banded pattern $S \subseteq \{1, \dots, n\} \times \{1, \dots, n\}$ be given. We prove our result by induction assuming the statement is correct for all generalized banded patterns $\widehat{S}$ which contain S as a proper subset. The case $S = (\{1, \dots, n\} \times \{1, \dots, n\}) \backslash \{(1,n),(n,1)\}$ reduces to the 3×3 problem. Let $(i_0, j_0) \notin S$, $i_0 < j_0$, be such that $\widehat{S} = S \cup \{(i_0,j_0),(j_0,i_0)\}$ is also generalized banded. Let $\{A_{ij}, (i,j) \in S\}$, F_c, and F be as in the statement of the proposition. Consider the partial matrix $\{B_{ij}, (i,j) \in \widehat{S}\}$, where $B_{ij} = F_{ij}$ for $(i,j) \in \widehat{S}$. Let $\widehat{F_c}$ denote the central completion of this latter partial matrix. By the induction hypothesis, since clearly F is a completion of $\{B_{ij}, (i,j) \in \widehat{S}\}$ we have that

$$\text{Ran}F^{1/2} \subseteq \text{Ran}\widehat{F}_c^{1/2}. \tag{2.6.7}$$

Observe that the matrices $\widehat{F_c}$ and F_c differ only on the positions (i,j) and (j,i), where $1 \le i \le i_0$ and $j_0 \le j \le n$. Moreover, defining $\widetilde{S} = (\{1, \dots, n\} \times$

$\{1,\ldots,n\})\backslash\{(i,j),(j,i)\}, 1 \leq i \leq i_0, j_0 \leq j \leq n)$ and the partial matrix $\{C_{ij},(i,j) \in \widetilde{S}\}$, where $C_{ij} = (\widehat{F_c})_{ij}$ for $(i,j) \in \widetilde{S}$, we have that F_c is also the central completion of this latter partial matrix. We can now use the 3×3 case to conclude that

$$\mathrm{Ran}\widehat{F}_c^{1/2} \subseteq \mathrm{Ran}F_c^{1/2}, \tag{2.6.8}$$

since $\widehat{F_c}$ can be viewed as a completion of $\{C_{ij},(i,j) \in \widetilde{S}\}$.

Now (2.6.5) is a consequence of (2.6.7), (2.6.8), and the remark preceding the proof. □

Proposition 2.6.5 *Let $S \subseteq \{1,\ldots,n\} \times \{1,\ldots,n\}$ be a generalized banded pattern and let $K = \{A_{ij} : (i,j) \in S\}$ be a partially positive semidefinite operator matrix. Let F_c denote the central completion and F an arbitrary positive semidefinite completion of K, and let*

$$F_c = V_c^* V_c = W_c^* W_c$$

be the lower-upper and upper-lower Cholesky factorizations of F_c. Then, if $F - F_c = \Omega^* + \Omega$, where $\Omega = (\Omega_{ij})_{ij=1}^n$ with $\Omega_{ij} = 0$ whenever $i \geq j$ or $(i,j) \in S$, there exists an operator matrix $Q = (Q_{ij})_{i,j=1}^n : \overline{\mathrm{Ran}V_c} \to \overline{\mathrm{Ran}W_c}$, with

$$\Omega = W_c^* Q V_c \tag{2.6.9}$$

and $Q_{ij} = 0$ whenever $i \geq j$ or $(i,j) \in S$.

Proof. We prove the proposition by induction in a similar way as Proposition 2.6.4. The 3×3 case is straightforward to check. (Using (2.5.24) for $i = 1$, (2.5.25), and (2.5.26), we obtain that only the $(1,3)$ entry of Q is nonzero, and equals Γ_{13}.)

Consider a generalized banded pattern $S \subseteq \{1,\ldots,n\} \times \{1,\ldots,n\}$ and assume that the proposition is true for all generalized banded patterns $\widehat{S}$ which contain S as a proper subset. The case

$$S = (\{1,\ldots,n\} \times \{1,\ldots,n\}\backslash\{(1,n),(n,1)\}$$

reduces to the 3×3 problem. Let $(i_0,j_0) \notin S$, $i_0 < j_0$, be such that $\widehat{S} = S \cup \{(i_0,j_0),(j_0,i_0)\}$ is also generalized banded. Let $K = \{A_{ij} : (i,j) \in S\}$, F_c, F, W_c, and V_c be as in the statement of the proposition. Consider the partial matrix $\{B_{ij} : (i,j) \in \widehat{S}\}$, where $B_{ij} = F_{ij}$ for $(i,j) \in \widehat{S}$. Let $\widehat{F_c}$ denote the central completion of this latter partial matrix. By the induction hypothesis,

$$\widehat{\Omega} = \widehat{W}_c^* \widehat{Q} \widehat{V}_c, \tag{2.6.10}$$

where $\widehat{\Omega}$ and $\widehat{Q}$ are upper triangular with support outside the band $\widehat{S}$, $\widehat{\Omega}^* + \widehat{\Omega} = F - \widehat{F}_c$, $\widehat{Q} : \overline{\mathrm{Ran}\widehat{V}_c} \to \overline{\mathrm{Ran}\widehat{W}_c}$, and $\widehat{F}_c = \widehat{V}_c^* \widehat{V}_c = \widehat{W}_c^* \widehat{W}_c$ are lower-upper and upper-lower Cholesky factorizations of $\widehat{F}_c$. By Proposition 2.6.4 and the remark preceding the proof of Proposition 2.6.4 we have that $\mathrm{Ran}\widehat{V}_c^* \subseteq \mathrm{Ran}V_c^*$ and $\mathrm{Ran}\widehat{W}_c^* \subseteq \mathrm{Ran}W_c^*$. But this yields that there exist an upper triangular α and a lower triangular β such that $\widehat{V_c} = \alpha V_c$ and $\widehat{W_c} = \beta W_c$. Now, taking $Q_1 = \beta^* \widehat{Q} \alpha$, we obtain from (2.6.10) that

$$\widehat{\Omega} = W_c^* Q_1 V_c, \tag{2.6.11}$$

and clearly Q_1 is upper triangular with support outside $\widehat{S}$.

As in the proof of Proposition 2.6.4, F_c is also the central completion of the partial matrix $\{C_{ij} : (i,j) \in \widetilde{S}\}$, where $\widetilde{S} = (\{1,\dots,n\} \times \{1,\dots,n\}) \setminus \{(i,j),(j,i), 1 \leq i \leq i_0, j_0 \leq j \leq n\}$ and $C_{ij} = (\widehat{F_c})_{ij}$ for $(i,j) \in \widetilde{S}$. By the 3×3 case we may conclude that

$$\widetilde{\Omega} = W_c^* Q_2 V_c, \tag{2.6.12}$$

where $\widetilde{\Omega}$ and Q_2 are upper triangular with support outside the band $\widetilde{S}$, $\widetilde{\Omega}^* + \widetilde{\Omega} = F - \widehat{F_c}$, $Q_2 : \overline{\text{Ran}} V_c \to \overline{\text{Ran}} W_c$.

Since $F - F_c = (F - \widehat{F_c}) + (\widehat{F_c} - F_c)$, we have that $\Omega = \widehat{\Omega} + \widetilde{\Omega}$, and thus (2.6.11) and (2.6.12) imply relation (2.6.9) with $Q = Q_1 + Q_2$, which clearly is of the desired form. □

We are now ready to prove the parametrization result.

Proof of Theorem 2.6.3. Write $F_c = C + C^*$, where C is upper triangular with $C_{ii} = \frac{1}{2} F_{ii}$, $i = 1,\dots,n$, and define for a contraction $G = (G_{ij})_{i,j=1}^n : \overline{\text{Ran}} W_c \to \overline{\text{Ran}} V_c$ with $G_{ij} = 0$ whenever $i > j$ or $(i,j) \in S$,

$$\mathcal{S}(G) = C - W_c^* (I + GU)^{-1} G V_c. \tag{2.6.13}$$

Since $U_{ij} = 0$ for $(i,j) \notin S$ with $i > j$, one easily sees that GU is strictly upper triangular, and so $(I + GU)^{-1}$ exists and is upper triangular. Since W_c^* and V_c are both also upper triangular, one readily obtains that

$$(\mathcal{S}(G))_{ij} = C_{ij}, (i,j) \in S. \tag{2.6.14}$$

Further, using (2.6.13) and the unitarity of U it is straightforward to check that $\mathcal{S}(G) + \mathcal{S}(G)^* = \mathcal{T}(G)$. This together with (2.6.14) yields that $\mathcal{T}(G)$ is a completion of K and since $\|G\| \leq 1$ the operator matrix $\mathcal{T}(G)$ is positive semidefinite.

Assume that for two contractions G_1 and G_2 (of the required form) we have that $\mathcal{T}(G_1) = \mathcal{T}(G_2)$. Then also $\mathcal{S}(G_1) = \mathcal{S}(G_2)$ and since W_c^* and V_c^* are injective on $\overline{\text{Ran}} W_c$ and $\overline{\text{Ran}} V_c$, respectively, (2.6.13) implies that $(I + G_1 U)^{-1} G_1 = (I + G_2 U)^{-1} G_2$. Thus $G_1 (I + U G_2) = (I + G_1 U) G_2$ which yields $G_1 = G_2$.

Conversely, let F be an arbitrary positive semidefinite completion of K. Consider $\Omega = (\Omega_{ij})_{i,j=1}^n$ such that $\Omega_{ij} = 0$ whenever $i \leq j$ or $(i,j) \in S$, and $F_c - F = \Omega + \Omega^*$. Then by Proposition 2.6.5 there exists an operator $Q = (Q_{ij})_{ij}^n : \overline{\text{Ran}} W_c \to \overline{\text{Ran}} V_c$ with $Q_{ij} = 0$ whenever $i > j$ or $(i,j) \notin S$, and $\Omega = W_c^* Q V_c$. Since UQ is strictly upper triangular, we can define

$$G = Q(I - UQ)^{-1},$$

which will give that $\Omega = W_c^* (I + GU)^{-1} G V_c$. Since $F = F_c - \Omega - \Omega^*$, and taking into account (2.6.13) we obtain that $F = \mathcal{T}(G)$. Since $F = \mathcal{T}(G)$ is positive semidefinite, (2.6.2) implies that G is a contraction. This completes our proof. □

2.6.2 The positive semidefinite operator-valued chordal case

A positive semidefinite scalar matrix $A = (a_{ij})$ (with i, j in some set of indices I) is always a *Gramian*; that is, there exist vectors ξ_i in a Hilbert space $\mathcal{K}$ such that $a_{ij} = \langle \xi_i, \xi_j \rangle$. Moreover, if $J \subset I$, and $A' = A(J)$, then the vectors ξ_i for $i \in J$ can be used to represent A' as a Gramian. These facts are easily extended to operator matrices: if $A = (A_{ij})_{i,j \in I}$ is a positive semidefinite (block operator) matrix, with entries $A_{ij} \in \mathcal{L}(\mathcal{H})$, then there exists an essentially unique Hilbert space, denoted by $\mathcal{K}(A)$, together with isometries $\omega_i = \omega_i(A) : \mathcal{H} \to \mathcal{K}(A)$, $i \in I$, such that $A_{ij} = \omega_i^* \omega_j$. (If ι_i denotes the canonical embedding of $\mathcal{H}$ into the ith coordinate of $\bigoplus_{i\in I} \mathcal{H}$, we may take, for instance, $\mathcal{K}(A)$ to be the subspace of $\bigoplus_{i\in I} \mathcal{H}$ spanned by all $\omega_i(A)\mathcal{H}$, with $\omega_i(A) = A^{1/2}\iota_i$.) The space $(\mathcal{K}(A), \omega_i(A))$ is called the *Kolmogorov decomposition* of A. If $J \subset I$, and $A' = A(J)$, then one can embed isometrically $\mathcal{K}(A')$ into $\mathcal{K}(A)$, such that $\omega_i(A)$ is just $\omega_i(A')$ followed by this embedding.

Suppose we are given the partially positive semidefinite matrix

$$A = \begin{pmatrix} A_{11} & A_{12} & ? \\ A_{21} & A_{22} & A_{23} \\ ? & A_{32} & A_{33} \end{pmatrix}, \tag{2.6.15}$$

which means that

$$A' = \begin{pmatrix} A_{11} & A_{12} \\ A_{21} & A_{22} \end{pmatrix} \geq 0, \quad A'' = \begin{pmatrix} A_{22} & A_{23} \\ A_{32} & A_{33} \end{pmatrix} \geq 0.$$

Proposition 2.5.2 has consequences for the Kolmogorov decomposition of a 3×3 matrix. First, if $A' = A(\{1,2\})$ and $A'' = A(\{2,3\})$, then a standard procedure for the Kolmogorov decomposition of 2×2 matrices implies that we may take

$$\mathcal{K}(A') = \overline{\mathrm{Ran}} A_{22} \oplus \mathcal{D}_{\Gamma_{12}^*}, \quad \mathcal{K}(A'') = \overline{\mathrm{Ran}} A_{22} \oplus \mathcal{D}_{\Gamma_{23}},$$

$$\omega_1(A')\xi = \Gamma_{12}^* A_{11}\xi \oplus \mathcal{D}_{\Gamma_{12}^*} A_{11}\xi, \quad \omega_2(A')\xi = A_{22}\xi \oplus 0, \tag{2.6.16}$$

$$\omega_2(A'')\xi = A_{22}\xi \oplus 0, \quad \omega_3(A'')\xi = \Gamma_{23} A_{33}\xi \oplus \mathcal{D}_{\Gamma_{23}} A_{33}\xi.$$

Further, the Kolmogorov decomposition of $\widetilde{A}$ corresponding to a given contraction $\Gamma_{13} : \mathcal{D}_{\Gamma_{23}} \to \mathcal{D}_{\Gamma_{12}^*}$ can be defined as follows:

$$\mathcal{K}(\widetilde{A}) = \mathcal{K}(A') \oplus \mathcal{D}_{\Gamma_{13}} = \overline{\mathrm{Ran}} A_{22} \oplus \mathcal{D}_{\Gamma_{12}^*} \oplus \mathcal{D}_{\Gamma_{13}}, \tag{2.6.17}$$

$$\omega_1(\widetilde{A})\xi = \omega_1(A')\xi \oplus 0, \quad \omega_2(\widetilde{A})\xi = \omega_2(A')\xi) \oplus 0, \tag{2.6.18}$$

$$\omega_3(\widetilde{A})\xi = \Gamma_{23} A_{33}\xi \oplus \Gamma_{13} D_{\Gamma_{23}} A_{33}\xi \oplus D_{\Gamma_{13}} D_{\Gamma_{23}} A_{33}\xi. \tag{2.6.19}$$

Corollary 2.6.6 *With the above notation, the following are equivalent:*

(i) $\widetilde{A}$ *is the central completion of* A*;*

(ii) $\mathcal{K}(\widetilde{A}(\{1,2\})) \ominus \mathcal{K}(\widetilde{A}(\{2\})) \perp \mathcal{K}(\widetilde{A}(\{2,3\})) \ominus \mathcal{K}(\widetilde{A}(\{2\}))$.

Proof. Obviously (ii) is equivalent to

$$\omega_3(\widetilde{A})\xi - P_{\mathcal{K}(\widetilde{A}(\{2\}))}\xi \perp \mathcal{K}(\widetilde{A}(\{1,2\}))$$

for all $\xi \in \mathcal{H}$. With respect to the decomposition (2.6.17) of $\mathcal{K}(\widetilde{A})$ we have

$$\mathcal{K}(\widetilde{A}(\{2\})) = \overline{\operatorname{Ran}} A_{22} \oplus \{0\} \oplus \{0\},$$
$$\mathcal{K}(\widetilde{A}(\{1,2\})) = \overline{\operatorname{Ran}} A_{22} \oplus \mathcal{D}_{\Gamma_{12}^*} \oplus \{0\}.$$

So (2.6.19) implies that (ii) is equivalent to $\Gamma_{13} D_{\Gamma_{23}} A_{33}\xi = 0$ for all $\xi \in \mathcal{H}$. Since the range of $D_{\Gamma_{23}} A_{33}$ is dense in $\mathcal{D}_{\Gamma_{23}}$, the last assertion is equivalent to $\Gamma_{13} = 0$. □

Let us consider Theorem 2.2.4 for the case of a partially positive semidefinite operator-valued matrix. One can consider an admissible sequence Σ for the graph $G = (V, E)$ of the partial matrix, $G = G_0 \subset G_1 \subset \cdots \subset G_k = K_n$. All the completions are in one-to-one correspondence with contractions $\gamma_1, \ldots, \gamma_k$ on Hilbert spaces, the domain and the range of each γ_s depending only on the previous γ_r, $r < s$. We call the parameters $\gamma_1, \ldots, \gamma_k$ the *completion parameters along* Σ. In particular, by considering at each step $\gamma_r = 0$, one obtains the *central completion* along Σ .

Thus for each admissible order on $E(G^c)$ we obtain a different parametrization. For further use, note that it follows immediately that, if $S \subset V$, and $G' = G|S$, then the restriction of an admissible order on $E(G^c)$ to $E(G'^c)$ is admissible. There are particular classes of admissible orders on $E(G)$, namely those given by Lemma 2.2.5. Suppose now that $v \in V$ is a simplicial vertex of G. Denote $G' = G|V \setminus \{v\}$, and take an admissible order Σ' on $E(G'^c)$. Then the order on $E(G^c)$ obtained by taking first the edges that do not contain v in the same order as in Σ', and then the edges that contain v (in any order) is admissible. We call such an *order adapted to* v. It has then obviously the following property: if $\widetilde{A}$ is the central completion of A with respect to Σ, then $\widetilde{A}(V \setminus \{v\})$ is the central completion of $A(V \setminus \{v\})$, with respect to Σ'.

We intend to prove in full generality that for every chordal graph G and every partial positive semidefinite block operator matrix, the central completion does not depend on the admissible order. First, we need a definition and a lemma, the proof of which is an easy exercise.

Let $G = (V, E)$ be a graph. If $C \subset V$ is a clique, we will define the graph G/C by "collapsing" C to a single vertex:

- the vertices of G/C are $V \setminus C$ together with a single vertex $\widehat{C}$ that replaces the clique C;
- the edges of G/C are the edges of V between vertices in $V \setminus C$, while, if $v \in V \setminus C$, then $(v, \widehat{C})$ is an edge of G/C if there exists $a \in C$ such that (v, a) is an edge of G.

The proof of the following lemma is elementary, and is omitted.

Lemma 2.6.7 *If G is chordal and C is a clique in G, then G/C is chordal as well.*

The following result is the main one of this subsection. It is referred to as the "inheritance principle."

Theorem 2.6.8 *Let A be a partially positive semidefinite matrix the graph G of which is chordal. Then all admissible orders on $E(G^c)$ yield the same central completion.*

Proof. We will use a double induction procedure, according to the order n of the matrix and to the number of missing edges k. So let us suppose that the statement is true for partially positive semidefinite matrices of order at most $n-1$, as well as for matrices of order n such that the number of edges of G^c is at most $k-1$. For any n and $k=1$ there is nothing to prove, since we have a single admissible order.

Consider then a partially positive semidefinite matrix A of order n, and such that $E(G^c)$ has k elements. Let Σ and Σ' be two admissible orders on $E(G^c)$, with first edges (i_1, j_1) and (i_1', j_1') respectively. We will now consider several possible cases.

Case 1. Suppose that there exists a simplicial vertex v of G such that $v \notin \{i_1, j_1, i_1', j_1'\}$. By Lemma 2.2.5, there exists an admissible order $\widehat{\Sigma}$ on $E(G^c)$ which is adapted to v, and such that its first elements coincide with the restriction of Σ to the elements in $E(G^c)$ that do not contain v. In particular, (i_1, j_1) is the first element of both Σ and $\widehat{\Sigma}$. Suppose $\widehat{G}$ is the graph with vertices V and edges obtained by adding (i_1, j_1) to $E(G)$, and $\widehat{A}$ is the corresponding partially positive semidefinite matrix which is the one-step central completion of A. Then $\widehat{G^c}$ has $k-1$ edges, and we may apply the induction hypothesis to conclude that the central completions of $\widehat{A}$ with respect to the two orders Σ and $\widehat{\Sigma}$ restricted to $\widehat{G^c}$ coincide. Obviously, these are the central completions of A with respect to Σ and $\widehat{\Sigma}$, which therefore coincide; we will denote this completion by $\widetilde{A}$.

Similarly, one can find an order $\widehat{\Sigma}'$ on $E(G^c)$ which is adapted to v, and such that the central completions of A with respect to Σ' and $\widehat{\Sigma}'$ coincide; denote it by $\widetilde{A}'$. Also, one can assume that the order of the edges which contain v is the same in $\widehat{\Sigma}$ and in $\widehat{\Sigma}'$.

Consider now $A'' = A(V \setminus \{v\})$. Since $\widehat{\Sigma}$ is adapted to v, it follows immediately that $\widetilde{A}(V \setminus \{v\})$ is the central completion of A'' with respect to $\widehat{\Sigma}$. Similarly, $\widetilde{A}'(V \setminus \{v\})$ is the central completion of A'' with respect to $\widehat{\Sigma}'$. These are both central completions of an $(n-1) \times (n-1)$ matrix, so they must coincide by the induction hypothesis. If we denote this $(n-1) \times (n-1)$ completed matrix by $\widetilde{A}''$, then $\widetilde{A}$ and $\widetilde{A}'$ are both central completions of $\widetilde{A}''$ with respect to the same order (on the edges containing v), and so they must coincide. This finishes the proof of Case 1.

Case 2. Suppose now that there exist two simplicial vertices v, v' such that $v \notin (i_1, j_1)$ and $v' \notin (i_1', j_1')$ (in particular, this is true if G has at least

3 simplicial vertices). Consider a perfect scheme for G which starts with the vertex v followed by v'. We have to consider two subcases.

Subcase 2.1. If $G|V\setminus\{v,v'\}$ is not the complete graph, then an associate order on $E(G^c)$ starts with an edge (ι,κ) disjoint from (v,v'); denote it by $\breve{\Sigma}$. Considering now the two admissible orders Σ and $\breve{\Sigma}$, we see that the simplicial vertex v does not belong to either one of their first edges (i_1,j_1) or to (ι,κ). We can then apply Case 1 to conclude that their associate central completions are the same.

Similarly, the vertex v' does not belong to the first edges of Σ' and $\breve{\Sigma}$, so their associate central completions coincide. So the central completions corresponding to Σ and Σ' coincide, which settles Subcase 2.1.

Subcase 2.2. If $G|V\setminus\{v,v'\}$ is the complete graph, then we must have $v'\in\{i_1,j_1\}$, say $v'=j_1$; and $v\in\{i_1',j_1'\}$, say $v=j_1'$. The edges in $E(G^c)$ are either edges joining v' or v to elements in $V\setminus\{v,v'\}$, or (v,v') itself.

Now, it is easily checked that any graph of this type is chordal, whence it follows that any order on $E(G^c)$ that leaves (v,v') as the last edge is admissible. Consider then such an order Σ_1 that begins with the edge (i_1,v') followed by (i_1',v), and another order Σ_1' that begins with (i_1',v) followed by (i_1,v'), and afterward coincides with Σ_1. Since Σ and Σ_1 have the same first edge, their central completions coincide by the induction hypothesis; we will denote this central completion by B. Similarly, the central completions of Σ' and Σ_1' coincide, say to B'.

Note now that, according to Proposition 2.5.2, whenever we have to complete one entry corresponding to the edge (i,j), this depends only on specified entries corresponding to edges with the property that if one of their vertices is neither i nor j, then it must be connected both to i and j. If we make the second step of completion according to Σ_1, then (i_1',v) does not have this property, since v is not connected to v'. Therefore the choice of $B_{i_1,v'}$ is not affected by the previous choice of $B_{i_1',v}$ and is therefore the same as in the case when we start by the edge (i_1,v'); that is, it is equal to $B'_{i_1,v'}$.

One shows similarly that $B_{i_1',v}=B'_{i_1',v}$. Therefore, after the first two steps the partially completed central matrices B and B' coincide; since from that moment on the orders Σ_1 and Σ_1' coincide, it follows that $B=B'$. This settles Subcase 2.2.

Case 3. The remaining case is when G has only two simplicial vertices v and v', and for either Σ or Σ' the first edge is (v,v'). In particular, the graph G' obtained by adding (v,v') to G is chordal. We will show that in this case the only missing edge in G is (v,v'); thus $\Sigma=\Sigma'$, and there is nothing to prove. Denote by C the clique $\mathrm{Adj}(v)\cap\mathrm{Adj}(v')$, and by C_v and $C_{v'}$ the connected components of $G|V\setminus C$ that contain v and v', respectively.

If $C_v=C_{v'}$, then a minimal path in C_v connecting v and v' has length at least 3. Adding the edge (v,v') to this path, one obtains a cycle of length at least 4 in G': a contradiction. We may then assume that $C_v\neq C_{v'}$.

Suppose then that C_v contains elements different from v. The set $C_1=C\cup\{v\}$ is a clique in the graph $G|(C_v\cup C)$. If $G_1=(G|(C_v\cup C))/C_1$, then G_1 is chordal (by Lemma 2.6.7). From Lemma 1.2.3, it must have a simplicial vertex u nonadjacent to v, and thus different from $\widehat{C}_1$. Then u is also simplicial in G, and it is different from v' (since C_v and $C_{v'}$ are disjoint): again a contradiction. A similar argument works if $C_{v'}$ contains elements different from v'.

Finally, if $C_v = \{v\}$ and $C_{v'} = \{v'\}$, consider the graph $G_2 = (G|(V \setminus \{v, v'\}))/C$. According to Lemma 2.6.7, G_2 is chordal. If it has a vertex different from $\widehat{C}$, again by Lemma 1.2.3 it should also have a simplicial vertex u different from $\widehat{C}$. It follows then that u is also simplicial in G, and it is different from v, v', which contradicts the assumption. Therefore the only possibility is that $V = C \cup \{v, v'\}$, which means that the only missing edge in G is (v, v'). □

Central completions are usually characterized by an extremal property. We have already discussed the maximum determinant property in the case of scalar positive definite matrices (with specified entries corresponding to a general chordal graph). This type of maximum principle cannot be formulated in the general case of a chordal pattern. We can, however, characterize by an extremal property the central completion; we have to use orthogonality in the Kolmogorov decomposition.

We call a completion $\widetilde{A}$ of A *maximal orthogonal* if, whenever $x, y \in V$ are nonadjacent vertices and $S \subset V$ is a minimal separator for x and y, we have

$$\mathcal{K}(\widetilde{A}(S \cup \{x\})) \ominus \mathcal{K}(\widetilde{A}(S)) \perp \mathcal{K}(\widetilde{A}(S \cup \{y\})) \ominus \mathcal{K}(\widetilde{A}(S)). \tag{2.6.20}$$

Theorem 2.6.9 *The following are equivalent for a completion $\widetilde{A}$:*
(i) $\widetilde{A}$ is central;
(ii) $\widetilde{A}$ is maximal orthogonal.

Proof. (i) $\Longrightarrow$ (ii). Note first that (2.6.20) is true if x is simplicial and S is its adjacency set. Indeed, in this case we may take the admissible order Σ adapted to x, such that the first edge that contains x is (x, y). When we arrive at (x, y) in the completion procedure, the set of vertices adjacent to both x and y is exactly S, and then Proposition 2.5.2 implies (2.6.20).

We will use induction with respect to the order n of the matrix. Take then $x, y \in V$ nonadjacent, and suppose first that there exists a simplicial vertex in G different from both x and y. Then S cannot contain v, since it would not be minimal anymore (if a path joining x and y passes through v, it passes before and after v through points in S, which are connected, since any minimal separator is a clique).

Let then $V' = V \setminus \{v\}$, and $G' = G|V'$. Using the order Σ adapted to x from above, one sees that the central completion of $\widetilde{A}$ compressed to V' coincides with the central completion of $\widetilde{A}(V')$.

We can apply the induction hypothesis to x, y and S in G'; denoting by $\mathcal{K}'$ the Kolmogorov decomposition of the central completion of $\widetilde{A}(V')$, we obtain that

$$\mathcal{K}'(\widetilde{A}(S\cup\{x\}))\ominus\mathcal{K}'(\widetilde{A}(S))\perp\mathcal{K}'(\widetilde{A}(S\cup\{y\}))\ominus\mathcal{K}'(\widetilde{A}(S)).$$

But, using again the admissible order Σ above, one sees that the central completion of $\widetilde{A}$ compressed to V' coincides with the central completion of $\widetilde{A}(V')$. Then $\mathcal{K}'\subset\mathcal{K}$, and actually we have

$$\mathcal{K}(\widetilde{A}(S\cup\{x\}))\ominus\mathcal{K}(\widetilde{A}(S))\perp\mathcal{K}(\widetilde{A}(S\cup\{y\}))\ominus\mathcal{K}(\widetilde{A}(S)),$$

which proves the claim.

Suppose now that x, y are the only simplicial vertices of G. Take S' to be the adjacency set of x. Using again the order Σ adapted to x, the central completion of $\widetilde{A}(V\setminus\{x\})$ is the compression of the central completion of $\widetilde{A}(V)$, and thus its Kolmogorov decomposition is embedded into $\mathcal{K}$. Also, note that S is still a minimal separator, in $G|V\setminus\{x\}$, for any pair $\{y,z\}$ with $z\in S'\setminus S$. Applying then the induction hypothesis to all such pairs, it follows that

$$\mathcal{K}(\widetilde{A}(S\cup\{y\}))\ominus\mathcal{K}(\widetilde{A}(S))\perp\mathcal{K}(\widetilde{A}(S\cup S'))\ominus\mathcal{K}(\widetilde{A}(S)),$$

or, equivalently,

$$\mathcal{K}(\widetilde{A}(S\cup\{y\}))\ominus\mathcal{K}(\widetilde{A}(S))\perp\mathcal{K}(\widetilde{A}(S\cup S')).$$

This implies that, if $\chi\in\mathcal{K}(\{y\})$ and P_S is the projection onto $\mathcal{K}(\widetilde{A}(S))$, then

$$\chi-P_S\chi\perp\mathcal{K}(\widetilde{A}(S\cup S')). \tag{2.6.21}$$

On the other hand, as noted above,

$$\mathcal{K}(\widetilde{A}(S'\cup\{x\}))\ominus\mathcal{K}(\widetilde{A}(S'))\perp\mathcal{K}(\{w\})$$

for all $w\notin S'\cup\{x\}$. This implies that, for any $\xi\in\mathcal{K}(\{x\})$, if $P_{S'}$ is the orthogonal projection onto $\mathcal{K}(\widetilde{A}(S'))$, then

$$\xi-P_{S'}\xi\perp\mathcal{K}(V\setminus\{x\}). \tag{2.6.22}$$

In particular, $\xi-P_{S'}\xi\perp\chi-P_S\chi$ for any $\chi\in\mathcal{K}(\{y\})$. But (2.6.21) implies that $\chi-P_S\chi\perp P_{S'}\xi$. Therefore $\xi\perp\chi-P_S\chi$, which proves the result.

(ii) $\Longrightarrow$ (i). We use induction with respect to the order n of the matrix.

Consider v a simplicial vertex. The maximal orthogonality condition applied to v, its adjacency set R, and any other vertex shows that

$$\mathcal{K}(\widetilde{A}(R\cup\{v\}))\ominus\mathcal{K}(\widetilde{A}(R))\perp\mathcal{K}(\widetilde{A}(\{x\})) \tag{2.6.23}$$

for all $x\in V\setminus\{v\}$.

Let now Σ be an admissible order adapted to v, and $G'=G|V\setminus\{v\}$. Suppose that x, y are two nonadjacent vertices in G', and S' is a minimal separator in G'. If there is a minimal path that connects x to y in G and avoids S', this path must go through two points of R, one before and one after v. But these points are connected, since R is a clique, and the path is no longer minimal. The contradiction obtained shows that S' is a separator of x, y in G also, and then the maximal orthogonality property is true for $\widetilde{A}(V \setminus \{v\})$. Thus this compression is central by the induction hypothesis.

We have now to add to this central completion of order $n-1$ the edges of G^c that contain v. Suppose that, in the order of Σ, these edges are $(v, u_1), \dots, (v, u_s)$, with completion parameters $\gamma_1, \dots, \gamma_s$.

Take $x = u_1$ in Theorem 2.6.8. From Corollary 2.6.6 it follows that $\gamma_1 = 0$; moreover, since, by Theorem 2.6.8,

$$\begin{aligned}\mathcal{K}(\widetilde{A}(R \cup \{u_1, v\})) = [\mathcal{K}(\widetilde{A}(R))] \oplus [\mathcal{K}(\widetilde{A}(R \cup \{v\})) \ominus \mathcal{K}(\widetilde{A}(R))] \\ \oplus [\mathcal{K}(\widetilde{A}(R \cup \{u_1\})) \ominus \mathcal{K}(\widetilde{A}(R))],\end{aligned}$$

we obtain

$$\mathcal{K}(\widetilde{A}(R \cup \{u_1, v\})) \ominus \mathcal{K}(\widetilde{A}(R \cup \{u_1\})) = \mathcal{K}(\widetilde{A}(R \cup \{v\})) \ominus \mathcal{K}(\widetilde{A}(R)).$$

Therefore, applying (2.6.23) for $x = u_2$, we have

$$\mathcal{K}(\widetilde{A}(R \cup \{u_1, v\})) \ominus \mathcal{K}(\widetilde{A}(R \cup \{u_1\})) \perp \mathcal{K}(\widetilde{A}(\{u_2\})).$$

Now, a similar argument implies $\gamma_2 = 0$ and

$$\mathcal{K}(\widetilde{A}(R \cup \{u_1, u_2\} \cup \{v\})) \ominus \mathcal{K}(\widetilde{A}(R \cup \{u_1, u_2\})) = \mathcal{K}(\widetilde{A}(R \cup \{v\})) \ominus \mathcal{K}(\widetilde{A}(R)).$$

It is clear now that the procedure can be continued, giving $\gamma_r = 0$ for all $0 \le r \le s$. Thus the completion $\widetilde{A}$ is central. □

Finally, let us show how the structure of $\widetilde{A}^{-1}$, when $\widetilde{A}$ is invertible, can be deduced directly from the maximal orthogonality property. Write $\widetilde{A} = (\langle \xi_i, \xi_j \rangle)_{i,j=1}^n$ for some linearly independent vectors ξ_i in a Hilbert space K, and denote $S_i = \text{Adj}(i)$, $K(S) = \mathcal{K}(\widetilde{A}(S))$ (which is in this case the linear span of ξ_i for $i \in S \subset V$), and

$$\eta_i = \xi_i - P_{K(S_i)} \xi_i = \xi_i + \sum_{r \in S_i} b_{ir} \xi_r. \tag{2.6.24}$$

We claim that $\eta_i \perp \xi_j$ for any $j \neq i$. This is immediate by definition for $j \in S_i$. Suppose now that $j \notin S_i \cup \{i\}$. Since S_i separates i from j, it contains a minimal separator S_{ij}; the maximal orthogonality property (2.6.20) implies that

$$\xi_i - P_{K(S_{ij})} \xi_i \perp \xi_j. \tag{2.6.25}$$

On the other hand, for any $k \in S_i \setminus S_{ij}$, S_{ij} separates k from j. Moreover, it is a minimal separator: since k is adjacent to i, any smaller minimal separator would also be a separator for i and j. Therefore (2.6.20) implies that $(\xi_k - P_{K(S_{ij})} \xi_k) \perp \xi_j$, whence $K(S_i) \ominus K(S_{ij}) \perp \xi_j$. Using (2.6.25), we obtain that

$$\eta_i = \xi_i - P_{K(S_i)} \xi_i = \xi_i - P_{K(S_i)} \xi_i - (P_{K(S_i) \ominus K(S_{ij})}) \xi_i$$

is indeed orthogonal to ξ_j.

If we define, for all i, $b_{ii} = 1$, then

$$0 = \langle \eta_i, \xi_j \rangle = \left\langle \sum_{r \in S_i \cup \{i\}} b_{ir} \xi_r, \xi_j \right\rangle = \sum_{r \in S_i \cup \{i\}} b_{ir} \langle \xi_r, \xi_j \rangle = \sum_{r \in S_i \cup \{i\}} b_{ir} a_{rj} \tag{2.6.26}$$

holds for any $i \neq j$. But the vectors ξ being linearly independent, we have $\eta_i \neq 0$, and therefore $d_i = \langle \eta_i, \eta_i \rangle > 0$. If we denote $B = (b_{ij})$ and D the diagonal matrix with entries $1/d_i$, then (2.6.26) implies that DB is the inverse of $\widetilde{A}$; while the definition formula (2.6.24) shows that it may have nonzero entries only in positions corresponding to E.

2.7 THE HAMBURGER MOMENT PROBLEM AND SPECTRAL FACTORIZATION ON THE REAL LINE

We first consider in this section the *Hamburger moment problem for positive measures on* $\mathbb{R}$, in both the scalar and operator-valued cases. The power moments of a positive measure on $\mathbb{R}$ form a sequence with the property that all Hankel matrices built on them are positive semidefinite, but unlike the situation for the trigonometric moment problem, in general they do not uniquely determine the measure. The truncated Hamburger moment problem asks necessary and sufficient conditions for m real number numbers (or Hermitian operators) to be the first m power moments of a positive measure on $\mathbb{R}$. We present a solution to the problem in the matrix-valued case, but again, the situation is different from that of the unit circle: the positive semidefinitess of a Hankel matrix is not sufficient by itself for the existence of a solution. The particular conditions make the operator-valued problem extra challenging, as pointed out in Example 2.7.8.

It is natural, by duality, to also expect a factorization result for matrix- and operator-valued polynomials that are positive semidefinite on the real line. We present in this section a finite dimensional result with a proof that relies on the algebraic Riccati equation. It is a continuation of the method based on Schur complements for proving the Fejér-Riesz factorization in Section 2.4. In Exercise 2.9.36 we present an operator-valued factorization result the proof of which relies on the results in Section 2.4 and a conversion from the unit circle to the real line.

The *power moments of a positive measure* μ *on* $\mathbb{R}$ are defined by

$$\widehat{\mu}(k) = \int_{-\infty}^{\infty} \lambda^k d\mu(\lambda), \quad k \geq 0,$$

assuming the integrals converge.

Let $(s_n)_{n\geq 0}$ be a sequence of real numbers. Consider for each $n \geq 0$ the Hankel matrix

$$K_n = \begin{pmatrix} s_0 & s_1 & \cdots & s_n \\ s_1 & s_2 & \cdots & s_{n+1} \\ \vdots & \vdots & \iddots & \vdots \\ s_n & s_{n+1} & \cdots & s_{2n} \end{pmatrix}. \tag{2.7.1}$$

The condition that for a real sequence $(s_n)_{n\geq 0}$ all matrices (2.7.1) are positive semidefinite is equivalent to: for every $n \geq 0$ and sequence $(\xi_k)_{k=0}^n$ of complex numbers,

$$\sum_{j,k=0}^{n} s_{j+k}\overline{\xi}_j \xi_k \geq 0. \tag{2.7.2}$$

Theorem 2.7.1 is a fundamental result that characterizes moment sequences of positive measures on $\mathbb{R}$.

Theorem 2.7.1 *Given a sequence $(s_n)_{n\geq 0}$ of real numbers, there exists a positive measure μ on $\mathbb{R}$ such that*

$$\widehat{\mu}(n) = s_n \text{ for } n \geq 0 \tag{2.7.3}$$

if and only if the sequence $(s_n)_{n\geq 0}$ is such that all Hankel matrices (2.7.1) are positive semidefinite for $n \geq 0$.

We will need the following classical operator theoretic result due to von Neumann, for which we do not include a proof (just a reference). An antilinear map $C : \mathcal{H} \to \mathcal{H}$ (i.e., $C(\alpha v + \beta w) = \overline{\alpha} Cv + \overline{\beta} CW$) is called a *conjugation* if it is norm preserving and $C^2 = I$.

Theorem 2.7.2 *Let A be a symmetric (unbounded) operator such that there exists a conjugation C with $C : \mathcal{D}(A) \to \mathcal{D}(A)$ such that $AC = CA$. Then A admits selfadjoint extensions.*

Proof. A proof may be found in [481, Theorem X.3]. □

Proof of Theorem 2.7.1 Assume μ is a positive measure on $\mathbb{R}$ for which (2.7.3) holds. Then

$$\sum_{j,k=0}^{n} s_{j+k}\overline{\xi}_j \xi_k = \int_{-\infty}^{\infty} \left| \sum_{j=0}^{n} \xi_j \lambda^j \right|^2 d\mu(\lambda) \geq 0,$$

implying that all Hankel matrices (2.7.1) are positive semidefinite for $n \geq 0$.

Assume conversely that $(s_n)_{n\geq 0}$ is such that all Hankel matrices (2.7.1) are positive semidefinite for $n \geq 0$. Let P denote the set of all polynomials on $\mathbb{R}$ with complex coefficients and define the nonnegative sesquilinear form on P by

$$\left\langle \sum_{k=0}^{n} \xi_k \lambda^k, \sum_{j=0}^{n} \xi_j \lambda^j \right\rangle = \sum_{j,k=0}^{n} s_{j+k}\overline{\xi}_j \xi_k.$$

Define $Q = \{\psi : \psi \in P, \langle \psi, \psi \rangle = 0\}$ and let $\mathcal{H}$ be the Hilbert space obtained by completing P/Q in the inner product $\langle \cdot, \cdot \rangle$. Consider the shift operator $S : P \to P$ defined by

$$S\left(\sum_{j=0}^{n} \xi_j \lambda^j \right) = \sum_{j=0}^{n} \xi_j \lambda^{j+1}.$$

It is easy to see that S is symmetric and $S : Q \to Q$ since the Schwarz inequality implies that

$$\langle S\psi, S\psi \rangle = \left| \left\langle S^2\psi, \psi \right\rangle \right| \leq \left\langle S^2\psi, S^2\psi \right\rangle^{\frac{1}{2}} \langle \psi, \psi \rangle^{\frac{1}{2}}.$$

Thus, S lifts to a symmetric operator $\widehat{S}$ on $\mathcal{H}$ with domain P/Q. If C denotes complex conjugation on P, then C also lifts to a map $\widehat{C} : P/Q \to P/Q$. It can easily be checked that $\widehat{C}$ extends to a conjugation on $\mathcal{H}$ and $\widehat{S}\widehat{C} = \widehat{C}\widehat{S}$. By Theorem 2.7.2, $\widehat{S}$ admits selfadjoint extensions, and let $\widetilde{S}$ be one of them. Let $E(\cdot)$ be the spectral measure of the operator $\widetilde{S}$, and put $\mu(\cdot) = \langle [1], E(\cdot)[1] \rangle$, where $[1]$ is the equivalence class of the function 1. Then

$$\int_{-\infty}^{\infty} \lambda^n d\mu = \left\langle 1, \widetilde{S}^n 1 \right\rangle = \langle 1, \lambda^n \rangle = s_n.$$

□

Remark 2.7.3 The major difference between the trigonometric and the Hamburger moment problems is that, in general, the power moments do not uniquely determine a positive measure on $\mathbb{R}$. Let, for instance,

$$s_n = \int_0^{\infty} \lambda^n e^{-\sqrt[4]{\lambda}} d\lambda, \ n = 0, 1, 2, \ldots.$$

One may verify that

$$\int_0^{\infty} \lambda^n \sin(\sqrt[4]{\lambda}) e^{-\sqrt[4]{\lambda}} d\lambda = 4 \int_0^{\infty} \lambda^{4n+3} (\sin \lambda) e^{-\lambda} d\lambda = 0, \quad n \geq 0.$$

The above relations imply the existence of nonzero real measures on $\mathbb{R}$ with all their power moments equal to 0. Now, for any $\alpha \in [-1, 1]$, the positive measure σ defined by

$$d\sigma(\lambda) = f(\lambda) d\lambda, \ f(\lambda) = \begin{cases} 0, & \lambda < 0, \\ (1 + \alpha \sin(\sqrt[4]{\lambda})) e^{-\sqrt[4]{\lambda}}, & \lambda \geq 0, \end{cases}$$

has the same power moments $(s_n)_{n \geq 0}$.

It is natural to consider next the so-called *truncated Hamburger moment problem*, namely, finding necessary and sufficient conditions for the given real numbers $s_0, s_1, \ldots, s_m$ to be the first $m+1$ power moments of a positive measure μ on $\mathbb{R}$, all moments of which are finite. The problem is equivalent by Theorem 2.7.1 to the existence of an extension $\{s_j\}_{j \geq 0}$ of $s_0, s_1, \ldots, s_m$ such that the Hankel matrices (2.7.1) are positive semidefinite for all $n \geq 0$. Contrary to the one-variable truncated trigonometric moment problem in Section 1.3, positive semidefiniteness of all prescribed moment matrices is necessary, but in general not sufficient, as shown by the following example.

Example 2.7.4 Let $s_0 = s_1 = 0, s_2 = 1$. Then we have

$$K_1 = \begin{pmatrix} s_0 & s_1 \\ s_1 & s_2 \end{pmatrix} = \begin{pmatrix} 0 & 0 \\ 0 & 1 \end{pmatrix} \geq 0.$$

The existence of a solution to the Hamburger problem with the above data implies that for some $s_3, s_4 \in \mathbb{R}$ we have that

$$K_2 = \begin{pmatrix} 0 & 0 & 1 \\ 0 & 1 & s_3 \\ 1 & s_3 & s_4 \end{pmatrix} \geq 0,$$

which is impossible.

Note that if we weaken the condition on s_2 to $\widehat{\sigma}(2) = \int_{-\infty}^{\infty} \lambda^2 d\sigma(\lambda) \leq s_2$, a solution does exist: any constant $\sigma(\lambda)$ will work, yielding $\widehat{\sigma}(j) = 0$, $j \geq 0$. It is also possible to add a range condition, as in the following theorem.

Theorem 2.7.5 *Let $s_0, \ldots, s_m$ be given real numbers.*

(i) When $m = 2n$, then there exist real numbers $s_{2n+1}, s_{2n+2}, \ldots$ such that $K_k := (s_{i+j})_{i,j=0}^k \geq 0$, $k \geq 0$, if and only if $K_n \geq 0$ and

$$\begin{pmatrix} s_{n+1} \\ \vdots \\ s_{2n} \end{pmatrix} \in \mathrm{Ran} K_{n-1}. \tag{2.7.4}$$

Equivalently, there exists a solution to the truncated Hamburger moment problem with given data $s_0, \ldots, s_{2n}$ if and only if $K_n \geq 0$ and (2.7.4) holds.

(ii) When $m = 2n + 1$, the necessary and sufficient conditions for the existence of a solution to the truncated Hamburger moment problem with given data $s_0, \ldots, s_{2n+1}$ are $K_n \geq 0$ and

$$\begin{pmatrix} s_{n+1} \\ \vdots \\ s_{2n+1} \end{pmatrix} \in \mathrm{Ran} K_n. \tag{2.7.5}$$

Proof. Clearly, the conditions are necessary (the range inclusions come from observing that $K_{n+1} \geq 0$, and thus also the principal submatrix of K_{n+1} obtained by removing its one but last row and column in case (i) or its last row and column in case (ii), is positive semidefinite).

For the sufficiency, consider first the case $m = 2n$. The condition (2.7.4) implies the existence of a vector $v = (v_i)_{i=0}^{n-1}$ such that

$$\begin{pmatrix} s_{n+1} \\ \vdots \\ s_{2n} \end{pmatrix} = K_{n-1} v.$$

Now define the real numbers s_{2n+1} and s_{2n+2} via

$$\begin{pmatrix} s_{2n+1} \\ s_{2n+2} \end{pmatrix} = \begin{pmatrix} s_n & \cdots & s_{2n-1} \\ s_{n+1} & \cdots & s_{2n} \end{pmatrix} \begin{pmatrix} v_0 \\ \vdots \\ v_{n-1} \end{pmatrix}.$$

Then K_{n+1} has the property that its last column is a linear combination of the first n columns. Since $K_n \geq 0$, and the Schur complement of K_{n+1} supported on the last row and column is 0, we have that $K_{n+1} \geq 0$ as well.

As K_{n+1} now satisfies the conditions of the theorem, one can repeat the process and obtain $s_{2n+3}, s_{2n+4}, \ldots$ to make $K_{n+2}, K_{n+3}, \ldots$ positive semidefinite.

Consider next $m = 2n + 1$. By condition (2.7.5), there exists a vector $v = (v_i)_{i=0}^n$ such that

$$\begin{pmatrix} s_{n+1} \\ \vdots \\ s_{2n+1} \end{pmatrix} = K_n v.$$

Defining

$$s_{2n+2} = \begin{pmatrix} s_{n+1} & \cdots & s_{2n+1} \end{pmatrix} \begin{pmatrix} v_0 \\ \vdots \\ v_n \end{pmatrix},$$

by arguments similar to the ones in the first part of the proof, we have that $K_{n+1} \geq 0$ and

$$\begin{pmatrix} s_{n+2} \\ \vdots \\ s_{2n+2} \end{pmatrix} \in \operatorname{Ran} K_{n+1}.$$

We are now in the situation (i) with $m = 2n + 2$, thus by the first part of the sufficiency proof, there exist $s_{2n+3}, s_{2n+4}, \ldots$ to make $K_{n+2}, K_{n+3}, \ldots$ positive semidefinite. □

We consider next the operator-valued Hamburger moment problem. Let $\mathcal{H}$ be a Hilbert space. First, a function $\sigma : \mathcal{B}(\mathbb{R}) \to \mathcal{L}(\mathcal{H})$ ($\mathcal{B}(\mathbb{R})$ is the collection of all Borel measurable sets in $\mathbb{R}$) is called a positive operator-valued measure on $\mathbb{R}$ if for each $h \in \mathcal{H}$, $\langle \sigma(\Delta)h, h\rangle$ defines a positive measure on $\mathbb{R}$. For a measurable function $f : \mathbb{R} \to \mathbb{C}$ its integral $\int_{-\infty}^{\infty} f(\lambda)d\sigma(\lambda) \in \mathcal{L}(\mathcal{H})$ is defined by the formula

$$\left\langle \int_{-\infty}^{\infty} f(\lambda)d\sigma(\lambda)h, k \right\rangle = \int_{-\infty}^{\infty} f(\lambda)\, \langle d\sigma(\lambda)h, k\rangle$$

for all $h, k \in \mathcal{H}$, provided all integrals on the right-hand side converge.

The *power moments of a positive operator-valued measure* σ are defined by

$$\widehat{\sigma}(k) = \int_{-\infty}^{\infty} \lambda^k d\sigma(\lambda), \quad k \geq 0,$$

provided the integrals converge.

Let $(S_n)_{n\geq 0}$ be a sequence of Hermitian operators in $\mathcal{L}(\mathcal{H})$ and define the Hankel operator matrices

$$K_n = \begin{pmatrix} S_0 & S_1 & \cdots & S_n \\ S_1 & S_2 & \cdots & S_{n+1} \\ \vdots & \vdots & \iddots & \vdots \\ S_n & S_{n+1} & \cdots & S_{2n} \end{pmatrix}, \quad n \geq 0. \tag{2.7.6}$$

If there exists a positive operator-valued measure σ on $\mathbb{R}$ such that $S_n = \widehat{\sigma}(n)$ for $n \geq 0$, then all matrices K_n are positive semidefinite. Indeed, we have in this case for every $h_0, \dots, h_n \in \mathcal{H}$,

$$\begin{aligned} 0 &\leq \int_{-\infty}^{\infty} \left\langle d\sigma(\lambda) \sum_{j=0}^{n} h_j \lambda^j, \sum_{k=0}^{n} h_k \lambda^k \right\rangle \\ &= \sum_{j,k=0}^{n} \int_{-\infty}^{\infty} \lambda^{j+k} \left\langle d\sigma(\lambda) h_j, h_k \right\rangle = \sum_{k,j=0}^{n} \left\langle S_{j+k} h_j, h_k \right\rangle, \end{aligned}$$

implying that $K_n \geq 0$. Conversely, if $(S_n)_{n\geq 0}$ is such that all $K_n \geq 0$ for $n \geq 0$, then by a construction similar to the one in the proof of Theorem 2.7.1, but in which the polynomials have operator coefficients, one obtains a positive operator-valued measure on $\mathbb{R}$ such that $\widehat{\sigma}(n) = S_n$ for all $n \geq 0$.

Let $S_0, \dots, S_m \in \mathcal{L}(H)$ be given Hilbert space operators. We look for a positive $\mathcal{L}(H)$-valued measure σ on $\mathbb{R}$ such that

$$\int_{-\infty}^{\infty} \lambda^k d\sigma(\lambda)$$

exists for all $k \geq 0$ and

$$\widehat{\sigma}(k) = S_k, \quad \text{for } k = 0, 1, \dots, m. \tag{2.7.7}$$

For a solution to exist it is necessary that the Hankel matrices $K_n = (S_{i+j})_{i,j=0}^{n}$, where $2n \leq m$, are positive semidefinite, implying in particular that $S_j^* = S_j$, $j = 0, 1, \dots, m$. The existence of a solution is equivalent to the existence of an extension $(S_k)_{k\geq 0}$ of $(S_k)_{k=0}^{m}$ such that all matrices K_n for $n \geq 0$ are positive semidefinite. As already known from the scalar case, the positive semidefiniteness condition of K_n with $2n \leq m$ is in general not sufficient. By Lemma 2.4.4, we also need that

$$\operatorname{Ran} \begin{pmatrix} S_n \\ \vdots \\ S_{2n-1} \end{pmatrix} \subseteq \operatorname{Ran} \begin{pmatrix} S_0 & \cdots & S_{n-1} \\ \vdots & \iddots & \vdots \\ S_{n-1} & \cdots & S_{2n-2} \end{pmatrix}^{\frac{1}{2}}. \tag{2.7.8}$$

The following theorem gives a solution to the matrix-valued truncated Hamburger moment problem. It is only valid in the finite dimensional case as shown by Example 2.7.8. Note that in the finite dimensional case $\operatorname{Ran} M^{\frac{1}{2}} = \operatorname{Ran} M$ for $M \geq 0$.

Theorem 2.7.6 *Let $S_0, \ldots, S_{2n-1}$ be given Hermitian matrices. There exists a solution to the Hamburger moment problem (2.7.7) with $m = 2n-1$ if and only if*

$$K_{n-1} \geq 0 \tag{2.7.9}$$

and

$$\text{Ran} \begin{pmatrix} S_n \\ \vdots \\ S_{2n-1} \end{pmatrix} \subseteq \text{Ran}K_{n-1}. \tag{2.7.10}$$

If, in addition, S_{2n} is given, then a solution to (2.7.7) with $m = 2n$ exists if and only if K_n is positive semidefinite and

$$\text{Ran} \begin{pmatrix} S_{n+1} \\ \vdots \\ S_{2n} \end{pmatrix} \subseteq \text{Ran}K_{n-1}. \tag{2.7.11}$$

In both cases we can find σ of the following form:

$$\sigma(\lambda) = \sum_{i=1}^{k} T_i \delta_{\lambda_i},$$

where $k = \text{rank}K_{\lfloor \frac{m}{2} \rfloor}$, $\lambda_1, \ldots, \lambda_k \in \mathbb{R}$, $T_1, \ldots, T_k \geq 0$ are of rank 1, and for $\lambda \in \mathbb{R}$, δ_λ denotes the Dirac mass at λ.

Remark 2.7.7 The measure $d\sigma(\lambda)$ constructed in Theorem 2.7.6 is one for which all moments $\int_{-\infty}^{\infty} \lambda^k d\sigma(\lambda)$, $k = 0, 1, \ldots$, automatically exist. Thus it follows from the existence of the moments $S_0, \ldots, S_{2n-1}$ satisfying (2.7.9) and (2.7.10), that a measure exists with well defined subsequent moments.

Notice that the finite dimensionality is important as the following example shows.

Example 2.7.8 Let

$$S_0 = \begin{pmatrix} 1 & & & \\ & \frac{1}{2} & & \\ & & \frac{1}{3} & \\ & & & \ddots \end{pmatrix}, \quad S_1 = \begin{pmatrix} 1 & & & \\ & \frac{1}{\sqrt{2}} & & \\ & & \frac{1}{\sqrt{3}} & \\ & & & \ddots \end{pmatrix},$$

acting on $l^2(\mathbb{N})$. Clearly $S_0 \geq 0$ and $\text{Ran}S_1 \subseteq \text{Ran}S_0^{\frac{1}{2}}$. However, no solution to the truncated Hamburger moment problem exists for these data. Indeed, if S_2 is such that

$$\begin{pmatrix} S_0 & S_1 \\ S_1 & S_2 \end{pmatrix} \geq 0,$$

then $S_2 \geq I$. But there is no choice for S_3 and S_4 possible such that

$$\begin{pmatrix} S_0 & S_1 & S_2 \\ S_1 & S_2 & S_3 \\ S_2 & S_3 & S_4 \end{pmatrix} \geq 0$$

as $\text{Ran}S_2 \not\subseteq \text{Ran}S_0^{\frac{1}{2}}$ when $S_2 \geq I$.

In order to prove Theorem 2.7.6 we first prove the following result on block Hankel matrices.

Theorem 2.7.9 *Let $S_0, \ldots, S_{2n-1} \in \mathcal{H}_p$ be Hermitian $p \times p$ matrices such that*

$$K_{n-1} = (S_{i+j})_{i,j=0}^{n-1} \geq 0$$

and

$$\text{Ran} \begin{pmatrix} S_n \\ \vdots \\ S_{2n-1} \end{pmatrix} \subseteq \text{Ran} \begin{pmatrix} S_0 & \cdots & S_{n-1} \\ \vdots & \iddots & \vdots \\ S_{n-1} & \cdots & S_{2n-2} \end{pmatrix}. \tag{2.7.12}$$

Put $k = \text{rank} K_{n-1}$ Then there exist a $p \times k$ matrix C_0 and a $k \times k$ real diagonal matrix D such that

$$S_i = C_0 D^i C_0^*, \quad i = 0, \ldots, 2n-1. \tag{2.7.13}$$

Moreover, if S_{2n} is so that $K_n \geq 0$, then $C_0 D^{2n} C_0^* \leq S_{2n}$. In fact, $S_{2n} - C_0 D^{2n} C_0^*$ is the Schur complement of K_n supported in its last row and column.

We need the following auxiliary result.

Lemma 2.7.10 *Let $A \geq 0$ and $B = B^*$ be $p \times p$ matrices so that* $\text{Ran} B \subseteq \text{Ran} A$. *Let $k = \text{rank} A$. Then there exist a $k \times k$ real diagonal D and a full rank $p \times k$ matrix C such that*

$$A = CC^* \text{ and } B = CDC^*. \tag{2.7.14}$$

Proof. Decompose $\mathbb{C}^p = \text{Ran} A \oplus \text{Ker} A$. Then with respect to this decomposition we have

$$A = \begin{pmatrix} \widetilde{A} & 0 \\ 0 & 0 \end{pmatrix} \text{ and } B = \begin{pmatrix} \widetilde{B} & 0 \\ 0 & 0 \end{pmatrix},$$

where we used that $\text{Ran} B \subseteq \text{Ran} A$. As $\widetilde{A} > 0$, consider the Hermitian matrix $\widetilde{A}^{-\frac{1}{2}} \widetilde{B} \widetilde{A}^{-\frac{1}{2}}$. Applying the spectral theorem we may write

$$\widetilde{A}^{-\frac{1}{2}} \widetilde{B} \widetilde{A}^{-\frac{1}{2}} = UDU^*$$

for some $k \times k$ unitary U and $k \times k$ real diagonal D. Letting now

$$C = \begin{pmatrix} \widetilde{A}^{\frac{1}{2}} \\ 0 \end{pmatrix} U, \tag{2.7.15}$$

we get (2.7.14). □

Proof of Theorem 2.7.9. Consider $A = K_{n-1}$ and $B = (S_{i+j+1})_{i,j=0}^{n-1}$. Then $A \geq 0, B = B^*$, and $\text{Ran} B \subseteq \text{Ran} A$, and thus we can apply Lemma 2.7.10 to obtain an injective matrix C and a real diagonal matrix D such that

$$A = CC^* \text{ and } B = CDC^*.$$

Write

$$C = \begin{pmatrix} C_0 \\ \vdots \\ C_{n-1} \end{pmatrix}.$$

Then we get that

$$S_{i+j+1} = C_i C_{j+1}^* = C_i D C_j^*, \ \ 0 \leq i \leq n-1, \ \ 0 \leq j \leq n-2. \tag{2.7.16}$$

Thus

$$C_i(C_{j+1}^* - DC_j^*) = 0, \ \ 0 \leq i \leq n-1, \ \ 0 \leq j \leq n-2,$$

which yields that

$$C(C_{j+1}^* - DC_j^*) = 0, \ \ 0 \leq j \leq n-2.$$

As C is injective, we get that

$$C_{j+1}^* = DC_j^*, \ \ 0 \leq j \leq n-2,$$

which yields $C_j = C_0 D^j, \quad j = 0, \ldots, n-1$. Thus

$$(S_{i+j})_{i,j=0}^{n-1} = \begin{pmatrix} C_0 \\ C_0 D \\ \vdots \\ C_0 D^{n-1} \end{pmatrix} \begin{pmatrix} C_0^* & DC_0^* & \cdots & D^{n-1}C_0^* \end{pmatrix},$$

and consequently we obtain that $S_i = C_0 D^i C_0^*, i = 0, \ldots, 2n-2$. Similarly, from $B = CDC^*$ we obtain in addition that $S_{2n-1} = C_0 D^{2n-1} C_0^*$.

Finally, let S_{2n} be given such that $K_n \geq 0$. Then $L := (C_0 D^{i+j} C_0^*)_{i,j=0}^n$ is a positive semidefinite matrix with rankK_{n-1} = rank$L$. As $K_{n-1} = (C_0 D^{i+j} C_0^*)_{i,j=0}^{n-1}$ we get that the Schur complement of L supported in the last row and column equals 0. But then we obtain that $L \leq K_n$ and thus $C_0 D^{2n} C_0^* \leq S_{2n}$ follows. □

The following lemma will allow us to reduce the even case in Theorem 2.7.6 to the odd case.

Lemma 2.7.11 *Let $S_0, \ldots, S_{2n}$ be $p \times p$ Hermitian matrices such that $K_n \geq 0$ and*

$$\operatorname{Ran} \begin{pmatrix} S_{n+1} \\ \vdots \\ S_{2n} \end{pmatrix} \subseteq \operatorname{Ran}(K_{n-1}). \tag{2.7.17}$$

Then there exists a $p \times p$ Hermitian matrix S_{2n+1} such that

$$\operatorname{Ran} \begin{pmatrix} S_{n+1} \\ \vdots \\ S_{2n+1} \end{pmatrix} \subseteq \operatorname{Ran}(K_n). \tag{2.7.18}$$

Proof. Let R be the Schur complement of K_n supported in the last row and column. Let C_0 and D be as in Theorem 2.7.9. Note that $S_{2n} - R = C_0 D^{2n} C_0^*$. As

$$\operatorname{rank} \begin{pmatrix} S_0 & \cdots & S_n \\ \vdots & \iddots & \vdots \\ S_n & \cdots & S_{2n} - R \end{pmatrix} = \operatorname{rank} K_{n-1},$$

we get that

$$\begin{pmatrix} S_n \\ \vdots \\ S_{2n-1} \\ S_{2n} - R \end{pmatrix} = \begin{pmatrix} S_0 & \cdots & S_{n-1} \\ \vdots & \iddots & \vdots \\ S_{n-1} & \cdots & S_{2n-2} \\ S_n & \cdots & S_{2n-1} \end{pmatrix} \begin{pmatrix} Y_0 \\ \vdots \\ Y_{n-1} \end{pmatrix} \tag{2.7.19}$$

for some $Y_0, \ldots, Y_{n-1}$. We claim that

$$S_{2n+1} := C_0 D^{2n+1} C_0^* + Y_{n-1}^* R + R Y_{n-1}$$

has the desired properties. Clearly, S_{2n+1} is Hermitian, so it remains to check (2.7.18).

Observe that $\begin{pmatrix} C_0^* & \cdots & D^{n-1} C_0^* \end{pmatrix}$ is surjective. Thus there exist $A_0, \ldots,$ A_{n-1} such that

$$C_0^* A_0 + \cdots + D^{n-1} C_0^* A_{n-1} = I_k.$$

But then

$$\begin{pmatrix} S_{n+1} \\ \vdots \\ S_{2n} - R \end{pmatrix} = K_{n-1} \begin{pmatrix} A_0 \\ \vdots \\ A_{n-1} \end{pmatrix} D^{n+1} C_0^*.$$

This together with (2.7.17) implies that

$$\begin{pmatrix} 0 \\ \vdots \\ 0 \\ R \end{pmatrix} = K_{n-1} \begin{pmatrix} X_0 \\ \vdots \\ X_{n-1} \end{pmatrix} \tag{2.7.20}$$

for some matrices $X_0, \ldots, X_{n-1}$. Note that

$$S_n X_0 + \cdots + S_{2n-1} X_{n-1} = \begin{pmatrix} Y_0^* & \cdots & Y_{n-1}^* \end{pmatrix} K_{n-1} \begin{pmatrix} X_0 \\ \vdots \\ X_{n-1} \end{pmatrix} = Y_{n-1}^* R.$$

But then

$$K_n \begin{pmatrix} X_0 \\ \vdots \\ X_{n-1} \\ 0 \end{pmatrix} = \begin{pmatrix} 0 \\ \vdots \\ 0 \\ R \\ Y_{n-1}^* R \end{pmatrix}.$$

Next,

$$K_n \begin{pmatrix} A_0 \\ \vdots \\ A_{n-1} \\ 0 \end{pmatrix} D^{n+1}C_0^* = \begin{pmatrix} C_0D^{n+1}C_0^* \\ \vdots \\ C_0D^{2n+1}C_0^* \end{pmatrix}$$

and

$$K_n[\begin{pmatrix} 0 \\ \vdots \\ 0 \\ Y_{n-1} \end{pmatrix} - \begin{pmatrix} A_0 \\ \vdots \\ A_{n-1} \\ 0 \end{pmatrix} D^n C_0^* Y_{n-1}] = \begin{pmatrix} 0 \\ \vdots \\ 0 \\ RY_{n-1} \end{pmatrix}.$$

Adding the last three equations yields (2.7.18) □

We are now ready to prove the Hamburger moment result.

Proof of Theorem 2.7.6. At the beginning of the section we showed the necessity of the conditions $K_{n-1} \geq 0$ and (2.7.10). Similarly, when S_{2n} is given, the necessity of $K_n \geq 0$ is clear.

For the sufficiency, consider first the case $m = 2n - 1$. Apply Theorem 2.7.9 to obtain $C_0 = \begin{pmatrix} c_1 & \cdots & c_k \end{pmatrix}$, with $c_i \in \mathbb{C}^p$, and $D = \text{diag}(\lambda_i)_{i=1}^k$ so that (2.7.13) holds. Now put $T_i = c_ic_i^*, i = 1, \ldots, k$, and define σ as in the statement of Theorem 2.7.6. One can directly verify (2.7.7).

Consider next the case $m = 2n$. We can apply then Lemma 2.7.11 to obtain a Hermitian S_{2n+1}. Now we are in the case with $m = 2n + 1$. Note that (2.7.9) and (2.7.10) hold with n replaced by $n + 1$. Thus we can apply the first part of this proof to obtain a solution. □

Corollary 2.7.12 *Let $S_0, \ldots, S_{2n}$ be given $p \times p$ Hermitian matrices. When either*

(i) $K_n > 0$ *or*

(ii) $K_n \geq 0$ *and* $\text{rank}\, K_{n-1} = \text{rank}\, K_n$

holds, then the Hamburger moment problem $\widehat{\sigma}(k) = S_k$, $k = 0, \ldots, 2n$, admits solutions. When $S_0, \ldots, S_{2n}$ are real numbers and the moment problem admits a solution, then either (i) or (ii) must hold.

Proof. First we assume that (ii) holds. As $K_n \geq 0$ we have that (2.7.12) is satisfied. Thus Theorem 2.7.9 yields the existence of matrices C_0 and a real diagonal D such that $S_j = C_0D^jC_0^*$, $j = 0, \ldots, 2n - 1$. By the same reasoning as in the last paragraph of the proof of Theorem 2.7.9, it follows that the Schur complement of K_n supported in the last row and column equals $S_{2n} - C_0D^{2n}C_0^*$. As rank K_{n-1} = rank K_n we have that this Schur complement equals 0, which thus implies that $S_{2n} = C_0D^{2n}C_0^*$. Putting $S_j = C_0D^jC_0^*$, $j > 2n$, we have a solution to the Hamburger problem.

Next assume that (i) holds. Choose S_{2n+1} to be any $p \times p$ Hermitian matrix, and put

$$S_{2n+2} = \begin{pmatrix} S_{n+1} & \cdots & S_{2n+1} \end{pmatrix} K_n^{-1} \begin{pmatrix} S_{n+1} \\ \vdots \\ S_{2n+1} \end{pmatrix}.$$

Now $K_{n+1} = (S_{i+j})_{i,j=0}^{n+1} \geq 0$ and rank $K_n =$ rank K_{n+1}. We can now apply part (ii) (with n replaced by $n+1$) to obtain the existence of a solution to the Hamburger moment problem.

Now let S_j, $j = 0, \ldots, 2n$, be real scalars, and suppose that there is a solution $(S_j)_{j=0}^{\infty}$ to the Hamburger problem. If $K_n > 0$ we are done, so let us assume that $\text{Ker}K_n$ is nontrivial and take $0 \neq (v_j)_{j=0}^{n} \in \text{Ker}K_n$. Let $k = \max\{j : v_j \neq 0\}$. If $v_n \neq 0$ then $K_n v = 0$ implies that the last column of K_n is a linear combination of the other columns and $\text{rank}K_n = \text{rank}K_{n-1}$ follows easily (as by the symmetry the last row is also a linear combination of the others). If $k < n$, consider $K_{2n-k} = (S_{i+j})_{i,j=0}^{2n-k}$. As $K_n v = 0$ and K_n is a principal submatrix of the positive semidefinite matrix K_{2n-k}, it follows that $K_{2n-k} \left(\begin{smallmatrix} v \\ 0 \end{smallmatrix} \right) = 0$ (use, for instance, Lemma 2.4.4). This implies that the kth column of K_{2n-k} is a linear combination of the columns $0, \ldots, k-1$ of K_{2n-k}; in fact, we have

$$\begin{pmatrix} S_k \\ \vdots \\ S_{2n} \end{pmatrix} = -\frac{v_0}{v_k} \begin{pmatrix} S_0 \\ \vdots \\ S_{2n-k} \end{pmatrix} - \cdots - \frac{v_{k-1}}{v_k} \begin{pmatrix} S_{k-1} \\ \vdots \\ S_{2n-1} \end{pmatrix}.$$

Looking at the bottom $n+1$ rows in this equality, we see that

$$\begin{pmatrix} S_n \\ \vdots \\ S_{2n} \end{pmatrix} = -\frac{v_0}{v_k} \begin{pmatrix} S_{n-k} \\ \vdots \\ S_{2n-k} \end{pmatrix} - \cdots - \frac{v_{k-1}}{v_k} \begin{pmatrix} S_{n-1} \\ \vdots \\ S_{2n-1} \end{pmatrix},$$

which gives that the last column of K_n is a linear combination of the other columns in K_n, and thus $\text{rank}K_{n-1} = \text{rank}K_n$ follows. □

We next consider the factorization of positive semidefinite matrix-valued polynomials in a real variable. Our main result is the following so-called spectral factorization on the real line.

Theorem 2.7.13 *Let $P_0, \ldots, P_{2n} \in \mathcal{H}_p$ be Hermitian $p \times p$ matrices with $P_{2n} > 0$ such that*

$$P(x) := P_0 + P_1 x + \cdots + P_{2n-1} x^{2n-1} + P_{2n} x^{2n} \geq 0 \quad \text{for all } x \in \mathbb{R}.$$

Then there exists a $p \times p$ matrix polynomial $G(x) = G_0 + \cdots + G_n x^n$ such that

$$P(x) = G(x)^* G(x), \quad x \in \mathbb{R}. \tag{2.7.21}$$

One may choose $G(x)$ so that $\det G(x)$ *has all its roots in the closed lower (upper) half-plane. Such a factorization may be found as follows. Let $\mathcal{F}$ be the convex set*

$$\mathcal{F} = \Big\{ F = (F_{ij})_{i,j=0}^{n} \in \mathbb{C}^{(n+1)p \times (n+1)p} : F \geq 0 \text{ and}$$

$$\sum_{\max\{0,k-n\} \leq i \leq \min\{n,k\}} F_{i,k-i} = P_k, k = 0, \ldots, 2n \Big\},$$

which is nonempty if and only if $P(x) \geq 0$, $x \in \mathbb{R}$. There is a unique $F \in \mathcal{F}$, F_{opt} say, such that $\text{Im}F_{n,n-1}(= \frac{1}{2i}(F_{n,n-1} - F_{n-1,n}))$ *is maximal (minimal) with respect to the Loewner ordering. Factorizing F_{opt} as*

$$F_{\text{opt}} = \begin{pmatrix} G_0^* \\ \vdots \\ G_n^* \end{pmatrix} \begin{pmatrix} G_0 & \cdots & G_n \end{pmatrix}, \tag{2.7.22}$$

each G_i being a $p \times p$ matrix (which is possible), one obtains the factorization (2.7.21) *with* $\det G(x)$ *having all its roots in the closed upper (lower) half-plane.*

Remark 2.7.14 For the maximization of $\text{Im}F_{n,n-1}$ it should be noted that some condition on P_{2n} is required. Indeed, if for instance $P_{2n} = 0$, then automatically $P_{2n-1} = 0$, and any $F \in \mathcal{F}$ will have $F_{ij} = 0$ when $i = n$ or $j = n$. So in this case one cannot recognize F_{opt} from looking at $\text{Im}F_{n,n-1}$.

Remark 2.7.15 In the scalar-valued case, the polynomial G is chosen so that all its roots are in the closed lower half-plane. This corresponds exactly to minimizing the imaginary part of the sum of the roots, which is exactly what maximizing $\text{Im}F_{n,n-1}$ does.

Remark 2.7.16 In Exercise 2.9.36 we point out how the spectral factorization is done in the operator-valued case. In that case, the maximization is done differently from the maximization of $\text{Im}F_{n,n-1}$.

The proof of Theorem 2.7.13 relies on the following result on algebraic Riccati equations. An *algebraic Riccati equation* is a quadratic matrix equation of the form

$$XDX - XA - A^*X - C = 0, \tag{2.7.23}$$

where $D = D^*$, $C = C^*$, and A are given, and a Hermitian solution X is to be found. Notice that when $D = 0$ one obtains a *Lyapunov equation*, while when $A = 0$ one obtains a *Stein equation*. For the proof of the factorization problem we will be in a situation where we know that a matrix X exists so that

$$\mathcal{T}(X) := XDX - XA - A^*X - C \leq 0. \tag{2.7.24}$$

The following result shows that under appropriate conditions on A, C, and D, inequality (2.7.24) implies the existence of a solution X of the Riccati equation (2.7.23) which has the additional property that $A - DX$ has all its eigenvalues in the closed left half-plane. We call a square matrix A *stable* if all its eigenvalues lie in the open left half-plane; in other words, $\lambda \in \sigma(A)$ implies that $\mathrm{Re}\lambda < 0$.

Theorem 2.7.17 *Let $A, D \geq 0, C^* = C$ be such that there exists a Hermitian X_0 so that $A - DX_0$ is stable. Assume that there exists a Hermitian solution of the inequality $\mathcal{T}(X) \leq 0$. Then there exists a (necessarily unique) Hermitian solution X_+ of $\mathcal{T}(X) = 0$ such that $X_+ \geq X$ for every Hermitian solution X of $\mathcal{T}(X) \leq 0$. In particular, X_+ is the maximal Hermitian solution of $\mathcal{T}(X) = 0$. Moreover, all the eigenvalues of $A - DX_+$ are in the closed left half-plane.*

We will use the following simple observation.

Lemma 2.7.18 *Let A be stable and $X = X^*$ so that $XA + A^*X \leq 0$. Then $X \geq 0$.*

Proof. Let S be invertible so that $A = SJS^{-1}$ with $J + J^* < 0$. This can always be done, for example, by taking J to be a direct sum of Jordan blocks of the type

$$\begin{pmatrix} \lambda & \epsilon & & \\ & \ddots & \ddots & \\ & & \ddots & \epsilon \\ & & & \lambda \end{pmatrix},$$

with ϵ small enough ($< \min\{|\mathrm{Re}\lambda| : \lambda \in \sigma(A)\}$.) Then

$$0 \geq S^*(XSJS^{-1} + S^{*-1}J^*S^*X)S = (S^*XS)J + J^*(S^*XS).$$

Suppose that $S^*XSv = \lambda v$ for some $v \neq 0$. Since S^*XS is Hermitian, we have $\lambda \in \mathbb{R}$. Then

$$0 \geq v^*((S^*XS)J + J^*(S^*XS))v = \lambda v^*(J + J^*)v.$$

As $v^*(J + J^*)v < 0$, we get that $\lambda \geq 0$. Thus $S^*XS \geq 0$, so $X \geq 0$. □

Proof of Theorem 2.7.17. Let $X = X^*$ be so that $\mathcal{T}(X) \leq 0$. Then

$$XDX - XA - A^*X - C' = 0, \tag{2.7.25}$$

where C' is a Hermitian matrix such that $C' \leq C$.

Let $X_0 = X_0^*$ be such that $A - DX_0$ is stable. Starting with X_0, we shall define a nonincreasing sequence of Hermitian matrices $\{X_\nu\}_{\nu=0}^\infty$ satisfying $X_\nu \geq X$ as well as the equalities

$$X_{\nu+1}(A - DX_\nu) + (A - DX_\nu)^*X_{\nu+1} = -X_\nu DX_\nu - C, \quad \nu = 0, 1, \ldots. \tag{2.7.26}$$

The sequence $\{X_\nu\}_{\nu=0}^\infty$ will also have the property that $A - DX_\nu$ is stable for all ν. We know that $A - DX_0$ is stable and, assuming inductively that we have already defined $X_\nu = X_\nu^*$ with $A - DX_\nu$ stable, it follows from $\sigma(A - DX_\nu) \cap \sigma(-(A - DX_\nu)^*) = \emptyset$ that (2.7.26) has a unique solution $X_{\nu+1}$ which is necessarily Hermitian.

We are now ready to show that $A - DX_{\nu+1}$ is stable. To this end note the following identity, which holds for any Hermitian matrices Y and $\widehat{Y}$:

$$Y(A - D\widehat{Y}) + (A - D\widehat{Y})^*Y + \widehat{Y}D\widehat{Y} - (Y - \widehat{Y})D(Y - \widehat{Y})$$

$$= Y(A - DY) + (A - DY)^*Y + YDY. \tag{2.7.27}$$

By assumption there exists a Hermitian solution X of (2.7.25). Letting $Y = X$ and $\widehat{Y} = X_\nu$ in (2.7.27), we get

$$X(A - DX_\nu) + (A - DX_\nu)^*X + X_\nu DX_\nu - (X - X_\nu)D(X - X_\nu) = -C'. \tag{2.7.28}$$

Subtract (2.7.28) from (2.7.26):

$$(X_{\nu+1} - X)(A - DX_\nu) + (A - DX_\nu)^*(X_{\nu+1} - X)$$

$$= -(X - X_\nu)D(X - X_\nu) - (C - C').$$

As the right-hand side in this equation is negative semidefinite and $A - DX_\nu$ is stable, it follows from Lemma 2.7.18 that $X_{\nu+1} \geq X$.

Next, use (2.7.27) again with $Y = X_{\nu+1}, \widehat{Y} = X_\nu$ and apply (2.7.26) to get

$$X_{\nu+1}(A - DX_{\nu+1}) + (A - DX_{\nu+1})^*X_{\nu+1} + X_{\nu+1}DX_{\nu+1}$$

$$= -C - (X_{\nu+1} - X_\nu)D(X_{\nu+1} - X_\nu). \tag{2.7.29}$$

Subtracting (2.7.28) with X_ν replaced by $X_{\nu+1}$, we obtain

$$(X_{\nu+1} - X)(A - DX_{\nu+1}) + (A - DX_{\nu+1})^*(X_{\nu+1} - X)$$

$$= -(X_{\nu+1} - X_\nu)D(X_{\nu+1} - X_\nu) - (X_{\nu+1} - X)D(X_{\nu+1} - X) - C + C' \tag{2.7.30}$$

Assume that $(A - DX_{\nu+1})x = \lambda x$ for some λ with $\mathrm{Re}\lambda \geq 0$ and some $x \neq 0$. Then

$$(\overline{\lambda} + \lambda)x^*(X_{\nu+1} - X)x = x^*Wx, \tag{2.7.31}$$

where $W \leq 0$ is the right-hand side of (2.7.30). As $X_{\nu+1} - X \geq 0$, (2.7.31) implies $x^*Wx = 0$ which, using the definition of W, in turn implies

$$x^*(X_{\nu+1} - X_\nu)D(X_{\nu+1} - X_\nu)x = 0$$

But $D \geq 0$, so $D(X_{\nu+1} - X_\nu)x = 0$. Now

$$(A - DX_\nu)x = (A - DX_{\nu+1})x = \lambda x.$$

This contradicts the stability of $A - DX_\nu$. Hence $A - DX_{\nu+1}$ is stable as well.

Next we show that $\{X_\nu\}_{\nu=0}^\infty$ is nonincreasing. Consider (2.7.29) with ν replaced by $\nu - 1$, and subtract from it (2.7.26) to get

$$(X_\nu - X_{\nu+1})(A - DX_\nu) + (A - DX_\nu)^*(X_\nu - X_{\nu+1})$$
$$= -(X_\nu - X_{\nu-1})D(X_\nu - X_{\nu-1}). \qquad (2.7.32)$$

As $A - DX_\nu$ is stable, $X_\nu - X_{\nu+1}$ is uniquely determined by (2.7.32), and by Lemma 2.7.18, it is positive semidefinite. So $\{X_\nu\}_{\nu=0}^\infty$ is a nonincreasing sequence of Hermitian matrices bounded below by X. Hence the limit $X_+ = \lim_{\nu\to\infty} X_\nu$ exists. Passing to the limit in (2.7.26) when $\nu \to \infty$ shows that X_+ is a Hermitian solution of $\mathcal{T}(X) = 0$. Since $A - DX_\nu$ is stable for all $\nu = 0, 1, \ldots$, the matrix $A - DX_+$ has all its eigenvalues in the closed left half-plane. Also $X_+ \geq X$ for every Hermitian solution of (2.7.24). □

We now prove the factorization result.

Proof of Theorem 2.7.13. We first show that $\mathcal{F} \neq \emptyset$ if and only if $P(x) \geq 0$, $x \in \mathbb{R}$.

First, suppose that $F \in \mathcal{F} \neq \emptyset$, and let $v \in \mathbb{C}^p$ and $x \in \mathbb{R}$. Put $g := \mathrm{col}(x^j v)_{j=0}^n$. Then $0 \leq g^* F g = v^* P(x) v$. As v and x we chosen arbitrarily, it follows that $P(x) \geq 0$, $x \in \mathbb{R}$.

Next, suppose that $\mathcal{F} = \emptyset$. Introduce the tridiagonal block matrix

$$F_0 = \begin{pmatrix} P_0 & \frac{1}{2}P_1 & & \\ \frac{1}{2}P_1 & P_2 & \ddots & \\ & \ddots & \ddots & \frac{1}{2}P_{2n-1} \\ & & \frac{1}{2}P_{2n-1} & P_{2n} \end{pmatrix} \in \mathbb{C}^{(n+1)p\times(n+1)p}, \qquad (2.7.33)$$

and let

$$\mathcal{L} = \Big\{ L = (L_{ij})_{i,j=0}^n \in \mathbb{C}^{(n+1)p\times(n+1)p} :$$
$$\sum_{\max\{0,k-n\}\leq i\leq\min\{n,k\}} L_{i,k-i} = 0, k = 0, \ldots, 2n \Big\}.$$

Then $\mathcal{F} = \emptyset$ is equivalent to the statement that $(F_0 + \mathcal{L}) \cap \mathrm{PSD}_{(n+1)p} = \emptyset$. By the convexity of the two subsets of $\mathcal{H}_{(n+1)p}$ involved, we obtain by the Hahn-Banach separation theorem that there exists a Hermitian matrix W such that

$$\langle F_0 + L, W\rangle < 0 \text{ for all } L \in \mathcal{L}, \text{ and } \langle M, W\rangle \geq 0 \text{ for all } M \in \mathrm{PSD}_{(n+1)p}.$$

The latter condition means that $W \geq 0$. As $\mathcal{L}$ is a linear subspace, the former condition can only be satisfied when $\langle F_0, W\rangle < 0$ and $\langle L, W\rangle = 0$ for all $L \in \mathcal{L}$. The condition $W \in \mathcal{L}^\perp$ is easily to be seen to be equivalent to W being a block Hankel matrix $(S_{i+j})_{i,j=0}^n$. By Theorem 2.7.9 it follows that $S_j = C_0 D C_0^*$, $j = 0, \ldots, 2n-1$, and $S_{2n} \geq C_0 D^{2n} C_0^*$, for some real diagonal matrix $D = \mathrm{diag}(\lambda_i)_{i=1}^k$ and some matrix $C_0 = \begin{pmatrix} c_1 & \cdots & c_k \end{pmatrix}$. But then

$$0 > \langle F_0, W\rangle = \sum_{i=1}^k c_i^* P(\lambda_i) c_i + \mathrm{tr}(S_{2n} - C_0 D^{2n} C_0^*) P_{2n}.$$

As the latter term is nonnegative, we get that $\sum_{i=1}^{k} c_i^* P(\lambda_i) c_i < 0$, and thus $P(x) \not\geq 0$ for some $x \in \mathbb{R}$.

In the remainder of the proof, for notational convenience and without loss of generality we will assume that $P_{2n} = I$ (otherwise, factorize $P_{2n}^{-\frac{1}{2}} P(x) P_{2n}^{-\frac{1}{2}} = G(x)^* G(x)$, then $P(x) = P_{2n}^{\frac{1}{2}} G(x)^* G(x) P_{2n}^{\frac{1}{2}}$). Let

$$Y = \begin{pmatrix} Y_{1,1} & \cdots Y_{1,n} \\ \vdots & \vdots \\ Y_{n,1} & \cdots Y_{n,n} \end{pmatrix} \in \mathbb{C}^{np \times np}$$

be a Hermitian matrix. Notice that any matrix

$$F = F_0 + \begin{pmatrix} 0_{pn \times p} & iY \\ 0_{p \times p} & 0_{p \times np} \end{pmatrix} + \begin{pmatrix} 0_{p \times np} & 0_{p \times p} \\ -iY & 0_{np \times p} \end{pmatrix} \geq 0$$

satisfies the constraints that $\sum_{l=0}^{k} F_{k-l,l} = P_k, k = 0, \ldots, 2n$, and that any matrix $F \geq 0$ satisfying them may be written in this form. (In the inequality above we emphasized the sizes of the zeros in the matrices as they are unusual. In the remainder of the proof we will not do this, but the reader should be aware that the diagonal entries in these block matrices are not necessarily square.) In other words, any matrix F as above belongs to $\mathcal{F}$, as defined in Theorem 2.7.13. So the optimization problem may be rewritten as

$$\begin{aligned} Y_{\text{opt}} = \arg\max \quad & \text{tr} Y_{nn} \\ \text{such that} \quad & \widetilde{F_0} + \begin{pmatrix} 0 & iY \\ 0 & 0 \end{pmatrix} + \begin{pmatrix} 0 & 0 \\ -iY & 0 \end{pmatrix} \geq 0. \end{aligned} \tag{2.7.34}$$

Let

$$A = i \begin{pmatrix} 0 & I_p & & \\ & 0 & \ddots & \\ & & \ddots & I_p \\ & & & 0 \end{pmatrix} \in \mathbb{C}^{np \times np}, \quad B = i \begin{pmatrix} 0 \\ \vdots \\ 0 \\ I_p \end{pmatrix} \in \mathbb{C}^{np \times p}.$$

Then

$$\begin{pmatrix} 0 & iY \\ 0 & 0 \end{pmatrix} = \begin{pmatrix} YA & YB \\ 0 & 0 \end{pmatrix}.$$

Further, let us divide F_0 into four blocks

$$F_0 = \begin{pmatrix} H_{11} & H_{12} \\ H_{21} & H_{22} \end{pmatrix}$$

with $H_{11} \in \mathcal{H}_{np}$ and $H_{22} = P_{2n} = I_p$. Then the constraint of (2.7.34) can be recast as

$$\begin{pmatrix} H_{11} & H_{12} \\ H_{21} & H_{22} \end{pmatrix} + \begin{pmatrix} YA + A^*Y & YB \\ B^*Y & 0 \end{pmatrix} \geq 0. \tag{2.7.35}$$

As $\mathcal{F} \neq \emptyset$ we have that (2.7.35) holds for some $Y = Y^*$. By taking the Schur complement supported in the first n block rows and columns, we get that the algebraic Riccati inequality

$$A^*Y + YA - (YB + H_{12})(B^*Y + H_{21}) + H_{11} \geq 0 \tag{2.7.36}$$

holds for the above choice of $Y = Y^*$. Note that $A - BH_{21} - BB^*X$ equals i times the companion matrix

$$\begin{pmatrix} 0 & I_p & & \\ & \ddots & \ddots & \\ & & 0 & I_p \\ -L_0 & \cdots & -L_{n-2} & -L_{n-1} \end{pmatrix},$$

where the coefficients of the corresponding polynomial $Iz^n + \sum_{j=0}^{n-1} L_j z^j$ can be chosen arbitrarily. When we restrict to Hermitian X the freedom we lose is only in the coefficient L_{n-1}. Therefore, we can always choose a Hermitian X_0 so that $A - BH_{21} - BB^*X_0$ is stable. By Theorem 2.7.17 we obtain that the algebraic Riccati equation

$$A^*Y + YA - (YB + H_{12})(B^*Y + H_{21}) + H_{11} = 0 \tag{2.7.37}$$

has a unique solution Y_+ such that $A - B(H_{21} + B^*Y_+)$ has all its eigenvalues in the closed left half-plane. In addition, if Y satisfies (2.7.36) then $Y \leq Y_+$.

We claim that $Y_{\text{opt}} = Y_+$. First of all, Y_{opt} satisfies (2.7.36), so $Y_{\text{opt}} \leq Y_+$. Next, we must have that

$$\text{tr}[(Y_{\text{opt}})_{nn}] = \text{tr}[(Y_+)_{nn}]. \tag{2.7.38}$$

Otherwise, we would have $\text{tr}[(Y_{\text{opt}})_{nn}] < \text{tr}[(Y_+)_{nn}]$, but this contradicts the optimality of Y_{opt} (Y_+ would yield a better objective function). Notice now that $(Y_{\text{opt}})_{nn} \leq (Y_+)_{nn}$ and (2.7.38) imply $(Y_{\text{opt}})_{nn} = (Y_+)_{nn}$. Thus

$$Y_+ - Y_{\text{opt}} = \begin{pmatrix} \Delta & 0 \\ 0 & 0 \end{pmatrix} =: E \geq 0, \tag{2.7.39}$$

where $\Delta \in \mathbb{C}^{(n-1)m \times (n-1)m}$. It remains to show that $\Delta = 0$. Since Y_+ satisfies (2.7.37), we get

$$A^*E + EA + Y_{\text{opt}}BB^*E + EBB^*Y_{\text{opt}} + H_{12}B^*E + EBH_{21} + EBB^*E \geq 0.$$

Due to the structure of E (see (2.7.39)) and the definition of B, $EB = 0$ and $B^*E = 0$. Thus we get that $EA + A^*E \geq 0$ (compare also with (2.4.41)). Since the $(0,0)$-entry of this inequality equals 0, the first row and first column must be equal to 0. So $E_{01} = \cdots = E_{0,n-1} = E_{n-1,0} = \cdots = E_{10} = 0$. But then its $(2,2)$-entry is also equal to 0, yielding by recursion that $E = 0$.

Further, the optimal matrix F_{opt} has a low rank:

$$\begin{aligned} \text{rank}\, F_{\text{opt}} &= \text{rank}\left(\begin{pmatrix} H_{11} & H_{12} \\ H_{21} & H_{22} \end{pmatrix} + \begin{pmatrix} Y_{\text{opt}}A + A^*Y_{\text{opt}} & Y_{\text{opt}}B \\ B^*Y_{\text{opt}} & 0 \end{pmatrix} \right) \\ &= \text{rank}\, H_{22} = \text{rank}\, P_{2n} = p \end{aligned} \tag{2.7.40}$$

due to the fact that the Schur complement supported in the first n rows and columns equals 0; see (2.7.37).

If we factorize F_{opt} as

$$F_{\text{opt}} = \begin{pmatrix} G_0^* \\ \vdots \\ G_n^* \end{pmatrix} \begin{pmatrix} G_0 & \cdots & G_n \end{pmatrix}$$

with $G_n = I_p$, the matrix $A_+ = A - B(H_{21} + B^*Y_+)$ equals

$$A_+ = A - B\begin{pmatrix} G_0 & \cdots & G_{n-1} \end{pmatrix},$$

which is exactly the companion matrix C_G in (2.3.6) multiplied by i. Since for all eigenvalues λ of A_+ we have that $\mathrm{Re}\lambda \leq 0$, we get that $-iA_+$ has all eigenvalues in the closed upper half-plane. But then Lemma 2.3.3 yields that the polynomial

$$\det(x^n + G_{n-1}x^{n-1} + \cdots + G_1 x + G_0) = \det G(x)$$

has all its roots in $\{z : \mathrm{Im}\, z \geq 0\}$. □

2.8 LINEAR PREDICTION

A stochastic process is a sequence $\{X_k\}_{k\in\mathbb{Z}}$ of random variables together with a family of probability distributions specifying probabilities for the joint occurrence of values for finite subsets $\{X_k, k \in L\}$, $L \subset \mathbb{Z}$ finite. The expectations $E(X_i X_j^*)$ are called *autocorrelation coefficients* or *covariances.* The process is termed *stationary in the wide sense* if for all i, j the quantity $E(X_i X_j^*)$ depends on $i - j$ only. In this case we denote

$$c_{i-j} := E(X_i X_j^*), i, j \in \mathbb{Z}.$$

Notice that $c_{-k}^* = c_k$, $k \in \mathbb{Z}$. When correlation coefficients $c_k, k \in L$, are given we can use them to try to estimate the value of X_j, based on observed values $X_i, i \in \widetilde{L}$. We will treat the case of *forward linear prediction*, the case when we predict the value of X_j based on a linear combination of observed values for $X_{j-1}, X_{j-2}, \ldots, X_{j-r}$. To find the "best" prediction scheme, we want to choose the linear combination so that the average square of the error is minimized. In other words, we seek $a_1, \ldots, a_r$ so that the quantity

$$E\left[\left(X_j + \sum_{k=1}^{r} a_k X_{j-k}\right)\left(X_j + \sum_{k=1}^{r} a_k X_{j-k}\right)^*\right] \tag{2.8.1}$$

is minimal. If we introduce the inner product,

$$\langle X_j, X_i \rangle = E(X_j X_i^*),$$

the minimum of (2.8.1) is attained at the linear combination of $X_{j-1}, \ldots,$ X_{j-r} that is closest to X_j in norm. Thus the best choice is characterized by having $X_j + \sum_{k=1}^{r} a_k X_{j-k}$ orthogonal to the subspace spanned by $X_{j-1}, \ldots, X_{j-r}$. In other words,

$$E\left[\left(X_j + \sum_{k=1}^{r} a_k X_{j-k}\right) X_j^*\right] = 0,\ j = 1, \ldots, r.$$

This leads to the equation

$$\begin{pmatrix} c_0 & \cdots & c_{-r} \\ \vdots & \ddots & \vdots \\ c_r & \cdots & c_0 \end{pmatrix}\begin{pmatrix} 1 \\ a_1 \\ \vdots \\ a_r \end{pmatrix} = \begin{pmatrix} \gamma \\ 0 \\ \vdots \\ 0 \end{pmatrix}.$$

It should be noticed that this equation corresponds exactly to (2.3.2). The minimal prediction error (2.8.1) is exactly the quantity γ.

2.9 EXERCISES

1 Let $A \in \mathcal{L}(\mathcal{H}), B \in \mathcal{L}(\mathcal{K}, \mathcal{H}), C \in \mathcal{L}(\mathcal{K})$, be Hilbert space operators, and assume that A is positive definite.

(a) Show that for $\lambda \in \mathbb{R}$ large enough, we have that

$$\begin{pmatrix} A & B \\ B^* & C + \lambda I \end{pmatrix}$$

is positive definite.

(b) Use part (a) to argue that in the positive definite completion problem the principal submatrix for which all diagonal entries are fully specified determines the existence of a positive definite completion.

2 Let $T = \left(\begin{smallmatrix} M & N \\ P & Q \end{smallmatrix}\right)$ be an operator matrix. Prove that when M is invertible, then T is invertible if and only if $Q - PM^{-1}N$ is.

3 Prove that if X and Y are matrices such that $0 < X \leq Y$, then $\det X \leq \det Y$. Moreover, $\det X = \det Y$ if and only if $X = Y$.

4 Let $K = (A_{i-j})_{i,j=1}^n$ be a partial operator Toeplitz matrix, with A_j specified for $|j| \leq p < n$. Assuming that $(A_{i-j})_{i,j=1}^p > 0$, state and prove an analogue of Theorem 2.1.2 for the parametrization of all positive definite Toeplitz completions of K. (Hint: see the remark after the proof of Theorem 2.1.2.)

5 Let

$$K = \begin{pmatrix} a & B & ? \\ B^* & C & D \\ ? & D^* & e \end{pmatrix}$$

be a partially positive definite scalar matrix.

(a) Show that the maximum of the determinant over all positive definite completions of K equals

$$\frac{\det \begin{pmatrix} a & B \\ B^* & C \end{pmatrix} \det \begin{pmatrix} C & D \\ D^* & e \end{pmatrix}}{\det C}.$$

(b) Use (a) to derive that for every $n \times n$ positive definite matrix A and every $j \leq k+1$ we have

$$\det A \leq \frac{\det(A|\{1, \ldots, k\}) \det(A|\{j, \ldots, n\})}{\det(A|\{j, \ldots, k\})}$$

(the so-called *Fischer-Hadamard inequality*). In particular,

$$\det A \leq A_{11}A_{22} \cdots A_{nn},$$

where $A = (A_{ij})_{i,j=1}^n$.

6 Let $\mathcal{A}$ be a C*-algebra with a unit e. A positive definite element of $\mathcal{A}$ is an element of the form a^*a with a invertible in $\mathcal{A}$, while a positive semidefinite element is an element of the form a^*a, with $a \in \mathcal{A}$.

(a) Show that a partially positive definite matrix of the form

$$\begin{pmatrix} * & * & ? \\ * & * & * \\ ? & * & * \end{pmatrix} \tag{2.9.1}$$

with given entries in $\mathcal{A}$ has a positive definite completion in $\mathcal{A}^{3\times 3}$.

(b) Show that the statement in (a) is no longer true when "positive definite" is replaced by "positive semidefinite." (Hint: let $\mathcal{A}$ be the C*-algebra consisting of continuous functions on $\overline{\mathbb{D}}$, and compose a partially positive semidefinite matrix using the functions $1, z, \overline{z}, |z|, |z|^2$.)

7 Prove Proposition 2.2.6. (Hint: For $m \geq 4$, consider the partial matrix

$$K_m = \begin{pmatrix} 1 & 1-\epsilon & 0 & 0 & \cdots & -1+\epsilon \\ 1-\epsilon & 1 & 1-\epsilon & 0 & \cdots & 0 \\ 0 & 1-\epsilon & 1 & 1-\epsilon & \cdots & 0 \\ \vdots & \vdots & \vdots & \vdots & \vdots & \vdots \\ -1+\epsilon & 0 & 0 & 0 & \cdots & 1 \end{pmatrix}.$$

Show that for $0 < \epsilon$ sufficiently small, K_m is partially positive definite but does not admit a positive definite completion. Proceed then as in the proof of Proposition 1.2.9.)

8 Consider a partially positive semidefinite matrix A with underlying graph G.

(a) Show that if G is an n-cycle with $n \geq 4$, then there always exists a completion of A with at most one negative eigenvalue.

(b) Show that for any $n \in \mathbb{N}$ there exists a graph G such that for any choice of a partially positive semidefinite matrix A with underlying graph G, there exists a completion with at most n negative eigenvalues, and such that for some partially positive semidefinite matrix A with underlying graph G all completions have at least n negative eigenvalues.

9 Find the band extension of the partially positive semidefinite matrix

$$\begin{pmatrix} 1 & \frac{1}{2} & ? & ? \\ \frac{1}{2} & 1 & \frac{1}{2} & ? \\ ? & \frac{1}{2} & 1 & \frac{1}{2} \\ ? & ? & \frac{1}{2} & 1 \end{pmatrix}.$$

10 Consider the partial matrix

$$\begin{pmatrix} 1 & a & ? & d \\ \overline{a} & 1 & b & ? \\ ? & \overline{b} & 1 & c \\ \overline{d} & ? & \overline{c} & 1 \end{pmatrix}$$

with $|a|, |b|, |c|, |d| \leq 1$.

(a) Find necessary and sufficient conditions on a, b, c, and d for the matrix to admit a positive semidefinite completion.

(b) Use part (a) to show that the partial matrix

$$\begin{pmatrix} 1 & a & ? & b \\ \overline{a} & 1 & b & ? \\ ? & \overline{b} & 1 & \overline{a} \\ \overline{b} & ? & \overline{a} & 1 \end{pmatrix}$$

with $|a|, |b| \leq 1$ always admits a positive semidefinite completion.

(c) Let $n \geq 4$, $0 \leq \theta_1, \ldots, \theta_n \leq \pi$, and assume that at most one of $\theta_1, \ldots, \theta_n$ is greater than $\frac{\pi}{2}$. Then

$$\begin{pmatrix} 1 & \cos\theta_1 & ? & & \cos\theta_n \\ \cos\theta_1 & 1 & \cos\theta_2 & ? & \\ ? & \cos\theta_2 & 1 & \ddots & ? \\ & ? & \ddots & \ddots & \cos\theta_{n-1} \\ \cos\theta_n & & ? & \cos\theta_{n-1} & 1 \end{pmatrix}$$

has a positive semidefinite completion of and only if

$$2 \max_{k=1,\ldots,n} \theta_k \leq \sum_{k=1}^{n} \theta_k.$$

Using part (a), prove this statement for $n = 4$. To prove this for general n is quite challenging.

11 Consider the partially positive definite matrix

$$K = \begin{pmatrix} 1 & \frac{1}{2} & \frac{1}{2} & \frac{1}{2} \\ \frac{1}{2} & 1 & ? & ? \\ \frac{1}{2} & ? & 1 & ? \\ \frac{1}{2} & ? & ? & 1 \end{pmatrix}.$$

(a) Find two distinct admissible sequences for the graph of K.

(b) Find the unique positive definite completion of K which maximizes its determinant, and find the right regular G-factorization of its inverse.

12 Let a_{jk}, $|j-k| \le m$, $1 \le j,k \le n$, be given such that

$$(a_{rs})_{r,s=j}^{j+m} \ge 0, \ j = 1, \ldots, n-m.$$

Show that the problem of finding vectors $x_1, \ldots, x_n \in \mathbb{C}^n$ such that $\langle x_k, x_j \rangle = a_{jk}$, $|j-k| \le m$ is equivalent to a positive semidefinite completion problem. Moreover, show that finding vectors as above so that the parallelepiped determined by $x_1, \ldots, x_n$ has maximal volume corresponds to finding a maximal determinant positive definite completion.

13 Let A and B be positive semidefinite operators, and assume that $\lambda_{\max}(A) \le \lambda_{\max}(B)$. Let $\mathcal{M}$ denote the set of positive semidefinite completions of $\left(\begin{smallmatrix} A & ? \\ ? & B \end{smallmatrix}\right)$. Show that

$$\bigcup_{M \in \mathcal{M}} \sigma(M) = [0, \lambda_{\max} A + \lambda_{\max}(B)] \setminus (\lambda_{\max}(A), \lambda_{\min}(B)).$$

14 Let $\mu : \mathcal{B}(\mathbb{T}) \to \mathcal{L}(\mathcal{H})$ be an operator-valued positive measure. Prove that for every $j \in \mathbb{Z}$ there exists $\widehat{\mu}(j) \in \mathcal{L}(\mathcal{H})$ such that for every $h \in \mathcal{H}$, $\langle \widehat{\mu}(j)h, h \rangle$ is the jth moment of the measure μ_h, as defined in Section 2.3. (Hint: show first that given any $f : \mathcal{H} \times \mathcal{H} \to \mathbb{C}$ such that $|f(h,k)| \le M\|h\|\|k\|$, and f is linear in h and antilinear in k, then there exists $T \in \mathcal{L}(\mathcal{H})$ such that $f(h,k) = \langle Th, k \rangle$. Define $\mu_{h,k}(\sigma) = \langle \mu(\sigma)h, k \rangle$ and apply the above to $f(h,k) = \widehat{\mu}_{h,k}(j)$. The set $(\mu_{h,k})_{h,k \in \mathcal{H}}$ is called the *spectral family* associated with μ.)

15 Let $\mu : \mathcal{B}(\mathbb{T}) \to \mathcal{L}(\mathcal{H})$ be an operator-valued positive measure that is absolutely continuous with respect to the Lebesgue measure. Prove that there exists a weakly measurable function $F : \mathbb{T} \to \mathcal{L}(\mathcal{H})$ with $F(e^{it}) \ge 0$ a.e. on $\mathbb{T}$, such that for each $h \in \mathcal{H}$, $d\langle \mu(t)h, h \rangle = \langle F(e^{it})h, h \rangle dt$. Moreover, $F(e^{it}) = \sum_{j=-\infty}^{\infty} \widehat{\mu}(j) e^{ijt}$, with the series converging weakly in the $L^1(\mathbb{T})$ norm. Here $\widehat{\mu}(j)$ is defined in Exercise 2.9.14; we also write $\widehat{F}(j) = \widehat{\mu}(j)$ in this case.

16 Consider the set-up of Lemma 2.3.3.

(a) Show that $zI - C_G$ is invertible if and only if $G(z)$ is invertible.

(b) Show that an eigenvector of C_G with eigenvalue z must be of the form

$$\begin{pmatrix} v \\ zv \\ \vdots \\ z^{n-1}v \end{pmatrix},$$

where $0 \ne v \in \mathcal{H}$ is so that $G(z)v = 0$.

17 Prove that the operator-valued polynomial $Y(z)$ defined in the statement of Theorem 2.3.1 is invertible for all $z \in \mathbb{C}_\infty \setminus \mathbb{D}$.

18 Apply Theorem 2.3.1 to the Toeplitz matrix

$$\begin{pmatrix} 4 & 2 & 1 \\ 2 & 4 & 2 \\ 1 & 2 & 4 \end{pmatrix}.$$

(Answer: $F(z) = \frac{20}{|z-2|^2}$.)

19 Let $f(z) = \frac{f_0}{2} + \sum_{i=1}^{\infty} f_i z_i \in H^\infty (= H^\infty(\mathcal{L}(\mathbb{C})))$ have a positive definite real part. Put

$$T_n = \begin{pmatrix} f_0 & \cdots & \overline{f_n} \\ \vdots & \ddots & \vdots \\ f_n & \cdots & f_0 \end{pmatrix}.$$

Show that

$$\lim_{n\to\infty} \frac{\det T_n}{\det T_{n-1}} = e^{\mathcal{E}(2\mathrm{Re} f)},$$

where $\mathcal{E}$ is defined in (1.4.2).

20 In Theorem 2.3.1 the operators C_j, $|j| \le n$, are fully specified. One can also consider the case when these operators are only partially defined. As an example, show the following.

Let $C_j = (C_{pq}^{(j)})_{p,q=1}^3$, $j = -2, \dots, 2$, and suppose that

$$C_{11}^{(0)} = C_{11}^{(0)*},\ C_{22}^{(0)} = C_{22}^{(0)*},\ C_{31}^{(0)} = C_{13}^{(0)*},\ C_{23}^{(0)} = C_{32}^{(0)*},$$

$$C_{33}^{(0)} = C_{33}^{(0)*},\ C_{12}^{(1)} = C_{21}^{(-1)*},\ C_{13}^{(1)} = C_{31}^{(-1)*},\ C_{31}^{(1)} = C_{13}^{(-1)*},$$

$$C_{32}^{(1)} = C_{23}^{(-1)*},\ C_{33}^{(1)} = C_{33}^{(-1)*},\ C_{33}^{(2)} = C_{33}^{(-2)*}$$

be given. Show that $C_j = (C_{pq}^{(j)})_{p,q=1}^3$, $j = -2, \dots, 2$, exists such that $(C_{i-j})_{i,j=0}^2 > 0$ if and only if

$$\begin{pmatrix} C_{11}^{(0)} & C_{13}^{(0)} & C_{13}^{(-1)} \\ C_{31}^{(0)} & C_{33}^{(0)} & C_{33}^{(-1)} \\ C_{31}^{(1)} & C_{33}^{(1)} & C_{33}^{(0)} \end{pmatrix} > 0, \quad \begin{pmatrix} C_{22}^{(0)} & C_{23}^{(0)} & C_{21}^{(-1)} & C_{23}^{(-1)} \\ C_{32}^{(0)} & C_{33}^{(0)} & C_{31}^{(-1)} & C_{33}^{(-1)} \\ C_{12}^{(1)} & C_{13}^{(1)} & C_{11}^{(0)} & C_{13}^{(0)} \\ C_{32}^{(1)} & C_{33}^{(1)} & C_{31}^{(0)} & C_{33}^{(0)} \end{pmatrix} > 0,$$

$$\begin{pmatrix} C_{33}^{(0)} & C_{31}^{(-1)} & C_{33}^{(-1)} & C_{33}^{(-2)} \\ C_{13}^{(1)} & C_{11}^{(0)} & C_{13}^{(0)} & C_{13}^{(-1)} \\ C_{31}^{(1)} & C_{31}^{(0)} & C_{33}^{(0)} & C_{33}^{(-1)} \\ C_{33}^{(2)} & C_{31}^{(1)} & C_{33}^{(1)} & C_{33}^{(0)} \end{pmatrix} > 0.$$

21 Let P be positive definite such that $P - SPS^*$ is positive definite. Show that the spectrum of S lies in $\mathbb{D}$. (Hint: see the argument after (2.3.15).)

22 This is an exercise about a basic proof for Szegő's theorem. For this, let $T_n > 0$, $\langle \cdot, \cdot \rangle$ be the associated inner product and let p_i, $i = 1, \ldots, n$, be associated orthogonal polynomials (e.g., given by (2.3.29)). Let λ be a root of $p_n(z)$.

(a) For $r(z)$ of degree less than n, show that $\langle r(z), r(z) \rangle = \langle zr(z), zr(z) \rangle$ and $\langle p_n(z), r(z) \rangle = 0$.

(b) Let λ be a root of $p_n(z)$, and write $p_n(z) = (z - \lambda) r(z)$. Show that

$$\langle r(z), r(z) \rangle = \langle p_n(z), p_n(z) \rangle + |\lambda|^2 \langle r(z), r(z) \rangle. \tag{2.9.2}$$

(c) Conclude from (2.9.2) that $|\lambda| < 1$.

23 The following is a generalization of Szegő's result. *Let* $T_n = (c_{i-j})_{i,j=0}^n$ *be a Hermitian Toeplitz matrix so that* $\det T_i \neq 0$, $i = 0, \ldots, n$. *Let* $p_i(z)$ *be defined via* (2.3.29). *Then* $p_n(z)$ *has no roots on the unit circle, and the number of roots outside the unit circle is equal to the number of negative eigenvalues of* $(\det T_n)(\det T_{n-1}) T_n$.

(a) Prove the result for $n = 1$.

(b) Prove the result for $n = 2$.

(c) Prove the general result.

24 The *Schur-Cohn test* states the following: $p(z) = p_0 + \cdots + p_n z^n$ is stable (no roots in $\overline{\mathbb{D}}$) if and only if

$$\begin{pmatrix} p_0 & & \\ \vdots & \ddots & \\ p_{n-1} & \cdots & p_0 \end{pmatrix} \begin{pmatrix} \overline{p}_0 & \cdots & \overline{p}_{n-1} \\ & \ddots & \vdots \\ & & \overline{p}_0 \end{pmatrix} - \begin{pmatrix} \overline{p}_n & & \\ \vdots & \ddots & \\ \overline{p}_1 & \cdots & \overline{p}_n \end{pmatrix} \begin{pmatrix} p_n & \cdots & p_1 \\ & \ddots & \vdots \\ & & p_n \end{pmatrix}$$

is positive definite.

(a) Prove the "only if" part.

(b) Prove the "if" part. (Hint: first prove that the inverse of the above matrix is Toeplitz.)

25 Let $\mathcal{H}, \mathcal{K}_1$, and $\mathcal{K}_2$ be Hilbert spaces.

(a) Prove that $T = \binom{A}{X} : \mathcal{H} \to \mathcal{K}_1 \oplus \mathcal{K}_2$ is a contraction if and only if $\|A\| \leq 1$ and $X = \Gamma D_A$, where $\Gamma : \mathcal{D}_A \to \mathcal{K}_2$ is also a contraction.

(b) Prove that $T = \begin{pmatrix} A & D_{A^*}\Gamma \end{pmatrix}$ is an isometry if and only if both A and Γ are isometries and T is a co-isometry if and only if Γ is a co-isometry.

(c) Prove that $T = \begin{pmatrix} A \\ \Gamma D_A \end{pmatrix}$ is an isometry if and only if Γ is an isometry and T is a co-isometry if and only if both A and Γ are co-isometries.

(Hint: The condition $\|T\| \leq 1$ is equivalent to both $T^*T \leq I$ and $TT^* \leq I$; use then Lemma 2.4.2 or Corollary 2.4.3.)

26 Show that with T as in (2.4.9) we have that $S(T; \mathcal{M}^\perp) = C^{\frac{1}{2}}(I - G^*G)C^{\frac{1}{2}}$.

27 An alternative way to introduce the Schur complement is as follows: Let $T : \mathcal{H} \to \mathcal{H}$ be positive semidefinite and $\mathcal{M}$ a subspace of $\mathcal{H}$. Write T as in (2.4.9). Define now $S : \mathcal{M} \to \mathcal{M}$ as the unique operator satisfying

$$\langle Sf, f \rangle = \inf \left\{ \left\langle \begin{pmatrix} A & B \\ B^* & C \end{pmatrix} \begin{pmatrix} f \\ g \end{pmatrix}, \begin{pmatrix} f \\ g \end{pmatrix} \right\rangle \; : \; g \in \mathcal{M}^\perp \right\}.$$

Show that S is well defined and equals the Schur complement of T supported on $\mathcal{M}$.

28 Show that $S(A + B; \mathcal{M}) \geq S(A; \mathcal{M}) + S(B; \mathcal{M})$. In addition, give an example where equality does not hold.

29

(a) Let $T : \mathcal{H} \to \mathcal{H}$ be positive semidefinite, and $\mathcal{M}$ a subspace of $\mathcal{H}$. Write T as in (2.4.9). Suppose that C is invertible. Show that $S(T; \mathcal{M}) = A - BC^{-1}B^*$.

(b) Using the observation in (a) one can also define the Schur complement for general 2×2 block matrices (not necessarily positive semidefinite, or even selfadjoint) as long as an invertibility condition is satisfied. This leads to the following definition: Let

$$T = \begin{pmatrix} A & B \\ C & D \end{pmatrix} : \begin{matrix} \mathcal{H}_1 \\ \oplus \\ \mathcal{H}_2 \end{matrix} \to \begin{matrix} \mathcal{K}_1 \\ \oplus \\ \mathcal{K}_2 \end{matrix} \tag{2.9.3}$$

be an operator matrix, and suppose that D is invertible. Define the Schur complement of T supported on $\mathcal{H}_1$ as $S(T; \mathcal{H}_1) = A - BD^{-1}C$. State and prove an analogue of Proposition 2.4.7 for this notion of Schur complement.

(c) Suppose that T in (b) is of finite rank (forcing $\dim \mathcal{H}_2 = \dim \mathcal{K}_2 < \infty$). Show that $\operatorname{rank} T = \operatorname{rank}(A - BD^{-1}C) + \operatorname{rank} D$.

30 For positive semidefinite matrices A and B of the same size, define their *parallel sum* $A : B$ via

$$\langle (A : B)x, x \rangle = \inf\{\langle Ay, y \rangle + \langle Bz, z \rangle \; : \; x = y + z\}.$$

(a) Show that $A : B = B : A \geq 0$.

(b) Show that $A : (B : C) = (A : B) : C$.

(c) Show that $(S^*AS) : (S^*BS) = S^*(A : B)S$.

(d) Show that $A : B$ is the Schur complement supported in the first row and column of the block matrices

$$\begin{pmatrix} A & A \\ A & A+B \end{pmatrix}, \begin{pmatrix} B & B \\ B & A+B \end{pmatrix}.$$

(e) Show that when A and B are positive definite, then $A : B = (A^{-1} + B^{-1})^{-1}$.

(f) Show that $A_1 \geq A_2$, $B_1 \geq B_2$ imply $A_1 : B_1 \geq A_2 : B_2$.

(g) Show that $(A_1 + A_2) : (B_1 : B_2) \geq A_1 : B_1 + A_2 : B_2$.

(h) Show that $\mathrm{Ran} A : B = \mathrm{Ran} A \cap \mathrm{Ran} B$.

31 Determine the extreme points of the following sets.

(a) $\{X \; : \; 0 \leq X \leq I\} \subset \mathcal{H}_n$.

(b) $\{X \; : \; 0 \leq X \leq I \text{ and } 0 \leq X \leq B\} \subset \mathcal{H}_n$.

(c) Do parts (a) and (b) with I replaced by a positive semidefinite A.

32 Let

$$r(A) = \sup\{x^*Ax : x^*x = 1\}$$

denote the *numerical radius* of an operator $A \in \mathcal{B}(\mathcal{H})$.

(a) Show that $r(A) \leq 1$ if and only if there exists a $Z = Z^*$ so that

$$\begin{pmatrix} I - Z & A \\ A^* & I + Z \end{pmatrix} \geq 0.$$

(Hint: Observe that $r(A) \leq 1$ if and only if $2I - e^{i\theta}A - e^{-i\theta}A^* \geq 0$ for all θ. Now use Theorem 2.4.16 to factorize this degree 1 trigonometric polynomial.)

(b) Show that there exists a positive definite X such that $X + A^*X^{-1}A = I$ if and only if $r(A) \leq \frac{1}{2}$.

33 Let $\Lambda = R(1, n) = \{0, 1\} \times \{0, \ldots, N\}$ and $S = \Lambda - \Lambda$. Use the Fejér-Riesz theorem (Theorem 2.4.16) to prove that $\mathbb{T}\mathrm{Pol}_S^+ = \mathbb{T}\mathrm{Pol}_\Lambda^2$, thus providing an alternative proof for Theorem 1.3.19.

(Hint: Let $p(e^{iz}, e^{iw}) = \overline{P_1(e^{iz})}e^{-iw} + P_0(e^{iz}) + P_1(e^{iz})e^{iw} \in \mathbb{T}\mathrm{Pol}_S^+$. Show that

$$P(e^{iz}) = \begin{pmatrix} P_0(e^{iz}) & 2P_1(e^{iz}) \\ 2\overline{P_1(e^{iz})} & P_0(e^{iz}) \end{pmatrix} \geq 0,$$

and apply Theorem 2.4.16 to obtain a factorization $P(e^{iz}) = Q(e^{iz})^*Q(e^{iz})$, where $Q = \begin{pmatrix} Q_1 & Q_2 \\ Q_3 & Q_4 \end{pmatrix}$, each Q_i being an analytic polynomial of degree at most n. Conclude that

$$2p(e^{iz}, e^{iw}) = |Q_1(e^{iz}) + e^{iw}Q_2(e^{iz})|^2 + |Q_3(e^{iz}) + e^{iw}Q_4(e^{iz})|^2.)$$

34 In this example we will show how a Fejér-Riesz factorization can be obtained by solving a so-called discrete algebraic Riccati equation (DARE); see (2.9.5) below. Let

$$\alpha = \begin{pmatrix} 0 & \cdots & 0 & 0 \\ I & \cdots & 0 & 0 \\ \vdots & \ddots & \vdots & \vdots \\ 0 & \cdots & I & 0 \end{pmatrix}, \quad \beta = \begin{pmatrix} I \\ 0 \\ \vdots \\ 0 \end{pmatrix}.$$

(a) Show that $A(X)$ in Theorem 2.4.19 equals

$$\begin{pmatrix} R & S \\ S^* & X \end{pmatrix} - \begin{pmatrix} \beta^*X\beta & \beta^*X\alpha \\ \alpha^*X\beta & \alpha^*X\alpha \end{pmatrix},$$

where

$$R = Q_0, \; S = \begin{pmatrix} Q_{-1} & \cdots & Q_{-n} \end{pmatrix}.$$

(b) Show that the set $\mathcal{G}$ as in Theorem 2.4.19 is nonempty if there exists an $X = X^*$ such that $R - \beta^*X\beta > 0$ and

$$X - \alpha^*X\alpha - (S^* - \alpha^*X\beta)(R - \beta^*X\beta)^{-1}(S - \beta^*X\alpha) \geq 0. \quad (2.9.4)$$

(c) Assume $R - \beta^*X\beta > 0$. Show that the rank of $A(X)$ equals the rank of $R - \beta^*X\beta$ if and only if equality holds in (2.9.4).

(d) Suppose that X has been found so that equality holds in (2.9.4) and so that $R - \beta^*X\beta > 0$. Write

$$A(X) = \begin{pmatrix} H_0^* \\ \vdots \\ H_n^* \end{pmatrix} \begin{pmatrix} H_0 & \cdots & H_n \end{pmatrix},$$

where $H_0 = (R - \beta^*X\beta)^{\frac{1}{2}}$. Show that $Q(z) = H(z)^*H(z)$, where $H(z) = \sum_{i=0}^n H_i z^i$. Moreover, show that the roots of $H(z)$ are exactly the eigenvalues of $\alpha - \beta(R - \beta^*X\beta)^{-1}(S - \beta^*X\alpha)$.

In Matlab a routine is available to solve the discrete algebraic Riccati equation, which in addition finds the solution that yields the outer (or co-outer) factorization.

35 Consider the discrete algebraic Riccati equation

$$X - \alpha^* X\alpha - (S^* - \alpha^* X\beta)(R - \beta^* X\beta)^{-1}(S - \beta^* X\alpha) + Q = 0, \quad (2.9.5)$$

with

$$\alpha = \begin{pmatrix} 0 & 1 \\ 1 & -\frac{3}{2} \end{pmatrix}, B = \begin{pmatrix} 1 & 0 \\ 0 & 1 \end{pmatrix}, Q = \begin{pmatrix} 3 & 0 \\ 0 & -1 \end{pmatrix}, R = \begin{pmatrix} 2 & 0 \\ 0 & 0 \end{pmatrix}, S = \begin{pmatrix} 0 & 0 \\ 0 & 1 \end{pmatrix}.$$

Show that

$$X = \begin{pmatrix} -3 & 1 \\ 1 & -\frac{17\pm 3\sqrt{21}}{10} \end{pmatrix}$$

gives two solutions. Compute also the eigenvalues of $\alpha - \beta(R - \beta^* X\beta)^{-1}(S - \beta^* X\alpha)$ for these two solutions.

36 Let $P_0, \ldots, P_{2n} \in \mathcal{B}(\mathcal{K})$ be bounded selfadjoint Hilbert space operators such that $P(x) = P_0 + P_1 x + \cdots + P_{2n-1}x^{2n-1} + P_{2n}x^{2n} \geq 0$ for all $x \in \mathbb{R}$.

(a) Use the maps Φ and Ψ from Exercise 1.6.38 in combination with Theorem 2.4.16 to show that there exists a polynomial $G(x) = G_0 + \cdots + G_n x^n$ such that

$$P(x) = G(x)^* G(x), x \in \mathbb{R}.$$

(b) Show that the factorization in (a) may be found as follows. Let $\mathcal{F}$ be the convex set

$$\mathcal{F} = \Big\{ F = (F_{ij})_{i,j=0}^n \in \mathcal{B}(\mathcal{K}^{n+1}) : F \geq 0 \text{ and}$$

$$\sum_{\max\{0,k-n\} \leq i \leq \min\{n,k\}} F_{i,k-i} = P_k, k = 0, \ldots, 2n \Big\},$$

which is nonempty exactly since $P(x) \geq 0$, $x \in \mathbb{R}$. Choose $z \in \mathbb{C}_+$ and let $\pi_n(z) = \text{col}(z^j)_{j=0}^n$. There is a unique $F \in \mathcal{F}$, F_{opt} say, for which $(\pi_n(z) \otimes I_{\mathcal{K}})^* F(\pi_n(z) \otimes I_{\mathcal{K}})$ is maximal (resp. minimal) in the Loewner ordering. Factoring F_{opt} as

$$F_{\text{opt}} = \begin{pmatrix} G_0^* \\ \vdots \\ G_n^* \end{pmatrix} \begin{pmatrix} G_0 & \ldots & G_n \end{pmatrix}, \quad (2.9.6)$$

with $G_i \in \mathcal{B}(\mathcal{K}), i = 0, \ldots, n$ (which is possible), yields a desired factorization.

(c) Show that one may also find the factorization by taking $z \in \mathbb{C}_-$ and maximize (minimize) $(\pi_n(z) \otimes I_{\mathcal{K}})^* F(\pi_n(z) \otimes I_{\mathcal{K}})$ among all members of $\mathcal{F}$ to obtain F_{opt}.

Note: in this exercise we do not need the finite dimensionality of the underlying spaces, which is crucial in Theorem 2.7.13.

37 State and prove the analogue of Proposition 2.5.1 for describing the general structure of a column contraction.

38 An undirected graph $G = (V, E)$ is called a *proper interval graph* if there exists a set of real intervals $(I_v)_{v \in V}$, such that none of them is a a subset of another one, and that $(v_i, v_j) \in E$ if and only if $I_{v_i} \cap I_{v_j} \neq \emptyset$.

(a) Prove that a pattern is generalized banded if and only if its graph is a proper interval graph.

(b) Prove that every proper interval graph is chordal.

(c) Give an example of a chordal graph which is not a proper interval graph. (Hint: let $G = (V, E)$, with $V = \{1, 2, 3, 4\}$ and $E = \{(1, 2), (1, 3), (1, 4)\}$.)

39 Given a partially positive semidefinite banded matrix $K = \{A_{ij} : |j - i| \leq p, 1 \leq i, j \leq k\}$, with A_{ii} Fredholm and A_{ij}, $i \neq j$, compact, prove that the minimal possible dimension for the kernel of a positive semidefinite completion of K is

$$\sum_{i=1}^{k-p} \dim \operatorname{Ker} \begin{pmatrix} A_{ii} & \cdots & A_{i,i+p} \\ \vdots & & \vdots \\ A_{i+p,i} & \cdots & A_{i+p,i+p} \end{pmatrix}$$

$$- \sum_{i=1}^{k-p-1} \dim \operatorname{Ker} \begin{pmatrix} A_{i+1,i+1} & \cdots & A_{i+1,i+p} \\ \vdots & & \vdots \\ A_{i+p,i+1} & \cdots & A_{i+p,i+p} \end{pmatrix}.$$

(Hint: prove the result first for the 3×3 partial matrix of the form (2.4.5); it is immediate that the dimension of the kernel of any positive semidefinite completion is at least that large, and show then this dimension is realized for the central completion. Use induction for proving the general result.)

40 Let G_1 and G_2 be Hilbert space contractions on appropriate spaces so that the partial matrix

$$\begin{pmatrix} I & G_1 & ? \\ G_1^* & I & G_2 \\ ? & G_2^* & I \end{pmatrix} \tag{2.9.7}$$

is well defined. Let $g = \max\{\|G_1\|, \|G_2\|\}$. Show that the central completion M of this partially positive semidefinite operator matrix satisfies

$$q_- I \leq M \leq q_+ I,$$

where

$$q_{\pm} = \frac{1}{2}g^2 + 1 \pm \frac{1}{2}\sqrt{g^4 + 8g^2}.$$

(Hint: Do first the scalar case where $G_1 = G_2$.)

41 Let $s_0 = s_1 = 1$ and $s_2 = 2$. Find a positive measure σ supported on two points of $\mathbb{R}$, such that $\widehat{\sigma}(k) = s_k$ for $k = 0, 1, 2$. (Hint: Find $\rho_1, \rho_2 > 0$ and $t_1, t_2 \in \mathbb{R}$ such that $\rho_1 t_1^k + \rho_2 t_2^k = s_k$ for $k = 0, 1, 2$.)

42 Given a positive matrix-valued Borel measure σ on $\mathbb{R}$ we consider integrals of the form

$$I(f) = \int_{\mathbb{R}} f d\sigma.$$

A *cubature* (or *quadrature*) formula for $I(f)$ is an expression of the from

$$C(f) = \sum_{j=1}^{k} \rho_j f(x_j),$$

with fixed nodes $x_1, \ldots, x_k \in \mathbb{R}$ and fixed weights $\rho_1, \ldots, \rho_k > 0$, so that $C(f)$ is a "good" approximation for $I(f)$ on a class of functions of interest. Use Theorem 2.7.9 to obtain a quadrature formula so that $I(f) = C(f)$ for polynomials f of degree $\leq 2n - 1$ based on the existence of the moments $S_k = \int_{-\infty}^{\infty} \lambda^k d\sigma(\lambda)$, $k = 0, \ldots, 2n$.

43 Let $T \in \mathcal{B}(\mathcal{H})$. The *rank k numerical range* of T is defined via

$$\Lambda_k(T) = \left\{ \lambda \in \mathbb{C} \ : \ \text{there exists a unitary } U \text{ so that } U^*TU = \begin{pmatrix} \lambda I_k & * \\ * & * \end{pmatrix} \right\}.$$

In this exercise we will show that $\Lambda_k(T)$ is convex.

(a) Show that $\Lambda_k(\alpha I + \beta T) = \alpha + \beta \Lambda_k(T)$.

(b) Show that it suffices to prove $1, -1 \in \Lambda_k(T)$ implies $0 \in \Lambda_k(T)$.

(c) Suppose that $1, -1 \in \Lambda_k(T)$. Then there exist k dimensional subspaces M_1, M_2 of $\mathcal{H}$ so that $P_{M_1} T \mid_{M_1} = I$, $P_{M_1} T \mid_{M_2} = -I$, where P_{M_i} is the orthogonal projection onto $M_i, i = 1, 2$. Show that $M_1 \cap M_2 = (0)$.

(d) Show that for the convexity of $\Lambda_k(T)$ it suffices to show that $0 \in \Lambda_k(X)$, where

$$X = P_{M_1 + M_2} T \mid_{M_1 + M_2} .$$

(e) Show that for a matrix X and an invertible S, $0 \in \Lambda_k(X)$ if and only if $0 \in \Lambda_k(S^*TS)$. (Hint use a QR-factorization of S.)

(f) Show that for the convexity of $\Lambda_k(T)$ it suffices to show that

$$0 \in \Lambda_k \begin{pmatrix} I_k & X \\ Y & -I_k \end{pmatrix}$$

for all $X, Y \in \mathbb{C}^{k \times k}$.

(g) Show that $0 \in \Lambda_k \left(\begin{smallmatrix} I & X \\ Y & -I \end{smallmatrix}\right)$ if and only if there exists a Z so that

$$I_k + XZ + Z^*Y - Z^*Z = 0. \tag{2.9.8}$$

(h) Show that for all X, Y there exists a Z such that (2.9.8) holds if and only for all M and $R > 0$ there exists a $H = H^*$ such that

$$I_k + MH + HM^* - HRH = H. \tag{2.9.9}$$

(Hint: Put $X = MR^{-\frac{1}{2}}, Y = R-\frac{1}{2}(M^* - I)$ and find Z solving (2.9.8). Then observe that $H = Z^*R^{-\frac{1}{2}}$ is Hermitian and solves (2.9.9). For the converse, first consider the case when $X - Y^*$ is nonsingular. Let $J = (X - Y^*)^{-1}, M = XJ$, and $R = J^*J$, solve for H in (2.9.9), and put $Z = JH$. If $X - Y^*$ is singular, use $(X + \frac{1}{m}I_k) - Y^*$ in the previous argument, and argue that the corresponding Z_m converges to the desired solution Z.)

(i) Finally, use Theorem 2.7.17 to show that (2.9.9) always has a solution. (Hint: Take $A = M - \frac{1}{2}H$ and $D = R$ in Theorem 2.7.17.)

44 Let A be a stable, and C a positive semidefinite matrix of the same size. Show that

$$X = \int_0^\infty e^{At}Ce^{A^*t}dt,$$

exists, is positive semidefinite, and satisfies the Lyapunov equation $XA + A^*X = -C$.

45 Let $S_0, \dots, S_{2n+1}$ be such that $K_n = (S_{i+j})_{i,j=0}^n$ and $\widetilde{K}_n = (S_{i+j+1})_{i,j=0}^n$ are positive semidefinite. Show that S_{2n+2} and S_{2n+3} may be constructed such that K_{n+1} and $\widetilde{K}_{n+2}$ are positive semidefinite. (This problem occurs when one is interested in solving the truncated matrix-valued Stieltjes moment problem where a measure is to be found supported on $[0, \infty)$ (as opposed to $\mathbb{R}$ in the Hamburger problem).)

2.10 NOTES

Section 2.1

Banded partially positive definite operator matrices were considered first in [265] as part of a general approach called the *band method*, which originated in [194] and was further developed in [265]–[267], [563], and [199]. Theorems 2.1.1 and 2.1.2 for the operator case are taken from [265], while Theorem 2.1.4 is from [268]. Remark 2.1.3 and Exercise 2.9.4 are based on Theorem 1.1 in [265].

Section 2.2

Lemma 2.2.5 was first proven in [285]. Theorem 2.2.4 was originally proved in the scalar positive semidefinite case in [285] by the same method as presented in Section 2.2. The operator-valued case was considered by a different method in [452]. The construction of the distinguished completions in Theorem 2.2.8 and the maximum entropy property in Theorem 2.2.10 are taken from [47]. In the scalar case, an explicit construction of the maximum determinant positive definite completion can be found in [85]. In the scalar case, Corollary 2.2.11 is a particular case of Corollary 1.4.5, previously known from [219] and [285]. In [85], a formula for the maximum over the determinants of all positive definite completions of a partially positive definite matrix A with a chordal graph was found in terms of the determinants of the fully specified submatrices of A. In [52], a formula was derived for an arbitrary positive definite completion of such a partial matrix in terms of some free parameters. The condition given in part (c) of Exercise 2.9.10 appears in [87] (an equivalent result can be found in [220]). This condition is referred to as a "cycle condition" as the graph of the underlying pattern in this exercise is a simple cycle. For an arbitrary graph G, these cycle conditions on the cycles of G are thus necessary for a real partially positive semidefinite matrix A with graph G to admit a positive semidefinite completion. In [83] the authors characterize the graphs G for which these cycle conditions on A with graph G are also sufficient for the existence of a positive semidefinite completion of A. For further results on positive semidefinite completions of partial matrices with nonchordal graphs see [84]. In [306] the cycle condition is further studied, also in the context of Toeplitz partial matrices. The answer to Exercise 2.9.6 can be found in [453]. Exercise 2.9.8 is based on [54], where the question is asked how to determine for a given graph G the minimal number K such that any partial positive semidefinite matrix with graph G always allows a completion with at most k negative eigenvalues. Exercise 2.9.12 is based on [88, Theorem 5.1] and 2.9.13 is based on [327].

Section 2.3

Theorems 2.3.1 and 2.3.8 were first obtained in [266] and [199], respectively, as part of the "band method" approach mentioned earlier. The Yule-Walker equations go back to [587] and [558], where they were used in the context of autoregressive models. The Christoffel-Darboux formula in the scalar case is a classical result (see [537]) while its matrix-valued version was proven in [174]. The formulas in Theorem 2.3.6 for computing the inverse of a Toeplitz matrix are known in the literature as the Gohberg-Semencul formulas; they were derived in [275] for the scalar case and in [261] for the matrix-valued case. In the earlier paper [547] it was already recognized that if T_{n-1} and T_n are invertible then T_n^{-1} is determined completely by its first row and column; in fact, the last four equations in [547] are computationally equivalent to the Gohberg-Semencul formula. See also [346] for a full account. For a discussion on the maximum entropy principle as in Theorem 2.3.9 see [268]; see also [414]. For the connection of the moment problem to Padé approximation, see for instance [113].

Basic properties of analytic function with values in Banach spaces may be found in [354]; among others, in [354, Theorem 1.37] states that analyticity in the weak sense (as defined in Section 2.3) implies analyticity in the strong sense. We recommend the reader [320] for an introduction to the theory of Hardy spaces. For further studies, see [189] and [368]. The operator-valued Bochner-Phillips generalization of the classical (scalar) Wiener's theorem, which states that a pointwise invertible Wiener class function is invertible in the Wiener class, may be found in [99, Theorem 1]. The fundamental theory of vector- and operator-valued Hardy spaces can be found in [534]. Theorem 2.3.11 dates back to [537]. The recent book [523] is a very comprehensive collection of results on the theory of orthogonal polynomials on the unit circle. Matrix-valued orthogonal polynomials started to be investigated in [174] and [177] and operator-valued ones in [142] (see also [53]). The proof for Szegő's theorem outlined in Exercise 2.9.22 can be found in [200]. Exercise 2.9.23 is a result by M.G. Krein (see [200]). The statement in Exercise 2.9.19 is known as Szegő's first limit theorem; the book [105] is an excellent source for a complete account on Szegő type limit theorems. The Carathéodory-Toeplitz problem with partial data, of which Exercise 2.9.20 is an example, was treated in [23] and [570].

Section 2.4

The notion of Schur complement (as defined in Exercise 2.9.29) was developed in early works of Frobenius and Schur. It is a simple concept that turned out to be very useful during the years. For positive semidefinite operators the Schur complement may be defined via the maximality property we used; this view goes back to Krein [374, Section 1.1]. The definition using the subspaces, as we did as well, goes back to Ando [34]. The book [588] gives a wonderful recent overview of the Schur complement including its origin and rediscoveries (e.g., in engineering it was called the shorted operator). Our presentation in Section 2.4 follows [187]. Lemma 2.4.2 originates in [183]. The Fejér-Riesz factorization theorem was first generalized to the case of matrix-valued polynomials in [494] and [307], and later for operator-valued polynomials with a compactness condition in [259], and in the general form in [495] (see also [497]). A recent account on the operator-valued Fejér-Riesz theorem may be found in [500]. Its proof presented in Section 2.4 is taken from [187] and it is based on the Schur complement approach started in [186] and [187]. A multivariable operator-valued version of the Fejér-Riesz factorization theorem can also be found in [187]. Exercise 2.9.30 is based on T. Ando's Section 5.3 in the book [588]. The notion of parallel sum goes back to [28]. Parts of this exercise are based on results from [29]. The answers to parts of Exercise 2.9.31 may be found in [35] and [36]. The extreme points of the interval $[0, A]$ are, in the $n \times n$ case, exactly all the Schur complements supported on finite dimensional subspaces of $\mathbb{C}^n$; see [30]. In the infinite dimensional case there are in general other extreme points as well, as was shown in [279]. Exercise 2.9.32 is based on [33] (part (a)) and [207] (part (b)).

Section 2.5

Section 2.5 is based on [141] (see also [143]). We briefly review here the history of these parameters. Schur proved in [513], [514] that there exists a one-to-one correspondence between the set of all analytic functions in $\mathbb{D}$ with their absolute value bounded by 1 and the set of all complex numbers $(\gamma_n)_{n\geq 0}$ such that $|\gamma_n| \leq 1$ for all $n \geq 0$ and if $|\gamma_{n_0}| = 1$, then $\gamma_{n_0+1} = \gamma_{n_0+2} = \cdots = 0$. These so-called Schur parameters have found during the years a wide range of applications in mathematics and applied fields such as geophysics, signal processing, and control theory (see [225] and [53]). In particular, the Schur parameters are closely related to the reflection coefficients (or Szegő parameters) of orthogonal polynomials on the unit circle (see [53] and [523]). A well-known result, originating from Rakhmanov's work [477], [478], states that if a measure has a strictly positive derivative almost everywhere on $\mathbb{T}$, then the corresponding Szegő parameters converge to 0. The papers [420], [434], and [407] provide an alternative proof. Rakhmanov's result was recently generalized to the matrix-valued case in [550]. In [124], the operator-valued correspondent of Schur parameters, called choice sequences, was introduced and used to parameterize the set of all commutant liftings of a contraction. For two Hilbert spaces $\mathcal{H}_1$ and $\mathcal{H}_2$, an $(\mathcal{H}_1, \mathcal{H}_2)$-*choice sequence* is a sequence $(\Gamma_n)_{n\geq 0}$ of contractions such that $\Gamma_0 \in \mathcal{L}(\mathcal{H}_1, \mathcal{H}_2)$ and for each $n \geq 1$, $\Gamma_n \in \mathcal{L}(\mathcal{D}_{\Gamma_{n-1}}, \mathcal{D}_{\Gamma_{n-1}^*})$. In [140] an explicit parametrization of positive semidefinite Toeplitz operator matrices in terms of choice sequences was given (see Theorem 2.5.8). For multivariable Toeplitz matrices a comprehensive Schur parameter analysis does not exist yet. In [251] an initial analysis is done for two-level Toeplitz matrices. For the non-Toeplitz case the choice triangle was introduced and used for the parametrization of arbitrary positive semidefinite operator matrices in [141], where Theorem 2.5.5 and Corollary 2.5.7 are taken from (see also [143]). Let us also mention that in [238], [237] and related papers alternative versions of some of the above mentioned results appear.

Section 2.6

Section 2.6 is based on [66] and [63]. It was first observed in [40] that every partially positive definite generalized banded matrix admits positive semidefinite completions and that the set of all such completions is in a one-to-one correspondence with the completions of $\{\Gamma_{ij} : (i,j) \in S,\ i < j\}$ to a choice triangle. The characterization of the central completion in terms of the zeros in the matrix representation of U, as well as Theorem 2.6.3, were proven in [66].

Consider a given partially positive definite scalar matrix A the graph G of which is chordal. Let G_j, $1 \leq j \leq t$, be an admissible sequence for G. Consider the positive definite completion of A obtained as in the proof of Theorem 2.2.4 along this chordal sequence, by considering at each step the maximum determinant completion of (2.2.5). It has been asked in [339], whether this way we always obtain at the end the maximum determinant completion of A. The question was positively answered independently in [85] and [52]. The band case was treated earlier in [201]. The results about the central completion in the general operator-valued case for a partially positive semidefinite matrix and its extremal property were answered in [63].

Section 2.7

The study of the Hamburger moment problem started with [299]. In this section we study the truncated moment problem. In [529] it was shown that a full solution to the truncated moment problem yields a solution to the full moment problem. There exists a very vast literature on various aspects of the Hamburger moment problem, among which we mention the well-known books [16], [520], [17], [376], [331], and [481]. There are numerous necessary and sufficient conditions for the Hamburger moment problem to be determined, for example, when the power moments uniquely determine the positive measure (see [16], [520], [17], and [481]). In [32] the operator-valued Hamburger moment problem is considered. We remark that in the literature, real sequences $(s_n)_{n \geq 0}$ for which all Hankel matrices (2.7.1) are positive definite (or equivalently (2.7.2) holds) are called positive semidefinite on $(-\infty, \infty)$. This is a different type of positive semidefitness from the one considered in Chapters 1 and 3, as it is expressed in terms of Hankel matrices instead of Toeplitz ones. Theorem 2.7.2 is due to von Neumann; a proof may be found in [481], where it is Theorem X.3. Theorem 2.7.1 and its proof are taken from [481]. The example in Remark 2.7.3 is from [528, page J. 105]; other examples may be found in [520, Chapter 1, Section 8]. Theorem 2.7.6 is similar to a result in the revised version of [101] (unpublished); see also [191] and the recent paper [196]. The proof of Theorem 2.7.9 is inspired by a similar argument in an unpublished paper [287]; see also [288]. The proof of Lemma 2.7.11 is taken from [209]. The scalar truncated Hamburger moment problem was first solved in [224] for the case (ii) in Corollary 2.7.12, and then for the case $K_n > 0$ in [16]. Both solutions are based on the fact that K_n in the situations considered is a convex combination of rank 1 Hankel matrices of the form $H_n = (t^{i+j})_{i,j=0}^n$ (see also Exercise 1.6.39). In [157] the truncated Hamburger moment problem was solved in the odd case, that is, Theorem 2.7.5 (ii). In [473] a solution is provided based on hyponormal operator theory. The matrix Stieltjes moment problem is treated in [100], where among others a solution to Exercise 2.9.45 may be found. The second part of Section 2.7, starting with Theorem 2.7.13, is based on [293]. The factorization result in Theorem 2.7.13 was earlier obtained for the operator-valued case in [497] and [416]. A more general version, concerning polynomials with selfadjoint coefficients (without necessarily a positive semidefiniteness constraint) was considered in [387]. For engineering applications various related factorizations are treated; see, for instance, [479], [548], [527], [590]. The books [269] and [270] provide a more elaborate background. Theorem 2.7.13 was proven in [293], in the operator-valued case with a maximality condition as it appears in Exercise 2.9.36. Other extremal properties of the outer factor G, in the scalar case, appear in [422]. The book [384] is an excellent source of information on the Riccati equation. Numerical algorithms to solve the Riccati equations go back to [391]. Exercise 2.9.35 is taken from [367]. For related topics we also mention [385] (Lyapunov and Stein equations), [486], and [324] (Lyapunov equation). The proof of Theorem 2.7.17 is almost identical to that of Theorem 9.1.2 in [384]. Exercise 2.9.43 is based on the papers [130] and [574]. An alternative proof of the convexity of the higher rank numerical range may be found in [403]. Exercise 2.9.44 is a well-known result by Lyapunov (see, e.g., [384], [385], or [324]).

Section 2.8

Finally, Section 2.8 follows the paper [271]. The results may also be applied to cyclostationary processes, which arise in models for phenomena that involve periodicities; see [491, Section 3.3]. The book [242] serves as a reference on cyclostationary processes.

Chapter Three

Multivariable moments and sums of Hermitian squares

In one variable one can formulate several very closely related problems that become vastly different in the multivariable case, as follows. Given are complex numbers $c_0 = \overline{c_0}, \ldots, c_n = \overline{c_{-n}}$. Consider the following problems.

1. *Carathéodory interpolation problem*: find an analytic function $\phi : \mathbb{D} \to \mathbb{C}$ such that $\operatorname{Re} \phi(z) \geq 0, z \in \mathbb{D}$ (that is, ϕ belongs to the *Carathéodory class*), $\phi(0) = c_0$, and $\frac{\phi^{(k)}(0)}{k!} = 2c_k$, $k \in \{1, \ldots, n\}$.

2. *Moment problem*: find a positive measure σ on $\mathbb{T}$ with moments $\widehat{\sigma}(k) := \int_{\mathbb{T}} z^k d\sigma = c_k$, $k \in \{-n, \ldots, n\}$.

3. *Bernstein-Szegő measure moment problem*: find a polynomial $p(z) = \sum_{k=0}^{n} p_k z^k$ with $p(z) \neq 0$, $z \in \overline{\mathbb{D}}$ ($p(z)$ is *stable*), so that the Fourier coefficients $\widehat{f}(k)$ of $f(z) := \frac{p_0}{|p(z)|^2}$ satisfy $\widehat{f}(k) = c_k$, $k \in \{-n, \ldots, n\}$.

4. *Ordered group moment problem*: consider the function $f : \{-n, \ldots, n\} \to \mathbb{C}$ given by $f(k) = c_k$, $|k| \leq n$. Find a function F defined on the ordered group $\mathbb{Z}$ such that $F|\{-n, \ldots, n\} = f$, and $(F(k-l))_{k,l \in J}$ is positive semidefinite for all finite subsets J of $\mathbb{Z}$ (i.e., F is positive semidefinite).

5. *Free group moment problem*: consider the (multiplicative) free group $\mathbb{F}$ with one generator a, and let $\phi : \{a^k \ : \ |k| \leq n\} \to \mathbb{C}$ be defined by $\phi(a^k) = c_k, k \in \{-n, \ldots, n\}$. Find $\Phi : \mathbb{F} \to \mathbb{C}$ such that $\Phi|\{a^k : |k| \leq n\} = \phi$ and Φ is positive definite (i.e., $(\Phi(s_k^{-1} s_l))_{k,l \in J}$ is positive semidefinite for all finite subsets $\{s_k \ : \ k \in J\}$ of $\mathbb{F}$).

There is a tight connection between these five problems. The Riesz-Herglotz representation of functions in the Carathéodory class connects Problems 1 and 2, so that they are only solvable simultaneously. Next, Bochner's theorem provides the equivalence of 2 and 4, and 4 and 5 are trivially equivalent (via the bijection $g : \mathbb{Z} \to \mathbb{F}$, $g(n) = a^n$). Finally, if $p(z)$ is a solution to 5, then $f(z)d\mu(z)$, where μ is the normalized Lebesgue measure, is a solution to 2. If 2 has a strictly positive absolutely continuous solution $f(z)d\mu$ (with $f(z) > 0$ on $\mathbb{T}$), then a combination of the maximum entropy principle and the Riesz-Fejér factorization lemma shows that it has in fact a solution of the form $\frac{p_0}{|p(z)|^2} d\mu$, with p a stable polynomial, thus providing a solution for

3. In fact, as we have seen in the previous chapters, 1,2,4, and 5 are solvable if and only if $T \geq 0$ and 3 is solvable if and only if $T > 0$, where T is the finite Toeplitz matrix $T = (c_{i-j})_{i,j=0}^{n}$.

We will see in this chapter that these five closely related problems become vastly different when one considers the multivariable case. In fact, Problems 4 and 5 require now different data sets from the other three problems. Furthermore, 1 and 2 are no longer simultaneously solvable. Finally, while a solution to 3 trivially still provides a solution to 2, it is no longer true that the existence of a strictly positive absolutely continuous solution to 2 implies the existence of a solution to 3.

In a similar way, various natural sums of Hermitian squares problems that one may consider in the multivariable case all come down to the classical factorization problem in one variable as discussed in Section 1.1. In particular, there are three sums of (Hermitian) squares problems in several variables that we can solve:

6. Every strictly positive operator-valued trigonometric polynomial in commutative variables is a sum of Hermitian squares.
7. Every positive semidefinite trigonometric polynomial in several noncommutative unitary variables is a sum of Hermitian squares.
8. Matrix positive real polynomials in noncommutative variables are sums of squares (that is, Hilbert's 17th problem holds for real polynomials in several noncommutative variables).

In addition to these, we also establish necessary and sufficient conditions for a trigonometric polynomial in two variables to have a stable factorization.

This area of research on multivariable problems is still very active. In this chapter we will give our view on the current state of the art, and in almost all cases we will consider the operator-valued versions.

3.1 POSITIVE CARATHÉODORY INTERPOLATION ON THE POLYDISK

In this section we shall address the d-variable Carathéodory interpolation problem, a multivariable generalization of Problem 1 above.

We first need to introduce the Herglotz-Agler class of functions. We will use the notation

$$z^k = (z_1, \ldots, z_d)^{(k_1,\ldots,k_d)} := z_1^{k_1} \cdots z_d^{k_d},$$

$$\phi^{(k)} = \frac{\partial^{k_1}}{\partial z_1^{k_1}} \cdots \frac{\partial^{k_d}}{\partial z_d^{k_d}} \phi,$$

$$k! = (k_1, \ldots, k_d)! := k_1! \cdots k_d!\,.$$

Let ϕ be a holomorphic $\mathcal{L}(\mathcal{E})$-valued function defined on the polydisk $\mathbb{D}^d = \{(z_1, \ldots, z_d) : |z_i| < 1, i = 1, \ldots, d\}$, and let $\phi(z) = \sum_{k \in \mathbb{N}_0^d} \phi_k z^k$, $z = (z_1, \ldots, z_d)$, $\phi_k \in \mathcal{L}(\mathcal{E})$, be its series expansion. For a commuting collection of strict contractions $R = (R_1, \ldots, R_d) \in \mathcal{L}(\mathcal{K})^d$, we may define $\phi(R) = \phi(R_1, \ldots, R_d)$ to be an operator in $\mathcal{L}(\mathcal{E}) \otimes \mathcal{L}(\mathcal{K}) \equiv \mathcal{L}(\mathcal{E} \otimes \mathcal{K})$ by $\phi(R) = \sum_{k \in \mathbb{N}_0^d} \phi_k \otimes R^k$, where $(R_1, \ldots, R_d)^{(k_1, \ldots, k_d)} := R_1^{k_1} \cdots R_d^{k_d}$. We say that ϕ is *Herglotz-Agler* if $\operatorname{Re} \phi(R) \geq 0$ for all commuting collections of strict contractions $R = (R_1, \ldots, R_d)$. Here $\operatorname{Re} A = \frac{1}{2}(A + A^*)$ denotes the real part of the operator A. We shall denote the class of Herglotz-Agler $\mathcal{L}(\mathcal{E})$-valued functions on $\mathbb{D}^d$ by $\mathcal{A}_d(\mathcal{E})$. By taking $R_j = z_j I$ one sees immediately that $\phi \in \mathcal{A}_d(\mathcal{E})$ implies that $\operatorname{Re} \phi(z) \geq 0$, $z \in \mathbb{D}^d$. The converse holds when $d = 1, 2$, but not when $d \geq 3$; see Example 3.1.8.

Theorem 3.1.1 gives a characterization of the elements ϕ in the Herglotz-Agler class that are normalized to satisfy $\phi(0) = \frac{1}{2}I$. One can always normalize a function in the Herglotz-Agler class so that $\phi(0) = \frac{1}{2}I$. Indeed, with ϕ as above and $\operatorname{Re}\phi \geq 0$, we have that for each $k \in \mathbb{N}_0^d$ that

$$\begin{pmatrix} \phi_0 + \phi_0^* & \phi_k^* \\ \phi_k & \phi_0 + \phi_0^* \end{pmatrix} \geq 0.$$

By Lemma 2.4.4 we have that $\phi_k = (\phi_0 + \phi_0^*)^{1/2} g_k (\phi_0 + \phi_0^*)^{1/2}$ for some contraction g_k. Put $g_0 = \frac{1}{2}I$ and $g(z) = \sum_{k \in \mathbb{N}_0^d} g_k z^k$, where all g_k act on $\overline{\operatorname{Ran}(\phi_0 + \phi_0^*)}$. Then g is in the Herglotz-Agler class if and only if ϕ is, and one can retrieve ϕ via the formula $\phi = (\phi_0 + \phi_0^*)^{1/2}(g(z) - g_0)(\phi_0 + \phi_0^*)^{1/2} + \phi_0$.

Theorem 3.1.1 *A holomorphic ϕ with $\phi(0) = \frac{1}{2}I$ belongs to $\mathcal{A}_d(\mathcal{E})$ if and only if there exist a Hilbert space $\mathcal{H} = \mathcal{H}_1 \oplus \cdots \oplus \mathcal{H}_d$, an isometry $V : \mathcal{E} \to \mathcal{H}$, and a unitary $U : \mathcal{H} \to \mathcal{H}$ such that*

$$\phi(z) = \frac{1}{2}I + V^* U (I - Z(z)U)^{-1} Z(z) V, \ z \in \mathbb{D}^d, \tag{3.1.1}$$

where $(Z(z))(\oplus_{i=1}^d h_i) := \oplus_{i=1}^d (z_i h_i)$.

Proof. We will only provide a proof for the "if" part. The proof for the converse can be found in [8].

Suppose ϕ is given by (3.1.1), with V, U, and $Z = Z(z)$ as above. Then we get that

$$\begin{aligned} \phi(z) + \phi(z)^* &= I + V^*U(I - ZU)^{-1}ZV + V^*Z^*(I - ZU)^{*-1}U^*V \\ &= V^*(I - Z^*U^*)^{-1}[(I - Z^*U^*)(I - UZ) \\ &\quad + Z^*U^*(I - UZ) + (I - Z^*U^*)UZ](I - UZ)^{-1}V \\ &= V^*(I - Z^*U^*)^{-1}[I - Z^*Z](I - UZ)^{-1}V. \end{aligned}$$

Plugging in a d-tuple R of commuting strict contraction yields that $\phi(R) + \phi(R)^*$ equals

$$(V^* \otimes I)(I - Z(R)(U \otimes I))^{-1}(I - Z(R)^*Z(R))(I - Z(R)(U \otimes I))^{*-1}(V \otimes I),$$

which is easily seen to be positive semidefinite. □

We next state our interpolation problem. Let $\leq$ be the partial order on $\mathbb{N}_0^d$ defined by $(k_1, \ldots, k_d) \leq (l_1, \ldots, l_d)$ if and only if $k_i \leq l_i$, $i = 1, \ldots, d$. A subset K of $\mathbb{N}_0^d$ will be called *lower inclusive* if $k \in K$ and $l \leq k$ imply $l \in K$. Given a finite lower inclusive subset K of $\mathbb{N}_0^d$ and operators $C_k \in \mathcal{L}(\mathcal{E})$, $k \in K$, *the Carathéodory interpolation* problem asks for a $\phi \in \mathcal{A}_d(\mathcal{E})$ so that $\phi_k = C_k$, $k \in K$, where $\phi(z) = \sum_{k \in \mathbb{N}_0^d} \phi_k z^k$.

Before we can state our main result, we first have to recall the notion of words. Let $A_1, \ldots, A_d$ be operators on $\mathcal{H}$. An expression of the form

$$A_1^{n_{11}} \cdots A_d^{n_{1d}} \cdots A_1^{n_{p1}} \cdots A_d^{n_{pd}}$$

is called a *word* in $(A_1, \ldots, A_d)$ of *multilength* $(\sum_{j=1}^p n_{j1}, \ldots, \sum_{j=1}^p n_{jd})$. Typically, factors with an exponent 0 are left out, and subexpressions $A_i^p A_i^q$ are contracted to A_i^{p+q}. Words are equal when after removing factors with a zero exponent and contracting as above, the expressions we have are equal. So, for example, there are exactly three different words in (A, B) of multilength $(2, 1)$: A^2B, ABA, and BA^2. Notice that even though it may happen that these three operators are equal (e.g., when A and B commute), the words are considered to be different. For $k \in \mathbb{N}_0^d$ we denote the set of all words of multilength k in $(A_1, \ldots, A_d)$ by $W_k(A_1, \ldots, A_d)$.

Let K be a finite subset of $\mathbb{N}_0^d$ with cardinality card$K$. We let $\mathcal{E}^{|K|}$ denote the Hilbert space of card$K$-tuples $\xi = (\xi_k)_{k \in K}$ with $\|\xi\|^2 := \sum_{k \in K} \|\xi_k\|^2 (< \infty$, since K is finite). Notice that instead of indexing the coordinates of the tuples by $1, \ldots, \mathrm{card}K$, we prefer to index them with the elements of K. This will be convenient when we define operators on $\mathcal{E}^{\mathrm{card}K}$. If $F_k \in \mathcal{L}(\mathcal{E})$, $k \in K$, the notation $F = \mathrm{col}(F_k)_{k \in K}$ stands for the operator $F : \mathcal{E} \to \mathcal{E}^{\mathrm{card}K}$ defined by $F\xi = (F_k \xi)_{k \in K}$. Next, for $T \in \mathcal{L}(\mathcal{K})$ we denote by M_T the conjugacy operator $M_T(X) = TXT^*$ on $\mathcal{L}(\mathcal{K})$. Note that $X \geq 0$ implies that $M_T(X) \geq 0$. Finally, denote by e_i, $i = 1, \ldots, d$, the ith standard vector in $\mathbb{N}_0^d$, and let δ_{kl} denote the Kronecker delta function on $\mathbb{N}_0^d$.

Theorem 3.1.2 *Given a nonempty finite lower inclusive set $K \subset \mathbb{N}_0^d$ and operators $C_k \in \mathcal{L}(\mathcal{E})$, $k \in K$, with $C_0 = \frac{1}{2}I$, the following are equivalent.*

(i) There exists a $\phi(z) = \sum_{k \in \mathbb{N}_0^d} \phi_k z^k \in \mathcal{A}_d(\mathcal{E})$ such that $\phi_k = C_k$, $k \in K$.

(ii) There exist positive semidefinite operators $G_1, \ldots, G_d$ on $\mathcal{E}^{\mathrm{card}K}$ such that
$\prod_{j \neq i}(I - M_{T_j})(G_i) \geq 0$, $i = 1, \ldots, d$, and

$$X + X^* = G_1 + \cdots + G_d. \tag{3.1.2}$$

Here $X = (C_{k-j})_{k,j \in K}$, $C_k = 0$ for $k \notin K$, and $T_j = (t_{k,l}^{(j)})_{k,l \in K}$, where $t_{k,l}^{(j)} = I$ if $k = l + e_j$ and $t_{k,l}^{(j)} = 0$ otherwise.

(iii) There exist positive semidefinite operators $\Gamma_1, \ldots, \Gamma_d$ on $\mathcal{E}^{\mathrm{card}K}$ such that

$$EC^* + CE^* = \Gamma_1 - T_1\Gamma_1T_1^* + \cdots + \Gamma_d - T_d\Gamma_dT_d^*, \tag{3.1.3}$$

where

$$C = \text{col}(C_k)_{k\in K}, \quad E = \text{col}(\delta_{0k})_{k\in K},$$

and T_j is as in (ii).

(iv) *There exist a Hilbert space $\mathcal{H} = \mathcal{H}_1 \oplus \cdots \oplus \mathcal{H}_d$, an isometry $V : \mathcal{E} \to \mathcal{H}$, and a unitary $U \in \mathcal{L}(\mathcal{H})$ such that $C_k = V^*(\sum_{w\in W_k(UP_1,\ldots,UP_d)} w)V$, $k \in K \setminus \{0\}$, where P_i is the orthogonal projection in $\mathcal{L}(\mathcal{H})$ with image $\mathcal{H}_i$, $i = 1, \ldots, d$.*

It should be noted that in the one-variable case (ii) and (iii) simply reduce to the statement that the Toeplitz matrix $X + X^*$ is positive semidefinite.

Example 3.1.3 Let $\Lambda = \{0, 1\}^2$ and $c_{00} = 1$, $c_{10} = c_{01} = c_{11} = 0.9$. Then there does not exist a solution to the Carathéodory interpolation problem, as

$$X + X^* = \begin{pmatrix} 1 & 0.9 & 0.9 & 0.9 \\ 0.9 & 1 & 0 & 0.9 \\ 0.9 & 0 & 1 & 0.9 \\ 0.9 & 0.9 & 0.9 & 1 \end{pmatrix}$$

is not positive semidefinite. We will see in Section 3.3 that for this set of data we can solve the Bernstein-Szegő measure problem. In addition, the measure $0.1d\mu + 0.9\delta_{(1,1)}$, where $d\mu$ is the normalized Lebesgue measure on $\mathbb{T}^2$ and δ_p is the Dirac mass at the point p, yields the moments $c_{00} = 1$, $c_{10} = c_{01} = c_{11} = 0.9$, and thus the moment problem is solvable for this data set.

Proof of Theorem 3.1.2. For the equivalence of (ii) and (iii) observe that $T_j^{\text{card}K} = 0$, and thus for all $j = 1, \ldots, d$ we have that $I - M_{T_j}$ is invertible on $\mathcal{L}(\mathcal{E}^{\text{card}K})$. But then the equalities $\prod_{j=1}^d (I - M_{T_j})(X + X^*) = EC^* + CE^*$ and $\Gamma_i = \prod_{j\neq i}(I - M_{T_j})(G_i)$, $i = 1, \ldots, d$, yield the equivalence.

Next, the equivalence of (i) and (iv) follows directly from writing (3.1.1) in its series expansion. A tedious but straightforward computation will show that for $k \in \mathbb{N}_0^d \setminus \{0\}$ we have that $\phi_k = V^*(\sum_{w\in W_k(UP_1,\ldots UP_d)} w)V$.

It remains to show the equivalence of (i) and (iii). First suppose that (i) holds, and write ϕ as in (3.1.1). Identify $\mathcal{E}^{\text{card}K}$ with $H^2(\mathbb{D}^d, \mathcal{E}, K) := \{f(z) = \sum_{k\in K} f_k z^k, z \in \mathbb{D}^d : f_k \in \mathcal{E}\}$. We may view $T_j : H^2(\mathbb{D}^d, \mathcal{E}, K) \to H^2(\mathbb{D}^d, \mathcal{E}, K)$ as the restriction of the multiplication operator with symbol z_j, namely,

$$(T_j f)(z) = P_K(z_j f(z)),$$

where P_K is the projection $P_K(\sum g_k z^k) = \sum_{k\in K} g_k z^k$. Likewise, X (as defined in (ii)) may be viewed as the restriction of the multiplication operator with symbol ϕ, namely,

$$(Xf)(z) = P_K(\phi(z)f(z)).$$

Define

$$\Omega = [\Omega_1 \ \cdots \ \Omega_d] : \mathcal{H}_1 \oplus \cdots \oplus \mathcal{H}_d \to H^2(\mathbb{D}^d, \mathcal{E}, K)$$

by $\Omega(\xi)(z) = P_K(V^*U(I - Z(z)U)^{-1}\xi)$. Since K is finite, this indeed defines a bounded operator. Then one easily checks that

$$[T_1\Omega_1 \ \cdots \ T_d\Omega_d]\,(\xi)(z) = P_K(V^*U(I - Z(z)U)^{-1}Z(z)\xi). \tag{3.1.4}$$

Letting $\Gamma_i = \Omega_i\Omega_i^*$ we get that

$$CE^* + EC^* = \Gamma_1 - T_1\Gamma_1T_1^* + \cdots + \Gamma_d - T_d\Gamma_dT_d^*.$$

For the last equality, let $h \in H^2(\mathbb{D}^d, \mathcal{E}, K)$ and $x \in \mathcal{E}$ (which we may also view as the constant function in $H^2(\mathbb{D}^d, \mathcal{E})$ with value x). Then, using (3.1.1) and (3.1.4), we get that

$$\begin{aligned} \langle (X^*h)(0), x\rangle_{\mathcal{E}} &= \langle X^*h, x\rangle_{H^2(\mathbb{D}^d,\mathcal{E},K)} = \langle h, Xx\rangle_{H^2(\mathbb{D}^d,\mathcal{E},K)} \\ &= \left\langle h, \frac{1}{2}x + P_K(V^*U(I - Z(z)U)^{-1}Z(z)Vx)\right\rangle_{H^2(\mathbb{D}^d,\mathcal{E},K)} \\ &= \frac{1}{2}\langle h(0), x\rangle_{\mathcal{E}} + \langle V^*(\Omega_1^*T_1^*h \oplus \cdots \oplus \Omega_d^*T_d^*h), x\rangle_{\mathcal{E}}, \end{aligned}$$

and thus we have that

$$(X^*h)(0) = \frac{1}{2}h(0) + V^*(\Omega_1^*T_1^*h \oplus \cdots \oplus \Omega_d^*T_d^*h). \tag{3.1.5}$$

Moreover, again using (3.1.4), we get that

$$\begin{aligned} \langle h, \Omega\xi\rangle_{H^2(\mathbb{D}^d,\mathcal{E},K)} &= \langle h, P_K(V^*U(I - ZU)^{-1}(I - ZU + ZU)\xi)\rangle_{H^2(\mathbb{D}^d,\mathcal{E},K)} \\ &= \langle h, V^*U\xi\rangle_{H^2(\mathbb{D}^d,\mathcal{E},K)} + \langle h, [T_1\Omega_1 \ \cdots \ T_d\Omega_d]\,U\xi\rangle_{H^2(\mathbb{D}^d,\mathcal{E},K)} \\ &= \langle U^*Vh(0), \xi\rangle_{\mathcal{E}} + \langle U^*(\Omega_1^*T_1^*h \oplus \cdots \oplus \Omega_d^*T_d^*h), \xi\rangle_{\mathcal{H}_1\oplus\cdots\oplus\mathcal{H}_d}. \end{aligned}$$

Thus

$$\Omega_1^*h \oplus \cdots \oplus \Omega_d^*h = U^*Vh(0) + U^*(\Omega_1^*T_1^*h \oplus \cdots \oplus \Omega_d^*T_d^*h). \tag{3.1.6}$$

Equations (3.1.5) and (3.1.6) yield that

$$\begin{pmatrix} U^* & U^*V \\ V^* & \frac{1}{2}I \end{pmatrix} \begin{pmatrix} \Omega_1^*T_1^*h \oplus \cdots \oplus \Omega_d^*T_d^*h \\ h(0) \end{pmatrix} = \begin{pmatrix} \Omega_1^*h \oplus \cdots \oplus \Omega_d^*h \\ (X^*h)(0) \end{pmatrix}. \tag{3.1.7}$$

Notice that when

$$\begin{pmatrix} U^* & U^*V \\ V^* & \frac{1}{2}I \end{pmatrix} \begin{pmatrix} x \\ y \end{pmatrix} = \begin{pmatrix} a \\ b \end{pmatrix}$$

we get that

$$\begin{aligned} \|a\|^2 - \|x\|^2 &= \langle U^*x + U^*Vy, U^*x + U^*Vy\rangle - \|x\|^2 \\ &= 2\operatorname{Re}\langle U^*x, U^*Vy\rangle + \langle y, y\rangle = 2\operatorname{Re}\langle V^*x + \frac{1}{2}y, y\rangle = 2\operatorname{Re}\langle b, y\rangle, \end{aligned}$$

where we used that U is unitary and that V is isometric. Thus from (3.1.7) we obtain that

$$2\operatorname{Re}\langle (X^*h)(0), h(0)\rangle = \|\Omega_1^*h\|^2 + \cdots + \|\Omega_d^*h\|^2 - (\|\Omega_1^*S_1^*h\|^2 + \cdots + \|\Omega_d^*S_d^*h\|^2).$$

It is straightforward to check that this is equivalent to the statement that

$$CE^* + EC^* = (I - M_{T_1})(\Gamma_1) + \cdots + (I - M_{T_d})(\Gamma_d).$$

Next, suppose that Γ_i exist as (iii). Equation (3.1.3) implies that

$$2\operatorname{Re}\langle (X^*h)(0), h(0)\rangle = \|\Gamma_1^{1/2}h\|^2 - \|\Gamma_1^{1/2}T_1^*h\|^2 + \cdots + \|\Gamma_d^{1/2}h\|^2 - \|\Gamma_d^{1/2}T_d^*h\|^2$$

for all $h \in H^2(\mathbb{D}^d, \mathcal{E}, K)$. But then

$$\left\|\frac{1}{2}h(0) + (X^*h)(0)\right\|^2 + \sum_{i=1}^{d}\left\|\Gamma_i^{1/2}T_i^*h\right\|^2$$

$$= \left\|\frac{1}{2}h(0) - (X^*h)(0)\right\|^2 + \sum_{i=1}^{d}\left\|\Gamma_i^{1/2}h\right\|^2.$$

Thus the map

$$\begin{pmatrix} \Gamma_1^{1/2}T_1^*h \oplus \cdots \oplus \Gamma_d^{1/2}T_d^*h \\ \frac{1}{2}h(0) + (X^*h)(0) \end{pmatrix} \mapsto \begin{pmatrix} \Gamma_1^{1/2}h \oplus \cdots \oplus \Gamma_d^{1/2}h \\ \frac{1}{2}h(0) - (X^*h)(0) \end{pmatrix} \tag{3.1.8}$$

defines an isometry from $\{\Gamma_1^{1/2}T_1^*h \oplus \cdots \oplus \Gamma_d^{1/2}T_d^*h \oplus (\frac{1}{2}h(0) + (X^*h)(0)) : h \in \mathcal{M}\}$ into $\mathcal{M} \oplus \cdots \oplus \mathcal{M} \oplus \mathcal{E}$, where $\mathcal{M} = H^2(\mathbb{D}^d, \mathcal{E}, K)$. Extend the isometry to a unitary operator $\mathcal{U}$ on $\mathcal{H}_1 \oplus \cdots \oplus \mathcal{H}_d \oplus \mathcal{E}$. By composing $\Gamma_i^{1/2}$ with the embeddings of $\mathcal{M}$ into $\mathcal{H}_i$, we obtain mappings $\Phi_i : \mathcal{M} \to \mathcal{H}_i$ $(i = 1, \dots, d)$ such that

$$\mathcal{U}^* : \begin{pmatrix} \Phi_1 T_1^*h \oplus \cdots \oplus \Phi_d T_d^*h \\ \frac{1}{2}h(0) + (X^*h)(0) \end{pmatrix} = \begin{pmatrix} \Phi_1 h \oplus \cdots \oplus \Phi_d h \\ \frac{1}{2}h(0) - (X^*h)(0) \end{pmatrix}.$$

Decompose

$$\mathcal{U}^* = \begin{pmatrix} \mathcal{U}_{11}^* & \mathcal{U}_{21}^* \\ \mathcal{U}_{12}^* & \mathcal{U}_{22}^* \end{pmatrix} : (\mathcal{H}_1 \oplus \cdots \oplus \mathcal{H}_d) \oplus \mathcal{E} = (\mathcal{H}_1 \oplus \cdots \oplus \mathcal{H}_d) \oplus \mathcal{E}.$$

Let $\xi \in \mathcal{E}$, and consider it as a constant function. Then, since $T_i^*\xi = 0$ and $(X^*\xi)(0) = \frac{1}{2}\xi$, we get that $\mathcal{U}_{22}^*(\xi) = 0$. But, since this holds for all $\xi \in \mathcal{E}$ we get that $\mathcal{U}_{22} = 0$. Let now $V = -\mathcal{U}_{12}$ and $U = \mathcal{U}_{11} - \mathcal{U}_{12}\mathcal{U}_{21}$. Then it is not hard to check that V is an isometry, U is unitary, and $V^*U = \mathcal{U}_{21}$. Furthermore,

$$\begin{pmatrix} U^* & U^*V \\ V^* & \frac{1}{2}I \end{pmatrix} \begin{pmatrix} \Phi_1 T_1^*h \oplus \cdots \oplus \Phi_d T_d^*h \\ h(0) \end{pmatrix} = \begin{pmatrix} \Phi_1 h \oplus \cdots \oplus \Phi_d h \\ (X^*h)(0) \end{pmatrix}. \tag{3.1.9}$$

Define, as before,

$$\Omega = [\Omega_1 \ \cdots \ \Omega_d] : \mathcal{H}_1 \oplus \cdots \oplus \mathcal{H}_d \to H^2(\mathbb{D}^d, \mathcal{E}, K)$$

by $\Omega(\xi)(z) = P_K(V^*U(I - Z(z)U)^{-1}\xi)$. Then (3.1.6) holds. Combining (3.1.6) and (3.1.9) we get that

$$U^*((\Omega_1^* - \Phi_1)T_1^*h \oplus \cdots \oplus (\Omega_d^* - \Phi_d)T_d^*h) = (\Omega_1^* - \Phi_1)h \oplus \cdots \oplus (\Omega_d^* - \Phi_d)h, h \in \mathcal{M}.$$

Thus

$$\Xi := \sum_{i=1}^{d} (\Omega_i^* - \Phi_i)^*(\Omega_i^* - \Phi_i) - \sum_{i=1}^{d} T_i(\Omega_i^* - \Phi_i)^*(\Omega_i^* - \Phi_i)T_i^* = 0.$$

We claim that

$$\Omega_i^* = \Phi_i,\ i = 1, \dots, d. \tag{3.1.10}$$

For this, fix an $N \geq \operatorname{card} K$. Then we have that $T_i^{N+1} = 0$, $i = 1, \dots, d$. Define $S_T(X) = \sum_{i=0}^{N} T^i X T^{i*}$. Clearly,

$$S_T(X - TXT^*) = X - T^{N+1} X T^{(N+1)*}.$$

Moreover, if $T^{N+1} = 0$, then $S_T(X - TXT^*) = X$. Working out the equality $S_{T_1}(\dots(S_{T_d}(\Xi))\dots) = 0$ yields

$$\sum_{i=1}^{d} S_{T_1}(\dots(S_{T_{i-1}}(S_{T_{i+1}}(\dots(S_{T_d}((\Omega_i^* - \Phi_i)^*(\Omega_i^* - \Phi_i))\dots))))) = 0.$$

Thus we obtain a sum of positive semidefinite operators summing to 0, yielding that every term must equal 0. In particular $(\Omega_i^* - \Phi_i)^*(\Omega_i^* - \Phi_i) = 0$, $i = 1, \dots, d$. This yields (3.1.10).

Next, let

$$\phi(z) = \frac{1}{2} I + V^* U (I - Z(z)U)^{-1} Z(z) V.$$

Using (3.1.10) we obtain that for $\xi \in \mathcal{E}$ and $h \in \mathcal{M}$ we have that

$$\begin{aligned} \langle h, T_\phi \xi \rangle &= \left\langle h, \left(\frac{1}{2} I + V^* U (I - Z(z)U)^{-1} Z(z) V \right) \xi \right\rangle \\ &= \left\langle h, \frac{1}{2} \xi \right\rangle + \langle V^*(\Phi_1 T_1^* h \oplus \cdots \oplus \Phi_d T_d^* h), \xi \rangle \\ &= \left\langle \frac{1}{2} h(0), \xi \right\rangle + \left\langle -\frac{1}{2} h(0) + (X^* h)(0), \xi \right\rangle = \langle (X^* h)(0), \xi \rangle = \langle h, X \xi \rangle. \end{aligned}$$

And thus

$$\langle T_\phi^* h, S_1^{n_1} \cdots S_d^{n_d} \xi \rangle = \langle T_\phi^* S_1^{n_1 *} \cdots S_d^{n_d *} h, \xi \rangle$$

$$= \langle X^* T_1^{n_1 *} \cdots T_d^{n_d *} h, \xi \rangle = \langle X^* h, S_1^{n_1} \cdots S_d^{n_d} \xi \rangle.$$

But now it follows that ϕ has the required properties. □

The process of extracting an isometry out of the given data as was done in the proof of (iii) ⇒ (i) is sometimes referred to as a "lurking isometry" technique.

The condition in Theorem 3.1.2(iii) may be checked numerically by semidefinite programming, as the following example illustrates.

Example 3.1.4 Let $K = \{0, 1, 2\}^2$ and $c_{00} = 1/2, c_{01} = 0, c_{02} = 0, c_{10} = 1/2\sqrt{2}, c_{11} = 1/2, c_{12} = -1/4\sqrt{2}, c_{20} = 1/2, c_{21} = 1/2\sqrt{2}$, and $c_{22} = -1/4$.

In order to build the matrices, we order K using the lexicographical order. Using LMIlab we find the matrices

$$\begin{pmatrix} .7437 & .1813 & -.1282 & .5259 & .5000 & -.2629 & .3718 & .6165 & -.1218 \\ .1813 & .1282 & -.0906 & .1282 & .1813 & -.0641 & .0906 & .1922 & .0000 \\ -.1282 & -.0906 & .0641 & -.0906 & -.1282 & .0453 & -.0641 & -.1359 & -.0000 \\ .5259 & .1282 & -.0906 & .3718 & .3536 & -.1859 & .2629 & .4359 & -.0861 \\ .5000 & .1813 & -.1282 & .3536 & .3782 & -.1768 & .2500 & .4442 & -.0609 \\ -.2629 & -.0641 & .0453 & -.1859 & -.1768 & .0930 & -.1315 & -.2180 & .0431 \\ .3718 & .0906 & -.0641 & .2629 & .2500 & -.1315 & .1859 & .3082 & -.0609 \\ .6165 & .1922 & -.1359 & .4359 & .4442 & -.2180 & .3082 & .5320 & -.0861 \\ -.1218 & .0000 & -.0000 & -.0861 & -.0609 & .0431 & -.0609 & -.0861 & .0305 \end{pmatrix},$$

and

$$\begin{pmatrix} .2563 & -.1813 & .1282 & .1813 & -.0000 & -.0906 & .1282 & .0906 & -.1282 \\ -.1813 & .1282 & -.0906 & -.1282 & .0000 & .0641 & -.0906 & -.0641 & .0906 \\ .1282 & -.0906 & .0641 & .0906 & -.0000 & -.0453 & .0641 & .0453 & -.0641 \\ .1813 & -.1282 & .0906 & .3718 & -.1723 & .0578 & .2629 & .0641 & -.1768 \\ -.0000 & .0000 & -.0000 & -.1723 & .1218 & -.0861 & -.1218 & -.0000 & .0609 \\ -.0906 & .0641 & -.0453 & .0578 & -.0861 & .0930 & .0408 & -.0320 & .0022 \\ .1282 & -.0906 & .0641 & .2629 & -.1218 & .0408 & .1859 & .0453 & -.1250 \\ .0906 & -.0641 & .0453 & .0641 & -.0000 & -.0320 & .0453 & .0320 & -.0453 \\ -.1282 & .0906 & -.0641 & -.1768 & .0609 & .0022 & -.1250 & -.0453 & .0945 \end{pmatrix}$$

for Γ_1 and Γ_2, respectively. The corresponding U and V are

$$V = \begin{pmatrix} 0.8543 \\ -0.1178 \\ 0.4141 \\ -0.2913 \end{pmatrix}, \quad U = \begin{pmatrix} 0.7061 & 0.0036 & 0.6952 & 0.1341 \\ -0.0012 & 0.7054 & 0.1318 & -0.6964 \\ 0.6947 & 0.1346 & -0.7066 & 0.0014 \\ 0.1368 & -0.6959 & 0.0006 & -0.7050 \end{pmatrix}.$$

The projections P_1 and P_2 are the projections onto $\text{Span}\{e_1, e_2\}$ and $\text{Span}\{e_3, e_4\}$, respectively, where $e_1, \ldots, e_4$ is the standard basis in $\mathbb{C}^4$. Of course, many choices for U, V, P_1, and P_2 are possible. In fact, the example was constructed using

$$V = \begin{pmatrix} 1 \\ 0 \end{pmatrix}, \quad U = \begin{pmatrix} 1/2\sqrt{2} & 1/2\sqrt{2} \\ 1/2\sqrt{2} & -1/2\sqrt{2} \end{pmatrix},$$

$$P_1 = \begin{pmatrix} 1/2 & 1/2 \\ 1/2 & 1/2 \end{pmatrix}, \quad P_2 = \begin{pmatrix} 1/2 & -1/2 \\ -1/2 & 1/2 \end{pmatrix}.$$

Example 3.1.5 Let $\Lambda = \{0,1\}^2$ and $c_{00} = 1$, $c_{10} = 0.3$, $c_{01} = 0.7$, and $c_{11} = 0.8$. One may check that the matrix

$$\begin{pmatrix} 1 & 0.7 & 0.3 & 0.8 \\ 0.7 & 1 & \overline{x} & 0.3 \\ 0.3 & x & 1 & 0.7 \\ 0.8 & 0.3 & 0.7 & 1 \end{pmatrix}$$

is positive definite when $x = 0$. Using Matlab's LMILab, we solve for Γ_1 and Γ_2 as in Theorem 3.1.2:

$$\Gamma_1 = \begin{pmatrix} 0.3000 & 0.4000 & -0.1000 & 0.0952 \\ 0.4000 & 0.5523 & -0.1523 & 0.1125 \\ -0.1000 & -0.1523 & 0.0523 & -0.0173 \\ 0.0952 & 0.1125 & -0.0173 & 0.0603 \end{pmatrix}$$

and

$$\Gamma_2 = \begin{pmatrix} 0.7000 & 0.3000 & 0.4000 & 0.7048 \\ 0.3000 & 0.1477 & 0.1523 & 0.2875 \\ 0.4000 & 0.1523 & 0.2477 & 0.4173 \\ 0.7048 & 0.2875 & 0.4173 & 0.7397 \end{pmatrix}.$$

In order to see that these matrices are positive semidefinite one may note that $(-1,1,1,0)^T$ belongs to the kernel of both. Furthermore, after we omit the third column and row in both matrices the determinants of the leading principal submatrices in exact arithmetic are

$$\frac{3}{10}, \frac{569}{100000}, \frac{106167}{976562500}, \frac{7}{10}, \frac{1339}{100000}, \frac{993207}{3906250000},$$

yielding the positive semidefiniteness of both Γ_1 and Γ_2. Next, one may easily check that

$$EC^* + CE^* - \Gamma_1 + T_1\Gamma_1T_1^* - \Gamma_2 + T_2\Gamma_2T_2^* = 0,$$

where C, E, T_1, and T_2 are as in Theorem 3.1.2. Thus the Carathéodory interpolation problem is solvable for this data set. We will see in Section 3.3 that for this data set, there is no solution to the Bernstein-Szegő measure problem.

Notice that the equivalence of (iii) and (iv) in Theorem 3.1.2 may be interpreted as a multivariable version of the Naimark dilation theorem on a finite index set. Recall that the Naimark dilation theorem states that a sequence of operators $C_k \in \mathcal{L}(\mathcal{E})$, $k = 1, 2, \dots$, may be represented as $C_k = V^*U^kV$, with $V : \mathcal{E} \to \mathcal{H}$ an isometry and $U : \mathcal{H} \to \mathcal{H}$ a unitary, if and only if for all $k \in \mathbb{N}_0$ the lower triangular Toeplitz operator matrix $(C_{i-j})_{i,j=0}^k$, where $C_0 = \frac{1}{2}I$ and $C_{-1} = C_{-2} = \cdots = 0$, has a positive semidefinite real part. The same is true for a finite collection of operators $C_k \in \mathcal{L}(\mathcal{E})$, $k = 1, 2, \dots, n$. Theorem 3.1.2 now extends the classical result to the case of a finite lower inclusive subset of $\mathbb{N}_0^d$.

Corollary 3.1.6 *Consider the sequence of operators $\{C_k\}_{k\in\mathbb{N}_0^2}$ on a Hilbert space $\mathcal{E}$ with $C_0 = I$. Define $C_{-k} = C_k^*$, $k \in \mathbb{N}_0^2$, and $C_k = 0$, $k \in \mathbb{Z}^2 \setminus (\mathbb{N}_0^2 \cup -\mathbb{N}_0^2)$. Then the sequence $\{C_k\}_{k\in\mathbb{N}_0^2}$ is positive semidefinite in the sense that for every finite $K \subset \mathbb{Z}^2$ the operator matrix $(C_{k-l})_{k,l\in K}$ is positive semidefinite if and only if there exists a Hilbert space $\mathcal{H} = \mathcal{H}_1 \oplus \mathcal{H}_2$, an isometry $V : \mathcal{E} \to \mathcal{H}$, and a unitary $U \in \mathcal{L}(\mathcal{H})$ such that*

$$C_k = V^* \left(\sum_{w\in W_k(UP_1,UP_2)} w \right) V, \ k \in K \setminus \{0\}, \tag{3.1.11}$$

where P_i is the orthogonal projection in $\mathcal{L}(\mathcal{H})$ with image $\mathcal{H}_i$, $i = 1, 2$.

3.1.1 Carathéodory interpolation for holomorphic functions

In this subsection we explore the Carathéodory interpolation problem in the class $\mathcal{M}_d(\mathcal{E})$ of holomorphic $\mathcal{L}(\mathcal{E})$-valued functions ϕ defined on $\mathbb{D}^d$ with positive real part. Thus,

$$\mathcal{M}_d(\mathcal{E}) = \{\phi : \mathbb{D}^d \to \mathcal{L}(\mathcal{E}) : \phi \text{ analytic and } \operatorname{Re}\phi(z) \geq 0, \ z \in \mathbb{D}^d\}.$$

As observed before, $\mathcal{A}_d(\mathcal{E}) \subseteq \mathcal{M}_d(\mathcal{E})$, and the inclusion is strict when $d \geq 3$ (see Example 3.1.8, where $\mathcal{E} = \mathbb{C}^2$, or Example 4.6.8, where $\mathcal{E} = \mathbb{C}$). We shall confine ourselves to the situation where $K^{(d)}(n)$ is the subset of $\mathbb{N}_0^d$ consisting of points $x = (x_1, \dots, x_d)$ satisfying $|x| := \sum_{i=1}^d x_i \leq n$.

We have the following necessary condition. Denote

$$K_j^{(d)} = \{(x_1, \dots, x_d) \in \mathbb{N}_0^d \, : \, x_1 + \cdots + x_d = j\}.$$

Further, for $j = 1, \dots, d$ let $S_j : H^2(\mathbb{D}^d, \mathbb{C}) \to H^2(\mathbb{D}^d, \mathbb{C})$ be the multiplication operator with symbol z_j, namely,

$$(S_j f)(z) = z_j f(z).$$

For $k = (k_1, \dots, k_d)$ we let S^k denote the operator $S_1^{k_1} \cdots S_d^{k_d}$ on $H^2(\mathbb{D}^d, \mathbb{C})$.

Theorem 3.1.7 *Let $n \in \mathbb{N}$ and $K^{(d)}(n) = \bigcup_{j=0}^n K_j^{(d)}$. Given are operators $C_k \in \mathcal{L}(\mathcal{E})$, $k \in K^{(d)}(n)$, with $C_0 = \frac{1}{2}I$. If there exists $f(z) = \sum_{k \in \mathbb{N}_0^d} f_k z^k \in \mathcal{M}_d(\mathcal{E})$ with $f_k = C_k$, $k \in K$, then the operator*

$$\Gamma := (\Gamma_{p-q})_{p,q=0}^n \tag{3.1.12}$$

has positive semidefinite real part. Here $\Gamma_p = \sum_{k \in K_p^{(d)}} C_k \otimes S^k$. When $n = 1$ the converse is also valid.

Proof of Theorem 3.1.7. First suppose that $f \in \mathcal{M}_d(\mathcal{E})$ exists with $f_k = C_k$, $k \in K^{(d)}(n)$. We can find unitary liftings $U_i : \mathcal{K} \to \mathcal{K}$ of S_i such that $U_1, \dots, U_d$ commute; for instance, take U_i to be the multiplication with z_i on $L^2(\mathbb{T}^d, \mathbb{C})$. Let now $g(w) = \sum_{j=0}^\infty w^j (\sum_{k \in K_j^{(d)}} f_k \otimes U^k), w \in \mathbb{D}$, where $U^k = U_1^{k_1} \cdots U_d^{k_d}$. Then $g \in \mathcal{M}_1(\mathcal{E} \otimes \mathcal{K})$, and thus by the classical one-variable result we get that $\Gamma + \Gamma^* \geq 0$.

Let now $n = 1$ and suppose that $\Gamma + \Gamma^* \geq 0$. We let $e_1, \dots, e_d$ denote the standard basis in $\mathbb{N}_0^d$. Since $\Gamma + \Gamma^* \geq 0$ it follows that $\| \sum_{i=1}^d z_i C_{e_i} \| < 1$ for all $z = (z_1, \dots, z_d) \in \mathbb{D}^d$. But then

$$f(z) := \frac{1}{2}I + \left(I - \sum_{i=1}^d z_i C_{e_i} \right)^{-1} \sum_{i=1}^d z_i C_{e_i}, z \in \mathbb{D}^d$$

has the required properties.

We observe that even in the case $n = 1$ the condition on C_k, $k \in K^{(d)}(n)$, in Theorem 3.1.7 is much weaker than the condition in Theorem 3.1.2. In fact, below we provide an example for which the Carathéodory interpolation problem is solvable in $\mathcal{M}_d(\mathcal{E})$ but not in $\mathcal{A}_d(\mathcal{E})$.

Example 3.1.8 First we observe the following. Let

$$A_1 = \begin{pmatrix} 1 & 0 \\ 0 & -1 \end{pmatrix}, \quad A_2 = \begin{pmatrix} 0 & 1 \\ 1 & 0 \end{pmatrix}, \quad A_3 = \begin{pmatrix} 1 & 0 \\ 0 & 1 \end{pmatrix},$$

and let T_i, $i = 1, 2, 3$, be the pairwise commuting contractions

$$T_1 = \begin{pmatrix} 0 & 0 \\ A_1 & 0 \end{pmatrix}, \quad T_2 = \begin{pmatrix} 0 & 0 \\ A_2 & 0 \end{pmatrix}, \quad T_3 = \begin{pmatrix} 0 & 0 \\ A_3 & 0 \end{pmatrix}. \tag{3.1.13}$$

Then

$$\|A_1 \otimes T_1 + A_2 \otimes T_2 + A_3 \otimes T_3\| = 3 > \alpha := \max_{z_1,z_2,z_3 \in \mathbb{T}} \|z_1 A_1 + z_2 A_2 + z_3 A_3\|. \tag{3.1.14}$$

Indeed, the left-hand side equals the norm of

$$A_1 \otimes A_1 + A_2 \otimes A_2 + A_3 \otimes A_3 = \begin{pmatrix} 2 & 0 & 0 & 1 \\ 0 & 0 & 1 & 0 \\ 0 & 1 & 0 & 0 \\ 1 & 0 & 0 & 2 \end{pmatrix},$$

which equals 3 (note that $\begin{pmatrix} 1 & 0 & 0 & 1 \end{pmatrix}^T$ is an eigenvector with eigenvalue 3), while the right-hand side of (3.1.14) can only equal 3 if we can find unit vectors $x, y \in \mathbb{C}^2$ such that

$$|x^* A_1 y| = |x^* A_2 y| = |x^* A_3 y| = 1.$$

As A_1, A_2, A_3 are contractions, this can only happen when $x, A_1 y, A_2 y, A_3 y$ are multiples of one another, which is impossible. Thus the right-hand side of (3.1.14) equals $\alpha < 3$ (in fact, $\alpha = \sqrt{6}$).

Let now $C_{(0,0,0)} = \frac{1}{2} I$, $C_{(1,0,0)} = \frac{1}{\alpha} A_1$, $C_{(0,1,0)} = \frac{1}{\alpha} A_2$, and $C_{(0,0,1)} = \frac{1}{\alpha} A_3$. The right-hand inequality in (3.1.14) implies that the condition $\Gamma + \Gamma^* \geq 0$ in Theorem 3.1.7 is satisfied. Thus the Carathéodory interpolation problem with data C_k, $k \in K^{(3)}(1)$ has a solution in $\mathcal{M}_3(\mathcal{E})$. The left-hand inequality in (3.1.14) implies, however, that the Carathéodory interpolation problem with data C_k, $k \in K^{(3)}(1)$ does not have a solution in $\mathcal{A}_3(\mathcal{E})$. Indeed, if a solution $\phi \in \mathcal{A}_3(\mathcal{E})$ exists, then the single-variable function

$$f(z) := \phi(zT_1, zT_2, zT_3)$$

should have the property that $\operatorname{Re} f(z) \geq 0$ for $z \in \mathbb{D}$. In particular,

$$\frac{df}{dz}(0) = \frac{1}{\alpha}(A_1 \otimes T_1 + A_2 \otimes T_2 + A_3 \otimes T_3)$$

should be a contraction, which contradicts (3.1.14).

Though the Carathéodory interpolation problems in $\mathcal{M}_d(\mathcal{E})$ and $\mathcal{A}_d(\mathcal{E})$ are in general different, there are also cases when the problems in the two classes are only solvable simultaneously. This happens for instance when $\mathcal{E} = \mathbb{C}$ and $n = 1$.

Proposition 3.1.9 *Given are complex numbers c_k, $k \in K^{(d)}(1)$, with $c_0 = \frac{1}{2}$. The following are equivalent.*

(i) There exist $f(z) = \sum_{k \in \mathbb{N}_0^d} f_k z^k \in \mathcal{A}_d(\mathbb{C})$ with $f_k = c_k$, $k \in K^{(d)}(1)$.

(ii) There exist $f(z) = \sum_{k \in \mathbb{N}_0^d} f_k z^k \in \mathcal{M}_d(\mathbb{C})$ with $f_k = c_k$, $k \in K^{(d)}(1)$.

(iii) $\sum_{i=1}^d |c_{e_i}| \leq 1$, where $e_1, \ldots, e_d$ is the standard basis in $\mathbb{N}_0^d$.

Proof. We will prove (i) $\Rightarrow$ (ii) $\Rightarrow$ (iii) $\Rightarrow$ (i).

Since $\mathcal{A}_d(\mathbb{C}) \subset \mathcal{M}_d(\mathbb{C})$, the implication (i) $\Rightarrow$ (ii) follows.

Suppose that (ii) holds. Choose complex numbers α_i, $i = 1, \ldots, d$, of modulus 1, such that $\alpha_i c_{e_i} \geq 0$, and let $g(w) = f(\alpha_1 w, \ldots, \alpha_d w)$, $w \in \mathbb{D}$. Then $\operatorname{Re} g(w) \geq 0$ for $w \in \mathbb{D}$, and by the one-variable results we get that $|g'(0)| \leq 1$. But this yields $\sum_{i=1}^d |c_{e_i}| \leq 1$.

Suppose that (iii) holds. Consider the functions $g_1(z_1) = \frac{1}{2}|c_{e_1}| + c_{e_1} z_1$, $\ldots$, $g_{d-1}(z_{d-1}) = \frac{1}{2}|c_{e_{d-1}}| + c_{e_{d-1}} z_{d-1}$, $g_d(z_d) = \frac{1}{2}(1 - |c_{e_1}| - \cdots - |c_{e_{d-1}}|) + c_{e_d} z_d$. By the one-variable result, there exist for $i = 1, \ldots, d$ functions $f_i \in \mathcal{A}_1(\mathbb{C})$ such that $f_i(0) = g_i(0)$ and $f_i'(0) = g_i'(0)$. But then $f(z) = f_1(z_1) + \cdots + f_d(z_d)$ is a function satisfying (i). $\square$

Notice that the proof of (iii) $\Rightarrow$ (i) in Proposition 3.1.9 goes through for operators C_{e_i}. That is, if $\sum_{i=1}^d \|C_{e_i}\| \leq 1$, then there exists a solution to the Carathéodory interpolation problem in $\mathcal{A}_d(\mathcal{E})$ with given data $C_0 = \frac{1}{2}I, C_{e_1}, \ldots, C_{e_d}$. Clearly, the condition $\sum_{i=1}^d \|C_{e_i}\| \leq 1$ is not necessary in general as one can start with a Hilbert space $\mathcal{H}$ of dimension greater than or equal to $d \geq 2$, a unitary $U : \mathcal{H} \to \mathcal{H}$, and nontrivial orthogonal projections $P_1, \ldots, P_d$ on $\mathcal{H}$ with $\oplus_{i=1}^d \operatorname{Ran} P_i = \mathcal{H}$, and put $C_{e_i} = UP_i$. Then $\frac{1}{2}I + (I - \sum_{i=1}^d z_i C_{e_i})^{-1} \sum_{i=1}^d z_i C_{e_i} \in \mathcal{A}_d(\mathcal{H})$ but $\sum_{i=1}^d \|C_{e_i}\| = d > 1$.

3.2 INVERSES OF MULTIVARIABLE TOEPLITZ MATRICES AND CHRISTOFFEL-DARBOUX FORMULAS

We need the following notation. For $z = (z_1, \ldots, z_d) \in \mathbb{C}^d$ and $k = (k_1, \ldots, k_d) \in \mathbb{Z}^d$, we let $z^k = z_1^{k_1} \cdots z_d^{k_d}$, where for negative k_i we have $z_i \neq 0$. When $n = (n_1, \ldots, n_d) \in \mathbb{N}_0^d$ we let $\underline{n}$ denote the set $\underline{n} = \underline{n_1} \times \cdots \times \underline{n_d}$, where $\underline{p} = \{0, \ldots, p\}$. Let $n \in \mathbb{N}_0^d$. When $P(z) = \sum_{k \in \underline{n}} P_k z^k$, with $P_k \in \mathcal{L}(\mathcal{H})$, we say that P is an operator-valued polynomial *of degree at most* n. We say that P is *stable* when $P(z)$ is invertible for $z \in \overline{\mathbb{D}}^d$. With P we associate its *adjoint* P^*, which is given by $P^*(z) = \sum_{k \in \underline{n}} P_k^* z^{-k}$. We will also associate with P its *reverse* $\overleftarrow{P}$ given by $\overleftarrow{P}(z) = z^n P(\frac{1}{\bar{z}})^*$. In other words, $\overleftarrow{P}(z) = \sum_{k \in \underline{n}} P_k^* z^{n-k}$. Note that the definition depends on the choice of n. This choice will always be made clear (and is typically n).

Theorem 3.2.1 *Let $n \in \mathbb{N}_0^d$, and let*

$$P(z) = \sum_{k \in \underline{n}} P_k z^k \text{ and } R(z) = \sum_{k \in \underline{n}} R_k z^k$$

be operator-valued stable polynomials of degree at most n such that

$$P(z)P(z)^* = R(z)^* R(z), z \in \mathbb{T}^d.$$

Put

$$F(z) = P(z)^{*-1} P(z)^{-1} = R(z)^{-1} R(z)^{*-1}, \tag{3.2.1}$$

and let $F(z) = \sum_{j\in\mathbb{Z}^d} F_j z^j$, $z \in \mathbb{T}^d$, *be its Fourier expansion. Put* $\Lambda = \underline{n} \setminus \{n\}$. *Then*

$$0 < \left[(F_{k-l})_{k,l\in\Lambda}\right]^{-1} \leq AA^* - B^*B, \tag{3.2.2}$$

where

$$A = (P_{k-l})_{k,l\in\Lambda}, \quad B = (R_{k-l})_{\substack{k\in n+\Lambda \\ l\in\Lambda}}$$

and $P_k = R_k = 0$ *whenever* $k \notin \Lambda$. *When* $d = 1$ *the second inequality in* (3.2.2) *is an equality.*

Let us start by illustrating Theorem 3.2.1 in a low dimensional case.

Example 3.2.2 Let $d = 2$ and $n_1 = n_2 = 2$. Thus $\Lambda = (\{0,1,2\}\times\{0,1,2\}) \setminus \{(2,2)\}$, which we will order lexicographically, giving

$$\Lambda = \{(0,0),(0,1),(0,2),(1,0),(1,1),(1,2),(2,0),(2,1)\}.$$

Then

$$A = \begin{pmatrix} P_{00} & & & & & & & \\ P_{01} & P_{00} & & & & & & \\ P_{02} & P_{01} & P_{00} & & & & & \\ P_{10} & & & P_{00} & & & & \\ P_{11} & P_{10} & & P_{01} & P_{00} & & & \\ P_{12} & P_{11} & P_{10} & P_{02} & P_{01} & P_{00} & & \\ P_{20} & & & P_{10} & & & P_{00} & \\ P_{21} & P_{20} & & P_{11} & P_{10} & & P_{01} & P_{00} \end{pmatrix}$$

and

$$B = \begin{pmatrix} R_{22} & R_{21} & R_{20} & R_{12} & R_{11} & R_{10} & R_{02} & R_{01} \\ & R_{22} & R_{21} & & R_{12} & R_{11} & & R_{02} \\ & & R_{22} & & & R_{12} & & \\ & & & R_{22} & R_{21} & R_{20} & R_{12} & R_{11} \\ & & & & R_{22} & R_{21} & & R_{12} \\ & & & & & R_{22} & & \\ & & & & & & R_{22} & R_{21} \\ & & & & & & & R_{22} \end{pmatrix}.$$

If we let $(F_{k-l})_{k,l\in\Lambda}$ be the matrix

$$\begin{pmatrix} 0.1532 & 0.0186 & 0.0005 & 0.0084 & 0.0155 & 0.0231 & 0.0118 & -0.0079 \\ 0.0186 & 0.1532 & 0.0186 & 0.0005 & 0.0084 & 0.0155 & 0.0017 & 0.0118 \\ 0.0005 & 0.0186 & 0.1532 & 0.0016 & 0.0005 & 0.0084 & 0.0002 & 0.0017 \\ 0.0084 & 0.0005 & 0.0016 & 0.1532 & 0.0186 & 0.0005 & 0.0084 & 0.0155 \\ 0.0155 & 0.0084 & 0.0005 & 0.0186 & 0.1532 & 0.0186 & 0.0005 & 0.0084 \\ 0.0231 & 0.0155 & 0.0084 & 0.0005 & 0.0186 & 0.1532 & 0.0016 & 0.0005 \\ 0.0118 & 0.0017 & 0.0002 & 0.0084 & 0.0005 & 0.0016 & 0.1532 & 0.0186 \\ -0.0079 & 0.0118 & 0.0017 & 0.0155 & 0.0084 & 0.0005 & 0.0186 & 0.1532 \end{pmatrix},$$

then $AA^* - B^*B - \left[(F_{k-l})_{k,l\in\Lambda}\right]^{-1}$ equals the positive semidefinite matrix

$$\begin{pmatrix} 0 & 0 & 0 & 0 & 0 & 0 & 0 & 0 \\ 0 & 0.1193 & 0.0566 & 0 & -0.0166 & 0.0954 & 0 & 0 \\ 0 & 0.0566 & 0.2564 & 0 & 0.0003 & 0.0339 & 0 & 0 \\ 0 & 0 & 0 & 0.0813 & -0.0441 & 0 & -0.0249 & -0.0309 \\ 0 & -0.0166 & 0.0003 & -0.0441 & 0.0494 & -0.0129 & 0.0528 & 0.0262 \\ 0 & 0.0954 & 0.0339 & 0 & -0.0129 & 0.0795 & 0 & 0 \\ 0 & 0 & 0 & -0.0249 & 0.0528 & 0 & 0.1575 & -0.0183 \\ 0 & 0 & 0 & -0.0309 & 0.0262 & 0 & -0.0183 & 0.0622 \end{pmatrix}.$$

The zeros in the matrix are no coincidence. We will explain this further in Corollary 3.2.6.

Assume that the operator matrix $(A_{ij})_{i,j=1}^2 : \mathcal{H}_1 \oplus \mathcal{H}_2 \to \mathcal{H}_1 \oplus \mathcal{H}_2$ and the operator A_{22} are invertible. Then by Exercise 2.9.2, $S = A_{11} - A_{12}A_{22}^{-1}A_{21}$ is invertible and

$$\begin{pmatrix} A_{11} & A_{12} \\ A_{21} & A_{22} \end{pmatrix}^{-1} = \begin{pmatrix} S^{-1} & * \\ * & * \end{pmatrix}. \tag{3.2.3}$$

We will be interested in looking at (3.2.3) in the following way. Suppose that we have identified the inverse of a block matrix, and we are interested in the inverse of the (1,1) block. Then taking a Schur complement in the inverse of the complete block matrix gives us a formula for the inverse of this $(1,1)$ block. Below is one way one can use this observation.

Lemma 3.2.3 *Let lower-upper and upper-lower factorizations of the inverse of a block matrix be given, as follows:*

$$\begin{pmatrix} B_{11} & B_{12} \\ B_{21} & B_{22} \end{pmatrix}^{-1} = \begin{pmatrix} P_{11} & 0 \\ P_{21} & P_{22} \end{pmatrix} \begin{pmatrix} Q_{11} & Q_{12} \\ 0 & Q_{22} \end{pmatrix} \tag{3.2.4}$$

$$= \begin{pmatrix} R_{11} & R_{12} \\ 0 & R_{22} \end{pmatrix} \begin{pmatrix} T_{11} & 0 \\ T_{21} & T_{22} \end{pmatrix}, \tag{3.2.5}$$

and suppose that R_{22} and T_{22} are invertible. Then

$$B_{11}^{-1} = P_{11}Q_{11} - R_{12}T_{21}. \tag{3.2.6}$$

Proof. Apply (3.2.3) to the equality

$$\begin{pmatrix} B_{11} & B_{12} \\ B_{21} & B_{22} \end{pmatrix}^{-1} = \begin{pmatrix} P_{11}Q_{11} & R_{12}T_{22} \\ R_{22}T_{21} & R_{22}T_{22} \end{pmatrix}.$$

□

Corollary 3.2.4 *Consider a positive definite operator matrix $(B_{ij})_{i,j=1}^3$ of which the lower-upper and upper-lower block Cholesky factorizations of its inverse are given, as follows:*

$$[(B_{ij})_{i,j=1}^3]^{-1} = \begin{pmatrix} P_{11} & 0 & 0 \\ P_{21} & P_{22} & 0 \\ P_{31} & P_{32} & P_{33} \end{pmatrix} \begin{pmatrix} P_{11}^* & P_{21}^* & P_{31}^* \\ 0 & P_{22}^* & P_{32}^* \\ 0 & 0 & P_{33}^* \end{pmatrix} \tag{3.2.7}$$

$$= \begin{pmatrix} R_{11}^* & R_{21}^* & R_{31}^* \\ 0 & R_{22}^* & R_{32}^* \\ 0 & 0 & R_{33}^* \end{pmatrix} \begin{pmatrix} R_{11} & 0 & 0 \\ R_{21} & R_{22} & 0 \\ R_{31} & R_{32} & R_{33} \end{pmatrix}, \tag{3.2.8}$$

with R_{22} and R_{33} invertible. Then

$$B_{11}^{-1} = P_{11}P_{11}^* - R_{21}^*R_{21} - R_{31}^*R_{31} \le P_{11}P_{11}^* - R_{31}^*R_{31}. \tag{3.2.9}$$

Proof. By Lemma 3.2.3 we have that

$$B_{11}^{-1} = P_{11}P_{11}^* - \begin{pmatrix} R_{21}^* & R_{31}^* \end{pmatrix} \begin{pmatrix} R_{21} \\ R_{31} \end{pmatrix}$$

$$= P_{11}P_{11}^* - R_{21}^*R_{21} - R_{31}^*R_{31} \le P_{11}P_{11}^* - R_{31}^*R_{31}. \tag{3.2.10}$$

□

Before we prove the main result, we need to introduce some notation. We let $L^\infty(\mathbb{T}^d, \mathcal{L}(\mathcal{H}))$ denote the Lebesgue space of essentially bounded $\mathcal{L}(\mathcal{H})$-valued measurable functions on $\mathbb{T}^d$, and we let $L^2(\mathbb{T}^d, \mathcal{H})$ and $H^2(\mathbb{T}^d, \mathcal{H})$ denote the Lebesgue and Hardy spaces of square integrable $\mathcal{H}$-valued functions on $\mathbb{T}^d$, respectively. As usual we view H^2 as a subspace of L^2. For $L(z) = \sum_{i\in\mathbb{Z}^d} L_i z^i \in L^\infty(\mathbb{T}^d, \mathcal{H})$ we will consider its multiplication operator $M_L : L^2(\mathbb{T}^d, \mathcal{H}) \to L^2(\mathbb{T}^d, \mathcal{H})$ given by

$$(M_L(f))(z) = L(z)f(z).$$

The Toeplitz operator $T_L : H^2(\mathbb{T}^d, \mathcal{H}) \to H^2(\mathbb{T}^2, \mathcal{H})$ is defined as the compression of M_L to $H^2(\mathbb{T}^d, \mathcal{H})$. For $\Lambda \subset \mathbb{Z}^d$ we let S_Λ denote the subspace $\{F \in L^2(\mathbb{T}^d, \mathcal{H}) : F(z) = \sum_{k\in\Lambda} F_k z^k\}$ consisting of those functions with Fourier support in Λ. In addition, we let P_Λ denote the orthogonal projection onto S_Λ. So, for instance, $P_{\mathbb{N}_0^d}$ is the orthogonal projection onto $H^2(\mathbb{T}^d, \mathcal{H})$ and $T_L = P_{\mathbb{N}_0^d} M_L P_{\mathbb{N}_0^d}$.

Proof of Theorem 3.2.1. Clearly we have that $M_{F^{-1}} = M_P M_{P^*} = M_{R^*} M_R$. With respect to the decomposition $L^2(\mathbb{T}^d, \mathcal{H}) = H^2(\mathbb{T}^d, \mathcal{H})^\perp \oplus H^2(\mathbb{T}^d, \mathcal{H})$ we get that

$$M_F = \begin{pmatrix} * & * \\ * & T_F \end{pmatrix}, \ M_P = \begin{pmatrix} * & 0 \\ * & T_P \end{pmatrix}, \ M_{P^{-1}} = \begin{pmatrix} * & 0 \\ * & T_{P^{-1}} \end{pmatrix}, \tag{3.2.11}$$

$$M_R = \begin{pmatrix} * & 0 \\ * & T_R \end{pmatrix}, \ M_{R^{-1}} = \begin{pmatrix} * & 0 \\ * & T_{R^{-1}} \end{pmatrix}, \tag{3.2.12}$$

where we used that $M_{P^{\pm 1}}[H^2(\mathbb{T}^d, \mathcal{H})] \subset H^2(\mathbb{T}^d, \mathcal{H})$ and $M_{R^{\pm 1}}[H^2(\mathbb{T}^d, \mathcal{H})] \subset H^2(\mathbb{T}^d, \mathcal{H})$, which follows as $P^{\pm 1}$ and $R^{\pm 1}$ are analytic in $\overline{\mathbb{D}}^d$. It now follows that $T_F = (T_P)^{*-1}(T_P)^{-1}$, and thus

$$(T_F)^{-1} = T_P(T_P)^*. \tag{3.2.13}$$

Next, decompose $H_2 = S_\Lambda \oplus S_\Theta \oplus S_{n+\mathbb{N}_0^d}$, where $\Lambda = \underline{n} \setminus \{n\}$ and $\Theta = \mathbb{N}_0^d \setminus (\Lambda \cup (n + \mathbb{N}_0^d))$, and write T_P and T_R with respect to this decomposition:

$$T_P = \begin{pmatrix} P_{11} & & \\ P_{21} & P_{22} & \\ P_{31} & P_{32} & P_{33} \end{pmatrix}, \quad T_R = \begin{pmatrix} R_{11} & & \\ R_{21} & R_{22} & \\ R_{31} & R_{32} & R_{33} \end{pmatrix}. \tag{3.2.14}$$

As the Fourier support of P and R lies in $\underline{n}$, and as $P(z)P(z)^* = R(z)^*R(z)$ on $\mathbb{T}^d$, it is not hard to show that

$$T_P T_P^* P_{n+\mathbb{N}_0^d} = T_R^* T_R P_{n+\mathbb{N}_0^d}, \tag{3.2.15}$$

which yields that

$$P_{31}P_{31}^* + P_{32}P_{32}^* + P_{33}P_{33}^* = R_{33}^* R_{33},$$

$$P_{21}P_{31}^* + P_{22}P_{32}^* = R_{32}^* R_{33}, \ P_{11}P_{31}^* = R_{31}^* R_{11}.$$

Thus we can factor $T_P T_P^*$ as

$$T_P T_P^* = \begin{pmatrix} \widetilde{R}_{11}^* & \widetilde{R}_{21}^* & R_{31}^* \\ & \widetilde{R}_{22}^* & R_{32}^* \\ & & R_{33}^* \end{pmatrix} \begin{pmatrix} \widetilde{R}_{11} & & \\ \widetilde{R}_{21} & \widetilde{R}_{22} & \\ R_{31} & R_{32} & R_{33} \end{pmatrix} \tag{3.2.16}$$

for some $\widetilde{R}_{11}, \widetilde{R}_{21}$, and $\widetilde{R}_{22}$. Combining now (3.2.13) with the two factorizations of $T_P T_P^*$ given via (3.2.16), we get by Corollary 3.2.4 that

$$[(F_{k-l})_{k,l \in \Lambda}]^{-1} \leq P_{11} P_{11}^* - R_{31} R_{31}^*.$$

This proves the claim. □

A more detailed analysis of where $T_P T_P^*$ and $T_R^* T_R$ coincide, other than indicated in (3.2.15), exposes some of the zero entries in $[(F_{k-l})_{k,l \in \Lambda}]^{-1}$. As we will see, the argument works only in the case of two variables ($d = 2$).

Lemma 3.2.5 *Let $d = 2$ and let n, P, R, F, and Λ be as in Theorem 3.2.1. Then*

$$[(F_{k-l})_{k,l \in \Lambda}]^{-1}$$

has zeros in locations $(k, l) = ((k_1, k_2), (l_1, l_2))$, where $(k_1, l_2) = (n_1, n_2)$ or $(k_2, l_1) = (n_2, n_1)$.

Proof. Let us decompose $H^2(\mathbb{T}^d, \mathcal{H})$ as

$$S_{\underline{n_1-1} \times \underline{n_2-1}} \oplus S_{\underline{n_1-1} \times \{n_2\}} \oplus S_{\{n_1\} \times \underline{n_2-1}} \oplus S_{n_1+\mathbb{N} \times \underline{n_2-1}} \oplus S_{\underline{n_1-1} \times n_2+\mathbb{N}} \oplus S_{n+\mathbb{N}_0^2}. \tag{3.2.17}$$

Writing $T_P T_P^* - T_R^* T_R$ with respect to this decomposition we get that this operator is of the form

$$T_P T_P^* - T_R^* T_R = \begin{pmatrix} * & * & * & * & * & 0 \\ * & * & 0 & 0 & * & 0 \\ * & 0 & * & * & 0 & 0 \\ * & 0 & * & * & 0 & 0 \\ * & * & 0 & 0 & * & 0 \\ 0 & 0 & 0 & 0 & 0 & 0 \end{pmatrix}.$$

For instance, to explain the zeros in the $(2, 3)$ and $(2, 4)$ positions, note that

$$P_{\underline{n_1-1} \times \{n_2\}} T_P T_P^* P_{(n_1+\mathbb{N}_0) \times \underline{n_2-1}} = P_{\underline{n_1-1} \times \{n_2\}} T_R^* T_R P_{(n_1+\mathbb{N}_0) \times \underline{n_2-1}}.$$

The fact that the last column is zero (and by symmetry, so is the last row) comes from observation (3.2.15). As a general observation, notice that if operator matrices G and H coincide on certain locations, then so will the Schur complement expressions $G - KL^{-1}K^*$ and $H - KL^{-1}K^*$. Therefore, taking the Schur complement in $T_P T_P^*$ and $T_R^* T_R$ with respect to the last row and column (which is where they completely coincide), we see that the resulting operators Σ_P and Σ_R, satisfy

$$\Sigma_P - \Sigma_R = \begin{pmatrix} * & * & * & * & * \\ * & * & 0 & 0 & * \\ * & 0 & * & * & 0 \\ * & 0 & * & * & 0 \\ * & * & 0 & 0 & * \end{pmatrix}. \tag{3.2.18}$$

The matrix $[(F_{k-l})_{k,l\in\Lambda}]^{-1}$ is the Schur complement of $T_P T_P^*$ supported in Λ, which is the same as the Schur complement in Σ_P supported in the first three rows and columns (as $\Lambda = (\underline{n_1-1}\times\underline{n_2-1})\cup(\underline{n_1-1}\times\{n_2\})\cup(\{n_1\}\times\underline{n_2-1})$). Using Lemma 2.4.9 one gets that $\Sigma_R = G^*G$, where

$$G = (R_{k-l})_{k,l\in\mathbb{N}_0^2\setminus n+\mathbb{N}_0^2}.$$

With respect to the decomposition (3.2.17) we have that Σ_R is of the form

$$\Sigma_R = \begin{pmatrix} * & * & * & * & * \\ * & * & 0 & 0 & * \\ * & 0 & * & * & 0 \\ * & 0 & * & * & 0 \\ * & * & 0 & 0 & * \end{pmatrix}.$$

By (3.2.18) it follows that Σ_P must have the same form. But now, if we take the Schur complement in Σ_P supported in the first three rows and columns, we get that $[(F_{k-l})_{k,l\in\Lambda}]^{-1}$ equals

$$\begin{pmatrix} * & * & * \\ * & * & 0 \\ * & 0 & * \end{pmatrix} - \begin{pmatrix} * & * \\ 0 & * \\ * & 0 \end{pmatrix}\begin{pmatrix} * & 0 \\ 0 & * \end{pmatrix}^{-1}\begin{pmatrix} * & 0 & * \\ * & * & 0 \end{pmatrix} = \begin{pmatrix} * & * & * \\ * & * & 0 \\ * & 0 & * \end{pmatrix}.$$

This proves the result. □

Corollary 3.2.6 *Let $d = 2$ and let A, B, and F be as in Theorem 3.2.1. Then*

$$AA^* - B^*B - [(F_{k-l})_{k,l\in\Lambda}]^{-1}$$

has zeros in the locations $(k,l) = ((k_1,k_2),(l_1,l_2))$, where $k = 0$ or $l = 0$ or $(k_1,l_2) = (n_1,n_2)$ or $(k_2,l_1) = (n_2,n_1)$.

Proof. As AA^*, B^*B and $[(F_{k-l})_{k,l\in\Lambda}]^{-1}$ all have zeros in positions $(k,l) = ((k_1,k_2),(l_1,l_2))$, where $(k_1,l_2) = (n_1,n_2)$ or $(k_2,l_1) = (n_2,n_1)$, then so does $AA^* - B^*B - [(F_{k-l})_{k,l\in\Lambda}]^{-1}$. It therefore remains to show that $AA^* - B^*B - [(F_{k-l})_{k,l\in\Lambda}]^{-1}$ has zeros in locations (k,l) where k or l is zero. We focus on the case when $l = 0$. We will show this by showing that $[(F_{k-l})_{k,l\in\Lambda}]^{-1}$ and $AA^* - B^*B$ coincide in the first column ($l = 0$). It follows from (3.2.1). Indeed, as $F(z)P(z) = P(z)^{*-1}$ and $P(z)$ is stable, we have that

$$\sum_{l\in\underline{n}} F_{-l}P_l = P_0^{*-1},\ \sum_{l\in\underline{n}} F_{k-l}P_l = 0,\ k\in\underline{n}\setminus\{0\}. \tag{3.2.19}$$

In addition, since $F(z)R(z)^* = R(z)^{-1}$, we also have that

$$\sum_{l\in\underline{n}} F_l R_l^* = R_0^{-1},\ \sum_{l\in\underline{n}} F_{l-k}R_l^* = 0,\ k\in-\underline{n}\setminus\{0\}. \tag{3.2.20}$$

Replacing l by $n-l$, and k by $n-k$ in (3.2.20) we obtain

$$\sum_{l\in\underline{n}} F_{n-l}R_{n-l}^* = R_0^{-1},\ \sum_{l\in\underline{n}} F_{k-l}R_{n-l}^* = 0,\ k\in\Lambda. \tag{3.2.21}$$

Now (3.2.19) implies that $\sum_{l\in\Lambda} F_{-l}P_l = P_0^{*-1} - F_{-n}P_n$, and thus

$$\sum_{l\in\Lambda} F_{-l}P_lP_0^* = I - F_{-n}P_nP_0^*. \tag{3.2.22}$$

Equation (3.2.21) with $k = 0$ and multiplied on the right with R_n implies that

$$\sum_{l\in\Lambda} F_{-l}R_{n-l}^*R_n = -F_{-n}R_0^*R_n. \tag{3.2.23}$$

Using that $P(z)P(z)^* = R(z)^*R(z)$, and thus $P_nP_0^* = R_0^*R_n$, we get by combining (3.2.22) and (3.2.23) that

$$\sum_{l\in\Lambda} F_{-l}P_lP_0^* - \sum_{l\in\Lambda} F_{-l}R_{n-l}^*R_n = I. \tag{3.2.24}$$

Next, combining (3.2.19) and (3.2.21), we get that

$$\sum_{l\in\Lambda} F_{k-l}P_lP_0^* - \sum_{l\in\Lambda} F_{k-l}R_{n-l}^*R_n = -F_{n-l}P_nP_0^* + F_{n-l}R_0^*R_n = 0,\ k\in\Lambda\backslash\{0\}. \tag{3.2.25}$$

Combining (3.2.24) and (3.2.25) yields that

$$[(F_{k-l})_{k,l\in\Lambda}](AA^* - B^*B)$$

has a first column equal to $\begin{pmatrix} I & 0 & \cdots & 0\end{pmatrix}^T$, and thus the first columns of $[(F_{k-l})_{k,l\in\Lambda}]^{-1}$ and $AA^* - B^*B$ coincide. This proves the claim. □

Next, let us derive a two-variable Christoffel–Darboux type formula; a two-variable generalization of Proposition 2.3.4.

Theorem 3.2.7 *Let*

$$P(z) = \sum_{k_1=0}^{n_1}\sum_{k_2=0}^{n_2} P_{k_1,k_2} z_1^{k_1} z_2^{k_2},\ R(z) = \sum_{k_1=0}^{n_1}\sum_{k_2=0}^{n_2} R_{k_1,k_2} z_1^{k_1} z_2^{k_2}$$

be stable operator-valued polynomials (with underlying Hilbert space $\mathcal{H}$), and let $P(z)P(z)^* = R(z)^*R(z)$, $z \in \mathbb{T}^2$. Then there exist operator-valued polynomials G_i and F_i (with underlying Hilbert space $\mathcal{H}$) such that

$$P(z)P(w)^* - \overleftarrow{R}(z)\overleftarrow{R}(w)^*$$

$$(1 - z_1\overline{w_1})\sum_{j=0}^{n_1-1} G_j(z)G_j(w)^* + (1 - z_2\overline{w_2})\sum_{j=0}^{n_2-1} F_j(z)F_j(w)^*. \tag{3.2.26}$$

The polynomials F_i and G_i can be constructed directly from Schur complements of the Toeplitz operator with symbol $F := P^{*-1}P^{-1} = R^{-1}R^{*-1}$. The G_i can be chosen to have degree at most (i, n_2), $i = 0,\dots,n_1-1$, and the F_j can be chosen to have degree at most (n_1, j), $j = 0,\dots,n_2-1$.

Before we prove this result, let us prove a useful corollary. We will use this corollary in Section 4.6 to prove the so-called von Neumann inequality in one and two variables.

Corollary 3.2.8 *Let P and R be as in Theorem 3.2.7, and let T_1, T_2 be commuting contractions on a Hilbert space $\mathcal{K}$. Define*

$$P(T_1, T_2) = \sum_{k_1=0}^{n_1} \sum_{k_2=0}^{n_2} P_{k_1,k_2} \otimes T_1^{k_1} T_2^{k_2},$$

and define $\overleftarrow{R}(T_1, T_2)$ analogously. Then

$$P(T_1, T_2)P(T_1, T_2)^* \geq \overleftarrow{R}(T_1, T_2)\overleftarrow{R}(T_1, T_2)^*. \tag{3.2.27}$$

Proof. Rewrite the right-hand side of (3.2.26) as

$$\sum_{j=0}^{n_1-1} G_j(z)(1 - z_1\overline{w_1})G_j(w)^* + \sum_{j=0}^{n_2-1} F_j(z)(1 - z_2\overline{w_2})F_j(w)^*.$$

Substituting T_1 for z_1, T_1^* for $\overline{w_1}$, T_2 for z_2, and T_2^* for $\overline{w_2}$, and using that $I - T_1T_1^* \geq 0$ and $I - T_2T_2^* \geq 0$, the corollary follows easily. □

In order to prove Theorem 3.2.7, we start by deriving a d-variable general formula for $P(z)P(w)^* - \overleftarrow{R}(z)\overleftarrow{R}(w)^*$. We denote

$$\ell_\Lambda(z) = \mathrm{row}(z^k)_{k\in\Lambda}, \Lambda \subset \mathbb{Z}^d.$$

Proposition 3.2.9 *Let $P(z) = \sum_{k\in\underline{n}} P_k z^k$ and $R(z) = \sum_{k\in\underline{n}} R_k z^k$ be operator-valued stable polynomials of degree at most $n \in \mathbb{N}_0^d$ so that*

$$P(z)P(z)^* = R(z)^*R(z), z \in \mathbb{T}^d. \tag{3.2.28}$$

Put

$$F(z) = P(z)^{*-1}P(z)^{-1} = R(z)^{-1}R(z)^{*-1}, \tag{3.2.29}$$

and let $F(z) = \sum_{j\in\mathbb{Z}^d} F_j z^j$, $z \in \mathbb{T}^d$, be its Fourier expansion. We let $T = (F_{k-l})_{k,l\in\underline{n}}$. Then

$$P(z)P(w)^* - \overleftarrow{R}(z)\overleftarrow{R}(w)^* = \ell_{\underline{n}}(z)(S(T^{-1}; \underline{n}\setminus\{n\}) - S(T^{-1}; \underline{n}\setminus\{0\}))\ell_{\underline{n}}(w)^*. \tag{3.2.30}$$

We need a simple observation.

Lemma 3.2.10 *Let the operator matrix $M = (A_{ij})_{i,j=1}^2 : \mathcal{H}_1 \oplus \mathcal{H}_2 \to \mathcal{H}_1 \oplus \mathcal{H}_2$ be positive definite. If*

$$\begin{pmatrix} A_{11} & A_{12} \\ A_{21} & A_{22} \end{pmatrix} \begin{pmatrix} L_{11} \\ L_{21} \end{pmatrix} = \begin{pmatrix} L_{11}^{*-1} \\ 0 \end{pmatrix},$$

then

$$\begin{pmatrix} A_{11} & A_{12} \\ A_{21} & A_{22} \end{pmatrix}^{-1} = \begin{pmatrix} L_{11} & 0 \\ L_{21} & L_{22} \end{pmatrix} \begin{pmatrix} L_{11}^* & L_{21}^* \\ 0 & L_{22}^* \end{pmatrix}$$

for some invertible L_{22}. In fact, $L_{22}L_{22}^* = A_{22}^{-1}$. Similarly, if

$$\begin{pmatrix} U_{12}^* & U_{22}^* \end{pmatrix} \begin{pmatrix} A_{11} & A_{12} \\ A_{21} & A_{22} \end{pmatrix} = \begin{pmatrix} 0 & U_{22}^{-1} \end{pmatrix},$$

then

$$\begin{pmatrix} A_{11} & A_{12} \\ A_{21} & A_{22} \end{pmatrix}^{-1} = \begin{pmatrix} U_{11} & U_{12} \\ 0 & U_{22} \end{pmatrix} \begin{pmatrix} U_{11}^* & 0 \\ U_{12}^* & U_{22}^* \end{pmatrix}$$

for some invertible U_{11}. In fact, $U_{11}U_{11}^* = A_{11}^{-1}$.

Proof. Immediate. □

Proof of Proposition 3.2.9. As $F(z)P(z) = P(z)^{*-1}$ and $P(z)$ is stable, we have that (3.2.19) holds. Letting $L_{11} = P_0$ and $L_{21} = \mathrm{col}(P_k)_{k\in\underline{n}\setminus\{0\}}$, we get from (3.2.19) that

$$T\begin{pmatrix} L_{11} \\ L_{21} \end{pmatrix} = \begin{pmatrix} L_{11}^{*-1} \\ 0 \end{pmatrix}.$$

Applying Lemma 3.2.10 we get that

$$T^{-1} = \begin{pmatrix} L_{11} & 0 \\ L_{21} & L_{22} \end{pmatrix} \begin{pmatrix} L_{11}^* & L_{21}^* \\ 0 & L_{22}^* \end{pmatrix}, \tag{3.2.31}$$

where $L_{22}L_{22}^* = [(F_{k-l})_{k,l\in\underline{n}\setminus\{0\}}]^{-1}$. Next, since $R(z)F(z) = R(z)^{*-1}$, we also have that

$$\sum_{l\in\underline{n}} R_l F_{-l} = R_0^{*-1}\ ,\ \sum_{l\in\underline{n}} R_l F_{k-l} = 0,\ k \in \underline{n}\setminus\{0\}. \tag{3.2.32}$$

Letting $U_{22}^* = R_0$ and $U_{12}^* = \mathrm{row}(R_{n-k})_{k\in\underline{n}\setminus\{n\}}$, we get that

$$\begin{pmatrix} U_{12}^* & U_{22}^* \end{pmatrix} T = \begin{pmatrix} 0 & U_{22}^{*-1} \end{pmatrix}.$$

Applying Lemma 3.2.10 we get that

$$T^{-1} = \begin{pmatrix} U_{11} & U_{12} \\ 0 & U_{22} \end{pmatrix} \begin{pmatrix} U_{11}^* & 0 \\ U_{12}^* & U_{22}^* \end{pmatrix}, \tag{3.2.33}$$

where $U_{11}U_{11}^* = [(F_{k-l})_{k,l\in\underline{n}\setminus\{n\}}]^{-1}$.

Next observe that

$$\ell_{\underline{n}}(z)\begin{pmatrix} L_{11} \\ L_{21} \end{pmatrix} = P(z)\ ,\ \ell_{\underline{n}}(z)\begin{pmatrix} U_{12} \\ U_{22} \end{pmatrix} = \overleftarrow{R}(z).$$

Now equating the right-hand sides of (3.2.31) and (3.2.33), and multiplying the equation by $\ell_{\underline{n}}(z)$ on the left and by $\ell_{\underline{n}}(w)^*$ on the right, we get that

$$P(z)P(w)^* + \ell_{\underline{n}}(z)\begin{pmatrix} 0 & 0 \\ 0 & L_{22}L_{22}^* \end{pmatrix}\ell_{\underline{n}}(w)^* \tag{3.2.34}$$

$$= \ell_{\underline{n}}(z)\begin{pmatrix} U_{11}U_{11}^* & 0 \\ 0 & 0 \end{pmatrix}\ell_{\underline{n}}(w)^* + \overleftarrow{R}(z)\overleftarrow{R}(w)^*.$$

Observing, using (3.2.3), that

$$\begin{pmatrix} 0 & 0 \\ 0 & L_{22}L_{22}^* \end{pmatrix} = S(T^{-1}; \underline{n}\setminus\{n\}),\ \begin{pmatrix} U_{11}U_{11}^* & 0 \\ 0 & 0 \end{pmatrix} = S(T^{-1}; \underline{n}\setminus\{0\}),$$

(3.2.30) follows. □

Proof of Theorem 3.2.7. We will use (3.2.30). Note that in the two-variable case we have that

$$\underline{n}\setminus\{n\} = (\underline{n_1 - 1}\times\underline{n_2})\cup(\underline{n_1}\times\underline{n_2 - 1}), \underline{n}\setminus\{0\} = (\underline{n_1}\setminus\{0\}\times\underline{n_2})\cup(\underline{n_1}\times\underline{n_2}\setminus\{0\}).$$

Next we observe that the Schur complement $S(T^{-1}; \underline{n} \setminus \{n\})$ is equal to the matrix $[(F_{k-l})_{k,l \in \underline{n} \setminus \{n\}}]^{-1}$ padded with zeros. It follows from Lemma 3.2.5 that $[(F_{k-l})_{k,l \in \underline{n} \setminus \{n\}}]^{-1}$ has zeros in locations $(k,l) = ((k_1,k_2),(l_1,l_2))$, where $(k_1,l_2) = (n_1,n_2)$ and $(k_2,l_1) = (n_2,n_1)$. But then we may apply Corollary 2.4.15 and obtain that

$$S(T^{-1}; \underline{n} \setminus \{n\}) = S(T^{-1}; \underline{n_1 - 1} \times \underline{n_2}) \tag{3.2.35}$$

$$+ S(T^{-1}; \underline{n_1} \times \underline{n_2 - 1}) - S(T^{-1}; \underline{n_1 - 1} \times \underline{n_2 - 1}).$$

Similarly,

$$S(T^{-1}; \underline{n} \setminus \{0\}) = S(T^{-1}; \underline{n_1} \setminus \{0\} \times \underline{n_2}) \tag{3.2.36}$$

$$+ S(T^{-1}; \underline{n_1} \times \underline{n_2} \setminus \{0\}) - S(T^{-1}; \underline{n_1} \setminus \{0\} \times \underline{n_2} \setminus \{0\}).$$

Next, let Z_1 and Z_2 be the shift matrices so that

$$\ell_{\underline{n}}(z_1,z_2) Z_1 = \frac{1}{z_1}(\ell_{\underline{n}}(z_1,z_2) - \ell_{\underline{n}}(0,z_2)),$$

$$\ell_{\underline{n}}(z_1,z_2) Z_2 = \frac{1}{z_2}(\ell_{\underline{n}}(z_1,z_2) - \ell_{\underline{n}}(z_1,0))$$

for all z_1, z_2. Using that $S(T^{-1}; \Lambda)$ is equal to the operator matrix $[(F_{k-l})_{k,l \in \Lambda}]^{-1}$ restricted to the rows and columns corresponding to Λ, one sees that $S(T^{-1}; \nu + \Lambda)$ consists of the same matrix $[(F_{k-l})_{k,l \in \nu+\Lambda}]^{-1} = [(F_{k-l})_{k,l \in \Lambda}]^{-1}$ but now restricted to the rows and columns corresponding to $\nu + \Lambda$. Using this it is not hard to see that

$$S(T^{-1}; \underline{n_1} \setminus \{0\} \times \underline{n_2}) = Z_1 S(T^{-1}; \underline{n_1 - 1} \times \underline{n_2}) Z_1^*,$$

$$S(T^{-1}; \underline{n_1} \times \underline{n_2} \setminus \{0\}) = Z_2 S(T^{-1}; \underline{n_1} \times \underline{n_2 - 1}) Z_2^*,$$

$$S(T^{-1}; \underline{n_1} \setminus \{0\} \times \underline{n_2} \setminus \{0\}) = Z_1 Z_2 S(T^{-1}; \underline{n_1} \times \underline{n_2 - 1}) Z_2^* Z_1^*.$$

Putting these observations together with (3.2.30) we get that

$$P(z)P(w)^* - \overleftarrow{R}(z)\overleftarrow{R}(w)^* = (1 - z_1\overline{w_1})\ell_{\underline{n}}(z) S(T^{-1}; \underline{n_1 - 1} \times \underline{n_2}) \ell_{\underline{n}}(w)^*$$

$$+ (1 - z_2\overline{w_2})\ell_{\underline{n}}(z) S(T^{-1}; \underline{n_1} \times \underline{n_2 - 1}) \ell_{\underline{n}}(w)^*$$

$$-(1 - z_1 z_2 \overline{w_1 w_2})\ell_{\underline{n}}(z) S(T^{-1}; \underline{n_1 - 1} \times \underline{n_2 - 1}) \ell_{\underline{n}}(w)^*.$$

One may write

$$1 - z_1 z_2 \overline{w_1 w_2} = 1 - z_1\overline{w_1} + (1 - z_2\overline{w_2}) z_1 \overline{w_1}.$$

Rewriting, now yields that

$$P(z)P(w)^* - \overleftarrow{R}(z)\overleftarrow{R}(w)^*$$
$$= (1 - z_1\overline{w_1})\ell_{\underline{n}}(z)[S(T^{-1}; \underline{n_1 - 1} \times \underline{n_2}) - S(T^{-1}; \underline{n_1 - 1} \times \underline{n_2 - 1})]\ell_{\underline{n}}(w)^*$$

$$+(1 - z_2\overline{w_2})\ell_{\underline{n}}(z)[S(T^{-1}; \underline{n_1} \times \underline{n_2 - 1}) - S(T^{-1}; \underline{n_1} \setminus \{n_1\} \times \underline{n_2 - 1})]\ell_{\underline{n}}(w)^*.$$

As $S(M; K) - S(M; \widetilde{K}) \geq 0$ whenever $\widetilde{K} \subset K$, we have that

$$S(T^{-1}; \underline{n_1 - 1} \times \underline{n_2}) - S(T^{-1}; \underline{n_1 - 1} \times \underline{n_2 - 1}) = CC^*,$$

$$S(T^{-1}; \underline{n_1} \times \underline{n_2 - 1}) - S(T^{-1}; \underline{n_1} \setminus \{0\} \times \underline{n_2 - 1}) = DD^*$$

for some triangular operator matrices C and D. But then writing

$$\ell_{\underline{n}}(z)CC^*\ell_{\underline{n}}(w)^* = \sum_{j=0}^{n_1-1} G_j(z)G_j(w)^* \ , \ \ell_{\underline{n}}(z)DD^*\ell_{\underline{n}}(w)^* \sum_{j=0}^{n_2-1} F_j(z)F_j(w)^*,$$

where the coefficients of the polynomials G_j and F_j are in the jth row of C and D, respectively, we get the result. □

In the case of three or more variables we can derive the following.

Theorem 3.2.11 *Let $P(z) = \sum_{k \in \underline{n}} P_k z^k$ and $R(z) = \sum_{k \in \underline{n}} R_k z^k$ be operator-valued stable polynomials of degree at most $n \in \mathbb{N}_0^d$ so that*

$$P(z)P(z)^* = R(z)^*R(z), z \in \mathbb{T}^d.$$

Then there exist analytic functions G_i and F_i such that

$$P(z)P(w)^* - \overleftarrow{R}(z)\overleftarrow{R}(w)^* = \prod_{i=3}^{d}(1 - z_i\overline{w_i}) \tag{3.2.37}$$

$$\times[(1 - z_1\overline{w_1}) \sum_{j=0}^{n_1-1} G_j(z)G_j(w)^* + (1 - z_2\overline{w_2}) \sum_{j=0}^{n_2-1} F_j(z)F_j(w)^*].$$

The functions F_i and G_i can be constructed directly from Schur complements of the Toeplitz operator associated with $F := P^{*-1}P^{-1} = R^{-1}R^{*-1}$. The G_i can be chosen to be polynomials in (z_1, z_2) of degree at most (i, n_2), $i = 0, \ldots, n_1 - 1$, and the F_j can be chosen to be polynomials in (z_1, z_2) of degree at most (n_1, j), $j = 0, \ldots, n_2 - 1$.

Of course, by interchanging the roles of the variables $z_1, \ldots, z_d$ one may derive other variations of (3.2.37).

Proof. Define

$$p_{z_1,z_2}(z_3, \ldots, z_d) = P(z_1, \ldots, z_d), \ r_{z_1,z_2}(z_3, \ldots, z_d) = R(z_1, \ldots, z_d).$$

Let $M_1 = M_1(z_1, z_2)$ and $M_2 = M_2(z_1, z_2) : H^2(\mathbb{T}^{d-2}, \mathcal{H}) \to H^2(\mathbb{T}^{d-2}, \mathcal{H})$ be the Toeplitz operators with symbols p_{z_1,z_2} and r_{z_1,z_2}, respectively. With respect to the standard basis $\{\widetilde{z}^k \ : k \in \mathbb{N}_0^{d-2}\}$, where $\widetilde{z} = (z_3, \ldots, z_d)$, we have the matrix representations

$$M_1(z_1, z_2) = \sum_{k_1=0}^{n_1} \sum_{k_2=0}^{n_2} z_1^{k_1} z_2^{k_2} (P_{(k_1,k_2,k-l)})_{k,l \in \mathbb{N}_0^{d-2}},$$

$$M_2(z_1,z_2)=\sum_{k_1=0}^{n_1}\sum_{k_2=0}^{n_2} z_1^{k_1}z_2^{k_2}(R_{(k_1,k_2,k-l)})_{k,l\in\mathbb{N}_0^{d-2}}.$$

The operator-valued polynomials $M_1 = M_1(z_1, z_2), M_2 = M_2(z_1, z_2)$ satisfy the conditions of Theorem 3.2.7, and thus we obtain the existence of polynomials f_j, $j = 0, \dots, n_2-1$, of degree at most (n_1, j), and g_i, $i = 0, \dots, n_1-1$, of degree at most (i, n_2), such that

$$\begin{aligned}
&M_1(z_1,z_2)M_1(w_1,w_2)^* - \overleftarrow{M_2}(z_1,z_2)\overleftarrow{M_2}(w_1,w_2)^*\\
&= (1-z_1\overline{w_1})\sum_{j=0}^{n_1-1} g_j(z_1,z_2)g_j(w_1,w_2)^*\\
&+ (1-z_2\overline{w_2})\sum_{j=0}^{n_2-1} f_j(z_1,z_2)f_j(w_1,w_2)^*.
\end{aligned} \tag{3.2.38}$$

Let S_i, $i = 3, \dots, d$, be the multiplication operator on $H^2(\mathbb{T}^{d-2}, \mathcal{H})$ with symbol z_i, $i = 3, \dots, d$, and denote $l(\widetilde{z}) = (\widetilde{z}^k)_{k\in\mathbb{N}_0^{d-2}}$. Then, denoting $z = (z_1, z_2, \widetilde{z})$ and $w = (w_1, w_2, \widetilde{w})$, we have

$$l(\widetilde{z})M_1(z_1,z_2)M_1(w_1,w_2)^*l(\widetilde{w})^* = \sum_{k\in\mathbb{N}_0^{d-2}} \widetilde{z}^k\overline{\widetilde{w}}^k P(z)P(w)^*.$$

Also,

$$\begin{aligned}
&l(\widetilde{z})(S_3\dots S_d)M_1(z_1,z_2)M_1(w_1,w_2)^*(S_3\dots S_d)^*l(\widetilde{w})^*\\
&= \sum_{k\in\mathbb{N}_0^{d-2}} \widetilde{z}\overline{\widetilde{w}}\widetilde{z}^k\overline{\widetilde{w}}^k P(z)P(w)^*.
\end{aligned}$$

Taking the difference, we obtain

$$\sum_{k\in\mathbb{N}_0^{d-2}} \widetilde{z}^k\overline{\widetilde{w}}^k P(z)P(w)^* - \sum_{k\in\mathbb{N}_0^{d-2}} \widetilde{z}\overline{\widetilde{w}}\widetilde{z}^k\overline{\widetilde{w}}^k P(z)P(w)^* = P(z)P(w)^*.$$

Similar equations hold above if we replace M_1 and P by $\overleftarrow{M_2}$ and $\overleftarrow{R}$, respectively.

Next, put $G_i(z) = l(\widetilde{z})g_i(z_1, z_2), F_j(z) = l(\widetilde{z})f_j(z_1, z_2)$, which are analytic for $\widetilde{z} = (z_3, \dots, z_d) \in \mathbb{D}^{d-2}$. Now if we multiply (3.2.38) on the left with $S_3 \cdots S_d$ and on the right with $S_3^* \cdots S_d^*$, and subtract the result from (3.2.38) we obtain (3.2.37). □

3.3 TWO-VARIABLE MOMENT PROBLEM FOR BERNSTEIN-SZEGŐ MEASURES

A classical Bernstein-Szegő measure is a measure of the form $\frac{1}{|p(e^{i\theta})|^2}d\theta$, where p is a stable polynomial. In this section we look at two-variable matrix-valued generalizations. We let e_{ij} denote the column vector with $I_{\mathcal{H}}$ in the (i,j) position and zeros elsewhere. Note that rows and columns are indexed by pairs of nonnegative integers; recall the discussion in Section 1.3. The main result of this section is the following.

Theorem 3.3.1 *Let bounded linear operators $c_{ij} \in \mathcal{L}(\mathcal{H})$, $(i,j) \in \Lambda := \{-n,\ldots,n\} \times \{-m,\ldots,m\} \setminus \{(n,m),(-n,m),(n,-m),(-n,-m)\}$ be given. There exist stable polynomials*

$$p(z,w) = \sum_{\substack{i\in\{0,\ldots,n\}\\ j\in\{0,\ldots,m\}}} p_{ij} z^i w^j \in \mathcal{L}(\mathcal{H}), r(z,w) = \sum_{\substack{i\in\{0,\ldots,n\}\\ j\in\{0,\ldots,m\}}} r_{ij} z^i w^j \in \mathcal{L}(\mathcal{H}) \tag{3.3.1}$$

with $p_{00} > 0$ and $r_{00} > 0$ such that

$$p(z,w)^{*-1} p(z,w)^{-1} = \sum_{(i,j)\in\mathbb{Z}^2} c_{ij} z^i w^j = r(z,w)^{-1} r(z,w)^{*-1}, \ (z,w) \in \mathbb{T}^2, \tag{3.3.2}$$

for some $c_{ij} \in \mathcal{L}(\mathcal{H}), (i,j) \notin \Lambda$, if and only if

(i) $\Phi_1 \Phi^{-1} \Phi_2^* = \Phi_2^* \Phi^{-1} \Phi_1$;

(ii) when we put

$$c_{-n,m} = \operatorname{row}(c_{k-l})_{\substack{k=(0,m-1)\\ l\in\{1,\ldots,n\}\times\{0,\ldots,m-1\}}} \Phi^{-1} \operatorname{col}(c_{k-l})_{\substack{k\in\{0,\ldots,n-1\}\times\{1,\ldots,m\}\\ l=(n-1,0)}},$$

then the operators

$$(c_{k-l})_{k,l\in\{0,\ldots,n\}\times\{0,\ldots,m\}\setminus\{(n,m)\}} \text{ and } (c_{k-l})_{k,l\in\{0,\ldots,n\}\times\{0,\ldots,m\}\setminus\{(0,0)\}}$$

are positive definite.

Here

$$\Phi = (c_{k-l})_{k,l\in\{0,\ldots,n-1\}\times\{0,\ldots,m-1\}},$$

$$\Phi_1 = (c_{k-l})_{k\in\{0,\ldots,n-1\}\times\{0,\ldots,m-1\},\ l\in\{1,\ldots,n\}\times\{0,\ldots,m-1\}},$$

$$\Phi_2 = (c_{k-l})_{k\in\{0,\ldots,n-1\}\times\{0,\ldots,m-1\},\ l\in\{0,\ldots,n-1\}\times\{1,\ldots,m\}}.$$

There is a unique choice for $c_{n,m}$ that results in $p_{n,m} = 0$, namely,

$$c_{n,m} = (c_{k-l})_{\substack{k=(n,m)\\ l\in\{0,\ldots,n\}\times\{0,\ldots,m\}\setminus\{(0,0),(n,m)\}}}$$

$$\times [(c_{k-l})_{k,l\in\{0,\ldots,n\}\times\{0,\ldots,m\}\setminus\{(0,0),(n,m)\}}]^{-1} (c_{k-l})_{\substack{k\in\{0,\ldots,n\}\times\{0,\ldots,m\}\setminus\{(0,0),(n,m)\}\\ l=(0,0)}}.$$

When (i) and (ii) are satisfied, the coefficients of the polynomials $p(z,w)$ and $r(z,w)$ can be found via the equations

$$(c_{k-l})_{k,l\in\{0,\ldots,n\}\times\{0,\ldots,m\}} \operatorname{col}(p_{ij})_{i\in\{0,\ldots,n\},j\in\{0,\ldots,m\}} = e_{00} p_{00}^{*-1}, \tag{3.3.3}$$

$$(c_{k-l})_{k,l\in\{0,\ldots,n\}\times\{0,\ldots,m\}} \operatorname{col}(r_{n-i,m-j})_{i\in\{0,\ldots,n\},j\in\{0,\ldots,m\}} = e_{nm} r_{00}^{*-1}. \tag{3.3.4}$$

If one requires that $p_{00} > 0$ and $r_{00} > 0$, the solutions in (3.3.3) and (3.3.4) are unique.

Equations (3.3.3) and (3.3.4) are two-variable generalizations of the Yule-Walker equations (2.3.2).

Notice that (i) is equivalent to the statement that $\Phi^{-1}\Phi_1$ and $\Phi^{-1}\Phi_2^*$ commute. When conditions (i) and (ii) are met, the polynomial p may be constructed by a Yule-Walker type of equation. Alternatively, the Fourier coefficients c_{ij} may be constructed by an iterative process.

In the matrix-valued case we can present the result in simpler terms, as follows.

Theorem 3.3.2 *Let $p \times p$ matrices c_{kl}, $(k,l) \in \{0,\dots,n\} \times \{0,\dots,m\}$ be given. There exist stable $p \times p$ matrix-valued polynomials*

$$p(z,w) = \sum_{k=0}^{n}\sum_{l=0}^{m} p_{kl} z^k w^l, \quad r(z,w) = \sum_{k=0}^{n}\sum_{l=0}^{m} r_{kl} z^k w^l$$

with $p_{00} > 0, r_{00} > 0$, such that the function

$$f(z,w) := p(z,w)^{*-1} p(z,w)^{-1} = r(z,w)^{-1} r(z,w)^{*-1}$$

has Fourier coefficients $\widehat{f}(k,l) = c_{kl}, (k,l) \in \{0,\dots,n\} \times \{0,\dots,m\}$, if and only if there exist $p \times p$ matrices c_{kl}, $(k,l) \in \{1,\dots,n\} \times \{-m,\dots,-1\}$ such that

1. *the matrix $\Gamma := (c_{t-s})_{t,s \in \underline{n} \times \underline{m}}$ is positive definite;*
2. *the matrix*

$$(c_{t-s})_{t \in \{1,\dots,n\} \times \underline{m}, s \in \underline{n} \times \{1,\dots,m\}}$$

has rank equal to nmp.

In this case one finds the block column and block row

$$\mathrm{col}(p_{st}p_{00}^*)_{s,t \in \underline{n} \times \underline{m}}, \quad \mathrm{row}(r_{00}^* r_{n-s,m-t})_{s,t \in \underline{n} \times \underline{m}},$$

as the first block column and last block row of the inverse of Γ, respectively. Equivalently, the coefficients of $p(z,w)$ may be obtained by taking the first block column of L, where $LL^* = \Gamma^{-1}$ is the lower-upper Cholesky factorization of Γ^{-1}, and the coefficients of $r(z,w)$ may be obtained by taking the last block row of U^*, where $UU^* = \Gamma^{-1}$ is the upper-lower Cholesky factorization of Γ^{-1}.

Example 3.3.3 Let $\Lambda = \{0,1\}^2$ and $c_{00} = 1$, $c_{10} = c_{01} = c_{11} = 0.9$. Then for $x = 0.81$ we have that

$$X + X^* = \begin{pmatrix} 1 & 0.9 & 0.9 & 0.9 \\ 0.9 & 1 & x & 0.9 \\ 0.9 & \overline{x} & 1 & 0.9 \\ 0.9 & 0.9 & 0.9 & 1 \end{pmatrix}$$

is positive definite and satisfies the rank condition in Theorem 3.3.2. Thus the Bernstein-Szegő measure problem has a solution. In Example 3.1.3 we have seen that there is no solution to the Carathéodory interpolation problem for this data set.

Using the above characterization the Bernstein-Szegő problem can be solved numerically. Below we give two examples.

Example 3.3.4 For the given data

$$c_{00} = 8, c_{01} = 4, c_{02} = 1, c_{03} = .25, c_{04} = 0.01, c_{12} = 2, c_{13} = 0.5,$$

$$c_{14} = 0.03, c_{15} = 0.006, c_{24} = 1, c_{25} = 0.1, c_{26} = 0.01, c_{27} = 0.001,$$

we arrive at the polynomial

$$\begin{aligned} p(z,w) = \tfrac{1}{\sqrt{0.1925}} \quad & (0.1925 - 0.1215w + 0.0450w^2 - 0.0158w^3 + 0.0049w^4 \\ & -0.0521zw^2 + 0.0486zw^3 - 0.0239zw^4 + 0.0083zw^5 \\ & -0.0157z^2w^4 + 0.0157z^2w^5 - 0.0089z^2w^6 + 0.0034z^2w^7). \end{aligned}$$

After computing the Fourier coefficients of $1/|p(w,z)|^2$ (by using 2D-fft and 2D-ifft with grid size 64; see also Example 1.4.10) we compute an error of

$$\sqrt{\sum \left| c_{ij} - \widehat{\frac{1}{|p|^2}}(i,j) \right|^2} = 1.1026 \times 10^{-9}.$$

Example 3.3.5 For the data

$$c_{00} = 1, \ c_{01} = .4, \ c_{02} = .1, \ c_{03} = .04, \ c_{10} = .2,$$

$$c_{11} = .05, \ c_{12} = .02, \ c_{13} = .005, \ c_{20} = .1, \ c_{21} = .05, \ c_{22} = .01,$$

$$c_{23} = .003, \ c_{30} = .04, \ c_{31} = .015, \ c_{32} = .002, \ c_{33} = .0005,$$

we find the polynomial

$$\begin{aligned} p(z,w) = \tfrac{1}{\sqrt{1.2646}} \quad & (1.2646 - .5572w + .1171w^2 - .0429w^3 - .2612z \\ & +.1791zw - .0791zw^2 + .0324zw^3 - .0607z^2 \\ & -.0171z^2w + .0336z^2w^2 - .0143z^2w^3 - .0132z^3 \\ & -.0107z^3w - .0058z^3w^2 + .0037z^3w^3). \end{aligned}$$

The error we find here is 2.0926×10^{-11}.

We will need the following auxiliary result.

Lemma 3.3.6 *Let A be a positive definite $r \times r$ operator matrix with entries $A_{ij} \in \mathcal{L}(\mathcal{H})$. Suppose that for some $1 \leq j < k \leq r$ we have that $(A^{-1})_{kl} = 0$, $l = 1, \ldots, j$. Write $A^{-1} = L^*L$, where L is a lower triangular operator matrix with positive definite diagonal entries. Then L satisfies $L_{kl} = 0$, $l = 1, \ldots, j$. Moreover, if $\widetilde{A}$ is the $(r-1) \times (r-1)$ matrix obtained from A by removing the kth row and column, and $\widetilde{L}$ is the lower triangular factor of $\widetilde{A}^{-1}$ with positive definite diagonal entries, then*

$$L_{il} = \widetilde{L}_{il}, \ i = 1, \ldots, k-1; \ l = 1, \ldots, j, \tag{3.3.5}$$

and

$$L_{i+1,l} = \widetilde{L}_{il}, \ i = k, \ldots, r-1; \ l = 1, \ldots, j. \tag{3.3.6}$$

In other words, the first j columns of L and $\widetilde{L}$ coincide after the kth row (which contains zeros in columns $1, \ldots, j$) in L has been removed.

Proof. Since the first j columns of a lower Cholesky factor of a matrix M are linear combinations of the first j columns of M, the first statement follows.

Let $M = (M_{ij})_{i,j=1}^3$ be a positive definite operator matrix, and without loss of generality assume $(N_{ij})_{i,j=1}^3 = M^{-1}$ satisfies $N_{13} = 0$. Then

$$\begin{pmatrix} M_{11} & M_{12}^* & M_{13}^* \\ M_{21} & M_{22} & M_{23}^* \\ M_{31} & M_{32} & M_{33} \end{pmatrix}^{-1} = \begin{pmatrix} N_{11} & N_{12}^* & 0 \\ N_{21} & N_{22} & N_{23}^* \\ 0 & N_{32} & N_{33} \end{pmatrix}.$$

Consider

$$N = \begin{pmatrix} A_{11} & 0 & 0 \\ A_{21} & A_{22} & 0 \\ 0 & A_{32} & A_{33} \end{pmatrix} \begin{pmatrix} A_{11} & A_{21}^* & 0 \\ 0 & A_{22} & A_{32}^* \\ 0 & 0 & A_{33} \end{pmatrix}.$$

As a consequence,

$$\begin{pmatrix} M_{11} & M_{12}^* \\ M_{21} & M_{22} \end{pmatrix} \begin{pmatrix} A_{11} \\ A_{12} \end{pmatrix} = \begin{pmatrix} I \\ 0 \end{pmatrix},$$

which implies $\begin{pmatrix} A_{11} \\ A_{12} \end{pmatrix}$ is the first column of the lower triangular Cholesky factor of M. □

In addition, we need the following characterization of stability.

Proposition 3.3.7 . *Let*

$$p(z,w) = \sum_{\substack{i \in \{0,\dots,n\} \\ j \in \{0,\dots,m\}}} p_{ij} z^i w^j \in \mathcal{L}(\mathcal{H}).$$

Then $p(z,w)$ is stable if and only if $p(z,w)$ is invertible for all $z \in \overline{\mathbb{D}}$ and $w \in \mathbb{T}$ and for all $z \in \mathbb{T}$ and $w \in \overline{\mathbb{D}}$.

Proof. Since $p(z,w)$ is invertible for all $z \in \overline{\mathbb{D}}$ and $w \in \mathbb{T}$ we can write

$$p(z,w)^{-1} = \sum_{k=-\infty}^{\infty} g_k(z) w^k, \; z \in \mathbb{T}, w \in \mathbb{T},$$

where $g_k(z)$ is analytic for $z \in \overline{\mathbb{D}}$. The second condition implies that $g_k(z) = 0$ for $k < 0$. Thus $p(z,w)^{-1}$ is analytic for all $(z,w) \in \overline{\mathbb{D}}^2$. Thus $p(z,w)$ is invertible for all $(z,w) \in \overline{\mathbb{D}}^2$, and hence $p(z,w)$ is stable. □

We will need the notions of left and right stable factorizations of operator-valued trigonometric polynomials. A polynomial $A(z) = A_0 + \cdots + A_n z^n$ is called *stable* if $A(z)$ is invertible for $z \in \overline{\mathbb{D}}$. We say that $B(z) = B_0 + \cdots + B_{-n} z^{-n}$ is *antistable* if $B(1/\overline{z})^*$ is stable. Let $Q(z) = \sum_{i=-n}^{n} Q_i z^i$ be a matrix-valued trigonometric polynomial that is positive definite on $\mathbb{T}$, that is, $Q(z) > 0$ for $|z| = 1$. In particular, since the values of $Q(z)$ on the unit circle are Hermitian, we have $Q_i = Q_{-i}^*$, $i = 0, \dots, n$. The positive matrix function $Q(z)$ allows a *left stable factorization*, that is, we may write

$$Q(z) = M(z) M(1/\overline{z})^*, \quad z \in \mathbb{C} \setminus \{0\},$$

with $M(z)$ a stable matrix polynomial of degree n. When we require that $M(0)$ is lower triangular with positive definite diagonal entries, the stable factorization is unique. Indeed, this follows from Theorem 2.4.16. We shall refer to this unique factor $M(z)$ as *the left stable factor* of $A(z)$. Similarly, we define right variations of the above notions. In particular, if $N(z)$ is such that $A(z) = N(1/\overline{z})^* N(z), z \in \mathbb{C} \setminus \{0\}$, $N(z)$ is stable, and $N(0)$ is lower triangular with positive definite diagonal elements, then $N(z)$ is called the *right stable factor* of $A(z)$.

We are now ready to prove Theorem 3.3.1.

Proof of Theorem 3.3.1. Observe that $\Phi_i \Phi^{-1}$ and $\Phi^{-1}\Phi_i$, $i = 1, 2$, have the following companion type forms:

$$\Phi_1 \Phi^{-1} = \begin{pmatrix} * & \cdots & * & * \\ I & 0 & & \\ & \ddots & \ddots & \\ & & I & 0 \end{pmatrix}, \quad \Phi^{-1}\Phi_1 = \begin{pmatrix} 0 & & & * \\ I & \ddots & & * \\ & \ddots & 0 & \vdots \\ & & I & * \end{pmatrix}, \tag{3.3.7}$$

$$\Phi_2^* \Phi^{-1} = (Q_{ij})_{i,j=0}^{n-1}, \quad Q_{ij} = \begin{pmatrix} * & \cdots & * & * \\ \delta_{ij} I & 0 & & \\ & \ddots & \ddots & \\ & & \delta_{ij} I & 0 \end{pmatrix}, \tag{3.3.8}$$

$$\Phi^{-1}\Phi_2^* = (R_{ij})_{i,j=0}^{n-1}, \quad R_{ij} = \begin{pmatrix} 0 & & & * \\ \delta_{ij} I & \ddots & & * \\ & \ddots & 0 & \vdots \\ & & \delta_{ij} I & * \end{pmatrix}, \tag{3.3.9}$$

where $\delta_{ij} = 1$ when $i = j$ and $\delta_{ij} = 0$ otherwise. Consequently, if $S = (S_{ij})_{i,j=0}^{n-1}$ satisfies

$$\Phi_1 \Phi^{-1} S = S \Phi^{-1} \Phi_1, \tag{3.3.10}$$

then S is block Toeplitz (i.e., $S_{ij} = S_{i+1,j+1}$ for all $0 \leq i, j \leq n-2$). Next, if $S = (S_{ij})_{i,j=0}^{n-1}$ satisfies

$$\Phi_2^* \Phi^{-1} S = S \Phi^{-1} \Phi_2^*, \tag{3.3.11}$$

then each S_{ij} is Toeplitz. It follows from (i) that all expressions of the form

$$S = \Psi_{i_1} \Phi^{-1} \Psi_{i_2} \Phi^{-1} \cdots \Phi^{-1} \Psi_{i_k}, \tag{3.3.12}$$

where $i_j \in \{1, 2\}, \Psi_1 = \Phi_1$ and $\Psi_2 = \Phi_2^*$, satisfy (3.3.10) and (3.3.11). Thus all expressions S in (3.3.12) are doubly Toeplitz. In particular, $\Phi_1 \Phi^{-1} \Phi_2^* = \Phi_2^* \Phi^{-1} \Phi_1$ is doubly Toeplitz. Upon closer inspection we have that

$$\Phi_1 \Phi^{-1} \Phi_2^* = (\Gamma_{i-j})_{i=-1,j=0}^{n-2,\ n-1}, \quad \Gamma_k = (c_{k,r-s})_{r=1,s=0}^{m,\ m-1}, \tag{3.3.13}$$

where we use that $c_{-n,m}$ satisfies the equation under (ii). Notice that due to (3.3.13) we have that

$$\begin{pmatrix} c_{-1,1} & \cdots & c_{-n,1} \\ \vdots & & \vdots \\ c_{-1,m} & \cdots & c_{-n,m} \end{pmatrix} = \begin{pmatrix} I & 0 & \cdots & 0 \end{pmatrix} \Phi_1 \Phi^{-1} \Phi_2^* \begin{pmatrix} e_0 & & 0 \\ & \ddots & \\ 0 & & e_0 \end{pmatrix}, \tag{3.3.14}$$

with $e_0 = \begin{pmatrix} 1 & 0 & \cdots & 0 \end{pmatrix}^*$. Due to (ii) and the 3×3 positive definite matrix completion problem, we can choose $c_{n,m} = c_{-n,-m}^*$ such that the matrix $\Gamma = (c_{k-l})_{k,l \in \{0,\dots,n\} \times \{0,\dots,m\}}$ is positive definite. View $\Gamma = (C_{i-j})_{i,j=0}^n$, where $C_k = (c_{k,r-s})_{r,s=0}^m$, and extend Γ to $(C_{i-j})_{i,j=0}^\infty$ according to the rule (see (2.3.5))

$$C_r^* = C_{-r} = \begin{pmatrix} C_{-1} & \cdots & C_{-n} \end{pmatrix} [(C_{i-j})_{i,j=0}^{n-1}]^{-1} \begin{pmatrix} C_{-r+1} \\ \vdots \\ C_{-r+n} \end{pmatrix}, \quad r \geq n+1.$$

Equivalently, if we let

$$\begin{pmatrix} Q_0 \\ \vdots \\ Q_n \end{pmatrix} = \Gamma^{-1} \begin{pmatrix} I \\ 0 \\ \vdots \\ 0 \end{pmatrix},$$

and we factor $Q_0 = LL^*$ with L lower triangular, and put $P_j = Q_j L^{*-1}$, $j = 0, \dots, n$, then $P(z) := P_0 + \cdots + z^n P_n$ is stable and

$$\sum_{j=-\infty}^{\infty} C_j z^j = P(z)^{*-1} P(z)^{-1}, \quad z \in \mathbb{T}. \tag{3.3.15}$$

Due to (3.3.14) it follows from Lemma 3.3.6 that P_j is of the form

$$P_j = \begin{pmatrix} \begin{pmatrix} p_{j0} \\ p_{j1} \\ \vdots \\ p_{jm} \end{pmatrix} & \begin{matrix} 0 \\ \widetilde{P}_j \end{matrix} \end{pmatrix}, \quad j = 0, \dots, n.$$

But then it follows that $\widetilde{P}(z) := \widetilde{P}_0 + \cdots + z^n \widetilde{P}_n$ is stable, and that

$$\sum_{j=-\infty}^{\infty} \widetilde{C}_j z^j = \widetilde{P}(z)^{*-1} \widetilde{P}(z)^{-1}, \quad z \in \mathbb{T},$$

where $\widetilde{C}_j$ is obtained from C_j by leaving out the first row and column.

Similarly, if we let

$$\begin{pmatrix} R_{-n} \\ \vdots \\ R_0 \end{pmatrix} = \Gamma^{-1} \begin{pmatrix} 0 \\ \vdots \\ 0 \\ I \end{pmatrix},$$

and we factor $R_0 = UU^*$ with U upper triangular, and put $S_j = R_j U^{*-1}$, $j = -n, \ldots, 0$, then $S(z) := S_0 + \cdots + z^{-n} S_{-n}$ is antistable and

$$\sum_{j=-\infty}^{\infty} C_j z^j = S(z)^{*-1} S(z)^{-1}, \quad z \in \mathbb{T}.$$

Due to (3.3.14) it follows from Lemma 3.3.6 that S_j is of the form

$$S_j = \begin{pmatrix} \widetilde{S}_j & \begin{pmatrix} \widetilde{p}_{j,-m} \\ \vdots \\ \widetilde{p}_{j,-1} \end{pmatrix} \\ 0 & \widetilde{p}_{0j} \end{pmatrix}, \quad j = -n, \ldots, 0. \tag{3.3.16}$$

But then it follows that $\widetilde{S}(z) := \widetilde{S}_0 + \cdots + z^{-n} \widetilde{S}_{-n}$ is antistable, and that

$$\sum_{j=-\infty}^{\infty} \widehat{C}_j z^j = \widetilde{S}(z)^{*-1} \widetilde{S}(z)^{-1}, \quad z \in \mathbb{T},$$

where $\widehat{C}_j$ is obtained from C_j by leaving out the last row and column.

Due to the block Toeplitz property of C_j, $j = -n, \ldots, n$, we have that $\widetilde{C}_j = \widehat{C}_j$, $j = -n, \ldots, n$. As $\widetilde{S}(z)$ and $\widetilde{P}(z)$ follow the one-variable construction with $(\widetilde{C}_{i-j})_{i,j=0}^n = (\widehat{C}_{i-j})_{i,j=0}^n$, we have by Theorem 2.3.1 (use (2.3.3)) that

$$\widetilde{P}(z)^{*-1} \widetilde{P}(z)^{-1} = \widetilde{S}(z)^{*-1} \widetilde{S}(z)^{-1}, \ z \in \mathbb{T},$$

and thus $\widetilde{C}_j = \widehat{C}_j$, $j \in \mathbb{Z}$. Thus C_j is Toeplitz for all j. Denote $C_j = (c_{j,r-s})_{r,s=0}^m$. As $\sum_{j=-\infty}^{\infty} C_j z^j > 0$, $z \in \mathbb{T}$, we have that the infinite block Toeplitz matrix $(C_{i-j})_{i,j=-\infty}^{\infty}$ is positive definite. We may regroup this infinite block Toeplitz matrix with Toeplitz entries as $(T_{i-j})_{i,j=0}^m$, where

$$T_j = (c_{r-s,j})_{r,s=-\infty}^{\infty}.$$

Taking (3.3.15), and performing a regrouping and extracting the first column from P_i, one arrives at

$$\begin{pmatrix} T_0 & \cdots & T_{-m} \\ \vdots & \ddots & \vdots \\ T_m & \cdots & T_0 \end{pmatrix} \begin{pmatrix} \Pi_0 \\ \vdots \\ \Pi_m \end{pmatrix} = \begin{pmatrix} Q_0 \\ 0 \\ \vdots \\ 0 \end{pmatrix},$$

where $\Pi_j = (p_{r-s,j})_{r,s=-\infty}^{\infty}$, $Q_0 = \Pi_0^{-1} = (q_{r-s,0})_{r,s=-\infty}^{\infty}$, $p_{ij} = 0$ for $i < 0$ or $i > n$, $j < 0$ or $j > m$, and

$$q(z) = \sum_{i=-\infty}^{0} q_{i0} z^i := \left(\sum_{i=0}^{n} p_{i0} z^i \right)^{*-1}.$$

Note that $q(z)$ is indeed anti-analytic as $\sum_{i=0}^n p_{i0} z^i$ is stable. Theorem 2.3.1 now yields that

$$\Pi(w) := \Pi_0 + \cdots + \Pi_m w^m$$

is invertible for all $w \in \overline{\mathbb{D}}$. As $\Pi(w)$ is Toeplitz, its symbol is invertible on $\mathbb{T}$, and thus $p(z,w) = \sum_{i=0}^{n} \sum_{j=0}^{m} p_{ij} z^i w^j$ is invertible for all $|w| \le 1$ and $|z| = 1$. By reversing the roles of z and w one can prove in a similar way that $p(z,w)$ is invertible for all $|z| \le 1$ and $|w| = 1$. Combining these two statements yields by Proposition 3.3.7 that $p(z,w)$ is stable. In addition, we obtain that

$$\Pi(w)^{*-1}\Pi(w)^{-1}$$

has Fourier coefficients $T_{-m}, \dots, T_m$. But then it follows that

$$p(z,w)^{*-1}p(z,w)^{-1}$$

has Fourier coefficients c_{ij}. Similarly, $r(z,w) := \sum_{\substack{i\in\{0,\dots,n\}\\ j\in\{0,\dots,m\}}} \widetilde{p}^*_{-i,-j} z^i w^j$ is stable and $r(z,w)^{-1}r(z,w)^{*-1}$ has Fourier coefficients c_{ij}. This proves one direction of the theorem.

For the converse, let p and r as in (3.3.1) be stable and suppose that (3.3.2) holds. Denote $f(z,w) = \sum_{(i,j)\in\mathbb{Z}^2} c_{ij} z^i w^j$. Write $f(z,w) = \sum_{i=-\infty}^{\infty} f_i(z) w^i$. Then $T_k(z) := (f_{i-j}(z))_{i,j=0}^{k} > 0$ for all $k \in \mathbb{N}_0$ and all $z \in \mathbb{T}$. Next, write

$$p(z,w) = \sum_{i=0}^{m} p_i(z) w^i, \quad r(z,w) = \sum_{i=0}^{m} r_i(z) w^i,$$

and put $p_i(z) = r_i(z) \equiv 0$ for $i > m$. By the inverse formula for block Toeplitz matrices (Theorem 2.3.6) we have that for $k \ge m-1$ and $z \in \mathbb{T}$

$$T_k(z)^{-1} = \begin{pmatrix} p_0(z) & & 0 \\ \vdots & \ddots & \\ p_k(z) & \cdots & p_0(z) \end{pmatrix} \begin{pmatrix} \overline{p}_0(1/\overline{z})^* & \cdots & p_k(1/\overline{z})^* \\ & \ddots & \vdots \\ 0 & & p_0(1/\overline{z})^* \end{pmatrix}$$
$$- \begin{pmatrix} r_{k+1}(1/\overline{z})^* & & 0 \\ \vdots & \ddots & \\ r_1(1/\overline{z})^* & \cdots & r_{k+1}(1/\overline{z})^* \end{pmatrix} \begin{pmatrix} r_{k+1}(z) & \cdots & r_1(z) \\ & \ddots & \vdots \\ 0 & & r_{k+1}(z) \end{pmatrix}$$
$$=: E_k(z). \tag{3.3.17}$$

Next, we have that for $k \ge m-1$, the left outer factors $M_k(z)$ and $M_{k+1}(z)$ of $E_k(z)$ and $E_{k+1}(z)$, respectively, satisfy

$$M_{k+1}(z) = \begin{pmatrix} p_0(z) & 0 \\ \operatorname{col}(p_l(z))_{l=1}^{k+1} & M_k(z) \end{pmatrix}. \tag{3.3.18}$$

Indeed, if we define $M_{k+1}(z)$ by this equality, then writing out the product $M_{k+1}(z)M_{k+1}(1/\overline{z})^*$ and comparing it to $E_{k+1}(z)$, it is straightforward to see that $M_{k+1}(z)M_{k+1}(1/\overline{z})^* = E_{k+1}(z)$. Since both $p_0(z)$ and $M_k(z)$ are stable, $M_{k+1}(z)$ is stable as well. Moreover, since $p_0(0) > 0$ and $M_k(0)$ is lower triangular with positive diagonal entries, the same holds for $M_{k+1}(0)$. Thus $M_{k+1}(z)$ must be the stable factor of $E_{k+1}(z)$. Let $C_k = (c_{k,r-s})_{r,s=0}^{m}$ as before. Then we have that $T_m(z) = \sum_{k=-\infty}^{\infty} C_k z^k = M_m(1/\overline{z})^{*-1}M_m(z)$.

Writing $M_m(z) = P_0 + \cdots + z^n P_n$, it follows from the one-variable result that

$$\begin{pmatrix} C_0 & \cdots & C_{-n} \\ \vdots & & \vdots \\ C_n & \cdots & C_0 \end{pmatrix} \begin{pmatrix} P_0 \\ \vdots \\ P_n \end{pmatrix} = \begin{pmatrix} P_0^{*-1} \\ 0 \\ \vdots \\ 0 \end{pmatrix}.$$

Due to the zeros in $P_1, \ldots, P_n$ (see (3.10.1)), it follows from Theorem 2.2.3 that (3.3.14) holds. By a similar argument, reversing the roles of z and w, we obtain that

$$\begin{pmatrix} \widetilde{C}_0 & \cdots & \widetilde{C}_{-n} \\ \vdots & & \vdots \\ \widetilde{C}_n & \cdots & \widetilde{C}_0 \end{pmatrix} \begin{pmatrix} S_{-n} \\ \vdots \\ S_0 \end{pmatrix} = \begin{pmatrix} 0 \\ \vdots \\ 0 \\ S_0^{*-1} \end{pmatrix},$$

where $\widetilde{C}_k = (c_{r-s,k})_{r,s=0}^n$ and S_j has the form as in (3.3.16). Using the zero structure of $S_{-1}, \ldots, S_{-n}$ one obtains equality (3.3.14) with $\Phi_1 \Phi^{-1} \Phi_2^*$ replaced by $\Phi_2^* \Phi^{-1} \Phi_1$. But then it follows that

$$\begin{pmatrix} I & 0 & \cdots & 0 \end{pmatrix} \Phi_1 \Phi^{-1} \Phi_2^* \begin{pmatrix} e_0 & & 0 \\ & \ddots & \\ 0 & & e_0 \end{pmatrix} \tag{3.3.19}$$

$$= \begin{pmatrix} I & 0 & \cdots & 0 \end{pmatrix} \Phi_2^* \Phi^{-1} \Phi_1 \begin{pmatrix} e_0 & & 0 \\ & \ddots & \\ 0 & & e_0 \end{pmatrix}.$$

From (3.3.7)–(3.3.9) it is easily seen that $\Phi_2^* \Phi^{-1} \Phi_1$ and $\Phi_1 \Phi^{-1} \Phi_2^*$ have the same block entries anywhere else, so combining this with (3.3.19) gives that $\Phi_2^* \Phi^{-1} \Phi_1 = \Phi_1 \Phi^{-1} \Phi_2^*$. This yields (i) and the equality for $c_{-n,m}$ in (ii). The positive definiteness of the matrices in (ii) follows as they are restriction of the multiplication operator with symbol f, which takes on positive definite values on $\mathbb{T}^2$. □

Example 3.3.8 Let $\Lambda = \{0,1\}^2$ and $c_{00} = 1$, $c_{10} = 0.3$, $c_{01} = 0.7$ and $c_{11} = 0.8$. One may check that the matrix

$$\begin{pmatrix} 1 & 0.7 & 0.3 & 0.8 \\ 0.7 & 1 & \overline{x} & 0.3 \\ 0.3 & x & 1 & 0.7 \\ 0.8 & 0.3 & 0.7 & 1 \end{pmatrix}$$

is positive definite when $x = 0$. By Theorem 1.3.18 a positive measure exists with the prescribed moments. When $x = \frac{c_{10}\overline{c_{01}}}{c_{00}} = 0.21$ the matrix has determinant equal to $-\frac{589}{250000}$ and is therefore not positive definite. Hence by Theorem 3.3.2 it follows that no solution to Bernstein-Szegő measure problem exists for this data set. Finally, the Carathédory interpolation problem is solvable for this data set; see Example 3.1.5.

3.4 FEJÉR-RIESZ FACTORIZATION AND SUMS OF HERMITIAN SQUARES

In this section we obtain the following two-variable variation of the Fejér-Riesz factorization result.

Theorem 3.4.1 *Let*

$$f(z,w) = \sum_{k=-n}^{n} \sum_{l=-m}^{m} f_{kl} z^k w^l \in \mathbb{C}^{p\times p}$$

be positive definite for $(z,w) \in \mathbb{T}^2$. *Let* $c_{rs}, (r,s) \in \mathbb{Z}^2$ *denote the Fourier coefficients of* $f(z,w)^{-1}$. *The following are equivalent:*

(i) there exists stable polynomials $p(z,w), r(z,w) \in \mathbb{C}^{p\times p}$ *of degree* (n,m) *such that* $f(z,w) = p(z,w)p(z,w)^* = r(z,w)^* r(z,w)$, $|z| = |w| = 1$;

(ii) the matrix

$$(c_{t-s})_{t\in\{1,\ldots,n\}\times\underline{m}, s\in\underline{n}\times\{1,\ldots,m\}}$$

has rank equal to nmp.

In the case one of (i)–(ii) (and thus all of (i)–(ii)) hold, one may find $p(z,w)$ *and* $r(z,w)$ *by letting*

$$p(z,w) = q_{00}^{-1/2} \left(\sum_{(k,l)\in\underline{n}\times\underline{m}} q_{kl} z^k w^l \right), \tag{3.4.1}$$

$$r(z,w) = \left(\sum_{(k,l)\in\underline{n}\times\underline{m}} s_{kl} z^k w^l \right) s_{00}^{-1/2}, \tag{3.4.2}$$

where

$$\mathrm{col}(q_u)_{u\in\underline{n}\times\underline{m}} = \left[(c_{u-v})_{u,v\in\underline{n}\times\underline{m}}\right]^{-1} \mathrm{col}(\delta_u)_{u\in\underline{n}\times\underline{m}}, \tag{3.4.3}$$

$$\mathrm{row}(s_u)_{u\in\underline{n}\times\underline{m}} = \mathrm{row}(\delta_u)_{u\in\underline{n}\times\underline{m}} \left[(c_{u-v})_{u,v\in\underline{n}\times\underline{m}}\right]^{-1}, \tag{3.4.4}$$

with δ *the Kronecker delta.*

We first need the following uniqueness result.

Proposition 3.4.2 *Let* $Q(z,w), \widetilde{Q}(z,w) \in \mathbb{C}^{p\times p}$ *be matrix-valued trigonometric polynomials with Fourier support in* $\{-n,\ldots,n\}\times\{-m,\ldots,m\}$ *that are positive definite for* $(z,w)\in\mathbb{T}^2$. *Let* c_{kl} *and* $\widetilde{c}_{kl}$, $k,l\in\mathbb{Z}$, *denote the Fourier coefficients of* Q^{-1} *and* $\widetilde{Q}^{-1}$, *respectively. Suppose that* $c_{kl} = \widetilde{c}_{kl}$, $(k,l)\in\{-n,\ldots,n\}\times\{-m,\ldots,m\}$. *Then* $Q \equiv \widetilde{Q}$.

Proof. For a positive definite matrix function F we define its entropy by

$$\mathcal{E}(F) := \frac{1}{(2\pi)^d} \int_{[0,2\pi]^d} \log\det F(e^{it_1}, \dots, e^{it_d}) dt_1 \dots dt_d.$$

Then as in Corollary 1.4.8 one shows that $\mathcal{E}$ is strictly concave on the cone $\mathcal{C}$ of positive definite $p \times p$ matrix-valued functions. If we consider the linear space $\mathcal{W}$ of all matrix-valued functions with Hermitian values with no Fourier support in $\{-n, \dots, n\} \times \{-m, \dots, m\}$, then for a fixed F_0 we have that the unique maximizer G of $\mathcal{E}$ over the set $(F_0 + \mathcal{W}) \cap \mathcal{C}$ is characterized by the property that $G^{-1} \in \mathcal{W}^\perp$ (as in Theorem 1.4.9). In other words, G^{-1} has Fourier support in $\{-n, \dots, n\} \times \{-m, \dots, m\}$.

If we now apply these observations to $F_0 = Q^{-1}$, we see that both Q and $\widetilde{Q}$ lie in $(F_0 + \mathcal{W}) \cap \mathcal{C}$, and that both Q^{-1} and $\widetilde{Q}^{-1}$ lie in $\mathcal{W}^\perp$. The uniqueness of the maximizer (due to the strict concavity) now implies that $Q \equiv \widetilde{Q}$. □

Proof of Theorem 3.4.1. (i) ⇒ (ii). Letting $c_u = \widehat{f^{-1}}(u)$, we get by Theorem 3.3.2 that (ii) holds.

For (ii) ⇒ (i), observe that the coefficients c_u satisfy (ii) of Theorem 3.3.2. Introduce now the stable $p(z,w)$ and $r(z,w)$ as in Theorem 3.3.2 (which is equivalent to (3.4.1), (3.4.2), (3.4.3), and (3.4.4)), obtaining that $\widehat{p^{-1}p^{*-1}}(u) = \widehat{r^{*-1}r^{-1}} = c_u = \widehat{\frac{1}{f}}(u)$, $u \in \{-n, \dots, n\} \times \{-m, \dots, m\}$. Consequently, $p^{-1}p^{*-1} = r^{*-1}r^{-1}$ and f^{-1} have matching Fourier coefficients for indices $\{-n, \dots, n\} \times \{-m, \dots, m\}$. By the uniqueness result in Proposition 3.4.2 we get $p^{-1}p^{*-1} = r^{*-1}r^{-1} \equiv f^{-1}$. □

The criterion in Theorem 3.4.1 allows for a numerical algorithm to obtain the factor p, when it exists. Let us illustrate this in the following example.

Example 3.4.3 Let

$$f(z,w) = \sum_{i=-2}^{2} \sum_{j=-2}^{2} z^i w^j \left(\sum_{r=0}^{2-|i|} \sum_{s=0}^{2-|j|} 2^{-2(r+s)-|i|-|j|} \right).$$

Computing the Fourier coefficients of the reciprocal of f (using Matlab; truncating the Fourier series at index 64), we get:

$$c_{00} = 1.6125, c_{01} = c_{10} = -0.6450, c_{02} = c_{20} = -0.0806, c_{1,-2} = 0.0322,$$
$$c_{1,-1} = 0.2580, c_{11} = 0.2580, c_{12} = c_{21} = 0.0322, c_{2,-2} = 0.0040,$$
$$c_{2,-1} = 0.0322, c_{22} = 0.0040,$$

where only the first four decimal digits show. In order to check Theorem 3.4.1(ii) (where $n = m = 2, q = 0$) we compute

$$\begin{pmatrix} c_{00} & c_{0,-1} & c_{-1,1} & c_{-1,0} & c_{-1,-1} & c_{-2,1} & c_{-2,0} & c_{-2,-1} \\ c_{01} & c_{00} & c_{-1,2} & c_{-1,1} & c_{-1,0} & c_{-2,2} & c_{-2,1} & c_{-2,0} \\ c_{1,-1} & c_{1,-2} & c_{00} & c_{0,-1} & c_{0,-2} & c_{-1,0} & c_{-1,-1} & c_{-1,-2} \\ c_{10} & c_{1,-1} & c_{01} & c_{00} & c_{0,-1} & c_{-1,1} & c_{-1,0} & c_{-1,-1} \\ c_{11} & c_{10} & c_{02} & c_{01} & c_{00} & c_{-1,2} & c_{-1,1} & c_{-1,0} \\ c_{2,-1} & c_{2,-2} & c_{10} & c_{1,-1} & c_{1,-2} & c_{00} & c_{0,-1} & c_{0,-2} \\ c_{20} & c_{2,-1} & c_{11} & c_{10} & c_{1,-1} & c_{01} & c_{00} & c_{0,-1} \\ c_{21} & c_{20} & c_{12} & c_{11} & c_{10} & c_{02} & c_{01} & c_{00} \end{pmatrix}^{-1}$$

$$
= \begin{pmatrix}
0.9375 & 0.3750 & 0.0000 & 0.4688 & 0.1875 & 0.0000 & 0.2344 & 0.0938 \\
0.3750 & 0.9375 & 0.0000 & 0.1875 & 0.4688 & 0.0000 & 0.0938 & 0.2344 \\
0.0000 & 0.0000 & 0.9375 & 0.4688 & 0.2344 & 0.3750 & 0.1875 & 0.0938 \\
0.4688 & 0.1875 & 0.4688 & 1.3477 & 0.5625 & 0.1875 & 0.5625 & 0.2344 \\
0.1875 & 0.4688 & 0.2344 & 0.5625 & 1.1719 & 0.0938 & 0.2344 & 0.4922 \\
0.0000 & 0.0000 & 0.3750 & 0.1875 & 0.0938 & 0.9375 & 0.4688 & 0.2344 \\
0.2344 & 0.0938 & 0.1875 & 0.5625 & 0.2344 & 0.4688 & 1.1719 & 0.4922 \\
0.0938 & 0.2344 & 0.0938 & 0.2344 & 0.4922 & 0.2344 & 0.4922 & 0.9961
\end{pmatrix},
$$

whose zeros imply the low rank condition Theorem 3.4.1(ii). Computing $p(z,w)(= r(z,w))$ gives $p(z,w) = \sum_{k,l=0}^{2} 2^{-k-l} z^k w^l$.

We remark that our result is quite different from results regarding writing positive trigonometric polynomials as sums of Hermitian squares of polynomials. For example, the positive function $|z-4|^2 + |w-2|^2$ cannot be written as $|p(z,w)|^2$, where p is a polynomial (i.e., p has finite Fourier support). One may, however, write $|z-4|^2 + |w-2|^2 = |p(z,w)|^2$ when one allows p to be a Wiener function with infinite Fourier support $\{0\} \times \{0,1,2,\dots\} \cup \{1\} \times \{\dots,-2,-1,0\}$, and in that case p can be chosen so that $p(z,w)$ is invertible for $z \in \overline{\mathbb{D}}$ and $w \in \mathbb{T}$.

We end this section with showing that any positive definite operator-valued trigonometric polynomial is a sum of Hermitian squares.

Theorem 3.4.4 *Let $Q(z) = \sum_{k \in \underline{n}-\underline{n}} Q_k z^k$ be strictly positive definite for $z \in \mathbb{T}^d$. Then there exists a multivariable polynomial $P(z) = \sum_{k \in \underline{N}} P_k z^k$, with $P_k : \mathcal{H} \to \mathcal{H}^M$, such that $Q(z) = P(z)^* P(z)$, $z \in \mathbb{T}^d$.*

Proof. For $N \in \mathbb{N}_0^d$, denote $\ell_N(z) = \text{col}(z^k I_{\mathcal{H}})_{k \in \underline{N}}$. Notice that it suffices to show that $Q(z) = \ell_N(z)^* A \ell_N(z)$, $z \in \mathbb{T}$, for some positive semidefinite $A : \mathcal{H}^M \to \mathcal{H}^M$, where $M = \prod_{i=1}^d (N_i + 1)$. Indeed, then we may define $P(z)$ via $P(z) = A^{\frac{1}{2}} \ell_N(z)$, to get the desired equality $Q(z) = P(z)^* P(z)$, $z \in \mathbb{T}^d$.

Notice that $Q(z) \geq \delta I_{\mathcal{H}} > 0$, $z \in \mathbb{T}^d$, implies that for any $N \in \mathbb{N}_0^d$ we have that

$$(Q_{k-l})_{k,l \in \underline{N}} \geq \delta I_{\mathcal{H}^M},$$

where $Q_k = 0$, $k \notin \underline{n} - \underline{n}$. In addition, notice that

$$\ell_N(z)^* (Q_{k-l})_{k,l \in \underline{N}} \ell_N(z) = \prod_{i=1}^d (N_i + 1) \sum_{k \in \underline{n}-\underline{n}} \frac{\prod_{i=1}^d (N_i + 1 - |k_i|)}{\prod_{i=1}^d (N_i + 1)} Q_k z^k.$$

For all $N \in \mathbb{N}_0^d$, put

$$\mu_{k,N} := \begin{cases} \frac{\prod_{i=1}^d (N_i + 1 - |k_i|)}{\prod_{i=1}^d (N_i + 1)} & \text{when } k \in \underline{n} - \underline{n}; \\ 1 & \text{otherwise.} \end{cases}$$

As $\lim_{N \to \infty} \mu_{k,N} = 1$, we may choose N large enough so that

$$\sum_{k \in \underline{n}-\underline{n}} \|1 - \frac{1}{\mu_{k,N}}\| \, \|Q_k\| < \delta.$$

Put $\widetilde{Q}_k = \frac{1}{\mu_{k,N}} Q_k$. Then

$$\widetilde{Q}(z) = \sum_{k \in \underline{n}-\underline{n}} \widetilde{Q}_k z^k = Q(z) - \sum_{k \in \underline{n}-\underline{n}} \left(1 - \frac{1}{\mu_{k,N}}\right) Q_k z^k \geq \delta I_{\mathcal{H}} - \delta I_{\mathcal{H}} = 0,$$

and thus

$$A := (\widetilde{Q}_{k-l})_{k,l \in \underline{N}} \geq 0.$$

As

$$Q(z) = \sum_{k \in \underline{n}-\underline{n}} \mu_{k,N} \widetilde{Q}_k z^k = \frac{1}{\prod_{i=1}^d (N_i + 1)} \ell_N(z)^* A \ell_N(z), z \in \mathbb{T}^d,$$

we get by the first paragraph of the proof the desired factorization. □

It is an open problem whether Theorem 3.4.4 holds when $Q(z)$ is positive semidefinite on $\mathbb{T}^d$.

3.5 COMPLETION PROBLEMS FOR POSITIVE SEMIDEFINITE FUNCTIONS ON AMENABLE GROUPS

In this section, G is a locally compact group, meaning that its unit element e has a compact neighborhood. Let $S^{-1} = S \subset G$ and $e \in S$. Such a set S is called *symmetric*. Let $\mathcal{H}$ be a Hilbert space. A function $f : G \to \mathcal{L}(\mathcal{H})$ is referred to as *positive semidefinite* if for every $n \in \mathbb{N}$ and every $g_1, \ldots, g_n \in G$ the operator matrix $(f(g_i^{-1} g_j))_{i,j=1}^n$ is positive semidefinite.

A function $f : S \to \mathcal{L}(\mathcal{H})$ is called *partially positive semidefinite* if for every $g_1, \ldots, g_n \in G$ such that $g_i^{-1} g_j \in S$, for all $1 \leq i,j \leq n$, implies that the operator matrix $(f(g_i^{-1} g_j))_{1 \leq i,j \leq n}$ is positive semidefinite. A function $f_{\text{ext}} : G \to \mathcal{L}(\mathcal{H})$ is called a *positive semidefinite completion* of f if f_{ext} is positive semidefinite and $f_{\text{ext}}|S = f$.

If $S \subset G$ is symmetric, we define the *Cayley graph* of G with respect to S, denoted $\Gamma(G, S)$, as the graph whose vertices are the elements of G, while $\{x, y\}$ is an edge if and only if $x^{-1} y \in S$. An infinite graph is called *chordal* if its restriction to every finite set of vertices is chordal.

Let G be a locally compact group, and let L_G^∞ denote the space of all measurable functions $f : G \to \mathbb{C}$ such that $\|f\|_\infty := \text{ess sup}_{g \in G} |f(g)| < \infty$. A linear functional $m : L_G^\infty \to \mathbb{C}$ is called a *state* on G if it maps nonnegative functions to nonnegative numbers and $m(1) = 1$. A state m which satisfies

$$m(f) = m(f_x)$$

for all $x \in G$, where $f_x(y) = f(yx)$, is called a *right invariant mean*. In the case there exists a right invariant mean on G, G is called *amenable*. We will write $m^x(f(x))$ for $m(f)$ in case we would like to make the underlying variable explicit. There exist several other equivalent characterizations of amenability. Examples of amenable groups are: all Abelian groups, all compact groups (see Exercise 3.10.26), all solvable groups, the Heisenberg group (upper triangular 3×3 matrices with unit diagonal and integer strict upper triangular entries), $\mathbb{Z}_2$ free product with $\mathbb{Z}_2$, and so on. An indication that a group is nonamenable is that it contains a subgroup isomorphic to $\mathbb{F}_2$, the free group with two generators.

The main result of this section is the following.

Theorem 3.5.1 *Suppose G is amenable and $S \subset G$, is symmetric. If $\Gamma(G, S)$ is chordal, then any positive semidefinite function ϕ on S admits a positive semidefinite completion Φ on G.*

Before we can prove Theorem 3.5.1 we first need to prove a generalization of Theorem 2.2.4 to infinite graphs. Consider an undirected graph Γ with vertex set G and edge set E. Necessarily, we have that $(g, h) \in E$ if and only if $(h, g) \in E$. We also require $(g, g) \in E$ for all $g \in G$. Let $F \subseteq G \times G$. We say that $K : F \to \mathcal{L}(\mathcal{H})$ is a *positive semidefinite kernel* if $(K(g_i, g_j))_{i,j=1}^n \geq 0$ for all finite sets $\{g_1, \ldots, g_n\} \subset G$ with $(g_i, g_j) \in F$, $1 \leq i, j \leq n$. Given a positive semidefinite kernel $K : E \to \mathcal{L}(\mathcal{H})$, we say that $\tilde{K} : G \times G \to \mathcal{L}(\mathcal{H})$ is a *positive semidefinite completion* of K if $\tilde{K}$ is a positive semidefinite kernel and $\tilde{K}|E = K$. We now have the following result.

Theorem 3.5.2 *Let the undirected graph Γ with vertex set G and edge set E, be chordal. Then every positive semidefinite kernel $K : E \to \mathcal{L}(\mathcal{H})$ has a positive semidefinite completion.*

Proof. Let $K : E \to \mathcal{L}(\mathcal{H})$ be a positive semidefinite kernel. Let $\mathcal{N}$ denote all finite subsets of G, and let inclusion be the partial order on $\mathcal{N}$. For $D \in \mathcal{N}$ consider $K|(E \cap (D \times D))$. By Theorem 2.2.4 there exists a positive semidefinite completion $\tilde{K}^{(D)}$ of $K|(E \cap (D \times D))$; that is, $\tilde{K}^{(D)} : D \times D \to \mathcal{L}(\mathcal{H})$ is positive semidefinite and $\tilde{K}^{(D)}|(E \cap (D \times D)) = K|(E \cap (D \times D))$. We extend $\tilde{K}^{(D)}$ to $G \times G$ by defining $\tilde{K}^{(D)}(g, h) = 0$ whenever $g \notin D$ or $h \notin D$. Notice that for all $g, h \in G$ we have that

$$\begin{pmatrix} K(g,g) & \tilde{K}^{(D)}(g,h) \\ \tilde{K}^{(D)*}(g,h) & K(h,h) \end{pmatrix} \geq 0.$$

But then $\tilde{K}^{(D)}(g, h)$ lies in the set $\mathcal{A}_{g,h} := K(g, g)^{1/2} \mathcal{G} K(h, h)^{1/2}$, where $\mathcal{G}$ is the unit ball of contractions. This set is compact in the weak operator norm. For each $D \in \mathcal{N}$, define $Q_D := (\tilde{K}^{(D)}(g, h))_{g,h \in G}$. Due to the compactness of the Cartesian product of the sets $\mathcal{A}_{g,h}$, the net $\{Q_D : D \in \mathcal{N}\}$ has a limit point $(\tilde{K}(g, h))_{g,h \in G}$, say. It is now straightforward to check that $\tilde{K}$ gives the desired completion. □

We are now ready to prove Theorem 3.5.1.

Proof of Theorem 3.5.1. Consider the partially positive semidefinite function $k : G \times G \to \mathcal{L}(\mathcal{H})$, defined only for pairs (x, y) for which $x^{-1}y \in S$, by the formula

$$k(x, y) = \phi(x^{-1}y).$$

Since the pattern of specified values for this kernel is chordal by assumption, it follows from Theorem 3.5.2 that k can be extended to a positive semidefinite kernel $K : G \times G \to \mathcal{L}(\mathcal{H})$. Note that $K(x, y)$ has no reason to depend only on $x^{-1}y$.

For any $x, y \in G$, the operator matrix $\begin{pmatrix} \phi(e) & K(x,y) \\ K(x,y)^* & \phi(e) \end{pmatrix}$ is positive semidefinite, whence it follows that all operators $K(x, y)$, $x, y \in G$, are bounded by the common constant $\|\phi(e)\|$.

Fix then $\xi, \eta \in \mathcal{H}$ and $x \in G$. The function $F_{x;\xi,\eta} : G \to \mathbb{C}$ defined by

$$F_{x;\xi,\eta}(y) = \langle K(yx, y)\xi, \eta\rangle \tag{3.5.1}$$

is in L^∞_G. Define then $\Phi : G \to \mathcal{L}(\mathcal{H})$ by

$$\langle \Phi(x)\xi, \eta\rangle = m(F_{x;\xi,\eta}). \tag{3.5.2}$$

We claim that Φ is a positive semidefinite function. Indeed, take arbitrary vectors $\xi_1, \dots, \xi_n \in \mathcal{H}$. We have

$$\sum_{i,j=1}^{n} \langle \Phi(g_i^{-1} g_j)\xi_i, \xi_j\rangle = \sum_{i,j=1}^{n} m(F_{g_i^{-1} g_j;\xi_i,\xi_j}) = \sum_{i,j=1}^{n} m^y \left(\langle K(y g_i^{-1} g_j, y)\xi_i, \xi_j\rangle\right).$$

Take one of the terms in the last sum; the mean m is applied to the function $y \mapsto \langle K(yg_i^{-1}g_j, y)\xi_i, \xi_j\rangle$. The right invariance of m implies that we may apply the change of variable $z = yg_i^{-1}$, $y = zg_i$, and thus

$$m^y \left(\langle K(yg_i^{-1}g_j, y)\xi_i, \xi_j\rangle\right) = m^z \left(\langle K(zg_j, g_i z)\xi_i, \xi_j\rangle\right).$$

Therefore

$$\begin{aligned} \sum_{i,j=1}^{n} \langle \Phi(g_i^{-1} g_j)\xi_i, \xi_j\rangle &= \sum_{i,j=1}^{n} m(\langle K(zg_j, g_i z)\xi_i, \xi_j\rangle) \\ &= m\Big(\sum_{i,j=1}^{n} \langle K(zg_j, g_i z)\xi_i, \xi_j\rangle\Big). \end{aligned}$$

But the positivity of K implies that for each $z \in G$,

$$\sum_{i,j=1}^{n} \langle K(zg_j, g_i z)\xi_i, \xi_j\rangle \geq 0.$$

Since m is a positive functional, it follows that indeed Φ is positive semidefinite. On the other hand, for $x \in S$, the function $F_{x;\xi,\eta}$ is constant, equal to $\langle \phi(x)\xi, \eta\rangle$. Therefore Φ is indeed the desired completion of ϕ. □

Remark 3.5.3 The chordality of $\Gamma(G, S)$ means that for every finite cycle $[x_1, \dots x_n, x_1]$, $n \geq 4$, at least one $\{x_i, x_{i+2}\}$ (with $x_{n+1} = x_1$ and $x_{n+2} = x_2$) is an edge. Denoting $\xi_k = x_k x_{k+1}^{-1}$, the condition is equivalent to: $\xi_1, \dots, \xi_n \in S$, $\xi_1\xi_2 \dots \xi_n = e$, $n \geq 4$, implies that there exists an $i \in \{1, \dots, m\}$ such that $\xi_i\xi_{i+1} \in S$ (here $\xi_{n+1} = \xi_1$).

We conjecture the following reciprocal of Theorem 3.5.1.

Conjecture 3.5.4 *For every symmetric subset S of an amenable group G such that $\Gamma(G,S)$ is not chordal, there exists a positive semidefinite function $k : S \to \mathcal{L}(\mathcal{H})$ which does not admit a positive semidefinite completion to G.*

The following are some immediate corollaries of Theorem 3.5.1.

Corollary 3.5.5 *Let a be a positive real number and let $k : (-a,a) \to \mathcal{L}(\mathcal{H})$ be a positive semidefinite function. Then k admits a positive semidefinite completion to $\mathbb{R}$.*

Proof. Let $G = \mathbb{R}$ (with addition) and $S = (-a,a)$. The result follows immediately from Theorem 3.5.1. □

Corollary 3.5.6 *Let a be a positive real number and let $k : (-a,a) \to \mathcal{L}(\mathcal{H})$ be a positive semidefinite function with a countable support. Then k admits a positive semidefinite completion to $\mathbb{R}$ which also has a countable support.*

Proof. Apply Theorem 3.5.1 with G the subgroup of $\mathbb{R}$ generated by the support of k and $S = G \cap (-a,a)$. □

Let $\alpha_1, \alpha_2, \dots, \alpha_r \in \mathbb{R}$ and $a > 0$ be given. Consider in $\mathbb{Z}^r$ the set

$$S = \{(m_1,\dots,m_r) \in \mathbb{Z}^r \ : \ |\alpha_1 m_1 + \cdots + \alpha_r m_r| < a\}. \tag{3.5.3}$$

Corollary 3.5.7 *Let S be as in* (3.5.3)*. Then every positive semidefinite $k : S \to \mathcal{L}(\mathcal{H})$ admits a positive semidefinite completion to $\mathbb{Z}^r$.*

Proof. Apply Theorem 3.5.1 with $G = \mathbb{Z}^r$ and S as in (3.5.3). □

3.6 MOMENT PROBLEMS ON FREE GROUPS

In this section we consider the moment problem on the group $\mathbb{F} = \mathbb{F}_m$, the free group with m generators $a_1, \dots, a_m$. Denote by e the unit and by $\mathcal{A}$ the set of generators of $\mathbb{F}$. Elements in $\mathbb{F}$ are therefore words $s = b_1 \dots b_n$, with letters $b_i \in \mathcal{A} \cup \mathcal{A}^{-1}$; each word is usually written in its *reduced form*, obtained after canceling all products aa^{-1}. The *length of a word* s, denoted by $|s|$, is the number of generators which appear in (the reduced form of) s. By definition $|e| = 0$. For a positive integer n, define S_n to be the set of all words of length $\leq n$ in $\mathbb{F}$ and $S'_n \subset S_n$ the set of words of length exactly n; the number of elements in S'_n is $2m(2m-1)^{n-1}$.

A function $\Phi : \mathbb{F} \to \mathcal{L}(\mathcal{H})$ is *positive semidefinite* if for every $s_1, \dots, s_k \in \mathbb{F}$ the operator matrix $A(\Phi; (s_1, \dots, s_k)) := [\Phi(s_i^{-1} s_j)]_{i,j=1}^k$ is positive semidefinite. In general, for a finite set $S \subset \mathbb{F}$, we will use the notation $A(\Phi; S) := [\Phi(s^{-1}t)]_{s,t \in S}$. There is here a slight abuse of notation, since the matrix on the right-hand side of the equality depends on the order of the elements of S, and changing the order amounts to permuting the rows and columns of $A(\Phi; S)$; however, the reader can easily check that this ambiguity is irrelevant in all instances where this notation is used below.

We will also consider positive semidefinite functions defined on subsets of $\mathbb{F}$. If $\Sigma \subset \mathbb{F}$ such that $\Sigma = \Sigma^{-1}$, then a function $\phi : \Sigma \to \mathcal{L}(\mathcal{H})$ is called *positive semidefinite* if for every $s_1, \dots, s_k \in \mathbb{F}$ such that $s_i^{-1}s_j \in \Sigma$ for every $i, j = 1, \dots, k$, the operator matrix $A(\Phi; \{s_1, \dots, s_k\})$ is positive semidefinite. Obviously, if $\Phi : \mathbb{F} \to \mathcal{L}(\mathcal{H})$ is positive semidefinite, then $\Phi|\Sigma$ is positive semidefinite for all $\Sigma \subset \mathbb{F}$.

The *moment problem on free groups* is now the following: "Given $\Sigma \subset \mathbb{F}$ such that $\Sigma = \Sigma^{-1}$, and $\phi : \Sigma \to \mathcal{L}(\mathcal{H})$ positive semidefinite, can we find a positive semidefinite $\Phi : \mathbb{F} \to \mathcal{L}(\mathcal{H})$ such that $\Phi|\Sigma = \phi$?" We will show that the answer is affirmative for sets that arise naturally when we put a total lexicographical order on $\mathbb{F}$.

To define the order, we first need the notion of "common beginning" of two elements of $\mathbb{F}$. If s, t are two words in reduced form, we will call u the *common beginning* of s and t and write $u = \mathrm{CB}(s, t)$ if $s = us_1$ and $t = ut_1$ and $|t_1^{-1}s_1| = |s_1| + |t_1|$; in other words, the first letters of s_1 and t_1 are different. For more than two words, we define the common beginning by induction:

$$\mathrm{CB}(s_1, \dots, s_p) = \mathrm{CB}(\mathrm{CB}(s_1, \dots, s_{p-1}), s_p).$$

It can be seen that $\mathrm{CB}(s_1, \dots s_p)$ does not depend on the order of the elements $s_1, \dots, s_p$, and is formed by their first common letters. We include for completeness the proof of the following lemma.

Lemma 3.6.1 *If $s \neq e$, then $|s^2| > |s|$; in particular, $s^2 \neq e$.*

Proof. When we write the word s^2, it may happen that some of the letters at the beginning of s cancel with some at the end of s. Let us group the former in s_1 and the latter in s_3; thus $s = s_1s_2s_3$, with s_i in reduced form, $s_3s_1 = e$, and the reduced form of s^2 is $s_1s_2^2s_3$. We have then $|s| = |s_1| + |s_2| + |s_3|$ and $|s^2| = |s_1| + 2|s_2| + |s_3|$. If $|s^2| \leq |s|$, we must have $s_2 = e$. Therefore $s = s_1s_3 = e$. □

Given a total order $\preceq$ on $\mathcal{A} \cup \mathcal{A}^{-1}$, we can extend it to $\mathbb{F}$ as follows. For two elements $s_1, s_2 \in \mathbb{F}$, we have $s_1 \preceq s_2$ whenever one of the following holds:

(i) $|s_1| < |s_2|$;

(ii) $|s_1| = |s_2|$, and, if $t = \mathrm{CB}(s_1, s_2)$ and a_i is the first letter of $t^{-1}s_i$ for $i = 1, 2$, then $a_1 \preceq a_2$.

This order will be called a *lexicographical order* on $\mathbb{F}$. We will write $s \prec t$ if $s \preceq t$ and $s \neq t$. One can see that lexicographic orders are in one-to-one correspondence with their restrictions to $\mathcal{A} \cup \mathcal{A}^{-1}$. For the remainder of this section a lexicographic order $\preceq$ on $\mathbb{F}$ will be fixed. Note that the order is a total order with the additional property that every element but e has a unique predecessor.

As for the moment problem we are interested in sets Σ such that $\Sigma = \Sigma^{-1}$, it is natural to introduce the following. Let $\sim$ be the equivalence relation on

$\mathbb{F}$ obtained by having the equivalence classes $\widehat{s} = \{s, s^{-1}\}$. We will denote also by $\preceq$ the order relation on $\widehat{\mathbb{F}} = \mathbb{F}/\sim$ defined by $\widehat{s} \preceq \widehat{t}$ if and only if $\min\{s, s^{-1}\} \preceq \min\{t, t^{-1}\}$. If $\nu \in \widehat{\mathbb{F}}$, then ν^- and ν^+ will be the predecessor and the successor of ν with respect to $\preceq$, respectively. Define $\Sigma_\nu = \bigcup_{\nu' \preceq \nu} \nu$.

The main result in this section is the following.

Theorem 3.6.2 *If $\phi : \Sigma_\nu \to \mathcal{L}(\mathcal{H})$ is a positive semidefinite function, then ϕ has a positive semidefinite extension $\Phi : \mathbb{F} \to \mathcal{L}(\mathcal{H})$.*

Let us illustrate these definitions for a particular case.

Example 3.6.3 If we have the free group with two generators a_1 and a_2, and we take the order $a_1 \prec a_2 \prec a_1^{-1} \prec a_2^{-1}$ on $\mathcal{A} \cup \mathcal{A}^{-1}$, it induces the following total order of $\mathbb{F}$:

$$e \prec a_1 \prec a_2 \prec a_1^{-1} \prec a_2^{-1} \prec a_1^2 \prec a_1 a_2 \prec a_1 a_2^{-1} \prec a_2 a_1 \prec a_2^2 \prec a_2 a_1^{-1}$$

$$\prec a_1^{-1} a_2 \prec a_1^{-2} \prec a_1^{-1} a_2^{-1} \prec a_2^{-1} a_1 \prec a_2^{-1} a_1^{-1} \prec a_2^{-2} \prec a_1^3 \prec \cdots .$$

On $\widehat{\mathbb{F}}$ this induces the ordering

$$\{e\} \prec \{a_1, a_1^{-1}\} \prec \{a_2, a_2^{-1}\} \prec \{a_1^2, a_1^{-2}\} \prec \cdots .$$

When we choose $\nu = \{a_1^2, a_1^{-2}\}$ we have

$$\Sigma_\nu = \{e, a_1, a_1^{-1}, a_2, a_2^{-1}, a_1^2, a_1^{-2}\}.$$

Suppose now that we are given a positive semidefinite function ϕ on Σ_ν. To find Φ on $\mathbb{F}$ so that $\Phi|\Sigma_\nu = \phi$, we first find a ψ positive semidefinite on Σ_{ν^+} such that $\psi|\Sigma_\nu = \phi$. As long as one can keep on going this way, one eventually defines all of Φ. Note that $\nu^+ = \{a_1 a_2, a_2^{-1} a_1^{-1}\}$, and

$$\Sigma_{\nu^+} = \{e, a_1, a_1^{-1}, a_2, a_2^{-1}, a_1^2, a_1^{-2} a_1 a_2, a_2^{-1} a_1^{-1}\}.$$

Thus we arrive (in the scalar-valued case $\mathcal{H} = \mathbb{C}$) at the following positive semidefinite completion problem:

$$\begin{pmatrix} 1 & c_1 & c_1^* & c_2 & c_2^* & c_{11} & c_{11}^* & x_{12} & x_{12}^* & \cdots \\ c_1^* & 1 & c_{11}^* & ? & ? & c_1 & ? & c_2 & ? & \cdots \\ c_1 & c_{11} & 1 & x_{12} & ? & ? & c_1^* & ? & ? & \cdots \\ c_2^* & ? & x_{12}^* & 1 & ? & ? & ? & ? & ? & \cdots \\ c_2 & ? & ? & ? & 1 & ? & ? & ? & c_1 & \cdots \\ c_{11}^* & c_1^* & ? & ? & ? & 1 & ? & ? & ? & \cdots \\ c_{11} & ? & c_1 & ? & ? & ? & 1 & ? & ? & \cdots \\ x_{12}^* & c_2^* & ? & ? & ? & ? & ? & 1 & ? & \cdots \\ x_{12} & ? & ? & ? & c_1 & ? & ? & ? & 1 & \cdots \\ \vdots & \vdots & \vdots & \vdots & \vdots & \vdots & \vdots & \vdots & \vdots & \ddots \end{pmatrix}.$$

Here $\phi(e) = 1$, $\phi(a_1) = c_1$, $\phi(a_2) = c_2$, $\phi(a_1^2) = c_{11}$, are given, and $\psi(a_1 a_2) = x_{12}$ is to be found such that the above (infinite) partial matrix remains partial positive semidefinite. The part of the matrix that is shown corresponds to the part with index set Σ_{ν^+}. As we will see, in each step the problem reduces to a completion problem where the underlying pattern is chordal, and thus the results from Chapter 2 imply that one can make the completion. While the same entries appear in different places in the matrix, it turns out that in each step the same entries lie in a shifted version of the same partial matrix, thus reducing it in each step to a single entry matrix completion problem. In the above particular case one only needs to worry about finding x_{12} such that

$$\begin{pmatrix} 1 & c_1^* & c_2 \\ c_1 & 1 & x_{12} \\ c_2^* & x_{12}^* & 1 \end{pmatrix}$$

remains positive semidefinite. This submatrix appears more than once (allowing permutations) in the above matrix, for example, by taking the submatrix in rows and columns 1,3,4, but also by taking rows and columns 1,2,8 or 1,5,9.

In order to prepare for the proof of Theorem 3.6.2 we first need some elementary results about chordal graphs.

Lemma 3.6.4 *If $G = (V, E)$ is chordal, then for any two nonadjacent vertices $v_1, v_2 \in V$ the set $\mathrm{Adj}(v_1) \cap \mathrm{Adj}(v_2)$ is a clique.*

Proof. Denote $\mathrm{Adj}(v_1) \cap \mathrm{Adj}(v_2) = \{w_1, \ldots, w_k\}$. If $k \leq 1$, there is nothing to prove. Suppose that w_i, w_j, $i \neq j$, are such that w_i is not adjacent to w_j. Then $[v_1, w_i, v_2, w_j]$ is a four cycle without a chord, yielding that G is not chordal. This yields the result. □

A *tree* is a connected graph $G = (V, E)$ with no cycles, or equivalently, any graph in which the number of edges exceeds by one the number of vertices. If G is a tree, and $v, w \in V$, there is a unique path $P(v, w)$ joining v and w which passes at most once through each vertex. We will call it the *minimal path* joining v and w, and define $d(v, w)$ to be its length (the number of edges it contains). For $n \geq 1$, we will denote by $\widehat{G}_n$ the graph that has the same vertices as G, while $E(\widehat{G}_n) = \{(v, w) : d(v, w) \leq n\}$ (in particular, $G = \widehat{G}_1$).

Lemma 3.6.5 *If G is a tree, then $\widehat{G}_n$ is chordal for any $n \geq 1$.*

Proof. Take a minimal cycle C of length > 3 in $\widehat{G}_n$. Suppose x, y are elements of C at a maximal distance. If $d(x, y) \leq n$, then C is actually a clique, which is a contradiction. Thus x and y are not adjacent in $\widehat{G}_n$. Suppose v, w are the two vertices of $\widehat{G}_n$ adjacent to x in the cycle C. Now $P(x, v)$ has to pass through a vertex which is on $P(x, y)$, since otherwise the union of these two paths would be the minimal path connecting y and v, and it would have length strictly larger than $d(x, y)$. Denote by v_0 the element of $P(x, v) \cap P(x, y)$ which has the largest distance to x; since $d(y, v) = d(y, v_0) + d(v_0, v) \leq d(y, x) = d(y, v_0) + d(v_0, x)$, it follows that $d(v_0, v) \leq d(v_0, x)$.

Similarly, if w_0 is the element of $P(x,w) \cap P(x,y)$ which has the largest distance to x, it follows that $d(w_0,w) \le d(w_0,x)$.

Suppose now that $d(v_0,x) \le d(w_0,x)$. Then

$$\begin{aligned} d(v,w) &= d(v,v_0) + d(v_0,w_0) + d(w_0,w) \\ &\le d(x,v_0) + d(v_0,w_0) + d(w_0,w) = d(x,w) \le n, \end{aligned}$$

since w is adjacent to x. Then $(v,w) \in E$, and C is not minimal: a contradiction. Thus $\widehat{G}_n$ is chordal. □

Corollary 3.6.6 *Suppose G is a tree, and $x,y \in V(G)$, $d(x,y) = n+1$. Then the set*

$$\{z \in V(G) : \max(d(z,x), d(z,y)) \le n\}$$

is a clique in $\widehat{G}_n$.

Proof. We use Lemmas 3.6.4 and 3.6.5. □

There is a graph γ naturally associated to $\mathbb{F}$: the *Cayley graph* corresponding to the set of generators $\mathcal{A}$. Namely, $V(\gamma)$ are the elements of $\mathbb{F}$, while s and t are connected by an edge if and only if $|s^{-1}t| = 1$. Moreover, γ is easily seen to be a tree: any cycle would correspond to a nontrivial relation satisfied by the generators of the group. The distance d between vertices of γ is defined as $d(s,t) = |s^{-1}t|$.

As a consequence of Lemma 3.6.5, $\widehat{\gamma}_n$ is chordal for any $n \ge 1$. We will introduce a sequence of intermediate graphs γ_ν, with $\nu \in \widehat{\mathbb{F}}$, as follows. We have $V(\gamma_\nu) = \mathbb{F}$ for all ν, while $(s,t) \in E(\gamma_\nu)$ if and only if $\widehat{s^{-1}t} = \nu'$ for some $\nu' \preceq \nu$. Obviously $\gamma_\nu \subset \gamma_{\nu'}$ for $\nu \preceq \nu'$, and $E(\gamma_{\nu^+})$ is obtained by adding to $E(\gamma_\nu)$ all edges (s,t) with $\widehat{s^{-1}t} = \nu^+$. Each γ_ν is invariant with respect to translations, and $\widehat{\gamma}_n = \gamma_{\nu_n}$, where ν_n is the last element in $\widehat{S}_n$.

The next proposition is the main technical ingredient of the section.

Proposition 3.6.7 *With the above notation, γ_ν is chordal for all ν.*

Proof. If γ_ν is not chordal, and $n = |\nu|$, we may assume that ν is the last element in $\widehat{\mathbb{F}}$ of length n with this property. Since $\widehat{\gamma}_\nu$ is chordal, $\nu \ne \nu_n$, and thus $|\nu^+| = n$.

Suppose then that γ_ν contains the cycle $(s_1,\ldots,s_q)$, with $q \ge 4$. At least one of (s_1,s_3) or (s_2,s_4) must be an edge of γ_{ν^+}, since otherwise (s_1,s_2,s_3,s_4) is a part of a cycle of length ≥ 4 in γ_{ν^+}. We may assume that $(s_1,s_3) \in V(\gamma_{\nu^+})$; if we denote $t = s_1^{-1}s_3$, then, since $(s_1,s_3) \notin V(\gamma_\nu)$, we must have $\widehat{t} = \nu^+$.

Suppose that $q > 4$. Then $(s_1,s_4) \notin E(\gamma_\nu)$. If $(s_1,s_4) \in V(\gamma_{\nu^+})$, then $\widehat{s_1^{-1}s_4} = \nu^+$, and thus either $s_1^{-1}s_4 = t$ or $s_1^{-1}s_4 = t^{-1}$. The first equality is impossible since it implies $s_3 = s_4$. As for the second, it would lead to $t^2 = s_4^{-1}s_3$. But $|t| = |\nu^+| = n$, and thus, by Lemma 3.6.1, $|s_4^{-1}s_3| = |t^2| > n$. This contradicts $(s_4,s_3) \in V(\gamma_\nu)$; consequently, we must have $q = 4$.

Performing, if necessary, a translation, we may suppose that the four cycle is (e, s, t, r), and that $(e, t) \in E(\gamma_{\nu^+})$ (thus $t \in \nu^+$, and $t \prec t^{-1}$). Thus, e, s, r, t are all different, $(e, s), (s, t), (r, t), (e, r)$ are edges of G_ν while $(e, t), (s, r)$ are not, and (e, t) is an edge of G_{ν^+}. These assumptions imply that

$$|t| = n, \quad |s| \leq n, \quad |r| \leq n, \quad |t^{-1}s| \leq n, \quad |t^{-1}r| \leq n, \quad |r^{-1}s| \geq n. \tag{3.6.1}$$

Let us denote $u = \mathrm{CB}(s, r, t)$. We have $r = ur_1$, $t = ut_1$, $s = us_1$, and at least two among the elements $\mathrm{CB}(s_1, r_1)$, $\mathrm{CB}(r_1, t_1)$, $\mathrm{CB}(s_1, t_1)$ are equal to e.

Suppose first that this happens with (s_1, t_1) and (r_1, t_1). If $v = \mathrm{CB}(s_1, r_1)$, then $s_1 = vs_2$, $r_1 = vr_2$. The inequalities (3.6.1) imply

$$|u| + |t_1| = n, \tag{3.6.2}$$

$$|u| + |v| + |s_2| \leq n, \quad |u| + |v| + |r_2| \leq n, \tag{3.6.3}$$

$$|t_1| + |v| + |s_2| \leq n, \quad |t_1| + |v| + |r_2| \leq n, \tag{3.6.4}$$

$$|s_2| + |r_2| \geq n. \tag{3.6.5}$$

Replacing $|t_1| = n - |u|$ from (3.6.2), one obtains from (3.6.4) that $|v| + |s_2| \leq |u|$ and $|v| + |r_2| \leq |u|$. Then (3.6.3) imply $|v| + |s_2| \leq n/2$, $|v| + |r_2| \leq n/2$. Comparing the last two inequalities with (3.6.5), one obtains $|v| = 0$. This means that all pairs $(s_1, r_1), (r_1, t_1)$, and (s_1, t_1) have as common beginning e, and so we may as well assume from the start that (s_1, r_1) and (s_1, t_1) have as common beginning e (the case (s_1, r_1) and (r_1, t_1) is symmetrical, and can be treated likewise; note that the whole situation is symmetric in s and r).

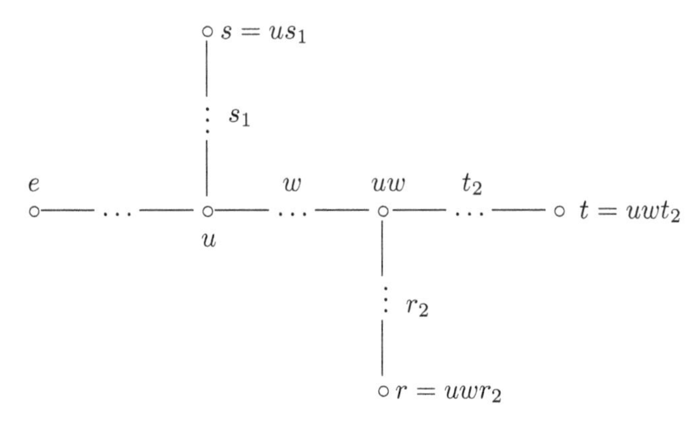

Figure 3.1

We will assume in the sequel that $w = \mathrm{CB}(t_1, r_1)$; thus $r_1 = wr_2$ and $t_1 = wt_2$ (see Figure 3.1). We have then

$$|u| + |w| + |t_2| = n, \quad |u| + |w| + |r_2| \leq n, \quad |s_1| + |w| + |t_2| \leq n, \tag{3.6.6}$$

whence $|s_1| \le |u|$, $|r_2| \le |t_2|$. But, since (s,r) is not an edge of γ_ν, we have $|s_1| + |w| + |r_2| \ge n$. Comparing this last inequality with the first inequality in (3.6.6), it follows that

$$|s_1| = |u|, \quad |r_2| = |t_2|.$$

Now, $s^{-1}t$ is a word of length n different from t; thus $s_1^{-1}w \ne uw$, and $|s_1^{-1}w| = |uw|$.

On the other hand, $t \in \nu^+$ begins with uw, and $t \preceq t^{-1}$. Suppose $t' \in \mathbb{F}$, with $|t'| = n$, begins with t_1', with $|t_1'| = |uw|$. If $t_1' \prec uw$, then the definition of lexicographic order implies that $t' \prec t$ and $\widehat{t'} \prec \widehat{t}$. Applying this argument to $t' = s^{-1}r$, $t_1' = s_1^{-1}w$, it follows that, if $s_1^{-1}w \prec uw$, then $\widehat{s^{-1}r} \prec \nu^+$. Then $\widehat{s^{-1}r} \prec \nu$, and therefore $(s,r) \in E(\gamma_\nu)$: a contradiction. Therefore $uw \prec s_1^{-1}w$, and $t \prec s^{-1}t$.

Since, however, $(s,t) \in \gamma_\nu$, it follows that $\widehat{t^{-1}s} \preceq \nu$, and thus

$$t^{-1}s \preceq t \preceq t^{-1}. \tag{3.6.7}$$

Also, $(e,s) \in E(\gamma_\nu)$ implies $|s| = |u| + |s_1| = 2|u| \le n$, and thus $|u| \le n/2$. Therefore $|t_2^{-1}w^{-1}| = |wt_2| \ge n/2$.

Now $t_2^{-1}w^{-1} = \mathrm{CB}(t^{-1}s, t^{-1})$, and (3.6.7) implies that $t_2^{-1}w^{-1}$ is also the beginning of t. Thus $\mathrm{CB}(t,t^{-1})$ has length $\ell \ge n/2$. Writing then $t^2 = (t^{-1})^{-1}t$, $\mathrm{CB}(t,t^{-1})$ cancels, and we obtain $|t^2| \le 2n - 2\ell \le n = |t|$. By Lemma 3.6.1, this would imply $t = e$: a contradiction, since the elements e, s, r, t of the assumed four cycle must be distinct. The proposition is proved. □

If $\nu \in \widehat{\mathbb{F}}$, then there exist cliques in γ_{ν^+} which are not cliques in γ_ν: we may start with any edge of γ_{ν^+} which is not an edge of γ_ν and take a maximal clique (in γ_{ν^+}) which contains it. Such a clique is necessarily finite, since the length of edges is bounded by $|\nu^+|$.

Corollary 3.6.8 *Suppose $\nu \in \widehat{\mathbb{F}}$, and C is a clique in γ_{ν^+} which is not a clique in γ_ν. Then C contains a single edge in γ_{ν^+} which is not an edge in γ_ν.*

Proof. Suppose C contains (s,t), (s',t'), with $s^{-1}t = s'^{-1}t' = v \in \nu^+$. Obviously $s \ne s'$ and $t \ne t'$. If $s = t'$, then $s'^{-1}t = v^2$, and thus the edge (s',t) has length strictly larger than $n = |v|$ by Lemma 3.6.1. One shows similarly that $s' \ne t$. Then (s,t,t',s') is a four cycle in γ_ν, which contradicts Proposition 3.6.7. □

Let us now fix $\nu \in \widehat{\mathbb{F}}$, $\nu \ne \{e\}$; thus $\nu = \{s_\nu, s_\nu^{-1}\}$, with $s_\nu \preceq s_\nu^{-1}$. Suppose that ϕ_{ν^-} is a positive semidefinite function on Σ_{ν^-}. Consider the maximal clique C_ν in γ_ν that contains (e, s_ν); by Corollary 3.6.8, (e, s_ν) is the unique edge of C_ν that is not in $E(\gamma_{\nu^-})$. Consequently, the following definitions make sense:

$$\begin{aligned} \mathrm{Domain}(\phi_{\nu^-}) &= \mathrm{Ran}(A(\phi_{\nu^-}; C_\nu \setminus \{s_\nu\})) \ominus \mathrm{Ran}(A(\phi_{\nu^-}; C_\nu \setminus \{e, s_\nu\})), \\ \mathrm{Ran}\,(\phi_{\nu^-}) &= \mathrm{Ran}(A(\phi_{\nu^-}; C_\nu \setminus \{e\})) \ominus \mathrm{Ran}(A(\phi_{\nu^-}; C_\nu \setminus \{e, s_\nu\})) \end{aligned} \tag{3.6.8}$$

It is immediate from the definition of positive semidefiniteness that $\Phi(e)$ is a positive semidefinite operator. A standard argument shows that, if $\mathcal{H}_0 = \overline{\Phi(e)\mathcal{H}}$, then there exists a positive semidefinite function $\Phi_0 : \mathbb{F} \to \mathcal{L}(\mathcal{H}_0)$ such that $\Phi_0(e) = I_{\mathcal{H}_0}$, and $\Phi(s)h = \Phi_0(s)\Phi(e)^{1/2}h$ for all $h \in \mathcal{H}$. We will suppose in the sequel that $\Phi(e) = I_{\mathcal{H}}$; equivalently, the operator V in the Naimark dilation theorem is an isometry.

Remember now that we have a fixed lexicographic order on $\mathbb{F}$, and consequently on $\widehat{\mathbb{F}}$. Define $\Sigma_\nu = \bigcup_{\nu' \preceq \nu} \nu$; the following lemma is then obvious, on closer inspection.

Lemma 3.6.9 *Suppose $\mu \in \mathbb{F}$, and $\phi : \Sigma_\mu \to \mathcal{L}(\mathcal{H})$ is positive semidefinite. Then the map $\Phi \mapsto (\Phi|\Sigma_\nu)_{\nu \in \widehat{\mathbb{F}}, \mu \preceq \nu}$ defines a one-to-one correspondence between*

1) positive semidefinite functions on $\mathbb{F}$ whose restriction to Σ_μ is ϕ and

2) sequences $(\phi_\nu)_{\nu \in \widehat{\mathbb{F}}, \mu \preceq \nu}$, ϕ_ν positive semidefinite function defined on Σ_ν, such that, if $\nu \preceq \nu'$, then $\phi_\nu = \phi_{\nu'}|\Sigma_\nu$.

Lemma 3.6.10 *With the above notation, if ϕ_{ν^-} is a positive semidefinite function on Σ_{ν^-}, then all positive semidefinite extensions ϕ_ν of ϕ_{ν^-} are in one-to-one correspondence with contractions* $\gamma_\nu : \text{Domain}(\phi_{\nu^-}) \to \text{Ran}(\phi_{\nu^-})$.

Proof. Note first that

$$\left(\phi_{\nu^-}(s^{-1}t)\right)_{\{s,t\} \subset C_\nu, \{s,t\} \in E(\gamma_{\nu^-})}$$

is a partially positive matrix whose corresponding pattern is the graph induced by γ_{ν^-} on C_ν. By applying Proposition 2.5.2, we obtain that there exists a positive semidefinite completion of this matrix, and, moreover, that all completions are in one to one correspondence with contractions $\gamma_\nu : \text{Domain}(\phi_{\nu^-}) \to \text{Ran}\,(\phi_{\nu^-})$. Denote the completed entry by $B(\gamma_\nu)$.

Now, any maximal clique in γ_ν that is not a clique in γ_{ν^-} is a translate of C_ν, and the corresponding partially positive semidefinite matrix determined by ϕ_{ν^-} is the same. If we define then

$$\phi_\nu(s) = \begin{cases} \phi_{\nu^-}(s) & \text{for } s \in \Sigma_{\nu^-}, \\ B(\gamma_\nu) & \text{for } s = s_\nu, \\ B(\gamma_\nu)^* & \text{for } s = s_\nu^{-1}, \end{cases}$$

we obtain a positive semidefinite function on Σ_ν that extends ϕ_{ν^-}. It is easy to see that this correspondence is one-to-one. □

In particular, such extensions exist, and thus any positive semidefinite function on Σ_{ν^-} can be extended to a positive semidefinite function on Σ_ν. As we know from Chapter 2, among them there exists a distinguished one, the central extension, obtained by taking $\gamma_\nu = 0$.

Proof of Theorem 3.6.2. Now, combining Lemmas 3.6.9 and 3.6.10, we obtain the result. □

Note that in this correspondence $\Phi|\Sigma_\mu$ depends only on $\gamma_{\mu'}$ with $\nu \preceq \mu' \preceq \mu$, and thus the domain and range of γ_μ are well defined by the previous $\gamma_{\mu'}$. Then at step μ one can choose γ_μ to be an arbitrary element of the corresponding operator unit ball.

The most important consequence is an extension result for positive semidefinite functions on S_n. It should be noted that, contrary to Σ_ν, the sets S_n do not depend on the particular lexicographic order $\preceq$.

Proposition 3.6.11 *Every positive semidefinite function ϕ on S_n has a positive semidefinite extension Φ on $\mathbb{F}$. If ϕ is radial (that is, $\phi(s)$ depends only on $|s|$), then one can choose also Φ radial.*

Proof. The first part is an immediate consequence of Theorem 3.6.2, obtained by taking $\Sigma_\nu = S_n$. Suppose Φ is positive semidefinite; then $\widetilde{\Phi}$ is also positive semidefinite. The second part of the proposition follows. □

Another consequence is the parametrization of all positive semidefinite functions.

Theorem 3.6.12 *There is a one-to-one correspondence between the set of all positive semidefinite functions $\Phi : \mathbb{F} \to \mathcal{L}(\mathcal{H})$ and the set of sequences of contractions $(\gamma_\mu)_{\mu \neq \{e\}}$, where $\gamma_\mu : \mathrm{Domain}(\Phi|\Sigma_{\mu^-}) \to \mathrm{Ran}(\Phi|\Sigma_{\mu^-})$.*

We may also rephrase Theorem 3.6.2 by using the Naimark dilation theorem.

Corollary 3.6.13 *Suppose $\phi : S_n \to \mathcal{L}(\mathcal{H})$ is a positive semidefinite function. Then there exist a representation π of $\mathbb{F}$ on a Hilbert space $\mathcal{K}$ and an operator $V : \mathcal{H} \to \mathcal{K}$, such that, for all $s, t \in \mathbb{F}$ $|s^{-1}t| \leq n$, we have*

$$\phi(s^{-1}t) = V^* \pi(s^{-1}t) V.$$

There is an alternate way of looking at Theorem 3.6.2. Denote by $\mathfrak{M}_n(\mathcal{H})$ the block operator matrices indexed by elements in S_n; they have order $N(n)$ equal to the cardinality of S_n. To any function ϕ on S_{2n} corresponds then a matrix $M(\phi) \in \mathfrak{M}_n(\mathcal{H})$, defined by $M(\phi)_{s,t} = \phi(s^{-1}t)$. We will call an element $M = (M_{s,t})_{s,t \in S_n} \in \mathfrak{M}_n(\mathcal{H})$ Toeplitz if its entries depend only on $s^{-1}t$. The space of Toeplitz operator matrices in $\mathfrak{M}_n(\mathcal{H})$ will be denoted by $\mathcal{T}_n(\mathcal{H})$. When $\mathcal{H} = \mathbb{C}$, we also write $\mathfrak{M}_n$ and $\mathcal{T}_n$, instead of $\mathfrak{M}_n(\mathbb{C})$ and $\mathcal{T}_n(\mathcal{H})$, respectively. Basis elements in $\mathcal{T}_n$ are the matrices $\epsilon(\sigma)$, $\sigma \in S_n$, where $\epsilon(\sigma)_{s,t} = \delta_{s\sigma,t}$.

Lemma 3.6.14 *If $M \in \mathcal{T}_n(\mathcal{H})$, then there exists a uniquely defined $\phi : S_{2n} \to \mathcal{L}(\mathcal{H})$, such that $M = M_\phi$.*

Proof. If $s \in S_{2n}$, we may write $s = t^{-1}r$, with $t, r \in S_n$. We define then $\phi(s) = M_{t,r}$. The Toeplitz condition implies that this definition does not depend on the particular decomposition of s. In order to show that ϕ is indeed positive semidefinite, take $S \in \mathbb{F}$ such that for any $s, t \in S$,

$s^{-1}t \in S_{2n}$. By adding elements to S, if necessary, we may assume that the maximum of the lengths $|s^{-1}t|$ for $s, t \in S$, is actually equal to $2n$. Take then $s_0, t_0 \in S$ such that $|s_0^{-1}t_0| = n$, and let $\widehat{s}$ be the "middle" of the path (in the Cayley graph of $\mathbb{F}$) going from s_0 to t_0; that is, $\widehat{s}$ is uniquely defined by the conditions $|s_0^{-1}\widehat{s}| = |\widehat{s}^{-1}t_0| = n$. Using the tree structure of the Cayley graph, it is then easy to see that $|\widehat{s}^{-1}s| \leq n$ for any $s \in S$. So

$$A(\phi : S) = [\phi(t^{-1}r)]_{t,r\in\widehat{s}^{-1}S}$$

is a submatrix of

$$M = [\phi(t^{-1}r)]_{t,r\in S_n},$$

and is thus positive semidefinite. The uniqueness of ϕ is immediate. □

We call a unital homomorphism $\pi : G \to \mathcal{L}(\mathcal{H})$, where $\mathcal{H}$ is a Hilbert space, a *representation* of the group G.

Corollary 3.6.15 *If $T = (T_{s,t})_{s,t\in S_n} \in \mathcal{T}_n$, then there exist a representation π of $\mathbb{F}$ and an operator V such that*

$$T_{s,t} = V^*\pi(s^{-1}t)V.$$

3.7 NONCOMMUTATIVE FACTORIZATION

In this section we present two applications to factorization of noncommutative polynomial problems. The first is a result based on Theorem 3.6.2; the second one proves that for polynomials in several noncommutative variables Hilbert's 17th problem is true.

We continue to use the notation from the previous section. The elements of $\mathbb{F}$, the free group with m generators $X_1, \ldots, X_m$, can be considered "monomials" in the indeterminates $X_1, \ldots, X_m$ and $X_1^{-1}, \ldots, X_m^{-1}$; the monomial $X(s)$ corresponding to $s \in \mathbb{F}$ is obtained by replacing a_i by X_i and a_i^{-1} by X_i^{-1}. It is then possible to consider also polynomials in these indeterminates; that is, formal finite sums

$$p(X) = \sum_s A_s X(s), \tag{3.7.1}$$

where we assume the coefficients A_s to be operators on a fixed Hilbert space $\mathcal{C}$. We denote by $\deg(p)$ (the *degree* of p) the maximum length of the words appearing in the sum in (3.7.1). We introduce an involution $p(X)^* = \sum_s A_s^* X(s^{-1})$.

If $U_1, \ldots, U_m$ are unitary (not necessarily commuting) operators acting on a Hilbert space $\mathcal{X}$, then, for $s \in \mathbb{F}$, $U(s)$ is the operator on $\mathcal{X}$ obtained by replacing X_i with U_i, and X_i^{-1} by $U_i^{-1} = U_i^*$. Then $p(U) \in \mathcal{L}(\mathcal{C} \otimes \mathcal{X})$ is defined by $p(U) = \sum_s A_s \otimes U(s)$. We say that such a polynomial is *positive semidefinite* if, for any choice of unitary operators $U_1, \ldots U_m$, the operator $p(U)$ is positive semidefinite.

We will need the following theorem, which is known as the Arveson extension theorem. Let A be a C^*-algebra. A linear subspace $\mathcal{T} \subset A$ is called an *operator system* if $x \in \mathcal{T}$ implies $x^* \in \mathcal{T}$. A linear map $l : A \to B$ between two C^*-algebras is called *positive* if $l(a) \geq 0$ for every $a \geq 0$ ($a \geq 0$ means there exists $x \in A$ such that $a = x^*x$). A linear map $l : A \to B$ is called *completely positive* if for every $n \geq 1$, the transformation $l_n : A \otimes \mathbb{C}^{n\times n} \to B \otimes \mathbb{C}^{n\times n}$ defined by $l((a_{ij})_{i,j=1}^n) = (l(a_{ij}))_{i,j=1}^n$ is positive.

Theorem 3.7.1 *Let A and B be C^*-algebras and $\mathcal{T} \subset A$ be an operator system. Then every completely positive map $l : \mathcal{T} \to B$ admits a completely positive extension $L : A \to B$.*

Proof. A proof of this result may be found in [44]. □

Theorem 3.7.2 *If p is a positive semidefinite polynomial, then there exist an auxiliary Hilbert space $\mathcal{E}$ and operators $B_s : \mathcal{C} \to \mathcal{E}$, defined for words s of length $\leq \deg(p)$ such that, if $q(X) = \sum_s B_s X(s)$, then*

$$p(X) = q(X)^* q(X). \tag{3.7.2}$$

Proof. As noted above, the proof uses a crucial point of Theorem 3.6.2 (or rather Corollary 3.6.15).

Denote $d = \deg(p)$, and consider the space of Toeplitz matrices $\mathcal{T}_d \subset \mathfrak{M}_d$; it is a finite dimensional operator system. We define a map $\psi : \mathcal{T}_d \to \mathcal{L}(\mathcal{C})$ by the formula $\psi(\epsilon(\sigma)) = A_\sigma$.

Suppose that an element $T \in \mathcal{T}_d \otimes \mathbb{C}^{k\times k}$ is a positive semidefinite matrix. By considering $T = [\tau_{s^{-1}t}]$ as an element of $\mathcal{T}_d(\mathbb{C}^k)$, we can apply Theorem 3.6.2 and obtain a representation $\pi_\Phi : \mathbb{F} \to \mathcal{L}(\mathcal{H})$ and an operator $V_\Phi : \mathbb{C}^k \to \mathcal{H}$, such that

$$\tau_{s^{-1}t} = V_\Phi^* \pi_\Phi(s^{-1}t) V_\Phi.$$

Therefore

$$\begin{aligned}(\psi \otimes 1_k)(T) &= (\psi \otimes 1_k)\left(\sum_s \epsilon(s) \otimes \tau_s\right) = \sum_s A_s \otimes \tau_s \\ &= (1_{\mathcal{C}} \otimes V_\Phi)^* \left(\sum_s A_h \otimes \pi_\Phi(s)\right)(1_{\mathcal{C}} \otimes V_\Phi).\end{aligned}$$

Since p is positive semidefinite, $\sum_s A_h \otimes \pi_\Phi(s)$ is a positive semidefinite operator, and therefore the same is true about $(\psi \otimes 1_k)(T)$. It follows then that ψ is completely positive. Applying Arveson's extension theorem (Theorem 3.7.1), we can extend ψ to a completely positive map $\widetilde{\psi} : \mathfrak{M}_n \to \mathcal{L}(\mathcal{C})$.

Suppose $E_{s,t} \in \mathfrak{M}_n$ has 1 in the (s,t) position and 0 everywhere else. The block operator matrix $(E_{s,t})_{s,t}$ is positive semidefinite. As $\widetilde{\psi}$ is completely positive it follows that $(\widetilde{\psi}(E_{s,t}))_{s,t} \in \mathcal{L}(\bigoplus_{i=1}^N \mathcal{C})$ is positive semidefinite, where $N = N(d)$. Therefore there exist operators $B_s : \mathcal{C} \to \bigoplus_{i=1}^N \mathcal{C}$, such that $B_s^* B_t = \widetilde{\psi}(E_{s,t})$. Consequently

$$A_x = \psi(\epsilon(x)) = \psi\left(\sum_{x=s^{-1}t} E_{s,t}\right) = \sum_{x=s^{-1}t} \widetilde{\psi}(E_{s,t}) = \sum_{x=s^{-1}t} B_s^* B_t.$$

If we define $q(X) = \sum_s B_s X(s)$, then the last equality is equivalent to $p(X) = q(X)^* q(X)$. □

Note that the factorization of p is usually not unique. The theorem produces a factor B which has degree at most equal to $\deg p$. Also, in case $\mathcal{C}$ is finite dimensional, the resulting space $\mathcal{E}$ can also be taken finite dimensional, with $\dim \mathcal{E} = N(d) \times \dim \mathcal{C}$.

One can rephrase the result of Theorem 3.7.2 as a decomposition into a sum of squares.

Corollary 3.7.3 *If p is a positive semidefinite polynomial, then there exist a finite number of polynomials $Q_1, \ldots, Q_N$, with coefficients in $\mathcal{L}(\mathcal{C})$, such that*

$$p(X) = \sum_{j=1}^{N} Q_j(X)^* Q_j(X). \tag{3.7.3}$$

Proof. Since $B_s : \mathcal{C} \to \bigoplus_{i=1}^N \mathcal{C}$, we can write $B_s = (B_s^{(1)} \ \ldots \ B_s^{(N)})^T$. We consider then $Q_j(X) = \sum_s B_s^{(j)} X(s)$, and obtain the required decomposition. □

Connections between extension problems for positive functions and factorizations exist in the noncommutative real case as well. We mention here a solution in the positive sense to Hilbert's 17th problem for real polynomials in noncommutative variables (which we present next). Here we consider polynomials that are real linear combinations of words in the indeterminates $X_1, \ldots, X_m$ and $X_1^T, \ldots, X_m^T$. The degree of the polynomial is the length of the longest word appearing in the polynomial. On these polynomials we have a "transpose" operator which replaces a term $aH_1 \cdots H_k$, where $a \in \mathbb{R}$ and $H_i \in \{X_1, \ldots, X_m, X_1^T, \ldots, X_m^T\}$, by $aH_k^T \cdots H_1^T$, where $(X_i^T)^T := X_i$. A polynomial is called *symmetric* when it equals its transpose. For instance, $2X_1X_2^T + 2X_2X_1^T$ is symmetric, while $2X_1X_2^T - 2X_2X_1^T$ is not. We call a polynomial $Q = Q(X_1, \ldots, X_m, X_1^T, \ldots, X_m^T)$ *matrix positive* if for all choices of r and all choices of $A_1, \ldots, A_m \in \mathbb{R}^{r\times r}$, the matrix $Q(A_1, \ldots, A_m, A_1^T, \ldots, A_m^T)$ is positive semidefinite. For instance, $Q(X_1, X_2, X_1^T, X_2^T) = X_1^T X_1 + 2X_2X_2^T$ is matrix positive, while $Q(X_1, X_1^T) = X_1^2 + X_1^{T2}$ is not (take, e.g., $A_1 = \left(\begin{smallmatrix} 0 & 1 \\ -1 & 0 \end{smallmatrix}\right)$). We say that a symmetric polynomial Q is a *sum of squares* if we can write

$$Q(X) = \sum_{i=1}^{k} h_i(X) h_i(X)^T,$$

where each h_i is a polynomial in $X_1, \ldots, X_m, X_1^T, \ldots, X_m^T$. We now have the following result.

Theorem 3.7.4 *Let $Q(X)$ be a symmetric polynomial of degree $2d$. Then $Q(X)$ is matrix positive if and only if it is a sum of squares.*

Proof. We represent $Q(X) = V(X)^T M V(X)$, with $M = M^T \in \mathbb{R}^{p \times p}$ and $V(X) = \text{col}(w(X))_{w \in W}$, where $W = \{w_1, \ldots, w_p\}$ is the collection of words in $X_1, \ldots, X_m, X_1^T, \ldots, X_m^T$ of length $\leq d$. The concatenation (or product) of two words w_1 and w_2 is simply denoted by $w_1 w_2$. For $w \in W$, its transpose w^T is defined in an obvious way. For instance, the transpose of the word $X_3 X_2^T X_5$ is $X_5^T X_2 X_3^T$. We (partially) order the words so that $|w_i| \leq |w_j|$ whenever $i \leq j$, where $|w|$ is the length of word w.

Let

$$\mathcal{B} = \{B = B^T \in \mathbb{R}^{p \times p} : V(X)^T B V(X) \equiv 0\}.$$

Then $\mathcal{B}$ is a subspace of $\mathbb{R}^{p \times p}$. In fact, $B = (B_{ij})_{i,j=1}^p \in \mathcal{B}$ if and only if for all words w of length $\leq 2d$ we have that

$$\sum_{w_i^T w_j = w} B_{ij} = 0.$$

To prove our theorem we need to show that Q is matrix positive if and only if $M \in \mathcal{B} + \text{PSD}$ (or, equivalently, that we can find a $B \in \mathcal{B}$ so that $M - B$ is positive semidefinite). The only if part of this statement is trivial, thus we focus on the if part.

First we show that $\mathcal{B}+\text{PSD}$ is closed. Let $A_i = B_i + P_i \to A$, where $B_i \in \mathcal{B}$ and $P_i \in \text{PSD}$. We will show that the sequence P_i is bounded. Let us write $P_i = (P_{kl}^{(i)})_{k,l=0}^d$, where the kth block column/row corresponds to those words of degree k. Decompose $B_i = (B_{kl}^{(i)})_{k,l=0}^d$ accordingly. By the definition of B we must have that $B_{00}^{(i)} = 0$ and $B_{dd}^{(i)} = 0$. Suppose that k is the largest index such that $P_{kk}^{(i)}$ is unbounded. Let us say $\|P_{kk}^{(i)}\| = \mu_k(i) \to \infty$ as $i \to \infty$. Then $B_{kk}^{(i)}$ is unbounded as well of the same order (as $B_{kk}^{(i)} + P_{kk}^{(i)}$ is bounded). By the definition of $\mathcal{B}$ this gives that there must be a $1 \leq l \leq \min\{k, d-k\}$ such that $B_{k-l,k+l}^{(i)}$ is unbounded of the same order. Choose l as small as possible. But then $P_{k-l,k+l}^{(i)}$ must be unbounded at least of order $\mu_k(i)$. As $P_i \in \text{PSD}$, and $P_{k+l,k+l}^{(i)}$ is bounded, we must have that $\|P_{k-l,k-l}^{(i)}\|$ tends to infinity at least as fast as $\mu_k(i)^2$. As above, this implies that there must be a $1 \leq l_2 \leq \min\{k-l, d-k+l\}$ so that $B_{k-l-l_2,k-l+l_2}^{(i)}$ is unbounded of the same order. Choose l_2 as small as possible. Since all P_i are positive semidefinite, and as $P_{k-l+l_2,k-l+l_2}^{(i)}$ either is bounded or tends to infinity at most at the rate of $\mu_k(i)$, we must have that $\|P_{k-l-l_2,k-l-l_2}^{(i)}\|$ tends to infinity at least as fast as $\mu_k(i)^3$. Continuing this way we obtain that $P_{00}^{(i)}$ must be unbounded, but this is impossible as $P_{00}^{(i)} = A_{00}^{(i)} \to A_{00}$.

Since $\{P_i : i \in \mathbb{N}\}$ is bounded, we can take a convergent subsequence $P_{k_i} \to P$. As PSD is closed, this gives $P \in \text{PSD}$. But then $B_{k_i} = A_{k_i} - P_{k_i} \to A - P \in \mathcal{B}$, as $\mathcal{B}$ is closed. This shows that $A \in \mathcal{B}+\text{PSD}$, and thus $\mathcal{B}+PSD$ is closed.

As $\mathcal{B} + \mathrm{PSD}$ is a closed convex cone, we get that

$$\mathcal{B} + \mathrm{PSD} = (\mathcal{B} + \mathrm{PSD})^{**} = (\mathcal{B}^\perp \cap \mathrm{PSD})^*.$$

Thus it suffices to show that when Q is matrix positive we have that

$$\mathrm{tr}(MZ) \geq 0 \text{ for all } Z \in \mathcal{Z},$$

where $\mathcal{Z}$ is some dense subset of $\mathcal{B}^\perp \cap \mathrm{PSD}$. We choose $\mathcal{Z} = \mathcal{B}^\perp \cap PD$. To see that $\mathcal{Z}$ is dense in $\mathcal{B}^\perp \cap \mathrm{PSD}$, first observe that $Z = (Z_{ij})_{i,j=1}^p \in \mathcal{B}^\perp$ if and only if

$$Z_{ij} = Z_{kl} \text{ whenever } w_i^T w_j = w_k^T w_l. \tag{3.7.4}$$

Notice that whenever (3.7.4) forces a diagonal entry (i,i) to equal another entry (k,l) in the matrix, the other entry is an off-diagonal entry with $k > i$ or $l > i$ (this is due to the ordering of the words in W as either $|w_k| > |w_i|$ or $|w_l| > |w_i|$). But then it is easy to see that when $Z \in \mathcal{B}^\perp \cap \mathrm{PSD}$ one can find an arbitrary close perturbation in $\mathcal{Z}$. Indeed, if k is the first index for which $(Z_{ij})_{i,j=1}^k$ is singular, one can add ϵ to the entry Z_{kk}. By requirement (3.7.4) this may affect some entries in rows and columns with index $> k$, and thus the matrix can remain in $Z \in \mathcal{B}^\perp \cap \mathrm{PSD}$ if we perturb some entries in rows and columns $k+1, \ldots, p$. Repeating this procedure ultimately yields a positive definite matrix in $\mathcal{B}^\perp$.

The fact that Q is matrix positive gives that

$$\mathrm{tr}(MZ) \geq 0 \text{ for all } Z \in \mathcal{Z}_q,$$

where $\mathcal{Z}_q = \{(z_i^T z_j)_{i,j=1}^p \ : \ \mathrm{col}(z_i)_{i=1}^p \in \mathcal{W}_q\}$ and

$$\mathcal{W}_q = \{V(A)x \ : \ (A_1, \ldots, A_n) \in (\mathbb{R}^{q \times q})^n, x \in \mathbb{R}^q\}.$$

Thus, it suffices to show that for q large enough we have that $\mathcal{Z} \subseteq \mathcal{Z}_q$. We will show that this inclusion holds for $q = p$.

Let $Z = (v_i^T v_j)_{i,j=1}^p \in \mathcal{Z}$, with $v_i \in \mathbb{R}^p$. As Z is positive definite, we get that $\{v_1, \ldots, v_p\}$ are linearly independent. Define now A_i, $\widehat{A}_i$, $i = 1, \ldots, m$, via

$$A_i v_j := v_k, \ \widehat{A}_i v_p := v_l,$$

whenever $X_i w_j = w_k$ and $X_i^T w_p = w_l$. This defines A_i and $\widehat{A}_i$ on the subspace $\mathcal{M}$ spanned by $\{v_i \ : \ |w_i| < d\}$. Write

$$A_i = \begin{pmatrix} A_{11}^{(i)} & A_{12}^{(i)} \\ A_{21}^{(i)} & A_{11}^{(i)} \end{pmatrix}, \ \widehat{A}_i = \begin{pmatrix} \widehat{A}_{11}^{(i)} & \widehat{A}_{12}^{(i)} \\ \widehat{A}_{21}^{(i)} & \widehat{A}_{11}^{(i)} \end{pmatrix},$$

with respect to the decomposition $\mathbb{R}^p = \mathcal{M} \oplus \mathcal{M}^\perp$. As $Z \in \mathcal{B}^\perp$ we have that $v_i^T v_j = v_k^T v_l$ whenever $w_i^T w_j = w_k^T w_l$. But then we get that

$$v_j^T A_{11}^{(i)T} v_s = (A_{11}^{(i)} v_j)^T v_s = (A_i v_j)^T v_s = v_k^T v_s = v_j^T v_p = v_j \widehat{A}_{11}^{(i)} v_s,$$

where $w_k = X_i w_j$ and $w_p = X_i^T w_s$. This yields that $A_{11}^{(i)T} = \widehat{A}_{11}^{(i)}$. If we now define $A_{12}^{(i)} = \widehat{A}_{21}^{(i)T}$, $\widehat{A}_{12}^{(i)} = A_{21}^{(i)T}$, and choose $A_{22}^{(i)} = \widehat{A}_{22}^{(i)T}$ arbitrarily, we get that $A_i^T = \widehat{A}_i$, $i = 1, \ldots, m$. It is now straightforward (using induction) to check that with the choice $x = v_1$ we have that $v_j = w_j(A)x$. This gives that $(v_i)_{i=1}^p \in \mathcal{W}_p$, and thus $Z \in \mathcal{Z}_p$. □

3.8 TWO-VARIABLE HAMBURGER MOMENT PROBLEM

We will consider a two-variable generalization to the matrix-valued case of the Hamburger moment problem in Section 2.7. While similar results also hold for more than two variables, it simplifies the presentation to restrict ourselves to the case of two variables. In the Notes we provide the reference where the more general results appear.

Given a finite lower inclusive set $\Gamma \subset \mathbb{N}_0^2$, the two-variable matrix-valued truncated Hamburger moment problem asks necessary and sufficient conditions for m Hermitian matrices, indexed by Γ, to be the corresponding moments of a matrix-valued positive measure on $\mathbb{R}^2$. We present a solution to the problem for a particular family of lower inclusive sets, which includes sets of the form

$$\Gamma = \{(k_1, k_2) \in \mathbb{N}_0^2 \colon k_1 + k_2 \leq 2n + 1\}. \tag{3.8.1}$$

The class of Borel measurable sets in $\mathbb{R}^2$ is denoted by $\mathcal{B}(\mathbb{R}^2)$. A function $\sigma \colon \mathcal{B}(\mathbb{R}^2) \to \mathcal{H}_p$ is called a *positive matrix-valued measure* on $\mathbb{R}^2$ if for each $x \in \mathbb{C}^p$, $\langle \sigma(\Delta)x, x\rangle$ defines a positive measure on $\mathbb{R}^2$ for all sets $\Delta \in \mathcal{B}(\mathbb{R}^2)$. For a measurable function $f \colon \mathbb{R}^2 \to \mathbb{C}$ its integral $\int_{\mathbb{R}^2} f(\xi_1, \xi_2) d\sigma(\xi_1, \xi_2)$ is defined by the formula

$$\left\langle \int_{\mathbb{R}^2} f(\xi_1, \xi_2) d\sigma(\xi_1, \xi_2) x, y \right\rangle = \int_{\mathbb{R}^2} f(\xi_1, \xi_2) \langle d\sigma(\xi_1, \xi_2) x, y\rangle$$

for all $x, y \in \mathbb{C}^p$, provided all integrals on the right-hand side converge.

The power moments of a positive matrix-valued measure σ on $\mathbb{R}^2$ are defined by the formulas

$$\hat{\sigma}(k_1, k_2) = \int_{\mathbb{R}^2} \xi_1^{k_1} \xi_2^{k_2} d\sigma(\xi_1, \xi_2), \qquad k_1, k_2 \in \mathbb{N}_0,$$

provided the integrals converge.

Given a finite lower inclusive set $\Gamma \subset \mathbb{N}_0^2$, let $\{S_\gamma\}_{\gamma \in \Gamma}$ be a given $\mathcal{H}_p$-valued sequence. We look for a positive $\mathcal{H}_p$-valued measure σ on $\mathbb{R}^2$ such that

$$\int_{\mathbb{R}^2} \xi_1^{k_1} \xi_2^{k_2} d\sigma(\xi_1, \xi_2)$$

exists for all $k_1, k_2 \in \mathbb{N}_0$ and

$$\hat{\sigma}(\gamma) = S_\gamma \text{ for all } \gamma \in \Gamma. \tag{3.8.2}$$

Given any lower inclusive set $\Lambda \subset \mathbb{N}_0^2$, we define $\Lambda + \Lambda = \{\lambda + \mu \colon \lambda, \mu \in \Lambda\}$ and $\Lambda + \Lambda + e_j = \{\lambda + \mu + e_j \colon \lambda, \mu \in \Lambda\}$, where e_j is the standard basis vector in $\mathbb{N}_0^2$ with a one in the jth entry, $j = 1, 2$. Consider the lower inclusive set

$$\Gamma = (\Lambda \cup \Lambda) \cup (\Lambda + \Lambda + e_1) \cup (\Lambda + \Lambda + e_2). \tag{3.8.3}$$

Note that when $\Lambda = \{(k_1, k_2) \in \mathbb{N}_0^2 \colon k_1 + k_2 \leq n\}$ we get Γ as in (3.8.1). In general, Γ will serve as an indexing set for the given Hermitian matrices in the following way. Index the rows and columns of a matrix Φ by Λ and then let the entry in the λ row and μ column, where $\lambda, \mu \in \Lambda$, be given by $S_{\lambda+\mu}$. That is,

$$\Phi = (S_{\lambda+\mu})_{\lambda,\mu\in\Lambda}.$$

Similarly index the rows and columns of a matrix Φ_j, and let the entry in the λ row and μ column, where $\lambda, \mu \in \Lambda$, be given by $S_{\lambda+\mu+e_j}$, $j = 1, 2$. That is,

$$\Phi_j = (S_{\lambda+\mu+e_j})_{\lambda,\mu\in\Lambda}, \quad j = 1, 2.$$

It should be noted that the matrices Φ, Φ_1, Φ_2 are dependent on the choice of Λ. However, we have decided not to highlight this in the notation as we believe that the choice of Λ will in all instances be clear from the context.

Let us consider the following example.

Example 3.8.1 Let $\Lambda = \{(0,0), (0,1), (1,0)\}$. Then

$$\Lambda + \Lambda = \{(0,0), (0,1), (0,2), (1,0), (1,2), (2,0)\},$$

$$\Lambda + \Lambda + e_1 = \{(1,0), (1,1), (1,2), (2,0), (2,2), (3,0)\}$$

and

$$\Lambda + \Lambda + e_2 = \{(0,1), (0,2), (0,3), (1,1), (1,3), (2,1)\}.$$

Hence

$$\Gamma = \{(0,0), (0,1), (0,2), (0,3), (1,0), (1,2), (1,3), (2,0), (2,1), (2,2), (3,0)\}$$

and we get the following matrices:

$$\Phi = \begin{pmatrix} S_{00} & S_{01} & S_{10} \\ S_{01} & S_{02} & S_{11} \\ S_{10} & S_{11} & S_{20} \end{pmatrix},$$

$$\Phi_1 = \begin{pmatrix} S_{10} & S_{11} & S_{20} \\ S_{11} & S_{12} & S_{21} \\ S_{20} & S_{21} & S_{30} \end{pmatrix},$$

and

$$\Phi_2 = \begin{pmatrix} S_{01} & S_{02} & S_{11} \\ S_{02} & S_{03} & S_{12} \\ S_{11} & S_{12} & S_{21} \end{pmatrix}.$$

The following theorem gives a solution to the two-variable matrix-valued truncated Hamburger moment problem, when the given Hermitian matrices are indexed by a particular family of lower inclusive sets, whose construction was given in (3.8.3).

Theorem 3.8.2 *Let $\Lambda \subset \mathbb{N}_0^2$ be a finite lower inclusive set and suppose the $\mathcal{H}_p$-valued sequence $\{S_\gamma\}_{\gamma\in\Gamma}$, where $\Gamma = (\Lambda+\Lambda)\cup(\Lambda+\Lambda+e_1)\cup(\Lambda+\Lambda+e_2)$, is given. Let $\Phi = (S_{\lambda+\mu})_{\lambda,\mu\in\Lambda}$, $\Phi_j = (S_{\lambda+\mu+e_j})_{\lambda,\mu\in\Lambda}$, $j = 1, 2$. There exists a solution to the two-variable Hamburger moment problem (3.8.2) if*

$$\Phi \geq 0 \tag{3.8.4}$$

and there exist commuting matrices Θ_1, Θ_2 such that

$$\Phi\Theta_j = \Phi_j, \quad j = 1, 2. \tag{3.8.5}$$

In that case, we can find σ of the following form:

$$\sigma = \sum_{i=1}^{k} T_i \delta_{(x_i,y_i)}, \tag{3.8.6}$$

where $(x_1, y_1), \ldots, (x_k, y_k)$ are different points in $\mathbb{R}^2$, and the matrices $T_1, \ldots, T_k$ are positive semidefinite and satisfy $\sum_{i=1}^k \text{rank}T_i = \text{rank } \Phi$.

Conversely, if σ is of the form (3.8.6), with $T_i \geq 0$, and (x_i, y_i) are different points, then there exists a finite lower inclusive set $\Lambda \subset \mathbb{N}_0^2$ with k points such that $\Phi \geq 0$, $\text{rank } \Phi = \sum_{i=1}^k \text{rank } T_i$, and there exist commuting Θ_1 and Θ_2 such that $\Phi\Theta_j = \Phi_j$, $j = 1, 2$.

Note that the choice of Λ in the converse part of Theorem 3.8.2 indeed depends on the given measure, unlike the one-variable case, as the following example will show.

Example 3.8.3 Let $\sigma = T_1\delta_{(1,0)} + T_2\delta_{(2,0)}$, where T_1 and T_2 are positive numbers. If we choose the finite lower inclusive set $\Lambda = \{(0,0), (0,1)\}$ then

$$\Phi = \begin{pmatrix} T_1 + T_2 & 0 \\ 0 & 0 \end{pmatrix}, \Phi_1 = \begin{pmatrix} T_1 + 2T_2 & 0 \\ 0 & 0 \end{pmatrix}, \Phi_2 = \begin{pmatrix} 0 & 0 \\ 0 & 0 \end{pmatrix}.$$

By this choice of Λ we get that $\text{rank}\Phi = 1 < 2 = \text{rank}T_1 + \text{rank}T_2$. The correct choice for Λ is $\{(0,0), (1,0)\}$.

Remark 3.8.4 As in the one-variable setting, the measure σ constructed in Theorem 3.8.2 is one for which all moments $\int_{\mathbb{R}^2} \xi_1^{k_1}\xi_2^{k_2} d\sigma(\xi_1, \xi_2)$, $k_1, k_2 \in \mathbb{N}_0$, automatically exist. Thus all subsequent moments are again well defined.

Remark 3.8.5 We will see in the proof of Theorem 3.8.2 that the points x_i and y_i are in fact the eigenvalues of $\Phi^{(-1/2)}\Phi_1\Phi^{(-1/2)}$ and $\Phi^{(-1/2)}\Phi_2\Phi^{(-1/2)}$, respectively, where both matrices are restricted to the range of Φ. The x_i and y_i should be paired up according to their common eigenvectors, which exist due to the commutativity requirement. Thus, in effect, one is able to read off the support of the measure immediately from the given data, and thus the result can be formulated as a solution to the K-moment problem, where the support is required to lie in a set $K \subseteq \mathbb{R}^2$.

In order to prove Theorem 3.8.2 we must introduce and prove a result for generalized Vandermonde matrices and prove an analogous result to Theorem 2.7.9 in the two-variable setting. Given a sequence of distinct points $w_1, \ldots, w_n$ in $\mathbb{R}^2$ and a lower inclusive set $\Lambda = \{\lambda_1, \ldots, \lambda_m\} \subset \mathbb{N}_0^2$ we define

$$V(w_1, \ldots, w_n; \Lambda) := (w_i^{\lambda_j})_{i=1,\ldots,n;\ j=1,\ldots,m}.$$

Consider the following example.

Example 3.8.6 If $w_1 = (1,1)$, $w_2 = (2,1)$ and $w_3 = (3,1)$ and $\Lambda = \{(0,0),(0,1),(1,0)\}$, then

$$V(w_1, w_2, w_3; \Lambda) = \begin{pmatrix} 1 & 1 & 1 \\ 1 & 1 & 2 \\ 1 & 1 & 3 \end{pmatrix}.$$

Note that Example 3.8.6 illustrates that, unlike square Vandermonde matrices in one variable, distinctness of points in two or more variables does not guarantee invertibility. A natural question is the following. Given distinct points $w_1, \dots, w_n$ in $\mathbb{R}^2$, can one construct a lower inclusive set $\Lambda \subset \mathbb{N}_0^2$, with $\operatorname{card}\Lambda = n$, such that $V(w_1, \dots, w_n; \Lambda)$ is invertible? The following theorem resolves this question in the affirmative.

Theorem 3.8.7 *Given distinct points $w_1 = (x_1, y_1), \dots, w_n = (x_n, y_n)$ in $\mathbb{R}^2$, there exists a lower inclusive set $\Lambda \subset \mathbb{N}_0^2$, with $\operatorname{card}\Lambda = n$, such that $V(w_1, \dots, w_n; \Lambda)$ is invertible. In fact, if we reorder $w_1, \dots, w_n$ via*

$$\begin{aligned} w_1^{(1)} &:= (x_1, y_1^{(1)}), \dots, w_{n_1}^{(1)} := (x_1, y_{n_1}^{(1)}), \\ w_1^{(2)} &:= (x_2, y_1^{(2)}), \dots, w_{n_2}^{(2)} := (x_2, y_{n_2}^{(2)}), \\ &\vdots \\ w_1^{(k)} &:= (x_k, y_1^{(k)}), \dots, w_{n_k}^{(k)} := (x_k, y_{n_k}^{(k)}), \end{aligned}$$

where $x_i \neq x_j$, when $i \neq j$, $n_1 \geq \dots \geq n_k \geq 1$ and $n_1 + \dots + n_k = n$, then one can choose

$$\Lambda = \bigcup_{i=1}^{k} (\{i-1\} \times \{0, \dots, n_i - 1\})$$

and obtain

$$\det V(w_1, \dots, w_n; \Lambda) = \prod_{1 \leq i < j \leq k} (x_j - x_i)^{n_i} \prod_{1 \leq p < q \leq n_i} (y_q^{(i)} - y_p^{(i)}) \neq 0. \quad (3.8.7)$$

Before we can prove Theorem 3.8.7 we must introduce some notation and results for divided differences. Given distinct points $x_0, \dots, x_n$ and a function $f \colon \mathbb{R} \to \mathbb{R}$, the zeroth divided difference for the function f with respect to x_0 is

$$f[x_0] := f(x_0).$$

The other divided differences can be defined inductively via

$$f[x_i, \dots, x_{i+j}] := \frac{f[x_{i+1}, \dots, x_{i+j}] - f[x_i, \dots, x_{i+j-1}]}{x_{i+j} - x_i}, \quad 0 \leq i \leq n - j.$$

When $f(x) = x^n$ then we will let $[x_i, \dots, x_{i+j}]^n = f[x_i, \dots, x_{i+j}]$. Divided differences arise naturally as coefficients of Newton interpolating polynomials. If we wish to construct the unique n degree interpolating polynomial for the given data $(x_0, f(x_0)), \ldots, (x_n, f(x_n))$, then we must solve the following system

$$\begin{pmatrix} 1 & & & 0 \\ 1 & x_1 - x_0 & & \\ \vdots & \vdots & \ddots & \\ 1 & x_n - x_0 & \cdots\cdots & \prod_{i=0}^{n-1}(x_n - x_i) \end{pmatrix} \begin{pmatrix} c_0 \\ \vdots \\ c_n \end{pmatrix} = \begin{pmatrix} f(x_0) \\ \vdots \\ f(x_n) \end{pmatrix}. \tag{3.8.8}$$

One easily sees that $c_0 = f[x_0], c_1 = f[x_0, x_1], \ldots, c_n = f[x_0, \ldots, x_n]$, and hence the interpolating polynomial has the form

$$p(x) = f[x_0] + f[x_0, x_1](x - x_0) + \cdots + f[x_0, \ldots, x_n] \prod_{i=0}^{n-1}(x - x_i). \tag{3.8.9}$$

From this point of view, we get the following results.

Lemma 3.8.8 *(i) If $\{i_0, \ldots, i_n\}$ is a permutation of the set $\{0, \ldots, n\}$ then $f[x_0, \ldots, x_n] = f[x_{i_0}, \ldots, x_{i_n}]$.*

(ii) If $f(x) = x^n$ then $[x_0, \ldots, x_n]^n = f[x_0, \ldots, x_n] = 1$.

Proof. (i) Since $f[x_0, \ldots, x_n]$ is the leading coefficient of the unique interpolating polynomial, it suffices to look at the leading coefficient of the interpolating polynomial, via Lagrange interpolation, which is given by

$$\sum_{i=1}^{n} \frac{f(x_i)}{\prod_{\substack{j=1 \\ i \neq j}}(x_i - x_j)}. \tag{3.8.10}$$

Since (3.8.10) is invariant upon permutation of $\{0, \ldots, n\}$, we get

$$f[x_0, \ldots, x_n] = f[x_{i_0}, \ldots, x_{i_n}].$$

(ii) As $p(x) = x^n$ is the unique interpolating polynomial for $(x_0, x_0^n), \ldots, (x_n, x_n^n)$ we get that $[x_0, \ldots, x_n]^n = f[x_0, \ldots, x_n] = 1$, as 1 is the leading coefficient of $p(x) = x^n$. □

We are now ready to prove the generalized Vandermonde result.

Proof of Theorem 3.8.7. First reorder $w_1, \ldots, w_n$ as in the statement of the theorem and denote $\Lambda_i = \{i-1\} \times \{0, \ldots, n-1\}$, $1 \leq i \leq k$. Then $\Lambda = \bigcup_{i=1}^{k} \Lambda_i$. We will prove (3.8.7) when $k = 3$. First observe that $V(w_1, \ldots, w_n; \Lambda)$ equals

$$\begin{pmatrix} V(w_1^{(1)}, \ldots, w_{n_1}^{(1)}; \Lambda_1) & V(w_1^{(1)}, \ldots, w_{n_1}^{(1)}; \Lambda_2) & V(w_1^{(1)}, \ldots, w_{n_1}^{(1)}; \Lambda_3) \\ V(w_1^{(2)}, \ldots, w_{n_2}^{(2)}; \Lambda_1) & V(w_1^{(2)}, \ldots, w_{n_2}^{(2)}; \Lambda_2) & V(w_1^{(2)}, \ldots, w_{n_2}^{(2)}; \Lambda_3) \\ V(w_1^{(3)}, \ldots, w_{n_3}^{(3)}; \Lambda_1) & V(w_1^{(3)}, \ldots, w_{n_3}^{(3)}; \Lambda_2) & V(w_1^{(3)}, \ldots, w_{n_3}^{(3)}; \Lambda_3) \end{pmatrix},$$

which can be rewritten as

$$\begin{pmatrix} V_{11} & x_1 V_{12} & x_1^2 V_{13} \\ V_{21} & x_2 V_{22} & x_2^2 V_{23} \\ V_{31} & x_3 V_{32} & x_3^2 V_{33} \end{pmatrix},$$

where $V_{ij} = V(y_1^{(i)}, \dots, y_{n_i}^{(i)}; \{0, \dots, n_j - 1\})$, $1 \le i, j \le 3$, Using column operations we see that $V(w_1, \dots, w_n; \Lambda)$ is column equivalent to

$$\begin{pmatrix} V_{11} & 0 & 0 \\ V_{21} & (x_2 - x_1)V_{22} & (x_2^2 - x_1^2)V_{23} \\ V_{31} & (x_3 - x_1)V_{32} & (x_3^2 - x_1^2)V_{33} \end{pmatrix}, \tag{3.8.11}$$

where we used the fact that the first n_j columns of $V_{i,1}$ form precisely $V_{i,j}$, $i = 1, 2, 3$ and $j = 2, 3$. Scaling the middle column in (3.8.11) and keeping in mind that for the determinant we are taking out a factor $(x_2 - x_1)^{n_2}(x_3 - x_1)^{n_3}$, we get the matrix

$$\begin{pmatrix} V_{11} & 0 & 0 \\ * & V_{22} & \frac{x_2^2 - x_1^2}{x_2 - x_1} V_{23} \\ * & V_{32} & \frac{x_3^2 - x_1^2}{x_3 - x_1} V_{33} \end{pmatrix}. \tag{3.8.12}$$

Rewriting (3.8.12) in divided difference notation yields

$$\begin{pmatrix} V_{11} & 0 & 0 \\ * & V_{22} & [x_1, x_2]^2 V_{23} \\ * & V_{32} & [x_1, x_3]^2 V_{33} \end{pmatrix}. \tag{3.8.13}$$

Using column operations (3.8.13) becomes

$$\begin{pmatrix} V_{11} & 0 & 0 \\ * & V_{22} & 0 \\ * & * & ([x_1, x_3]^2 - [x_1, x_2]^2) V_{33} \end{pmatrix}. \tag{3.8.14}$$

Scale the third block rows by $x_3 - x_2$ and note that

$$\frac{[x_1, x_3]^2 - [x_1, x_2]^2}{x_3 - x_2} = [x_2, x_1, x_3]^2 = [x_1, x_2, x_3]^2 = 1,$$

where Lemma 3.8.8(i) was used to deduce $[x_2, x_1, x_3]^2 = [x_1, x_2, x_3]^2$ and Lemma 3.8.8(ii) was used to get $[x_1, x_2, x_3]^2 = 1$. Note that a factor of $(x_3 - x_2)^{n_3}$ must now be taken into account for $\det V(w_1, \dots, w_n; \Lambda)$ and (3.8.14) becomes

$$\begin{pmatrix} V_{11} & 0 & 0 \\ * & V_{22} & 0 \\ * & * & V_{33} \end{pmatrix}.$$

Notice that V_{ii}, $i = 1, 2, 3$, are invertible Vandermonde matrices and so we are have the invertibility of $V(w_1, \dots, w_n, \Lambda)$. Moreover, for $i = 1, 2, 3$, since

$$\det V_{ii} = \prod_{1 \le p < q \le n_i} (y_q^{(i)} - y_p^{(i)}),$$

it is easy to arrive at (3.8.7), when $k = 3$, upon consideration of the row scaling that was performed. When proving Theorem 3.8.7 for $k > 3$, note that the coefficient of V_{ii} can always be scaled to be 1 by recognizing the appropriate divided difference and then use Lemma 3.8.8. □

We will now prove a two-variable analog of Lemma 2.7.10.

Lemma 3.8.9 *Let $A \geq 0$ and $B_j = B_j^*$ be $p \times p$ matrices so that there exist commuting matrices W_1 and W_2 satisfying $AW_j = B_j$, $j = 1, 2$. Put $k = \text{rank}A$. Then there exist $k \times k$ real diagonal matrix D_j and an injective $p \times k$ matrix C with the property that*

$$A = CC^* \text{ and } B_j = CD_jC^*, \quad j = 1, 2. \tag{3.8.15}$$

Proof. Decompose $\mathbb{C}^p = \text{Ran}A \oplus \text{Ker}A$. Then with respect to this decomposition we have

$$A = \begin{pmatrix} \tilde{A} & 0 \\ 0 & 0 \end{pmatrix}, \; W_j = \begin{pmatrix} \tilde{W}_j & 0 \\ * & * \end{pmatrix}, \text{ and } B_j = \begin{pmatrix} \tilde{B}_j & 0 \\ 0 & 0 \end{pmatrix}, \; j = 1, 2,$$

where we used that $AW_j = B_j = B_j^*$, $j = 1, 2$. Notice that $AW_j = B_j$ yields $\tilde{A}\tilde{W}_j = \tilde{B}_j$. Since $\tilde{A}$ is invertible, we get $\tilde{W}_j = \tilde{A}^{-1}\tilde{B}_j$. Since W_1 and W_2 commute, $\tilde{W}_1$ and $\tilde{W}_2$ commute. Hence $\tilde{A}^{-\frac{1}{2}}\tilde{B}_1\tilde{A}^{-\frac{1}{2}}$ and $\tilde{A}^{-\frac{1}{2}}\tilde{B}_2\tilde{A}^{-\frac{1}{2}}$ commute, so they must be simultaneously diagonalizable. Thus there exist a unitary matrix U and real diagonal matrices D_1 and D_2 so that

$$\tilde{A}^{-\frac{1}{2}}\tilde{B}_j\tilde{A}^{-\frac{1}{2}} = UD_jU^*, \quad j = 1, 2.$$

Put $C = \begin{pmatrix} \tilde{A}^{\frac{1}{2}} \\ 0 \end{pmatrix} U$ so then (3.8.15) holds. □

We will now prove a result that is analogous to Theorem 2.7.9.

Theorem 3.8.10 *Let $\Lambda \subset \mathbb{N}_0^2$ be a finite lower inclusive set and let $\Gamma = (\Lambda + \Lambda) \cup (\Lambda + \Lambda + e_1) \cup (\Lambda + \Lambda + e_2)$. Let the $p \times p$ Hermitian matrix-valued sequence $\{S_\gamma\}_{\gamma \in \Gamma}$ be given. Set $\Phi = (S_{\lambda+\mu})_{\lambda,\mu \in \Lambda}$, $\Phi_j = (S_{\lambda+\mu+e_j})_{\lambda,\mu \in \Lambda}$, $j = 1, 2$, and put $k = \text{rank}\,\Phi$. Suppose that $\Phi \geq 0$ and there are commuting $mp \times mp$ matrices Θ_1, Θ_2 such that $\Phi\Theta_j = \Phi_j$, $j = 1, 2$. Then there exist an $mp \times k$ matrix C_0 and real $k \times k$ diagonal matrices D_j, $j = 1, 2$, such that*

$$S_\gamma = C_0 D_1^{g_1} D_2^{g_2} C_0^*, \quad \gamma = (g_1, g_2) \in \Gamma. \tag{3.8.16}$$

Proof. Consider $A = \Phi$, $W_j = \Theta_j$, and $B_j = \Phi_j$, $j = 1, 2$. Then $A \geq 0, B_j = B_j^*$, $W_1W_2 = W_2W_1$, and $AW_j = B_j$, $j = 1, 2$, so we can apply Lemma 3.8.9 to obtain an injective matrix C and real diagonal matrices D_j, $j = 1, 2$ such that

$$\Phi = CC^* \text{ and } \Phi_j = CD_jC^*, \quad j = 1, 2. \tag{3.8.17}$$

Write C as a block column with entries C_λ, $\lambda \in \Lambda$; that is, $C = \text{col}(C_\lambda)_{\lambda \in \Lambda}$. When $\lambda, \mu + e_j \in \Lambda$ we get from (3.8.17) that

$$S_{\lambda+\mu+e_j} = C_\lambda C_{\mu+e_j}^* = C_\lambda D_j C_\mu^*. \tag{3.8.18}$$

Notice that (3.8.18) readily implies

$$C(C_{\mu+e_j}^* - D_jC_\mu^*) = 0. \tag{3.8.19}$$

Since C is injective, (3.8.19) yields

$$C_{\mu+e_j} = C_\mu D_j. \tag{3.8.20}$$

Consider $\lambda = (k_1, k_2) \in \Lambda$. If $k_1 \in \mathbb{N}$ then choose $\mu = (k_1 - 1, k_2) \in \Lambda$ (use that Λ is lower inclusive), so then (3.8.20) gives

$$C_{\mu+e_1} - C_\mu D_1.$$

Hence $C_{(k_1,k_2)} = C_{(k_1-1,k_2)}D_1$. Similarly, if $k_2 \in \mathbb{N}$ then choose $\mu = (k_1, k_2 - 1)$, so then (3.8.20) gives

$$C_{\mu+e_2} - C_\mu D_2,$$

and hence $C_{(k_1,k_2)} = C_{(k_1,k_2-1)}D_1$. Continuing this way we arrive at

$$C_{(k_1,k_2)} = C_{(0,0)}D_1^{k_1}D_2^{k_2}.$$

But then from $\Phi = CC^*$ and $\Phi_j = CD_jC^*$, $j = 1, 2$, we obtain that (3.8.16) holds with $C_0 = C_{(0,0)}$. □

We are now ready to prove the two-variable Hamburger moment result.

Proof of Theorem 3.8.2. Suppose (3.8.4) holds, and there exist commuting matrices Θ_1 and Θ_2 so that, (3.8.5) holds. Apply Theorem 3.8.10 to obtain $C_0 = \begin{pmatrix} c_1 & \cdots & c_m \end{pmatrix}$, $D_1 = \operatorname{diag}(x_i)_{i=1}^m$, and $D_2 = \operatorname{diag}(y_i)_{i=1}^m$ with $c_i \in \mathbb{C}^p$, $i = 1, \ldots, m$, where $m = \operatorname{rank} \Phi$. Then (3.8.16) holds. Without loss of generality, assume that $(x_1, y_1), \ldots, (x_k, y_k)$, $k \leq m$, are the distinct points among (x_i, y_i), $i = 1, \ldots, m$. Put σ as in the statement of Theorem 3.8.2 with $T_i = \sum_{\{j:\, (x_j,y_j)=(x_i,y_i)\}} c_j c_j^* \geq 0$. One can directly verify (3.8.2).

For the converse, let $\sigma = \sum_{i=1}^k T_i \delta_{(x_i,y_i)}$. Then we have that

$$S_{(k_1,k_2)} = \int_{\mathbb{R}^2} \xi_1^{k_1}\xi_2^{k_2} d\sigma(\xi_1, \xi_2) = \sum_{i=1}^k T_i x_i^{k_1} y_i^{k_2}, \quad k_1, k_2 \in \mathbb{N}.$$

Use Theorem 3.8.7 to obtain a lower inclusive set $\Lambda \subset \mathbb{N}_0^2$, with $\operatorname{card}\Lambda = k$, so that $V := V((x_1, y_1), \ldots, (x_k, y_k); \Lambda)$ is invertible. One can check that

$$\Phi = (V \otimes I)^T R(V \otimes I) \geq 0,$$

$$\Phi_1 = (V \otimes I)^T RX(V \otimes I), \ \Phi_2 = (V \otimes I)^T RY(V \otimes I),$$

where

$$R = \operatorname{diag}(\tilde{T}_i)_{i=1}^k, \ X = \operatorname{diag}(x_i)_{i=1}^k \otimes I, \ Y = \operatorname{diag}(y_i)_{i=1}^k \otimes I.$$

Choosing $\Theta_1 = (V \otimes I)^{-1}X(V \otimes I)$ and $\Theta_2 = (V \otimes I)^{-1}Y(V \otimes I)$ yield $\Theta_1\Theta_2 = \Theta_2\Theta_1$ and (3.8.5). □

3.9 BOCHNER'S THEOREM AND AN APPLICATION TO AUTOREGRESSIVE STOCHASTIC PROCESSES

In this section we return to the notation of G for a locally compact Abelian group. In the first part we present Bochner's theorem, the simplest version of which was already mentioned in Section 1.1. The scalar version is presented with proof, while only the necessary part of the operator version is fully exposed. In the second part of the section we find a characterization for the existence of a solution of a causal bivariate autoregressive filter design problem based on Bochner's theorem.

A *character* of G is a continuous homomorphism $\gamma : G \to \mathbb{T}$. The set of characters of G together with pointwise multiplication form an Abelian group. This group endowed with the uniform convergence on compact sets in G is called the *character group* or the *dual group* of G. It is known that the character group of a compact group is discrete, and that of a discrete group it is compact. For example, see the dual group of $\mathbb{T}^r$ is $\mathbb{Z}^r$, and the dual group of $\mathbb{R}_d$ ($\mathbb{R}_d$ with the discrete topology) is the Bohr compactification of $\mathbb{R}$.

Let μ be a finite Borel measure on G. The function $\widehat{\mu} : \Gamma \to \mathbb{C}$ defined by

$$\widehat{\mu}(\gamma) = \int_G \overline{\gamma(x)} d\mu(x)$$

is called the *Fourier transform* of μ. If μ is a finite Borel measure on Γ, then $\check{\mu} : G \to \mathbb{C}$ defined by

$$\check{\mu}(x) = \int_\Gamma \gamma(x) d\mu(\gamma)$$

is called the *inverse Fourier transform* of μ. We let $C_0(\Gamma)$ denote the class of continuous functions $h : \Gamma \to \mathbb{C}$ such that for all $\epsilon > 0$ there exists a compact $K \subset \Gamma$ such that $|f(x)| < \epsilon$ for $x \in \Gamma \setminus K$. For a Banach space $\mathcal{B}$, let $L^1_G(\mathcal{B})$ denote the Lebesgue space of all measurable functions $f : G \to \mathcal{B}$ such that $\int_G \|f(x)\| dm(x) < \infty$, where m is the normalized Haar measure on G. We write L^1_G for $L^1_G(\mathbb{C})$.

Theorem 3.9.1 *Every measure on Γ is uniquely determined by the inverse Fourier transform $\check{\mu}$.*

Proof. It is sufficient to prove that $\check{\mu} = 0$ implies $\mu = 0$. Let $f \in L^1_G$ be arbitrary, and by Fubini's theorem,

$$\int_\Gamma \widehat{f} d\mu = \int \Big[\int_G f(x)\overline{\gamma(x)} dx \Big] d\mu(x)$$

$$= \int_H f(x) \int_\Gamma \overline{\gamma(x)} d\mu(\gamma) dx = \int_H f(x) \check{\mu}(x^{-1}) dx = 0.$$

Thus $\int_\Gamma \widehat{f} d\mu = 0$ for all $f \in L^1_G$ and hence, by Exercise 3.10.42, $\int_\Gamma h d\mu = 0$ for all continuous $h \in C_0(\Gamma)$. Consequently, $\mu = 0$. □

A very important result concerning positive semidefinite functions is the following result known as Bochner's theorem.

Theorem 3.9.2 *If μ is a finite positive Borel measure on Γ, then $\check{\mu}$ is a continuous positive semidefinite function on G. Conversely, to every continuous positive semidefinite function f on G, there corresponds a uniquely determined finite positive Borel measure μ on Γ, such that $f = \check{\mu}$.*

Before we can prove Bochner's theorem, we need the notion of barycenter. Let X be a nonempty compact subset of a locally convex space E, and let μ be a positive Borel measure on X with $\mu(X) = 1$ (i.e., μ is a *probability measure* on X). We say that $x \in E$ is the *barycenter* of μ if $f(x) = \int_X f d\mu$ for every continuous linear functional on E. The following is a basic result.

Proposition 3.9.3 *Suppose that Y is a compact subset of a locally convex space E such that the closed convex hull of Y is compact. Then x is in the closed convex hull of Y if and only if x is the barycenter of a probability measure on Y.*

Proof. See, for instance, [90, Chapter 2, Proposition 5.3]. □

Proof of Theorem 3.9.2. If μ is a finite positive Borel measure on Γ, by Exercise 3.10.41, $\check{\mu}$ is uniformly continuous on G. The positive semidefinitess of $\check{\mu}$ follows directly from the computation

$$\sum_{i,j=1}^{n} \check{\mu}(x_i^{-1}x_j)\overline{c}_i c_j = \sum_{i,j=1}^{n} \overline{c}_i c_j \int_\Gamma \gamma(x_i^{-1}x_j)d\mu(\gamma)$$

$$= \int_\Gamma \left\| \sum_{i=1}^{n} c_i \overline{\gamma(x_i)} \right\|^2 d\mu(\gamma) \geq 0.$$

To prove the converse statement we may assume $f \in P^1(G)$, where $P^1(G)$ stands for positive semidefinite functions on G with the property that $f(e) \leq 1$. For every $h \in L^1_G$, the linear functional $L_h(g) := \int_G g(x)h(x)dx$, is continuous on L^∞_G in the weak-* topology. The set $P^1(G)$ is convex and compact in this topology and the set $\Gamma_0 = \Gamma \cup \{0\}$ of its extreme points is closed by Exercise 3.10.40. By Proposition 3.9.3 it follows that f is the barycenter of some positive measure μ_0 on Γ_0. Therefore

$$L_h(f) = \int_{\Gamma_0} L_h(\gamma_0)d\mu(\gamma_0)$$

for all $h \in L^1_G$. Applying Fubini's theorem, we obtain

$$\int_G f(x)h(x)dx = \int_{\Gamma_0}\int_G \gamma_0(x)h(x)dxd\mu_0(\gamma_0)$$

$$= \int_G \Big[\int_{\Gamma_0} \gamma_0(x)d\mu_0(\gamma)\Big]h(x)dx.$$

This being true for all $h \in L^1_G$, we must have

$$f(x) = \int_{\Gamma_0} \gamma_0(x)d\mu_0(\gamma_0), \quad x \in G.$$

Both sides of the above relation represent continuous functions, so denoting by μ the restriction of μ_0 to Γ, we have that μ is a positive measure on Γ

and

$$f(x) = \int_{\Gamma} \gamma(x) d\mu(\gamma) = \check{\mu}(x).$$

The uniqueness of μ follows from Theorem 3.9.1. □

The identification problem for wide sense stationary autoregressive stochastic processes is a classical signal processing problem. In this section we consider (wide sense) stationary processes $X_{m,n}$ depending on two discrete variables defined on a fixed probability space $(\Omega, \mathcal{A}, P)$. We shall assume that the random variables $X_{m,n}$ are *centered*, i.e., their means $E(X_{m,n})$ equal zero. Recall that the space $L^2(\Omega, \mathcal{A}, P)$ of square integrable random variables endowed with the *inner product of centered random variables*

$$\langle X, Y \rangle := E(Y^* X)$$

is a Hilbert space. A sequence $X = (X_{m,n})_{(m,n) \in \mathbb{Z}^2}$ is called a *stationary process* on $\mathbb{Z}^2$ if for $m, n, k, \ell \in \mathbb{Z}$ we have that

$$E(X^*_{mn} X_{k\ell}) = E(X^*_{m+p,n+q} X_{k+p,\ell+q}) =: R_X(m-k, n-\ell) \text{ for all } p, q \in \mathbb{Z}^2.$$

It is known that the function R_X, termed the *covariance function* of X, defines a positive semidefinite function on $\mathbb{Z}^2$, that is,

$$\sum_{i,j=1}^{k} \alpha_i \overline{\alpha}_j R_X(r_i - r_j, s_i - s_j) \geq 0$$

for all $k \in \mathbb{N}$, $\alpha_1, \ldots, \alpha_k \in \mathbb{C}, r_1, \ldots, r_k, s_1, \ldots, s_k \in \mathbb{Z}$. Bochner's theorem on positive semidefinite functions (Theorem 3.9.2) states that for such a function R_X there is a positive measure μ_X defined for Borel sets on the torus $\{(u, v) : u, v \in [0, 2\pi]\}$ such that

$$R_X(r, s) = \int e^{-i(ru+sv)} d\mu_X(u, v)$$

for all integers r and s. The measure μ_X is referred to as the *spectral distribution measure* of the process X.

A centered stationary stochastic process X is said to be AR($\underline{n} \times \underline{m}$) if there exist complex numbers $a_{kl}, (k,l) \in \underline{n} \times \underline{m} \setminus \{(0,0)\}$, such that for every t and s,

$$x_{ts} + \sum_{\substack{k,l \in \underline{n} \times \underline{m} \\ (k,l) \neq (0,0)}} a_{kl} x_{t-k,s-l} = e_{ts}, \qquad (t, s) \in \mathbb{Z}^2, \tag{3.9.1}$$

where $\{e_{kl} \, ; (k,l) \in \mathbb{Z}^2\}$ is a white noise zero mean process with variance σ^2. Here AR stands for auto-regressive. Let $H = \{(n, m) : n > 0 \text{ or } (n = 0 \text{ and } m > 0)\}$ be the standard half-space in $\mathbb{Z}^2$. The AR($\underline{n} \times \underline{m}$) process is said to be *causal* if there is a solution to (3.9.1) of the form

$$x_{ts} = \sum_{k,l \in H \cup \{(0,0)\}} \phi_{kl} e_{t-k,s-l}, \quad (t, s) \in \mathbb{Z}^2, \tag{3.9.2}$$

with $\sum_{k,l \in H \cup \{(0,0)\}} |\phi_{kl}| < \infty$. It is not difficult to see that the $AR(\underline{n} \times \underline{m})$ process X is causal if and only if the polynomial

$$p(z, w) = 1 + \sum_{\substack{k,l \in \underline{n} \times \underline{m} \\ (k,l) \neq (0,0)}} \overline{a}_{kl} z^k w^l$$

is stable. A causal AR($\underline{n} \times \underline{m}$) process is in fact *quarterplane causal*, which by definition means that there is a solution to (3.9.1) of the form

$$x_{ts} = \sum_{\substack{k,l \geq 0 \\ (k,l) \neq (0,0)}} \phi_{kl} e_{t-k,s-l}, \quad (t, s) \in \mathbb{Z}^2. \tag{3.9.3}$$

The *bivariate autoregressive filter design problem* is the following. "Given are covariances

$$c_{kl} = E(x_{kl}\overline{x}_{00}), \qquad (k, l) \in \underline{n} \times \underline{m}.$$

What conditions must the covariances satisfy in order that these are the covariances of a causal AR($\underline{n} \times \underline{m}$) process? And in that case, how does one compute the filter coefficients a_{kl}, $(k, l) \in \underline{n} \times \underline{m} \setminus \{(0, 0)\}$ and σ^2?"

The following characterization for the existence of a causal solution for an AR($\underline{n} \times \underline{m}$) can now be obtained.

Theorem 3.9.4 *Let c_{kl}, $(k, l) \in \underline{n} \times \underline{m}$, be given complex numbers. There exists a causal AR($\underline{n} \times \underline{m}$) process with the given covariances c_{kl} if and only if there exist complex numbers c_{kl}, $(k, l) \in \{1, \dots, n\} \times \{-m, \dots, 1\}$, such that*

(i) the $(n + 1)(m + 1) \times (n + 1)(m + 1)$ doubly indexed Toeplitz matrix $\Gamma = (c_{t-s})_{s,t \in \underline{n} \times \underline{m}}$ is positive definite;

(ii) the matrix $(c_{s-t})_{s \in \{1,\dots,n\} \times \underline{m},\ t \in \underline{n} \times \{1,\dots,m\}}$ has rank equal to nm.

In this case one finds the vector

$$\frac{1}{\sigma^2}[a_{nm} \cdots a_{n0} \quad \cdots \quad a_{0m} \cdots a_{01} \ 1]$$

as the last row of the inverse of Γ.

The theorem follows directly from the scalar case ($p = 1$) of Theorem 3.3.2. How to generalize this result to more than two variables is still an open problem.

3.10 EXERCISES

1 Let $K_1 \subset \mathbb{N}_0^d$ and $K_2 \subset \mathbb{N}_0^d$ be lower inclusive sets. Is $K_1 \cap K_2$ lower inclusive? What about $K_1 \cup K_2$?

2 Let $\Lambda \subset \mathbb{N}_0^d$ be lower inclusive. Show that

$$\Gamma = (\Lambda + \Lambda) \cup (\Lambda + \Lambda + e_1) \cup \cdots \cup (\Lambda + \Lambda + e_d)$$

is lower inclusive.

3 Show that the matrix function

$$F(z, v, w) = \frac{1}{1 + w^2 + (z + v)w} \begin{pmatrix} zw^2 + (zv + 1) + v & w(1 - zv) \\ w(1 - zv) & vw^2 + (zv + 1)w + z \end{pmatrix}$$

satisfies $F + F^*|H \geq 0$, where

$$H = \{(z, v, w) \in \mathbb{C}^3 \ : \ \text{Re } z > 0, \text{Re } w > 0, \text{Re } v > 0\}.$$

4 Show that α in Example 3.1.8 equals $\sqrt{6}$.

5 Let $p(z, w)$ be a stable polynomial of degree (n, m). We say that (z_0, w_0) is an *intersecting zero* of p if $p(z_0, w_0) = \overleftarrow{p}(z_0, w_0)$. We allow ∞ as a root, as follows. Write

$$p(z, w) = \sum_{k=0}^{n} \sum_{l=0}^{m} p_{kl} z^k w^l = \sum_{i=0}^{m} p_i(z) w^i.$$

We say that $p(\infty, \infty) = 0$ if $p_{nm} = 0$, and we say that $p(z, \infty) = 0$, $z \in \mathbb{C}$, when $p_m(z) = 0$.

(a) Give the definition of what it should mean that $p(\infty, w) = 0$, $w \in \mathbb{C}$.

(b) What does it mean that (z, ∞) is an intersecting zero of p?

(c) How about (∞, ∞) being an intersecting zero of p?

(d) Show that (z, w) is an intersecting zero if and only if $(1/\overline{z}, 1/\overline{w})$ is an intersecting zero. Here we use the convention that $\frac{1}{0} = \infty$ and $\frac{1}{\infty} = 0$.

(e) For a stable polynomial, show that the intersecting zeros lie in $(\mathbb{D} \times \mathbb{E}) \cup (\mathbb{E} \times \mathbb{D})$, where $\mathbb{E} = (\mathbb{C} \setminus \overline{\mathbb{D}}) \cup \{\infty\}$.

6 Check whether the following polynomials are stable.

(a) $p(z, w) = 1 + \frac{1}{2}(z + w) + \frac{1}{4}(zw + z^2 + w^2)$;

(b) $p(z, v, w) = 5 + \frac{1}{4}(z + v + w) + z^2 + v^2 + w^2$;

(c) $p(z, w) = 1 + z + zw + \frac{1}{2}w$;

(d) $p(z,w) = 4 + 2z + 2w + zw + z^2w + 2w^2 + zw^2 + z^2w^2$.

7 We say that a square matrix A is *centrotranspose symmetric* if $JAJ = A^T$, where J is the matrix with ones on the antidiagonal and zeros elsewhere.

(a) Show that an $(n+1)\times(n+1)$ matrix $A = (a_{i,j})_{i,j=0}^n$ is Toeplitz if and only if both A and $\widehat{A} := (a_{i,j})_{i,j=0}^{n-1}$ are centrotranspose symmetric.

(b) Formulate a result as under (i) for a block matrix $A = (A_{ij})_{i,j=1}^n$ to be double Toeplitz (i.e., each A_{ij} is Toeplitz and $A_{ij} = A_{i+1,j+1}$ for all $1 \le i,j \le n-1$).

8 Let $p(z,w)$ be a stable scalar polynomial. Determine the eigenvalues of $\Phi_1\Phi^{-1}$ and $\Phi_2^*\Phi^{-1}$ appearing in Theorem 3.3.1 (with $P = R = p$), and determine the corresponding eigenvectors. (Hint: consider the components of the intersecting zeros of p that lie inside the unit disk.)

9 Let $p(z,w)$ be a stable polynomial of degree (n_1,n_2), and consider $L^2(\mathbb{T}^2)$, with the inner product

$$\langle f,g\rangle = \int_{[0,2\pi]^2} f(e^{it},e^{ix})\overline{g(e^{it},e^{ix})}\frac{1}{|p(e^{it},e^{ix})|^2}dtdx.$$

Let M be the subspace

$$M = \left\{\frac{q}{p} \; : \; q \text{ is a polynomial of degree at most } (n_1-1,n_2-1)\right\}.$$

Find the matrix representations of the compressions of multiplication by z and w to the subspace M (i.e., find the matrix representation of $P_M M_z P_M$ and $P_M M_w P_M$, where P_M is the orthogonal projection onto M in the inner product given above, and M_z and M_w are the multiplication operators with symbol $\phi(z,w) = z$ and $\psi(z,w) = w$, respectively). (Hint: the answer is closely related to the operators $\Phi_1\Phi^{-1}$ and $\Phi_2^*\Phi^{-1}$ from Theorem 3.3.1.)

10 Let $q(z,w) = |z-w|^2$ and let p be a polynomial in z and w such that $|p|^2 \le q$ on $\mathbb{T}^d$. Show that $\overline{pH^2(\mathbb{D}^d)} \ne H^2(\mathbb{D}^d)$.

11 Let $p(z,w) = \frac{11}{6} + w^2 - \frac{11}{6}z + \frac{25}{3}zw - 11zw^2$. Find the intersecting zeros of p. Is p stable? Is $\overleftarrow{p}$ stable?

12 Let $f(z,w) = 1 + \frac{3w}{10} + \frac{4z}{10} + \frac{15zw}{100} + \frac{7z}{100w} + \frac{3}{10w} + \frac{4}{10z} + \frac{15}{100zw} + \frac{7w}{100z}$. Show that f cannot be factored as $f = p(z,w)\overline{p}(1/z,1/w)$, where p is a stable polynomial of degree $(1,1)$.

13 Let $p(z,w) = (1 - z/2)^2(1 - w/3)^2$. Show that $\Phi_1\Phi^{-1}$ and $\Phi_2^*\Phi^{-1}$ appearing in Theorem 3.3.1 equal

$$\begin{pmatrix} 0 & 0 & -\frac{1}{9} & 0 \\ 0 & 0 & 0 & -\frac{1}{9} \\ 1 & 0 & \frac{2}{3} & 0 \\ 0 & 1 & 0 & \frac{2}{3} \end{pmatrix}, \quad \begin{pmatrix} 1 & 1 & 0 & 0 \\ -\frac{1}{4} & 0 & 0 & 0 \\ 0 & 0 & 1 & 1 \\ 0 & 0 & -\frac{1}{4} & 0 \end{pmatrix},$$

respectively. In addition, show that they are similar to the Jordan matrices

$$\begin{pmatrix} \frac{1}{3} & 1 & 0 & 0 \\ 0 & \frac{1}{3} & 0 & 0 \\ 0 & 0 & \frac{1}{3} & 1 \\ 0 & 0 & 0 & \frac{1}{3} \end{pmatrix}, \quad \begin{pmatrix} \frac{1}{2} & 0 & 1 & 0 \\ 0 & \frac{1}{2} & 0 & 1 \\ 0 & 0 & \frac{1}{2} & 0 \\ 0 & 0 & 0 & \frac{1}{2} \end{pmatrix},$$

respectively.

14 Let $n = m = 1$ and put $c_{00} = 1$, $c_{01} = \frac{1}{4} = c_{1,-1}$, $c_{10} = 0 = c_{11}$.

(a) Build Γ as in Theorem 3.3.2 and show that $\Gamma > 0$.

(b) Show that Γ does not satisfy condition Theorem 3.3.2(2).

(c) Build a polynomial p of degree $(1,1)$ via the first column of Γ^{-1} as indicated in Theorem 3.3.2. (One may use, for instance, Maple, Mathematica, or Matlab.)

(d) Show that $p(z,w)$ is stable.

(e) Why does this not contradict the statement of Theorem 3.3.2?

15 Let C be a $(n_1 + n_2) \times (n_1 + n_2)$ contractive matrix, and let $p_0 \in \mathbb{C}$. Show that $p(z_1, z_2) = p_0 \det\left(I - \begin{pmatrix} z_1 I_{n_1} & 0 \\ 0 & z_2 I_{n_2} \end{pmatrix} C\right)$ has no roots in $\mathbb{D}^2$.

16 Let $p(z,w)$ be a stable polynomial of degree (n,m) with $p(0,0) > 0$, and let $f = \frac{1}{|p|^2}$ be its spectral density function. Write

$$p(z,w) = \sum_{i=0}^{m} p_i(z) w^i, \quad f(z,w) = \sum_{i=-\infty}^{\infty} f_i(z) w^i.$$

Put $p_i(z) \equiv 0$ for $i > m$. Show that the following hold.

(a) $T_k(z) := (f_{i-j}(z))_{i,j=0}^{k} > 0$ for all $k \in \mathbb{N}_0$ and all $z \in \mathbb{T}$.

(b) For all $k \geq m - 1$ and for all z in the domain of $T_k^{\pm 1}$:

$$T_k(z)^{-1} = \begin{pmatrix} p_0(z) & & 0 \\ \vdots & \ddots & \\ p_k(z) & \cdots & p_0(z) \end{pmatrix} \begin{pmatrix} \overline{p}_0(1/\overline{z})^* & \cdots & p_k(1/\overline{z})^* \\ & \ddots & \vdots \\ 0 & & p_0(1/\overline{z})^* \end{pmatrix}$$
$$- \begin{pmatrix} \overline{p}_{k+1}(1/z) & & 0 \\ \vdots & \ddots & \\ \overline{p}_1(1/z) & \cdots & \overline{p}_{k+1}(1/z) \end{pmatrix} \begin{pmatrix} p_{k+1}(z) & \cdots & p_1(z) \\ & \ddots & \vdots \\ 0 & & p_{k+1}(z) \end{pmatrix}$$

$=: E_k(z)$.

(Hint: Use the scalar case of Theorem 2.3.22.)

(c) for $k \geq m-1$, the left stable factors $M_k(z)$ and $M_{k+1}(z)$ of the positive trigonometric matrix polynomials $E_k(z)$ and $E_{k+1}(z)$, respectively, satisfy

$$M_{k+1}(z) = \begin{pmatrix} p_0(z) & 0 \\ \mathrm{col}(p_l(z))_{l=1}^{k+1} & M_k(z) \end{pmatrix}. \tag{3.10.1}$$

(d) Show that the spectrum of $\overleftarrow{M}_{m-1}$ is given given by

$$\Sigma(\overleftarrow{M}_{m-1}) = \{z \in \mathbb{D} \ : \ \exists w \text{ such that}$$

$$(z,w) \text{ is an intersecting zero of } p\}.$$

In particular, p has only a finite number of intersecting zeros. In addition, show that for $k \geq m$, $\Sigma(\overleftarrow{M}_k) = \Sigma(\overleftarrow{M}_{m-1}) \cup \{z \in \mathbb{C}_\infty : \overleftarrow{p_0}(z) = 0\}$.

(e) Find the spectrum of M_{m-1} and M_m.

17 Let Toeplitz matrices C_i, $-n \leq i \leq n$, be given so that

$$T_n = (C_{i-j})_{i,j=0}^n > 0.$$

Put

$$C_{n+1} = C_{-n-1}^* = \begin{pmatrix} C_n & \cdots & C_1 \end{pmatrix} T_n^{-1} \begin{pmatrix} C_1 \\ \vdots \\ C_n \end{pmatrix}. \tag{3.10.2}$$

The following question is of interest: "Under what conditions on C_i, $-n \leq i \leq n$, is C_{n+1} Toeplitz as well?"

Let e_j denote the j standard basis vector in $\mathbb{C}^{n+1}$. Put

$$\alpha_{11} = \begin{pmatrix} e_1^* C_n e_1 & \cdots & e_1^* C_1 e_1 \end{pmatrix}, \ \alpha_{12} = \begin{pmatrix} e_1^* C_n P & \cdots & e_1^* C_1 P \end{pmatrix},$$

$$\alpha_{21} = \begin{pmatrix} P^* C_n e_1 & \cdots & P^* C_1 e_1 \end{pmatrix}, \ \alpha_{22} = \begin{pmatrix} P^* C_n P & \cdots & P^* C_1 P \end{pmatrix},$$

$$\beta_{11} = (e_1^* C_{i-j} e_1)_{i,j=0}^{n-1}, \ \beta_{12} = (e_1^* C_{i-j} P)_{i,j=0}^{n-1},$$

$$\beta_{21} = (P^* C_{i-j} e_1)_{i,j=0}^{n-1}, \ \beta_{22} = (P^* C_{i-j} P)_{i,j=0}^{n-1},$$

$$\gamma_{11} = \begin{pmatrix} e_1^* C_1 e_1 \\ \vdots \\ e_1^* C_n e_1 \end{pmatrix}, \ \gamma_{12} = \begin{pmatrix} e_1^* C_1 P \\ \vdots \\ e_1^* C_n P \end{pmatrix},$$

$$\gamma_{21} = \begin{pmatrix} P^* C_1 e_1 \\ \vdots \\ P^* C_n e_1 \end{pmatrix}, \ \gamma_{22} = \begin{pmatrix} P^* C_1 P \\ \vdots \\ P^* C_n P \end{pmatrix}.$$

If

$$(\alpha_{21} - \alpha_{22}\beta_{22}^{-1}\beta_{21})(\beta_{11} - \beta_{12}\beta_{22}^{-1}\beta_{21})^{-1}(\gamma_{12} - \beta_{12}\beta_{22}^{-1}\gamma_{22}) = 0, \quad (3.10.3)$$

then C_{n+1} defined in (3.10.2) is Toeplitz.

Observe that the rank condition in Theorem 3.3.2 corresponds exactly to $\gamma_{12} - \beta_{12}\beta_{22}^{-1}\gamma_{22} = 0$.

18 Consider the partial positive semidefinite operator matrix

$$T = \begin{pmatrix} T_{11} & T_{12} & ? & T_{14} \\ T_{21} & T_{22} & T_{23} & ? \\ ? & T_{32} & T_{33} & T_{34} \\ T_{41} & ? & T_{43} & T_{44} \end{pmatrix},$$

where $T_{ij} \in \mathcal{L}(\mathcal{H}_j, \mathcal{H}_i)$. Let $\mathcal{M} = \{(a_{ij})_{i,j=1}^4 \in \mathbb{C}^{4\times 4} : a_{13} = a_{24} = a_{31} = a_{42} = 0\}$. Define $\phi_T : \mathcal{M} \to \mathcal{L}(\oplus_{i=1}^4 \mathcal{H}_i)$, by

$$\phi_T[(a_{ij})_{i,j=1}^4] = (a_{ij}T_{ij})_{i,j=1}^4,$$

where 0 times an unknown entry is defined as 0. Show that if ϕ_T is a positive map, then T has a positive semidefinite completion.

19 Let U be the Fourier matrix $U = \frac{1}{\sqrt{6}}(\omega^{kl})_{k,l=0}^5$, and write $U = (U_{ij})_{i,j=1}^3$ with $U_{ij} \in \mathbb{C}^{2\times 2}$. Consider the linear operator ϕ_U acting $\mathbb{C}^{3\times 3} \to \mathbb{C}^{6\times 6}$ (both endowed with the spectral norm) defined by

$$\phi_U[(a_{ij})_{i,j=1}^3] = (a_{ij}T_{ij})_{i,j=1}^3.$$

(a) Show that the norm of the operator ϕ_U is strictly less than 1.

(b) Show that $\sup_n \|\phi_u \otimes I_n\| = 1$.

20 Let U, U_{ij}, and ϕ_U be as in Exercise 3.10.19. Let α be the norm of ϕ_U. Consider the partial matrix

$$T = \begin{pmatrix} \alpha I_2 & ? & ? & U_{11} & U_{12} & U_{13} \\ ? & \alpha I_2 & ? & U_{21} & U_{22} & U_{23} \\ ? & ? & \alpha I_2 & U_{31} & U_{32} & U_{33} \\ U_{11}^* & U_{21}^* & U_{31}^* & \alpha I_2 & ? & ? \\ U_{12}^* & U_{22}^* & U_{32}^* & ? & \alpha I_2 & ? \\ U_{13}^* & U_{23}^* & U_{33}^* & ? & ? & \alpha I_2 \end{pmatrix}.$$

Write $T = (T_{ij})_{i,j=1}^6$ with entries T_{ij} in $\mathbb{C}^{2\times 2}$, where some are unknown. Let $\mathcal{M} = \{(a_{ij})_{i,j=1}^6 \in \mathbb{C}^{4\times 4} : a_{12} = a_{13} = a_{21} = a_{23} = a_{31} = a_{32}a_{45} = a_{46} = a_{54} = a_{56} = a_{64} = a_{65} = 0\}$. Define $\phi_T : \mathbb{C}^{6\times 6} \to \mathbb{C}^{12\times 12}$ by $\phi_T[(a_{ij})_{i,j=1}^6] = (a_{ij}T_{ij})_{i,j=1}^6$. Show that ϕ_T is positive but not completely positive.

21 In Theorem 3.4.4 provide an upper bound for N and the degree of the polynomial $P(z)$ in terms of δ, where $Q(z) \geq \delta > 0$, and the degree of Q.

22 For a function f of d variables and a $N \in \mathbb{N}_0$, define the mapping Ψ_N via

$$(\Psi_N f)(x_1, \ldots, x_d) = \frac{1}{2^{dN}} \left[\prod_{j=1}^{d} (1 + x_j^2)^N \right] f\left(\frac{i - x_1}{i + x_1}, \ldots, \frac{i - x_m}{i + x_m} \right).$$

(a) Show that if f is a trigonometric polynomial defined on $\mathbb{T}^d$ of degree at most $(N, \ldots, N)$, then $\Psi_N f$ is a polynomial of degree at most $(2N, \ldots, 2N)$ on $\mathbb{R}^d$.

(b) Show that

$$(\Phi_N g)(z_1, \ldots, z_d) = \frac{\prod_j (1 + z_j)^{2N}}{2^{dN} \prod_j z_j^N} g\left(i\frac{1 - z_1}{1 + z_1}, \ldots, i\frac{1 - z_d}{1 + z_d} \right)$$

defines the inverse of Ψ_N.

(c) Show that if $g = q^* q$, then $\Phi_N g = p^* p$, where

$$p(z_1, \ldots, z_d) = \frac{1}{2^{dN/2}} [\prod_j (1 + z_j)^N] q\left(i\frac{1 - z_1}{1 + z_1}, \ldots, i\frac{1 - z_d}{1 + z_d} \right).$$

(d) Show that if $f = p^* p$, then $\Psi_N f = q^* q$ for an appropriately chosen q. Find the formula for q in terms of p.

(e) Let $g(x_1, x_2) = x_1^4 x_2^2 + x_1^2 x_2^4 - x_1^2 x_2^2 + 1$. Show that g is positive on $\mathbb{R}^2$, but cannot be factored as $q^* q$ for some polynomial q.

(f) Compute $\Phi_2 g$ and show that $(\Phi_2 g)(-1, -1) = 0$ (the latter can also be seen directly).

23 Let

$$f(z_1, z_2, z_3) = a_0 + a\overline{z_1} z_2 + \overline{a} z_1 \overline{z_2} + b\overline{z_1} z_3 + \overline{b} z_1 \overline{z_3} + c\overline{z_2} z_3 + \overline{c} z_2 \overline{z_3}.$$

Use Exercise 1.6.9(b) to show that if $f(z_1, z_2, z_3) \geq 0$, $(z_1, z_2, z_3) \in \mathbb{T}^3$, then f is a sum of at most three Hermitian squares (where in fact the underlying polynomials are linear in (z_1, z_2, z_3)).

24 Let g be a polynomial of degree at most $(2N, \ldots, 2N)$, and assume that $g(x_1, \ldots, x_d) \geq \epsilon \prod_j (1 + x_j^2)^N$ for all $(x_1, \ldots, x_d) \in \mathbb{R}^d$. Use Theorem 3.4.4 and the maps Ψ_N and Φ_N in Exercise 22 to show that g can be written as a sum of squares of polynomials.

25 Let $A_0, \ldots, A_n \in \mathcal{L}(Y)$ be so that the Toeplitz operator matrix

$$\begin{pmatrix} A_0 & \frac{n+1}{n} A_1^* & \frac{n+1}{n-1} A_2^* & \cdots & (n+1) A_n^* \\ \frac{n+1}{n} A_1 & \ddots & \ddots & \ddots & \vdots \\ \frac{n+1}{n-1} A_2 & \ddots & \ddots & \ddots & \frac{n+1}{n-1} A_2^* \\ \vdots & \ddots & \ddots & \ddots & \frac{n+1}{n} A_1^* \\ (n+1) A_n & \cdots & \frac{n+1}{n-1} A_2 & \frac{n+1}{n} A_1 & A_0 \end{pmatrix}$$

is positive semidefinite. Show that for all $z \in \mathbb{T}$ we have

$$z^{-n}A_n^* + \cdots + z^{-1}A_1^* + A_0 + zA_1 + \cdots + z^n A_n \geq 0.$$

(Hint: use the ideas from the proof of Theorem 3.4.4.)

26 Give an example of a nonamenable Abelian group and prove that every compact group is amenable.

27 Consider the expression $Q(X_1, X_2) = X_1^2 + (X_1^T)^2 + X_2^T X_2$. Show that for real scalar X_1 and X_2 we have that $Q(X_1, X_2) \geq 0$, but that for real matrices X_1, X_2 we do not necessarily have that $Q(X_1, X_2) \geq 0$.

28 Given the measure

$$\sigma = \begin{pmatrix} 1 & 0 \\ 0 & 1 \end{pmatrix} \delta_{(0,0)} + \begin{pmatrix} 1 & 1 \\ 1 & 1 \end{pmatrix} \delta_{(2,1)} + \begin{pmatrix} 2 & 0 \\ 0 & 0 \end{pmatrix} \delta_{(3,0)}.$$

Let S_γ, $\gamma \in \mathbb{N}_0^2$, be defined by (3.8.2) with respect to the given measure σ.

(a) Compute S_{00}, S_{01}, S_{10}, and S_{11}.

(b) Is it possible to find a lower inclusive set $\Lambda \subset \mathbb{N}_0^2$ with $\text{card}\Lambda = 3$ so that $\Phi = (S_{\lambda+\mu})_{\lambda,\mu\in\Lambda}$ is invertible?

29 Given the measure

$$\sigma = \begin{pmatrix} 2 & 1 \\ 1 & 2 \end{pmatrix} \delta_{(0,0)} + \begin{pmatrix} 1 & 0 \\ 0 & 1 \end{pmatrix} \delta_{(0,1)}$$

$$+ \begin{pmatrix} 3 & 2 \\ 2 & 1 \end{pmatrix} \delta_{(1,0)} + \begin{pmatrix} 5 & 0 \\ 0 & 5 \end{pmatrix} \delta_{(1,1)} + \begin{pmatrix} 2 & 0 \\ 0 & 2 \end{pmatrix} \delta_{1,2},$$

use Theorem 3.8.7 to produce a lower inclusive set $\Lambda \subset \mathbb{N}_0^2$, with $|\Lambda| = 5$, so that $\Phi = (S_{\lambda+\mu})_{\lambda,\mu\in\Lambda}$ is invertible.

30 What are precise condition(s) on a measure $\sigma = \sum_{i=1}^k T_i \delta_{(x_i, y_i)}$, with $T_i \geq 0$, and (x_i, y_i) all different, so that there exists a lower inclusive set $\Lambda \subset \mathbb{N}_0^2$, with $\text{card}\Lambda = k$, such that $\Phi = (S_{\lambda+\mu})_{\lambda,\mu\in\Lambda}$ is invertible?

31 Given the data

$$S_{00} = \begin{pmatrix} 3 & 1 \\ 1 & 3 \end{pmatrix}, S_{01} = S_{02} = S_{03} = \begin{pmatrix} 0 & 0 \\ 0 & 0 \end{pmatrix},$$

$$S_{10} = \begin{pmatrix} 1 & 0 \\ 0 & 1 \end{pmatrix}, S_{11} = S_{12} = \begin{pmatrix} 0 & 0 \\ 0 & 0 \end{pmatrix}, S_{20} = \begin{pmatrix} 4 & 0 \\ 0 & 4 \end{pmatrix},$$

$$S_{21} = \begin{pmatrix} 0 & 0 \\ 0 & 0 \end{pmatrix}, S_{30} = \begin{pmatrix} 8 & 0 \\ 0 & 8 \end{pmatrix},$$

use Theorem 3.8.2 to determine whether the given data come from a measure. In the case that the data do indeed come from a measure use Theorem 3.8.2 to produce a representing measure.

32 Given the data

$$\Phi = \begin{pmatrix} s_{00} & s_{01} & s_{10} \\ s_{01} & s_{02} & s_{11} \\ s_{10} & s_{11} & s_{20} \end{pmatrix} = \begin{pmatrix} 4 & 0 & 0 \\ 0 & 3 & 1 \\ 0 & 1 & 3 \end{pmatrix},$$

$$\Phi_1 = \begin{pmatrix} s_{10} & s_{11} & s_{20} \\ s_{11} & s_{20} & s_{21} \\ s_{20} & s_{21} & s_{30} \end{pmatrix} = \begin{pmatrix} 0 & 1 & 3 \\ 1 & 2 & 0 \\ 3 & 0 & ? \end{pmatrix},$$

and

$$\Phi_2 = \begin{pmatrix} s_{01} & s_{02} & s_{11} \\ s_{02} & s_{03} & s_{12} \\ s_{11} & s_{12} & s_{21} \end{pmatrix} = \begin{pmatrix} 0 & 3 & 1 \\ 3 & 1 & 0 \\ 1 & 0 & ? \end{pmatrix},$$

is it possible to choose s_{21} and s_{30} so that there is a representing measure with 3 atoms for the given data?

33 Let $\sigma = \delta_{(0,0)} + \delta_{(0,1)} + \delta_{(1,0)} + \delta_{(1,1)}$, $\{S_\gamma\}$ be the moments of σ, and $\Lambda = \{(0,0), (0,1), (1,0)\}$. Compute Φ, Φ_1, Φ_2, Θ_1, and Θ_2 as in Theorem 3.8.2. Show that Θ_1 and Θ_2 do not commute. Why is this not in contradiction with Theorem 3.8.2?

34 Finish the proof of Theorem 3.8.7 by considering the case $k > 3$.

35 For complex numbers $\gamma_{00} = 1, \gamma_{01}, \gamma_{10}, \gamma_{02}, \gamma_{11}, \gamma_{20}$, introduce the matrix

$$M = \begin{pmatrix} \gamma_{00} & \gamma_{01} & \gamma_{10} \\ \gamma_{10} & \gamma_{11} & \gamma_{20} \\ \gamma_{01} & \gamma_{02} & \gamma_{11} \end{pmatrix}.$$

In this exercise we are interested in the *truncated complex moment problem*, i.e., we are seeking a positive Borel measure μ on $\mathbb{C}$ so that

$$\int_{\mathbb{C}} \overline{z}^i z^j d\mu = \gamma_{ij}, \ 0 \leq i + j \leq 2. \tag{3.10.4}$$

(a) Show that if a solution μ exists then $M \geq 0$. (Hint: compute $\int_{\mathbb{C}} v(z)^* v(z) d\mu$, where $v(z) = \begin{pmatrix} 1 & z & \overline{z} \end{pmatrix}$.)

(b) Show that if $M \geq 0$ then a solution exists. In fact, μ can be chosen to be an atomic measure supported on r points, where $r = \text{rank} M$. (Hint: In the case $r = 2$, consider a measure of the form

$$\mu = \rho \delta_{\gamma_{01} + w\sqrt{(1-\rho)/\rho}} + (1-\rho)\delta_{\gamma_{01} - w\sqrt{\rho/(1-\rho)}},$$

with $0 < \rho < 1$ and $w^2 = \gamma_{02} - \gamma_{01}^2$. When $r = 3$, choose a $z \in \mathbb{C}$ and choose $\alpha > 0$ so that $M - \alpha v(z)^* v(z)$ is positive semidefinite of rank 2, and apply the rank 2 case to (a scaled) $M - \alpha v(z)^* v(z)$.)

(c) Show that the measure μ can be chosen to have support in $\mathbb{T}$ if and only if $M \geq 0$ and $\gamma_{00} = \gamma_{11}$. (Hint for the "if" part, show that

$$\begin{pmatrix} \gamma_{00} & \gamma_{10} & \gamma_{20} \\ \gamma_{01} & \gamma_{00} & \gamma_{10} \\ \gamma_{02} & \gamma_{01} & \gamma_{00} \end{pmatrix} \geq 0$$

and apply Theorem 1.3.6.)

(d) Show that if the measure μ has support in $\overline{\mathbb{D}}$ then $\gamma_{11} \leq \gamma_{00}$.

(e) Solve the truncated complex moment problem for $\gamma_{00} = 1, \gamma_{01} = \frac{1}{2} + \frac{1}{4}i = \overline{\gamma_{10}}, \gamma_{02} = \frac{3}{8} = \overline{\gamma_{20}}, \gamma_{11} = \frac{5}{8}$. Can μ be chosen to have support in $\mathbb{T}$? in $\overline{\mathbb{D}}$?

(f) Same questions for $\gamma_{00} = 1, \gamma_{01} = \frac{1+i}{2} = \overline{\gamma_{10}}, \gamma_{02} = \frac{1}{8} + \frac{i}{2} = \overline{\gamma_{20}}, \gamma_{11} = \frac{3}{4}$.

36 Show that the truncated complex moment problem with data γ_{ij}, $0 \leq i+j \leq 2n$, can be converted to a two-variable Hamburger moment with data s_{ij}, $0 \leq i+j \leq 2n$, and conversely. Provide the appropriate transformation between γ_{ij} and s_{ij}.

37 Prove that for every $a > 0$, the function

$$\phi_a(x) = \begin{cases} 1 - \frac{|x|}{a} & \text{if } |x| \leq a, \\ 0 & \text{if } x > a. \end{cases}$$

is positive semidefinite on $\mathbb{R}$.

38 Prove that the following functions are all positive semidefinite on $\mathbb{R}$.

(a) e^{itx}, with $t \in \mathbb{R}$ fixed,

(b) $\cos x$,

(c) $(1 + x^2)^{-a}, \quad a \geq 0$,

(d) $e^{-|x|^\alpha}$, $0 \leq \alpha \leq 2$.

39 Let G_1 and G_2 be two arbitrary groups and let $G_1 \circ G_2$ denote their *free product.* That means each $g \in G_1 \circ G_2$ can be represented in the form

$$g = x_1^\epsilon y_1 \cdots x_k y_k^\delta, \tag{3.10.5}$$

where $x_i \in G_1 - \{e\}$, $y_i \in G_2 - \{e\}$, and $\epsilon, \delta \in \{0, 1\}$. For two functions $f_1 : G_1 \to \mathbb{C}$ and $f_2 : G_2 \to \mathbb{C}$ such that $f_1(e) = f_2(e) = 1$ we define their *free product function* $f_1 \circ f_2$ on $G_1 \circ G_2$ by setting

$$(f_1 \circ f_2)(g) = f_1(x_1^\epsilon) f_2(y_1) \cdots f_1(x_k) f_2(y_k^\delta)$$

if g has representation (3.10.5). Prove that if f_1 and f_2 are positive semidefinite on the groups G_1 and G_2, respectively, then the function $f = f_1 \circ f_2$ is positive semidefinite on the free product group $G = G_1 \circ G_2$.

40 Let G denote a locally compact Abelian group and denote by $P^1(G)$ the set of all positive semidefinite functions $f : G \to \mathbb{C}$ such that $f(e) \leq 1$.

(a) Prove that the set of extreme points of $P^1(G)$ equals the set $\Gamma_0 := \Gamma \cup \{0\}$.

(b) Show that Γ_0 is compact in the weak-* topology of L^∞_G.

(Hint: To show that extreme points of $P^1(G)$ lie in Γ_0, introduce for an extremal f the functions

$$f_c^y(x) = (1 + |c|^2)f(x) + \bar{c}f(yx) + cf(y^{-1}x), y \in G, c \in \mathbb{C},$$

which again are positive semidefinite. As

$$f = \frac{f_c^y + f_{-c}^y}{2(1 + |c|^2)}$$

is extremal, f_c^y must be a nonnegative multiple of f; thus $f_c^y(x) = k(c, y)f(x)$ with $k(c, y) \geq 0$. Put $\ell(y) = \frac{1}{4}[k(1, y) - ik(i, y)]$, and show that $f(xy) = \ell(y)f(x)$. From this conclude that $f(xy) = f(y)f(x)$ and continue.)

41 Let G be a locally compact Abelian group G and Γ its character group.

(a) Prove that for every positive finite Borel measure μ on G the function $\widehat{\mu}$ is uniformly continuous on Γ. (Hint: For $\delta > 0$, let $K \subset G$ be such that $\mu(G \setminus K) < \delta$. Now estimate $|\hat{\mu}(\gamma_1) - \hat{\mu}(\gamma_2)|$ by using $\int_G = \int_K + \int_{G \setminus K}$.)

(b) Prove that for every positive finite Borel measure μ on Γ the function $\check{\mu}$ is uniformly continuous on G.

42 For any locally compact Abelian group G prove that the set $\{\widehat{f} : f \in L^1_G\}$ is dense in $C_0(\Gamma)$ in the topology of uniform convergence on Γ. (Hint: Use the Stone-Weierstrass theorem.)

3.11 NOTES

Section 3.1

Section 3.1 follows the presentation of [571]. Theorem 3.1.1 is due to [8]. Parts of the proof of Theorem 3.1.2 are inspired by [79], [74], and [72]. The argument to arrive at an isometry as in (3.1.8) is referred to as the "lurking isometry" argument; this term was introduced by J. A. Ball. In Example 3.1.8 it is used that there exist T_i and A_i, satisfying (3.1.14), which is a result of [352]. The latter also follows from [450]. In [208] the Carathéodory-Fejér problem on the polydisk is considered. Corollary 3.1.6 is related to results in [369]. An open problem is to obtain necessary and sufficient conditions in Theorem 3.1.7 when $d \geq 3$ and $n \geq 2$. A possible approach to Exercise 3.10.4 may be found in [450, Section 2]. Exercise 3.10.3 is based on an example in [366].

Section 3.2

Section 3.2 is based on the papers [573] and [576]. Theorem 3.2.1 represents a multivariable generalization of the classical Gohberg-Semencul formula. In [431] an approximation algorithm is discussed to approximate the inverse of a two variable Toeplitz matrix. More recently, inversion algorithms for two-variable Toeplitz matrices have been proposed in [440] and [441], and indirectly in [551]. For two-level Toeplitz matrices with a Bernstein-Szegő symbol an exact formula for the inverse is given in [371], where this formula is also used to give good approximation for other positive symbols. For three or more variable Toeplitz matrices such a result does not exist. Theorem 3.2.7 is a two-variable Christoffel-Darboux formula, which was useful in solving the Bernstein-Szegő measure problem (see Theorem 2.3.1). Equation (3.2.28) may be viewed as the left and right stable factorization of a multivariable trigonometric polynomial. The main result in [424] yields a characterization for when a positive definite matrix-valued $Q(z)$ has a left (right) stable factorization. This result in [424] concerns factorization along commutative subspace lattices, and is based on the general factorization result in [388]. Exercise 3.10.6 is based on examples from [517] (parts (a), (b)) and [357] (parts (c), (d)).

Section 3.3

The Bernstein-Szegő measures received their name as Szegő first used them in [536] and Bernstein used them later in [92] for the real line case. Section 3.3 is based on [249] and [248]. The scalar version was done in the latter paper, while the operator-valued case was done in [249]. In [246], [175], [176] the first steps were made toward a general multivariable theory, followed soon after by [401] from the perspective of covariance extension. An open problem is how to generalize Theorems 3.3.1 and 3.3.2 to the three or more variable case. For instance, if p is a stable polynomial of degree $n \in \mathbb{N}_0^3$ and $c_k = \widehat{\frac{1}{|p|^2}}(k)$, $k \in \mathbb{Z}^3$, are there low rank conditions on $(c_{k-l})_{k,l}$ as in

Theorem 3.3.2(2)? The notion of intersecting zeros, as defined in Exercise 3.10.5, was introduced and developed in [248]. An earlier version of Theorem 3.2.7 can be found in [139], but there no explicit construction is given of the polynomials G_j and F_j. In [362] and [363] alternative proofs using reproducing kernels can be found for some of these (sometimes strengthened) results. In [364] scalar versions of (3.2.26) are obtained, where now $P = R$ is allowed to have roots on the bitorus; uniqueness of the representation (which in the case that P, R are stable never happens) is also discussed there. Partial results were obtained in [254], [516], [123], and [122]. In [250] the authors give an algorithm to construct polynomials with a prescribed set of intersecting zeros and discuss the relation with the Cayley-Bacharach theorem. Exercise 3.10.16 is based on [248, Example 2.3.2]. In the paper [251] the authors continued to investigate the properties of bivariate polynomials orthogonal on the bicircle with respect to a positive linear functional. The lexicographical and reverse lexicographical orderings are used to order the monomials. Recurrence formulas are derived between the polynomials of different degrees. These formulas link the orthogonal polynomials constructed using the lexicographical ordering with those constructed using the reverse lexicographical ordering. These results are then used to construct a class of two variable measures supported on the bicircle that are given by one over the magnitude squared of a stable polynomial (the Bernstein-Szegő measure problem). Some of the results in this section may be used in radar applications; see [1], [2]. Exercises 3.10.11 and 3.10.12 are based on [250, Example 3.11] and Example [577, Example 6.2], respectively. Exercise 3.10.7 is inspired by [251, Lemmas 2.1 and 2.2]. Exercise 3.10.9 is based on [78, Section 5]. Exercise 3.10.10 is in fact [292, Example 2].

Section 3.4

Section 3.4 is based on [248] and [186]. Theorem 3.4.1 is somewhat more general than the corresponding scalar-valued version appearing in [248]. The uniqueness result Proposition 3.4.2 was proven for the scalar case in [579] and [58]. This section generalizes the well-known Fejér-Riesz lemma (Theorem 1.1.5) to the two-variable case. An open question, however, is how to deal with the singular case, that is, when $f(z,w)$ is singular for some $|z| = |w| = 1$. In the one-variable case the Fejér-Riesz factorization of a necessarily singular trigonometric polynomial plays an important role in the construction of wavelets with compact support; see [164]. For an operator-valued generalization of Theorem 3.4.1 it is expected that the rank condition needs to be replaced by the conditions in Theorem 3.3.1; however, a proof does not exist yet. Finally, a three or more variable generalization of Theorem 3.4.1 is another open problem. Given now

$$f(z,w) = \sum_{(k,l)\in\Lambda_+-\Lambda_+} f_{kl} z^k w^l, \qquad \sum_{(k,l)\in\Lambda_+-\Lambda_+} |f_{kl}| < \infty,$$

and suppose that $f(z,w) > 0$ for $|z| = |w| = 1$, one may ask the question: when does there exists a stable Wiener function $p(z,w)$ with Fourier support in Λ_+ such that $f(z,w) = |p(z,w)|^2$, $(z,w) \in \mathbb{T}^2$? For the case when Λ_+ is the strip $\Lambda_+ = \{(n,m) : 0 < n \leq r \text{ or } (n = 0 \text{ and } m \geq 0)\}$ this question was answered affirmatively in [60]. Also, for the truncated strip

$$\Lambda_+ = \{(n,m) : 0 < n < r \text{ or } (n = 0 \text{ and } m \geq 0) \text{ or } (n = r \text{ and } m \leq s)\}$$

the answer is affirmative, as was observed in [490]; see also [197] for a more general version. It needs to be noted that in both these cases (as well as in the classical one-variable case) $\Lambda_+ - \Lambda_+ = \Lambda_+ \cup (-\Lambda_+)$. As already mentioned in the Notes of Chapter 2, the author gives in [186] a new proof to a result generalizing the Fejér-Riesz lemma to the operator-valued case. The new ideas allow him to also obtain the multivariable result Theorem 3.4.4, which applies to strictly positive trigonometric polynomials. The case when there are singularities on $\mathbb{T}^d$ is still open. The proof presented here is based on an idea from [247]. It is natural to relate this to the real line case where it is known that not every positive polynomial in several variables is a sum of squares. The latter is reminiscent to the articles [482], [530], [475], and [556], where the latter three are primarily dedicated to solving the multivariable Hamburger problem. For positive polynomials on compact semialgebraic sets, positive results do exist; see [512], [474], and the recent book [417]. These results have also led to semidefinite programming-based algorithms for polynomial optimization, a previously intractable problem. See, for example, [448], [515], [389], [396].

Section 3.5

The presentation of Section 3.5 follows that of [65] and was also inspired by a similar solution for Nehari's problem in [210]. For more examples of amenable groups see [179] or [280]. Theorem 3.5.2 was proven in [542] in case of a countable vertex set. It is interesting to note that [542] provides an example of a chordal graph with a countable number of vertices so that adding any edge destroys the chordality. For results on ordered groups we refer the reader to [502]. Krein's extension theorem ([373]) proves that every positive semidefinite continuous scalar function on a real interval $(-a, a)$ admits a continuous positive semidefinite extension to $\mathbb{R}$. A new proof was found in [42] (Artjomenko's proof) for this result without the continuity requirement (see also [510]). The corollaries 3.5.5–3.5.7 are taken from [50]. The ideas are based on [308], where such subspaces were already considered for $r = 2$. Corollary 3.5.6 can be viewed as a generalization of a result in [489] stating that every strictly positive definite matrix-valued almost periodic Wiener class function with spectrum in the real interval $(-a, a)$ can be extended to a (pointwise) strictly positive almost periodic function on $\mathbb{R}$ which also belongs to the Wiener class (see [526] for the scalar case). Corollary 3.5.7 can be viewed as a generalization of a result in [60] stating that every matrix-valued strictly positive definite function on a set of the form (3.5.3) for $r = 2$ which belongs to the Wiener class can be extended to a (pointwise) positive function on $\mathbb{Z}^2$ which also belongs to the Wiener class. Corollary 3.5.7 in the case that $\alpha_1 = 1$ and $\alpha_2 = \cdots = \alpha_r = 0$ was proved in [112] by a different approach.

Section 3.6

The presentation of Section 3.6 is based on [64]. Positive semidefinite functions on free groups are an important object of study in relation to group theory and C^*-algebras since the basic work of [257] and [211]. In the case of Abelian groups, Bochner's theorem ensures that such a function is the Fourier transform of a positive measure on the dual group, and much of the theory develops along this line. On non-Abelian groups Fourier analysis is no longer available, but the focus is now on the relation to group representations. The theory of positive semidefinite functions is more intricate; for some notable results, see [109], [110], [171], and [291]. In Section 3.6 we consider positive semidefinite functions on the most basic non-Abelian group, namely, the free group with an arbitrary number of generators. We obtain an analogue of Krein's theorem: a positive semidefinite function defined on the set of words of length bounded by a fixed constant can be extended to the whole group. This is rather surprising in view of the result concerning $\mathbb{Z}^2$ discussed in the previous sections. For general facts about Fourier analysis on free groups, see [223]. The extension property is closely related to a parametrization of all operator-valued positive semidefinite functions by means of sequences of contractions. These are an analogue of the *choice sequences* of contractions, a generalization of the *Schur or Szegő parameters* that has been developed by Foias and his coauthors in connection with intertwining liftings (see [225] and the references within) and the theory of orthogonal polynomials ([252] and [523]).

One should note that results in a close area of investigation have been obtained in [465], [466], and [467]. Most notably, in [467] a similar extension problem is proved for the free semigroup, together with a description of all solutions. The group case that we have considered requires however new arguments. Theorems 3.6.2 and 3.6.12 and their corollaries extend to the case when $\mathbb{F}$ is a free group with an infinite number of generators. The main ideas are similar, but the details are more cumbersome, since we have to use transfinite induction in several instances, along the lines of [542].

For definitions and results on group representation and Naimark's dilation theorem we refer the reader to [510] and [451]. Choi's theorem, initially proved in [128], can also be found in the third chapter of [451].

Haagerup has considered in [291] functions $s \mapsto e^{-t|s|}$, and has proved that they are positive semidefinite for $t > 0$. In [171] a larger class is defined: a function $u : \mathbb{F} \to \mathbb{C}$ is called a *Haagerup function* if $u(e) = 1$, $|u(s)| \leq 1$, $u(s^{-1}) = \overline{u(s)}$, and $u(st) = u(s)u(t)$ whenever $|st| = |s| + |t|$; it is proved therein that any Haagerup function is positive semidefinite. The results are generalized in [109], where the analogue operator-valued functions (called *quasimultiplicative*) are proved to be positive semidefinite and also the central extension of their restriction to S_n.

Section 3.7

The analogue of Theorem 3.7.2, [421, Theorem 0.1] deals with the case when the words appearing in the polynomial are of the form $s_+^{-1}t_+$, where

s_+, t_+ are elements in the free semigroup with m generators. The positivity condition replaces the indeterminates with unitary matrices, and the consequence is a corresponding decomposition.

In second part of Section 3.10, the latter is applied to noncommutative factorization problems, in the spirit of [421] and [310]. Theorem 3.7.4 is due to [310]. Some adjustments were in order. An alternative proof of this result may be found in [423]. It is worth mentioning that the connection between extension problems of positive semidefinite functions and factorization has been noted and used in the commutative case already in [503]. A more recent and comprehensive account on positive polynomials in noncommutative variables can be found in [312].

It is worth comparing the factorization obtained in Theorem 3.7.2 with the results of [310] and [421]. In [310] the author considers real polynomials in the indeterminates $X_1, \ldots, X_m$ and $X_1^T, \ldots, X_m^T$; p is called positive when any replacement of X_i with real matrices A_i and of X_i^T with the transpose of A_i leads to a positive semidefinite matrix. The main result is an analogue of decomposition 3.7.3 for such polynomials. Exercise 3.10.27 is based on [310, Section 1].

Section 3.8

Let $\Gamma = \{\gamma \in \mathbb{N}_0^d \colon 0 \leq |\gamma| \leq m\}$ and suppose a real-valued multisequence $\{s_\gamma\}_{\gamma \in \Gamma}$ and a closed set $K \subseteq \mathbb{R}^d$ are given. The *truncated K-moment problem* consists of answering the question of whether there exists a positive Borel measure μ on $\mathbb{R}^d$, with supp $\mu \subseteq K$, so that $s_\gamma = \int_{\mathbb{R}^d} \xi^\gamma d\mu(\xi)$, for all $\gamma \in \Gamma$. When m is even or odd we will call this the even case or odd case, respectively. When $K = \mathbb{R}^d$, the truncated Hamburger moment problem is recovered. In [302] a solution is provided that is based on checking positivity of polynomials on K (called the *Riesz-Haviland criterion*), while [158] provides a solution based on this idea of constructing a rank-preserving positive semidefinite extension (called *flat extension* in [158]); see also [343] for a variation where the given data is on a different lower inclusive set. Moreover, it was shown in [158] that solving the complex truncated K-moment problem for $K = \mathbb{C}^d$ is equivalent to solving the truncated K-moment problem for $K = \mathbb{R}^{2d}$. Minimality of representing measures and special cases relative to given data for $K = \mathbb{C}^d$ were investigated in [159]. Exercise 3.10.35 is based on [160]; see also [158, Theorem 6.1] for parts (a) and (b). See also [218] for more recent results. A complete solution to Exercise 3.10.36 can be found in [161].

Theorem 3.8.2 is a special case ($d = 2$) of the d-variable result in [360]. Also, [360] provides a solution to the odd case for the truncated multivariable Hamburger moment problem when Γ is as above. Given points in $\mathbb{R}^d$, the question whether one can produce a lower inclusive set which results in the corresponding Vandermonde matrix being invertible was answered, in the affirmative, in [511]. Furthermore, [511] provided a concrete algorithm for generating the desired lower inclusive set. It should be noted that (3.8.16) implies a factorization for a positive semidefinite two-variable Hankel matrix. In the one variable case such factorizations appear in [544], also in the indefinite case. Finally, we refer to [530] and [475] for approaches of the power moment problem by checking positivity of an extended sequence.

Section 3.9

The classical Bochner's theorem goes back to [97] and [98]; see also [510]. In [559], [560] and [308], the authors considered already elements of prediction and factorization for functions on ordered groups (in particular $\mathbb{Z}^2$); see also [419] for an overview. For more recent work see [411]. We refer to the book [46] as a general reference on ARMA (Auto Regressive Moving Averages) processes and related issues. Let us remark that in [418] the domain

$$S_{M,N} = \{(n,m) : n = -N, m \geq -M \text{ or}$$

$$-N+1 \leq n \leq N-1 \text{ or } n = N, m \leq M\}$$

is considered. In that paper reflection coefficients are developed as well as a two-dimensional Levinson algorithm. The papers [348], [401], [356] are useful sources for an explanation how the autoregressive filters are used in signal processing.

Exercises

Exercises 3.10.8 and 3.10.13 concern the eigenstructure of $\Phi_1\Phi^{-1}$ and $\Phi_2^*\Phi^{-1}$ appearing in Theorem 3.3.1. When p and $\overleftarrow{p}$ have common simple roots eigenvectors may be constructed directly from the intersecting zeros. When roots have a higher multiplicity it is still an open problem how to determine the Jordan structure of $\Phi_1\Phi^{-1}$ and $\Phi_2^*\Phi^{-1}$.

Exercise 3.10.15 is inspired by [381], where it was shown that any polynomial without roots in $\mathbb{D}^2$ is of the form described in this exercise.

Exercise 3.10.18 is based on [452, Remark 4.12]. Exercises 3.10.19 and 3.10.20 are based on [452, Remark 4.13].

Possible answers to Exercise 3.10.21 may be found in [186] and [247].

Exercises 3.10.22 and 3.10.24 are based on [186, Section 6].

Exercise 3.10.23 was inspired by [470].

Exercise 3.10.25 is based on [186, Theorem 7.1].

Exercise 3.10.40 combines Theorems 1.8.10 and 1.8.12(a) in [510].

Additional notes

In addition to the five multivariable generalizations of the classical Carathéodory problem mentioned in the introduction, one can also consider the Carathéodory interpolation problem in the so-called noncommutative Herglotz-Agler class, as introduced in [353]. This class concerns formal power series

$$f(z) = \sum_{w \in \mathbb{F}_m} f_w z^w$$

with coefficients $f_w \in \mathcal{L}(Y)$, $w \in \mathbb{F}_m$, where for the indeterminates $z = (z_1, \dots, z_m)$ and words $w = a_{i_1} \cdots a_{i_k}$ one puts $z^w = z_{i_1} \cdots z_{i_k}$, and $z^e = 1$. The Herglotz-Agler class $\mathcal{HA}_m^{\rm nc}(Y)$ consists of such series for which the associated series

$$f(C) := \sum_{w \in \mathbb{F}_m} f_w \otimes C^w,$$

converges in the operator norm and $\operatorname{Re} f(C) \geq 0$, where $(C_1, \dots C_m)$ is an arbitrary m-tuple of strict contractions on some Hilbert space, $\mathcal{E}$ say. Here, for $w = a_{i_1} \cdots a_{i_k}$, one puts $C^w = C_{i_1} \cdots C_{i_k}$, and $C^e = I_{\mathcal{E}}$.

The interpolation problem is now the following. Let $\Lambda \subset \mathbb{F}_m$ be an admissible set, that is, Λ is a finite set so that $a_k w, w a_k \in \mathbb{F}_m \setminus \Lambda$ for every $w \in \mathbb{F}_m \setminus \Lambda$ and every $k = 1, \dots, m$. Given a collection $\{c_w \in \mathcal{L}(Y) : w \in \Lambda\}$ with $c_e \geq 0$, find f in the noncommutative Herglotz-Agler class $\mathcal{HA}_m^{\rm nc}(Y)$ such that $f_e = \frac{c_e}{2}$ and $f_w = c_w$ for $w \in \Lambda \setminus \{e\}$.

In [353, Theorem 4.11] it was shown that a solution exists if and only if the polynomial

$$p(z) := \frac{c_e}{2} + \sum_{w \in \Lambda \setminus \{e\}} c_w z^w$$

satisfies $\operatorname{Re} p(T) \geq 0$ for every m-tuple T of Λ-jointly nilpotent contractive operators (i.e., $T^w = 0$ for all $w \in \mathbb{F}_m \setminus \Lambda$). In fact, it suffices to check that $\operatorname{Re} p(T) \geq 0$ for every m-tuple T of Λ-jointly nilpotent contractive $n \times n$ matrices, for all $n \in \mathbb{N}$. In the classical one-variable case, it suffices to take

$$T = \begin{pmatrix} 0 & \cdots & \cdots & 0 \\ 1 & 0 & \cdots & 0 \\ \vdots & \ddots & \ddots & \vdots \\ 0 & \cdots & 1 & 0 \end{pmatrix}.$$

Indeed, $\operatorname{Re} p(T) \geq 0$ is exactly the condition $T_n \geq 0$, where T_n is the Toeplitz matrix in Theorem 1.3.6.

Chapter Four

Contractive analogs

An operator $T \in \mathcal{L}(\mathcal{H}, \mathcal{H}')$ is called a *contraction* if $\|T\| \leq 1$, and a *strict contraction* if $\|T\| < 1$. In this chapter we deal with contractive completions of partial operator matrices. Since the norm of a submatrix is always less or equal to the norm of the matrix itself, every partial matrix which admits a contractive completion has to be *partially contractive* (or a *partial contraction*), that is, all its fully specified submatrices are contractions.

A contractive completion problem can always be reduced to a positive semidefinite one. That is because Lemma 2.4.4 implies that $\|T\| \leq 1$ if and only if $\begin{pmatrix} I & T \\ T^* & I \end{pmatrix} \geq 0$. In Section 4.1 we describe those patterns which have the property that every partially contractive operator matrix with that pattern of specified/unspecified entries admits a contractive completion. We answer a similar question for partially contractive Toeplitz contractions.

We start Section 4.2 with a parametrization of the set of all solutions to the 2×2 completion problem of a partial matrix with one unknown entry. Next we present the solution to a linearly constrained contractive completion problem of lower-triangular partial matrices. In the 2×2 case we parameterize the set of all solutions and give necessary and sufficient conditions for the existence of isometric, co-isometric, and unitary solutions.

In Sections 4.3, 4.6, and 4.5 we solve several operator-valued contractive interpolation problems. First, the solution to Nehari's problem is stated by the means of a Hankel operator. From there we derive the solution to the contractive Carathéodory-Fejér problem. Next we solve the Nevanlinna-Pick interpolation problem via a linearly fractional transform by a unitary matrix. Finally, we solve the problems of Nehari and Carathéodory-Fejér for operator-valued functions defined on compact groups with an ordered dual.

In Section 4.7 we continue by the methods started in Section 4.2 and solve a linearly constrained contractive interpolation problem in nest algebras. As an application, we obtain an operator theoretic Corona type theorem.

In Section 4.8, we first consider lower triangular operator matrices with all their specified entries being Hilbert-Schmidt operators. We prove bounds for both the operator and the Hilbert-Schmidt norms of a distinguished completion. We extend the method for obtaining joint L^∞/L^2 norm control solutions for the problems of Nehari and Carathéodory-Fejér.

Given a polynomial $p(z) = C_0 + zC_1 + \cdots + z_n C_n$ with coefficients in $\mathcal{L}(\mathcal{H})$, in Section 4.9 we estimate the quantity $\nu(p)$ representing the infimum over the norms of extensions of p which are in $L^\infty(\mathcal{L}(\mathcal{H}))$. As a consequence, we obtain estimates for Toeplitz completions of partially contractive lower triangular Toeplitz matrices.

In Section 4.10 we consider lower triangular partial matrices the specified entries of which are compact operators. In general, there are multiple compact completions which minimize the norm. We show that there exists a unique compact completion which minimizes the sequence of singular values with respect to the lexicographic order, leading to a unique so-called "superoptimal completion."

Given a function $G \in (H^\infty + C)(\mathbb{C}^{m\times n})$, it is shown that the Hankel operator H_G is compact. The distance from G to $H^\infty(\mathbb{C}^{m\times n})$ is $\|H_G\|$, but in general there are multiple such best approximations for G. Section 4.11 is devoted to showing the existence in this case of a unique "superoptimal" approximation. Finally, in Section 4.12 we show how the so-called model matching problem can be solved using Nehari's theorem.

4.1 CONTRACTIVE OPERATOR-MATRIX COMPLETIONS

The first half of this section is devoted to describing $m \times n$ patterns which have the property that every partially contractive operator-matrix with that particular pattern of specified/unspecified entries admits a contractive completion. In the second half, we consider partial contractions which are Toeplitz at the extent they are specified. We describe those patterns of specified consecutive diagonals which have the property that every partially contractive Toeplitz matrix with that particular pattern of specified/unspecified entries admits a contractive Toeplitz completion.

Let $\mathcal{K} = \mathcal{H}_1 \oplus \cdots \oplus \mathcal{H}_n$ and $\mathcal{K}' = \mathcal{H}'_1 \oplus \cdots \oplus \mathcal{H}'_n$, where $\mathcal{H}_1, \ldots, \mathcal{H}_n, \mathcal{H}'_1, \ldots, \mathcal{H}'_n$ are Hilbert spaces, and let

$$K = \begin{pmatrix} A_{11} & ? & ? & \cdots & ? \\ A_{21} & A_{22} & ? & \cdots & ? \\ \vdots & \vdots & \vdots & \vdots & \vdots \\ A_{n1} & A_{n2} & A_{n3} & \cdots & A_{nn} \end{pmatrix} \tag{4.1.1}$$

be a lower triangular partial matrix acting between $\mathcal{K}$ and $\mathcal{K}'$. Then K is partially contractive when all fully specified operator matrices

$$A^{(k)} = \begin{pmatrix} A_{k1} & \cdots & A_{kk} \\ \vdots & & \vdots \\ A_{n1} & \cdots & A_{nk} \end{pmatrix}$$

are contractive.

Theorem 4.1.1 *Every partially contractive lower triangular operator matrix admits a contractive completion.*

Proof. By the remarks above and Lemma 2.4.4, a partial matrix K admits a contractive completion if and only if $B = \begin{pmatrix} I & K \\ K^* & I \end{pmatrix}$ admits a positive semidefinite completion. If K is a partially contractive lower triangular operator matrix, then B is a banded partially positive semidefinite matrix, which admits a positive semidefinite completion (see the introduction to Section 2.6). Thus K admits a contractive completion. □

Remark 4.1.2 For a partial matrix as in (4.1.1), the quantity

$$d_\infty(K) = \max_{1 \le k \le n} \|A^{(k)}\|$$

is called the *Arveson distance* of K. By Theorem 4.1.1, it represents the minimum over the norm of all completions of K, and thus the distance between K and the set of all strictly upper triangular operator matrices.

A partial operator matrix $K = (A_{ij})_{i=1,j=1}^{m,n}$ acting between $\mathcal{K} = \mathcal{H}_1 \oplus \cdots \oplus \mathcal{H}_n$ and $\mathcal{K}' = \mathcal{H}_1' \oplus \cdots \oplus \mathcal{H}_m'$ is called *generalized lower triangular* if for every A_{pq} which is specified, all A_{ij}, with $p \le i \le m$ and $1 \le j \le q$ are specified as well.

Theorem 4.1.3 *All partially contractive generalized lower triangular matrices K admit contractive completions.*

Proof. Let K be a partially contractive generalized lower triangular matrix, and introduce that partial matrix $B = \begin{pmatrix} I & K \\ K^* & I \end{pmatrix}$. It is straightforward to check that B is a generalized banded partially positive semidefinite matrix. Thus B has a positive semidefinite completion $B_1 = \begin{pmatrix} I & K_1 \\ K_1^* & I \end{pmatrix}$, say. But then K_1 is a contractive completion of K. □

Let K be a partially contractive generalized lower-triangular matrix. If $B_c = \begin{pmatrix} I & F_c \\ F_c^* & I \end{pmatrix}$ is the central completion of $\begin{pmatrix} I & K \\ K^* & I \end{pmatrix}$ (as defined in Section 2.6), then F_c is called the *contractive central completion of* K. This completion is uniquely determined by Theorem 2.6.1.

An $m \times n$ pattern of specified/unspecified entries is said to have the *contractive completion property* if every partial contraction with that pattern of specified/unspecified entries admits a contractive completion. The pattern

$$\begin{pmatrix} * & * & ? \\ * & ? & * \end{pmatrix} \tag{4.1.2}$$

does not have the contractive completion property. Consider indeed the partial contraction

$$\begin{pmatrix} \frac{1}{\sqrt{2}} & \frac{1}{\sqrt{2}} & ? \\ \frac{1}{\sqrt{2}} & ? & \frac{1}{\sqrt{2}} \end{pmatrix}. \tag{4.1.3}$$

The only contractive completion of the partial matrix representing the first two columns of (4.1.3) is

$$\begin{pmatrix} \frac{1}{\sqrt{2}} & \frac{1}{\sqrt{2}} \\ \frac{1}{\sqrt{2}} & -\frac{1}{\sqrt{2}} \end{pmatrix},$$

but then the second row becomes $\begin{pmatrix} \frac{1}{\sqrt{2}} & -\frac{1}{\sqrt{2}} & \frac{1}{\sqrt{2}} \end{pmatrix}$, with norm equal to $\sqrt{\frac{3}{2}}$. In conclusion, (4.1.2) does not have the contractive completion property. Because the norm of a matrix and its transpose are equal, the transpose of (4.1.2) does not have the contractive completion property as well.

The following theorem describes all patterns that possess the contractive completion property.

Theorem 4.1.4 *The following statements are equivalent for every rectangular pattern of specified/unspecified entries.*

(i) The pattern has the contractive completion property.

(ii) The pattern does not contain any subpattern of the form (4.1.2) or its transpose.

(iii) The rows and columns of the pattern can be permuted such that the resulting pattern is the direct sum of generalized lower triangular patterns, bordered above and on the right by rows and columns of unspecified entries.

Proof. A first observation is that the direct sum of any generalized lower triangular patterns has the contractive completion property. The reason is that in this case the contractive completion can be realized by choosing all unspecified positions within a single direct summand by the use of Theorem 4.1.3, and all the other entries made to equal 0. Also, adding rows and columns with all their entries equal to 0 to an operator matrix does not modify its norm. Thus (iii)⇒(i). The implication (i)⇒(ii) is already clear.

Assume that a certain rectangular pattern verifies (ii) and for every column j of it let col_j denote the the set of all rows u such that the position (u, j) is specified. We want to point out first that for two distinct columns j and k, either one of col_j or col_k is included in the other, or that the two are disjoint. If we contradict the latter statement, one of the following things has to happen.

(a) There exist two distinct columns j and k such that $\mathrm{col}_j \cap \mathrm{col}_k \neq \emptyset$, and neither $\mathrm{col}_j \subseteq \mathrm{col}_k$, nor $\mathrm{col}_k \subseteq \mathrm{col}_j$. This means there exist rows $u \in \mathrm{col}_j \cap \mathrm{col}_k$, $v \in \mathrm{col}_j \setminus \mathrm{col}_k$, and $w \in \mathrm{col}_k \setminus \mathrm{col}_j$. Then the subpattern formed by the rows u, v, and w, and columns j and k is of the type (4.1.2) transposed, which gives a contradiction.

(b) There exist two distinct columns j and k such that neither $\mathrm{col}_j \subseteq \mathrm{col}_k$, nor $\mathrm{col}_k \subseteq \mathrm{col}_j$, and also a third column l, such that $\mathrm{col}_l \cap \mathrm{col}_j \neq \emptyset$ and $\mathrm{col}_l \cap \mathrm{col}_k \neq \emptyset$. This implies the existence of two distinct rows u and v such that $u \in \mathrm{col}_l \cap \mathrm{col}_j$, $u \notin \mathrm{col}_l \cap \mathrm{col}_k$, $v \notin \mathrm{col}_l \cap \mathrm{col}_j$, and $v \in \mathrm{col}_l \cap \mathrm{col}_k$. Then the subpattern formed by the rows u and v, and columns l, j, and k is of type (4.1.2), which gives a contradiction.

We can thus partition the columns $\{1, \ldots, n\}$ of the pattern as $X_1 \cup X_2 \cup \cdots \cup X_q$ such that $\{\mathrm{col}_j : j \in X_s\}$ is completely ordered by inclusion for each

$1 \leq s \leq q$, and for $s \neq t$, $j \in X_s$ and $k \in X_t$ imply that $\text{col}_j \cap \text{col}_k = \emptyset$. This implies (iii) and completes the proof of the theorem. □

Since the rows and columns of every partial matrix with only one unspecified entry can be permuted such that the unspecified entry is in the upper right-hand corner, Theorem 4.1.4 has the following simple but useful consequence.

Corollary 4.1.5 *Every partially contractive operator matrix with a single unspecified entry admits a contractive completion.*

4.1.1 The Toeplitz case

In this subsection we consider contractive completions of partially contractive Toeplitz matrices. We present a Toeplitz version of Theorem 4.1.4 for the situation when the specified diagonals form a consecutive sequence.

An $m \times n$ partial operator matrix $(A_{ij})_{i=1,j=1}^{m,n}$ is called a *partial Toeplitz matrix* if it is Toeplitz to the extent it is specified, namely that if A_{ij} is specified, then each A_{kl} with $k - l = i - j$ is specified as well and $A_{kl} = A_{ij}$. We number the diagonals of an $m \times n$ Toeplitz matrix by $-(n-1), \ldots, 0, \ldots, (m-1)$ from the upper right to the lower left. A partial Toeplitz matrix corresponds then to the situation of some specified and some unspecified diagonals. A *partial Toeplitz contraction* is a partial matrix that is both Toeplitz and a partial contraction. We say that an $m \times n$ pattern of specified/unspecified diagonals has the *Toeplitz contractive completion property* if every scalar partial Toeplitz contraction with that pattern of specified/unspecified entries admits a contractive Toeplitz completion. Assuming that the specified diagonals are consecutive, the following theorem describes the patterns which have the Toeplitz contractive completion property.

Theorem 4.1.6 *Let P be an $m \times n$ pattern whose specified d diagonals form a consecutive sequence. Then P has the Toeplitz contractive completion property if and only if one of the following is satisfied:*

(i) $m = 1$ or $n = 1$;

(ii) $d \leq 1$;

(iii) $d \geq m + n - 2$; or

(iv) $m = n$ and $d = 2n - 3$.

Equivalently, the only patterns which have the Toeplitz contractive completion property (up to transposition) are the following:

(i) a single row;

(ii) at most one specified diagonal;

(iii) $\begin{pmatrix} * & \cdots & * & ? \\ * & \cdots & * & * \\ \vdots & & \vdots & \vdots \\ * & \cdots & * & * \end{pmatrix}$; and

(iv)(a) $\begin{pmatrix} * & * & \cdots & * & ? \\ * & * & \cdots & * & * \\ \vdots & \vdots & & \vdots & \vdots \\ ? & * & \cdots & * & * \end{pmatrix}$ or

(iv)(b) $\begin{pmatrix} * & \cdots & * & ? & ? \\ * & \cdots & * & * & ? \\ \vdots & & \vdots & \vdots & \vdots \\ * & \cdots & * & * & * \end{pmatrix}$ in the square case.

Proof of Theorem 4.1.6 (sufficiency). Every partial Toeplitz contraction with pattern (i) or (ii) can be completed to a Toeplitz contraction by choosing all unspecified entries to be 0. Every partial Toeplitz contraction with pattern (iii) can be completed to a contraction by Corollary 4.1.5. All such completions are Toeplitz. Let

$$K = \begin{pmatrix} a_0 & a_{-1} & \cdots & a_{-n+1} & ? \\ a_1 & a_0 & \cdots & a_{-n+2} & a_{-n+1} \\ \vdots & \vdots & \ddots & \vdots & \vdots \\ a_{n-1} & a_{n-2} & \cdots & a_0 & a_{-1} \\ ? & a_{n-1} & \cdots & a_1 & a_0 \end{pmatrix}$$

be a partial Toeplitz contraction with pattern (iv)(a). Let M be the partial contraction obtained by deleting the first column of K. By Corollary 4.1.5, we can fix the unspecified entry as a_{-n}, such that the resulting matrix

$$F = \begin{pmatrix} a_{-1} & \cdots & a_{-n+1} & ? \\ a_0 & \cdots & a_{-n+2} & a_{-n+1} \\ \vdots & \ddots & \vdots & \vdots \\ a_{n-2} & \cdots & a_0 & a_{-1} \\ a_{n-1} & \cdots & a_1 & a_0 \end{pmatrix}$$

is contractive. In F^T reverse the order of its rows and columns. The so obtained matrix G is also a contraction and coincides with the completion of the partial matrix obtained by deleting the last row of K and by choosing its unspecified entry to equal a_{-n}. This implies that by choosing the upper right-hand corner entry in K to equal a_{-n}, we obtain a partial contraction B. Again by Corollary 4.1.5, B admits a contractive completion. This implies that pattern (iv)(a) has the Toeplitz contractive completion property. The fact that pattern (iv)(b) has the Toeplitz contractive completion property follows similarly. Its proof is left as an exercise (see Exercise 4.13.1).

Example 4.1.7 We would like to illustrate by an example that the pattern (iv)(b) only has the Toeplitz contractive completion property when the entries are scalar. It does not already have this property when the entries are assumed to be 2×2 matrices. Let U and V be two noncommuting unitary matrices; for example, let

$$U = \begin{pmatrix} 1 & 0 \\ 0 & -1 \end{pmatrix}, V = \begin{pmatrix} \frac{1}{\sqrt{2}} & -\frac{1}{\sqrt{2}} \\ \frac{1}{\sqrt{2}} & \frac{1}{\sqrt{2}} \end{pmatrix}.$$

Let α and β be two numbers in the interval $(0, 1)$ such that, denoting $D(\alpha, \beta) = \sqrt{1 - \alpha^2 - \beta^2}$, we have

$$\frac{\alpha\beta}{D(\alpha, \beta)} + \beta < \alpha. \tag{4.1.4}$$

For example, let $\alpha = 0.1$ and $\beta = 0.01$. Consider the partial Toeplitz matrix

$$A = \begin{pmatrix} D(\alpha, \beta)I & ? & ? \\ \beta V & D(\alpha, \beta)I & ? \\ \alpha U & \beta V & D(\alpha, \beta)I \end{pmatrix}.$$

Let

$$M(X) = \begin{pmatrix} \beta V & D(\alpha, \beta)I & X \\ \alpha U & \beta V & D(\alpha, \beta)I \end{pmatrix}$$

be any completion of the partial matrix obtained by deleting the first row of A. Then $I - M(X)M(X)^*$ equals

$$\begin{pmatrix} \alpha^2 - XX^* & -\alpha\beta VU^* - D(\alpha, \beta)(\beta V^* + X) \\ -\alpha\beta UV^* - D(\alpha, \beta)(\beta V + X^*) & 0 \end{pmatrix}.$$

Thus $\|M(X)\| \leq 1$ if and only if

$$X = -\frac{\alpha\beta}{D(\alpha, \beta)} VU^* - \beta V^* \tag{4.1.5}$$

verifies $\|X\| \leq \alpha$. The latter condition is automatically satisfied by X in (4.1.5), since U and V are unitary, and so (4.1.4) implies $\|X\| \leq \frac{\alpha\beta}{D(\alpha,\beta)} + \beta < \alpha$. Thus the matrix formed by the last two rows of A is a partial contraction.

Let

$$N(X) = \begin{pmatrix} D(\alpha, \beta)I & X \\ \beta V & D(\alpha, \beta)I \\ \alpha U & \beta V \end{pmatrix}$$

be any completion of the partial matrix obtained by deleting the last column of A. Then $I - N(X)^*N(X)$ equals

$$\begin{pmatrix} 0 & -D(\alpha, \beta)(X + \beta V^*) - \alpha\beta U^*V \\ -D(\alpha, \beta)(X^* + \beta V) - \alpha\beta V^*U & \alpha^2 I - X^*X \end{pmatrix}.$$

Thus $\|N(X)\| \leq 1$ if and only if

$$X = -\frac{\alpha\beta}{D(\alpha, \beta)} U^*V - \beta V^* \tag{4.1.6}$$

verifies $\|X\| \leq \alpha$. Again, U and V unitary together with (4.1.4) imply that X in (4.1.6) verifies $\|X\| < \alpha$. Thus the matrix formed by the first two columns of A is a partial contraction, so A is a partial contraction. Since U and V do not commute, there is no X which simultaneously verifies (4.1.5) and (4.1.6), so A does not admit a contractive Toeplitz completion.

Remark 4.1.8 Following Example 4.1.7 and Exercise 4.13.2, we remark that when the entries of a partial matrix are matrices of size at least 2×2 or operators, the pattern has the Toeplitz contractive completion property if and only it is of type (i), (ii), or (iii) in Theorem 4.1.6.

Before we proceed with the necessity part of the proof of Theorem 4.1.6 we need some preliminary results.

Lemma 4.1.9 *For $1 \leq q \leq r$, let $G(x, y)$ be the $q \times r$ matrix all of whose elements, with exception of one diagonal consisting of q elements, are equal to y, and these exceptional entries are equal to x (with $x, y \in \mathbb{C}$). Then $G(x, y)$ is a contraction if and only if the following conditions hold.*

(i) For $q \geq 2$, either

$$|x - y| < 1 \quad \text{and} \quad (r-2)|y|^2 + x\overline{y} + \overline{x}y \leq q^{-1}(1 - |x-y|^2)$$

or

$$|x - y| = 1 \quad \text{and} \quad (r-2)|y|^2 + x\overline{y} + \overline{x}y = 0.$$

(ii) For $q = 1$,

$$(r-1)|y|^2 + |x|^2 \leq 1.$$

In particular, $G(x, x)$ is a contraction if and only if $|x| \leq (qr)^{-\frac{1}{2}}$.

Proof. The case $q = 1$ is elementary, so assume $q \geq 2$. By a straightforward computation, the $q \times q$ matrix $I - G(x, y)G(x, y)^*$ equals

$$(1 - |x-y|^2)I - ((r-2)|y|^2 + x\overline{y} + \overline{x}y)e_q^* e_q, \qquad (4.1.7)$$

where e_q is the row with q entries all equal to 1. Since $q \geq 2$, when the matrix in (4.1.7) is positive semidefinite, we necessarily have $|x - y| \leq 1$. The case $|x - y| = 1$ is evident, so assume that $|x - y| < 1$. Then (4.1.7) is positive semidefinite if and only if $I - we_q^* e_q$ is such, where

$$w = (1 - |x-y|^2)^{-1}((r-2)|y|^2 + x\overline{y} + \overline{x}y).$$

Denoting by $\langle \cdot, \cdot \rangle$ the standard inner product on $\mathbb{C}^q$, for every $x \in \mathbb{C}^q$ we have that

$$\langle (I - we_q^* e_q)x, x \rangle = \langle x, x \rangle - w \left| \langle e_q^*, x \rangle \right|^2. \qquad (4.1.8)$$

It follows that (4.1.8) is nonnegative for every $x \in \mathbb{C}^q$ if and only if

$$1 - w \langle e_q^*, e_q^* \rangle \geq 0,$$

which means that $w \leq \frac{1}{q}$. □

Lemma 4.1.10 *For $1 \le r \le q$, $q \ge 2$, let $F_{q,r}$ be the $q \times r$ partial matrix whose every entry, with the exception of one unspecified entry in the upper right-hand corner, is equal to $[(q-1)r]^{-\frac{1}{2}}$. Then there is a unique contractive completion of $F_{q,r}$ which is obtained when the upper right-hand corner equals*

$$w_0 = -(r-1)[(q-1)r]^{-\frac{1}{2}}. \tag{4.1.9}$$

Proof. It follows from Lemma 4.1.9 that $F_{q,r}$ is a partial Toeplitz contraction. Let $F(x)$ be the $q \times r$ matrix obtained from $F_{q,r}$ by specifying the upper right-hand corner to be $x \in \mathbb{C}$. Then, denoting $w = [(q-1)r]^{-\frac{1}{2}}$, we have

$$I - F(x)F(x)^* = \begin{pmatrix} 1-(r-1)|w|^2 - |x|^2 & [-(r-1)|w|^2 - x\overline{w}]e_{q-1} \\ [-(r-1)|w|^2 - \overline{x}w]e_{q-1}^* & V \end{pmatrix},$$

where

$$V = I - r|w|^2 e_{q-1}^* e_{q-1} = I - (q-1)^{-1} e_{q-1}^* e_{q-1}$$

is a $(q-1) \times (q-1)$ matrix. One easily sees that e_{q-1}^* belongs to the kernel of V. Let U be a unitary $(q-1) \times (q-1)$ matrix whose first column is $e_{q-1}^*/\|e_{q-1}^*\|$. Then the 2×2 upper left-hand corner in

$$\begin{pmatrix} 1 & 0 \\ 0 & U^* \end{pmatrix} (I - F(x)F(x)^*) \begin{pmatrix} 1 & 0 \\ 0 & U \end{pmatrix}$$

is

$$\begin{pmatrix} 1-(r-1)|w|^2 - |x|^2 & (q-1)^{\frac{1}{2}}[-(r-1)|w|^2 - x\overline{w}] \\ (q-1)^{\frac{1}{2}}[-(r-1)|w|^2 - \overline{x}w] & 0 \end{pmatrix}. \tag{4.1.10}$$

The matrix (4.1.10) is positive semidefinite only if

$$-(r-1)|w|^2 - x\overline{w} = 0,$$

or

$$x = -(r-1)w.$$

So, if there exists a contractive completion of $F_{q,r}$, then it must be specified by (4.1.9). Since $F_{q,r}$ is partially contractive, the existence of a contractive completion of $F_{q,r}$ follows from Corollary 4.1.5. This completes the proof. □

Using Lemma 4.1.9 and a calculation similar to those in the proof of Lemma 4.1.10, one obtains the following statement.

Lemma 4.1.11 *For $2 \le q \le r-1$ let $H(x,y)$ be a $q \times r$ partial matrix all of whose entries except for the upper right-hand corner are specified. The specified entries are all equal to y, except for one diagonal consisting of q elements which does not intersect the last column in $H(x,y)$; the entries of this exceptional diagonal are equal to x. Here $x, y \in \mathbb{C}$ are such that*

$$|x-y| < 1, \quad y \ne 0,$$

$$(r-2)|y|^2 + x\overline{y} + \overline{x}y = (q-1)^{-1}(1 - |x-y|^2),$$

$$(r-3)|y|^2 + x\overline{y} + \overline{x}y \leq q^{-1}(1-|x-y|^2).$$

Then there is a unique z such that by specifying the upper right-hand corner in $H(x,y)$ to be z, one obtains a contraction. This number is given by the formula

$$z = [-(r-3)|y|^2 - x\overline{y} - \overline{x}y]\overline{y}^{-1}.$$

Proof of Theorem 4.1.6 (necessity). Let P be an $m \times n$ pattern of specified consecutive diagonals, which is of none of the types (i)–(iv). We show that not every partial Toeplitz contraction of pattern P admits a contractive Toeplitz completion. We assume first that $2 \leq m < n$ and either

$$P = \begin{pmatrix} & & ? \\ ? & & \end{pmatrix} \tag{4.1.11}$$

or

$$P = \begin{pmatrix} & ? & ? \\ & & ? \end{pmatrix}. \tag{4.1.12}$$

We will argue later why solving the problem for the above two patterns P leads to a complete solution. Consider first the case P as given by (4.1.12), and let $3 \leq m \leq n-2$. Let K be an $m \times n$ partial matrix with pattern P with the following properties.

(1) The specified entries in K, with the exception of one diagonal consisting of m elements which does not intersect the last two columns of K, are equal to

$$y = \left((m-1)(m-2) + \frac{1}{4}(n+1-m)^2\right)^{-\frac{1}{2}}.$$

(ii) The specified entries on the exceptional diagonal in K are equal to

$$x = \frac{m-n+1}{2}y.$$

One verifies that

$$|x-y| < 1,$$

$$(n-2)y^2 + 2xy = (m-2)^{-1}(1-|x-y|^2),$$

$$(n-3)y^2 + 2xy = (m-1)^{-1}(1-|x-y|^2),$$

$$(n-4)y^2 + 2xy < m^{-1}(1-|x-y|^2).$$

So by Lemma 4.1.11, K is a partial Toeplitz contraction which does not admit a contractive Toeplitz completion. Indeed, removing from (4.1.12) the last column or the first row yields two different partial matrices with one unknown corner entry. By Lemma 4.1.11 each of these corner entries is uniquely determined when completing to a contraction. As these two unique values are not the same, the partial matrix (4.1.12) does not have a contractive Toeplitz completion.

Assume that $m = n - 1 \geq 3$ (and P is given by (4.1.12)). Let K be a partial Toeplitz matrix with pattern P each entry of which is $((n-2)(n-1))^{-\frac{1}{2}}$. By Lemma 4.1.9, K is a partial contraction. If K were to admit a contractive Toeplitz completion, then by Lemma 4.1.10 the only possibility in the $(1, n-1)$, hence also the $(2, n)$ entry, would be

$$-(n-2)((n-2)(n-1))^{-\frac{1}{2}}.$$

However, this is impossible, because one easily checks that the n-dimensional row

$$\begin{pmatrix} w & w & \cdots & w & w_0 \end{pmatrix},$$

where $w = ((n-2)(n-1))^{-\frac{1}{2}}$, $w_0 = -(n-2)w$, is not a contraction for $n > 1$. It remains to consider (still assuming P has the form (4.1.12)) the case $m = 2$, $n > 2$. If $n = 3$, then we are done by letting the specified entries to be $\frac{\sqrt{2}}{2}$. If $n \geq 4$, then put

$$y = \left(n - 2 + \frac{1}{4}(n-5)^2\right)^{-\frac{1}{2}}, \quad x = -\frac{n-5}{2}y,$$

and verify that

$$|x - y| < 1,$$

$$(n-4)y^2 + 2xy = \frac{1}{2}(1 - |x-y|^2),$$

$$(n-2)y^2 + x^2 = 1.$$

Let K be the partial Toeplitz matrix with pattern P, all of whose specified entries are y with the exception of the $(1,1)$ and $(2,2)$ entries, which are x. By Lemma 4.1.9, K is a partial contraction. Arguing as in the proof of Lemma 4.1.11, we see that the only way K can be completed to a Toeplitz contraction is by putting

$$z = -(n-4)y - 2x$$

in the $(1, n-1)$ position. However, the $1 \times n$ matrix

$$\begin{pmatrix} y & x & y & \cdots & y & z \end{pmatrix}$$

is not a contraction, so K does not admit a contractive Toeplitz completion.

Consider now the case when P is given by (4.1.11). If $3 \leq m \leq n-2$, then we are done by arguing as in the case when P has the form (4.1.12). Assume that $m = n - 1 \geq 2$. Let K be the partial matrix with the pattern P all of whose specified entries are $((n-2)(n-1))^{-\frac{1}{2}}$. Then K is a partial contraction by Lemma 4.1.9. By Lemma 4.1.10 applied to the partial matrix obtained by deleting the last column of K, one obtains that K can be completed to a contraction by putting

$$z = -(n-2)[(n-2)(n-1)]^{-\frac{1}{2}}$$

in the $(1, n-1)$ position. The $(2, n)$ position of the matrix being the same, we obtain that the second row is not contractive, so K does not admit a contractive Toeplitz completion. Finally assume that $m = 2$ and $n \geq 4$. Let

$$y = \left(n - 2 + \frac{1}{4}(n-5)^2\right)^{-\frac{1}{2}}, \quad x = -\frac{n-5}{2}y,$$

and let K be the partial matrix with pattern P all of whose entries are y except for the $(1,2)$ and $(2,3)$ entries, which are x. As in the case when P was given by (4.1.12), one verifies that K is a partial contraction which is not completable to a Toeplitz contraction.

Let P be an $m \times n$ pattern of d specified consecutive diagonals, which is of none of the types (i)–(iv). Without loss of generality, we may assume $m \leq n$ (otherwise consider the transpose pattern with its rows and columns in reverse order). Then P must contain at least one subpattern of type (4.1.11) or type (4.1.12). Among such subpatterns of P, consider the one of size $q \times r$, $q < r$, with q maximal. Then $d = p + q - 1$ and the possible size of any fully specified subpattern of K is $s \times t$, where $d = s + t + 1$. Assume first $r - 1 = q \geq 3$ and let K be the partial matrix with pattern P with all its specified entries equal to $\frac{1}{\sqrt{(q-1)(q-2)}}$. It is easy to see that K is a partial contraction and K does not admit a contractive Toeplitz completion (since one of its submatrices does not). In case $q = 2$, K has only two fully specified submatrices, and we have the counterexamples for this situation described earlier. Assume second that $r-2 \geq q \geq 3$. Define then the partially Toeplitz matrix K with pattern P and having all diagonals equal to

$$y = \left((q-1)(q-2) + \frac{1}{4}(r+1-q)^2\right)^{-\frac{1}{2}},$$

except one which is equal to

$$x = \frac{q-r+1}{2}y.$$

The diagonal with x appears in the $q \times r$ submatrix as the one of length q which does not intersect its last two columns. It is a straightforward computation to show that for every s and t such that $s + t = p + q - 2$ and $t \geq s \geq 2$, we have that

$$(s-2)y^2 + 2xy \leq t^{-1}(1 - |x-y|^2).$$

One can elementarily show that the row matrix of one x and $p+q-3$ entries equal to y is a contraction. Thus by Lemma 4.1.9, K is a partial contraction which does not admit a contractive Toeplitz completion (since one of its submatrices does not). This completes the proof of Theorem 4.1.6. □

4.2 LINEARLY CONSTRAINED COMPLETION PROBLEMS

We start this section by describing all contractive completions of a partial operator-valued contraction of the form $\left(\begin{smallmatrix} * & * \\ * & ? \end{smallmatrix}\right)$. We continue with an $n \times n$ linearly constrained operator-valued contractive completion problem. We first establish necessary and sufficient conditions for a solution to exist. We concentrate then on the 2×2 problem, for which we describe the set of all solutions. Necessary and sufficient conditions are also found for the existence of isometric, co-isometric, and unitary solutions. As a consequence, similar conditions are found for the existence of such solutions in the unconstrained case.

By Exercise 2.10.25, $\left(\begin{smallmatrix} A & B \\ C & X \end{smallmatrix}\right) \in \mathcal{L}(\mathcal{H}_1 \oplus \mathcal{H}_2, \mathcal{K}_1 \oplus \mathcal{K}_2)$ is a partial contraction if and only if there exist contractions $G_1 \in \mathcal{L}(\mathcal{H}_2, \mathcal{D}_{A^*})$ and $G_2 \in \mathcal{L}(\mathcal{D}_A, \mathcal{K}_2)$ such that $B = D_{A^*}G_1$ and $C = G_2 D_A$.

Theorem 4.2.1 *The formula*

$$X = -G_2 A^* G_1 + D_{G_2^*} \Gamma D_{G_1} \tag{4.2.1}$$

establishes a one-to-one formula between the set of all operators $X \in \mathcal{L}(\mathcal{H}_2, \mathcal{K}_2)$ *such that*

$$\widetilde{A} = \begin{pmatrix} A & D_{A^*}G_1 \\ G_2 D_A & X \end{pmatrix}$$

is a contraction, and the set of all contractions $\Gamma \in \mathcal{L}(\mathcal{D}_{G_1}, \mathcal{D}_{G_2^*})$.

Proof. Let

$$T = \begin{pmatrix} A \\ G_2 D_A \end{pmatrix}.$$

By Exercise 2.10.25, $\widetilde{A}$ is a contraction if and only if it is of the form $\widetilde{A} = \begin{pmatrix} T & D_{T^*}G \end{pmatrix}$, where $G \in \mathcal{L}(\mathcal{H}_1 \oplus \mathcal{H}_2, \mathcal{D}_{T^*})$ is a contraction. Since

$$\begin{aligned} I - TT^* &= \begin{pmatrix} D_{A^*}^2 & -AD_A G_2^* \\ -G_2 D_A A^* & I - G_2 D_A^2 G_2^* \end{pmatrix} \\ &= \begin{pmatrix} D_A^* & 0 \\ -G_2 A^* & D_{G_2^*} \end{pmatrix} \begin{pmatrix} D_A^* & 0 \\ -G_2 A^* & D_{G_2^*} \end{pmatrix}^*, \end{aligned}$$

by Lemma 2.4.2, there exists a partial isometry U mapping $(\mathrm{Ker} D_{T^*})^\perp$ onto $\left(\mathrm{Ker}\begin{pmatrix} D_A^* & 0 \\ -G_2 A^* & D_{G_2^*} \end{pmatrix}\right)^\perp$ such that $D_{T^*} = \begin{pmatrix} D_A^* & 0 \\ -G_2 A^* & D_{G_2^*} \end{pmatrix} U$. Since

$$\left(\mathrm{Ker}\begin{pmatrix} D_A^* & 0 \\ -G_2 A^* & D_{G_2^*} \end{pmatrix}\right)^\perp \subseteq \mathcal{D}_{A^*} \oplus \mathcal{D}_{G_2^*}$$

and $r_1 \oplus r_2 \in \mathcal{D}_{A^*} \oplus \mathcal{D}_{G_2^*}$ with $\begin{pmatrix} D_A^* & 0 \\ -G_2 A^* & D_{G_2^*} \end{pmatrix}\begin{pmatrix} r_1 \\ r_2 \end{pmatrix} = 0$ implies $r_1 = 0$ and $r_2 = 0$, we have that

$$\left(\mathrm{Ker}\begin{pmatrix} D_A^* & 0 \\ -G_2 A^* & D_{G_2^*} \end{pmatrix}\right)^\perp = \mathcal{D}_{A^*} \oplus \mathcal{D}_{G_2^*}.$$

Thus

$$D_{T^*}G = \begin{pmatrix} D_A^* & 0 \\ -G_2A^* & D_{G_2^*} \end{pmatrix} UG,$$

where UG is an arbitrary contraction in $\mathcal{L}(\mathcal{H}_2, \mathcal{D}_{A^*} \oplus \mathcal{D}_{G_2^*})$. Again by Exercise 2.10.25, there exist contractions $G_1' \in \mathcal{L}(\mathcal{D}_{G_1'}, \mathcal{D}_{G_2^*})$, and $\Gamma \in \mathcal{L}(\mathcal{D}_{G_1'}, \mathcal{D}_{G_2^*})$ such that

$$UG = \begin{pmatrix} G_1' \\ \Gamma D_{G_1'} \end{pmatrix}.$$

We have then $D_{A^*}G_1 = D_{A^*}G_1'$, and since D_{A^*} is injective on its range, $G_1' = G_1$. This implies $\widetilde{A}$ is a contraction if and only if

$$\widetilde{A} = \begin{pmatrix} A & D_{A^*}G_1 \\ G_2D_A & -G_2A^*G_1 + D_{G_2^*}\Gamma D_{G_1} \end{pmatrix},$$

$\Gamma \in \mathcal{L}(\mathcal{D}_{G_1}, \mathcal{D}_{G_2^*})$ being an arbitrary contraction. □

It can be verified by computation that the contractive central completion is always defined by the choice of $\Gamma = 0$ in (4.2.1); for this it is better to permute the rows of the matrix making it in the form $\left(\begin{smallmatrix} * & ? \\ * & * \end{smallmatrix}\right)$.

For $1 \leq j \leq i \leq n$ let $A_{ij} : \mathcal{H}_j \to \mathcal{K}_i$ be given bounded linear operators acting between Hilbert spaces. Further, let also be given the operators

$$S = \begin{pmatrix} S_1 \\ \vdots \\ S_n \end{pmatrix} : \mathcal{H} \to \bigoplus_{i=1}^n \mathcal{K}_i, \quad T = \begin{pmatrix} T_1 \\ \vdots \\ T_n \end{pmatrix} : \mathcal{H} \to \bigoplus_{i=1}^n \mathcal{H}_i. \tag{4.2.2}$$

We want to find contractive completions of the following problem:

$$\begin{pmatrix} A_{11} & ? & \cdots & ? \\ A_{21} & A_{22} & \cdots & ? \\ \vdots & \vdots & & \vdots \\ A_{n1} & A_{n2} & \cdots & A_{nn} \end{pmatrix} \begin{pmatrix} S_1 \\ \vdots \\ S_n \end{pmatrix} = \begin{pmatrix} T_1 \\ \vdots \\ T_n \end{pmatrix}; \tag{4.2.3}$$

that is, we want to find A_{ij}, $1 \leq i < j \leq n$, such that $\widetilde{A} = (A_{ij})_{i,j=1}^n$ is a contraction satisfying the linear constraint $\widetilde{A}S = T$.

We first need the following observation.

Lemma 4.2.2 *Let $\widetilde{A}$, S, and T be bounded linear operators acting between corresponding Hilbert spaces. Then*

$$\begin{pmatrix} I & T & \widetilde{A} \\ T^* & S^*S & S^* \\ \widetilde{A}^* & S & I \end{pmatrix} \geq 0 \tag{4.2.4}$$

if and only if $||\widetilde{A}|| \leq 1$ and $\widetilde{A}S = T$.

Proof. By Theorem 2.4.1, (4.2.4) holds if and only if the Schur complement of the matrix supported on the first two rows and columns is positive semidefinite. This condition is equivalent to

$$\begin{pmatrix} I & T \\ T^* & S^*S \end{pmatrix} - \begin{pmatrix} \widetilde{A} \\ S^* \end{pmatrix} \begin{pmatrix} \widetilde{A}^* & S \end{pmatrix} = \begin{pmatrix} I - \widetilde{A}\widetilde{A}^* & T - \widetilde{A}S \\ T^* - S^*\widetilde{A}^* & 0 \end{pmatrix} \geq 0,$$

and the latter inequality is satisfied if and only if $\|\widetilde{A}\| \leq 1$ and $\widetilde{A}S = T$. □

Theorem 4.2.3 *Let $A_{ij} : \mathcal{H}_j \to \mathcal{K}_i$, $1 \leq j \leq i \leq n$, $S_i : \mathcal{H} \to \mathcal{H}_i$, $i = 1, \ldots, n$, and $T_j : \mathcal{H} \to \mathcal{K}_j$ be given linear operators acting between Hilbert spaces, and let S and T be as in (4.2.2). Then there exist contractive completions $\widetilde{A}$ of $K = \{A_{ij}, 1 \leq j \leq i \leq n\}$ satisfying the linear constraint $\widetilde{A}S = T$ if and only if*

$$\begin{pmatrix} I - A^{(i)}A^{(i)^*} & T^{(i)} - A^{(i)}S^{(i)} \\ T^{(i)*} - S^{(i)*}A^{(i)^*} & S^*S - S^{(i)*}S^{(i)} \end{pmatrix} \geq 0 \tag{4.2.5}$$

for $i = 1, \ldots, n$, where

$$A^{(i)} = \begin{pmatrix} A_{i1} & \cdots & A_{ii} \\ \vdots & & \vdots \\ A_{ni} & \cdots & A_{ni} \end{pmatrix}, \quad S^{(i)} = \begin{pmatrix} S_1 \\ \vdots \\ S_i \end{pmatrix}, \quad T^{(i)} = \begin{pmatrix} T_i \\ \vdots \\ T_n \end{pmatrix} \tag{4.2.6}$$

for $i = 1, \ldots, n$.

Proof. By Lemma 4.2.2 there exists a contractive completion $\widetilde{A}$ of K satisfying the linear constrained $\widetilde{A}S = T$ if and only if there exists a positive semidefinite completion of the partial matrix

$$\begin{pmatrix} I & 0 & \cdots & 0 & T_1 & A_{11} & ? & \cdots & ? \\ 0 & I & \cdots & 0 & T_2 & A_{12} & A_{22} & \cdots & ? \\ \vdots & \vdots & & \vdots & \vdots & \vdots & \vdots & & \vdots \\ 0 & 0 & \cdots & I & T_n & A_{1n} & A_{2n} & \cdots & A_{nn} \\ T_1^* & T_2^* & \cdots & T_n^* & S^*S & S_1^* & S_2^* & \cdots & S_n^* \\ A_{11}^* & A_{12}^* & \cdots & A_{1n}^* & S_1 & I & 0 & \cdots & 0 \\ ? & A_{21}^* & \cdots & A_{2n}^* & S_2 & 0 & I & \cdots & 0 \\ \vdots & \vdots & & \vdots & \vdots & \vdots & \vdots & & \vdots \\ ? & ? & \cdots & A_{nn}^* & S_n & 0 & 0 & \cdots & I \end{pmatrix}. \tag{4.2.7}$$

As it is known, the existence of a positive semidefinite completion of (4.2.7) is equivalent to the positive semidefiniteness of all principal submatrices of (4.2.7) formed with known entries. This condition is equivalent to

$$\begin{pmatrix} I & T^{(i)} & A^{(i)} \\ T^{(i)*} & S^*S & S^{(i)*} \\ A^{(i)*} & S^{(i)} & I \end{pmatrix} \geq 0$$

for $i = 1, \ldots, n$. By (a minor variation of) Lemma 2.2.1, the latter condition is equivalent with that of the Schur complements supported on the first two rows and columns being positive semidefinite, and that is exactly (4.2.5). □

For the particular case $n = 2$ we obtain the following.

Theorem 4.2.4 *The linearly constrained problem*

$$\begin{pmatrix} A & B \\ C & X \end{pmatrix} \begin{pmatrix} S_1 \\ S_2 \end{pmatrix} = \begin{pmatrix} T_1 \\ T_2 \end{pmatrix} \tag{4.2.8}$$

admits a contractive solution if and only if the following conditions are satisfied:

(i) $\|(A \quad B)\| \leq 1$;

(ii) $AS_1 + BS_2 = T_1$;

(iii) $\begin{pmatrix} S^*S - T^*T & S_1^* - T_1^*A - T_2^*C \\ S_1 - A^*T_1 - C^*T_2 & I - A^*A - C^*C \end{pmatrix} \geq 0.$

Proof. We first interchange the rows of the matrix. Then Theorem 4.2.3 implies that problem (4.2.8) admits a contractive solution if and only if

$$\begin{pmatrix} I & 0 & T_2 & C \\ 0 & I & T_1 & A \\ T_2^* & T_1^* & S^*S & S_1^* \\ C^* & A^* & S_1 & I \end{pmatrix} \geq 0 \tag{4.2.9}$$

and

$$\begin{pmatrix} I & T_1 & A & B \\ T_1^* & S^*S & S_1^* & S_2^* \\ A^* & S_1 & I & 0 \\ B^* & S_2 & 0 & I \end{pmatrix} \geq 0. \tag{4.2.10}$$

By Theorem 2.4.1, (4.2.9) is equivalent to conditions (i) and (ii), while (4.2.10) is equivalent to (iii). □

Problem (4.2.8) being a particular case of the contractive completion problem in Theorem 4.2.1, in case a solution to (4.2.8) exists, there exist contractions $G_1 \in \mathcal{L}(\mathcal{H}_1, \mathcal{D}_{A^*})$ and $G_2 \in \mathcal{L}(\mathcal{D}_A, \mathcal{K}_2)$ such that $B = D_{A^*}G_1$ and $C = G_2 D_A$. If $\widetilde{A} = \left(\begin{smallmatrix} A & B \\ C & X \end{smallmatrix}\right)$ is a contractive solution to (4.2.8), there exists a contraction $\Gamma \in \mathcal{L}(\mathcal{D}_{G_1}, \mathcal{D}_{G_2^*})$ such that $X = -G_2A^*G_1 + D_{G_2^*}\Gamma D_{G_1}$. The equation $CS_1 + XS_2 = T_2$ implies that $\Gamma|\overline{\operatorname{Ran}}(D_{G_1}S_2) = \Gamma_0$, where $\Gamma_0 : \overline{\operatorname{Ran}}(D_{G_1}S_2) \to \mathcal{D}_{G_2^*}$ is uniquely determined by

$$D_{G_2^*}\Gamma_0 D_{G_1} S_2 = T_2 - CS_1 + G_2A^*G_1S_2. \tag{4.2.11}$$

The contractivity of Γ_0 is ensured by the existence of a solution to problem (4.2.8).

We are ready to state our parametrization result.

Theorem 4.2.5 *Consider the completion problem (4.2.8), where the data satisfy the conditions of Theorem 4.2.4. Then there exists a one-to-one correspondence between the set of all contractive solutions of the problem and the set of all contractions* $G : \operatorname{Ker}(S_2^* D_{G_1}|\mathcal{D}_{G_1}) \to \mathcal{D}_{\Gamma_0}$, *where* $G_1 \in \mathcal{L}(\mathcal{H}_1, \mathcal{D}_{A^*})$ *is such that* $B = D_{A^*}G_1$ *and* Γ_0 *is defined by (4.2.11).*

Proof. Following the discussion preceding the theorem, there exists a one-to-one correspondence between the set of all contractive solutions of (4.2.8) and the set of all contractive extension of Γ_0 to $\mathcal{D}_{G_1}$. By Exercise 2.10.25, the set of all contractive extensions of Γ_0 to $\mathcal{D}_{G_1}$ is in a one-to-one correspondence with the set of all contractions $G : \mathcal{D}_{G_1} \ominus \overline{\text{Ran}}(D_{G_1}S_2) \to \mathcal{D}_{\Gamma_0^*}$. Since $\mathcal{D}_{G_1} \ominus \overline{\text{Ran}}(D_{G_1}S_2) = \text{Ker}(S_2^* D_{G_1}|\mathcal{D}_{G_1})$, our result follows. □

Corollary 4.2.6 *Under the conditions of Theorem 4.2.4, the completion problem (4.2.8) admits a unique solution if and only if* $\mathcal{D}_{G_1} \subseteq \overline{\text{Ran}}(S_1)$ *or* Γ_0 *is a co-isometry.*

Proof. Following Theorem 4.2.5, the existence of a unique contractive solution is equivalent to either $\text{Ker}(S_2^* D_{G_1}|\mathcal{D}_{G_1}) = \{0\}$ or $\mathcal{D}_{\Gamma_0} = \{0\}$. The first condition is equivalent to $\mathcal{D}_{G_1} \subseteq \overline{\text{Ran}}(S_1)$ while the second one is equivalent to Γ_0 being a co-isometry. □

Theorem 4.2.7 *Consider problem* (4.2.8) *where the data satisfy the conditions of Theorem 4.2.4. Then:*

(i) Problem (4.2.8) *admits an isometric solution if and only if* G_2 *is an isometry,* $S^*S = T^*T$, *and* $\dim\mathcal{D}_{G_1} \leq \dim\mathcal{D}_{G_2^*}$.

(ii) Problem (4.2.8) *admits a co-isometric solution if and only if* G_1 *is a co-isometry and* $\dim\mathcal{D}_{\Gamma_0^*} \leq \dim(\text{Ker}G_1 \cap \text{Ker}S_2^*)$.

(iii) Problem 4.2.8 admits a unitary solution if and only if G_1 *is a co-isometry,* G_2 *is an isometry,* $S^*S = T^*T$, *and* $\dim\mathcal{D}_{G_1} = \dim\mathcal{D}_{G_2^*}$.

Proof. We first prove (i). It is clear that the conditions $S^*S = T^*T$ and G_2 being an isometry are necessary; for the latter one may use Exercise 2.9.25. Assume now that these two conditions are satisfied and let $\widetilde{A}$ be a solution corresponding to the parameter Γ satisfying $\Gamma|\overline{\text{Ran}}(D_{G_1}S_2) = \Gamma_0$. We have then $S^*(I - \widetilde{A}^*\widetilde{A})S = S^*S - T^*T = 0$. A straightforward computation shows that in this case

$$I - \widetilde{A}^*\widetilde{A} = \begin{pmatrix} D_{G_1}D_\Gamma^2 D_{G_1} & 0 \\ 0 & 0 \end{pmatrix}$$

and thus $D_\Gamma D_{G_1} S_2 = 0$. This implies that $\widetilde{A}$ is an isometry if and only if Γ_0 is an isometry. By the proof of Theorem 4.2.5, Γ_0 can be chosen to be an isometry if and only if $G : \mathcal{D}_{G_1} \ominus \overline{\text{Ran}}(D_{G_1}S_2) \to \mathcal{D}_{\Gamma_0^*}$ can be chosen to be an isometry. The operator G can be chosen to be an isometry if $\dim\text{Ker}(S_2^* D_{G_1}|\mathcal{D}_{G_1}) \leq \dim\mathcal{D}_{\Gamma_0^*}$, and so it remains to be shown that the last inequality is equivalent to $\dim\mathcal{D}_{G_1} \leq \dim\mathcal{D}_{G_2^*}$. But this follows immediately from the equalities

$$\dim\text{Ker}(S_2^* D_{G_1}|\mathcal{D}_{G_1}) = \dim\mathcal{D}_{G_1} - \dim\overline{\text{Ran}}(D_{G_1}S_2),$$

$$\dim D_{\Gamma_0^*} = \dim D_{G_2^*} - \dim\overline{\text{Ran}}(D_{G_1}S_2),$$

and thus (i) follows.

Similarly, problem (4.2.8) admits a co-isometric solution if G_1 is a co-isometry and Γ_0 defined by (4.2.11) admits a co-isometric extension $\Gamma : \mathcal{D}_{G_1} \to \mathcal{D}_{G_2^*}$. The latter is equivalent to the existence of a co-isometry $G : \mathrm{Ker}(S_2^* D_{G_1}|\mathcal{D}_{G_1}) \to \mathcal{D}_{\Gamma_0^*}$, that is, $\dim D_{\Gamma_0^*} \leq \dim\mathrm{Ker}(S_2^* D_{G_1}|\mathcal{D}_{G_1})$. When G_1 is a co-isometry, D_{G_1} is the orthogonal projection onto $\mathrm{Ker}G_1$ and so part (ii) follows.

The proof of (iii) is similar to that of (i), but we must add the condition that G_1 is a co-isometry, and that there exists a unitary G acting $\mathrm{Ker}(S_2^* D_{G_1}|\mathcal{D}_{G_1}) \to \mathcal{D}_{\Gamma_0^*}$, which is equivalent to $\dim\mathcal{D}_{G_1} = \dim\mathcal{D}_{G_2^*}$. □

In the particular case $S = 0$ and $T = 0$, Theorem 4.2.7 implies the following result, which can also be derived directly from Exercise 2.10.25 and the proof of Theorem 4.2.1 (see Exercise 4.13.15) .

Theorem 4.2.8 *Consider the partial contraction*

$$\begin{pmatrix} A & D_{A^*}G_1 \\ G_2 D_A & X \end{pmatrix}. \tag{4.2.12}$$

(i) The partial matrix (4.2.12) admits an isometric completion if and only if the following conditions hold:

(a) G_2 *is an isometry;*

(b) $\dim\mathcal{D}_{G_1} \leq \dim\mathrm{Ker}G_2^*$.

When both (a) and (b) hold, the formula

$$X = -G_2 A^* G_1 + \Gamma D_{G_1}$$

represents a one-to-one correspondence between the set of all operators $X \in \mathcal{L}(\mathcal{H}_2, \mathcal{K}_2)$ *that provide an isometric completion of (4.2.12) and the set of all isometries* $\Gamma : \mathcal{D}_{G_1} \to \mathrm{Ker}G_2^*$.

(ii) The partial contraction (4.2.12) admits a co-isometric completion if and only if the following conditions hold:

(c) G_1 *is a co-isometry;*

(d) $\dim\mathcal{D}_{G_2^*} \leq \dim\mathrm{Ker}G_1$.

When both (c) and (d) hold, the formula

$$X = -G_2 A^* G_1 + D_{G_2^*}\Gamma$$

represents a one-to-one correspondence between the set of all operators $X \in \mathcal{L}(\mathcal{H}_2, \mathcal{K}_2)$ *that provide a co-isometric completion of (4.2.12) and the set of all co-isometries* $\Gamma : \mathrm{Ker}G_1 \to \mathcal{D}_{G_2^*}$.

(iii) The partial contraction (4.2.12) admits a unitary completion if and only if the conditions (a), (c), and

(e) $\dim\mathrm{Ker}G_1 = \dim\mathrm{Ker}G_2^*$

hold. When all (a), (c), and (e) hold, the formula

$$X = -G_2 A^* G_1 + \Gamma$$

represents a one-to-one correspondence between the set of all operators $X \in \mathcal{L}(\mathcal{H}_2, \mathcal{K}_2)$ *that provide a unitary completion of (4.2.12) and the set of all unitary operators* $\Gamma : \mathrm{Ker} G_1 \to \mathrm{Ker} G_2^*$.

4.3 THE OPERATOR-VALUED NEHARI AND CARATHÉODORY PROBLEMS

In this section we solve the classical interpolation problem of Nehari. We start by a solution to the operator-valued case (which is also referred to as Page's problem) based on contractive completions of operator matrices. Second, we present an alternate proof in the scalar case based on a factorization of type $f = f_1 f_2$, $f \in H^1(\mathbb{T})$, $f_1, f_2 \in H^2(\mathbb{T})$. We characterize the finite rank and compact Hankel operators with a scalar symbol. We finally show that if $G \in (H^\infty + C)(\mathbb{C}^{m\times n})$ and if $m = 1$ or $n = 1$, then there exists a unique best approximant $Q \in H^\infty(\mathbb{C}^{m\times n})$ to G and $\|(G-Q)(z)\|$ is constant a.e. on $\mathbb{T}$. A solution to the contractive Carathéodory-Fejér operator-valued problem is obtained as a consequence of our first approach for the solution of Nehari's problem.

Let $\mathcal{H}$ be a separable Hilbert space and let P_+ denote the orthogonal projection of $L^2(\mathcal{H})$ onto $H^2(\mathcal{H})$ and let $P_- = I - P_+$. Recall the multiplication operator M_F with symbol $F \in L^\infty(\mathcal{L}(\mathcal{H}))$ defined in Section 2.3, and for $F \in H^\infty(\mathcal{L}(\mathcal{H}))$ the Toeplitz operator $T_F : H^2(\mathcal{L}(\mathcal{H})) \to H^2(\mathcal{L}(\mathcal{H}))$ defined in Section 2.4 by $(T_F g)(e^{it}) = F(e^{it})g(e^{it})$. By Proposition 2.4.17 we have that $\|T_F\| = \|F\|_\infty$. It is straightforward that with respect to the decomposition $H^2(\mathcal{H}) = \bigoplus_{k=0}^\infty e^{ikt}\mathcal{H}$, T_F has the matrix representation

$$T_F = \begin{pmatrix} C_0 & 0 & 0 & 0 & \cdots \\ C_1 & C_0 & 0 & 0 & \cdots \\ C_2 & C_1 & C_0 & 0 & \cdots \\ \vdots & \ddots & \ddots & \ddots & \vdots \end{pmatrix} \tag{4.3.1}$$

The multiplication by the free variable e^{it} on $L^2(\mathcal{H})$ is called the *bilateral shift operator*, while the multiplication on $H^2(\mathcal{H})$ by z is called the *unilateral shift operator* and is usually denoted by S. We have then

$$S\left(\sum_{n=0}^\infty z^n h_n\right) = \sum_{n=0}^\infty z^{n+1} h_n, \quad S^*\left(\sum_{n=0}^\infty z^n h_n\right) = \sum_{n=0}^\infty z^n h_{n+1}.$$

Consider the function $k(e^{it}) = \sum_{n=-\infty}^{-1} C_n e^{int}$ and define the (possibly unbounded) operator $H_k : H^2(\mathcal{H}) \to [H^2(\mathcal{H})]^\perp$ by $H_k f = P_- M_k f$. The operator H_k is referred to as the *Hankel operator with symbol* k. An equivalent way of defining for a sequence $\{C_n\}_{n=-\infty}^{-1}$ of operators in $\mathcal{L}(\mathcal{H})$ the Hankel operator is via the formula

$$H(\{C_n\}_{n=-\infty}^{-1}) = \begin{pmatrix} \ddots & & & & \\ \ddots & \ddots & & & \\ C_{-3} & \ddots & \ddots & & \\ C_{-2} & C_{-3} & \ddots & \ddots & \\ C_{-1} & C_{-2} & C_{-3} & \ddots & \ddots \end{pmatrix}, \tag{4.3.2}$$

where the matrix is acting from $\bigoplus_{k=0}^{\infty} \mathcal{H}$ to $\bigoplus_{k=-\infty}^{-1} \mathcal{H}$.

The following is known as *Nehari's problem* (sometimes as Page's problem in the operator-valued case). Given a sequence of operators $\{C_n\}_{n=-\infty}^{-1}$ in $\mathcal{L}(\mathcal{H})$, find necessary and sufficient conditions for the existence of a function $F \in L^{\infty}(\mathcal{L}(\mathcal{H}))$ with $\|F\|_{\infty} \leq 1$, of the form $F(e^{it}) = \sum_{n=-\infty}^{-1} C_n e^{int} + \sum_{n=0}^{\infty} D_n e^{int}$.

Theorem 4.3.1 *For a given sequence of operators $\{C_n\}_{n=-\infty}^{-1}$ in $\mathcal{L}(\mathcal{H})$, Nehari's problem admits a solution if and only if $\|H(\{C_n\}_{n=-\infty}^{-1})\| \leq 1$.*

Proof. When Nehari's problem admits a solution $F(e^{it}) = \sum_{n=-\infty}^{\infty} F_n e^{int}$, then as the Hankel operator (4.3.2) is the upper-right corner of the multiplication operator M_F represented by the doubly infinite Toeplitz matrix $(F_{i-j})_{i,j=-\infty}^{\infty}$, we must have that $\|H(\{C_n\}_{n=-\infty}^{-1})\| \leq \|M_F\| = \|F\|_{\infty} \leq 1$.

For the sufficiency part, assume that $\|H(\{C_n\}_{n=-\infty}^{-1})\| \leq 1$, and consider the contractive completion problem for the following operator from $\mathcal{H} \oplus (\bigoplus_{k=0}^{\infty} \mathcal{H})$ to $(\bigoplus_{k=-\infty}^{-1} \mathcal{H}) \oplus \mathcal{H}$. This problem can be viewed as a problem of the form

$$\begin{pmatrix} \ddots & & & & \\ C_{-3} & \ddots & & & \\ C_{-2} & C_{-3} & \ddots & & \\ C_{-1} & C_{-2} & C_{-3} & \ddots & \\ ? & C_{-1} & C_{-2} & C_{-3} & \ddots \end{pmatrix}. \tag{4.3.3}$$

The above problem can be partitioned as $\left(\begin{smallmatrix} C & B \\ X & A \end{smallmatrix}\right)$, where $\begin{pmatrix} C & B \end{pmatrix} = H(\{C_n\}_{n=-\infty}^{-1})$, and $\left(\begin{smallmatrix} B \\ A \end{smallmatrix}\right) = H(\{C_n\}_{n=-\infty}^{-1})$. Since $\|H(\{C_n\}_{n=-\infty}^{-1})\| \leq 1$, by Corollary 4.1.5 there exists $D_0 \in \mathcal{L}(\mathcal{H})$ which defines a contractive completion of (4.3.3). We continue by induction, $D_n \in \mathcal{L}(\mathcal{H})$ being defined so to realize a contractive completion of the partial matrix

$$\begin{pmatrix} C_{-2} & \ddots & & & & & \\ C_{-1} & C_{-2} & \ddots & & & & \\ D_0 & C_{-1} & C_{-2} & \ddots & & & \\ \ddots & \ddots & \ddots & \ddots & \ddots & & \\ D_{n-1} & \ddots & \ddots & \ddots & \ddots & \ddots & \\ ? & D_{n-1} & \ddots & D_0 & C_{-1} & C_{-2} & \ddots \end{pmatrix}.$$

Finally, define $F(e^{it}) = \sum_{n=-\infty}^{-1} C_n e^{int} + \sum_{n=0}^{\infty} D_n e^{int}$. Then $\|F\|_\infty \leq \|M_F\| \leq 1$, and F is a solution to Nehari's problem. $\square$

Let us give a construction for the solution to the Nehari problem in case the Hankel is strictly contractive. For a countable set K we let $l^2_{\mathcal{H}}(K)$ denote the Hilbert space of sequences $\eta = (\eta_j)_{j\in K}$ satisfying $\|\eta\| := \sqrt{\sum_{j\in K} \|\eta_j\|^2_{\mathcal{H}}} < \infty$. We shall typically write Hankels in a Toeplitz like format by reversing the order of the columns of our Hankel matrices. For example, in the one-variable case our Hankels typically act $l^2(-\mathbb{N}_0) \to l^2(\mathbb{N}_0)$ as opposed to the usual convention of acting $l^2(\mathbb{N}_0) \to l^2(\mathbb{N}_0)$.

Theorem 4.3.2 *Let $\Gamma_i \in \mathcal{L}(\mathcal{H}), i \geq 0$, be bounded linear Hilbert space operators such that the Hankel*

$$H := \begin{pmatrix} & \Gamma_1 & \Gamma_0 \\ & \ddots & \Gamma_1 \\ & & \end{pmatrix} : l^2_{\mathcal{H}}(-\mathbb{N}_0) \to l^2_{\mathcal{H}}(\mathbb{N}_0) \tag{4.3.4}$$

is a strict contraction. Solve for operators $\Delta_0, D_{-1}, D_{-2}, \dots, B_0, B_1, \dots$, satisfying the equation

$$\begin{pmatrix} I & H \\ H^* & I \end{pmatrix} \begin{pmatrix} B \\ D \end{pmatrix} = \begin{pmatrix} 0 \\ \Delta \end{pmatrix}, \tag{4.3.5}$$

where

$$B = \begin{pmatrix} B_0 \\ B_1 \\ B_2 \\ \vdots \end{pmatrix} : \mathcal{H} \to l^2_{\mathcal{H}}(\mathbb{N}_0), \quad D = \begin{pmatrix} \vdots \\ D_{-2} \\ D_{-1} \\ I_{\mathcal{H}} \end{pmatrix} : \mathcal{H} \to l^2_{\mathcal{H}}(-\mathbb{N}_0),$$

$$\Delta = \begin{pmatrix} \vdots \\ 0 \\ 0 \\ \Delta_0 \end{pmatrix} : \mathcal{H} \to l^2_{\mathcal{H}}(-\mathbb{N}_0).$$

For $j = -1, -2, \ldots$, put

$$\Gamma_j = -\Gamma_{j+1}D_{-1} - \Gamma_{j+2}D_{-2} - \cdots = -\sum_{k=1}^{\infty} \Gamma_{j+k}D_{-k}. \tag{4.3.6}$$

Then $f(z) = \sum_{j=-\infty}^{\infty} \Gamma_j z^j$ belongs to $L^\infty(\mathcal{L}(\mathcal{H}))$ and $\|f\|_\infty < 1$.

Alternatively, the Fourier coefficients Γ_j of f may be constructed as follows. Solve for operators $\alpha_0, A_1, A_2, \ldots, C_0, C_{-1}, \ldots$, satisfying the equation

$$\begin{pmatrix} I & H \\ H^* & I \end{pmatrix} \begin{pmatrix} A \\ C \end{pmatrix} = \begin{pmatrix} \alpha \\ 0 \end{pmatrix}, \tag{4.3.7}$$

where

$$A = \begin{pmatrix} I_{\mathcal{H}} \\ A_1 \\ A_2 \\ \vdots \end{pmatrix} : \mathcal{H} \to l^2_{\mathcal{H}}(\mathbb{N}_0), \quad C = \begin{pmatrix} \vdots \\ C_{-2} \\ C_{-1} \\ C_0 \end{pmatrix} : \mathcal{H} \to l^2_{\mathcal{H}}(-\mathbb{N}_0),$$

$$\alpha = \begin{pmatrix} \alpha_0 \\ 0 \\ 0 \\ \vdots \end{pmatrix} : \mathcal{H} \to l^2_{\mathcal{H}}(\mathbb{N}_0).$$

For $j = -1, -2, \ldots$, put now

$$\Gamma_j^* = -\Gamma_{j+1}^* A_1 - \Gamma_{j+2}^* A_2 - \cdots = -\sum_{k=1}^{\infty} \Gamma_{j+k}^* A_k. \tag{4.3.8}$$

Proof. Let

$$\widetilde{H} = \begin{pmatrix} & \Gamma_2 & \Gamma_1 \\ & \ddots & \Gamma_2 \\ & & \end{pmatrix}.$$

Then it follows from (4.3.5) that

$$\begin{pmatrix} I & \widetilde{H} \\ \widetilde{H}^* & I \end{pmatrix} \begin{pmatrix} B \\ \widetilde{D} \end{pmatrix} = -\begin{pmatrix} \Gamma \\ 0 \end{pmatrix},$$

where

$$\Gamma = \begin{pmatrix} \Gamma_0 \\ \Gamma_1 \\ \vdots \end{pmatrix}, \quad \widetilde{D} = \begin{pmatrix} \vdots \\ D_{-2} \\ D_{-1} \end{pmatrix}.$$

But then it follows that (4.3.6) is equivalent to the equation

$$\Gamma_j = \begin{pmatrix} 0 & Z_{j+1} \end{pmatrix} \begin{pmatrix} I & \widetilde{H} \\ \widetilde{H}^* & I \end{pmatrix}^{-1} \begin{pmatrix} \Gamma \\ 0 \end{pmatrix}, \; j \leq -1, \tag{4.3.9}$$

where

$$Z_k = \begin{pmatrix} \cdots & \Gamma_{k+1} & \Gamma_k \end{pmatrix}.$$

Let now

$$H_k = \begin{pmatrix} & \Gamma_{k+1} & \Gamma_k \\ & \ddots & \Gamma_{k+1} \\ & & \end{pmatrix}, \ k \in \mathbb{Z}.$$

Note that $H_0 = H$ and $H_1 = \widetilde{H}$. Then (4.3.9) implies that for $k \leq -1$

$$\begin{pmatrix} I & H_k \\ H_k^* & I \end{pmatrix}^{-1} = \begin{pmatrix} P_k & Q_k \\ Q_k^* & R_k \end{pmatrix},$$

where

$$Q_k = \begin{pmatrix} \cdots & Q_{12}^{(k)} & Q_{11}^{(k)} \\ \cdots & Q_{22}^{(k)} & Q_{21}^{(k)} \\ & \vdots & \vdots \end{pmatrix}$$

has the property that

$$Q_{ij}^{(k)} = 0, \ i + j \leq -k + 1. \tag{4.3.10}$$

Indeed, viewing Γ_j as the unknown, equation (4.3.9) is obtained from applying Theorem 2.2.3 to the partial matrices

$$\begin{pmatrix} I_{\mathcal{H}} & 0 & Z_{j+1} & ? \\ 0 & I & H_{j+2} & W_{j+1} \\ z_{j+1}^* & H_{j+2}^* & I & 0 \\ ? & W_{j+1}^* & 0 & I_{\mathcal{H}} \end{pmatrix}, \ j = -1, -2, \ldots,$$

inductively. Here

$$W_j = \begin{pmatrix} \Gamma_j \\ \Gamma_{j+1} \\ \vdots \end{pmatrix} : \mathcal{H} \to l^2_{\mathcal{H}}(j + \mathbb{N}_0).$$

Theorem 2.2.3 thus yields that $\|H_j\| < 1$, $j \in \mathbb{Z}$, and that $Q^{(k)}$ has the zero structure as described in (4.3.10). Next, let

$$S = I_{\mathcal{H}} - \begin{pmatrix} 0 & Z_0 \end{pmatrix} \begin{pmatrix} I & \widetilde{H} \\ \widetilde{H}^* & I \end{pmatrix}^{-1} \begin{pmatrix} 0 \\ Z_0^* \end{pmatrix}.$$

By the zero structure of $Q_{(k)}$ we get that for every $j \leq 0$ the Schur complement of

$$\begin{pmatrix} I & H_j \\ H_j^* & I \end{pmatrix} = \begin{pmatrix} I_{\mathcal{H}} & 0 & Z_j \\ 0 & I & H_{j-1} \\ Z_j^* & H_{j-1}^* & I \end{pmatrix}$$

supported in the upper left corner equals the positive definite operator S, which is independent of j. But then it also follows that $\|(\Gamma_{i-j})_{i,j\in\mathbb{Z}}\| < 1$, and thus $f(z) = \sum_{j=-\infty}^{\infty} \Gamma_j z^j$ satisfies $\|f\|_\infty < 1$.

For the alternative construction of Γ_j, use that (4.3.7) implies that

$$\begin{pmatrix} I & \widetilde{H} \\ \widetilde{H}^* & I \end{pmatrix} \begin{pmatrix} \widetilde{A} \\ C \end{pmatrix} = - \begin{pmatrix} 0 \\ \widehat{\Gamma} \end{pmatrix},$$

where

$$\widehat{\Gamma} = \begin{pmatrix} \vdots \\ \Gamma_1^* \\ \Gamma_0^* \end{pmatrix}, \quad \widetilde{A} = \begin{pmatrix} A_1 \\ A_2 \\ \vdots \end{pmatrix}.$$

But (4.3.8) is equivalent to the equality

$$\Gamma_j^* = \begin{pmatrix} \widehat{Z}_{j+1} & 0 \end{pmatrix} \begin{pmatrix} I & \widetilde{H} \\ \widetilde{H}^* & I \end{pmatrix}^{-1} \begin{pmatrix} 0 \\ \widehat{\Gamma} \end{pmatrix},$$

with

$$\widehat{Z}_k = \begin{pmatrix} \Gamma_k & \Gamma_{k+1} & \cdots \end{pmatrix}.$$

This yields the same sequence of operators Γ_k, $k \leq -1$, as in (4.3.9). □

The following is an immediate consequence of Theorem 4.3.1.

Corollary 4.3.3 *For every* $k \in L^\infty(\mathcal{L}(\mathcal{H}))$, $\operatorname{dist}(k, H^\infty(\mathcal{L}(\mathcal{H}))) = \|H_k\|$. *Moreover, the distance is attained.*

Proof. If $k \in L^\infty(\mathcal{L}(\mathcal{H}))$ and $f \in H^\infty(\mathcal{L}(\mathcal{H}))$, then $H_{k-f} = H_k$, consequently,

$$\|H_k\| = \|H_{k-f}\| \leq \|M_{k-f}\| = \|k - f\|_\infty,$$

and thus $\|H_k\| \leq \operatorname{dist}(k, H^\infty(\mathcal{L}(\mathcal{H})))$. If $k \in L^\infty(\mathcal{L}(\mathcal{H}))$, by Theorem 4.3.1 there exists $K \in L^\infty(\mathcal{L}(\mathcal{H}))$ such that the Fourier coefficients of k and K coincide for $n \leq 0$ and $\|K\| = \|H_k\|$. This implies that $k - K \in H^\infty(\mathcal{L}(\mathcal{H}))$, so $\operatorname{dist}(k, H^\infty(\mathcal{L}(\mathcal{H}))) = \|H_k\|$. □

In order to present an alternative proof for the scalar version of Theorem 4.3.1, let $L^p(\mathbb{T})$ and $H^p(\mathbb{T})$, $1 \leq p \leq \infty$, denote the Lebesgue and Hardy spaces of scalar-valued functions defined on $\mathbb{T}$. We need the following well-known factorization result.

Theorem 4.3.4 *For every* $f \in H^1(\mathbb{T})$ *there exist* $f_1, f_2 \in H^2(\mathbb{T})$, *such that* $f = f_1 f_2$, *and* $\|f_1\|_2^2 = \|f_2\|_2^2 = \|f\|_1$.

Second proof for the scalar version of Theorem 4.3.1. Let $\{c_n\}_{n=-\infty}^{-1}$ be a sequence of complex numbers such that $\|H(\{c_n\}_{n=-\infty}^{-1})\| \leq 1$, and define the linear functional L on $H_0^1(\mathbb{T}) = \{f \in H^1(\mathbb{T}) : f(0) = 0\}$ by $L(f) = \sum_{n=1}^{\infty} c_{-n} d_n$, where $f(e^{it}) = \sum_{n=1}^{\infty} d_n e^{int}$. By Theorem 4.3.4, for every given $f \in H_0^1(\mathbb{T})$ there exist $f_1 \in H^2(\mathbb{T})$ and $f_2 \in H^2(\mathbb{T})$, with $f = f_1 f_2$, $f_1(0) = 0$, and $\|f_1\|_2^2 = \|f_2\|_2^2 = \|f\|_1$. Define $\widetilde{f_1}(e^{it}) = f(e^{-it})$, and $\widehat{f_2}(e^{it}) = \overline{f_2(e^{-it})}$. A straightforward computation shows that $L(f) = \left\langle H(\{c_n\}_{n=-\infty}^{-1})\widetilde{f_1}, \widehat{f_2}\right\rangle_{L^2(\mathbb{T})}$; thus

$$|L(f)| \leq \|H(\{c_n\}_{n=-\infty}^{-1})\| \cdot \|f_1\|_2 \cdot \|f_2\|_2 = \|H(\{c_n\}_{n=-\infty}^{-1})\| \cdot \|f\|_1.$$

By the Hahn-Banach theorem, L can be extended to a linear functional $\widetilde{L}$ on $L^1(\mathbb{T})$ with $\|\widetilde{L}\| \leq 1$. So there exists $g \in L^\infty(\mathbb{T})$, such that $\widetilde{L}(f) = \frac{1}{2\pi}\int_0^{2\pi} f(e^{it})g(e^{it})dt$, and $\|g\|_\infty \leq 1$. For each $n \geq 1$,

$$c_{-n} = L(e^{int}) = \widetilde{L}(e^{int}) = \frac{1}{2\pi}\int_0^{2\pi} g(e^{it})e^{int}dt;$$

thus c_{-n} is the Fourier coefficient of g corresponding to $(-n)$, so g is a solution to Nehari's problem. □

Let $r = \frac{p}{q}$ be a rational function where p and q are polynomials. If $\frac{p}{q}$ is in its lowest terms, the degree of r is by definition

$$\deg r = \max\{\deg p, \deg q\},$$

where $\deg p$ and $\deg q$ are the degrees of the polynomials p and q. It is easy to see that $\deg r$ is the sum of the multiplicities of the poles of r (including a possible pole at infinity).

For proving the next result we identify sequences $a = \{a_j\}_{j\geq 0}$ of complex numbers with formal power series

$$a(z) = \sum_{j=0}^{\infty} a_j z^j.$$

Consider the shift operator S and its adjoint S^* defined on the space of formal series in the following way:

$$(Sa)(z) = \sum_{j=0}^{\infty} a_j z^{j+1}, \quad (S^*a)(z) = \sum_{j=0}^{\infty} a_{j+1} z^j.$$

The space of all formal series $a(z) = \sum_{j=0}^{\infty} a_j z^j$ forms an algebra with respect to the usual multiplication. It is easy to see that $a(z) = \sum_{j=0}^{\infty} a_j z^j$ is invertible in the space of formal series if and only if $a_0 \neq 0$.

Next we describe the Hankel operators of finite rank without any assumption on the boundedness of the operator. The result is known as *Kronecker's theorem.*

Theorem 4.3.5 *Given a sequence of complex numbers $\{c_n\}_{n=-\infty}^{-1}$, the Hankel operator $H(\{c_n\}_{n=-\infty}^{-1})$ has a finite rank if and only if the power series*

$$c(z) = \sum_{n=0}^{\infty} c_{-n-1} z^n$$

defines a rational function. In this case

$$\text{rank } H(\{c_n\}_{n=-\infty}^{-1}) = \deg zc(z).$$

Proof. Assuming that $\operatorname{rank} H(\{c_n\}_{n=-\infty}^{-1}) = k$, its first $k+1$ rows are linearly dependent. It means that there exists a family $\{\alpha_j\}_{j=0}^{k}$ of complex numbers, not all of them 0, such that

$$\alpha_0 c + \alpha_1 S^* c + \cdots + \alpha_k S^{*k} c = 0. \tag{4.3.11}$$

It is easy to see that

$$S^k S^{*l} c = S^{k-l} c - S^{k-l} \sum_{j=0}^{l-1} c_j z^j, \quad l \le k. \tag{4.3.12}$$

It follows easily from (4.3.11) and (4.3.12) that

$$0 = S^k \sum_{l=0}^{k} c_l S^{*l} c = \sum_{l=0}^{k} \alpha_l S^k S^{*l} c = \sum_{l=0}^{k} \alpha_l S^{k-l} c - p, \tag{4.3.13}$$

where p has the form $p(z) = \sum_{j=0}^{k-1} p_j z^j$. Put

$$q(z) = \sum_{j=0}^{k} \alpha_{k-j} z^j. \tag{4.3.14}$$

Then p and q are polynomials, and it follows from (4.3.13) that $qc = p$. Note that $\alpha_k \ne 0$, since otherwise the rank of $H(\{c_n\}_{n=-\infty}^{-1})$ would be less than k. Thus we can divide by q within the space of formal power series. Hence, $c(z) = \frac{p(z)}{q(z)}$ is a rational function. Clearly,

$$\deg zc(z) \le \max\{\deg zp(z), \deg q(z)\} = k.$$

Conversely, suppose that $c(z) = \frac{p(z)}{q(z)}$, where p and q are polynomials such that $\deg p \le k-1$ and $\deg q \le k$. Consider the complex numbers α_j defined by (4.3.14). We have then

$$\sum_{j=0}^{k} \alpha_j S^{k-j} c = p.$$

Therefore

$$S^{*k} \sum_{j=0}^{k} \alpha_j S^{k-j} c = \sum_{j=0}^{k} \alpha_j S^{*j} c = 0,$$

which means that the first $k+1$ rows of $H(\{c_n\}_{n=-\infty}^{-1})$ are linearly dependent. Let $m \le k$ be the largest number for which $c_m \ne 0$. Then $S^{*m} c$ is a linear combination of the $S^{*j} c$ with $0 \le j \le m-1$:

$$S^{*m} c = \sum_{j=0}^{m-1} d_j S^{*j} c, \quad d_j \in \mathbb{C}.$$

We prove by induction that any row of $H(\{c_n\}_{n=-\infty}^{-1})$ is a linear combination of the first m rows. Let $l > m$. We have

$$S^{*l} c = S^{*(l-m)} S^{*m} c = \sum_{j=0}^{m-1} d_j S^{*(l-m+j)} c. \tag{4.3.15}$$

Since $l - m + j < l$ for $0 \leq j \leq m - 1$, by the inductive hypothesis each of the terms on the right-hand side of (4.3.15) is a linear combination of the first m rows. Therefore, rank $H(\{c_n\}_{n=-\infty}^{-1}) \leq m$, and this completes the proof. □

For Hankel operators H_f on $H^2(\mathbb{T})$ Kronecker's theorem can immediately be reformulated as follows.

Corollary 4.3.6 *Let $f \in L^\infty(\mathbb{T})$. Then H_f has finite rank if and only if $P_- f$ is a rational function. In this case*

$$\text{rank } H_f = \deg P_- f.$$

Let $C(\mathbb{T})$ denote the set of all continuous complex-valued functions on $\mathbb{T}$, while $C(\mathbb{C}^{m \times n})$ denotes the set of all continuous functions from $\mathbb{T}$ to $\mathbb{C}^{m \times n}$. The linear span $H^\infty(\mathbb{T}) + C(\mathbb{T})$, is abbreviated as $H^\infty + C$, while $H^\infty(\mathbb{C}^{m \times n}) + C(\mathbb{C}^{m \times n})$ as $(H^\infty + C)(\mathbb{C}^{m \times n})$. Next, let A denote the uniform closure of the set of all analytic polynomials on $\mathbb{T}$. These spaces play a great role in the theory of Hankel operators. First of all, following Exercises 4.13.21 and 4.13.22, the natural mapping embedding $C(\mathbb{T})/A$ into its bidual $L^\infty(\mathbb{T})/H^\infty(\mathbb{T})$ is $i(f + A) = f + H^\infty(\mathbb{T})$. Since the natural map is an isometry, it follows that its image is a closed subspace of $L^\infty(\mathbb{T})/H^\infty(\mathbb{T})$. Hence, the inverse image of this latter subspace under the natural homomorphism of $L^\infty(\mathbb{T})$ onto $L^\infty(\mathbb{T})/H^\infty(\mathbb{T})$ is closed, therefore $H^\infty + C$ is a closed subspace of $L^\infty(\mathbb{T})$.

The following result, known as Hartman's theorem, is a description of the scalar-valued Hankel operators that are compact. A similar characterization holds for matrix-valued symbols; see Exercise 4.13.23.

Theorem 4.3.7 *Let $f \in L^\infty(\mathbb{T})$. Then H_f is compact if and only if $f \in H^\infty + C$.*

Proof. Let first $f = f_1 + f_2$ with $f_1 \in H^\infty(\mathbb{T})$ and $f_2 \in C(\mathbb{T})$. Let p_n be a sequence of trigonometric polynomials such that $p_n \to f_2$ uniformly on $\mathbb{T}$. Then $H_f = H_{f_2} = \lim_{n \to \infty} H_{p_n}$ (the limit is in the operator norm). It is clear that each H_{p_n} is finite rank; thus H_f is compact, being the norm limit of finite rank operators.

Assume now $f(e^{it}) = \sum_{n=-\infty}^{\infty} c_n e^{int} \in L^\infty(\mathbb{T})$ such that H_f is compact. Let $p_n(e^{it}) = \sum_{k=-n}^{-1} c_k e^{ikt}$. Since H_f is compact, it is the norm limit of H_{p_n}. Thus for each $\epsilon > 0$ there exists $N \in \mathbb{N}$ such that $\|H_{p_n - f}\| < \epsilon$ for $n \geq N$. By Nehari's theorem (Theorem 4.3.1), there exists $q_n \in H^\infty(\mathbb{T})$ such that $\|p_n - f - q_n\| \leq \epsilon$ for $n \geq N$. Since $H^\infty + C$ is closed, we have that $f \in H^\infty + C$. □

Given $T \in \mathcal{L}(\mathcal{E}, \mathcal{F})$, we refer to the unit vectors $e \in \mathcal{E}$ and $f \in \mathcal{F}$ as a *maximizing pair of unit vectors for T* if $Te = \|T\| f$. It is well known that such vectors always exist when T is compact.

Theorem 4.3.8 *Assume $G \in L^\infty(\mathbb{C}^{m\times n})$ is such that H_G admits a maximizing pair of unit vectors v and w. Let $Q \in H^\infty(\mathbb{C}^{m\times n})$ be at minimal distance from G (which exists because of Corollary 4.3.3). Then*

$$(G-Q)v = \|H_G\| w \tag{4.3.16}$$

and

$$\|w(z)\|_{\mathbb{C}^m} = \|v(z)\|_{\mathbb{C}^n} \quad \text{a.e. on } \mathbb{T}. \tag{4.3.17}$$

Furthermore,

$$\|G(z)-Q(z)\| = \|H_G\| \quad \text{a.e. on } \mathbb{T}. \tag{4.3.18}$$

Proof. From Nehari's theorem and the optimality of Q, it immediately follows that

$$\begin{aligned}\|H_G\| = \|H_G v\| = \|H_{G-Q}v\| &= \|P_-(G-Q)v\| \\ \le \|(G-Q)v\| \le \|G-Q\|_\infty &= \|H_G\|.\end{aligned}$$

Equality must hold throughout; in particular

$$\|P_-(G-Q)v\| = \|(G-Q)v\|,$$

implying $(G-Q)v \perp H^2(\mathbb{C}^m)$, so

$$H_G v = P_-(G-Q)v = (G-Q)v,$$

which is (4.3.16). Moreover,

$$\|(G-Q)v\| = \|G-Q\|_\infty \|v\|.$$

But then

$$\int \|w(z)\|^2 = \|G-Q\|_\infty^2 \int \|v(z)\|^2.$$

Combining this with the observation that

$$\|w(z)\| \le \|G-Q\|_\infty \|v(z)\| \text{ a.e.}$$

yields (4.3.17). Now $v(z)$ is a maximizing vector for $G(z)-Q(z)$ for $z \in \mathbb{T}$ a.e., so (4.3.18) follows. □

Corollary 4.3.9 *Let $G \in (H^\infty + C)(\mathbb{C}^{m\times n})$. If either $m=1$ or $n=1$, then there exists a unique best approximation $Q \in H^\infty(\mathbb{C}^{m\times n})$ to G and $\|(G-Q)(z)\|$ is constant a.e. on $\mathbb{T}$.*

Proof. Suppose $n=1$. Since H_G is compact by Theorem 4.3.7, it has a maximizing pair of unit vectors v and w. Since $v \in H^2(\mathbb{T})$, it is known that $v(z) \neq 0$ a.e. on $\mathbb{T}$. Next observe that $\frac{w}{v}$ is independent of the choice of maximizing pair of unit vectors for H_G. Indeed, if v_1 and w_1 is another one, we get the following relationship between the kth Fourier coefficients. For $k \in \mathbb{N}_0$ we have

$$\begin{aligned}\widehat{v_1^* v}(-k) = \langle z^k v, v_1\rangle &= \frac{1}{\|H_G\|}\langle z^k v, H_G^* w_1\rangle = \frac{1}{\|H_G\|}\langle H_G z^k v, w_1\rangle \\ &= \frac{1}{\|H_G\|}\langle P_-(z^k H_G v), w_1\rangle = \langle z^k w, w_1\rangle = \widehat{w_1^* w}(-k).\end{aligned}$$

Similarly, for $k \in \mathbb{N}_0$ we have $\widehat{v_1^* v}(-k) = \widehat{w_1^* w}(-k)$, and thus

$$v_1^* v = w_1^* w. \tag{4.3.19}$$

As $\|v_1^* v\| = \|v_1\|\|v\| = 1$ (where we use that they are scalars), and $\|w\| = 1 = \|w_1\|$, we get from (4.3.19) that $\|w_1^* w\| = \|w_1\|\|w\|$. But this can only happen if w_1 is a scalar multiple of w, and by (4.3.19) this scalar multiple must be $\frac{v_1}{v}$. But now we get

$$Q = G - \frac{\|H_G\| w}{v} = G - \frac{\|H_G\| w_1}{v_1},$$

yielding the uniqueness. The case $m = 1$ follows from a similar consideration of G^T. □

The following is referred to in the literature as the (contractive) Carathéodory-Fejér problem. Given a sequence $\{C_k\}_{k=0}^n$ of operators in $\mathcal{L}(\mathcal{H})$, find necessary and sufficient conditions for the existence of a $F \in H^\infty(\mathcal{L}(\mathcal{H}))$ such that $F(z) = \sum_{k=0}^{n} C_k z^k + O(z^{n+1})$ and $\|F\|_\infty \leq 1$.

The (truncated) Toeplitz operator built on a sequence of operators $\{C_k\}_{k=0}^n$ is defined by the formula

$$T(\{C_k\}_{k=0}^n) = \begin{pmatrix} C_0 & 0 & 0 & \cdots & 0 \\ C_1 & C_0 & 0 & \cdots & 0 \\ C_2 & C_1 & C_0 & \cdots & 0 \\ \vdots & \vdots & \ddots & \ddots & \vdots \\ C_n & C_{n-1} & C_{n-2} & \cdots & C_0 \end{pmatrix}. \tag{4.3.20}$$

Theorem 4.3.10 *For a given sequence of operators $\{C_k\}_{k=0}^n$ in $\mathcal{L}(\mathcal{H})$, the Carathéodory-Fejér problem admits a solution if and only if $\|T(\{C_k\}_{k=0}^n)\| \leq 1$.*

Proof. If the Carathéodory-Fejér problem admits a solution F, then the operator $T(\{C_k\}_{k=0}^n)$ is the upper left corner of the matrix representing the operator T_F. Thus $\|T(\{C_k\}_{k=0}^n)\| \leq \|T_F\| = \|F\|_\infty \leq 1$.

For the sufficiency part, assume that $\|T(\{C_k\}_{k=0}^n)\| \leq 1$ and define the sequence of operators $\{D_k\}_{k=-\infty}^{-1}$ in $\mathcal{L}(\mathcal{H})$ by: $D_k = 0$ for $k \leq -(n+2)$ and $D_k = C_{k+n+1}$ for $-(n+1) \leq k \leq -1$. In this case the matrix representation Hankel operator $H(\{D_k\}_{k=-\infty}^{-1})$ coincides with that of $T(\{C_k\}_{k=0}^n)$ bordered with zeros. Thus these operators have the same norm, so there exists $G \in L^\infty(\mathcal{L}(\mathcal{H}))$,

$$G(e^{it}) = \sum_{k=0}^{n} C_k e^{i(k-n-1)t} + \sum_{k=0}^{\infty} G_k e^{ikt},$$

and $\|G\|_\infty \leq 1$. Consequently, $F(z) = \sum_{k=0}^n C_k z^k + \sum_{k=n+1}^\infty G_k e^{ikt} \in H^\infty(\mathcal{L}(\mathcal{H}))$ has $\|F\|_\infty \leq 1$, and so it is a solution of the contractive Carathéodory-Fejér problem. □

4.4 NEHARI'S PROBLEM IN TWO VARIABLES

In one variable the Hankel operators appear as operators acting H^2 to $(H^2)^\perp$. For functions in two variables there are the following two natural generalizations, resulting in the "big" Hankel and the "small" Hankel.

One can identify $L^2(\mathbb{T}^2)$ with $L^2(\mathbb{T}) \otimes L^2(\mathbb{T})$. Now consider the decompositions

$$L^2(\mathbb{T}^2) = H^2(\mathbb{T}^2) \oplus H^2(\mathbb{T}^2)^\perp = (H^2(\mathbb{T}) \otimes H^2(\mathbb{T})) \oplus (H^2(\mathbb{T})^\perp \otimes H^2(\mathbb{T})) \oplus (H^2(\mathbb{T}) \otimes H^2(\mathbb{T})^\perp) \oplus (H^2(\mathbb{T})^\perp \otimes H^2(\mathbb{T})^\perp)$$

The *big Hankel* operators act $H^2(\mathbb{T}^2) \to H^2(\mathbb{T}^2)^\perp$, while the *small Hankel operators* act $H^2(\mathbb{T}) \otimes H^2(\mathbb{T}) \to H^2(\mathbb{T})^\perp \otimes H^2(\mathbb{T})^\perp$. Differently put, if we write

$$L^2(\mathbb{T}^2) = \left\{ \sum_{i,j=-\infty}^{\infty} f_{ij} z^i w^j \ : \ \sum_{i,j=-\infty}^{\infty} \|f_{ij}\|^2 < \infty \right\},$$

then a big Hankel operator acts

$$\left\{ \sum_{i,j=0}^{\infty} f_{ij} \ : \ \sum_{i,j=-0}^{\infty} \|f_{ij}\|^2 < \infty \right\} \to \left\{ \sum_{i,j \in \mathbb{Z}^2 \setminus \mathbb{N}_0^2} f_{ij} \ : \ \sum_{i,j \in \mathbb{Z}^2 \setminus \mathbb{N}_0^2} \|f_{ij}\|^2 < \infty \right\},$$

while the small Hankel operator acts

$$\left\{ \sum_{i,j=0}^{\infty} f_{ij} \ : \ \sum_{i,j=-0}^{\infty} \|f_{ij}\|^2 < \infty \right\} \to \left\{ \sum_{i,j=-\infty}^{-1} f_{ij} \ : \ \sum_{i,j=-\infty}^{-1} \|f_{ij}\|^2 < \infty \right\}.$$

As it turns out, a bounded big Hankel operator does not always have a bounded symbol. The following result shows this.

Theorem 4.4.1 *Consider $\phi(z_1, z_2) = \log(1 - \frac{z_2}{z_1})$, and define the big Hankel operator*

$$\Gamma : H^2(\mathbb{T}^2) \to H^2(\mathbb{T}^2)^\perp, \ \Gamma(f) = P_{H^2(\mathbb{T}^2)^\perp}(\phi f).$$

Then $\|\Gamma\| \leq \pi$, but there does not exist a $\psi \in L^\infty(\mathbb{T}^2)$ such that $\Gamma(f) = P_{H^2(\mathbb{T}^2)^\perp}(\psi f)$.

We first need the following proposition.

Proposition 4.4.2 *Let $M = (m_{ij})_{i,j=-\infty}^{\infty}$, where $m_{ij} = \frac{1}{i-j}$, $i \neq j$, and $m_{ii} = 0, i \in \mathbb{Z}$. Then $\|M\| \leq \pi$ (viewed as an operator on l^2).*

Proof. Notice that M is skewadjoint, so $\|M^2\| = \|M\|^2$. The operator M^2 has a matrix $(b_{ij})_{i,j \in \mathbb{Z}}$, where

$$b_{ij} = \sum_{k=-\infty}^{\infty} m_{ik} m_{kj} = \sum_{k \neq i,j} \frac{1}{(i-k)(k-j)} = \sum_{n \neq 0, j-i} \frac{-1}{n(n-j+i)}.$$

Thus

$$b_{ii} = -\sum_{k\neq 0} \frac{1}{k^2} = -2\sum_{k=1}^{\infty} \frac{1}{k^2} = -\frac{\pi^2}{3}.$$

If $j - i = r \neq 0$, then

$$\begin{aligned} b_{ij} &= -\sum_{n>r} \frac{1}{n(n-r)} - \sum_{n=1}^{r-1} \frac{1}{n(n-r)} - \sum_{n<0} \frac{1}{n(n-r)} \\ &= -2\sum_{n=r+1}^{\infty} \frac{1}{r}\left(\frac{1}{n-r} - \frac{1}{n}\right) - \sum_{n=1}^{r-1} \frac{1}{r}\left(\frac{1}{n-r} - \frac{1}{n}\right) \\ &= -\frac{2}{r}\sum_{k=1}^{r} \frac{1}{k} + \frac{2}{r}\sum_{n=1}^{r-1} \frac{1}{n} = -\frac{2}{r^2}. \end{aligned}$$

Thus $M^2 = -\frac{\pi^2}{3} I - \sum_{k\neq 0} \frac{2}{k^2} U^k$, where U is the bilateral shift ($Ue_i = e_{i+1}$, where e_i is the ith standard basis vector in l^2). So

$$\|M^2\| \leq \frac{\pi^2}{3} + \sum_{k\neq 0} \frac{2}{k^2} = \frac{\pi^2}{3} + 4\sum_{n=1}^{\infty} \frac{1}{n^2} = \pi^2.$$

□

Proof of Theorem 4.4.1. Let ψ be so that $\Gamma(f) = P_{H^2(\mathbb{T}^2)^\perp}(\psi f)$ for all f. Then, there exist analytic $b_i(z_2)$ such that

$$\psi(z_1, z_2) = \phi(z_1, z_2) + \sum_{i=0}^{\infty} b_i(z_2) z_1^i = \sum_{i=1}^{\infty} \frac{1}{i} \frac{z_2^i}{z_1^i} + \sum_{i=0}^{\infty} b_i(z_2) z_1^i,$$

and thus

$$\sup_{|z_1|=|z_2|=1} \psi(z_1, z_2) = \sup_{|z_2|=1} \left\| \begin{pmatrix} b_0(z_2) & z_2 & \frac{z_2^2}{2} & \cdots \\ b_1(z_2) & b_0(z_2) & z_2 & \ddots \\ \vdots & \ddots & \ddots & \ddots \end{pmatrix} \right\|.$$

If we multiply this Toeplitz matrix with the unitary matrix $\operatorname{diag}(z_2^i)_{i=0}^{\infty}$ on the left, and with its adjoint $\operatorname{diag}(z_2^{-i})_{i=0}^{\infty}$ on the right, we do not change the norm. This yields

$$\begin{aligned} \sup_{|z_1|=|z_2|=1} \psi(z_1, z_2) &= \sup_{|z_2|=1} \left\| \begin{pmatrix} b_0(z_2) & 1 & \frac{1}{2} & \cdots \\ z_2 b_1(z_2) & b_0(z_2) & 1 & \ddots \\ \vdots & \ddots & \ddots & \ddots \end{pmatrix} \right\| \\ &=: \sup_{|z_2|=1} \|\Psi(z_2)\|. \end{aligned}$$

Notice that $\Psi(z_2)$ is analytic, and thus $\|\Psi(0)\| \leq \sup_{|z_2|=1} \|\Psi(z_2)\|$. Next, notice that

$$\Psi(0) - b_0(0)I = \begin{pmatrix} 0 & 1 & \frac{1}{2} & \cdots \\ 0 & 0 & 1 & \ddots \\ \vdots & \ddots & \ddots & \ddots \end{pmatrix},$$

which is the one-variable Toeplitz matrix with unbounded symbol $\log(1-z)$. Thus $\Psi(0) - b_0(0)I$ is not bounded, and therefore $\Psi(0)$ is unbounded. This yields that ψ cannot be in $L^\infty(\mathbb{T}^2)$.

To show that Γ is bounded, observe that

$$\overline{\text{Ran}\Gamma} \subseteq \text{Span}\{z_1^{-i} z_2^j \ : \ j \geq i \geq 1\} = \bigoplus_{k=0}^{\infty} M_k,$$

where

$$M_k = \text{Span}\{z_1^{-i} z_2^{i+k} \ : \ i \in \mathbb{N}\}.$$

Writing

$$\Gamma = (H_{kj})_{k,j=0}^{\infty} : \bigoplus_{j=0}^{\infty} \text{Span}\{z_1^i z_2^{j-i} \ : \ i = 0, \ldots, j\} \to \bigoplus_{k=0}^{\infty} M_k,$$

it is straightforward to check that $H_{kj} = 0$, $k \neq j$, and

$$H_{kk} = \begin{pmatrix} 1 & \frac{1}{2} & \cdots & \frac{1}{k+1} \\ \frac{1}{2} & & \iddots & \frac{1}{k+2} \\ \vdots & \iddots & \iddots & \vdots \\ \frac{1}{k+1} & \iddots & & \vdots \\ \vdots & & & \vdots \end{pmatrix}.$$

As each H_{kk} is a submatrix of the matrix M in Proposition 4.4.2, we have that $\|H_{kk}\| \leq \pi$, $k \in \mathbb{N}_0$. Thus $\|\Gamma\| \leq \pi$ follows. $\square$

We now come to the main result in this section, which concerns the little Hankel operator.

Theorem 4.4.3 *Let $\gamma_{ij} \in \mathcal{L}(\mathcal{H}, \mathcal{K})$, $i, j \geq 0$, be given so that the little Hankel operator $h_\gamma : l^2_{\mathcal{K}}(-\mathbb{N}_0 \times -\mathbb{N}_0) \to l^2_{\mathcal{H}}(\mathbb{N}_0 \times \mathbb{N}_0)$ defined via*

$$h_\gamma = \begin{pmatrix} & \Gamma_1 & \Gamma_0 \\ & \ddots & \Gamma_1 \\ & & \end{pmatrix}, \ \Gamma_j = \begin{pmatrix} & \gamma_{j1} & \gamma_{j0} \\ & \ddots & \gamma_{j1} \\ & & \end{pmatrix},$$

is a strict contraction. Put

$$\Phi = P_{\mathbb{N}_0\times\mathbb{N}} \oplus P_{-\mathbb{N}\times-\mathbb{N}_0} \begin{pmatrix} I & h_\gamma \\ h_\gamma^* & I \end{pmatrix} P^*_{\mathbb{N}_0\times\mathbb{N}} \oplus P^*_{-\mathbb{N}\times-\mathbb{N}_0}$$

$$= P_{\mathbb{N}\times\mathbb{N}_0} \oplus P_{-\mathbb{N}_0\times-\mathbb{N}} \begin{pmatrix} I & h_\gamma \\ h_\gamma^* & I \end{pmatrix} P^*_{\mathbb{N}\times\mathbb{N}_0} \oplus P^*_{-\mathbb{N}_0\times-\mathbb{N}},$$

$$\Phi_1 = P_{\mathbb{N}_0\times\mathbb{N}_0} \oplus P_{-\mathbb{N}\times-\mathbb{N}} \begin{pmatrix} I & h_\gamma \\ h_\gamma^* & I \end{pmatrix} P^*_{\mathbb{N}\times\mathbb{N}_0} \oplus P^*_{-\mathbb{N}_0\times-\mathbb{N}}$$

$$= P_{\mathbb{N}_0\times\mathbb{N}} \oplus P_{-\mathbb{N}\times-\mathbb{N}_0} \begin{pmatrix} I & h_\gamma \\ h_\gamma^* & I \end{pmatrix} P^*_{\mathbb{N}\times\mathbb{N}} \oplus P^*_{-\mathbb{N}_0\times-\mathbb{N}_0},$$

$$\Phi_2 = P_{\mathbb{N}_0\times\mathbb{N}_0} \oplus P_{-\mathbb{N}\times-\mathbb{N}} \begin{pmatrix} I & h_\gamma \\ h_\gamma^* & I \end{pmatrix} P^*_{\mathbb{N}_0\times\mathbb{N}} \oplus P^*_{-\mathbb{N}\times-\mathbb{N}_0}$$

$$= P_{\mathbb{N}\times\mathbb{N}_0} \oplus P_{-\mathbb{N}_0\times-\mathbb{N}} \begin{pmatrix} I & h_\gamma \\ h_\gamma^* & I \end{pmatrix} P^*_{\mathbb{N}\times\mathbb{N}} \oplus P^*_{-\mathbb{N}_0\times-\mathbb{N}_0},$$

where the projection $P_K : l^2(M) \to l^2(K)$, $K \subseteq M$, *is defined by* $P_K((\eta_j)_{j\in M}) = (\eta_j)_{j\in K}$. *Suppose that*

$$\Phi_1\Phi^{-1}\Phi_2^* = \Phi_2^*\Phi^{-1}\Phi_1. \tag{4.4.1}$$

Then there exist $\gamma_{ij} \in \mathcal{L}(\mathcal{H})$, $(i,j) \in (\mathbb{Z}\times -\mathbb{N}) \cup (-\mathbb{N}\times\mathbb{Z})$, *such that the operator matrix*

$$(\gamma_{i-j,k-l})_{i,j,k,l\in\mathbb{Z}} : l^2_{\mathcal{K}}(\mathbb{Z}\times\mathbb{Z}) \to l^2_{\mathcal{H}}(\mathbb{Z}\times\mathbb{Z})$$

is a strict contraction. Equivalently, the essentially bounded function $f(z) = \sum_{i,j\in\mathbb{Z}} \gamma_{ij} z^i w^j$ *satisfies* $\|f\|_\infty < 1$.

Proof. We start by applying Theorem 4.3.2 to construct Γ_j, $j \le -1$, via (4.3.6) or, equivalently, (4.3.8), yielding the strict contraction

$$(\Gamma_{i-j})_{i,j\in\mathbb{Z}} : l^2_{l^2_{\mathcal{K}}(-\mathbb{N}_0)}(\mathbb{Z}) \to l^2_{l^2_{\mathcal{H}}(\mathbb{N}_0)}(\mathbb{Z}).$$

The main step in the proof is to show that (4.4.1) implies that Γ_j, $j \le -1$, are also Hankel; that is, they are of the form

$$\Gamma_j = \begin{pmatrix} & \gamma_{j1} & \gamma_{j0} \\ & \ddots & \gamma_{j1} \\ & & \end{pmatrix}, \quad j \le -1,$$

for some operators γ_{ij}, $j \ge 0$, $i \le -1$. To show this we need to prove the following claim.

Claim. *Equation* (4.4.1) *implies that the operators* D_j, $j \le -1$, *in* (4.3.5) *are of the form*

$$D_j = \begin{pmatrix} & \vdots & \vdots & \vdots \\ \cdots & * & * & * \\ \cdots & * & * & * \\ \cdots & 0 & 0 & * \end{pmatrix} : l^2_{\mathcal{K}}(-\mathbb{N}_0) \to l^2_{\mathcal{K}}(-\mathbb{N}_0).$$

Similarly, (4.4.1) *implies that* A_j *in* (4.3.7) *is of the form*

$$A_j = \begin{pmatrix} * & 0 & 0 & \cdots \\ * & * & * & \cdots \\ * & * & * & \cdots \\ \vdots & \vdots & \vdots & \end{pmatrix} : l^2_{\mathcal{H}}(\mathbb{N}_0) \to l^2_{\mathcal{H}}(\mathbb{N}_0).$$

Proof of Claim. It is not hard to see that $\Phi_i\Phi^{-1}$ and $\Phi^{-1}\Phi_i$, $i = 1, 2$, have a certain companion type form (variations of the ones in the proof of Theorem 4.4.3). For instance,

$$\Phi_1\Phi^{-1} = \begin{pmatrix} \widehat{S} & Q \\ 0 & S \end{pmatrix}, \quad \Phi^{-1}\Phi_1 = \begin{pmatrix} Z & \widehat{Q} \\ 0 & \widehat{Z} \end{pmatrix},$$

where $\widehat{S}$ and $\widehat{Z}$ have an infinite companion form

$$\widehat{S} = \begin{pmatrix} * & * & \cdots & \\ I & & & \\ & I & & \\ & & \ddots & \end{pmatrix}, \quad \widehat{Z} = \begin{pmatrix} \ddots & & \vdots \\ & I & * \\ & & I & * \end{pmatrix},$$

the operators S and Z are shifts

$$S = \begin{pmatrix} \ddots & & \vdots \\ & I & 0 \\ & & I & 0 \end{pmatrix}, \quad Z = \begin{pmatrix} 0 & 0 & \cdots & \\ I & & & \\ & I & & \\ & & \ddots & \end{pmatrix},$$

and Q and $\widehat{Q}$ are zero except for the first block row and last block, respectively:

$$Q = \begin{pmatrix} * & * & \cdots \\ 0 & 0 & \cdots \\ \vdots & \vdots & \end{pmatrix}, \quad \widehat{Q} = \begin{pmatrix} & \vdots & \vdots \\ \cdots & 0 & * \\ \cdots & 0 & * \end{pmatrix}.$$

But then, viewing $R := \Phi_2^*\Phi^{-1}\Phi_1 = \Phi_1\Phi^{-1}\Phi_2^*$ in the four possible ways $(\Phi_2^*\Phi^{-1})\Phi_1$, $\Phi_2^*(\Phi^{-1}\Phi_1)$, $(\Phi_1\Phi^{-1})\Phi_2^*$, $\Phi_1(\Phi^{-1}\Phi_2^*)$ one easily deduces that

$$\Phi_1\Phi^{-1}\Phi_2^* = \Phi_2^*\Phi^{-1}\Phi_1 = P_{\mathbb{N}_0\times\mathbb{N}} \oplus P_{-\mathbb{N}\times-\mathbb{N}_0} \begin{pmatrix} I & h_\gamma \\ h_\gamma^* & I \end{pmatrix} P^*_{\mathbb{N}\times\mathbb{N}_0} \oplus P^*_{-\mathbb{N}_0\times-\mathbb{N}}.$$

Multiplying the above equation on the left by $0 \oplus P_{-\mathbb{N}\times\{0\}}$ and on the right by $0 \oplus P^*_{\{0\}\times-\mathbb{N}}$ gives that $YW^{-1}U = X$, where U, W, X, and Y are defined via

$$Y = 0 \oplus P_{-\mathbb{N}\times\{0\}} \begin{pmatrix} I & h_\gamma \\ h_\gamma^* & I \end{pmatrix} P^*_{\mathbb{N}_0\times\mathbb{N}_0} \oplus P^*_{-\mathbb{N}\times-\mathbb{N}},$$

$$W = P_{\mathbb{N}_0\times\mathbb{N}_0} \oplus P_{-\mathbb{N}\times-\mathbb{N}} \begin{pmatrix} I & h_\gamma \\ h_\gamma^* & I \end{pmatrix} P^*_{\mathbb{N}_0\times\mathbb{N}_0} \oplus P^*_{-\mathbb{N}\times-\mathbb{N}},$$

$$U = P_{\mathbb{N}_0\times\mathbb{N}_0} \oplus P_{-\mathbb{N}\times-\mathbb{N}} \begin{pmatrix} I & h_\gamma \\ h_\gamma^* & I \end{pmatrix} 0 \oplus P^*_{\{0\}\times-\mathbb{N}},$$

and

$$X = 0 \oplus P_{-\mathbb{N}\times\{0\}} \begin{pmatrix} I & h_\gamma \\ h_\gamma^* & I \end{pmatrix} 0 \oplus P^*_{\{0\}\times-\mathbb{N}}.$$

View the operator

$$M = \begin{pmatrix} I & h_\gamma P^*_{(-\mathbb{N}_0\times-\mathbb{N}_0)\setminus\{(0,0)\}} \\ P_{(-\mathbb{N}_0\times-\mathbb{N}_0)\setminus\{(0,0)\}} h_\gamma^* & I \end{pmatrix}$$

after permutation as the operator matrix

$$\begin{pmatrix} * & Y & X \\ * & W & U \\ * & * & * \end{pmatrix}$$

acting on

$$[0 \oplus l^2(-\mathbb{N}\times\{0\})] \oplus [l^2(\mathbb{N}_0\times\mathbb{N}_0) \oplus l^2(-\mathbb{N}\times-\mathbb{N})] \oplus [0 \oplus l^2(\{0\}\times-\mathbb{N})].$$

Then the equality $YW^{-1}U = X$ together with Theorem 2.2.3 gives that

$$(0 \oplus P_{-\mathbb{N}\times\{0\}})M^{-1}(0 \oplus P^*_{\{0\}\times-\mathbb{N}}) = 0.$$

This exactly yields the required zeros in D_j, $j \leq -1$.

The proof of the zeros in A_j, $j \geq 1$, is similar. This proves the claim. □

Following the claim, we may now write D_j and A_j as

$$D_j = \begin{pmatrix} \widetilde{D}_j & q_j \\ 0 & \delta_j \end{pmatrix},\ j \leq -1;\quad A_j = \begin{pmatrix} \alpha_j & 0 \\ r_j & \widehat{A}_j \end{pmatrix},\ j \geq 1. \tag{4.4.2}$$

Write

$$\Gamma_j = \begin{pmatrix} & \gamma_{j0} \\ \widetilde{\Gamma}_j & \gamma_{j1} \\ & \vdots \end{pmatrix},\quad \Gamma_j = \begin{pmatrix} \cdots & \gamma_{j1} & \gamma_{j0} \\ & \widehat{\Gamma}_j & \end{pmatrix}.$$

Note that $\widetilde{\Gamma}_j = \widehat{\Gamma}_j$, $j \geq 0$. Observe that due to (4.4.2), equation (4.3.5) implies

$$\begin{pmatrix} I & \widetilde{h}_\gamma \\ \widetilde{h}_\gamma^* & I \end{pmatrix} \begin{pmatrix} B \\ \widetilde{D} \end{pmatrix} = \begin{pmatrix} 0 \\ \widetilde{\Delta} \end{pmatrix},$$

with

$$\widetilde{h}_\gamma = (\widetilde{\Gamma}_{i-j})_{i\in\mathbb{N}_0, j\in-\mathbb{N}_0},\quad \widetilde{D} = \begin{pmatrix} \vdots \\ \widetilde{D}_{-2} \\ \widetilde{D}_{-1} \\ I \end{pmatrix},\quad \widetilde{\Delta} = \begin{pmatrix} \vdots \\ 0 \\ 0 \\ \widetilde{\Delta}_0 \end{pmatrix},$$

where $\widetilde{\Delta}_0$ is obtained from Δ_0 by removing the last row and column; that is $\Delta_0 = \begin{pmatrix} \widetilde{\Delta}_0 & * \\ * & * \end{pmatrix}$. Moreover, if we define

$$\widetilde{\Gamma}_j = -\sum_{k=1}^{\infty} \widetilde{\Gamma}_{j+k}\widetilde{D}_{-k},\quad j \leq -1,$$

then we have that $\widetilde{\Gamma}_j$ corresponds to Γ_j without the last column for $j \leq -1$ as well. In other words,

$$\Gamma_j = \begin{pmatrix} \widetilde{\Gamma}_j & * \end{pmatrix}, \quad j \leq -1.$$

Likewise, due to the form of A_j, we have that $\widehat{A}_j$ may be constructed from (4.3.7) with Γ_j replaced by $\widehat{\Gamma}_j$. Moreover, if we define

$$\widehat{\Gamma}_j^* = -\sum_{k=1}^{\infty} \widehat{\Gamma}_{j+k}^* \widehat{A}_j, \quad j \leq -1,$$

then we have that $\Gamma_j = \begin{pmatrix} * \\ \widehat{\Gamma}_j \end{pmatrix}$, $j \leq -1$. But since $\widehat{\Gamma}_j = \widetilde{\Gamma}_j$, $j \geq 0$, we obtain from Theorem 4.3.2 that

$$\widetilde{\Gamma}_j = -\sum_{k=1}^{\infty} \widetilde{\Gamma}_{j+k} \widetilde{D}_{-k} = -\sum_{k=1}^{\infty} \widehat{\Gamma}_{j+k} \widetilde{D}_{-k} = -\sum_{k=1}^{\infty} \widehat{A}_j^* \widehat{\Gamma}_{j+k} = \widehat{\Gamma}_j, \ j \leq -1.$$

Since

$$\Gamma_j = \begin{pmatrix} * \\ \widehat{\Gamma}_j \end{pmatrix} = \begin{pmatrix} \widetilde{\Gamma}_j & * \end{pmatrix}, \ j \leq -1,$$

it now follows that Γ_j, $j \leq -1$, is Hankel.

The last step in the proof is to recognize that

$$\|(\Gamma_{i-j})_{i,j \in \mathbb{Z}}\| < 1$$

implies that the Hankel $(H_{i-j})_{i \in \mathbb{N}_0, j \in -\mathbb{N}_0}$ is a strict contraction, where

$$H_i = (\gamma_{p-q,i})_{p,q \in \mathbb{Z}}, i \geq 0.$$

But now it follows that $H_i = (\gamma_{p-q,i})_{p,q \in \mathbb{Z}}$, $i \leq -1$, exist so that $(H_{i-j})_{i,j=-\infty}^{\infty}$ is a strict contraction. □

4.5 NEHARI AND CARATHÉODORY PROBLEMS FOR FUNCTIONS ON COMPACT GROUPS

Let G be a connected compact Abelian group. As G is connected, its character group Γ can be totally ordered by a semigroup, say, P (see Section 3.5 for definition), and let $\mathcal{H}$ be a Hilbert space. Let $L_G^p(\mathcal{H})$, $1 \leq p < \infty$, denote the set of all measurable functions $f : G \to \mathcal{H}$ such that $\int_G \|f(x)\|^p dm(x) < \infty$, where m is the normalized Haar measure on G. Then $L_G^2(\mathcal{H})$ is a Hilbert space under the inner product

$$\langle f, g \rangle = \int_G \langle f(x), g(x) \rangle dm(x).$$

For $f \in L_G^1 = L_G^1(\mathbb{C})$ let $\widehat{f}$ denote its Fourier transform, defined by

$$\widehat{f}(\gamma) = \int_G \overline{\gamma(x)} f(x) dm(x), \ \gamma \in \Gamma.$$

A function f in $L^2_G = L^2_G(\mathbb{C})$ is called *analytic* if $\widehat{f}(\gamma) = 0$ for all $\gamma \notin P$. A function f in $L^2_G = L^2_G(\mathbb{C})$ is called *analytic* if $\widehat{f}(\gamma) = 0$ for all $\gamma \in P$. A function $f \in L^2_G(\mathcal{H})$ is called *analytic (antianalytic)* if for every $h \in \mathcal{H}$, the function $f_h(x) = \langle h, f(x)\rangle$ is a scalar analytic (antianalytic) function on G. The set of all analytic (antianalytic) functions in $L^2_G(\mathcal{H})$ forms a closed subspace of $L_G(\mathcal{H})$ denoted by $H^2_G(\mathcal{H})$ ($[H^2_G(\mathcal{H})]^\perp$). If $f \in L^2_G(\mathcal{H})$, by the linearity of the Fourier transform, for every $\gamma \in \Gamma$ there exists $\widehat{f}_\gamma \in \mathcal{H}$ such that $\widehat{f}_h(\gamma) = \langle h, \widehat{f}_\gamma\rangle$, for every $h \in \mathcal{H}$. Then $\|f\|_2^2 = \sum_{\gamma\in\Gamma} \|f_\gamma\|^2$ and $f \in H^2_G(\mathcal{H})$ if and only if $\widehat{f}(\gamma) = 0$ for every $\gamma \notin P$.

We denote by $L^\infty_G(\mathcal{L}(\mathcal{H}))$ the set of all functions $F : G \to \mathcal{L}(\mathcal{H})$ with the property that for every $h, k \in \mathcal{H}$, $\langle F(x)h, k\rangle$ is a measurable function and with $\|F\|_\infty := \text{ess sup}\{\|F(g)\|,\ g \in G\} < \infty$. A function $F \in L^\infty_G(\mathcal{L}(\mathcal{H}))$ is called analytic (antianalytic) if for every $h, k \in \mathcal{H}$, the function $\langle F(x)h, k\rangle$ is a scalar analytic (antianalytic) function on G. The set of all such analytic operator-valued functions is denoted by $H^\infty_G(\mathcal{L}(\mathcal{H}))$. If $F \in L^\infty_G(\mathcal{L}(\mathcal{H}))$, the operator $M_F : L^2_G(\mathcal{H}) \to L^2_G(\mathcal{H})$, defined by $(M_F g)(x) = F(x)g(x)$ for $g \in L^2_G(\mathcal{H})$, is called the *multiplication operator with symbol* F. It is not hard to see (by an argument as in Proposition 2.3.7) that $\|M_F\| = \|F\|_\infty$. In case $F \in H^\infty_G(\mathcal{L}(\mathcal{H}))$, the operator $T_F : H^2_G(\mathcal{H}) \to H^2_G(\mathcal{H})$ defined by $(T_F g)(x) = F(x)g(x)$ is called the Toeplitz operator with symbol F and also verifies $\|T_F\| = \|F\|_\infty$; see Proposition 2.4.17 for the special case when $G = \mathbb{Z}$.

By $L^2_G(\mathcal{L}(\mathcal{H}))$ we denote the set of all weakly measurable functions $F : G \to \mathcal{L}(\mathcal{H})$ such that for every $h, k \in \mathcal{H}$, $\langle F(x)h, k\rangle \in L^2_G(\mathbb{C})$. If $F \in L^2_G(\mathcal{L}(\mathcal{H}))$, then for every $\gamma \in \Gamma$ there exists $F_\gamma \in \mathcal{L}(\mathcal{H})$ such that $F(x) = \sum_{\gamma\in\Gamma} \gamma(x)F_\gamma$ (the convergence is in the weak sense). For $K \in L^1_G(\mathcal{L}(\mathcal{H}))$ and $h, k \in \mathcal{H}$, define $K_{h,k}(x) = \langle K(x)h, k\rangle$. By the linearity of the Fourier transform, for every $\gamma \in \Gamma$ there exists $\widehat{K}(\gamma) \in \mathcal{L}(\mathcal{H})$ such that $\langle \widehat{K}(\gamma)h, k\rangle = \widehat{K}_{h,k}(\gamma)$ for $h, k \in \mathcal{H}$. It can be easily seen that $\widehat{K}(\gamma) \neq 0$ for only at most countably many $\gamma \in \Gamma$.

Let $k \in L^2_G(\mathcal{L}(\mathcal{H}))$ be of the form $k(x) = \sum_{\gamma\in(-P)} \gamma(x)k_\gamma$. Define the (possibly unbounded) operator $H_k : H^2_G(\mathcal{H}) \to [H^2_G(\mathcal{H})]^\perp$ by $H_k f = P_- M_k f$, where P_- is the orthogonal projection of $L^2_G(\mathcal{H})$ onto $[H^2_G(\mathcal{H})]^\perp$. The operator H_k is referred to as the *Hankel operator* with symbol k.

The following is the main result of this section.

Theorem 4.5.1 *Let G be a compact Abelian group so that its character group Γ has a total order induced by a semigroup P. Let $k \in L^2_G(\mathcal{L}(\mathcal{H}))$ of the form $k(x) = \sum_{\gamma\in(-P)} \gamma(x)k_\gamma$ be such that H_k is bounded. Then there exists $K \in L^\infty(\mathcal{L}(\mathcal{H}))$, such that $\|K\|_\infty = \|H_k\|$, and $\widehat{K}(\gamma) = k_\gamma$, for every $\gamma \in (-P)$.*

Before proving Theorem 4.5.1, we need some lemmas. The first one can be viewed as an alternative to the necessary part of the operator-valued version of Bochner's Theorem.

Lemma 4.5.2 *Let $F \in L^1_G(\mathcal{L}(\mathcal{H}))$, $F(g) \geq 0$ for $g \in G$ a.e. Then $\widehat{F} : \Gamma \to \mathcal{L}(\mathcal{H})$ is positive semidefinite.*

Proof. Let $\{\gamma_i\}_{i=1}^n \subset \Gamma$ and $\{h_i\}_{i=1}^n \subset \mathcal{H}$. Then

$$\sum_{i,j=1}^n \langle \widehat{F}(\gamma_i - \gamma_j)h_i, h_j\rangle = \sum_{i,j=1}^n \int_G \gamma_i(g)\overline{\gamma_j(g)}\langle F(g)h_i, h_j\rangle dm(g)$$
$$= \int_G \langle F(g) \sum_{i=1}^n \gamma_i(g)h_i, \sum_{i=1}^n \gamma_i(g)h_i\rangle dm(g) \geq 0.$$

This implies that the operator matrix $\{\widehat{F}(\gamma_i - \gamma_j)\}_{i,j=1}^n$ is positive semidefinite, thus $\widehat{F}$ is a positive semidefinite function. □

Lemma 4.5.3 *Let $\Phi : \Gamma \to \mathcal{L}(\mathcal{H})$ be a positive semidefinite function with finite support on the character group Γ of a compact Abelian group G. Then $\sum_{\gamma\in\Gamma} \Phi(\gamma) \geq 0$.*

Proof. Let $h \in \mathcal{H}$ and define $\Phi_h(\gamma) = \langle \Phi(\gamma)h, h\rangle$. Then Φ is a positive semidefinite scalar function and we have that

$$\left\langle \sum_{\gamma\in\Gamma} \Phi(\gamma)h, h \right\rangle = \sum_{\gamma\in\Gamma} \Phi_h(\gamma) = \check{\Phi}_h(e) \geq 0, \tag{4.5.1}$$

where for the last inequality we used the fact the inverse Fourier transform of an $L^1(\Gamma)$ positive semidefinite function is a (pointwise) positive function in L^1_G. Since (4.5.1) holds for every $h \in \mathcal{H}$, we have that $\sum_{\gamma\in\Gamma} \Phi(\gamma) \geq 0$. □

Lemma 4.5.4 *Let $k \in L^2_G(\mathcal{L}(\mathcal{H}))$, $k(x) = \sum_{\gamma\in(-P)} \gamma(x)k_\gamma$ be such that $\|H_k\| \leq 1$. Then for every $\{\gamma_i\}_{i=1}^n \subset \Gamma$, $\gamma_1 \leq \gamma_2 \leq \cdots \leq \gamma_n$, the operator matrix*

$$\begin{pmatrix} k_{\gamma_1-\gamma_i} & k_{\gamma_2-\gamma_i} & \cdots & k_e \\ k_{\gamma_1-\gamma_{i+1}} & k_{\gamma_2-\gamma_{i+1}} & \cdots & k_{\gamma_i-\gamma_{i+1}} \\ \vdots & \vdots & & \vdots \\ k_{\gamma_1-\gamma_n} & k_{\gamma_2-\gamma_n} & \cdots & k_{\gamma_i-\gamma_n} \end{pmatrix} \tag{4.5.2}$$

is a contraction for every $i = 1, \ldots, n$. (For $i = 1$ the matrix reduces to a column while for $i = n$ it reduces to a row).

Proof. The matrix (4.5.2) represents the compression of H_k to the subspaces $\{f \in H^2_G(\mathcal{H}) : \widehat{f}(\gamma) = 0 \text{ for } \gamma \notin \{\gamma_i - \gamma_1, \ldots, \gamma_i - \gamma_{i-1}, 0\}\}$ and $\{f \in [H^2_G(\mathcal{H})]^\perp : \widehat{f}(\gamma) = 0, \text{ for } \gamma \notin \{0, \gamma_i - \gamma_{i+1}, \ldots, \gamma_i - \gamma_n\}\}$. □

For G compact, the space of continuous functions $f : G \to \mathbb{C}$ endowed with the supremum norm $\|f\|_\infty$ is denoted by $C(G)$.

Proposition 4.5.5 *Let G be a connected compact Abelian group and let $L : C(G) \to \mathcal{L}(\mathcal{H})$ be such that $\|L(p)\| \leq \|p\|_1$, for every $p \in C(G)$. Then there exists a function $Q \in L_G^\infty(\mathcal{L}(\mathcal{H}))$, such that $\|Q\|_\infty \leq 1$, and*

$$\langle L(p)h, k\rangle = \int_G p(x)\langle Q(x)h, k\rangle dm(x) \tag{4.5.3}$$

for every $p \in C(G)$ and $h, k \in \mathcal{H}$.

Proof. For fixed $h, k \in \mathcal{H}$, $\alpha_{h,k}(p) = \langle L(p)h, k\rangle$ is a functional on $C(G)$ such that $|\alpha_{h,k}(p)| \leq \|h\|\|k\|\|p\|_1$. By the Hahn-Banach theorem, α can be extended to a bounded linear functional on L_G^1 of norm $\leq \|h\||k\|$. Thus there exists $f_{h,k} \in L_G^\infty$, $\|f_{h,k}\|_\infty \leq \|h\|\|k\|$ such that

$$\langle L(p)h, k\rangle = \int_G p(x) f_{h,k}(x) dm(x), \tag{4.5.4}$$

for every $p \in C(G)$. Since $\mathcal{H}$ is separable, (4.5.4) implies that for almost every $x \in G$, the mapping $\langle h, k\rangle \in \mathcal{H} \times \mathcal{H} \to f_{h,k}(x)$ is linear in h, antilinear in k, and $|f_{h,k}(x)| \leq \|h\|\|k\|$. Thus for such x there exists $Q(x) \in \mathcal{L}(\mathcal{H})$ such that $f_{h,k}(x) = (Q(x)h, k)$. We can extend this function to a function defined on G, $\|Q\|_\infty \leq 1$, and then (4.5.4) implies (4.5.3) and the proof is complete. □

Note that the tensor product $\mathbb{C}^{n\times n} \otimes \mathbb{C}^{m\times m}$ can alternatively be identified either with $n \times n$ matrices with entries in $\mathbb{C}^{m\times m}$ or with $m \times m$ matrices with entries in $\mathbb{C}^{n\times n}$. The isomorphism between these identifications is called the *canonical shuffle*. Thus the canonical shuffle is the map $A \mapsto P^*AP \in \mathbb{C}^{nm\times nm}$, where P is the permutation defined via $Pe_{(i-1)*m+j} = e_{(j-1)*n+i}$, $i \in \{1, \dots, \}, j \in \{1, \dots, m\}$. Here e_k, $k = 1, \dots, nm$, are the standard basis vectors in $\mathbb{C}^{nm}$. Actually, they will be used below in a more general sense, namely, when the entries are operators rather than scalars (i.e., P is replaced by $P \otimes I$).

Finally, we will need the following generalized version of Schur's theorem. Let $M = (M_{ij})_{i,j=1}^n$ and $N = (N_{ij})_{i,j=1}^n$ be matrices. The *Schur product* of M and N is defined as $M \odot N = (M_{ij}N_{ij})_{i,j=1}^n$.

Proposition 4.5.6 *Let M and N be positive semidefinite matrices of the same size, M operator-valued, and N scalar. Then the Schur product $M \odot N$ is also positive semidefinite.*

Proof. Let $M \geq 0$ and $N \geq 0$, where M has entries in $\mathcal{L}(\mathcal{H})$. If $v = \begin{pmatrix} v_1 & \cdots & v_n \end{pmatrix}^T \in \mathbb{C}^n$, then

$$M \odot vv^* = \operatorname{diag}(v_i I_{\mathcal{H}})_{i=1}^n M (\operatorname{diag}(v_i I_{\mathcal{H}})_{i=1}^n)^*,$$

and thus $M \geq 0$ implies $M \odot vv^* \geq 0$. Thus the statement is true when N has rank 1. For the general case, write now N as $N = N_1 + \cdots + N_k$ with $N_1, \dots, N_k$ rank 1 positive semidefinite matrices. Then $M \geq 0$ yields that

$$M \odot N = \sum_{i=1}^k M \odot N_i \geq 0.$$

□

Proof of Theorem 4.5.1. If $H_k = 0$, then we can choose $K = 0$ and the result follows immediately. If $\|H_k\| > 0$, we can, without loss of generality, assume that $\|H_k\| = 1$.

Let $\mathcal{S}$ be the set of all 2×2 matrix-valued functions of the form $\begin{pmatrix} p & q \\ \overline{q} & p \end{pmatrix}$, where $p \in C(G)$ and q is an antianalytic polynomial on G. For $\begin{pmatrix} p & q \\ \overline{q} & p \end{pmatrix} \in \mathcal{S}$, define $\Phi : \Gamma \to \mathcal{L}(\mathcal{H} \oplus \mathcal{H})$ by

$$\Phi(e) = \begin{pmatrix} \widehat{p}(e)I & \widehat{q}(e)k_e \\ \overline{\widehat{q}(e)}k_e^* & \widehat{p}(e)I \end{pmatrix}, \Phi(\gamma) = \begin{pmatrix} 0 & 0 \\ \overline{\widehat{q}(\gamma)}k_\gamma^* & 0 \end{pmatrix} \text{ for } \gamma \in P \setminus \{e\},$$

and $\Phi(\gamma) = \Phi(-\gamma)^*$ for $\gamma \in (-P) \setminus \{e\}$. We first prove that Φ is a positive semidefinite function on Γ whenever

$$\begin{pmatrix} p(x) & q(x) \\ \overline{q(x)} & p(x) \end{pmatrix} \geq 0 \text{ for all } x \in G.$$

We have to show that for any $\{\gamma_i\}_{i=1}^n \subset \Gamma$, $\gamma_1 \leq \gamma_2 \leq \cdots \leq \gamma_n$, the operator matrix $\{\Phi(\gamma_i - \gamma_j)\}_{i,j=1}^n$ is positive semidefinite. By canonically reshuffling the matrix $\{\Phi(\gamma_i - \gamma_j)\}_{i,j=1}^n$ one obtains the operator matrix

$$\begin{pmatrix} \widehat{p}(e)I_n & C \\ C^* & \widehat{p}(e)I_n \end{pmatrix},$$

where I_n is the identity operator on $\bigoplus_{i=1}^n \mathcal{H}$ and

$$C = \begin{pmatrix} \widehat{q}(e)k_e & 0 & \cdots & 0 \\ \widehat{q}(\gamma_1 - \gamma_2)k_{\gamma_1-\gamma_2} & \widehat{q}(e)k_e & \cdots & 0 \\ \vdots & \vdots & & \vdots \\ \widehat{q}(\gamma_1 - \gamma_n)k_{\gamma_1-\gamma_n} & \widehat{q}(\gamma_2 - \gamma_n)k_{\gamma_2-\gamma_n} & \cdots & \widehat{q}(e)k_e \end{pmatrix}.$$

Let $A = \{A_{ij}\}_{i,j=1}^n$ be the lower triangular partial operator matrix defined by

$$(A_{ij})_{i,j=1}^n = \begin{cases} k(\gamma_i - \gamma_j) & \text{for } i \geq j, \\ \text{unspecified} & \text{for } i < j. \end{cases} \tag{4.5.5}$$

By Lemma 4.5.4, all fully specified submatrices of A are contractions. By Theorem 4.1.1, A admits a contractive extension B. Then $\begin{pmatrix} I & B \\ B^* & I \end{pmatrix} \geq 0$.

Since

$$\begin{pmatrix} p(x) & q(x) \\ \overline{q(x)} & p(x) \end{pmatrix} \geq 0$$

for all $x \in G$, by Lemma 4.5.2, the matrix

$$Z = \begin{pmatrix} \{\widehat{p}(\gamma_i - \gamma_j)\}_{i,j=1}^n & \{\widehat{q}(\gamma_i - \gamma_j)\}_{i,j=1}^n \\ (\{\widehat{q}(\gamma_i - \gamma_j)\}_{i,j=1}^n)^* & \{\widehat{p}(\gamma_i - \gamma_j\}_{i,j=1}^n \end{pmatrix}$$

is positive semidefinite. The matrix

$$\begin{pmatrix} \widehat{p}(e)I_n & C \\ C^* & \widehat{p}(e)I_n \end{pmatrix}$$

is the Schur product of $\left(\begin{smallmatrix} I & B \\ B^* & I \end{smallmatrix}\right)$ and Z (the unspecified positions in A correspond to zeros in Z). By the generalized version of Schur's theorem (Proposition 4.5.6), it follows that

$$\begin{pmatrix} \widehat{p}(e)I_n & C \\ C^* & \widehat{p}(e)I_n \end{pmatrix} \geq 0,$$

implying that Φ is a positive semidefinite function on Γ.

By Lemma 4.5.3, the function $l : \mathcal{S} \to \mathcal{L}(\mathcal{H} \oplus \mathcal{H})$, defined by

$$l \begin{pmatrix} p & q \\ \overline{q} & p \end{pmatrix} = \sum_{\gamma \in \Gamma} \Phi(\gamma) = \begin{pmatrix} \widehat{p}(e)I & \sum_{\gamma \in \Gamma} \widehat{q}(\gamma)k_\gamma \\ \sum_{\gamma \in \Gamma} \overline{\widehat{q}(\gamma)}k_\gamma^* & \widehat{p}(e)I \end{pmatrix},$$

is positive.

We will prove that l is completely positive. For $n \geq 1$ denote $\mathcal{S}_n = \mathcal{S} \otimes \mathbb{C}^{n \times n}$. For

$$\left\{ \begin{pmatrix} p_{ij} & q_{ij} \\ \overline{q}_{ij} & p_{ij} \end{pmatrix} \right\}_{i,j=1}^n \in \mathcal{S}_n,$$

define the functions $U_{ij} : \Gamma \to \mathcal{L}(\mathcal{H} \oplus \mathcal{H})$, for $i, j = 1, \ldots, n$, by

$$U_{ij}(e) = \begin{pmatrix} \widehat{p}_{ij}(e)I & \widehat{q}_{ij}(e)k_e \\ \overline{\widehat{q}_{ij}(e)}k_e^* & \widehat{p}_{ij}(e)I \end{pmatrix}, U_{ij}(\gamma) = \begin{pmatrix} 0 & 0 \\ \overline{\widehat{q}_{ij}(\gamma)}k_\gamma^* & 0 \end{pmatrix} \text{ for } \gamma \in (-P) - \{e\},$$

and $U_{ij}(\gamma) = U_{ij}(-\gamma)^*$ for $\gamma \in P - \{e\}$. Let $U : \Gamma \to \mathcal{L}(\mathcal{H} \oplus \mathcal{H}) \otimes \mathbb{C}^{n \times n}$, $U(\gamma) = \{U_{ij}(\gamma)\}_{i,j=1}^n$. We prove that U is positive semidefinite on Γ whenever

$$\left\{ \begin{pmatrix} p_{ij}(x) & q_{ij}(x) \\ \overline{q}_{ij}(x) & p_{ij}(x) \end{pmatrix} \right\}_{i,j=1}^n \geq 0 \text{ for all } x \in G.$$

We have to show that for any $\{\gamma_k\}_{k=1}^m \subset \Gamma$, $\gamma_1 \leq \gamma_2 \leq \cdots \leq \gamma_m$, the operator matrix $\{U(\gamma_l - \gamma_k)\}_{k,l=1}^m$ is positive semidefinite. Applying the canonical shuffle to $\{U(\gamma_l - \gamma_k)\}_{k,l=1}^m$ one obtains the operator matrix

$$\left\{ \begin{pmatrix} \widehat{p}_{ij}(e)I_m & C_{ij} \\ C_{ij}^* & \widehat{p}_{ij}(e)I_m \end{pmatrix} \right\}_{i,j=1}^n \odot \left(\begin{pmatrix} I & B \\ B^* & I \end{pmatrix} \otimes J_n \right), \tag{4.5.6}$$

where I_m is the identity matrix on $\bigoplus_{i=1}^m \mathcal{H}$,

$$C_{ij} = \begin{pmatrix} \widehat{q}_{ij}(e)k_e & 0 & \cdots & 0 \\ \widehat{q}_{ij}(\gamma_1 - \gamma_2)k_{\gamma_1 - \gamma_2} & \widehat{q}_{ij}(e)k_e & \cdots & 0 \\ \vdots & \vdots & & \vdots \\ \widehat{q}_{ij}(\gamma_1 - \gamma_n)k_{\gamma_1 - \gamma_n} & \widehat{q}_{ij}(\gamma_2 - \gamma_n)k_{\gamma_2 - \gamma_n} & \cdots & \widehat{q}_{ij}(e)k_e \end{pmatrix},$$

where J_n is the $n \times n$ matrix with all entries equal to 1, and B is the contractive extension of the partial matrix A defined by (4.5.5). Then (4.5.6) and the generalized version of Schur's theorem (Proposition 4.5.6) imply that $\{U(\gamma_k - \gamma_l)\}_{k,l=1}^m$ is positive semidefinite, thus U is a positive semidefinite function. By Lemma 4.5.3, the function $l_n : \mathcal{S}_n \to \mathcal{L}(\mathcal{H} \oplus \mathcal{H}) \otimes \mathbb{C}^{n \times n}$, defined by

$$l_n\left(\left\{\begin{pmatrix} p_{ij} & q_{ij} \\ \overline{q}_{ij} & p_{ij} \end{pmatrix}\right\}_{i,j=1}^{n}\right) = \sum_{\gamma \in \Gamma} U(\gamma)$$

$$= \left\{\begin{pmatrix} \widehat{p}_{ij}(e)I & \sum_{\gamma \in \Gamma} \widehat{q}_{ij}(\gamma)k_\gamma \\ \sum_{\gamma \in \Gamma} \overline{\widehat{q}_{ij}(\gamma)}k_\gamma^* & \widehat{p}_{ij}(e)I \end{pmatrix}\right\}_{i,j=1}^{n},$$

is positive, implying that l is a completely positive function on $\mathcal{S}$. Since $\mathcal{S}$ is an operator system in $C(G) \otimes \mathbb{C}^{2\times 2}$, by Arveson's extension theorem (Theorem 3.7.1), l admits a (completely) positive extension $L : C(G) \otimes \mathbb{C}^{2\times 2} \to \mathcal{L}(\mathcal{H} \oplus \mathcal{H})$.

Let $p \in C(G)$. Since $\widehat{|p|}(e) = \int\limits_G |p(x)|dm(x) = \|p\|_1$,

$$L\begin{pmatrix} |p| & 0 \\ 0 & |p| \end{pmatrix} = \begin{pmatrix} \|p\|_1 I & 0 \\ 0 & \|p\|_1 I \end{pmatrix}.$$

Let

$$L\begin{pmatrix} 0 & p \\ 0 & 0 \end{pmatrix} = \begin{pmatrix} L_{11}(p) & L_{12}(p) \\ L_{21}(p) & L_{22}(p) \end{pmatrix}.$$

Since

$$\begin{pmatrix} |p(x)| & \pm p(x) \\ \pm \overline{p(x)} & |p(x)| \end{pmatrix} \geq 0 \text{ for } x \in G,$$

we have that

$$\begin{pmatrix} \|p\|_1 I & 0 \\ 0 & \|p\|_1 I \end{pmatrix} \pm \begin{pmatrix} L_{11}(p) + L_{11}(p)^* & L_{12}(p) + L_{21}(p)^* \\ L_{12}(p)^* + L_{21}(p) & L_{22}(p) + L_{22}(p)^* \end{pmatrix} \geq 0,$$

implying that

$$\|L_{12}(p) + L_{21}(p)^*\| \leq \|p\|_1. \tag{4.5.7}$$

Since

$$\begin{pmatrix} |p(x)| & \pm ip(x) \\ \mp i\overline{p(x)} & |p(x)| \end{pmatrix} \geq 0 \;\text{ for } x \in G,$$

we have that

$$\begin{pmatrix} \|p\|_1 I & 0 \\ 0 & \|p\|_1 I \end{pmatrix} \pm \begin{pmatrix} i(L_{11}(p) - L_{11}(p)^*) & i(L_{12}(p) - L_{21}(p)^*) \\ i(L_{12}(p)^* - L_{21}(p)) & i(L_{22}(p) - L_{22}(p)^*) \end{pmatrix} \geq 0,$$

implying that

$$\|L_{12}(p) - L_{21}(p)^*\| \leq \|p\|_1. \tag{4.5.8}$$

By (4.5.7) and (4.5.8) we get that $\|L_{12}(p)\| \leq \|p\|_1$.

By Proposition 4.5.5 there exists $Q \in L_G^\infty(\mathcal{L}(\mathcal{H}))$ such that $\|Q\|_\infty \leq 1$ and

$$\langle L_{12}(p)h, k\rangle = \int\limits_G p(x)\langle Q(x)h, k\rangle dm(x),$$

for every $h, k \in \mathcal{H}$. Let $\gamma_0 \in (-P)$ and define $p_0(x) = \gamma_0(x)$. Since L_{12} is an extension of the mapping that assigns to every antianalytic polynomial p the operator $\sum_{\gamma\in\Gamma} \widehat{p}(\gamma)k_\gamma$, we have that

$$\langle k_{\gamma_0}h, k\rangle = \langle L_{12}(p_0)h, k\rangle = \int_G \gamma_0(x)\langle Q(x)h, k\rangle dm(x). \tag{4.5.9}$$

Define $K(x) = Q(-x)$. Then (4.5.9) implies that $k_\gamma = \widehat{K}(\gamma)$, for every $\gamma \in (-P)$. Since $H_K = H_k$, we have that $\|K\|_\infty = \|M_K\| \geq \|H_K\| = 1$, and this completes the proof. □

Next we present two consequences of Theorem 4.5.1.

Corollary 4.5.7 *For every $k \in L^\infty_G(\mathcal{L}(\mathcal{H}))$, $\operatorname{dist}(k, H^\infty_G(\mathcal{L}(\mathcal{H}))) = \|H_k\|$. Moreover, the distance is attained.*

Proof. Similar to the proof of Corollary 4.3.3. □

Let $a \in P$ and consider the subspace $H^2_a(\mathcal{H}) = \{f \in H^2_G(\mathcal{H}) : f(x) = \sum_{0\leq\gamma\leq a} \gamma(x)h_\gamma\}$ of $H^2_G(\mathcal{H})$. Let P_a denote the orthogonal projection of $H^2_G(\mathcal{H})$ onto $H^2_a(\mathcal{H})$. If $k \in H^2_G(\mathcal{L}(\mathcal{H}))$, $k(x) = \sum_{0\leq\gamma\leq a} \gamma(x)k_\gamma$, define the (possibly unbounded) *generalized Toeplitz* operator $T_a = P_aM_k|H^2_a(\mathcal{H})$.

The following is a generalization of the (contractive) Carathéodory-Fejér problem.

Corollary 4.5.8 *Let $k \in H^2_G(\mathcal{L}(\mathcal{H}))$, $k(x) = \sum_{0\leq\gamma\leq a} \gamma(x)k_\gamma$ be such that T_a is bounded. Then there exists $K \in H^\infty_G(\mathcal{L}(\mathcal{H}))$ such that $\|K\|_\infty = \|T_a\|$ and $\widehat{K}(\gamma) = k_\gamma$ for $0 \leq \gamma \leq a$.*

Proof. Define $k_a(x) = \overline{a(x)}\sum_{0\leq\gamma\leq a} \gamma(x)k_\gamma$. Then $\|H_{k_a}\| = \|T_a\|$ and one can apply Theorem 4.5.1 to k_a. Thus there exists $K_a \in L^\infty_G(\mathcal{L}(\mathcal{H}))$ such that $\|K_a\|_\infty = \|H_{k_a}\| = \|T_a\|$ and $\widehat{K_a}(\gamma) = k_{\gamma+a}$ for every $\gamma \in (-P)$ (We consider $k_\gamma = 0$ for $\gamma \in (-P)$). Define $K(x) = a(x)K_a(x)$. Then $K \in H^\infty_G(\mathcal{L}(\mathcal{H}))$, and $\widehat{K}(\gamma) = k_\gamma$ for $0 \leq \gamma \leq a$. □

4.6 THE NEVANLINNA-PICK PROBLEM

This section is dedicated to solving the Nevanlinna-Pick interpolation problem. The solution to the problem is presented as a realization via a linearly fractional transform by a unitary operator matrix. This fact when generalized to several variables provides a solution to the multivariable Nevanlinna-Pick problem. Using Nehari's theorem (Theorem 4.3.1) one can solve the following interpolation problem, which is known as the *Nevanlinna-Pick problem.* Let $z_1, \ldots, z_n \in \mathbb{D}$, with $z_i \neq z_j$ for $i \neq j$, and $w_1, \ldots, w_n \in \mathbb{C}$ be given. Find, if possible $f \in H^\infty(\mathbb{T})$ such that

(1) $f(z_i) = w_i, \ i = 1, \ldots, n;$

(2) $\|f\|_\infty \leq 1$.

Clearly one needs that $|w_i| \leq 1$, $i = 1, \ldots, n$. In general, though, this is not sufficient. The necessary and sufficient condition is in terms of positive semidefiniteness of the so-called *Pick matrix*, which is the matrix Λ below. The statement is as follows.

Theorem 4.6.1 *Let different $z_1, \ldots, z_n \in \mathbb{D}$ and $w_1, \ldots, w_n \in \mathbb{C}$ be given. There exists an $f \in H^\infty(\mathbb{T})$ so that*

(1) $f(z_i) = w_i$, $i = 1, \ldots, n$,

(2) $\|f\|_\infty \leq 1$,

if and only if the $n \times n$ Hermitian matrix

$$\Lambda := \left(\frac{1 - w_i \overline{w_j}}{1 - z_i \overline{z_j}} \right)_{i,j=1}^{n}$$

is positive semidefinite. In that case f can be chosen to be a rational function with at most k poles where $k = \operatorname{rank} \Lambda$.

Proof. For $i = 1, \ldots, n$ let $b_i(z)$ denote the *Blaschke function*

$$b_i(z) = \frac{z - z_i}{1 - \overline{z_i} z},$$

which lies in $H^\infty(\mathbb{T})$ (since $|z_i| < 1$) and satisfies $|b_i(z)| = 1, z \in \mathbb{T}$. In addition, let $B(z) = \prod_{i=1}^n b_i(z)$. Put

$$g(z) = \sum_{j=1}^{n} \frac{w_j}{B_j(z_j)} B_j(z),$$

where $B_j = \frac{B}{b_j}$. Then $g(z_i) = w_i, i = 1, \ldots, n$, and $g \in H^\infty(\mathbb{T})$. Suppose now that $f \in H^\infty(\mathbb{T})$ is a solution to the Nevanlinna-Pick interpolation problem. Then $f(z_i) - g(z_i) = 0, i = 1, \ldots, n$, and thus $f = g - Bh$ for some $h \in H^\infty(\mathbb{T})$. Then, it follows that $1 \geq \|f\|_\infty = \|g - Bh\|_\infty = \|gB^{-1} - h\|_\infty$, and thus the Hankel operator $H_{gB^{-1}}$ has norm ≤ 1, and thus $H_{gB^{-1}}^* H_{gB^{-1}} \leq I$. If we let $\psi_j(z) = \frac{B(z)}{z - z_j}$, $j = 1, \ldots, n$, we therefore must have that for all complex numbers $c_1, \ldots, c_n$, that

$$\sum_{j,k=1}^{n} c_j \overline{c_k} \langle H_{gB^{-1}} \psi_j, H_{gB^{-1}} \psi_k \rangle \leq \sum_{j,k=1}^{n} c_j \overline{c_k} \langle \psi_j, \psi_k \rangle. \tag{4.6.1}$$

Notice that

$$H_{gB^{-1}} \psi_j = P_- \left(\frac{g}{z - z_j} \right) = \frac{g(z_j)}{z - z_j} = \frac{w_j}{z - z_j},$$

and thus

$$\langle H_{gB^{-1}} \psi_j, H_{gB^{-1}} \psi_k \rangle = w_j \overline{w_k} \langle \frac{1}{z - z_j}, \frac{1}{z - z_k} \rangle = w_j \overline{w_k} \frac{1}{1 - z_j \overline{z_k}}.$$

Also

$$\begin{aligned}\langle \psi_j, \psi_k \rangle &= \left\langle \frac{B(z)}{z - z_j}, \frac{B(z)}{z - z_k} \right\rangle \\ &= \left\langle \frac{1}{z - z_j}, \frac{1}{z - z_k} \right\rangle = \frac{1}{1 - z_j z_k},\end{aligned}$$

where we used that $|B| \equiv 1$ on $\mathbb{T}$. Thus (4.6.1) is equivalent to

$$\sum_{j,k=1}^{n} c_j \overline{c_k} \left(\frac{1 - w_j \overline{w_k}}{1 - z_j \overline{z_k}} \right) \geq 0,$$

for all $c_1, \ldots, c_n \in \mathbb{C}$. Thus $\Lambda \geq 0$.

For the converse, let $k = \operatorname{rank} \Lambda$, and factor Λ as

$$\Lambda = G^* G,$$

with $G = (g_1 \ldots g_n) \in \mathbb{C}^{k \times n}$. Then

$$\frac{1 - w_i \overline{w_j}}{1 - z_i \overline{z_j}} = \langle g_i, g_j \rangle, \quad 1 \leq i, j \leq n,$$

and thus

$$1 + \langle z_i g_i, z_j g_j \rangle = w_i \overline{w_j} + \langle g_i, g_j \rangle, \quad 1 \leq i, j \leq n.$$

But then there exists an isometry V such that

$$V \begin{pmatrix} 1 \\ z_i g_i \end{pmatrix} = \begin{pmatrix} w_i \\ g_i \end{pmatrix}, i = 1, \ldots, n. \tag{4.6.2}$$

Extend the isometry to a unitary U acting on all of $\mathbb{C}^{k+1} \simeq \mathbb{C} \oplus \mathbb{C}^k$. Write

$$U = \begin{pmatrix} A & B \\ C & D \end{pmatrix} : \begin{matrix} \mathbb{C} \\ \oplus \\ \mathbb{C}^k \end{matrix} \rightarrow \begin{matrix} \mathbb{C} \\ \oplus \\ \mathbb{C}^k \end{matrix}.$$

Let now

$$f(z) = A + Bz(I - zD)^{-1} C. \tag{4.6.3}$$

As $\|D\| \leq 1$, clearly $f(z)$ is well defined for $z \in \mathbb{D}$. Next

$$\begin{aligned}I - f(w)^* f(z) = I - \big(&A^* A + C^*(I - \overline{w} D^*)^{-1} \overline{w} B^* A \\ &+ A^* B z (I - zD)^{-1} C + C^*(I - \overline{w} D^*)^{-1} \overline{w} B^* B z (I - zD)^{-1} C\big).\end{aligned}$$

Now use that U is unitary, so

$$I - A^* A = C^* C, \quad B^* A = -D^* C, \quad B^* B = I - D^* D,$$

and thus

$$\begin{aligned}I - f(w)^* f(z) &= C^*(I - \overline{w} D^*)^{-1} [(I - \overline{w} D^*)(I - zD) + \overline{w} D^*(I - zD) \\ &\quad + (I - \overline{w} D^*) D z - \overline{w} z (I - D^* D)](I - zD)^{-1} C \\ &= C^*(I - \overline{w} D^*)^{-1} (1 - \overline{w} z)(I - zD)^{-1} C,\end{aligned} \tag{4.6.4}$$

which is positive semidefinite when $z = w \in \mathbb{D}$. This yields that $f \in H^\infty(\mathbb{T})$ and $\|f\|_\infty \leq 1$. Finally, by (4.6.2) we get that

$$\begin{cases} A + Bz_i g_i = w_i, \\ C + Dz_i g_i = g_i, \end{cases}$$

and thus $(I - z_i D)^{-1} C = g_i$, which yields that $f(z_i) = A + Bz_i(I - z_i D)^{-1} C = A + Bz_i g_i = w_i, i = 1, \ldots, n$. This proves the result. □

Remark 4.6.2 It should be noted that $H_{gB^{-1}} \mid (\text{Span}\{\psi_1, \ldots, \psi_n\})^\perp = 0$, so that $\Lambda \geq 0$ is equivalent to $\|H_{gB^{-1}}\| \leq 1$. Thus Nehari's theorem (Theorem 4.3.1) also yields the existence of a solution to the Nevanlinna-Pick interpolation problem as soon as $\Lambda \geq 0$. The sufficiency proof we have provided, though, gives the additional information that the solution can be chosen to be rational.

The expression (4.6.3) is called a realization of the function f. In Exercise 4.13.24 we will see how the norm of a Hankel operator can be computed via a realization of its symbol. For functions of several variables the following realization formula is used:

$$f(z_1, \ldots, z_d) = A + BZ(I - ZD)^{-1} C, \tag{4.6.5}$$

where

$$\begin{pmatrix} A & B \\ C & D \end{pmatrix} = \begin{pmatrix} A & B_1 & \cdots & B_d \\ C_1 & D_{11} & \cdots & D_{1d} \\ \vdots & \vdots & & \vdots \\ C_d & D_{d1} & \cdots & D_{dd} \end{pmatrix} : \begin{matrix} \mathbb{C} \\ \oplus \\ \mathcal{H}_1 \\ \oplus \\ \vdots \\ \oplus \\ \mathcal{H}_d \end{matrix} \to \begin{matrix} \mathbb{C} \\ \oplus \\ \mathcal{H}_1 \\ \oplus \\ \vdots \\ \oplus \\ \mathcal{H}_d \end{matrix} \tag{4.6.6}$$

and

$$Z = Z(z) = Z(z_1, \ldots, z_d) = \begin{pmatrix} z_1 I_{\mathcal{H}_1} & & 0 \\ & \ddots & \\ 0 & & z_d I_{\mathcal{H}_d} \end{pmatrix}. \tag{4.6.7}$$

If, as in the proof of Theorem 4.6.1 the operator matrix $\left(\begin{smallmatrix} A & B \\ C & D \end{smallmatrix}\right)$ is unitary, then it is not hard to check that $\|f\|_\infty \leq 1$ (see also Exercise 4.13.30); we will see this in the proof of the following theorem. If we restrict our attention to the class of functions f with the above type of realization (which is the Schur-Agler class) one can state the following solution to the *multivariable Nevanlinna-Pick interpolation problem.*

Theorem 4.6.3 *Let different* $z^{(i)} = (z_1^{(i)}, \ldots, z_d^{(i)}) \in \mathbb{D}^d, i = 1, \ldots, n,$ *and* $w_i \in \mathbb{C}, i = 1, \ldots, n,$ *be given. There exists a function with realization* (4.6.5) *where* $\left(\begin{smallmatrix} A & B \\ C & D \end{smallmatrix}\right)$ *is unitary such that* $f(z^{(i)}) = w_i,\ i = 1, \ldots, n,$ *if and only if there exist* $n \times n$ *positive semidefinite matrices* $\Lambda_1, \ldots, \Lambda_d$ *such that*

$$(1 - w_i \overline{w_j})_{i,j=1}^n = \sum_{k=1}^d \left[(1 - z_k^{(i)} \overline{z_k^{(j)}})_{i,j=1}^n \right] \odot \Lambda_k. \tag{4.6.8}$$

(Here $\odot$ *denotes the Schur product of matrices.)*

Proof. First assume that an interpolant f with realization (4.6.5) exists. Similarly to (4.6.4) we compute

$$I - f(w)^* f(z) = C^*(I - Z(w)D)^{*-1}(I - Z(w)^* Z(z))(I - Z(z)D)^{-1}C.$$

In particular if $z = w \in \mathbb{D}^d$, we get that

$$I - f(z)^* f(z) \geq 0,$$

and thus $\|f\|_\infty \leq 1$. Next, if $z = z^{(i)}$ and $w = z^{(i)}$, we get

$$1 - \overline{w_j} w_i = \sum_{k=1}^{d} \Gamma_k^{(j)*}(I - \overline{z_k^{(j)}} z_k^{(i)})\Gamma_k^{(i)},$$

where

$$(I - Z(z^{(i)})D^{-1})C = \begin{pmatrix} \Gamma_1^{(i)} \\ \vdots \\ \Gamma_d^{(i)} \end{pmatrix} : \mathbb{C} \to \begin{matrix} \mathcal{H}_1 \\ \oplus \\ \vdots \\ \oplus \\ \mathcal{H}_d \end{matrix}.$$

Letting

$$\Lambda_k = (\Gamma^{(j)*} \Gamma_k^{(i)})_{i,j=1}^n$$

we obtain condition (4.6.8).

Conversely, suppose that (4.6.8) holds. Factorize Λ_k as

$$\Lambda_k = G_k^* G_k, \ G_k = (g_1^{(k)} \dots g_n^{(k)}),$$

where $g_i^{(k)} \in \mathbb{C}^n$. Then (4.6.8) yields

$$1 - w_i \overline{w_j} = \sum_{k=1}^{d} (1 - z_k^{(i)} \overline{z_k^{(j)}}) g_j^{(k)*} g_i^{(k)},$$

and thus

$$1 + \sum_{k=1}^{d} \langle z_k^{(i)} g_i^{(k)}, z_k^{(j)} g_j^{(k)} \rangle = w_i \overline{w_j} + \sum_{k=1}^{d} \langle g_i^{(k)}, g_j^{(k)} \rangle.$$

Let now V be the isometry, defined on a subspace of $\mathbb{C} \oplus \mathbb{C}^n \oplus \cdots \oplus \mathbb{C}^n$, such that

$$V \begin{pmatrix} 1 \\ z_1^{(i)} g_i^{(1)} \\ \vdots \\ z_d^{(i)} g_i^{(d)} \end{pmatrix} = \begin{pmatrix} w_i \\ g_i^{(1)} \\ \vdots \\ g_i^{(d)} \end{pmatrix}, \quad i = 1, \dots, n,$$

and extend V to a unitary $\left(\begin{smallmatrix} A & B \\ C & D \end{smallmatrix}\right)$ of the form (4.6.6) defined on all of $\mathbb{C} \oplus \mathbb{C}^n \oplus \cdots \oplus \mathbb{C}^n$. Define now f via (4.6.5). Letting

$$g_i = \begin{pmatrix} g_i^{(1)} \\ \vdots \\ g_i^{(d)} \end{pmatrix},$$

and $Z(z) = Z(z_1, \ldots, z_d)$ be as in (4.6.7), we have

$$A + BZ(z^{(i)})g_i = w_i,$$

$$C + DZ(z^{(i)})g_i = g_i,$$

and thus $(I - DZ(z^{(i)}))^{-1}C = g_i$, yielding

$$f(z^{(i)}) = A + BZ(z^{(i)})g_i = w_i, \quad i = 1, \ldots, n.$$

□

Let N be a positive integer, $z^{(j)} = (z_1^{(j)}, \ldots, z_d^{(j)})$, $j = 1, \ldots, N$, be points in the polydisk $\mathbb{D}^d$, and $w^{(j)} \in \mathbb{C}$, $i = 1, \ldots, N$. The *d-variable Schur class* $\mathcal{S}_d$ consists of analytic functions $f : \mathbb{D}^d \to \mathbb{C}$ with $\|f\|_\infty \leq 1$. The Nevanlinna-Pick interpolation problem in the Schur class with data $(z^{(1)}, \ldots, z^{(N)}, w^{(1)}, \ldots, w^{(N)})$ asks for the existence of an $f \in \mathcal{S}_d$ such that $f(z^{(j)}) = w^{(j)}$, $j = 1, \ldots, N$. Using Theorem 3.2.11 we now obtain necessary conditions for the existence of a solution to the Nevanlinna-Pick interpolation problem in the d-variable Schur class.

Theorem 4.6.4 *If $f \in \mathcal{S}_d$ is a solution to the Nevanlinna-Pick interpolation problem with data $(z^{(1)}, \ldots, z^{(N)}, w^{(1)}, \ldots, w^{(N)})$, then for any integers r and s, $1 \leq r, s \leq d$, there exist positive semidefinite matrices A_{rs}, B_{rs} such that*

$$[1 - w^{(i)}w^{(j)*}]_{i,j=1}^N = \left[\prod_{k \neq r}(1 - z_k^{(i)}\overline{z_k^{(j)}})\right]_{i,j=1}^N \odot A_{rs} \tag{4.6.9}$$

$$+ \left[\prod_{k \neq s}(1 - z_k^{(i)}\overline{z_k^{(j)}})\right]_{i,j=1}^N \odot B_{rs}.$$

Before we can prove this result, we first need the following approximation result. Recall that $\overleftarrow{p}$ denotes the reverse of the polynomial p; see Section 3.2.

Theorem 4.6.5 *Let $f : \mathbb{D}^d \to \mathbb{C}$ be analytic, so that $\|f\|_\infty \leq 1$. Then for every compact $K \subset \mathbb{D}^d$ and every $\epsilon > 0$, there exists a stable polynomial p such that*

$$\sup_{z \in K} |f(z) - \frac{\overleftarrow{p}(z)}{p(z)}| < \epsilon.$$

In other words, every element of the Schur class on the polydisk is a limit (uniformly on compact subsets of $\mathbb{D}^d$) of a sequence of rational inner functions on $\mathbb{D}^d$.

Proof. Let K be compact in $\mathbb{D}^d$, and let $0 < \epsilon < \frac{1}{2}$. Choose a polynomial g such that $|g(z)| < 1$ for $z \in \overline{\mathbb{D}}^d$ and $\sup_{z \in K} |f(z) - g(z)| < \epsilon$. Such a choice is possible, as one can take a partial sum of the power series of $f(rz)$, where

$r \in (0,1)$ is chosen to be sufficiently close to 1. Next, let $m(z) = z^N$ be a monomial of sufficiently high degree so that (i) $b(z) := m(z)\overline{g}(1/z)$ is a polynomial, where $\overline{g}$ is obtained from g by taking the complex conjugates of the coefficients of g; (ii) $|m(z)| < \epsilon$ and $|g(z)b(z)| < \epsilon$ for $z \in K$. Put now $p(z) = 1 + b(z)$. On $\mathbb{T}^d$, we have that $|b(z)| = |m(z)\overline{g(z)}| < 1$, and thus $|b(z)| < 1$ for $z \in \overline{\mathbb{D}}^d$. This implies that p is stable. Next, for $z \in K$,

$$|f(z) - \frac{\overleftarrow{p}(z)}{p(z)}| \leq |f(z) - g(z)| + \left| g(z) - \frac{g(z) + m(z)}{1 + b(z)} \right|$$

$$= |f(z) - g(z)| + \left| \frac{m(z) - g(z)b(z)}{1 + b(z)} \right| < \epsilon + \frac{2\epsilon}{1 - \epsilon} < 5\epsilon,$$

proving the claim. □

We also need the following lemma.

Lemma 4.6.6 *Let p be a stable polynomial, and let $z^{(j)} \in \mathbb{D}^d$, $j = 1, \ldots, N$, be given. For all choices $r, s \in \{1, \ldots, d\}$ there exist positive semidefinite matrices A_{rs} and B_{rs} such that*

$$[1 - w^{(i)}w^{(j)*}]_{i,j=1}^N = \left[\prod_{k \neq r} (1 - z_k^{(i)} \overline{z_k^{(j)}}) \right]_{i,j=1}^N \odot A_{rs} \tag{4.6.10}$$

$$+ \left[\prod_{k \neq s} (1 - z_k^{(i)} \overline{z_k^{(j)}}) \right]_{i,j=1}^N \odot B_{rs},$$

where $w^{(j)} = \frac{\overleftarrow{p}(z^{(j)})}{p(z^{(j)})}$, $j = 1, \ldots, N$. Moreover, the entries of A_{rs} and B_{rs} are bounded in absolute value by $(1 - \delta^2)^{d-1}$, where $\delta = \max_{j,k} |z_k^{(j)}|$.

Proof. Apply Theorem 3.2.11 to the stable polynomial $p(= r)$, to obtain that

$$p(z)p(\widetilde{z})^* - \overleftarrow{p}(z)\overleftarrow{p}(\widetilde{z})^* \tag{4.6.11}$$

$$= \prod_{i \neq r} (1 - z_i \overline{\widetilde{z}_i}) \sum_{j=0}^{n_1 - 1} g_j(z) g_j(\widetilde{z})^* + \prod_{i \neq s} (1 - z_i \overline{\widetilde{z}_i}) \sum_{j=0}^{n_2 - 1} f_j(z) f_j(\widetilde{z})^*,$$

where g_j and f_j are analytic functions on $\mathbb{D}^d$. Divide now both sides by $p(z)p(\widetilde{z})^*$, to obtain that

$$1 - f(z)f(\widetilde{z}) = \prod_{i \neq r} (1 - z_i \overline{\widetilde{z}_i}) \sum_{j=0}^{n_1 - 1} \psi_j(z)\psi_j(\widetilde{z})^* + \prod_{i \neq s} (1 - z_i \overline{\widetilde{z}_i}) \sum_{j=0}^{n_2 - 1} \phi_j(z)\phi_j(\widetilde{z})^*,$$

where $f = \frac{\overleftarrow{p}}{p}$, $\psi_j = \frac{g_j}{p}$, and $\phi_j = \frac{f_j}{p}$. Letting

$$A_{rs} = \left(\sum_{l=0}^{n_1 - 1} \psi_l(z^{(i)}) \psi_l(z^{(j)})^* \right)_{i,j=1}^N, \quad B_{rs} = \left(\sum_{l=0}^{n_2 - 1} \psi_l(z^{(i)}) \phi_l(z^{(j)})^* \right)_{i,j=1}^N,$$

one obtains the desired positive semidefinite matrices.

Next, observe that the diagonal entries on the left-hand side of (4.6.10) are in absolute value less than one. This yields, using (4.6.10), that the diagonal entries of A_{rs} and B_{rs} are less than $(1-\delta^2)^{-1}$. As A_{rs} and B_{rs} are positive semidefinite, the same holds for the off-diagonal entries. □

Proof of Theorem 4.6.4. Fix $r, s \in \{1, \dots, d\}$. Suppose that f is a solution to the Nevanlinna-Pick problem. By Theorem 4.6.5 there exist stable polynomials p_m so that $f_m := \frac{\overleftarrow{p_m}}{p_m}$ converges to f (uniformly on compact subsets of $\mathbb{D}^d$). By Lemma 4.6.6 we have that for every m, positive semidefinite matrices $A_{rs}^{(m)}$ and $B_{rs}^{(m)}$ exist such that

$$[1 - f_m(z^{(i)}) f_m(z^{(j)})^*]_{i,j=1}^N = \left[\prod_{k \neq r} (1 - z_k^{(i)} \overline{z_k^{(j)}}) \right]_{i,j=1}^N \odot A_{rs}^{(m)} + \left[\prod_{k \neq s} (1 - z_k^{(i)} \overline{z_k^{(j)}}) \right]_{i,j=1}^N \odot B_{rs}^{(m)}. \quad (4.6.12)$$

As, by Lemma 4.6.6, the absolute values of the entries of matrices $A_{rs}^{(m)}$ and $B_{rs}^{(m)}$ are all bounded by the same number, there must be a convergent subsequence of $(A_{rs}^{(m)}, B_{rs}^{(m)})_{m \in \mathbb{N}}$ with limit (A_{rs}, B_{rs}) say. Taking the limit in (4.6.12) with respect to this subsequence now yields (4.6.9). □

We can now also prove the von Neumann inequality, which holds in one and two variables. In Example 4.6.8 we will see that the inequality fails, in general, for three and more variables.

Theorem 4.6.7 *Let $f : \mathbb{D}^2 \to \mathbb{C}$ be analytic such that $\|f\|_\infty < \infty$, and let $T_1, T_2 : \mathcal{H} \to \mathcal{H}$ be commuting strict contractions. Then*

$$\|f(T_1, T_2)\| \leq \|f\|_\infty. \quad (4.6.13)$$

Proof. By scaling, we can assume that $\|f\|_\infty = 1$. Let $r < 1$ be so that $r > \max\{\|T_1\|, \|T_2\|\}$. By Theorem 4.6.5 there exist stable polynomials p_n, $n \in \mathbb{N}$, such that $\frac{\overleftarrow{p_n}}{p_n}$, $n \in \mathbb{N}$, converges uniformly to f on the set $r\overline{\mathbb{D}}^2$. By Corollary 3.2.8 we have that $\|\frac{\overleftarrow{p_n}}{p_n}(T_1, T_2)\| \leq 1$. But then, by taking limits, we find that $\|f(T_1, T_2)\| \leq 1$. □

The classical von Neumann inequality is that for a polynomial p and a contraction T, we have that

$$\|p(T)\| \leq \|p\|_\infty.$$

This, of course, follows easily from Theorem 4.6.7.

The following example shows that the von Neumann inequality does not hold in three or more variables. Subsequently we show that there exists a Nevanlinna-Pick problem which is solvable in the d-variable Schur class but not in the Schur-Agler class.

Example 4.6.8 Let $T = (T_1, \dots, T_d)$ be a d-tuple ($d \geq 3$) of commuting (simultaneously) diagonalizable $N \times N$ matrices, $\|T_k\| < 1$, and let $p(z_1, \dots, z_d)$ be a scalar-valued polynomial with $\|p\|_\infty \leq 1$ such that the von Neumann inequality fails for T, that is, $\|p(T)\| > 1$. Below we show that this choice is possible. Once such a choice is made, let Y be a $N \times N$ nonsingular matrix such that $T_k = Y \Delta_k Y^{-1}$, where $\Delta_k = \operatorname{diag}\left(z_k^{(1)}, \dots, z_k^{(N)}\right)$, $k = 1, \dots, d$. Consider the Nevanlinna–Pick data

$$z^{(j)} = (z_1^{(j)}, \dots, z_d^{(j)}) \in \mathbb{D}^d, \quad w^{(j)} = p(z^{(j)}) \in \overline{\mathbb{D}}, \quad j = 1, \dots, N.$$

The interpolation problem is clearly solvable in the class $\mathcal{S}_d$ (for instance, p is a solution). However, for any analytic function $f(z) = \sum_{n \in \mathbb{Z}_+^d} c_n z^n$ on $\mathbb{D}^d$ satisfying $f(z^{(j)}) = w^{(j)}$, $j = 1, \dots, N$, one has

$$\begin{aligned} f(T) &= \sum_{n \in \mathbb{Z}_+^d} c_n T^n = Y \sum_{n \in \mathbb{Z}_+^d} c_n \Delta^n \, Y^{-1} \\ &= Y \operatorname{diag}\left(f(z^{(1)}), \dots, f(z^{(N)})\right) \, Y^{-1} \\ &= Y \operatorname{diag}\left(w^{(1)}, \dots, w^{(N)}\right) \, Y^{-1} \\ &= Y \operatorname{diag}\left(p(z^{(1)}), \dots, p(z^{(N)})\right) \, Y^{-1} \\ &= Y p(\Delta) Y^{-1} = p(T). \end{aligned}$$

Hence $\|f(T)\| = \|p(T)\| > 1$, and the interpolation problem is not solvable in the Schur-Agler class $\mathcal{SA}_d$.

It remains to show that p and T may be chosen as above. First we note that for

$$p(z) = \frac{1}{5}(z_1^2 + z_2^2 + z_3^2 - 2(z_1 z_2 + z_2 z_3 + z_3 z_1))$$

and

$$T_1 = \begin{pmatrix} 0 & 1 & 0 & 0 \\ 0 & 0 & 0 & 1 \\ 0 & 0 & 0 & 0 \\ 0 & 0 & 0 & 0 \end{pmatrix}, \quad T_2 = \begin{pmatrix} 0 & -\frac{1}{2} & \frac{\sqrt{3}}{2} & 0 \\ 0 & 0 & 0 & -\frac{1}{2} \\ 0 & 0 & 0 & \frac{\sqrt{3}}{2} \\ 0 & 0 & 0 & 0 \end{pmatrix},$$

$$T_3 = \begin{pmatrix} 0 & -\frac{1}{2} & -\frac{\sqrt{3}}{2} & 0 \\ 0 & 0 & 0 & -\frac{1}{2} \\ 0 & 0 & 0 & -\frac{\sqrt{3}}{2} \\ 0 & 0 & 0 & 0 \end{pmatrix},$$

we have that $\|p\|_\infty = 1$, (T_1, T_2, T_3) are commuting contractions, and $\|p(T_1, T_2, T_3)\| = \frac{6}{5}$. It remains to see that we can perturb (T_1, T_2, T_3) slightly to make them diagonalizable. For this let

$$Q = \begin{pmatrix} 0 & 0 & 1 & 0 \\ 0 & 0 & 0 & 0 \\ 0 & 0 & 0 & 1 \\ 0 & 0 & 0 & 0 \end{pmatrix}, \quad T_1(\epsilon) = \begin{pmatrix} 0 & 1 & \epsilon & 0 \\ 0 & \epsilon & 0 & 1 \\ 0 & 0 & 0 & \epsilon \\ 0 & 0 & 0 & 0 \end{pmatrix}.$$

It is now straightforward to check that

$$T_2 = -\frac{1}{2}T_1 + \frac{\sqrt{3}}{2}Q, \quad T_3 = -\frac{1}{2}T_1 - \frac{\sqrt{3}}{2}Q, \quad Q = a_\epsilon(T_1(\epsilon)),$$

where

$$a_\epsilon(x) = \frac{x}{\epsilon} + \frac{x^2}{\epsilon^4} - \frac{x^3}{\epsilon^3} - \frac{x^3}{\epsilon^5}.$$

If we now take $\widetilde{T}_1$ to be a diagonalizable perturbation of $T_1(\epsilon)$ and set

$$\widetilde{T}_2 = -\frac{1}{2}\widetilde{T}_1 + \frac{\sqrt{3}}{2}a_\epsilon(\widetilde{T}_1), \quad \widetilde{T}_3 = -\frac{1}{2}\widetilde{T}_1 - \frac{\sqrt{3}}{2}a_\epsilon(\widetilde{T}_1),$$

we obtain commuting diagonalizable $\widetilde{T}_1, \widetilde{T}_2, \widetilde{T}_3$. By a scaling (which is a small perturbation of 1) they can be made into strict contractions A_1, A_2, A_3, say. Keeping the perturbations small enough, we can always see to it that $\|p(A_1, A_2, A_3)\| > 1$.

The question whether any of the conditions of Theorem 4.6.4 are sufficient for the solvability of the Nevanlinna-Pick problem in the d-variable Schur class is open.

4.7 THE OPERATOR CORONA PROBLEM

We continue by the methods started in Section 4.2 to solve a linearly constrained interpolation problem in nest algebras. The result generalizes the well-known Arveson distance formula. Applications concern quasitriangular operators and a variant of the operator theoretic Corona theorem.

Let $\mathcal{H}$ be a Hilbert space and $\mathcal{N}$ a nest of orthogonal projections in $\mathcal{L}(\mathcal{H})$, i.e., $\mathcal{N}$ is a strongly closed, linearly ordered collection of orthogonal projections on $\mathcal{H}$, containing 0 and the identity. Recall that the *nest algebra* corresponding to $\mathcal{N}$ is defined as

$$\text{alg}\mathcal{N} := \{A \in \mathcal{L}(\mathcal{H}) \ : \ P^\perp AP = 0 \text{ for all } P \in \mathcal{N}\},$$

where $P^\perp = I - P$. The main result of this section is the following.

Theorem 4.7.1 *Let $\mathcal{N}$ be a nest on a Hilbert space $\mathcal{H}$ and $B \in \mathcal{L}(\mathcal{H})$, $S, T \in \mathcal{L}(\mathcal{K}, \mathcal{H})$ given operators. Then*

$$\inf\{\|A\| \ : \ A \in B + \text{alg}\mathcal{N}, AS = T\}$$

$$= \inf\left\{k \ : \begin{pmatrix} S^*P^\perp S & (T^* - S^*PB^*)P^\perp \\ P^\perp(T - BPS) & k^2 I - P^\perp BPB^*P^\perp \end{pmatrix} \geq 0 \text{ for all } P \in \mathcal{N}\right\}. \tag{4.7.1}$$

Proof. In order to obtain the inequality $\geq$, let $A \in B + \text{alg}\mathcal{N}$ satisfy $AS = T$ and denote $k = \|A\|$. By Lemma 4.2.2

$$\begin{pmatrix} I & S & A^*/k \\ S^* & S^*S & T^*/k \\ A/k & T/k & I \end{pmatrix} \geq 0.$$

Multiplying this inequality on the right with

$$\begin{pmatrix} P & -PS & -PA^*P^\perp \\ 0 & I & 0 \\ 0 & 0 & kP^\perp \end{pmatrix}$$

and on the left with its adjoint yields

$$\begin{pmatrix} P & 0 & 0 \\ 0 & S^*P^\perp S & (T^* - S^*PB^*)P^\perp \\ 0 & P^\perp(T - BPS) & k^2P^\perp - P^\perp BPB^*P^\perp \end{pmatrix} \geq 0$$

(use $P^\perp AP = P^\perp BP$). Thus $k = \|A\|$ is an element of the set on the right-hand side of (4.7.1), yielding the inequality $\geq$ in (4.7.1).

It remains to prove $\leq$. Denote the right-hand side of (4.7.1) by L. In case $L = 0$, we get that $P^\perp BP = 0$ for all $P \in \mathcal{N}$, implying $B \in \text{alg}\mathcal{N}$, and also that

$$\begin{pmatrix} S^*S & T^* \\ T & 0 \end{pmatrix} \geq 0$$

(take $P = 0$), yielding $T = 0$. But then it follows that $A = 0$ satisfies $A \in B + \text{alg}\mathcal{N}$ and $AS = T$, thus the left-hand side of (4.7.1) also equals 0.

Assume that $L > 0$. Without loss of generality we may assume that $L = 1$ (divide in (4.7.1) the operators A, B and T by L). In the case that $\mathcal{N}$ is a finite nest the result follows directly from Theorem 4.2.3.

Denote the left-hand side of (4.7.1) by $\text{cd}(B, \text{alg}\mathcal{N})$. The set $\{\text{alg}\mathcal{M}\}$ as $\mathcal{M}$ runs over finite subnests of $\mathcal{N}$ ordered by inclusion is a net of weakly closed subspaces of $\mathcal{L}(\mathcal{H})$ with intersection $\text{alg}\mathcal{N}$. Note that since $\text{alg}\mathcal{N} \subseteq \text{alg}\mathcal{M}$ we have that $L \geq \sup_{\mathcal{M}} \text{cd}(B, \text{alg}\mathcal{M})$. For each $\mathcal{M}$ let

$$C_{\mathcal{M}} := \{D \in \text{alg}\mathcal{M} \;:\; \|B + D\| \leq L \text{ and } (B + D)S = T\}.$$

Then $C_{\mathcal{M}}$ is a nonempty set that is compact in the weak operator topology. Moreover,

$$\{C_{\mathcal{M}} \;:\; \mathcal{M} \text{ is a finite subnest of } \mathcal{N}\}$$

has the finite intersection property, because $C_{(\cup\mathcal{M}_i)} \subseteq \cap C_{\mathcal{M}_i}$. Thus there exists an operator D in the intersection of all $C_{\mathcal{M}}$. Thus $D \in \text{alg}\mathcal{N}$ and $(B + D)S = T$, yielding that $A := B + D$ belongs to the set on the left-hand side of (4.7.1). Since $\|A\| \leq L$ we obtain the inequality $\leq$ in (4.7.1). □

Remark 4.7.2 In the case that $S = T = 0$, (4.7.1) yields

$$\begin{aligned} &\inf\{\, \|A\| \;:\; A \in B + \text{alg}\mathcal{N}\} \\ &\quad = \inf\{k \;:\; k^2 I - P^\perp BPB^*P^\perp \geq 0 \text{ for all } P \in \mathcal{N}\} \\ &\quad = \sup\{\|P^\perp BP\| \;:\; P \in \mathcal{N}\}. \end{aligned}$$

We obtain as a corollary the following existence result.

Corollary 4.7.3 *Under the hypothesis of Theorem 4.7.1, there exists an operator $A \in B + \text{alg}\mathcal{N}$ such that $AS = T$ if and only if there exists an $\epsilon > 0$ such that*

$$\epsilon \|P^\perp (T - BPS)x\| \le \|P^\perp Sx\| \tag{4.7.2}$$

for all $P \in \mathcal{N}$ and $x \in \mathcal{H}$. In that case, $\frac{1}{\epsilon} + \|B\|$ is an upper bound for (4.7.1).

Proof. It is easy to see that there exists a $k > 0$ such that

$$\begin{pmatrix} S^* P^\perp S & (T^* - S^* PB^*)P^\perp \\ P^\perp (T - BPS) & k^2 I - P^\perp BPB^* P^\perp \end{pmatrix} \ge 0 \tag{4.7.3}$$

if and only if there exists an $M > 0$ such that

$$\begin{pmatrix} S^* P^\perp S & (T^* - S^* PB^*)P^\perp \\ P^\perp (T - BPS) & M^2 I \end{pmatrix} \ge 0.$$

The latter holds if and only if (4.7.2) holds with $\epsilon = \frac{1}{M}$. Putting now $k = M + \|B\|$, we see that (4.7.3) holds, so $M + \|B\|$ is an upper bound for (4.7.1). □

Let $\mathcal{H}$ be a separable Hilbert space. An operator $A \in \mathcal{L}(\mathcal{H})$ is called *quasitriangular* if there is an increasing sequence P_n of finite dimensional projections such that $P_n \uparrow I$, and $\|P_n^\perp A P_n\| \to 0$. In this section we want to consider the algebra of all operators which are quasitriangular relative to a *fixed* sequence $\mathcal{P} = \{P_n\}$. We denote the set of quasitriangular operators by $\mathcal{QT}$. It is known that $\mathcal{QT} = \text{alg}\mathcal{P} + \mathcal{K}(\mathcal{H})$, where $\mathcal{K}(\mathcal{H})$ denotes the ideal of compact operators on $\mathcal{H}$.

Theorem 4.7.4 *Let $B \in \mathcal{L}(\mathcal{H})$, and $S, T \in \mathcal{L}(\mathcal{K}, \mathcal{H})$. Then*

$$\inf\{\|A\| \ : \ A \in B + \mathcal{QT}, AS = T\} = \limsup k_n, \tag{4.7.4}$$

where

$$k_n = \inf\left\{k \ : \begin{pmatrix} S^* P_n^\perp S & (T^* - S^* P_n B^*)P_n^\perp \\ P_n^\perp (T - BP_n S) & k^2 I - P_n^\perp BP_n B^* P_n^\perp \end{pmatrix} \ge 0\right\}.$$

Proof. Let $L = \limsup k_n$ and $\epsilon > 0$. Then there exists an index n_0 such that $k_n \le L + \epsilon$ for all $n \ge n_0$. By Theorem 4.7.1 there exists an operator $A \in B + \text{alg}\{0, I, P_{n_0}, P_{n_0+1}, \ldots\}$ such that $\|A\| \le L + \epsilon$ and $AS = T$. It is easy to see that A is a finite rank perturbation of an operator in $B + \text{alg}\mathcal{N}$. This proves the inequality $\le$ in (4.7.4).

In order to prove the opposite inequality, let $A \in B + \mathcal{QT}$ be such that $AS = T$. Write $A = B + C + K$, where $C \in \text{alg}\mathcal{P}$ and $K \in \mathcal{K}(\mathcal{H})$. Then $A \in (B + K) + \text{alg}\mathcal{P}$ and $AS = T$. By Theorem 4.7.1 we obtain that
$\|A\| \ge p$

$$= \inf\left\{q : \begin{pmatrix} S^* P_n^\perp S & (T^* - S^* P_n (B + K)^*)P_n^\perp \\ P_n^\perp (T - (B + K)P_n S) & q^2 I - P_n^\perp (B + K)P_n (B + K)^* P_n^\perp \end{pmatrix} \ge 0 \right.$$

$$\left. \text{for all } n \in \mathbb{N}\right\}.$$

Since $\lim_{n\to\infty} \|(I - P_n)K\| = 0$, we have that for any $\delta > 0$ there is an index n_1 such that $p \geq k_n - \delta$ for all $n \geq n_1$. This implies that $\|A\| \geq \limsup_{n\to\infty} k_n$, finishing the proof. □

An easy application of Theorem 4.7.1 yields the following.

Theorem 4.7.5 *Let $\mathcal{N}$ be a nest on a Hilbert space $\mathcal{H}$ and let $B_1, \ldots B_n \in \mathcal{L}(\mathcal{H})$, $S_1, \ldots, S_n, T \in \mathcal{L}(\mathcal{K}, \mathcal{H})$ be given operators. Then*
$\inf\{\| \begin{pmatrix} A_1 & \cdots & A_n \end{pmatrix} \| \; : \; A_i \in B_i + \mathrm{alg}\mathcal{N}, A_1S_1 + \cdots + A_nS_n = T\}$

$$= \inf \left\{ k \; : \; \begin{pmatrix} \sum_{i=1}^n S_i^* P^\perp S_i & \left(T^* - \sum_{i=1}^n (S_i^* P B_i^*)\right) P^\perp \\ P^\perp \left(T - \sum_{i=1}^n (B_i P S_i)\right) & k^2 I - P^\perp \sum_{i=1}^n (B_i P B_i^*) P^\perp \end{pmatrix} \geq 0 \right.$$

$$\left. \text{for all } P \in \mathcal{N} \right\}. \tag{4.7.5}$$

Proof. Consider the Banach algebra $\mathcal{C}^{n\times n} \otimes \mathcal{L}(\mathcal{H})$ of all $n \times n$ matrices over $\mathcal{L}(\mathcal{H})$, regarded as an algebra of operators on the Hilbert space direct sum $\mathcal{H}^{(n)} = \mathcal{H}\oplus\cdots\oplus\mathcal{H}$. Writing $P^{(n)} = P\oplus\cdots\oplus P$ (i.e., the $n\times n$ diagonal matrix with P in the diagonal entries), one easily checks that $\mathcal{N}^{(n)} := \{P^{(n)} \; : \; P \in \mathrm{alg}\mathcal{N}\}$ is a nest algebra on $\mathbb{C}^{n\times n} \otimes \mathcal{L}(\mathcal{H})$. Further, $(C_{ij}) \in \mathrm{alg}\mathcal{N}^{(n)}$ if and only if $C_{ij} \in \mathrm{alg}\mathcal{N}$ for all i and j. Let now

$$B = \begin{pmatrix} B_1 & B_2 & \cdots & B_n \\ 0 & 0 & \cdots & 0 \\ \vdots & \vdots & \cdots & \vdots \\ 0 & 0 & \cdots & 0 \end{pmatrix},$$

$$S = \begin{pmatrix} S_1 \\ S_2 \\ \vdots \\ S_n \end{pmatrix}, \quad T_1 = \begin{pmatrix} T \\ 0 \\ \vdots \\ 0 \end{pmatrix},$$

and apply Theorem 4.7.1 with these choices of B, S and T_1. This gives a solution $(A_{ij})_{i,j=1}^n$. Put $A_j = A_{1j}$, $j = 1, \ldots, n$. Using $\| \begin{pmatrix} A_{11} & \cdots & A_{1n} \end{pmatrix} \| \leq \|(A_{ij})_{i,j=1}^n\|$ and Theorem 4.7.1 yield the result. □

The analog of Corollary 4.7.3 is the following.

Corollary 4.7.6 *Under the hypothesis of Theorem 4.7.5, there exist operators $A_i \in B_i + \mathrm{alg}\mathcal{N}$ such that $\sum_{i=1}^n A_iS_i = T$ if and only if there exists an $\epsilon > 0$ such that*

$$\epsilon^2 \left\| P^\perp (T - \sum_{i=1}^n B_i P S_i) x \right\|^2 \leq \sum_{i=1}^n \|P^\perp S_i x\|^2$$

for all $P \in \mathcal{N}$ and $x \in \mathcal{H}$. In that case, $\frac{1}{\epsilon} + \| \begin{pmatrix} B_1 & \cdots & B_n \end{pmatrix} \|$ is an upper bound for (4.7.5).

For quasitriangular operators we obtain the following.

Theorem 4.7.7 *Let $B_1, \ldots, B_n \in \mathcal{L}(\mathcal{H})$, $S_1, \ldots, S_n, T \in \mathcal{L}(\mathcal{K}, \mathcal{H})$ be given operators. Then*

$$\inf \left\{ \left\| \begin{pmatrix} A_1 & \cdots & A_n \end{pmatrix} \right\| \, : \, A_i \in B_i + \mathcal{Q}T, A_1S_1 + \cdots + A_nS_n = T \right\} = \limsup_{m \to \infty} \, k_m,$$

where

$$k_m = \inf \left\{ k \, : \, \begin{pmatrix} \sum_{i=1}^n S_i^* P_m^\perp S_i & (T^* - \sum_{i=1}^n (S_i^* P_m B_i^*)) P_m^\perp \\ P_m^\perp (T - \sum_{i=1}^n (B_i P_m S_i)) & k^2 I - P_m^\perp \sum_{i=1}^n (B_i P_m B_i^*) P_m^\perp \end{pmatrix} \geq 0 \right\}. \quad (4.7.6)$$

The proof is straightforward, and therefore omitted.

We apply Theorem 4.7.5 to give an operator theoretic solution to the following *Corona-type problem.* "Given are functions $\psi, \phi_1, \ldots, \phi_n, g_1, \ldots, g_n \in L^\infty(\mathbb{T})$. When do there exist $f_1, \ldots, f_n \in H^\infty(\mathbb{T})$ such that

$$(f_1 + g_1)\phi_1 + \cdots + (f_n + g_n)\phi_n = \psi?\text{''}$$

In that case, find norm estimates for $\|f_i + g_i\|_\infty$, $i = 1, \ldots, n$.

Our solution is the following. (P_+ denotes the projection of $L^2(\mathbb{T})$ onto $H^2(\mathbb{T})$ and $P_- = I - P_+$.)

Theorem 4.7.8 *Let $\psi, \phi_1, \ldots, \phi_n, g_1, \ldots, g_n \in L^\infty(\mathbb{T})$ be given functions. Then there exist $f_1, \ldots, f_n \in H^\infty(\mathbb{T})$ such that $\sum_{i=1}^n (f_i + g_i)\phi_i = \psi$ if and only if there exists a $k \geq 0$ such that*

$$M_\psi^* M_\psi \leq k^2 \sum_{i=1}^n M_{\phi_i}^* M_{\phi_i} \quad (4.7.7)$$

and

$$\begin{pmatrix} \sum_{i=1}^n M_{\phi_i}^* P_- M_{\phi_i} & (M_\psi^* - \sum_{i=1}^n (M_{\phi_i}^* P_+ M_{g_i}^*)) P_- \\ P_-(M_\psi - \sum_{i=1}^n (M_{g_i} P_+ M_{\phi_i})) & k^2 I - P_- \sum_{i=1}^n (M_{g_i} P_+ M_{g_i}^*) P_- \end{pmatrix} \geq 0. \quad (4.7.8)$$

In that case $f_1, \ldots, f_n$ may be chosen such that

$$\| \begin{pmatrix} f_1 + g_1 & \cdots & f_n + g_n \end{pmatrix} \|_\infty := \sup_{z \in \mathbb{T}} \left(\sum_{i=1}^n |f_i(z) + g_i(z)|^2 \right)^{\frac{1}{2}} \leq k.$$

In particular, they may be chosen such that $\|f_i + g_i\|_\infty \leq k$ for $i = 1, \ldots, n$.

Proof. Let $P_\nu : L^2(\mathbb{T}) \to L^2(\mathbb{T})$, $\nu \in \mathbb{Z}$, denote the orthogonal projection given by $P_\nu(\sum_{j=-\infty}^{\infty} f_j z^j) = \sum_{j=-\infty}^{\nu} f_j z^j$. Put $\mathcal{N} = \{P_\nu : \nu \in \mathbb{Z}\} \cup \{0, I\}$, $B_i = M_{g_i}$, $S_i = M_{\phi_i}$, $i = 1, \ldots, n$, and $T = M_\psi$. If we let U denote the bilateral forward shift on $L^2(\mathbb{T})$ we obtain that $P_{\nu+1} = UP_\nu U^*$ and $M_f = UM_fU^*$ for any $f \in L^\infty(\mathbb{T})$. Therefore the positive semidefiniteness of the matrices in (4.7.5) for $P_\nu, \nu \in \mathbb{Z}$ is equivalent to the positive semidefiniteness of the operator matrix in (4.7.8). When $P = 0$ we obtain condition (4.7.7), and for $P = I$ the positive semidefiniteness of the operator matrix in (4.7.5) is trivially satisfied. Now we may apply Theorem 4.7.5 to obtain that the existence of $k \geq 0$ with the required properties is equivalent to the existence of $A_i \in M_{g_i} + \text{alg}\mathcal{N}$, $i = 1, \ldots, n$, such that $\sum_{i=1}^n A_i M_{\phi_i} = M_\psi$. Also, $A_1, \ldots, A_n$ may be chosen such that $\| \begin{pmatrix} A_1 & \cdots & A_n \end{pmatrix} \| \leq k$.

It is known that there exists an *expectation* (i.e., a norm one projection) $\pi : \mathcal{L}(L^2(\mathbb{T})) \to \{M_\zeta : \zeta \in L^\infty(\mathbb{T})\}$ such that $\pi(\text{alg}\mathcal{N}) \subseteq \{M_\zeta : \zeta \in H^\infty(\mathbb{T})\}$. Using that $\pi(AM_\zeta) = \pi(A)M_\zeta$ and $\pi(A_i - M_{g_i}) = M_{f_i}$, for some $f_i \in H^\infty(\mathbb{T})$, one obtains that

$$M_\psi = \pi(M_\psi) = \pi\left(\sum_{i=1}^n A_i M_{\phi_i}\right) = \sum_{i=1}^n M_{f_i+g_i} M_{\phi_i}$$

yielding $\sum_{i=1}^n (f_i + g_i)\phi_i = \psi$.

Since $\pi \otimes I : \mathcal{L}(L^2(\mathbb{T})) \otimes \mathbb{C}^{n\times n} \to \{M_\zeta : \zeta \in L^\infty(\mathbb{T})\} \otimes \mathbb{C}^{n\times n}$ is also an expectation, we obtain that

$$\left\| \begin{pmatrix} f_1 + g_1 & \cdots & f_n + g_n \end{pmatrix} \right\|_\infty = \left\| \begin{pmatrix} M_{f_1+g_1} & \cdots & M_{f_n+g_n} \\ 0 & \cdots & 0 \\ \vdots & & \vdots \\ 0 & \cdots & 0 \end{pmatrix} \right\|$$

$$= \left\| (\pi \otimes I) \begin{pmatrix} A_1 & \cdots & A_n \\ 0 & \cdots & 0 \\ \vdots & & \vdots \\ 0 & \cdots & 0 \end{pmatrix} \right\| \leq \left\| \begin{pmatrix} A_1 & \cdots & A_n \end{pmatrix} \right\| \leq k.$$

□

Corollary 4.7.9 *Under the hypothesis of Theorem 4.7.8, there exist* $f_1, \ldots, f_n \in H^\infty(\mathbb{T})$ *such that* $\sum_{i=1}^n (f_i + g_i)\phi_i = \psi$ *if and only if there exists an* $\epsilon \geq 0$ *such that*

$$\epsilon^2 \|M_\psi x\|^2 \leq \sum_{i=1}^n \|M_{\phi_i} x\|^2 \tag{4.7.9}$$

and

$$\epsilon^2 \|(P_-(M_\psi - \sum_{i=1}^n M_{g_i} P_+ M_{\phi_i}))x\|^2 \leq \sum_{i=1}^n \|P_- M_{\phi_i} x\|^2, \tag{4.7.10}$$

for any $x \in L^2(\mathbb{T})$. *In that case,* $1/\epsilon + \|g_i\|_\infty$ *is an upper bound for* $\|f_i\|_\infty$ *for* $i = 1, \ldots, n$.

Proof. In a way similar to the proof of Corollary 4.7.3, we prove that the existence of k which satisfies (4.7.8) is equivalent to the existence of an ϵ which satisfies (4.7.10). The rest is straightforward. □

4.8 JOINT OPERATOR/HILBERT-SCHMIDT NORM CONTROL EXTENSIONS

In this section we consider lower-triangular partial matrices such that all specified entries are Hilbert-Schmidt operators. We first prove estimates for both the operator and the Hilbert-Schmidt norms of the contractive central extension of the partial matrix. We extend the results to the problems of Nehari and Carathéodory-Fejér. This leads to $\sqrt{2}$ being the joint norm control constant for each of the problems under consideration.

Let $K = \left(\begin{smallmatrix} C & X \\ A & B \end{smallmatrix}\right)$ be a partial operator matrix such that $\| \left(\begin{smallmatrix} C \\ A \end{smallmatrix}\right) \| < 1$ and $\| \begin{pmatrix} A & B \end{pmatrix} \| \leq d < 1$. Then by Exercise 2.10.25, there exist contractions G_1 and G_2 such that $B = D_{A^*}G_1$ and $C = G_2 D_A$. In addition, assume that A, B, and C are Hilbert-Schmidt operators. Let F_c be the central completion of K, which corresponds to the choice of $X_c = -G_2 A^* G_1$ in (4.2.1). As $G_1 = D_{A^*}^{-1} B$, denoting by $\| \cdot \|_2$ the Hilbert-Schmidt norm of an operator, we have that

$$\|X_c\|_2 = \|G_2 A^* D_{A^*}^{-1} B\|_2 \leq \|G_2\| \|A^* D_{A^*}^{-1}\| \|B\|_2 \leq \frac{d}{\sqrt{1-d^2}} \|B\|_2, \quad (4.8.1)$$

where we used that $\|A^* D_{A^*}^{-1}\| \leq \frac{d}{\sqrt{1-d^2}}$ and that $f(x) = \frac{x}{\sqrt{1-x^2}}$ increases on $[0,1)$. Introducing $d_2(K) := (\|A\|_2^2 + \|B\|_2^2 + \|C\|_2^2)^{\frac{1}{2}}$, (4.8.1) implies that

$$\|F_c\|_2 = (\|A\|_2^2 + \|B\|_2^2 + \|C\|_2^2 + \|X_c\|_2^2)^{\frac{1}{2}} \leq \frac{1}{\sqrt{1-d^2}} d_2(K). \quad (4.8.2)$$

Consider the lower triangular partial operator matrix

$$K = \begin{pmatrix} A_{11} & ? & \cdots & ? \\ A_{21} & A_{22} & \cdots & ? \\ \vdots & \vdots & & \vdots \\ A_{n1} & A_{n2} & \cdots & A_{nn} \end{pmatrix} \quad (4.8.3)$$

and denote

$$A^{(i)} = \begin{pmatrix} A_{i1} & \cdots & A_{ii} \\ \vdots & & \vdots \\ A_{n1} & \cdots & A_{ni} \end{pmatrix}$$

for $i = 1, \ldots, n$. Recall the notation $d_\infty(K)$ for the Arveson distance of K. When $d_\infty(K) \leq 1$, Theorem 4.1.1 implies the existence of a contractive completion of K. Using the description in the introduction to Section 2.6, one can construct the contractive central completion F_c of K inductively as follows. First define A_{12} as the central completion of the contractive completion problem of the partial matrix representing the first two columns of K. Continue by defining at step j all missing entries in column $j+1$ by considering the central completion of the partial matrix representing the first $j+1$ columns of K.

Proposition 4.8.1 *The central completion F_c of any strictly contractive partial operator matrix K as in (4.8.3) is a strict contraction.*

Proof. If K is a strict contraction, then $\begin{pmatrix} I & K \\ K^* & I \end{pmatrix}$ is positive definite. By Corollary 2.6.2, $\begin{pmatrix} I & F_c \\ F_c^* & I \end{pmatrix}$ is positive definite, so F_c is a strict contraction. □

We assume in the rest of this section that all operators A_{ij}, $1 \leq j \leq i \leq n$ are Hilbert-Schmidt and denote $d_2(K) = (\sum_{1\leq j\leq i\leq n} \|A_{ij}\|_2^2)^{\frac{1}{2}}$. Assume first that $d_\infty(K) = d < 1$. At the jth step of the construction of the central completion F_c of K, consider the central completion of the problem $\begin{pmatrix} C & X \\ A & B \end{pmatrix}$ with

$$C = \begin{pmatrix} A_{11} & \cdots & A_{1j} \\ \vdots & & \vdots \\ A_{j1} & \cdots & A_{jj} \end{pmatrix}, \quad X = \begin{pmatrix} A_{1,j+1} \\ \vdots \\ A_{j,j+1} \end{pmatrix},$$

$$A = \begin{pmatrix} A_{j+1,1} & \cdots & A_{j+1,j} \\ \vdots & & \vdots \\ A_{n1} & \cdots & A_{nj} \end{pmatrix}, \quad B = \begin{pmatrix} A_{j+1,j+1} \\ \vdots \\ A_{n,j+1} \end{pmatrix}.$$

Since we have that $\| \begin{pmatrix} C \\ A \end{pmatrix} \| < 1$, and $\| \begin{pmatrix} A & B \end{pmatrix} \| \leq d$ the central completion X_c verifies (4.8.1). Similarly to (4.8.2), this leads to

$$\|F_c\|_2 \leq \frac{1}{\sqrt{1-d^2}} d_2(K). \tag{4.8.4}$$

Theorem 4.8.2 *Let K be a partial operator matrix as in (4.8.3) and let $\delta > 1$. Then K admits a completion F_c which simultaneously verifies*

$$\|F_c\| < \delta d_\infty(K) \quad \textit{and} \quad \|F_c\|_2 \leq \frac{\delta}{\sqrt{\delta^2 - 1}} d_2(K).$$

Proof. Let $\widetilde{F_c}$ be the central completion of $\frac{K}{\delta d_\infty(K)}$ and let $F_c = \delta d_\infty(K)\widetilde{F_c}$. Then certainly $\|F_c\| < \delta d_\infty(K)$, and applying (4.8.4) to $\widetilde{F_c}$ with $d = \frac{1}{\delta}$ we obtain $\|F_c\|_2 \leq \frac{\delta}{\sqrt{\delta^2-1}} d_2(K)$. □

The previous theorem with the choice of $\delta = \sqrt{2}$ leads to the so-called joint norm control constant.

Corollary 4.8.3 *Let K be a partial operator matrix as in (4.8.3) and let $\delta > 1$. Then K admits a completion F_c which simultaneously verifies*

$$\|F_c\| < \sqrt{2} d_\infty(K) \quad \textit{and} \quad \|F_c\|_2 \leq \sqrt{2} d_2(K).$$

We turn now to finding joint norm control estimates for the solutions of Nehari's problem. First assume that $\{C_n\}_{n=-\infty}^{-1}$ is a sequence of Hilbert-Schmidt operators in $\mathcal{L}(\mathcal{H})$. Denote $d_2(\{C_n\}_{n=-\infty}^{-1}) = (\sum_{n=-\infty}^{-1} \|C_n\|_2^2)^{\frac{1}{2}}$ and assume that $\|H(\{C_n\}_{n=-\infty}^{-1})\| = d < 1$. We recall the step-by-step construction in the proof of Theorem 4.3.1 (Nehari's problem). Consider in this process each D_k to define the central completion of the corresponding problem for $k = 0, \ldots, n$. One can easily prove by induction that for each $n \geq 0$, the contractive central completion of the problem

$$\begin{pmatrix} C_{-2} & \ddots & & & & \\ C_{-1} & C_{-2} & \ddots & & & \\ ? & C_{-1} & C_{-2} & \ddots & & \\ ? & D_0 & C_{-1} & C_{-2} & \ddots & \\ \vdots & \vdots & \ddots & \ddots & \ddots & \ddots \\ ? & D_{n-1} & \cdots & D_0 & C_{-1} & C_{-2} & \ddots \end{pmatrix}$$

is defined so that the column of "?" equals $\operatorname{col}(D_j)_{j=0}^n$. By (4.8.1), we have that

$$\left(\sum_{k=0}^{n} \|D_k\|_2^2\right)^{\frac{1}{2}} \leq \frac{d}{\sqrt{1-d^2}} d_2(\{C_n\}_{n=-\infty}^{-1}). \tag{4.8.5}$$

Letting n go to ∞ in the left-hand side, we obtain from (4.8.5) that Nehari's problem with data $\{C_n\}_{n=-\infty}^{-1}$ admits a solution $F_c(e^{it}) = \sum_{n=-\infty}^{-1} C_n e^{int} + \sum_{n=0}^{\infty} D_n e^{int}$ such that

$$\|F_c\|_2 = \left(\sum_{n=-\infty}^{-1} \|C_n\|_2^2 + \sum_{n=0}^{\infty} \|D_n\|_2^2\right)^{\frac{1}{2}} \leq \frac{1}{\sqrt{1-d^2}} d_2(\{C_n\}_{n=-\infty}^{-1}). \tag{4.8.6}$$

Theorem 4.8.4 *Let $\{C_n\}_{n=-\infty}^{-1}$ is a sequence of Hilbert-Schmidt operators in $\mathcal{L}(\mathcal{H})$ and let $\delta > 1$. Then there exists a function $F_c(e^{it}) = \sum_{n=-\infty}^{-1} C_n e^{int} + \sum_{n=0}^{\infty} D_n e^{int} \in L^\infty(\mathcal{L}(\mathcal{H}))$ which simultaneously verifies*

$$\|F_c\|_\infty \leq \delta \|H(\{C_n\}_{n=-\infty}^{-1})\| \quad \text{and} \quad \|F_c\|_2 \leq \frac{\delta}{\sqrt{\delta^2-1}} d_2(\{C_n\}_{n=-\infty}^{-1}).$$

Proof. The result can be derived from (4.8.6) by a trick similar to the one in the proof of Theorem 4.8.2. □

For the contractive Carathéodory-Fejér problem we can similarly derive the following joint norm control extension result. For a sequence $\{C_k\}_{k=0}^n$ of Hilbert-Schmidt operators in $\mathcal{L}(\mathcal{H})$ we denote $d_2(\{C_k\}_{k=0}^n) = (\sum_{k=0}^n \|C_k\|_2^2)^{\frac{1}{2}}$.

Theorem 4.8.5 *Let $\{C_k\}_{k=0}^n$ be a sequence of Hilbert-Schmidt operators in $\mathcal{L}(\mathcal{H})$ and let $\delta > 1$. Then there exists a function $F_c(z) = \sum_{k=0}^n C_k z^k + \sum_{k=n+1}^{\infty} D_k z^k \in H^\infty(\mathcal{L}(\mathcal{H}))$ which simultaneously verifies*

$$\|F_c\|_\infty \leq \delta \|T(\{C_k\}_{k=0}^n)\| \quad \text{and} \quad \|F_c\|_2 \leq \frac{\delta}{\sqrt{\delta^2-1}} d_2(\{C_k\}_{k=0}^n).$$

For both the Nehari and the contractive Carathéodory-Fejér problems, the solutions F_c obtained by considering at each step the central completion of the corresponding problem for D_n, $n \geq 0$, are referred to as the *contractive central solutions* of the problem. Both Theorems 4.8.4 and 4.8.5 contain estimates for the contractive central solutions.

4.9 AN L^∞ EXTENSION PROBLEM FOR POLYNOMIALS

Let $\mathcal{H}$ be a separable Hilbert space and $\{C_k\}_{k=0}^n$ be a sequence of operators in $\mathcal{L}(\mathcal{H})$. Let $p(z) = C_0 + C_1 z + \cdots + C_n z^n$. Define

$$\mathcal{S}(p) = \left\{ f \in L^\infty(\mathcal{L}(\mathcal{H})) : \ f(e^{it}) = \sum_{k=-\infty}^{\infty} D_k e^{ikt}, \ D_k = C_k, \ k = 0, \ldots, n \right\}.$$

In this section we are estimating the quantity $\nu(p) = \inf\{\|f\|_\infty : f \in \mathcal{S}(p)\}$; namely, we are looking for the smallest norm $L^\infty(\mathcal{L}(\mathcal{H}))$ extension of a polynomial, versus the smallest norm $H^\infty(\mathcal{L}(\mathcal{H}))$ extension in the Carathéodory-Fejér problem. So, by Theorem 4.3.10 we have that $\nu(p) \leq \|T(\{C_k\}_{k=0}^n)\|$, with $T(\{C_k\}_{k=0}^n)$ being the Toeplitz operator defined by (4.3.20). At the end of the section, we obtain as a consequence that every partially contractive lower diagonal Toeplitz matrix admits a Toeplitz completion of norm < 2. (It is already known from Subsection 4.1.1 that in general such a partial matrix does not admit a contractive Toeplitz completion.)

Let $\mathcal{C}_1(\mathcal{H})$ denote the set of all trace class operators on $\mathcal{H}$. Let $L^1(\mathcal{C}_1(\mathcal{H}))$ denote the set of all (weakly) measurable functions $f : \mathbb{T} \to \mathcal{C}_1(\mathcal{H})$ such that

$$\|f\|_{L^1(\mathcal{C}_1(\mathcal{H}))} = \frac{1}{2\pi} \int_0^{2\pi} \|f(e^{it})\|_1 dt < \infty,$$

where $\| \cdot \|_1$ denotes the trace norm.

It is known that $\mathcal{C}_1(\mathcal{H})^* = \mathcal{L}(\mathcal{H})$, and based on this, one can prove that $L^1(\mathcal{C}_1(\mathcal{H}))^* = L^\infty(\mathcal{L}(\mathcal{H}))$, with the duality map

$$\langle f, g \rangle = \frac{1}{2\pi} \int_0^{2\pi} \operatorname{tr}(f(e^{it}) g(e^{it})) dt, \quad f \in L^1(\mathcal{C}_1(\mathcal{H})), \ g \in L^\infty(\mathcal{L}(\mathcal{H})).$$

Using this duality, it is easy to see that $\nu(p)$ is a minimum. Indeed, suppose $\mathcal{P}_n(\mathcal{C}_1(\mathcal{H}))$ is the linear space of all analytic polynomials of degree $\leq n$, with coefficients in $\mathcal{C}_1(\mathcal{H})$; endow it with the norm inherited from $L^1(\mathcal{C}_1(\mathcal{H}))$. Our given $p(z)$ defines a linear functional ϕ_p on $\mathcal{P}_n(\mathcal{C}_1(\mathcal{H}))$ by the formula

$$\phi_p(q) = \sum_{k=0}^{n} \operatorname{tr}(B_k C_k), \tag{4.9.1}$$

where $q(z) = B_0 + zB_1 + \cdots + z^n B_n$. Since $L^1(\mathcal{C}_1(\mathcal{H}))^* = L^\infty(\mathcal{L}(\mathcal{H}))$, any function $f \in \mathcal{S}(p)$ corresponds to an extension of ϕ_p to the whole $L^1(\mathcal{C}_1(\mathcal{H}))$. It follows that

$$\nu(p) = \|\phi_p\|. \tag{4.9.2}$$

Moreover, by the Hahn-Banach theorem the infimum in the definition of $\nu(p)$ is actually a minimum.

In the scalar case, we may say more about the minimizing function by applying some classical duality arguments.

Theorem 4.9.1 *There exists a unique function $h \in \mathcal{S}(p)$ such that $\|h\|_\infty = \nu(p)$; moreover, $h = \nu(p)\frac{|q|}{q}$, for a certain $q \in \mathcal{P}_n$.*

Proof. The functional ϕ_p on $\mathcal{P}_n$ attains its norm in $q \in \mathcal{P}_n$; this means that $\phi_p(q) = \nu(p)\|q\|_1$. If $h \in \mathcal{S}(p)$ satisfies $\|h\|_\infty = \nu(p)$, the last relation becomes

$$\frac{1}{2\pi}\int_0^{2\pi} h(e^{it})q(e^{it})\,dt = \|h\|_\infty \|q\|_1.$$

Since q is a polynomial, with a finite number of zeros, this implies that $h(e^{it}) = \nu(p)\frac{|q|}{q}$ almost everywhere. The uniqueness of h is an immediate consequence. □

The above result shows thus that the (unique) minimizing function has constant modulus and is continuous outside a finite set.

For the given polynomial $p(z) = C_0 + C_1 z + \cdots + C_n z^n$, consider the partial matrix

$$A = \begin{pmatrix} C_0 & ? & ? & \cdots & ? \\ C_1 & C_0 & ? & \cdots & ? \\ C_2 & C_1 & C_0 & \cdots & ? \\ \vdots & \ddots & \ddots & \ddots & \vdots \\ C_n & C_{n-1} & C_{n-2} & \cdots & C_0 \end{pmatrix} \tag{4.9.3}$$

and define $\delta(p) = \max\{\|A_r\|,\ 0 \le r \le n\}$, where

$$A_r = \begin{pmatrix} C_r & C_{r-1} & \cdots & C_0 \\ C_{r+1} & C_r & \cdots & C_1 \\ \vdots & \ddots & \ddots & \cdots \\ C_n & C_{n-1} & \cdots & C_{n-r} \end{pmatrix}. \tag{4.9.4}$$

By Theorem 4.1.1, the quantity $\delta(p)$ represents the minimum of the norm of a completion of (4.9.3).

If we define $\widetilde{p}(z) = C_n + C_{n-1}z + \cdots + C_0 z^n$, then $\delta(\widetilde{p}) = \delta(p)$ and $\nu(\widetilde{p}) = \nu(p)$.

The following estimation is the main result of this section.

Theorem 4.9.2 *For every polynomial p, $\nu(p) \le 2\frac{n+1}{n+2}\delta(p)$.*

The following lemma is essential for the proof.

Lemma 4.9.3 *If $q(z) = B_0 + B_1 z + \cdots + B_n z^n$, with $B_k \in \mathcal{C}_1(\mathcal{H})$, then*

$$\|T(\{B_k\}_{k=0}^n)\|_1 \le (n+1)\|q\|_{L^1(\mathcal{C}_1(\mathcal{H}))},$$

where $\|T(\{B_k\}_{k=0}^n)\|_1$ is the trace norm of the Toeplitz operator defined by (4.3.20).

Proof. For every $s \in [0, 2\pi]$, define $x(s), y(s) \in H^2(\mathbb{T})$ by the formulas

$$x(s) = \sum_{k=0}^{n} e^{iks} e_k, \quad y(s) = \sum_{k=0}^{n} e^{-iks} e_k.$$

Then $x(s) \cdot y(s)(g) = \langle g, x(s) \rangle y(s)$ is a rank 1 operator (on H^2), and

$$T(s) = (x(s) \cdot y(s)) \otimes q(e^{is})$$

is, for each s, an operator on $H^2(\mathcal{H})$ (using the identification $H^2(\mathcal{H}) = H^2 \otimes \mathcal{H}$). All $B_k \in \mathcal{C}_1(\mathcal{H})$; therefore $q(e^{is}) \in \mathcal{C}_1(\mathcal{H})$, $T(s)$ is also trace class, and

$$\|T(s)\|_{\mathcal{C}_1(H^2(\mathcal{H}))} = \|(x(s) \cdot y(s))\|_{\mathcal{C}_1(H^2)} \|q(e^{is})\|_{\mathcal{C}_1(\mathcal{H})}.$$

The trace class norm of the rank 1 operator $(x(s) \cdot y(s))$ is easily computed to be equal to $n+1$, and thus

$$\|T(s)\|_{\mathcal{C}_1(H^2(\mathcal{H}))} = (n+1) \|q(e^{is})\|_{\mathcal{C}_1(\mathcal{H})}. \tag{4.9.5}$$

Note now that q is continuous as a function from $\mathbb{T}$ to $\mathcal{C}_1(\mathcal{H})$, and thus T is also continuous from $[0, 2\pi]$ to $\mathcal{C}_1(H^2(\mathcal{H}))$. If $\Gamma' = \int_0^{2\pi} T(s)\, ds$, and $k, l \in \mathbb{Z}$, $\xi, \eta \in \mathcal{H}$, then

$$\langle \Gamma'(e_k \otimes \xi), e_l \otimes \eta \rangle = \int_0^{2\pi} \langle e_k, x(s) \rangle \langle y(s), e_l \rangle \langle q(e^{is})\xi, \eta \rangle \, ds$$

$$= \int_0^{2\pi} e^{-i(k+l)s} \langle q(e^{is})\xi, \eta \rangle \, ds = \left\langle \int_0^{2\pi} q(e^{is})\xi, e^{i(k+l)s}\eta \right\rangle ds$$

$$= \langle \Gamma(q)(e_k \otimes \xi), e_l \otimes \eta \rangle.$$

It follows then from (4.9.5) that

$$\|\Gamma(q)\|_{\mathcal{C}_1(H^2(\mathcal{H}))} \leq \int_0^{2\pi} \|T(s)\|_{\mathcal{C}_1(H^2(\mathcal{H}))} \leq (n+1)\|q\|_{L^1(\mathcal{C}_1(\mathcal{H}))}$$

and the lemma follows. □

Proof of Theorem 4.9.2. Let $\mathcal{M}_n(\mathcal{H}) = \mathcal{M}_n \otimes \mathcal{L}(\mathcal{H})$ be the algebra set of $n \times n$ matrices with entries in $\mathcal{L}(\mathcal{H})$, and let $\mathcal{G} \subset \mathcal{M}_{n+1}(\mathcal{H})$ be the linear space spanned by all Toeplitz matrices $T(\{B_k\}_{k=0}^n)$ with all $B_k \in \mathcal{C}_1(\mathcal{H})$. We may define a linear map ψ on $\mathcal{G}$ by the formula

$$\psi(T(\{B_k\}_{k=0}^n)) = \mathrm{tr}(T(\{B_k\}_{k=0}^n)^* T(\{C_k\}_{k=0}^n)).$$

Obviously $\mathcal{G}$ can be considered as a subspace of $\mathcal{C}_1(H^2(\mathcal{H}))$, identifying $\mathcal{M}_{n+1}(\mathcal{H})$ with operators acting on the set of polynomials of degree $\leq n$ with coefficients in $\mathcal{H}$. By the Hahn-Banach theorem, ψ can be extended to a functional on the whole $\mathcal{C}_1(H^2(\mathcal{H}))$, without increasing the norm. Such extensions correspond to operators on $H^2(\mathcal{H})$ which with respect to the decomposition $H^2(\mathcal{H}) = \bigoplus_{k=0}^{\infty} \mathbb{C}e^{ikt} \otimes \mathcal{H}$ have their first $n+1$ diagonals fixed. But, since any operator on the set of polynomials of degree $\leq n$ can be extended with the same norm to the whole of $H^2(\mathcal{H})$, it is enough to consider completions of (4.9.3). By Theorem 4.1.1, it follows that $\|\psi\| \leq \delta(p)$. Therefore, we have, using Lemma 4.9.3,

$$|\psi(T(\{B_k\}_{k=0}^n))| \leq \|\psi\| \|T(\{B_k\}_{k=0}^n)\|_{\mathcal{C}_1} \leq \delta(p)(n+1)\|q\|_1.$$

Since $\psi(T(\{B_k\}_{k=0}^n)) = \sum_{k=0}^n (k+1)\mathrm{tr}(B_{n-k}^* C_{n-k})$, we obtain

$$\left|\sum_{k=0}^n (k+1)\mathrm{tr}(B_{n-k}^* C_{n-k})\right| \leq (n+1)\delta(p)\|q\|_1. \tag{4.9.6}$$

On the other hand, the same argument applied to $\widetilde{p}$ leads to

$$\left|\sum_{k=0}^n (n-k+1)\mathrm{tr}(B_{n-k}^* C_{n-k})\right| \leq (n+1)\delta(\widetilde{p})\|q\|_1 = (n+1)\delta(p)\|q\|_1. \tag{4.9.7}$$

Adding (4.9.6) and (4.9.7) gives that

$$\left|\sum_{k=0}^n \mathrm{tr}(B_k^* C_k)\right| \leq 2\frac{n+1}{n+2}\delta(p)\|q\|_1.$$

According to (4.9.1) the last relation says that the functional ϕ_p defined on $\mathcal{P}_n(\mathcal{C}_1(\mathcal{H}))$ satisfies $\|\phi_p\| \leq 2\frac{n+1}{n+2}\delta(p)$. Relation (4.9.2) completes then the proof of the theorem. □

Let $\delta_T(p)$ denote the minimum over all Toeplitz completions of the partial matrix A in (4.9.3). As known from Subsection 4.1.1, we have in general that $\delta_T(p) > \delta(p)$, meaning that among all completions of A of minimum norm there fails to exist a Toeplitz one. The following is a consequence of Theorem 4.9.2.

Theorem 4.9.4 *With the above notation, $\delta_T(p) \leq 2\frac{n+1}{n+2}\delta(p)$.*

Proof. For a given partial matrix A as in (4.9.3), let $p(z) = C_0 + C_1 z + \cdots + C_n z^n$, and let $F(e^{it}) = \sum_{k=-\infty}^{\infty} C_k e^{ikt} \in \mathcal{S}(p)$ be an extension of p with $\|F\|_\infty = \nu(p)$. Then

$$\begin{pmatrix} C_0 & C_{-1} & \cdots & C_{-n} \\ C_1 & C_0 & \cdots & C_{-n+1} \\ \vdots & \ddots & \ddots & \vdots \\ C_n & C_{n-1} & \cdots & C_0 \end{pmatrix}$$

is a Toeplitz completion of A and also a principal submatrix of M_F. Thus $\delta_T(p) \leq \nu(p) \leq 2\frac{n+1}{n+2}\delta(p)$. □

Theorem 4.9.4 has the following immediate consequence.

Corollary 4.9.5 *Every lower triangular partially contractive operator Toeplitz matrix admits a Toeplitz completion of norm less than 2.*

4.10 SUPEROPTIMAL COMPLETIONS

In this section we consider lower triangular partial matrices all specified entries of which are compact operators. Minimizing the norm (largest singular value) of a compact completion leads generally to several solutions. We show that when one minimizes the sequence of singular values with respect to a lexicographical order, the solution becomes unique. This leads to the so-called "superoptimal solution."

For a compact operator B acting on a Hilbert space we order its sequence of singular values $(s_1(B), s_2(B), \ldots)$ in nonincreasing order, that is,

$$s_1(B) \geq s_2(B) \geq \cdots .$$

A partial lower triangular (compact operator) matrix

$$K = \begin{pmatrix} A_{11} & ? & \cdots & ? \\ A_{21} & A_{22} & \cdots & ? \\ \vdots & \vdots & & \vdots \\ A_{k1} & A_{k2} & \cdots & A_{kk} \end{pmatrix} \tag{4.10.1}$$

is a matrix in which the lower triangular blocks A_{ij}, $i \geq j$, are specified compact operators acting from $\mathcal{H}_j$ into $\mathcal{K}_i$, where $\mathcal{H}_j$ and $\mathcal{K}_i$ are separable Hilbert spaces which we allow to be trivial, and in which the upper triangular blocks A_{ij}, $i < j$, are free variables in the ideal of compact operators acting $\mathcal{H}_j \to \mathcal{K}_i$. It is for notational convenience that we allow trivial spaces, since in this way we are able to view also a $m \times m$ strictly lower triangular partial matrix as being of the form (4.10.1). Indeed, this is done by putting $k = m + 1$, $\mathcal{H}_1 = (0)$ and $\mathcal{K}_n = (0)$. For instance, the partial matrix $(?) : \mathbb{C} \to \mathbb{C}$ is viewed as the 2×2 partial matrix

$$\begin{pmatrix} 0 & ? \\ 0 & 0 \end{pmatrix} : (0) \times \mathbb{C} \to \mathbb{C} \times (0).$$

Given a partial matrix K as in (4.10.1), recall from Section 4.1 the notation $d_\infty(K)$ for the Arveson distance of K (the minimum over the norms of any of the completions of K). We shall prove the following representation for the set of all completions of a partial matrix K with Arveson distance equal to $d_\infty(K)$.

Theorem 4.10.1 *Let K be a lower triangular partial matrix (4.10.1). Then there exist a partial lower triangular matrix D and unitary operators V and W such that the set of all completions A of K with $s_1(A) = d_\infty(K)$ is of the form*

$$\left\{ WBV^* \ : \ B = \begin{pmatrix} d_\infty(K) & 0 \\ 0 & C \end{pmatrix} \text{ with } C \text{ a completion of } D \text{ with } s_1(C) \leq d_\infty(K) \right\}. \tag{4.10.2}$$

Moreover, this correspondence is one-to-one.

We shall give an explicit construction for the unitary operators V and W, and the partial matrix D.

As a consequence of the above theorem we obtain a procedure to find the unique *superoptimal completion* A of K, that is, the completion A for which the sequence of singular values $(s_1(A), s_2(A), \ldots)$ is minimal in lexicographical order among all completions of K (thus, if B is another completion of K, we have that there exists an i such that $s_1(B) = s_1(A), \ldots, s_{i-1}(B) = s_{i-1}(A)$ and $s_i(B) > s_i(A)$). Indeed, after having found representation (4.10.2) for the set of all completions of K whose norm is minimal, one can proceed with finding a similar representation for the set of completions of D with minimal norm, and continue in that way. The completion obtained by taking the limit is the superoptimal one for K. When the underlying spaces are finite dimensional the procedure consists of a finite number of steps. We shall illustrate this process with an example.

We prove Theorem 4.10.1 by constructing the unitaries V and W and the partial matrix D.

Lemma 4.10.2 *Let K be the partial matrix* (4.10.1), *and let $i \in \{1, \ldots, k\}$ be such that*

$$s_1 \begin{pmatrix} A_{i1} & \ldots & A_{ii} \\ \vdots & & \vdots \\ A_{k1} & \ldots & A_{ki} \end{pmatrix} = d_\infty(K).$$

Let $v = \mathrm{col}(v_p)_{p=1}^{i}$ and $w = \mathrm{col}(w_p)_{p=i}^{k}$ be a maximizing pair of unit vectors such that

$$\begin{pmatrix} A_{i1} & \ldots & A_{ii} \\ \vdots & & \vdots \\ A_{k1} & \ldots & A_{ki} \end{pmatrix} \begin{pmatrix} v_1 \\ \vdots \\ v_i \end{pmatrix} = \begin{pmatrix} w_i \\ \vdots \\ w_k \end{pmatrix}. \tag{4.10.3}$$

Put $\widehat{v} = \binom{v}{0}$ and $\widehat{w} = \binom{0}{w}$. Then for all completions A of K with minimal possible norm we have that

$$A\widehat{v} = d_\infty(K)\widehat{w}. \tag{4.10.4}$$

Proof. If A is a completion with minimal possible norm we have that

$$M := \begin{pmatrix} d_\infty(K)I & A \\ A^* & d_\infty(K)I \end{pmatrix} \geq 0.$$

Since $\binom{-w}{v}$ belongs to the kernel of a principal submatrix of M (due to (4.10.3)) and M is positive semidefinite, we must have that $\binom{-\widehat{w}}{\widehat{v}}$ belongs to the kernel of M. This yields (4.10.4). □

Corollary 4.10.3 *Under the hypotheses of Lemma 4.10.2, we have for any completion $A = (A_{ij})_{i,j=1}^{k}$ of K of minimal possible norm that*

$$\begin{pmatrix} A_{p,p+1} & \ldots & A_{pi} \end{pmatrix} \begin{pmatrix} v_{p+1} \\ \vdots \\ v_i \end{pmatrix} = -\begin{pmatrix} A_{p1} & \ldots & A_{pp} \end{pmatrix} \begin{pmatrix} v_1 \\ \vdots \\ v_p \end{pmatrix}, \quad 1 \leq p \leq i-1, \tag{4.10.5}$$

and

$$\begin{pmatrix} w_i^* & \dots & w_{p-1}^* \end{pmatrix} \begin{pmatrix} A_{ip} \\ \vdots \\ A_{p-1,p} \end{pmatrix} = - \begin{pmatrix} w_p^* & \dots & w_k^* \end{pmatrix} \begin{pmatrix} A_{pp} \\ \vdots \\ A_{kp} \end{pmatrix}, \ i+1 \le p \le k. \tag{4.10.6}$$

Proof. Equations (4.10.5) and (4.10.6) follow directly from (4.10.4) and the dual equality $\widehat{w}^* A = d_\infty(K)\widehat{v}^*$, respectively. □

Notice that the right-hand sides of (4.10.5) and (4.10.6) only involve given parts of the partial matrix K, and that the left-hand sides of (4.10.5) and (4.10.6) only involve unspecified parts of K. With this knowledge we are now able to transform the minimal norm completion problem for K into the lower dimensional problem described in the right-hand side of (4.10.2).

Proof of Theorem 4.10.1 (construction of V, W, and D). Given K as in (4.10.1), choose i, v, and w such that (4.10.3) is satisfied. Let V be the matrix given by

$$\begin{pmatrix} v_1 & F_1 & \frac{\alpha_1}{\mu_1} v_1 & 0 & 0 & \cdots & 0 & 0 & 0 & \cdots & 0 \\ v_2 & 0 & \frac{1}{\mu_1} v_2 & F_2 & \frac{\alpha_2}{\mu_2} v_2 & \cdots & 0 & 0 & 0 & \cdots & 0 \\ \vdots & \vdots & \vdots & \vdots & \vdots & & \frac{\alpha_{i-1}}{\mu_{i-1}} v_{i-1} & \vdots & \vdots & & \vdots \\ v_i & 0 & \frac{1}{\mu_1} v_i & 0 & \frac{1}{\mu_2} v_i & \cdots & \frac{1}{\mu_{i-1}} v_i & F_i & 0 & \cdots & 0 \\ 0 & 0 & 0 & 0 & 0 & \cdots & 0 & 0 & I_{H_{i+1}} & \cdots & 0 \\ \vdots & \vdots & \vdots & \vdots & \vdots & & \vdots & \vdots & \vdots & & \vdots \\ 0 & 0 & 0 & 0 & 0 & \cdots & 0 & 0 & 0 & \cdots & I_{H_k} \end{pmatrix},$$

defined on

$$\mathbb{C} \oplus (\mathcal{H}_1 \ominus \operatorname{Span}\{v_1\}) \oplus \cdots \oplus \mathbb{C} \oplus (\mathcal{H}_i \ominus \operatorname{Span}\{v_i\}) \oplus \mathcal{H}_{i+1} \oplus \cdots \oplus \mathcal{H}_k,$$

where

F_j is the identical embedding of $\mathcal{H}_j \ominus \operatorname{Span}\{v_j\}$ in $\mathcal{H}_j$, $j = 1, \dots, i$,

$\alpha_j = -\frac{\|v_{j+1}\|^2 + \cdots + \|v_i\|^2}{\|v_j\|^2}$, $j = 1, \dots, i-1$,

$\mu_j = \sqrt{|\alpha_j|^2 \|v_j\|^2 + \|v_{j+1}\|^2 + \cdots + \|v_i\|^2}$, $j = 1, \dots, i-1$,

and in case $v_j = 0$ (making α_j and μ_j undefined) the column in V involving α_j and μ_j should be omitted. One may easily check that V is unitary.

Analogously, let W be the matrix given by

$$\begin{pmatrix} 0 & I_{K_1} & \cdots & 0 & 0 & 0 & 0 & \cdots & 0 & 0 \\ \vdots & \vdots & & \vdots & \vdots & \vdots & \vdots & & \vdots & \vdots \\ 0 & 0 & \cdots & I_{K_{i-1}} & 0 & 0 & 0 & \cdots & 0 & 0 \\ w_i & 0 & \cdots & 0 & G_i & \frac{1}{\nu_{i+1}} w_i & 0 & \cdots & \frac{1}{\nu_k} w_i & 0 \\ w_{i+1} & 0 & \cdots & 0 & 0 & \frac{\beta_{i+1}}{\nu_{i+1}} w_{i+1} & G_{i+1} & \cdots & \frac{1}{\nu_k} w_{i+1} & 0 \\ w_{i+2} & 0 & \cdots & 0 & 0 & 0 & 0 & \cdots & \frac{1}{\nu_k} w_{i+2} & 0 \\ \vdots & \vdots & & \vdots & \vdots & \vdots & \vdots & & \vdots & \vdots \\ w_k & 0 & \cdots & 0 & 0 & 0 & 0 & \cdots & \frac{\beta_k}{\nu_k} w_k & G_k \end{pmatrix},$$

defined on

$$\mathbb{C} \oplus \mathcal{K}_1 \oplus \cdots \oplus \mathcal{K}_{i-1} \oplus (\mathcal{K}_i \ominus \operatorname{Span}\{w_i\}) \oplus \mathbb{C} \oplus \cdots \oplus \mathbb{C} \oplus (\mathcal{K}_k \ominus \operatorname{Span}\{w_k\}),$$

where G_j is the identical embedding of $\mathcal{K}_j \ominus \operatorname{Span}\{w_j\}$ in $\mathcal{K}_j$, $j = i, \dots, k$,
$\beta_j = -\frac{\|w_i\|^2 + \cdots + \|w_{j-1}\|^2}{\|w_j\|^2}$, $j = i+1, \dots, k$,
$\nu_j = \sqrt{|\beta_j|^2 \|w_j\|^2 + \|w_{j-1}\|^2 + \cdots + \|w_i\|^2}$, $j = i+1, \dots, k$,
and in case $w_j = 0$ (making β_j and ν_j undefined) the column in W involving β_j and ν_j should be omitted. Again, the unitarity of W follows easily.

Let now $A = (A_{ij})_{i,j=1}^k$ be a completion of K with minimal possible norm, and view A_{ij}, $i < j$, as variables. Computing W^*AV, and making use of Corollary 4.10.3 one sees that W^*AV is of the form

$$\begin{pmatrix} d_\infty(K) & 0 \\ 0 & C \end{pmatrix},$$

in which the free variables in C are concentrated in the upper triangular part. This finishes the proof of Theorem 4.10.1. □

We proceed with an example to illustrate the procedure.

Example 4.10.4 Let

$$K = \begin{pmatrix} 1 & a_{12} & a_{13} & a_{14} \\ 1 & 1 & a_{23} & a_{24} \\ 1 & 1 & \frac{1}{4} & a_{34} \\ 1 & 1 & 0 & 1 \end{pmatrix},$$

with a_{ij}, $1 \le i < j \le 4$, the free variables. Clearly,

$$d_\infty(K) = s_1 \begin{pmatrix} 1 & 1 \\ 1 & 1 \\ 1 & 1 \end{pmatrix} = \sqrt{6},$$

yielding $i = 2$, and we have the maximizing pair of unit vectors

$$v = \begin{pmatrix} \frac{1}{2\sqrt{2}} \\ \frac{1}{2\sqrt{2}} \end{pmatrix}, w = \begin{pmatrix} \frac{1}{3\sqrt{3}} \\ \frac{1}{3\sqrt{3}} \\ \frac{1}{3\sqrt{3}} \end{pmatrix}.$$

Thus we get

$$V = \begin{pmatrix} \frac{1}{2\sqrt{2}} & -\frac{1}{2\sqrt{2}} & 0 & 0 \\ \frac{1}{2\sqrt{2}} & \frac{1}{2\sqrt{2}} & 0 & 0 \\ 0 & 0 & 1 & 0 \\ 0 & 0 & 0 & 1 \end{pmatrix}, \quad W = \begin{pmatrix} 0 & 1 & 0 & 0 \\ \frac{1}{3\sqrt{3}} & 0 & \frac{1}{2\sqrt{2}} & \frac{1}{6\sqrt{6}} \\ \frac{1}{3\sqrt{3}} & 0 & -\frac{1}{2\sqrt{2}} & \frac{1}{6\sqrt{6}} \\ \frac{1}{3\sqrt{3}} & 0 & 0 & -\frac{1}{3\sqrt{6}} \end{pmatrix},$$

since

$$\alpha_1 = -\frac{\frac{1}{2}}{\frac{1}{2}} = -1, \quad \mu_1 = \sqrt{\frac{1}{2} + \frac{1}{2}} = 1, \quad \beta_3 = -\frac{\frac{1}{3}}{\frac{1}{3}} = -1,$$

$$\nu_3 = \sqrt{\frac{1}{3} + \frac{1}{3}} = \frac{1}{3}\sqrt{6}, \quad \beta_4 = -\frac{\frac{1}{3} + \frac{1}{3}}{\frac{1}{3}} = -2, \quad \nu_4 = \sqrt{\frac{4}{3} + \frac{1}{3} + \frac{1}{3}} = \sqrt{2}.$$

So we get that W^*AV is equal to

$$\begin{pmatrix} \sqrt{6} & 0 & \frac{1}{3\sqrt{3}}(a_{34}+\frac{1}{4}) & \frac{1}{3\sqrt{3}}(a_{24}+a_{34}+1) \\ \frac{1}{2\sqrt{2}}(1+a_{12}) & \frac{1}{2\sqrt{2}}(-1+a_{12}) & a_{13} & a_{14} \\ 0 & 0 & \frac{1}{2\sqrt{2}}(a_{34}-\frac{1}{4}) & \frac{1}{2\sqrt{2}}(a_{24}-a_{34}) \\ 0 & 0 & \frac{1}{6\sqrt{6}}(a_{23}+\frac{1}{4}) & \frac{1}{6\sqrt{6}}(a_{24}+a_{34}-2) \end{pmatrix}$$

$$= \begin{pmatrix} \sqrt{6} & 0 & 0 & 0 \\ 0 & -\sqrt{2} & a_{13} & a_{14} \\ 0 & 0 & -\frac{1}{4\sqrt{2}} & \frac{1}{2\sqrt{2}}(a_{24}-a_{34}) \\ 0 & 0 & 0 & -\frac{1}{2\sqrt{6}} \end{pmatrix},$$

where we used that $A\widehat{v} = \sqrt{6}\widehat{w}$ and $\widehat{w}^*A = \sqrt{6}\widehat{v}^*$, which imply that $1+a_{12} = 0$, $a_{23}+\frac{1}{4} = 0$, and $a_{24}+a_{34}+1 = 0$. Thus the partial matrix D is

$$\begin{pmatrix} -\sqrt{2} & c_{12} & c_{13} \\ 0 & -\frac{1}{4\sqrt{2}} & c_{23} \\ 0 & 0 & -\frac{1}{2\sqrt{6}} \end{pmatrix},$$

with $c_{12} = a_{13}$, $c_{13} = a_{14}$, and $c_{23} = \frac{1}{2\sqrt{2}}(a_{24}-a_{34})$ the new free variables. It is clear that in this case we get the superoptimal completion by choosing $c_{ij} = 0$, $i < j$, so that the superoptimal completion of K is

$$\begin{pmatrix} 1 & -1 & 0 & 0 \\ 1 & 1 & -\frac{1}{4} & -\frac{1}{2} \\ 1 & 1 & \frac{1}{4} & -\frac{1}{2} \\ 1 & 1 & 0 & 1 \end{pmatrix}.$$

Given a partial matrix K as in (4.10.1) one may ask what the lowest possible singular values (in lexicographical ordering) of a completion of K are. Let us first look at what happens when we apply the algorithm in the proof of Theorem 4.10.1 to the 2×2 partial matrix

$$K(X) = \begin{pmatrix} A_{11} & X \\ A_{21} & A_{22} \end{pmatrix},$$

where X is such that $\|K(X)\| \leq d_\infty(K)$. Let us assume that in (4.10.3) we have $i = 1$ (the case when $i = 2$ is similar). Thus we have

$$\begin{pmatrix} A_{11} \\ A_{21} \end{pmatrix} v_1 = \begin{pmatrix} w_1 \\ w_2 \end{pmatrix}.$$

Then the unitaries V and W in the proof of Theorem 4.10.1 are given by

$$V = \begin{pmatrix} v_1 & F_1 & 0 \\ 0 & 0 & I_{H_2} \end{pmatrix}, \quad W = \begin{pmatrix} w_1 & G_1 & \frac{1}{\nu_2}w_1 & 0 \\ w_2 & 0 & \frac{\beta_2}{\nu_2}w_2 & G_2 \end{pmatrix}.$$

Computing $W^*K(X)V$, we get that

$$W^*K(X)V = \begin{pmatrix} d_\infty(K) & 0 & 0 \\ 0 & G_1^*A_{11}F_1 & ? \\ 0 & \frac{1}{\nu_2}(w_1^*A_{11}+\beta_2 w_2^*A_{21})v & ? \\ 0 & G_2^*A_{21}F_1 & G_2^*A_{22} \end{pmatrix}, \quad (4.10.7)$$

where the ?s are entries that depend on X and can be treated as unknowns. The first two columns of this matrix are unitarily equivalent to $\binom{A_{11}}{A_{21}}$, while the last row of (4.10.7) is unitarily equivalent to $G_2^*\left(A_{21} \quad A_{22}\right)$. Let us denote the singular values of $\binom{A_{11}}{A_{21}}$ by $d_\infty(K) = s_1^{(1)} \geq s_2^{(1)} \geq \cdots$, and the singular values of $\left(A_{21} \quad A_{22}\right)$ by $s_1^{(2)} \geq s_2^{(2)} \geq \cdots$. Then the singular values $t_1 \geq t_2 \geq \cdots$ of $G_2^*\left(A_{21} \quad A_{22}\right)$ satisfy $t_i \leq s_i^{(2)}$, $i \in \mathbb{N}$. Now it follows that the partial matrix (4.10.7) can be completed so that its norm is $d_\infty(K)$ and its second singular value is $\max\{s_2^{(1)}, t_1\}(\leq \max\{s_2^{(1)}, s_1^{(2)}\})$. Repeating this argument, one can see that the superoptimal completion of K has singular values that are bounded above by the merging of the sequences $s_1^{(1)} \geq s_2^{(1)} \geq \cdots$, and $s_1^{(2)} \geq s_2^{(2)} \geq \cdots$ into a single nonincreasing sequence (keeping multiple appearances). By keeping exact track of the multiplicity of the singular value $d_\infty(K)$, one can in fact make stronger statement. The general case is as follows.

Theorem 4.10.5 *Let K be the partial matrix (4.10.1), and for $j \leq i$ let m_{ij} equal the number of singular values of*

$$\begin{pmatrix} A_{i1} & \cdots & A_{ij} \\ \vdots & & \vdots \\ A_{k1} & \cdots & A_{kj} \end{pmatrix}$$

equal to $d_\infty(K)$. Put $m = \sum_{i=1}^k m_{ii} - \sum_{i=1}^{k-1} m_{i+1,i}$. Furthermore, let $\{s_j^{(i)}\}_j$ denote the singular values of

$$\begin{pmatrix} A_{i1} & \cdots & A_{ii} \\ \vdots & & \vdots \\ A_{k1} & \cdots & A_{ki} \end{pmatrix}, \quad i = 1, \ldots, k,$$

that are not equal to $d_\infty(K)$. Merge the lists $\{s_j^{(1)}\}_j, \ldots, \{s_j^{(k)}\}_j$ giving the sequence $t_1 \geq t_2 \geq t_3 \geq \cdots$ (keep multiple appearances of a value). Then there exists a completion A of K that satisfies

$$s_1(A) = d_\infty(K), \ldots, s_m(A) = d_\infty(K) \tag{4.10.8}$$

and

$$s_{i+m}(A) \leq t_i, i = 1, 2, \ldots . \tag{4.10.9}$$

Proof. If we apply the result in Exercise 2.9.39 to the partially positive semidefinite banded Fredholm operator

$$\mathcal{M} = \begin{pmatrix} d_\infty(K)I & K \\ K^* & d_\infty(K)I \end{pmatrix},$$

then it follows that the minimal possible dimension of the kernel of a positive semidefinite completion of $\mathcal{M}$ is m. Consequently, any completion of A of K with $s_1(A) = d_\infty(K)$ necessarily has m singular values equal to $d_\infty(K)$.

Moreover, there exist completions among them where this count is precise. This takes care of (4.10.8).

We use induction to obtain also (4.10.9). First let $k = 2$, and thus

$$K = \begin{pmatrix} A_{11} & ? \\ A_{21} & A_{22} \end{pmatrix}.$$

As A_{21} has the singular value $d_\infty(K)$ appearing m_{21} times (which could be 0), we can decompose A_{21} as

$$A_{21} = \begin{pmatrix} d_\infty(K) I_{m_{21}} & 0 \\ 0 & \widetilde{A_{21}} \end{pmatrix}.$$

Decomposing A_{11} and A_{22} accordingly, we can write

$$K = \begin{pmatrix} 0 & \widetilde{A_{11}} & ? \\ d_\infty(K) I_{m_{21}} & 0 & 0 \\ 0 & \widetilde{A_{21}} & \widetilde{A_{22}} \end{pmatrix}.$$

Note that the zeros in the first column and second row are due to the norm of the first row and second column being equal to $d_\infty(K)$. Considering the partial matrix

$$\widetilde{K} = \begin{pmatrix} \widetilde{A_{11}} & ? \\ \widetilde{A_{21}} & \widetilde{A_{22}} \end{pmatrix},$$

we have that $\begin{pmatrix} \widetilde{A_{11}} \\ \widetilde{A_{21}} \end{pmatrix}$ has the singular value $d_\infty(K)$ appearing $m_{11} - m_{21}$ times, and then singular values $s_1^{(1)} \geq s_2^{(1)} \geq \cdots$, and $\begin{pmatrix} \widetilde{A_{21}} & \widetilde{A_{22}} \end{pmatrix}$ has singular value $d_\infty(K)$ appearing $m_{22} - m_{21}$ times, and then singular values $s_1^{(2)} \geq s_2^{(2)} \geq \cdots$. By the observations in the paragraph before Theorem 4.10.5, we now obtain that $\widetilde{K}$ has a completion with the singular value $d_\infty(K)$ appearing $m_{11} + m_{22} - 2m_{21}$ times followed by singular values that are bounded above by the merging of the sequences $s_1^{(1)} \geq s_2^{(1)} \geq \cdots$ and $s_1^{(2)} \geq s_2^{(2)} \geq \cdots$. But then the partial matrix K has a completion with an additional m_{21} singular eigenvalues $d_\infty(K)$, and the result follows.

We now assume that the theorem has been proven for partial matrices of size at most $(k-1) \times (k-1)$, and we let K be as in (4.10.1). Consider now the partial matrix obtained from K by leaving out the last column (keep the last row). This is a $k \times (k-1)$ partial matrix. For this matrix we can find a completion B satisfying

$$s_1(B) = \cdots = s_q(B) = d_\infty(K),$$

in which $q = \sum_{i=1}^{k-1} m_{ii} - \sum_{i=1}^{k-2} m_{i+1,i}$, and such that

$$s_{j+q}(B) \leq u_j, \ j = 1, 2, \dots , \tag{4.10.10}$$

in which $u_1 \geq u_2 \geq \cdots$ is the list obtained by merging the lists $\{s_j^{(1)}\}_j, \dots,$ $\{s_j^{(k-1)}\}_j$. Write $B = \begin{pmatrix} \widehat{B} \\ C \end{pmatrix}$ in which $C = \begin{pmatrix} A_{k1} & \cdots & A_{k,k-1} \end{pmatrix}$. Next we consider the 2×2 problem $\begin{pmatrix} \widehat{B} & ? \\ C & A_{kk} \end{pmatrix}$. Applying the 2×2 case gives that we can find a completion A satisfying (4.10.8) and

$$s_{j+m} \leq v_j, \; j = 1, 2, \ldots , \tag{4.10.11}$$

in which the list $v_1 \geq v_2 \geq \cdots$ is obtained by merging the lists $\{u_j\}_j$ and $\{s_j^{(k)}\}_j$. Inequalities (4.10.10) and (4.10.11) yield (4.10.9). □

Corollary 4.10.6 *Let K be the partial matrix (4.10.1). Then there exists a unique superoptimal completion A of K. Moreover, the singular values of this completion satisfy (4.10.8) and (4.10.9).*

Proof. By repeating the procedure outlined in Theorem 4.10.1 and choosing in each step a completion C of D with $s_1(C) = d_\infty(D)$, we produce a sequence of operators $A_1, A_2, \ldots$ such that for each i the sequence $(s_1(A_i), \ldots, s_i(A_i))$ is as low as possible in lexicographical order among all completions. This sequence $A_1, A_2, \ldots$ has the property that the singular vectors of $A_i, A_{i+1}, A_{i+2}, \ldots$ corresponding to singular values $\mu > s_i(A_i)$ coincide. By Theorem 4.10.5 the singular values are subject to the upper bounds (4.10.9), and since $t_j \to 0$, the operators A_i converge to a limit A. The operator A is uniquely defined by the process and is the superoptimal completion of K. In addition, (4.10.8) and (4.10.9) hold. □

In general it is not easy to read off the singular values exactly from the given data, but clearly we can give a complete answer in the 2×2 case.

Proposition 4.10.7 *Let*

$$K = \begin{pmatrix} \alpha & ? \\ \beta & \gamma \end{pmatrix}.$$

Then the singular values (s_1, s_2) of the superoptimal completion A of K are

$$(s_1, s_2) = \begin{cases} (|\beta|, 0) & \text{if } \alpha = \gamma = 0, \\ (\sqrt{|\alpha|^2 + |\beta|^2}, \sqrt{|\alpha|^2 + |\beta|^2}\frac{|\gamma|}{|\alpha|}) & \text{if } 0 \neq |\alpha| \geq |\gamma|, \\ (\sqrt{|\gamma|^2 + |\beta|^2}, \sqrt{|\gamma|^2 + |\beta|^2}\frac{|\alpha|}{|\gamma|}) & \text{if } 0 \neq |\gamma| \geq |\alpha| \;. \end{cases}$$

Proof. The superoptimal completion is

$$\begin{pmatrix} 0 & 0 \\ \beta & 0 \end{pmatrix} \text{ if } \alpha = \gamma = 0, \text{ and } \begin{pmatrix} \alpha & -\frac{\overline{\beta}\gamma}{\overline{\alpha}} \\ \beta & \gamma \end{pmatrix} \text{ if } 0 \neq |\alpha| \geq |\gamma|.$$

Interchanging γ and α gives the remaining case. □

Note that the singular values of the optimal completion of K (and the superoptimal completion itself) do not necessarily depend continuously upon the specified entries in K. Indeed, the superoptimal singular values of

$$\begin{pmatrix} \epsilon & ? \\ 1 & \epsilon \end{pmatrix}$$

are $(\sqrt{1 + |\epsilon|^2}, \sqrt{1 + |\epsilon|^2})$ if $\epsilon \neq 0$ and $(1, 0)$ if $\epsilon = 0$.

The strongest statements about the superoptimal singular values using the above results can be obtained by first performing a reduction based on the following simple observation.

Lemma 4.10.8 *Suppose that K is a partial matrix of the form*

$$\begin{pmatrix} K_1 & \mathcal{O}_{12} & \cdots & \mathcal{O}_{1k} \\ \mathcal{O}_{21} & K_2 & \cdots & \mathcal{O}_{2k} \\ \vdots & \vdots & \ddots & \vdots \\ \mathcal{O}_{k1} & \mathcal{O}_{k2} & \cdots & K_k \end{pmatrix} \tag{4.10.12}$$

in which K_i are partial matrices and $\mathcal{O}_{ij}$ are partial matrices with only zero operators as given entries. Then the superoptimal completion of (4.10.12) is a diagonal operator whose main diagonal entries are the superoptimal completions of K_i.

Given now a lower triangular partial matrix K, one may, via block diagonal unitary equivalence and permutation equivalence, reduce K to the form (4.10.12) with each K_i a triangular partial matrix. Then one can apply the bounds (4.10.8) and (4.10.9) to each of the components K_i and merge the thus obtained lists to get bounds for the superoptimal singular values of K.

Clearly, a lower bound for the jth singular value $s_j(A)$ of the superoptimal completion of a partial triangular matrix K is

$$s_j(A) \geq \max\left\{ s_j\begin{pmatrix} A_{i1} & \cdots & A_{ii} \\ \vdots & & \vdots \\ A_{k1} & \cdots & A_{ki} \end{pmatrix} : i = 1, \ldots, k \right\}.$$

4.11 SUPEROPTIMAL APPROXIMATIONS OF ANALYTIC FUNCTIONS

Before arriving at the main subject of this section, we first state two results related to bounded analytic operator-valued functions. A function $\Theta \in H^\infty(\mathcal{L}(\mathcal{K},\mathcal{H}))$ is called *inner* if the values of $\Theta(e^{it})$ are isometries a.e. on $\mathbb{T}$, and *outer* if the closure of $\{\Theta u : u \in H^2(\mathcal{K})\}$ is dense in $H^2(\mathcal{H})$ (the operator T_Θ has a dense range).

The following result, called the *Beurling-Lax-Halmos theorem*, describes all closed subspaces of $H^2(\mathcal{H})$ which are invariant under the unilateral shift operator S on $H^2(\mathcal{H})$.

Theorem 4.11.1 *A closed subspace $\mathcal{L}$ of $H^2(\mathcal{H})$ is invariant under the unilateral shift operator if and only if there exist a Hilbert space $\mathcal{K}$ and an inner function $\Theta \in H^\infty(\mathcal{L}(\mathcal{K},\mathcal{H}))$ such that $\mathcal{L} = \Theta H^2(\mathcal{K})$.*

Proof. A proof of this result may be found in [534]. □

The next result is about the existence of the so-called *inner-outer factorization* of a bounded analytic operator-valued function.

Theorem 4.11.2 *For every $\Theta \in H^\infty(\mathcal{L}(\mathcal{K},\mathcal{H}))$ there exist a Hilbert space $\mathcal{E}$, an outer function $\Theta_o \in H^\infty(\mathcal{L}(\mathcal{K},\mathcal{E}))$, and an inner function $\Theta_i \in H^\infty(\mathcal{L}(\mathcal{E},\mathcal{H}))$, such that*

$$\Theta(e^{it}) = \Theta_i(e^{it})\Theta_o(e^{it}). \tag{4.11.1}$$

Proof. Consider the subspace $\overline{\Theta H^2(\mathcal{H})}$ of $H^2(\mathcal{K})$. Since it is invariant under the unilateral shift, there exist a Hilbert space $\mathcal{E}$ and an inner function $\Theta_i \in H^\infty(\mathcal{L}(\mathcal{E},\mathcal{H}))$ such that the subspace $\overline{\Theta H^2(\mathcal{H})}$ of $H^2(\mathcal{K})$ coincides with $\Theta_i H^2(\mathcal{E})$. Define $\Theta_o(e^{it}) = \Theta_e(e^{it})^*\Theta(e^{it})$. The ranges of $\Theta(e^{it})$ are included in the ranges of Θ_i, so the relation (4.11.1) holds. Since Θ_o maps $H^2(\mathcal{K})$ to $H^2(\mathcal{E})$, Θ_e is analytic. The operator T_Θ being isometric, it follows from (4.11.1) that Θ_e is outer. □

Given a function $G \in L^\infty(\mathbb{C}^{m\times n})$, Corollary 4.3.3 gives the distance between G and $H^\infty(\mathbb{C}^{m\times n})$. When $G \in (H^\infty + C)(\mathbb{C}^{m\times n})$ and $m = 1$ or $n = 1$, we know from Corollary 4.3.9 that G admits a unique $H^\infty(\mathbb{C}^{m\times n})$ approximant. When both $m, n > 1$, it is exceptional that G has a unique best approximant in $H^\infty(\mathbb{C}^{m\times n})$. The purpose of this section is to prove that for $G \in (H^\infty + C)(\mathbb{C}^{m\times n})$, there is a unique $Q \in H^\infty(\mathbb{C}^{m\times n})$ which minimizes not only $\|G - Q\|_\infty$, but also the the supremum of the singular values of $G - Q$. Thus such functions have a unique "superoptimal" $H^\infty(\mathbb{C}^{m\times n})$ approximant, which is similar to what we already know from Section 4.10 for completions of lower triangular operator matrices.

We start with the following example.

Example 4.11.3 Let

$$G(z) = \begin{pmatrix} 2z^{-1} & 0 \\ 0 & z^{-1} \end{pmatrix}, \quad z \in \mathbb{T}.$$

Certainly

$$\operatorname{dist}_{L^\infty}(G, H^\infty(\mathbb{C}^{2\times 2})) \le \|G - 0\|_\infty = 2,$$

while consideration of the $(1,1)$ entry gives

$$\operatorname{dist}_{L^\infty}(G, H^\infty(\mathbb{C}^{2\times 2})) \ge \operatorname{dist}_{L^\infty}(2z^{-1}, H^\infty) = 2.$$

So the distance is exactly 2, and the functions for which $\|G - Q\|_\infty$ is minimized are all functions

$$Q(z) = \begin{pmatrix} 0 & 0 \\ 0 & g(z) \end{pmatrix}, \tag{4.11.2}$$

where $g \in H^\infty$, $\|g - z^{-1}\|_\infty \le 2$.

The idea of finding a unique best approximant led to the introduction of s^∞ minimization, described in the following. Recall that $s_j(M)$ denote the singular values of M in a nonincreasing order. For $F \in L^\infty(\mathbb{C}^{m\times n})$ define, for $j \ge 0$,

$$s_j^\infty := \operatorname*{ess\,sup}_{z\in\mathbb{T}} s_j((F(z))$$

and

$$s^\infty(F) := (s_1^\infty(F), s_2^\infty(F), s_3^\infty(F), \ldots).$$

Of course, at most $\min\{m, n\}$ terms of this sequence are nonzero. Given $G \in L^\infty(\mathbb{C}^{m\times n})$, we seek $Q \in H^\infty(\mathbb{C}^{m\times n})$ which minimizes the sequence

$s^\infty(G-Q)$ with respect to the lexicographical ordering. As $s_1^\infty(G-Q) = \|G-Q\|_\infty$, this is a strengthening of the criterion in Corollary 4.3.3.

In Example 4.11.3, the L^∞-optimal approximations Q as in (4.11.2) satisfy

$$s^\infty(G-Q) = (2, \|z^{-1} - g\|_\infty, 0, 0, \ldots),$$

and among all such sequences the lexicographic minimum is $(2, 1, 0, 0, \ldots)$, attained for $g = 0$ (and thus $Q = 0$).

The following is the main result of this section.

Theorem 4.11.4 *Let $G \in (H^\infty + C)(\mathbb{C}^{m\times n})$. Then the minimum with respect to the lexicographic ordering of $s^\infty(G-Q)$ over all $Q \in H^\infty(\mathbb{C}^{m\times n})$ is attained at a unique function Q_0. The singular values $s_j(G(z) - Q_0(z))$ are constant a.e. on $\mathbb{T}$ for $j \geq 1$.*

Let $\Omega \subseteq \mathbb{C}$ be an open subset and let $H(\Omega)$ denote the set of all analytic functions on Ω. A family $\mathcal{F} \subseteq H(\Omega)$ is called *normal* if every sequence of functions in $\mathcal{F}$ contains a subsequence which converges uniformly on compact subsets of Ω to an element of $\mathcal{F}$. We say that $\mathcal{F}$ is *locally bounded* if for each $a \in \Omega$ there exist constants M and $r > 0$ such that for all $f \in \mathcal{F}$, $|f(z)| \leq M$ for $|z - a| < r$. The following is known as Montel's theorem.

Theorem 4.11.5 *A family $\mathcal{F} \subseteq H(\Omega)$ is normal if and only if it is locally bounded.*

Proof. A proof of this result may be found in [150]. □

The fact that the minimum of $s^\infty(G-Q)$ is always attained can be shown by repeatedly using Montel's theorem (see Exercise 4.13.39).

The rest of the proof of Theorem 4.11.4 requires several preparatory steps and will be completed only at the end of this section. First, we recall and introduce some new notation. P_+ denotes the orthogonal projection of $L^2(\mathbb{C}^n)$ onto $H^2(\mathbb{C}^n)$, and $P_- = I - P_+$. By Exercise 4.13.23, Theorem 4.3.7 holds with a similar proof for Hankel operators with symbol in $L^\infty(\mathbb{C}^{m\times n})$. A function $F \in H^\infty(\mathbb{C}^{m\times n})$ is called *co-outer* if F^T is outer. We extend the notion of Toeplitz operators to symbols $F \in L^\infty(\mathbb{C}^{m\times n})$; namely, for such an F, $T_F : H^2(\mathbb{C}^n) \to H^2(\mathbb{C}^m)$, is defined by $T_F f = P_+(Ff)$ for every $f \in H^2(\mathbb{C}^n)$.

A function $u \in L^\infty(\mathbb{T})$ is called *badly approximable* if a closest point of $H^\infty(\mathbb{T})$ to u is the zero function, that is, if

$$\|u\|_\infty = \operatorname{dist}(u, H^\infty(\mathbb{T})) = \|H_u\|.$$

A function $u \in L^\infty(\mathbb{T})$ such that both u and $\overline{u}$ belong to $H^\infty + C$ is called *quasicontinuous*. We have the following result.

Theorem 4.11.6 *If $u \in H^\infty + C$ is badly approximable, then $\overline{u} \in H^\infty + C$, so u is quasicontinuous.*

Proof. A proof of this result may be found in [458]. □

For a matrix $A \in \mathbb{C}^{n\times m}$ and any choice of indices $1 \le i_1 \le i_2 \le n$ and $1 \le j_1 \le j_2 \le n$ we denote by $A_{i_1 i_2, j_1 j_2}$ the corresponding minor of A, that is, the determinant of the 2×2 submatrix of A corresponding to rows i_1, i_2 and columns j_1, j_2. By a *minor of A on the first column* we mean a minor $A_{i_1 i_2, j_1 j_2}$ with $j_1 = 1$.

The Example 4.11.3 is about a diagonal matrix function, in which case the notion of superoptimal approximation is pretty easy to analyze. The general case is treated by diagonalization, and for this we need some facts about unitary-valued matrix functions on the circle.

Consider a $\mathbb{C}^n$-valued inner function v_1; that is, $v_1 \in H^\infty(\mathbb{C}^n)$ is such that $v_1(z)$ is a unit vector a.e. $z \in \mathbb{T}$. We wish to complete v_1 to a unitary-valued function, which means finding a $\mathbb{C}^{n\times(n-1)}$-valued function v_2 such that $V = \begin{pmatrix} v_1 & v_2 \end{pmatrix}$ is unitary a.e. on $\mathbb{T}$. For proving Theorem 4.11.4, we have to use such unitary-valued V with v_2 taken conjugate analytic such that certain minors of V are analytic.

Theorem 4.11.7 *Let ϕ be a $\mathbb{C}^n$-valued inner function. Then there exists a co-outer function $\phi_c \in H^\infty(\mathbb{C}^{n\times(n-1)})$ such that*

$$\Phi = \begin{pmatrix} \phi & \overline{\phi_c} \end{pmatrix}$$

is unitary-valued on $\mathbb{T}$ and all 2×2 minors of Φ on the first column are in H^∞.

Proof. We begin by constructing a single column of ϕ_c. We seek $\vartheta \in H^\infty(\mathbb{C}^n)$ such that $\overline{\vartheta}$ is pointwise orthogonal to ϕ, and $\phi \wedge \overline{\vartheta}$, the $\frac{n(n-1)}{2}$-component column vector function whose components are the 2×2 minors of $\begin{pmatrix} \phi & \overline{\vartheta} \end{pmatrix}$, belongs to H^∞. Let

$$L := \mathrm{Ker} T_{\phi^T} = \{x \in H^2(\mathbb{C}^n) : \phi^T x = 0\}.$$

Pick $C \in \mathbb{C}^n$ and let $\vartheta = P_L C$, the orthogonal projection of C onto L in $H^2(\mathbb{C}^n)$. Certainly $\phi^T \vartheta = 0$, that is, ϕ is pointwise orthogonal to $\overline{\vartheta}$. The statement that $\phi \wedge \overline{\vartheta}$ is analytic is equivalent to $P_+(\overline{\phi} \wedge \vartheta)$ being constant. By choice of ϑ, $\vartheta = C - g$ for some $g \in L^\perp = \overline{\mathrm{Ran}}(T_{\overline{\phi}})$. Hence

$$\begin{aligned} P_+(\overline{\phi} \wedge \vartheta) = P_+(\overline{\phi} \wedge (C - g)) &= P_+(\overline{\phi} \wedge C) - P_+(\overline{\phi} \wedge g) \\ &= \overline{\phi(0)} \wedge C - P_+(\overline{\phi} \wedge g). \end{aligned} \tag{4.11.3}$$

For any $g \in \mathrm{Ran}(T_{\overline{\phi}})$, say $g = T_{\overline{\phi}} f$, with $f \in H^2(\mathbb{C}^n)$, we have

$$P_+(\overline{\phi} \wedge g) = P_+(\overline{\phi} \wedge T_{\overline{\phi}} f) = P_+(\overline{\phi} \wedge \overline{\phi}) f = 0.$$

To see the above, note that a typical component of $P_+(\overline{\phi} \wedge T_{\overline{\phi}} f)$ is

$$P_+(\overline{\phi}_i T_{\overline{\phi}_j} f - \overline{\phi}_j T_{\overline{\phi}_i} f) = T_{\overline{\phi}_i} T_{\overline{\phi}_j} f - T_{\overline{\phi}_j} T_{\overline{\phi}_i} f = 0$$

since $T_{\overline{u}v} = T_{\overline{u}} T_v$ for any $u \in H^\infty(\mathbb{T})$ and $v \in L^\infty(\mathbb{T})$. It follows by continuity that $P_+(\overline{\phi} \wedge g) = 0$ for any $g \in \overline{\mathrm{Ran}}(T_{\overline{\phi}})$. Hence from (4.11.3),

$$P_+(\overline{\phi} \wedge \vartheta) = \overline{\phi(0)} \wedge C, \tag{4.11.4}$$

which is constant, so $\phi \wedge \overline{\vartheta}$ is analytic as desired.

Now we show that we can choose $C_1, \ldots, C_{n-1} \in \mathbb{C}^n$ such that the corresponding projections $\vartheta_1, \ldots, \vartheta_{n-1}$ on L are pointwise orthonormal. As L is a closed subspace of $H^2(\mathbb{C}^n)$ which is invariant under the multiplication by the invariant variable z, we have by Theorem 4.11.1 that it can be expressed as $\Theta H^2(\mathbb{C}^k)$ for some k, $1 \leq k \leq n$, and some inner function $\Theta \in H^\infty(\mathbb{C}^{n\times k})$. It is easy to see that

$$\dim\{x(z) : x \in L\} = n - 1$$

for a.e. $z \in \mathbb{T}$, and hence we must have $k = n-1$. Moreover, $P_L = \Theta P_+ \Theta^*$. Hence, for any $C \in \mathbb{C}^n$,

$$\vartheta = P_L C = \Theta P_+ \Theta^* C = \Theta\Theta(0)^* C.$$

From (4.11.4) it is clear that $\vartheta \neq 0$ (and so $\Theta(0)^* C \neq 0$) whenever C and $\overline{\phi(0)}$ are linearly independent. Hence $\Theta(0)^* : \overline{\phi(0)}^\perp \to \mathbb{C}^{n-1}$ is surjective, and so there exist $C_1, \ldots, C_{n-1} \in \overline{\phi(0)}^\perp$ such that $\Theta(0)^* C_1, \ldots, \Theta(0)^* C_{n-1}$ are the standard basis vectors $e_1, \ldots, e_{n-1}$ of $\mathbb{C}^{n-1}$. Let

$$\vartheta_j := P_L C_j = \Theta\Theta(0)^* C_j = \Theta e_j, \quad 1 \leq j \leq n-1.$$

Then

$$\Psi := \begin{pmatrix} \phi & \overline{\Theta} \end{pmatrix} = \begin{pmatrix} \phi & \overline{\vartheta}_1 & \cdots & \overline{\vartheta}_{n-1} \end{pmatrix}$$

is unitary-valued a.e. on $\mathbb{T}$. The analyticity of the 2×2 minors on the first column follows similarly as above for $n = 2$. □

Lemma 4.11.8 *Let*

$$\Phi := \begin{pmatrix} \phi & \overline{\phi}_c \end{pmatrix}$$

be unitary-valued on $\mathbb{T}$, *where* $\phi \in H^\infty(\mathbb{C}^{n\times 1})$, $\phi \in H^\infty(\mathbb{C}^{n\times(n-1)})$, *and* ϕ_c *is co-outer. Then*

$$\Phi^T H^2(\mathbb{C}^n) \cap \begin{pmatrix} 0 \\ L^2(\mathbb{C}^{n-1}) \end{pmatrix} = \begin{pmatrix} 0 \\ H^2(\mathbb{C}^{n-1}) \end{pmatrix}.$$

Proof. We show first that

$$\{x \in H^2(\mathbb{C}^n) : \ \phi^T x = 0\} = \phi_c H^2(\mathbb{C}^{n-1}). \tag{4.11.5}$$

Since Φ is unitary-valued, $\phi^T \phi_c = 0$ and hence in (4.11.5) the left-hand side contains the right-hand side. Conversely, suppose $\phi^T x = 0$ for some $x \in H^2(\mathbb{C}^n)$. For almost all $z \in \mathbb{T}$, $\overline{x}(z) \perp \phi(z)$, and since Φ is unitary-valued, $\overline{x}(z)$ is in the linear span of the columns of $\overline{\phi}_c(z)$. Hence we may write

$$\overline{x}(z) = \overline{\phi}_c(z)\overline{h}(z)$$

for some function $h : \mathbb{T} \to \mathbb{C}^{n-1}$. Then $x = \phi_c h$ and since $\phi_c^* \phi_c = I_{n-1}$, it follows that

$$h = \phi_c^* \in L^2(\mathbb{C}^{n-1}).$$

Fix i, $1 \leq i \leq n-1$. Since ϕ_c^T is outer, $\phi_c^T H^2(\mathbb{C}^n)$ is dense in $H^2(\mathbb{C}^{n-1})$ and hence there is a sequence $\{g_k\}$ in $H^2(\mathbb{C}^n)$ such that $\phi_c^T g_k \to e_i$ in $H^2(\mathbb{C}^{n-1})$, where e_i is the constant function equal to the ith standard basis vector of $\mathbb{C}^{n-1}$ at every point. Then

$$g_k^T x = g_k^T(\phi_c h) \to e_i^T h \quad \text{in} \quad L^1(\mathbb{T}).$$

Since $g_k^T x \in H^1(\mathbb{T})$, it follows that the negative Fourier coefficient of $e_i^T h$ vanish, $1 \leq i \leq n-1$. Thus $h \in H^2(\mathbb{C}^{n-1})$ and so $x \in \phi_c H^2(\mathbb{C})^{n-1}$ as required. This proves (4.11.5). Finally we have

$$\begin{aligned} \Phi^T H^2(\mathbb{C}^n) \cap \begin{pmatrix} 0 \\ L^2(\mathbb{C}^{n-1}) \end{pmatrix} &= \begin{pmatrix} \phi^T \\ \phi_c^* \end{pmatrix} H^2(\mathbb{C}^n) \cap \begin{pmatrix} 0 \\ L^2(\mathbb{C}^{n-1}) \end{pmatrix} \\ &= \begin{pmatrix} \phi^T \\ \phi_c^* \end{pmatrix} \phi_c H^2(\mathbb{C}^{n-1}) = \begin{pmatrix} 0 \\ H^2(\mathbb{C}^{n-1}) \end{pmatrix}. \end{aligned}$$

□

Lemma 4.11.9 *Let $V \in L^\infty(\mathbb{C}^{n\times n})$ and $W \in L^\infty(\mathbb{C}^{m\times m})$, both unitary a.e. on $\mathbb{T}$, of the form*

$$V = \begin{pmatrix} v_1 & \overline{v}_2 \end{pmatrix}, \quad W^T = \begin{pmatrix} w_1 & \overline{w}_2 \end{pmatrix},$$

where v_1, v_2, w_1, w_2 are H^∞ functions, v_1 and w_1 are column matrices, and v_2 and w_2 are co-outer. Then

$$WH^\infty(C^{m\times n})V \cap \begin{pmatrix} 0 & 0 \\ 0 & L^\infty(C^{(m-1)\times(n-1)}) \end{pmatrix} = \begin{pmatrix} 0 & 0 \\ 0 & H^\infty(C^{(m-1)\times(n-1)}) \end{pmatrix}.$$

Proof. Consider first $Q \in H^\infty(\mathbb{C}^{m\times n})$ such that

$$WQV = \begin{pmatrix} 0 & 0 \\ 0 & g \end{pmatrix},$$

for some $g \in L^\infty(\mathbb{C}^{(m-1)\times(n-1)})$. Since W is unitary-valued we have

$$QV = \begin{pmatrix} 0 & k \end{pmatrix},$$

for some $k \in L^\infty(\mathbb{C}^{m\times(n-1)})$. Then

$$V^T Q^T = \begin{pmatrix} 0 \\ k^T \end{pmatrix} \in \begin{pmatrix} 0 \\ L^\infty(\mathbb{C}^{(n-1)\times m}) \end{pmatrix}.$$

Applying Lemma 4.11.8 to each column of this equation we find $k \in H^\infty(\mathbb{C}^{m\times(n-1)})$. Now

$$\begin{pmatrix} 0 & Wk \end{pmatrix} = W \begin{pmatrix} 0 & k \end{pmatrix} = WQV = \begin{pmatrix} 0 & 0 \\ 0 & g \end{pmatrix}$$

and so

$$Wk \in WH^\infty(\mathbb{C}^{m\times(n-1)}) \cap \begin{pmatrix} 0 \\ L^\infty(\mathbb{C}^{(m-1)\times(n-1)}) \end{pmatrix}.$$

Apply Lemma 4.11.8 to each column in turn to deduce that $Wk \in H^\infty(\mathbb{C}^{(m-1)\times(n-1)})$. Thus

$$WQV \in \begin{pmatrix} 0 & 0 \\ 0 & H^\infty(C^{(m-1)\times(n-1)}) \end{pmatrix}.$$

For the reverse inclusion, let $g \in H^\infty(C^{(m-1)\times(n-1)})$. Applying Lemma 4.11.8 column after column implies the existence of $k \in H^\infty(\mathbb{C}^{m\times(n-1)})$ such that

$$Wk = \begin{pmatrix} 0 \\ g \end{pmatrix},$$

and gives a $Q \in H^\infty(\mathbb{C}^{m\times n})$ such that

$$V^T Q^T = \begin{pmatrix} 0 \\ k^T \end{pmatrix}.$$

Then

$$WQV = W\begin{pmatrix} 0 & k \end{pmatrix} = \begin{pmatrix} 0 & 0 \\ 0 & g \end{pmatrix}.$$

□

Lemma 4.11.10 *Let $m, n > 1$, let G be as in Theorem 4.11.4, and suppose that $t_0 := \|H_G\| \neq 0$. Let v and w be a pair of maximizing unit vectors for H_G. Then v, $\overline{zw} \in H^2$ admit the factorizations*

$$v = v_0 h, \quad \overline{zw} = \phi w_0 h, \tag{4.11.6}$$

for some scalar outer function h, some scalar inner ϕ, and inner columns v_0 and w_0. Moreover, there exist unitary-valued functions V and W of types $n \times n$ and $m \times m$, respectively, of the form

$$V = \begin{pmatrix} v_0 & \overline{\alpha} \end{pmatrix}, \quad W^T = \begin{pmatrix} w_0 & \overline{\beta} \end{pmatrix}, \tag{4.11.7}$$

where α and β are co-outer functions of types $n \times (n-1)$ and $m \times (m-1)$, respectively, and all minors of the first columns of V and W^T are in H^∞.

Proof. Theorem 4.3.8 implies that $\|v(z)\| = \|w(z)\|$ a.e. on $\mathbb{T}$, so the H^2 column vector functions v and $\overline{zw}$ have the same (scalar) outer factor h. This gives the inner-outer factorizations (4.11.6) (we can take out a common scalar inner factor ϕ from the components of the inner factor of $\overline{zw}$ if we wish). Unitary-valued functions V and W as in (4.11.7) exist by Theorem 4.11.7. □

Lemma 4.11.11 *Keep the notation of Lemma 4.11.10. Then every matrix-valued function $Q \in H^\infty(\mathbb{C}^{m\times n})$ which is at minimal distance from G satisfies*

$$W(G-Q)V = \begin{pmatrix} t_0 u_0 & 0 \\ 0 & F \end{pmatrix} \tag{4.11.8}$$

for some $F \in (H^\infty + C)(\mathbb{C}^{(m-1)\times(n-1)})$ and some quasicontinuous unimodular function u_0 given by

$$u_0 = \frac{\bar{z}\overline{\phi h}}{h}. \tag{4.11.9}$$

Proof. Let $Q \in H^\infty(\mathbb{C}^{m\times n})$ satisfy $\|G - Q\|_\infty = t_0$. By Theorem 4.3.8, $(G-Q)v = t_0 w$ and, in view of the factorizations (4.11.6), this means

$$(G-Q)v_0 h = t_0 \bar{z}\overline{\phi h_1}\,\overline{w}_0,$$

or, by (4.11.7) and (4.11.9),

$$(G-Q)V\begin{pmatrix}1 & 0 & \cdots & 0\end{pmatrix}^T = W^*\begin{pmatrix}t_0u_0 & 0 & \cdots 0\end{pmatrix}^T.$$

It follows that

$$W(G-Q)V = \begin{pmatrix} t_0u_0 & f \\ 0 & F\end{pmatrix}$$

for some $f \in L^\infty(\mathbb{C}^{1\times(n-1)})$ and $F \in L^\infty(\mathbb{C}^{(m-1)\times(n-1)})$. Certainly $|u_0| = 1$ a.e. on $\mathbb{T}$, and from Nehari's theorem (Theorem 4.3.1)

$$\|W(G-Q)V\|_\infty = \|G-Q\|_\infty = \|H_G\| = t_0.$$

It follows that $f = 0$ and so $W(G-Q)V$ is of the form (4.11.8).

We have $\|H_{u_0}\| \le \|u_0\|_\infty = 1$, while $\|H_{u_0}h\| = \|\bar{z}\overline{\phi h}\| = \|h\|$. Thus $\|H_{u_0}\| = 1 = \|u_0\|_\infty$, so u_0 is badly approximable. The equality of the $(1,1)$ entries of (4.11.8) implies

$$w_0^T(G-Q)v_0 = t_0u_0.$$

Since v_0 and w_0 are in $H^\infty(\mathbb{T})$ and $H^\infty + C$ is an algebra, it follows that $u_0 \in H^\infty + C$. Since u_0 is also badly approximable, Theorem 4.11.6 implies that $\overline{u}_0 \in H^\infty + C$, so u_0 is quasicontinuous.

For $1 < i \le m$ and $1 < j \le n$ take the 2×2 minor of (4.11.8) with indices $1i$, $1j$ to get

$$\sum_{r<s,k<l} W_{1i,rs}(G-Q)_{rs,kl}V_{kl,1j} = t_0u_0F_{i-1,j-1}. \tag{4.11.10}$$

By choice of W and V, $V_{kl,1j}$ and $W_{1i,rs} = (W^T)_{rs,1i}$ are in $H^\infty(\mathbb{T})$. Since $G - Q \in (H^\infty + C)(\mathbb{C}^{m\times n})$, all terms on the left-hand side of (4.11.10) are in $H^\infty + C$, and hence $u_0F \in (H^\infty + C)(\mathbb{C}^{(m-1)\times(n-1)})$. This implies

$$F = \overline{u}_0(u_0F) \in (H^\infty + C)(\mathbb{C}^{(m-1)\times(n-1)}).$$

□

Lemma 4.11.12 *Let $m, n > 1$, let G be as in Theorem 4.11.4, and let $Q \in H^\infty(\mathbb{C}^{m\times n})$ be at minimal distance from G, so that with the notation of Lemma 4.11.11,*

$$W(G-Q_1)V = \begin{pmatrix} t_0u_0 & 0 \\ 0 & F\end{pmatrix} \tag{4.11.11}$$

for some $F \in (H^\infty + C)(\mathbb{C}^{(m-1)\times(n-1)})$. Let

$$\mathcal{E} = \{G - Q : \ Q \in H^\infty(\mathbb{C}^{m\times n}), \|G-Q\|_\infty = t_0\}.$$

Then

$$W\mathcal{E}V = \begin{pmatrix} t_0u_0 & 0 \\ 0 & F + H^\infty(\mathbb{C}^{(m-1)\times(n-1)})\end{pmatrix} \cap B(t_0), \tag{4.11.12}$$

where $B(t_0)$ is the closed ball of radius t_0 in $L^\infty(\mathbb{C}^{m\times n})$.

Proof. Let

$$E_1 = G - Q_1 \in \mathcal{E}$$

and consider any element

$$E = E_1 - Q \in \mathcal{E}, \quad Q \in H^\infty(\mathbb{C}^{m\times n}).$$

By Lemma 4.11.11 there is some $g \in L^\infty(\mathbb{C}^{(m-1)\times(n-1)})$ such that

$$WEV = \begin{pmatrix} t_0u_0 & 0 \\ 0 & g \end{pmatrix},$$

and combining this equation with (4.11.11) we find

$$\begin{aligned} WQV &= W(G - Q_1)V - WEV \\ &= \begin{pmatrix} 0 & 0 \\ 0 & F - g \end{pmatrix} \in \begin{pmatrix} 0 & 0 \\ 0 & L^\infty(\mathbb{C}^{(m-1)\times(n-1)}) \end{pmatrix} \cap WH^\infty(\mathbb{C}^{m\times n})V. \end{aligned}$$

It follows from Lemma 4.11.9 that $WQV \in H^\infty(\mathbb{C}^{m\times n})$, say $F - g = q \in H^\infty(\mathbb{C}^{(m-1)\times(n-1)})$. Then

$$WEV = \begin{pmatrix} t_0u_0 & 0 \\ 0 & g \end{pmatrix} = \begin{pmatrix} t_0u_0 & 0 \\ 0 & F - q \end{pmatrix},$$

and one inclusion in (4.11.12) is proved.

Conversely, suppose $q \in H^\infty(\mathbb{C}^{(m-1)\times(n-1)})$ and $\|F - q\|_\infty \le t_0$. Again by Lemma 4.11.9, there exists $Q \in H^\infty(\mathbb{C}^{m\times n})$ such that

$$WQV = \begin{pmatrix} 0 & 0 \\ 0 & q \end{pmatrix},$$

and hence

$$W(E_1 - Q)V = \begin{pmatrix} t_0u_0 & 0 \\ 0 & F - q \end{pmatrix}.$$

Then

$$E_1 - Q = G - (Q_1 - Q) \in \mathcal{E},$$

and so

$$\begin{pmatrix} t_0u_0 & 0 \\ 0 & F - q \end{pmatrix} \in W\mathcal{E}V.$$

Hence equality holds in (4.11.12). □

Proof of Theorem 4.11.4. By Corollary 4.3.9 the conclusion holds when $\min\{m, n\} = 1$. Suppose that $\min\{m, n\} > 1$ and that the conclusion is true for any lesser value of $\min\{m, n\}$. Let $\mathcal{E}$ be the set of minimal norm errors $G - Q$ (as in Lemma 4.11.12). Since H_G is compact, it has a maximizing pair of unit vectors v and w. Then $\overline{zw} \in H^2(\mathbb{C}^m)$, and so the inner-outer factorizations (4.11.6) are defined (we may take $\phi = 1$). Let t_0, u_0, V, W, and F be as in Lemma 4.11.10. Consider any $E \in \mathcal{E}$. By Lemma 4.11.12 we have

$$WEV = \begin{pmatrix} t_0 u_0 & 0 \\ 0 & F - q \end{pmatrix}$$

for some $q \in H^\infty(\mathbb{C}^{(m-1)\times(n-1)})$, and furthermore, as E ranges over $\mathcal{E}$, q ranges over all $H^\infty(\mathbb{C}^{(m-1)\times(n-1)})$ functions satisfying $\|F - q\|_\infty \le t_0$. Since W and V are unitary-valued, we have that

$$s^\infty(E) = s^\infty(WEV) = s^\infty\left(\begin{pmatrix} t_0 u_0 & 0 \\ 0 & F - q \end{pmatrix}\right) = (t_0, s^\infty(F - q)).$$

Thus the lexicographic maximum of s^∞ over $\mathcal{E}$ is attained precisely when the corresponding $q \in H^\infty(\mathbb{C}^{(m-1)\times(n-1)})$ minimizes $s^\infty(F - q)$. By Lemma 4.11.11, $F \in (H^\infty + C)(\mathbb{C}^{(m-1)\times(n-1)})$. By the inductive hypothesis $s^\infty(F - q)$ is minimized by a unique $q \in H^\infty(\mathbb{C}^{(m-1)\times(n-1)})$, and for this q the singular values of $F(z) - q(z)$ are constant. From the relation

$$E = W^* \begin{pmatrix} t_0 u_0 & 0 \\ 0 & F - q \end{pmatrix} V^*$$

we see that there is a unique $E \in \mathcal{E}$ which maximizes $s^\infty(E)$ and that the singular values of $E(z)$ are constant a.e. on $\mathbb{T}$. This completes the proof. □

If Q_0 is the (unique) superoptimal approximation of G then the constant singular values of $G(z) - Q_0(z)$ are called the *superoptimal singular values* of G. An interesting problem is giving an intrinsic characterization of the superoptimal singular values of G. The following is a plausible try. For $G \in L^\infty(\mathbb{C}^{m\times n})$ and $0 \le k \le \min\{m, n\}$, let

$$\tau_k = \inf\{\|H_{G-F}\| : F \in L^\infty(\mathbb{C}^{m\times n}), \quad \text{rank}\, F(z) \le k \text{ a.e. on } \mathbb{T}\}.$$

Clearly, $\tau_0 = t_0$. It can be shown without much difficulty that $\tau_k \le t_k$. However, as shown by the following example, in general it is not true that $\tau_k = t_k$.

Example 4.11.13 Let

$$G(z) = \frac{1}{\sqrt{2}} \begin{pmatrix} \bar{z} & 0 \\ 0 & \alpha \bar{z} \end{pmatrix} \begin{pmatrix} 1 & \bar{z} \\ -1 & \bar{z} \end{pmatrix},$$

where $0 \le \alpha \le 1$. The superoptimal $H^\infty(\mathbb{C}^{2\times 2})$ approximation of G is the zero function, so that the superoptimal singular values of G are $t_0 = 1$ and $t_1 = \alpha$. Let

$$F(z) = \frac{1}{\sqrt{2}} \begin{pmatrix} \bar{z} & 0 \\ 0 & 0 \end{pmatrix} \begin{pmatrix} 1 & \bar{z} \\ -1 & \bar{z} \end{pmatrix}.$$

Clearly rank $F(z) = 1$ on $\mathbb{T}$, and

$$(G - F)(z) = \frac{\alpha}{\sqrt{2}} \begin{pmatrix} 0 & 0 \\ -1 & 0 \end{pmatrix} + A(z),$$

where

$$A(z) = \frac{\alpha}{\sqrt{2}} \begin{pmatrix} 0 & 0 \\ 0 & \bar{z} \end{pmatrix}.$$

Thus

$$\|H_{G-F}\| = \|H_A\| \leq \|A\|_\infty = \frac{\alpha}{\sqrt{2}}.$$

Hence

$$\tau_1(G) = \frac{\alpha}{\sqrt{2}} < t_1(G).$$

4.12 MODEL MATCHING

The generic model matching problem reduces to the Nehari problem, as we will explain in this section. With an element $f \in H^\infty(\mathbb{C}^{n\times m})$, we associate a filter $\Sigma_f : H^2(\mathbb{C}^m) \to H^2(\mathbb{C}^n)$, defined by

$$\Sigma_f(u) = f(z)u(z) = y(z).$$

We shall depict the filter as

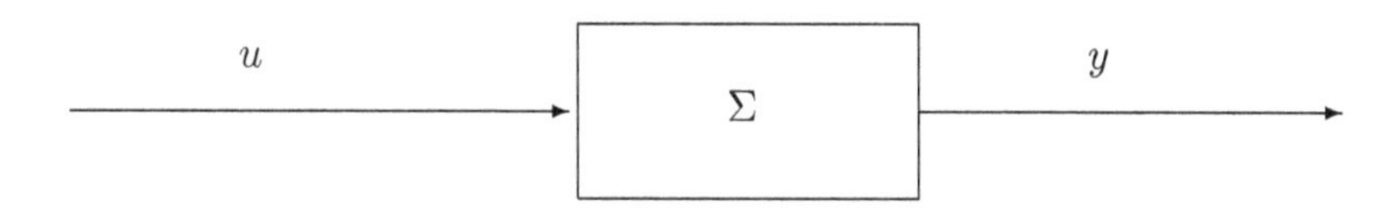

and call u the *input* and y the *output* of the filter. The concatenation of two filters results in the *product filter* $\Sigma_h\Sigma_f$. The difference filter $\Sigma_f - \Sigma_h$ may be depicted as in Figure 4.1.

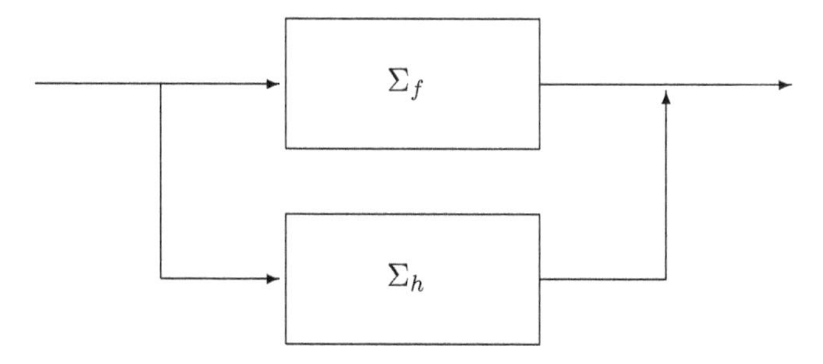

Figure 4.1

Note that $\Sigma_h\Sigma_f = \Sigma_{hf}$ and $\Sigma_h - \Sigma_f = \Sigma_{h-f}$. For a filter Σ_f we define its norm by

$$\|\Sigma_f\| = \sup_{u\neq 0} \frac{\|\Sigma_f(u)\|}{\|u\|}.$$

It is not hard to see that $\|\Sigma_f\| = \|f\|_\infty$.

The *model matching problem* for linear filters is the following. Given filters Σ_{f_1}, Σ_{f_2}, Σ_{f_3}, find a filter Σ_h such that the filter $\Sigma_{f_1} - \Sigma_{f_2}\Sigma_h\Sigma_{f_3}$ depicted in Figure 4.2 below has minimal possible norm. In other words, we would like the filter $\Sigma_{f_2}\Sigma_h\Sigma_{f_3}$ to best match the filter Σ_{f_1}.

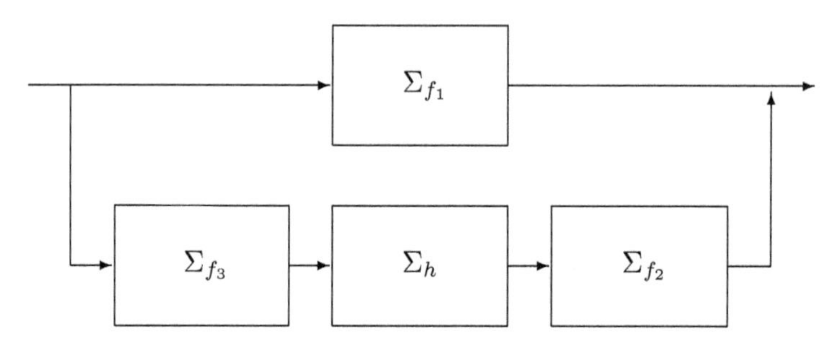

Figure 4.2

Equivalently, given f_1, f_2, f_3, find h such that $\|f_1 - f_2 h f_3\|_\infty$ is as small as possible.

We shall assume that f_2 and f_3^T allow inner-outer factorizations

$$f_2 = f_{2,i} f_{2,o}, \quad f_3^T = f_{3,i} f_{3,o}, \quad f_{2,i}^* f_{2,i} \equiv I, \quad f_{3,i}^* f_{3,i} \equiv I, \qquad (4.12.1)$$

where $f_{2,o}$ has an inverse $f_{2,o}^{-1} \in H^\infty(\mathbb{C}^{n\times n})$, and $f_{3,o}$ has an inverse $f_{3,o}^{-1} \in H^\infty(\mathbb{C}^{m\times m})$. We get that

$$\|f_1 - f_2 h f_3\|_\infty = \|f_{2,i}^*(f_1 - f_2 h f_3) f_{3,i}^{*T}\|_\infty = \|f_{2,i}^* f_1 f_{3,i}^{*T} - f_{2,o} h f_{3,o}^T\|_\infty,$$

where now the norm is in $L^\infty(\mathbb{C}^{n\times m})$ (as opposed to $H^\infty(\mathbb{C}^{n\times m})$). When we let h range through all of $H^\infty(\mathbb{C}^{p\times q})$, then $\widetilde{h} = f_{2,o} h f_{3,o}^T$ also runs through all of $H^\infty(\mathbb{C}^{p\times q})$, due to the invertibility of $f_{2,o}$ and $f_{3,o}$. Thus the problem has been reduced to finding the closest $H^\infty(\mathbb{C}^{n\times m})$ function to $f_{2,i}^* f_1 f_{3,i}^{*T} \in L^\infty(\mathbb{C}^{n\times m})$. Consequently, by Nehari's theorem (Theorem 4.3.1) the minimal norm of $\|f_1 - f_2 h f_3\|_\infty$ equals the norm of the Hankel operator with symbol $f_{2,i}^* f_1 f_{3,i}^{*T}$.

4.13 EXERCISES

1 Prove that the patterns of the form (iv)(b) in Theorem 4.1.6 have the Toeplitz contractive completion property.

2 Show that the patterns of the form (iv)(b) in Theorem 4.1.6 do not have the Toeplitz contractive completion property when the entries are assumed to be matrices of size 2×2 or higher.

3 Show that the partial Toeplitz contraction

$$\begin{pmatrix} ? & ? & ? \\ \frac{1}{\sqrt{2}} & ? & ? \\ \frac{1}{\sqrt{2}} & \frac{1}{\sqrt{2}} & ? \end{pmatrix}$$

does not admit a contractive Toeplitz completion.

4 Show directly that not every 4×4 lower-triangular partial Toeplitz contraction admits a contractive Toeplitz completion. Hint: consider the partial matrix

$$\begin{pmatrix} \frac{1}{\sqrt{6}} & ? & ? & ? \\ \frac{1}{\sqrt{6}} & \frac{1}{\sqrt{6}} & ? & ? \\ \frac{1}{\sqrt{6}} & \frac{1}{\sqrt{6}} & \frac{1}{\sqrt{6}} & ? \\ \frac{1}{\sqrt{6}} & \frac{1}{\sqrt{6}} & \frac{1}{\sqrt{6}} & \frac{1}{\sqrt{6}} \end{pmatrix}.$$

5 Prove that the operator-matrix $\left(\begin{smallmatrix} A & 0 \\ X & B \end{smallmatrix}\right)$ is a contraction if and only if $\|A\| \leq 1$, $\|B\| \leq 1$, and there exists a contraction $\Gamma : \mathcal{D}_A \rightarrow \mathcal{D}_{B^*}$ such that $X = D_{B^*}\Gamma D_A$.

6 Let

$$H(a_1, \ldots, a_n; x_1, \ldots, x_{n-1}) = \begin{pmatrix} a_1 & a_2 & \cdots & a_{n-1} & a_n \\ a_2 & a_3 & \cdots & a_n & x_1 \\ \vdots & \vdots & \iddots & \iddots & \vdots \\ a_{n-1} & a_n & \cdots & x_{n-3} & x_{n-2} \\ a_n & x_1 & \cdots & x_{n-2} & x_{n-1} \end{pmatrix} \tag{4.13.1}$$

be a partial Hankel matrix, where $x_1, \ldots, x_{n-1}$ are the unknowns.

(a) Prove that the partial Hankel matrix

$$H = H\left(0, -\frac{3}{4}, \frac{3}{8}, 0, x_1, x_2, x_3\right)$$

is contractive but does not admit a contractive Hankel completion.

(b) Prove the same about the partial matrix

$$H = H\left(0, \frac{7}{10}, 0, \frac{7}{10}, 0, \frac{7}{10}, x_1, x_2, x_3, x_4\right).$$

7 Prove that the partial Hankel matrix

$$H = H(0, \dots, 0, a_n; x_1, \dots, x_{n-1})$$

is contractive for every $|a_n| < 1$ and admits a contractive Hankel completion.

8 Let $\mathcal{T}_n \otimes \mathcal{T}_n$ in $\mathbb{C}^{n^2}$, where $\mathcal{T}_n$ is the nest algebra of upper triangular $n \times n$ matrices. For example, the pattern for $n = 2$ is

$$\begin{pmatrix} ? & ? & ? & ? \\ & ? & & ? \\ & & ? & ? \\ & & & ? \end{pmatrix}.$$

(a) Give the pattern for $n = 3$.

(b) Let $n = 2^k$. Show that when C is a partial contraction with pattern $\mathcal{T}_n \otimes \mathcal{T}_n$, then a completion can be found of norm at most $k + 1$. (Hint: use induction in k. Observe that when C has pattern $\mathcal{T}_n \otimes \mathcal{T}_n$, and one decomposes C as

$$C = \begin{pmatrix} C_{11} & C_{12} \\ C_{21} & C_{22} \end{pmatrix}$$

with C_{ij} of size $2^{k-1} \times 2^{k-1}$, then C_{11}, C_{22} both have a pattern $\mathcal{T}_{\frac{n}{2}} \otimes \mathcal{T}_{\frac{n}{2}}$ and C_{12} has a pattern which after permutation is block triangular.)

9 Show that there exist partial contractions with the patterns below such that all completions have norm $\geq \sqrt{\frac{9}{8}}$.

(a) $\begin{pmatrix} ? & * & * \\ * & ? & * \\ * & * & * \end{pmatrix}$.

(b) $\begin{pmatrix} ? & * & ? \\ * & ? & ? \\ * & * & ? \end{pmatrix}$.

10 In this exercise we show that the minimal possible norm of a completion of a partial contraction can be arbitrarily large.

(a) Show that for $Y, X_1, X_2, X_3 \in \mathcal{L}(\mathcal{H})$ we have that

$$\left\| \begin{pmatrix} X_1 & Y & Y \\ Y & X_2 & Y \\ Y & Y & X_3 \end{pmatrix} \right\| \geq \frac{3}{2}\|Y\|,$$

and that equality is achieved by choosing $X_1, X_2, X_3 = -\frac{1}{2}Y$.

(b) Let $A_0 = (1)$, and define the $3^n \times 3^n$ matrices A_n, $n \in \mathbb{N}_0$, via

$$A_{n+1} = \begin{pmatrix} 0 & A_n & A_n \\ A_n & 0 & A_n \\ A_n & A_n & 0 \end{pmatrix}, \quad n \in \mathbb{N}_0.$$

Consider the partial matrix T_n obtained by replacing all zeros in A_n by question marks. So, for instance,

$$T_1 = \begin{pmatrix} ? & 1 & 1 \\ 1 & ? & 1 \\ 1 & 1 & ? \end{pmatrix}.$$

Show that $\frac{1}{(\sqrt{2})^n} T_n$ is a partial contraction for which the norm of a minimal norm completion equals $\left(\sqrt{\frac{9}{8}}\right)^n$.

11 Show that if a partial contraction with pattern

$$\begin{pmatrix} ? & * & * \\ * & ? & * \\ * & * & ? \end{pmatrix}$$

does not have a contractive completion, then its minimal norm completion is unique.

12 Let $0 \leq A, B < I$. Show that $\left(\begin{smallmatrix} A & X^* \\ X & B \end{smallmatrix}\right)$ is a contraction if and only if $X = (I - A)^{-\frac{1}{2}} G (I - B)^{-\frac{1}{2}}$ for some contraction G.

13 Let $K = \{A_{ij} : 1 \leq j \leq i \leq n, \ (i,j) \in \mathcal{T}\}$ be a partially contractive generalized lower triangular matrix, and let F_c denote its contractive central completion. Let Φ_c and Ψ_c be upper and lower triangular operator matrices such that

$$\Phi_c^* \Phi_c = I - F_c^* F_c, \quad \Psi_c^* \Psi_c = I - F_c F_c^*.$$

Further, let $\omega_1 : \mathcal{D}_{F_c} \to \overline{\mathrm{Ran}}(\Phi_c)$ and $\omega_2 : \mathcal{D}_{F_c^*} \to \overline{\mathrm{Ran}}(\Psi_c)$ be unitary operator matrices such that

$$\Phi_c = \omega_1 D_{F_c}, \quad \Psi_c = \omega_2 D_{F_c^*},$$

and put

$$\tau_c = -\omega_1 F_c^* \omega_2^*.$$

Prove that each contractive completion of K is of the form

$$\mathcal{S}(G) = F_c - \Psi_c^* G (I + \tau_c G)^{-1} \Phi_c,$$

where $G = (G_{ij})_{i,j=1}^n : \overline{\mathrm{Ran}}(\Phi_c) \to \overline{\mathrm{Ran}}(\Psi_c)$ is a contraction with $G_{ij} = 0$ whenever $(i,j) \notin \mathcal{T}$. Moreover, the correspondence between the set of all contractive completions and all such contractions G is one-to-one.

Furthermore, $\mathcal{S}(G)$ is isometric (co-isometric, unitary) if and only if G is. (Hint: use the proof of Theorem 2.6.3.)

14 Let $K = \{A_{ij} : 1 \leq j \leq i \leq n, \ (i,j) \in \mathcal{T}\}$ be a partially contractive generalized lower triangular matrix. Prove that the following are equivalent.

(a) F is the contractive central completion of K.

(b) $\Delta_u(I - FF^*) \geq \Delta_u(I - \widetilde{F}\widetilde{F}^*)$ for all contractive completions $\widetilde{F}$ of K.

(c) $\Delta_r(I - F^*F) \geq \Delta_r(I - \widetilde{F}^*\widetilde{F})$ for all contractive completions $\widetilde{F}$ of K.

15 Provide a proof for Theorem 4.2.8 based on Exercise 2.10.25 and the proof of Theorem 4.2.1.

16 Suppose that $H(\{c_n\}_{n=-\infty}^{-1})$ has rank k. Define

$$A = [(c_{-i-j-1})_{i,j=0}^{k}]^{-1}(c_{-i-j-1})_{i=0,j=1}^{k-1,\ k}, \; B = \begin{pmatrix} 1 & 0 & \cdots & 0 \end{pmatrix}^T,$$

$$C = \begin{pmatrix} c_{-1} & \cdots & c_{-k} \end{pmatrix}.$$

(a) Show that A is well defined.

(b) Show that $CA^iB = c_{-i-1}, i \in \mathbb{N}_0$.

(c) Conclude that $c(z) = \sum_{n=0}^{\infty} c_{-1-n}z^{-n} = C(z-A)^{-1}B$.

This provides an alternative proof of the forward direction of Theorem 4.3.5. The procedure is known as Silverman's algorithm. The representation $C(z-A)^{-1}B$ is called a realization of the rational function $c(z)$.

17 Let $\{C_n\}_{n=-\infty}^{-1}$ be a sequence of operators in $\mathcal{L}(\mathcal{H})$ be such that $\|H(\{C_n\}_{n=-\infty}^{-1})\| < 1$ and $\sum_{n=-\infty}^{-1} \|C_n\| < \infty$. Define:

$$\begin{pmatrix} \vdots \\ \widehat{\alpha}_{-2} \\ \widehat{\alpha}_{-1} \end{pmatrix} = (I - H(\{C_n\}_{n=-\infty}^{-1})H(\{C_n\}_{n=-\infty}^{-1})^*)^{-1} \begin{pmatrix} \vdots \\ 0 \\ I \end{pmatrix},$$

$$\begin{pmatrix} \vdots \\ \widehat{\gamma}_1 \\ \widehat{\gamma}_0 \end{pmatrix} = H(\{C_n\}_{n=-\infty}^{-1})^* \begin{pmatrix} \vdots \\ \widehat{\alpha}_{-2} \\ \widehat{\alpha}_{-1} \end{pmatrix},$$

$$\begin{pmatrix} \vdots \\ \widehat{\delta}_{-2} \\ \widehat{\delta}_{-1} \end{pmatrix} = (I - H(\{C_n\}_{n=-\infty}^{-1})^* H(\{C_n\}_{n=-\infty}^{-1})^{-1} \begin{pmatrix} \vdots \\ 0 \\ I \end{pmatrix},$$

$$\begin{pmatrix} \vdots \\ \widehat{\beta}_1 \\ \widehat{\beta}_0 \end{pmatrix} = H(\{C_n\}_{n=-\infty}^{-1}) \begin{pmatrix} \vdots \\ \widehat{\delta}_{-2} \\ \widehat{\delta}_{-1} \end{pmatrix},$$

$$\alpha(e^{it}) = \sum_{n=-\infty}^{-1} \widehat{\alpha}_n \widehat{\alpha}_{-1}^{-\frac{1}{2}} e^{int}, \; \gamma(e^{it}) = \sum_{n=0}^{\infty} \widehat{\gamma}_n \widehat{\alpha}_{-1}^{-\frac{1}{2}} e^{int},$$

$$\beta(e^{it}) = \sum_{n=0}^{\infty} \widehat{\beta}_n \widehat{\delta}_{-1}^{-\frac{1}{2}} e^{int}, \quad \delta(e^{it}) = \sum_{n=-\infty}^{-1} \widehat{\delta}_n \widehat{\delta}_{-1}^{-\frac{1}{2}} e^{int}.$$

Prove that the formula

$$F(e^{it}) = (\mathcal{S}h)(e^{it}) = (\alpha(e^{it})h(e^{it}) + \beta(e^{it}))(\gamma(e^{it})h(e^{it}) + \delta(e^{it}))^{-1}$$

establishes a one-to-one correspondence between the set of all operator-valued functions

$$h(e^{it}) = \sum_{n=0}^{\infty} h_n e^{int}, \quad \|h\|_\infty < 1, \quad \sum_{n=0}^{\infty} \|h_n\| < \infty$$

and the set of all solutions of the form

$$F(e^{it}) = \sum_{n=-\infty}^{-1} C_n e^{int} + \sum_{n=0}^{\infty} D_n e^{int}, \quad \|F\|_\infty < 1, \quad \sum_{n=0}^{\infty} \|D_n\| < \infty$$

of Nehari's problem with data $\{C_n\}_{n=-\infty}^{-1}$.

18 Assume that in addition to the assumptions in Exercise 4.13.17, $\mathcal{H}$ is also finite dimensional. Prove that the solution

$$F_c(e^{it}) = \beta(e^{it})\delta(e^{it})^{-1}$$

is the unique one which maximizes the entropy integral

$$\mathcal{E}(F) = \frac{1}{2\pi} \int_0^{2\pi} \log \det(I - F(e^{it})^* F(e^{it})) dt$$

over the set of all strictly contractive solutions of the problem.

19 Let $\{C_k\}_{k=0}^{n}$ be a sequence of operators such that $\|T(\{C_k\}_{k=0}^{n})\| < 1$. Find a linear fractional type parametrization similar to the one in Exercise 4.13.17 for the set of all solutions of the form

$$F(e^{it}) = \sum_{k=0}^{n} C_k e^{ikt} + \sum_{k=n+1}^{\infty} D_k e^{ikt}, \quad \|F\|_\infty < 1, \quad \sum_{k=n+1}^{\infty} \|D_k\| < \infty$$

of the contractive Carathéodory-Fejér problem with data $\{C_k\}_{k=0}^{n}$.

20 In this exercise we will see that the map $G \to G_u$, where $G_u \in H^\infty$ is the unique approximant to G guaranteed by Corollary 4.3.9, is not continuous.

(a) Let $G(z) = \frac{1}{z^2}$. Show that $G_u(z) \equiv 0$ is the unique approximant.

(b) Let $G(z) = \frac{\alpha}{z} + \frac{1}{z^2}$. Show that the unique approximant is

$$G_u(z) = \alpha \frac{\sqrt{\alpha^2 + 4} - \alpha}{z(\sqrt{\alpha^2 + 4} - \alpha) + 2}.$$

(c) Conclude that the map $G \to G_u$ is not continuous. (Hint: look at $G_u(-1)$.)

21 Prove that the dual space $H_0^1(\mathbb{T})^*$ can be identified with $L^\infty(\mathbb{T})/H^\infty(\mathbb{T})$. (Hint: Prove that the mapping

$$[\Psi(\phi)](f + H^\infty(\mathbb{T})) = \frac{1}{2\pi}\int_0^{2\pi} \phi(e^{it})f(e^{it})dt$$

establishes an isometric isomorphism between the two spaces.)

22 Let A denote the uniform closure of the set of all analytic polynomials on the unit circle. Prove that the dual space $(C(\mathbb{T})/A)^*$ can be identified with $H_0^1(\mathbb{T})$. (Hint: for any $\phi \in H_0^1$, show that the mapping

$$\Phi(f + A) = \frac{1}{2\pi}\int_0^{2\pi} f(e^{it})\phi(e^{it})dt$$

is well defined and $\|\Phi\| = \|\phi\|_1$. In order to prove that each element of $(C(\mathbb{T})/A)^*$ can be realized this way, use the following theorem of F. and M. Riesz: "If μ is a regular Borel measure on $\mathbb{T}$ such that $\int_0^{2\pi} e^{int} d\mu = 0$ for all $n > 0$, then there exists $\phi \in H^1(\mathbb{T})$ such that $d\mu = \phi(e^{it})dt$.")

23 Let $F \in L^\infty(\mathbb{C}^{m\times n})$. Show that H_F is compact if and only if $F \in (H^\infty + C)(\mathbb{C}^{m\times n})$.

24 Let $f(z) = D + C(zI - A)^{-1}B$ with $\sigma(A) \subset \mathbb{D}$.

(a) Show that for $|z| \geq 1$, $f(z) = D + \sum_{k=1}^\infty CA^{k-1}Bz^{-k}$.

(b) Let

$$\mathcal{C} = \begin{pmatrix} C \\ CA \\ CA^2 \\ \vdots \end{pmatrix}, \quad \mathcal{O} = \begin{pmatrix} B & AB & A^2B & \cdots \end{pmatrix}$$

Show that the Hankel H_f equals $H_f = \mathcal{C}\mathcal{O}$.

(c) Show that $X = \mathcal{C}^*\mathcal{C}$ and $Y = \mathcal{O}\mathcal{O}^*$ are the solution to the Stein equations

$$X - A^*XA = C^*C, \quad Y - AYA^* = BB^*,$$

respectively.

(d) Show that

$$\|H_f\| = \sqrt{\rho(XY)},$$

where $\rho(M) = \max\{|\lambda| : \lambda \in \sigma(M)\}$ is the *spectral radius* of M.

(Hint: Use that $\mu - PQ, \mu \neq 0$, is invertible if and only if $\mu - QP$ is invertible, which can be seen by using the equivalence of both to the invertibility of $\left(\begin{smallmatrix} \mu I & P \\ Q & I \end{smallmatrix}\right)$.)

Notice that in this way the computation of the norm of a Hankel of a rational function is reduced to a finite matrix eigenvalue problem.

25 Let $\lambda_1, \dots, \lambda_n \in \mathbb{D}^2$ and let $W_1, \dots, W_n \in \mathcal{B}(\mathcal{L}(\mathcal{H}, \mathcal{K}))$. Prove that there exists Φ in the closed unit ball of $H^\infty(\mathbb{D}^2, \mathcal{L}(\mathcal{H}, \mathcal{K}))$, such that $\Phi(\lambda_i) = W_i$ for $1 \leq i \leq n$, if and only if there exist positive semidefinite $\mathcal{L}(\mathcal{K})$-valued functions Γ and Δ on $\{\lambda_1, \dots, \lambda_n\} \times \{\lambda_1, \dots, \lambda_n\}$ such that

$$I_{\mathcal{K}} - W_i^* W_j = (1 - \overline{\lambda}_i^{(1)} \lambda_j^{(1)})\Gamma(\lambda_i, \lambda_j) + (1 - \overline{\lambda}_i^{(2)} \lambda_j^{(2)})\Delta(\lambda_i, \lambda_j).$$

Here $\lambda_j = (\lambda_j^{(1)}, \lambda_j^{(2)})$, $j = 1, \dots, n$.

26 Let

$$p(z_1, \dots, z_d) = C_1 D_1(z_{i_1}) C_2 D_2(z_{i_2}) \cdots C_l D_l(z_{i_l}), \tag{4.13.2}$$

where C_k are strict contractions of compatible sizes, $D_k(z)$ are diagonal matrices with monomials in z on the diagonal, and $z_{i_k} \in \{z_1, \dots, z_d\}$, $k = 1, \dots, l$.

(a) Show that there exist polynomials $R_i(z) = R_i(z_1, \dots, z_d)$, $i = 0, \dots, d$, and a positive definite matrix R such that

$$I - p(z)p(w)^* = R + R_0(z)R_0(w)^* + \sum_{i=1}^{d} (1 - z_i \overline{w_i}) R_i(z) R_i(w)^*.$$

(b) Show that there exists an $\alpha < 1$ such that if $(T_1, \dots, T_d)$ are commuting contractions, then $\|p(T_1, \dots, T_d)\| \leq \alpha$. Here, when $p(z) = \sum_k B_k z^k$, we define $p(T) = \sum_k B_k \otimes T^k$ with $T^k = (T_1, \dots, T_d)^{(k_1, \dots, k_d)} := \prod_{i=1}^d T_i^{k_i}$.

27 Let $F(z)$ be matrix-valued and analytic on the open unit disk so that $\sup_{|z|<1} \|F(z)\| \leq 1$, and suppose that $F(z_i)x_i = y_i$, $i = 1, \dots, n$, where $z_1, \dots, z_n$ are different points in $\mathcal{D}$, and $x_1, \dots, x_n$ and $y_1, \dots, y_n$ are vectors of appropriate size. Show that the matrix

$$\left(\frac{x_i^* x_j - y_i^* y_j}{1 - \overline{z_i} z_j}\right)_{i,j=1}^n$$

is positive semidefinite. The converse also holds.

28 Given $z_i \in \mathbb{D}, w_i \in \mathbb{C}, i = 1, \dots, n$, so that $\Lambda := \left(\frac{1 - w_i \overline{w_j}}{1 - z_i \overline{z_j}}\right)_{i,j=1}^n > 0$. Let

$$\begin{pmatrix} a_1 \\ \vdots \\ a_n \end{pmatrix} = \Lambda^{-1} \begin{pmatrix} (1 - \overline{z_1})^{-1} \\ \vdots \\ (1 - \overline{z_n})^{-1} \end{pmatrix}.$$

Show that

$$f(z) := \frac{-(z-1)\sum_{i=1}^n w_i a_i \prod_{j\neq i}(z-z_j)}{\prod_{j=1}^n (z-z_j) - (z-1)\sum_{i=1}^n a_i \prod_{j\neq i}(z-z_j)}$$

solves the Nevanlinna-Pick interpolation problem (i.e., $f \in H^\infty, \|f\|_\infty \leq 1, f(z_i) = w_i, i = 1, \ldots, n$).

29 For the following data verify that condition (4.6.8) is satisfied, and construct a function that solves the Nevanlinna-Pick problem with these data. How much freedom is there for a solution?

(a) $z^{(1)} = (0,0), z^{(2)} = (\frac{1}{2}, 0)$ and $w_1 = 0, w_2 = \frac{1}{2}$.

(b) $z^{(1)} = (0,0), z^{(2)} = (\frac{1}{2}, \frac{1}{2})$ and $w_1 = 0, w_2 = \frac{1}{2}$.

30 Let

$$U = \begin{pmatrix} A & B \\ C & D \end{pmatrix} \in \mathcal{L}(\mathcal{H}_1 \oplus \mathcal{H}_2, \mathcal{K}_1 \oplus \mathcal{K}_2)$$

be a unitary operator and let $f : \mathbb{D} \to \mathcal{L}(\mathcal{K}_2, \mathcal{H}_2)$ be an analytic function such that $\sup_{z\in\mathbb{D}} \|f(z)\| \leq 1$. Prove that

$$(U_f)(z) = A + zBf(z)(I - zf(z)D)^{-1}C$$

is analytic in $\mathbb{D}$ and $\sup_{z\in\mathbb{D}} \|U_f(z)\| \leq 1$.

31 Prove the one-variable von Neumann inequality using Proposition 2.3.4.

32 Consider the function

$$f(z_1, z_2, z_3) = \frac{3z_1z_2z_3 - z_1z_2 - z_2z_3 - z_1z_3}{3 - z_1 - z_2 - z_3}.$$

(a) Show that f is inner.

(b) Define

$$S(z, w) = 3\left|zw - \frac{z}{2} - \frac{w}{2}\right|^2 + 3\left|1 - \frac{z}{2} - \frac{w}{2}\right|^2 + \frac{1}{2}|z - w|^2.$$

Show that

$$|3 - z_1 - z_2 - z_3|^2 - |3z_1z_2z_3 - z_1z_2 - z_2z_3 - z_1z_3|^2$$
$$= (1 - |z_1|^2)S(z_2, z_3) + (1 - |z_2|^2)S(z_1, z_3) + (1 - |z_3|^2)S(z_1, z_2).$$

(c) Use (b) to show that f is in the Schur-Agler class; that is, show that f has a representation (4.6.5) with $\left(\begin{smallmatrix} A & B \\ C & D \end{smallmatrix}\right)$ unitary.

33 Let f be as in (4.6.5) with $\left(\begin{smallmatrix} A & B \\ C & D \end{smallmatrix}\right)$ unitary, and let $T = (T_1, \ldots, T_d)$ be a d-tuple of commuting strict contractions (i.e., $\|T_i\| \leq 1$ and $T_iT_j = T_jT_i$ for all i and j). Define $f(T) = f(T_1, \ldots, T_d)$ as usual (e.g., one can write out f in its Taylor series $f(z) = \sum_{k\in\mathbb{N}_0^d} f_k z^k$, and define $f(T) = \sum_{k\in\mathbb{N}_0^d} f_k T^k$, where $T^k = T_1^{k_1} \cdots T_d^{k_d}$ for $k = (k_1, \ldots, k_d)$.

(a) Show that $\|f(T)\| \le 1$.

Hint: analogously to (4.6.4) show that

$$I - f(T)^* f(T) = (C \otimes I)^*(I - Z(T)(D \otimes I))^{-1} \\ \times (I - Z(T)^* Z(T))(I - Z(T)(D \otimes I))^{-1}(C \otimes I)$$

is positive semidefinite, where

$$Z(T) = \begin{pmatrix} I \otimes T_1 & & 0 \\ & \ddots & \\ 0 & & I \otimes T_d \end{pmatrix}.$$

(b) Let

$$f(z_1, z_2, z_3) = z_1^2 + z_2^2 + z_3^2 - 2(z_1 z_2 + z_2 z_3 + z_3 z_1).$$

Show that $\|f\|_\infty = 5$.

(c) Let

$$T_1 = \begin{pmatrix} 0 & 1 & 0 & 0 \\ 0 & 0 & 0 & 1 \\ 0 & 0 & 0 & 0 \\ 0 & 0 & 0 & 0 \end{pmatrix}, \quad T_2 = \begin{pmatrix} 0 & -\frac{1}{2} & \frac{\sqrt{3}}{2} & 0 \\ 0 & 0 & 0 & -\frac{1}{2} \\ 0 & 0 & 0 & \frac{\sqrt{3}}{2} \\ 0 & 0 & 0 & 0 \end{pmatrix},$$

$$\text{and } T_3 = \begin{pmatrix} 0 & -\frac{1}{2} & -\frac{\sqrt{3}}{2} & 0 \\ 0 & 0 & 0 & -\frac{1}{2} \\ 0 & 0 & 0 & -\frac{\sqrt{3}}{2} \\ 0 & 0 & 0 & 0 \end{pmatrix}.$$

Show that (T_1, T_2, T_3) is a 3-tuple of commuting contractions.

(d) Show that $\|f(T_1, T_2, T_3)\| = 6$.

(e) Conclude that $\frac{1}{5} f$ does not have a realization (4.6.5) with $\left(\begin{smallmatrix} A & B \\ C & D \end{smallmatrix}\right)$ unitary, even though $\|\frac{1}{5} f\|_\infty \le 1$.

34 Let

$$p(z_1, z_2, z_3) = z_1 z_2 z_3 - z_1^3 - z_2^3 - z_3^3.$$

Define $T_1, T_2, T_3 : \mathbb{C}^8 \to \mathbb{C}^8$, where $\mathbb{C}^8$ has an orthonormal basis

$$\{e, f_1, f_2, f_3, g_1, g_2, g_3, h\},$$

via $T_i e = f_i$, $T_i f_i = -g_i$, $T_i f_j = g_k$ (if $i \ne j$, with $k \ne i$ and $k \ne j$), $T_i g_j = \delta_{ij} h$, $T_i h = 0$.

(a) Show that $\|p\|_\infty < 4$.

(b) Show that T_1, T_2 and T_3 are commuting contractions.

(c) Show that $\|p(T_1, T_2, T_3)\| = 4$, thus violating von Neumann's inequality.

35 In this exercise we show that $\nu(p) \leq 2\sqrt{2}\delta(p)$ using a different method from that in Theorem 4.9.2.

(a) Let A be as in (4.9.3) and consider

$$B = \begin{pmatrix} ? & C_n & \cdots & \cdots & C_1 \\ ? & ? & C_n & \cdots & C_2 \\ \vdots & & \ddots & \ddots & \vdots \\ ? & ? & \cdots & ? & C_n \\ ? & ? & \cdots & \cdots & ? \end{pmatrix}.$$

Find X_{ij} and Y_{ij} to complete A and B to A_1 and B_1, say, both with norm $\leq \delta(p)$. Show that the bidiagonal operator matrix

$$M = \begin{pmatrix} \ddots & \ddots & & & \\ & B_1 & A_1 & & \\ & & B_1 & A_1 & \\ & & & \ddots & \ddots \end{pmatrix}$$

has norm $\leq 2\delta(p)$.

(b) Let

$$S = \begin{pmatrix} \ddots & \ddots & & & \\ & I_{\mathcal{H}} & 0 & & \\ & & I_{\mathcal{H}} & 0 & \\ & & & \ddots & \ddots \end{pmatrix}$$

be the bilateral shift (notice that the block sizes in M and S are different; in M each block A_1 and B_1 acts on $\mathcal{H}^{n+1}$). Show that

$$\frac{1}{n+1}(M + SMS^* + \cdots + S^n M S^{n*})$$

converges (as $n \to \infty$) to a Toeplitz operator of norm $\leq 2\delta(p)$, and conclude that $\nu(p) \leq 2\delta(p)$.

(c) Show that if $\|\begin{pmatrix} C & D \end{pmatrix}\| \leq 1$, then $\|\begin{pmatrix} C \\ D \end{pmatrix}\| \leq \sqrt{2}$.

(d) Use the above to show that $\delta(p) \leq \sqrt{2}\|A_k\|$, where $\deg p = 2k$ and A_k is defined by (4.9.4).

(e) Conclude that $\nu(p) \leq 2\sqrt{2}\|A_k\|$.

36 Define

$$\mathcal{P}_1 = \{a_0 + a_1 z : a_0, a_1 \in \mathbb{C}\}$$

and

$$\nu_1 = \max\left\{\frac{\nu(p)}{\delta(p)} : p \in \mathcal{P}_1, p \neq 0\right\}.$$

Show that $\nu_1 \leq \frac{2}{\sqrt{3}}$.

37 Find the contractive central and the superoptimal completions of the partially contractive matrix

$$\begin{pmatrix} \frac{1}{\sqrt{2}} & ? & ? & ? \\ \frac{1}{\sqrt{2}} & \frac{1}{\sqrt{2}} & ? & ? \\ 0 & 0 & \frac{1}{2} & ? \\ 0 & 0 & \frac{1}{2} & \frac{1}{2} \end{pmatrix}.$$

38 Find the superoptimal completion of the partial matrix

$$\begin{pmatrix} 1 & -1 & ? & ? \\ 0 & \frac{1}{2} & ? & ? \\ 1 & 1 & 1 & 0 \\ 1 & 0 & -1 & \frac{1}{2} \end{pmatrix}$$

.

39 Prove the part of Theorem 4.11.4 stating that the minimum of $s^\infty(G-Q)$ is always attained. (Hint: the result follows from a compactness argument by applying $\min\{m,n\}$ times Montel's theorem.)

4.14 NOTES

Section 4.1

Theorem 4.1.1 was first obtained in [43]. In the same paper, the author used Theorem 4.1.1 to find the distance from an operator to a nest algebra, a result known as Arveson's distance formula (see Remark 4.7.3). The contractive completion problem of operator-matrices of the form

$$\begin{pmatrix} A_{11} & ? \\ A_{21} & A_{22} \end{pmatrix}$$

is referred to as *Parrott's problem* (see [449]). Theorem 4.1.4 originates in [338]. Our presentation follows [338] and [48]. Motivated by considerations in [167], in [166] the question of determining some exact distance constants was raised. It was shown that every operator-valued partially contractive matrix of type (4.1.2) admits a completion of norm $\frac{3}{2\sqrt{2}}$ and the estimate is sharp. The following is by now a quite deep conjecture, initially raised in [165]: let $\mathcal{T}_n$ denote the algebra of the upper triangular $n \times n$ matrices. Given an $n^2 \times n^2$ matrix A, estimate the distance from A to $\mathcal{T}_n \otimes \mathcal{T}_n$ for large n (the main question is whether it approaches infinity). In Exercise 4.13.8 it is shown that the distance is bounded above by $\lceil \log_2 n \rceil + 1$ (see, e.g., [429, Section 5] for a solution). Other exercises involving the distance constant are the following. Exercise 4.13.9 is based on examples in [372], which are used to show that if an atomic-core commutative subspace lattice (CSL) algebra has a distance constant great than 1, it must be greater than or equal to $\sqrt{\frac{9}{8}}$. Exercise 4.13.10 is based on [167], where the authors construct commutative subspace lattice algebras without a finite distance constant. Exercise 4.13.11 is based on [294, Theorem 6.2]. The presentation of Subsection 4.1.1 follows the paper [340], where its results originate, except Example 4.1.7, which is new. Some adjustments were necessary. The paper [427] discusses the problem of completing to lower triangular Toeplitz contractions.

The question of completion of partially contractive Toeplitz completion becomes one about Hankels after reversing the columns (or rows). In [565], [303], and [156] this problem was studied. Let

$$H = H(a_1, \ldots, a_n; x_1, \ldots, x_{n-1})$$

be defined as in (4.13.1). The problem is to characterize when this partially contractive Hankel matrix H admits a contractive Hankel completion. For $n = 2$ the problem is trivial. For $n = 3$, it is a consequence of Theorem 4.1.6(iv)(b). For $n = 4$, there is not always a contractive Hankel completion, as shown by Exercise 4.13.6(a), taken from [156]. The same holds for $n \geq 5$, as seen in Exercise 4.13.6(b), taken from [565]. The cases $n = 3$ and $n = 4$ were intensively studied from the numerical point of view in [156].

Another interesting problem studied over the years is to find a real number

x such that the Hankel *triangle*

$$H_\Delta(a_1, \dots, a_n; x) = \begin{pmatrix} a_1 & a_2 & \cdots & a_{n-1} & a_n & x \\ a_2 & a_3 & \cdots & a_n & x & \\ \cdots & \cdots & \cdot^{\cdot^{\cdot}} & \cdot^{\cdot^{\cdot}} & & \\ a_{n-1} & a_n & \cdots & & & \\ a_n & x & & & & \\ x & & & & & \end{pmatrix}$$

is partially contractive; see [340], [565], [303], and [156]. The problem does not admit a solution already for $n = 2$; see for example $H_\Delta(\frac{1}{\sqrt{2}}, \frac{1}{\sqrt{2}}, x)$. In [313] another variation was studied where only some of the a_i above are prescribed.

Section 4.2

The first part of Section 4.2 is based on [68]. Theorem 4.2.1 was independently proved in [41], [169], and [519]. However, in [41], the defect spaces of an arbitrary contractive completion $\widetilde{A}$ were identified. Theorem 4.2.7 can also be obtained as consequence of these identifications. The rest of Section 4.2 is based on [66]. Exercise 4.13.13 is also a result from [66]. The particular case of problem (4.2.8) when $S^*S = T^*T$ was first considered in [229] and named the *strong Parrott problem.* It arose out of questions in the theory of intertwining dilations (see [225]). A variation appears in [358]. The idea of reducing problem (4.2.8) to a positive semidefinite completion problem is based on an observation in [543]. Exercise 4.13.13 is also a result from [66]. Exercise 4.13.12 is Corollary 3.3 in [428].

Section 4.3

Scalar Hankel matrices were introduced in [300], where determinants of finite such type matrices were studied. The factorization in Theorem 4.3.4 is a well-known result that can be found in most books on H^p spaces (see, e.g., [320], [497], [189], and [368]). Theorem 4.3.10 (in the scalar case) was first proved by Carathéodory and Fejér in [120], while Theorem 4.3.1 was proved by Nehari in [430]. The operator-valued version of Theorem 4.3.1 was first obtained in [444]. Its proof based on matrix completions in Section 4.3 originates in [449]. The second proof of the scalar version of Theorem 4.3.1 can be found in [438]. For both Theorems 4.3.1 and 4.3.10, there are several other proofs based on different approaches such as the *commutant lifting theorem*, *Schur's algorithm*, the band method, reproducing kernels, etc. ([509], [225], [53], [266], [260, Chapter XXXV], [190], etc.). In [272] the Nehari problem is applied in the context of prediction for two stochastic processes. Exercise 4.13.20 is based on [524, Section 3.1].

A detailed study of Hankel operators and related extension problem was done in the papers [4], [3], and [6], while in [454], [455], and [456] a completed characterization of Hankel operators belonging to the Schatten-von Neumann classes $\mathcal{C}_p$, $0 < p < \infty$, was given. The books [472] and [457] provide a comprehensive collection of results on Hankel operators.

Theorem 4.3.5 was first obtained in [377]. Its proof in Section 4.3 is taken from [457]. An alternative proof based on the scalar version of Theorem 4.11.1 can be found in [509] (see also [438] and [457]).

The proof of the closedness of $H^\infty + C$, as well as Exercises 4.13.21 and 4.13.22 on which its proof is based, are taken from [509] (see also [184]). Another proof of this fact can be found in [505]. Extensions and variations of this results may be found, e.g., in [282] and [492]. Many important properties of this class of functions are collected in [184]. Among them is the fact that $H^\infty + C$ is an algebra. The proof of Theorem 4.3.7 comes from [301]. In [443], it was obtained for the case of operator-valued functions. As proved there, for $F \in L^\infty(\mathcal{L}(\mathcal{H}))$, H_F is compact if and only if $F \in H^\infty(\mathcal{L}(\mathcal{H})) + C(\mathcal{K}(\mathcal{H}))$, where $C(\mathcal{K}(\mathcal{H}))$ denotes the set of all continuous functions defined on $\mathbb{T}$ with values compact linear operators on $\mathcal{H}$. Theorem 4.3.8 and Corollary 4.3.9 are taken from [460]. Exercise 4.13.17 is a result from [266].

For a variation of the Carathéodory-Fejer interpolation problem, but now on the ball $\{z \in \mathbb{C}^n : |z_1|^2 + \cdots + |z_d|^2 < 1\}$; see, for instance, [102] and [22] and references therein.

Section 4.4

Section 4.4 starts off with a result proved in [152] stating that not all bounded big Hankel operators $\Gamma : H^2(\mathbb{T}^2) \to L^2(\mathbb{T}^2) \ominus H^2(\mathbb{T}^2)$ have a bounded symbol. The authors of the latter paper also conjectured that when it has a bounded symbol, the norm of the Hankel is not comparable to the norm of any symbol. The conjecture was solved independently in [215] and [61]. Proposition 4.4.2 was taken from [178, Example 4.1]. Exercise 4.13.24 is based on [185].

In contrast to the result on big Hankels (Theorem 4.4.1), it was shown [216] that any bounded little Hankel has a bounded symbol. It is still unknown how the minimum infinity norms among all bounded symbols compares to the norm of the little Hankel. The result in [216] establishes the so-called weak factorization of functions in the two-variable biholomorphic Hardy space $H^1(\mathbb{R}^2_+ \times \mathbb{R}^2_+)$, where $\mathbb{R}^2_+$ denotes the upper half-plane. It is shown that there exists a constant $c > 0$ such that for each $h \in H^1(\mathbb{R}^2_+ \times \mathbb{R}^2_+)$, there exist sequences of functions $(f_j)_{j\in\mathbb{N}}$, $(g_j)_{j\in\mathbb{N}}$ in the Hardy space $H^2(\mathbb{R}^2_+ \times \mathbb{R}^2_+)$ such that

$$h = \sum_{j=1}^{\infty} f_j g_j \quad \text{and} \quad \sum_{j=1}^{\infty} \|f_j\|_2 \|g_j\|_2 \leq c\|h\|_1.$$

The question as to whether such a factorization holds has been an open problem for a number of years. (On other domains factorization results existed earlier, e.g., on the boundary of the multivariable ball; see [136].) In their proof, the authors rely on a dual formulation of the problem from [217]. The dual of the real variable Hardy space $H^1(\mathbb{R}^2_+ \times \mathbb{R}^2_+)$ as the product BMO space $\mathrm{BMO}(\mathbb{R}^2_+ \times \mathbb{R}^2_+)$, was characterized in [126] in terms of a Carleson-type condition on open sets. Recent articles on the multivariable Nehari problem include [468] and [383]. The remainder of the section was taken from [249], in which necessary and sufficient conditions are given for the solvability of the operator-valued two-variable autoregressive filter problem. The authors also give a condition by which the small Hankel of norm less than 1 implies the existence of a symbol $\|\phi\|_\infty < 1$.

Section 4.5

Section 4.5 follows the paper [51]; Theorem 4.5.1 and its consequences, Corollary 4.3.3 and 4.5.8 were taken from there. It is known that $\mathcal{L}(\mathcal{H})$ does not have the Radon-Nikodym property. This fact does not conflict with Proposition 4.5.5, since our notion of measurability is weaker than the ones usually considered in connection with vector measures. The notion of canonical shuffle is taken from [451], while the result mentioned in the proof of Lemma 4.5.3 is [510, Theorem 1.9.8].

Let G be a locally compact Abelian group and let Γ be its character group. Let Γ_d be the group Γ with the discrete topology, and let $\overline{G}$ be the character group of Γ_d. Then $\overline{G}$ is a compact Abelian group called the Bohr compactification of G ([502], Theorem 1.8.1). Theorem 4.5.1 can be applied to any such $\overline{G}$. When $\Gamma = \mathbb{R}_d$, then $G = \overline{\mathbb{R}}$ and the continuous functions on $\overline{\mathbb{R}}$ are referred to in the literature as almost periodic functions. Unfortunately, when k is an almost periodic function, there might not exist any almost periodic function K such that $\|K\|_\infty = \|H_k\|$. (As shown by a counterexample in [3], the latter fails even for scalar continuous functions on the unit circle.) This is why in [489]–[492], the (matrix-valued) function k was considered to belong to the Wiener class (i.e., $\sum_{\gamma<0} \|k_\gamma\| < \infty$). It was proved in that case that, if $\|H_k\| < 1$, there exists an almost periodic extension K of k, K in the Wiener class, such that $\|K\|_\infty < 1$. The authors successfully parameterized in the matrix-valued case the set of all such extensions K. However, Theorem 4.5.1 implies that if $k : \overline{\mathbb{R}} \to \mathcal{L}(\mathcal{H})$ is of the form $k(x) = \sum_{n=-\infty}^{0} e^{i\gamma_n x} k_n$, $\gamma_n \le 0$ for every $n \le 0$, and $\|H_k\| < \infty$, then k has an extension $K(x) = \sum_{n=-\infty}^{\infty} e^{i\gamma_n x} k_n$, $\gamma_n > 0$ for $n > 0$, with $\|K\|_\infty = \|H_k\|$. The convergence of the latter series is in the L^2 sense, which is weaker than the l^1 convergence in the Wiener algebra case. A result similar to Theorem 4.5.1 for the case of matrix-valued functions in the Wiener class and P as in (3.5.3) was proved in [492], together with a parametrization of the set of all strictly contractive solutions. Some of the Wiener algebra properties are useful in the study of localized frames; see [284], [71], [18]. The scalar versions of Corollaries 4.3.3 and 4.5.8 were proved in [182], while in [492] a proof based on the commutant lifting approach can be found for the matrix-valued version of Corollary 4.3.3.

Section 4.6

The Nevanlinna-Pick problem was independently studied in [463] and [435], where Theorem 4.6.1 was proved. For the latter theorem there exist several alternative proofs, as for example in [509]. The study of the multivariable version of the Nevanlinna-Pick problem was started in [7]. Representation (4.6.5) appears in [8] and [380]. It was shown in both these references that in two variables every Schur function has the representation (4.6.5); see also [75] for a discussion. In [10] it was shown that if there is a solution, then there is a solution which is a rational inner function. Theorem 4.6.5 is Theorem 5.5.1 in [504]. It was also shown in [504] that every scalar rational inner function is of the form $\frac{z^m \overleftarrow{p}}{p}$ with p a polynomial without roots in $\mathbb{D}^d$. A natural matrix-valued generalization would be to consider $\overleftarrow{R} D P^{-1}$, where R and P are matrix-valued polynomials that are invertible on $\mathbb{D}^d$ with $RR^* = P^*P$ on $\mathbb{T}^d$, and where D is a diagonal matrix of monomials; it is unknown, however, whether this is the correct generalization. Theorem 4.6.4 was taken from [283]. While Theorem 4.6.4 gives necessary conditions for the existence to the d-variable Nevanlinna-Pick problem, it is unclear whether these are also sufficient. In fact, the authors of [12] suggest that these conditions are in general not sufficient. Example 4.6.8 was also taken from [283]. The polynomial $p(z)$ in this example is due to [555], while the last part of this example is due to [322]. Exercise 4.13.34 provides another example, and is based on [153]. In [410] another construction of commuting diagonalizable contractions violating von Neumann's inequality is presented. The two-variable von Neumann inequality (Theorem 4.6.7; also known as *Ando's theorem*) is due to [31]. For a proof along the lines of the proof given here, please see [139]. A generalization of Ando's theorem where variables commute according to a graph without cycles may be found in [442]. Many open questions remain for the von Neumann inequality in 3 or more variables. In [321] it was shown that $\|p(A_1, \ldots, A_n)\| \leq \|p\|_\infty$, where p is a polynomial and $A_1, \ldots, A_n$ are 2×2 commuting contractions. For 3×3 commuting contractions the question is still open. An excellent account on the von Neumann inequality (including open problems) may be found in [464, Chapter 1]. A reference on multivariable holomorphic functional calculus is [539]. Treatments of the multivariable Nevanlinna-Pick theorem can also be found in [79], [137], [151], and [138]. There has been an explosion of interest in interpolation in several variables in general and on the bidisk in particular (see [11] and the references within). Other variations of the Nevanlinna-Pick problem are the spectral problem, where the norm condition is replaced by a condition on the spectral radius (see [89]), and the tangential Nevanlinna-Pick problem, where the operator/matrix values at the points are only prescribed in certain directions (see, e.g., [190], [76], [507], [227], and [125] for applications); see Exercise 4.13.27. The authors are not aware of multivariable tangential Nevanlinna-Pick interpolation results; this would be an interesting problem to pursue. Exercise 4.13.25 is taken from [11] and represents the operator-valued Pick problem on the bidisk. Unitary realizations of functions have been widely used in systems theory (see, e.g., [111], [78], and [79]). Exercise 4.13.29 comes from examples in [10].

Section 4.7

Section 4.7 follows the paper [67]. Theorem 4.7.1 unifies results of [43],

[386], [38], [39], and [355]. We refer to [178] as a reference on nest algebras and to [297] for results on quasitriangular operators. Remark 4.7.2 is called Arveson's distance formula (see [43]), representing the distance between an operator and a nest algebra. Remark 4.1.2 represents its version for finite nests.

The fact that $\mathcal{QT} = \text{alg}\mathcal{P} + \mathcal{K}(\mathcal{H})$ was proved in [43]. By taking $B = 0$ in Corollary 4.7.3 we recover a result in [386] (see also [355]). The Corona problem was solved by L. Carleson in [121] and states that given the functions $\phi_1, \ldots, \phi_n \in L^\infty(\mathcal{T})$, the existence of $f_1, \ldots, f_n \in H^\infty$ such that

$$f_1\phi_1 + \cdots + f_n\phi_n = 1$$

is equivalent to

$$|\phi_1(z)| + \cdots + |\phi_n(z)| \geq \epsilon > 0 \quad \text{for all } |z| < 1.$$

Carleson's result was generalized to the matrix-valued case in [239]. The proof for the existence of an expectation π necessary for the proof of Theorem 4.7.8 can be found in [43] and [178]. Theorem 4.7.8 and Corollary 4.7.9 are *Toeplitz-type Corona problems* as the necessary and sufficient conditions are on the level of the multiplication operators associated with the given functions. Their solutions are in the spirit of [43] (see also [535]). In the case when $\psi = 1$, $g_i = 0$, and $\phi_i \in H^\infty(\mathbb{T})$, $i = 1, \ldots, n$, the bound in Corollary 4.7.9 coincides with the bound found in [535]. In [496] the Corona problem is considered with a countable number number of functions, and in [488] a version with almost periodic functions is established.

Section 4.8

The results in Section 4.8 are based on results in [345], where the problem of joint norm control extensions was first addressed. There are several subsequent papers addressing the joint norm control issue and its applications; among them we mention [226], [228], [498], [233], [232], and [56]. In these papers the constants were improved locally, but not globally. It is still unknown whether $\sqrt{2}$ is the best absolute joint norm control constant. As explained in Section 4.8, the contractive central completion of a strictly contractive partial operator matrix is always strictly contractive, but the same is not true for Nehari's problem, not even in the scalar-valued case, when the condition $\sum_{n=-\infty}^{-1} |c_n| < \infty$ is not satisfied. In [49] such an example of a sequence $\{c_n\}_{n=-\infty}^{-1}$ was given for which $\|H(\{c_n\}_{n=-\infty}^{-1})\| < 1$ and the central solution f_c of Nehari's problem with data $\{c_n\}_{n=-\infty}^{-1}$ verifies $\|f_c\|_\infty = 1$. This is the reason why the first norm estimate in Theorem 4.8.4 is not strict, as opposed to the one in 4.8.2. The same has been shown to be true for the central solution of the commutant lifting interpolation problem in [241]. We notice that the solution F_c in Exercise 4.13.18 is always the central solution of the problem. This exercise is a result from [195] which was generalized in [268] for the general framework of the band method.

Section 4.9

The presentation of Section 4.9 follows the paper [62]. Theorems 4.9.2 and 4.9.4 were proved there. Theorem 4.9.2 answers in the scalar case a question raised in [376], Section IX.2.3, where Theorem 4.9.1 also originates. A proof for the duality $L^1(\mathcal{C}_1(\mathcal{H}))^* = L^\infty(\mathcal{L}(\mathcal{H}))$ can be found in [509]. Lemma 4.9.3 is an operator-valued generalization proven in [62] of a result in [454]. Another estimate for $\nu(p)$, namely,

$$\nu(p) \leq 3\|A_k\|,$$

where $\deg p = 2k$ and A_k is defined by (4.9.4), was obtained in [437] (see also [557]). Notice that Exercise 4.13.35, which was inspired by the paper [557], yields the better bound $\nu(p) \leq 2\sqrt{2}\|A_k\|$.

The estimate given by Theorem 4.9.2 is in general not sharp. For instance, if $p(z) = z$, then $\nu(p) = \delta(p) = 1$. It would be more reasonable to consider global estimates. To fix ideas, let us assume again $\dim \mathcal{H} = 1$ ($\mathcal{P}_n$ being then the space of the usual scalar-valued polynomials of degree at most n); thus $p(z) = a_0 + \cdots + a_n z^n$, with $a_k \in \mathbb{C}$. If we define

$$\nu_n = \max\left\{\frac{\nu(p)}{\delta(p)} \ : \ p \in \mathcal{P}_n,\ p \neq 0\right\},$$

we obtain from Theorem 4.9.2 that $\nu_n \leq 2\frac{n+1}{n+2}$. Again, we do not have equality in general. (Exercise 4.13.36, namely $\nu_1 \leq \frac{2}{\sqrt{3}} < \frac{4}{3}$, repeats a statement from [62].) For any natural n we have $\nu_n < 2$; an open conjecture is whether $\nu_n \to 2$ when $n \to \infty$.

The relation with the classical Carathéodory-Fejér problem can also be given in terms of global estimates. If

$$\mu_n = \max\left\{\frac{\mu(p)}{\nu(p)} \ : \ p \in \mathcal{P}_n,\ p \neq 0\right\},$$

then μ_n is of order $\log n$. Indeed, the natural projection onto $\mathcal{P}_n$ in the space of analytic functions of bounded mean oscillation has norm of order $\log n$. (A precise reference for this well-known fact is hard to find; however, it is obviously equivalent by duality to the same assertion for H^1, which can be found, for instance, in [591].) If f is an L^∞ extension of $\widetilde{p}$ of minimum norm, then $P_+ f \in$ BMOA, with norm equivalent to $\nu(\widetilde{p}) = \nu(p)$; if $g = P_{\mathcal{P}_n} f$, then the BMOA norm of g is equivalent to $\mu(p)$. In view of (4.9.1) and the definition of $\delta(p)$, it is worth noting that this estimate for μ_n is closely related to the fact that the norm of the triangular truncation in the space of $n \times n$ matrices is also of order $\log n$. For a detailed discussion on this last problem, including historical references, see [37].

Section 4.10

The results in Section 4.10 are taken from [567]. The problem was first considered for 2×2 compact operator matrices in [168]. The idea of superoptimal solutions for the Nevanlinna-Pick and Nehari problems was started in [586] and [460], respectively. Theorem 4.10.1 is reminiscent of the main technical result in [586]. The problem of minimizing the jth singular value over the set of all completions of a lower triangular block matrix was considered in [274], and will be treated in Section 5.6. The question of the lowest possible singular values in (the lexicographic order) was raised for the corresponding Nehari problem in [460]. Theorem 4.10.5 is generalizing a result in [168].

Section 4.11

Theorem 4.11.1 was first proved for $H^2(\mathbb{T})$ in [93]. The matrix-valued case was considered in [397], and the operator-valued case in [296] and [499]. The approach in [397] and [296] is based on the Wold decomposition of an isometry (see [578]). A proof of Theorem 4.11.1 can also be found in [534, Section 5.5]. In [77] (and subsequent papers) versions of the Beurling-Lax-Halmos theorem were used to derive several new results for classical interpolation results, such as Nevanlinna-Pick and Carathéodory-Fejér. The scalar version of Theorem 4.11.2 is a well-known results that can be found in most books on H^p spaces (see, e.g., [320], [189], and [368]). The operator-valued version of Theorem 4.11.2 can be found in [307] and [534]. Its proof in Section 4.11 is taken from [438]. Montel's theorem is a well-known result in complex analysis. Its proof can be found, for example, in [150]. The rest of Section 4.11 is based on [460] (the results are also included in [457]). The notion of superoptimal solution for Nehari's problem is motivated by considerations in engineering ([230], [382], [408], and [471]), where the notion was found useful in the context of H^∞ control theory. The proof of Theorem 4.11.6 can be found in [458], and the one for the inequalities $\tau_k \leq t_k$ in [460]. More recent accounts and additional results on superoptimality may be found in [546], [459]. In [295] a state space algorithm is presented to compute the superoptimal solution.

Section 4.12

A good source for more information on the model matching problem is [231, Chapters 6 and 8].

Exercises

The polynomial in Exercise 4.13.33(b) is from [555] and the matrices in part (c) are from [322].

Exercise 4.13.32 is based on an example in [361]. The equality in part (b) is an illustration of a result by [379], which was recently sharpened in [365]. This sharpened result says that for any stable polynomial $p(z_1, z_2, z_3)$ of degree at most $(1, 1, 1)$ we have that

$$|p|^2 - |\overleftarrow{p}|^2 = (1 - |z_1|^2) \sum_{i=1}^{4} |g_i|^2 + (1 - |z_2|^2) \sum_{i=1}^{4} |h_i|^2 + (1 - |z_3|^2) \sum_{i=1}^{2} |k_i|^2,$$

where g_i is of degree at most $(0, 1, 1)$, h_i of degree at most $(1, 0, 1)$, and k_i of degree at most $(1, 1, 0)$. Of course, one may interchange the variables to derive a different variation, where, for instance, there are only two g_i.

Exercise 4.13.26 is based on [390, Theorem 1]. In fact the three statements about p are equivalent; that is, p has a representation (4.13.2) if and only (a) holds and if and only if (b) holds. Parts of the proof of this statement were done earlier in [94]; see also [451, Corollary 18.2].

Chapter Five

Hermitian and related completion problems

In this chapter we consider various completion problems that are in one way or another closely related to positive semidefinite or contractive completion problems. For instance, as a variation on requiring that all eigenvalues of the completion are positive/nonnegative, one can consider the question how many eigenvalues of a Hermitian completion have to be positive/nonnegative. In the solution to the latter problem ranks of off-diagonal parts will play a role, which is why we also discuss minimal rank completions. Related is a question on real measures on the real line. As a variation of the contractive completion problem, we will consider the question how many singular values of a completion have to be smaller (or larger) than one? We will also consider completions in the class of normal matrices and the class of distance matrices. As applications we look at questions regarding Hermitian matrix expressions, a minimal representation problem for discrete systems, and the separability problem that appears in quantum information.

5.1 HERMITIAN COMPLETIONS

Hermitian completion problems concern partial Hermitian matrices for which completions are to be found that have certain requirements on the eigenvalues.

A $n \times n$ partial matrix is called *Hermitian* if entry (i, j) is a specified complex number a_{ij} if and only if entry (j, i) is a specified complex number a_{ji} and $a_{ji} = \overline{a}_{ij}$. We are interested in the possible inertia of Hermitian completions of partial Hermitian matrices. Recall that *the inertia* of a $n \times n$ Hermitian matrix M is a triple $i(M) = (i_+(M),\ i_0(M),\ i_-(M))$ of natural numbers in which $i_+(M),\ i_0(M),\ i_-(M)$ denotes the number of positive, zero, and negative eigenvalues of M, respectively. We shall also use

$$i_{\leq 0}(M) = i_-(M) + i_0(M),\ i_{\geq 0}(M) = i_0(M) + i_+(M),$$

which we shall refer to as the nonpositive and nonnegative inertia of M, respectively. Note that for an $n \times n$ Hermitian matrix M we have that

$$i_-(M) + i_0(M) + i_+(M) = n.$$

Some of the observations that we will use repeatedly are the following. First, we have for $A \in \mathcal{H}_n$ and $B \in \mathcal{H}_m$ that

$$i\begin{pmatrix} A & 0 \\ 0 & B \end{pmatrix} = i(A) + i(B).$$

Next, the Cauchy interlacing inequalities imply that when

$$M = \begin{pmatrix} A & * \\ * & * \end{pmatrix} \in \mathcal{H}_{n+m}, \ A \in \mathcal{H}_n,$$

then

$$i_-(A) \le i_-(M) \le i_-(A) + m, \ \ i_+(A) \le i_+(M) \le i_+(A) + m, \tag{5.1.1}$$

and

$$i_{\le 0}(A) \le i_{\le 0}(M) \le i_{\le 0}(A) + m, \ \ i_{\ge 0}(A) \le i_{\ge 0}(M) \le i_{\ge 0}(A) + m. \tag{5.1.2}$$

In addition, Sylvester's law of inertia states that

$$i(S^*MS) = i(M) \text{ for all } M \in \mathcal{H}_n, \ \ S \in \mathbb{C}^{n\times n}, \ \ \det S \ne 0.$$

As a consequence of Sylvester's law we obtain that if

$$M = \begin{pmatrix} A & B \\ B^* & C \end{pmatrix} \in \mathcal{H}_{n+m}$$

with C invertible, then

$$i(M) = i(C) + i(A - BC^{-1}B^*). \tag{5.1.3}$$

Indeed, this follows from applying Sylvester's law with

$$S = \begin{pmatrix} I & 0 \\ -C^{-1}B^* & I \end{pmatrix}.$$

Similarly, when A is invertible, we have that

$$i(M) = i(A) + i(C - B^*A^{-1}B). \tag{5.1.4}$$

We also have the following useful auxiliary results.

Lemma 5.1.1 *If $M = M^*$,*

$$M^{-1} = \begin{pmatrix} A & B \\ B^* & C \end{pmatrix}^{-1} = \begin{pmatrix} P & Q \\ Q^* & R \end{pmatrix},$$

with A invertible, then

$$i(M) = i(A) + i(R).$$

Proof. Observe that $C - B^*A^{-1}B$ is invertible, and that its inverse equals R. The lemma now follows directly from (5.1.4). □

Lemma 5.1.2 *Let $M = M^*$ be invertible, and*

$$M = \begin{pmatrix} A & B \\ B^* & C \end{pmatrix},$$

with A of size $n_1 \times n_1$ and C of size $n_2 \times n_2$. If $i_0(A) = n_2$, then

$$i(M) = (i_{\ge 0}(A), 0, i_{\le 0}(A)) = (i_+(A) + n_2, 0, i_-(A) + n_2).$$

Proof. As $i_{\leq 0}(M) \geq i_{\leq 0}(A)$, and $i_{\leq 0}(M) = i_-(M)$ (as M is invertible), we get that $i_-(M) \geq i_{\leq 0}(A) = i_-(A) + n_2$. Similarly, $i_+(M) \geq i_+(A) + n_2$. Finally,

$$i_+(M) + i_-(M) = \text{rank}(M) \leq \text{rank}(A) + \text{rank}\begin{pmatrix} B \\ C \end{pmatrix} + \text{rank}B^*$$
$$\leq i_+(A) + i_-(A) + 2n_2,$$

so we must have that $i_\pm(M) = i_\pm(A) + n_2$. □

Let us now start with a simple example.

Example 5.1.3 For the partial Hermitian matrix $M(x) = \begin{pmatrix} 1 & x \\ \bar{x} & 1 \end{pmatrix}$ we have the following possible inertia:

$$i(M(x)) = \begin{cases} (2,0,0) & |x| < 1, \\ (1,1,0) & |x| = 1, \\ (1,0,1) & |x| > 1. \end{cases}$$

For the partial matrix $N(x) = \begin{pmatrix} 1 & x \\ \bar{x} & -1 \end{pmatrix}$, the only possible inertia is $i(N(x)) = (1,0,1)$, regardless of the value of x.

In the example both partial matrices involve 2×2 matrices with specified diagonal entries. The following proposition addresses the general situation where two diagonal block matrices are specified and the remaining entries are to be chosen freely.

Proposition 5.1.4 *Let $A = A^* \in \mathcal{H}_{n_1}$ and $B = B^* \in \mathcal{H}_{n_2}$. Consider the $(n_1 + n_2 + n_3) \times (n_1 + n_2 + n_3)$ partial Hermitian matrix*

$$M = \begin{pmatrix} A & ? & ? \\ ? & B & ? \\ ? & ? & ? \end{pmatrix}.$$

Then all possible inertia $i(M)$ are described by the following inequalities:

$$\max\{i_+(A), i_+(B)\} \leq i_+(M) \leq \min\{i_+(A) + n_2 + n_3, i_+(B) + n_1 + n_3\}, \tag{5.1.5}$$

$$\max\{i_-(A), i_-(B)\} \leq i_-(M) \leq \min\{i_-(A) + n_2 + n_3, i_-(B) + n_1 + n_3\}, \tag{5.1.6}$$

$$-(i_-(A) + i_-(B) + n_3) \leq i_+(M) - i_-(M) \leq i_+(A) + i_+(B) + n_3, \tag{5.1.7}$$

$$i_-(M) + i_+(M) \leq n_1 + n_2 + n_3. \tag{5.1.8}$$

When the given matrices A and B are real, a Hermitian completion with the desired inertia can be chosen to be real as well.

To prove this result we will need the following lemma.

Lemma 5.1.5 *Let $M = \begin{pmatrix} A & X \\ X^* & B \end{pmatrix}$, where $A = A^* \in \mathcal{H}_n$ and $B = B^* \in \mathcal{H}_m$. Put*

$$\delta = \text{rank}\begin{pmatrix} A & X \end{pmatrix} - \text{rank}\begin{pmatrix} A \end{pmatrix}.$$

Then

$$i_+(M) \geq i_+(A) + \delta, \; i_-(M) \geq i_-(A) + \delta.$$

Proof. With a change of basis we can write $A = \begin{pmatrix} A_1 & 0 \\ 0 & 0 \end{pmatrix}$, where $A_1 = A_1^*$ is invertible. Writing $X = \begin{pmatrix} X_1 \\ X_2 \end{pmatrix}$ according to the same decomposition, we get that $\delta = \text{rank}X_2$. Next we perform an equivalence to write X_2 as

$$X_2 = \begin{pmatrix} I_\delta & 0 \\ 0 & 0 \end{pmatrix}.$$

Decomposing X_1 and B accordingly, we get that M is congruent to

$$\begin{pmatrix} A_1 & 0 & 0 & X_{11} & X_{12} \\ 0 & 0 & 0 & I_\delta & 0 \\ 0 & 0 & 0 & 0 & 0 \\ X_{11}^* & I_\delta & 0 & B_{11} & B_{12} \\ X_{12}^* & 0 & 0 & B_{21} & B_{22} \end{pmatrix}.$$

It is not hard to see that

$$i_+ \begin{pmatrix} A_1 & 0 & X_{11} \\ 0 & 0 & I_\delta \\ X_{11}^* & I_\delta & B_{11} \end{pmatrix} = i_+ \begin{pmatrix} A_1 & 0 & 0 \\ 0 & 0 & I_\delta \\ 0 & I_\delta & B_{11} \end{pmatrix} = i_+(A) + \delta.$$

But then the first statement follows from the interlacing inequalities. The second statement follows in a similar way. □

Proof of Proposition 5.1.4. Clearly, by the interlacing inequalities we have that $i_+(A) \leq i_+(M)$ and $i_{\leq 0}(A) \leq i_{\leq 0}(M)$. The latter inequality yields that $i_+(M) \leq n_1 + n_2 + n_3 - i_{\leq 0}(A) = n_2 + n_3 + i_+(A)$. Repeating the same argument with B instead of A yields that $i_+(B) \leq i_+(M) \leq n_1 + n_3 + i_+(B)$. Combining these inequalities yields (5.1.5). The inequalities (5.1.6) are obtained similarly (or can be obtained by applying (5.1.5) to $-M$).

For inequalities (5.1.7) we introduce the numbers

$$k_1 = \dim \text{Ker} \begin{pmatrix} A \\ X^* \end{pmatrix} \quad k_2 = \dim \text{Ker} \begin{pmatrix} X \\ B \end{pmatrix},$$

where X denotes the $(1,2)$ entry in a completion of M. Observe that

$$\text{rank} \begin{pmatrix} A \\ X^* \end{pmatrix} - \text{rank}A = i_0(A) - k_1, \quad \text{rank} \begin{pmatrix} X \\ B \end{pmatrix} - \text{rank}B = i_0(B) - k_2.$$

Let $N = \begin{pmatrix} A & X \\ X^* & B \end{pmatrix}$. Then, $i_0(N) \geq k_1 + k_2$. Applying now Lemma 5.1.5 to both A and B we get that

$$2i_-(N) \geq i_-(A) + i_-(B) + i_0(A) - k_1 + i_0(B) - k_2$$

$$\geq i_-(A) + i_-(B) + i_0(A) + i_0(B) - i_0(N).$$

Subtracting this inequality from the equality $i_-(N) + i_+(N) = n_1 + n_2 - i_0(N)$ yields

$$i_+(N) - i_-(N) \leq i_+(A) + i_+(B).$$

Next, since $i_+(M) \leq i_+(N) + n_3$ and $i_-(M) \geq i_-(N)$, we get

$$i_+(M) - i_-(M) \leq i_+(N) + n_3 - i_-(N) \leq i_+(A) + i_+(B) + n_3,$$

yielding the right-side inequality in (5.1.7). The left-side inequality in (5.1.7) follows in the same manner.

As (5.1.8) must obviously hold, we have proved the necessity of inequalities (5.1.5)–(5.1.8).

For the converse denote $i(A) = (p, q, r)$ and $i(B) = (u, v, w)$. By performing a congruence it suffices to assume that

$$A = \begin{pmatrix} I_p & 0 & 0 \\ 0 & -I_q & 0 \\ 0 & 0 & 0_r \end{pmatrix}, \quad B = \begin{pmatrix} I_u & 0 & 0 \\ 0 & -I_v & 0 \\ 0 & 0 & 0_w \end{pmatrix}.$$

It is not hard to see that for any $i(M)$ satisfying (5.1.5)–(5.1.8) one can choose the unknown matrices to be real diagonal to get the desired inertia. □

We now present the main result in this section.

Theorem 5.1.6 *Consider the partial Hermitian* $(n_1+n_2+n_3)\times(n_1+n_2+n_3)$ *matrix*

$$M(X) = \begin{pmatrix} A & B & X \\ B^* & C & D \\ X^* & D^* & E \end{pmatrix}, \tag{5.1.9}$$

where X *denotes the unknown. Introduce the following notation:*

$$i\begin{pmatrix} A & B \\ B^* & C \end{pmatrix} = (\pi, \delta, \nu)\ ,\ i\begin{pmatrix} C & D \\ D^* & E \end{pmatrix} = (\pi', \delta', \nu'),$$

$$r = \operatorname{rank}\begin{pmatrix} B^* & C & D \end{pmatrix},$$

$$p = \max\left\{\pi + r - \operatorname{rank}\begin{pmatrix} B^* & C \end{pmatrix},\ \pi' + r - \operatorname{rank}\begin{pmatrix} C & D \end{pmatrix}\right\},$$

$$q = \max\left\{\nu + r - \operatorname{rank}\begin{pmatrix} B^* & C \end{pmatrix},\ \nu' + r - \operatorname{rank}\begin{pmatrix} C & D \end{pmatrix}\right\}.$$

Then all possible inertias for $M(X)$ *are given via the following inequalities:*

$$p \le i_+(M) \le \min\{n_1 + \pi', n_3 + \pi\}, \tag{5.1.10}$$

$$q \le i_-(M) \le \min\{n_1 + \nu', n_3 + \nu\}, \tag{5.1.11}$$

$$r - \nu - \nu' \le i_+(M) - i_-(M) \le \pi + \pi' - r, \tag{5.1.12}$$

$$i_+(M) + i_-(M) \le n_1 + n_3 + r. \tag{5.1.13}$$

When the given data are real, a Hermitian completion with the desired inertia can always be chosen to be real as well.

Let us first illustrate the result with an example.

Example 5.1.7 Let $A = I_2 \oplus (-I_2) \oplus 0_2$, $E = 1 \oplus (-1) \oplus 0_4$, $C = 0$, and $B = 0_{6\times 1} = D^*$. Thus M is a 13×13 matrix. Denoting $p = i_+(M)$ and $n = i_-(M)$, we get the inequalities

$$2 \le p \le 7,\ 2 \le n \le 7,\ -3 \le p - n \le 3,\ p + n \le 12.$$

Note that none of the inequalities are redundant.

We will make use of the following lemma.

Lemma 5.1.8 *Given matrices A and B of sizes $n\times m$ and $n\times p$, respectively, there exist invertible matrices S, T_1, and T_2 of sizes $n\times n$, $m\times m$, and $p\times p$, respectively, such that*

$$S\begin{pmatrix} A & B \end{pmatrix}\begin{pmatrix} T_1 & 0 \\ 0 & T_2 \end{pmatrix} = \begin{pmatrix} I_a & 0 & 0 & 0 & 0 & 0 \\ 0 & I_b & 0 & 0 & I_b & 0 \\ 0 & 0 & 0 & 0 & 0 & I_c \\ 0 & 0 & 0 & 0 & 0 & 0 \end{pmatrix}, \tag{5.1.14}$$

where

$$\begin{aligned} a &= \operatorname{rank}\begin{pmatrix} A & B \end{pmatrix} - \operatorname{rank}(B), \\ c &= \operatorname{rank}\begin{pmatrix} A & B \end{pmatrix} - \operatorname{rank}(A), \\ b &= \operatorname{rank}(A) + \operatorname{rank}(B) - \operatorname{rank}\begin{pmatrix} A & B \end{pmatrix}. \end{aligned}$$

Proof. Let

$$M_2 = \text{RanA} \cap \text{RanB}, M_1 = \text{Ran}A \ominus M_2,$$

$$M_3 = \text{Ran}B \ominus M_2, M_4 = (M_1 \oplus M_2 \oplus M_3)^{\perp}.$$

Then we may write

$$\begin{pmatrix} A & B \end{pmatrix} = \begin{pmatrix} A_1 & 0 \\ A_2 & B_1 \\ 0 & B_2 \\ 0 & 0 \end{pmatrix} : \mathbb{C}^m \oplus \mathbb{C}^p \to \bigoplus_{i=1}^{4} M_i. \tag{5.1.15}$$

Next we decompose

$$\mathbb{C}^m = (\text{Ker}A_1)^{\perp} \oplus (\text{Ker}A_1 \ominus (\text{Ker}A_1 \cap \text{Ker}A_2)) \oplus (\text{Ker}A_1 \cap \text{Ker}A_2)$$

and

$$\mathbb{C}^p = (\text{Ker}B_1 \cap \text{Ker}B_2) \oplus (\text{Ker}B_2 \ominus (\text{Ker}B_1 \cap \text{Ker}B_2)) \oplus (\text{Ker}B_2)^{\perp}.$$

Writing (5.1.15) with respect to these decompositions, we get

$$\begin{pmatrix} A_{11} & 0 & 0 & 0 & 0 & 0 \\ A_{21} & A_{22} & 0 & 0 & B_{11} & B_{12} \\ 0 & 0 & 0 & 0 & 0 & B_{22} \\ 0 & 0 & 0 & 0 & 0 & 0 \end{pmatrix}, \tag{5.1.16}$$

where A_{11}, A_{22}, B_{11}, and B_{22} are invertible. Now, using row operations, one can eliminate A_{21} and B_{12}. Finally, it remains to convert A_{11}, A_{22}, B_{11}, and B_{22} to identities of the appropriate size. It should be noticed that all the steps above indeed lead to appropriate invertible matrices S, T_1, and T_2. Finally, it is straightforward to check the formulas for a, b, and c. □

Proof of Theorem 5.1.6. First we prove the necessity of (5.1.10) – (5.1.13). Note that

$$\gamma := \operatorname{rank}\begin{pmatrix} A & B & X \\ B^* & C & D \end{pmatrix} - \operatorname{rank}\begin{pmatrix} A & B \\ B^* & C \end{pmatrix}$$

$$\geq \operatorname{rank}\begin{pmatrix} B^* & C & D \end{pmatrix} - \operatorname{rank}\begin{pmatrix} B^* & C \end{pmatrix},$$

and

$$\gamma' := \operatorname{rank}\begin{pmatrix} B^* & C & D \\ X^* & D^* & E \end{pmatrix} - \operatorname{rank}\begin{pmatrix} C & D \\ D^* & E \end{pmatrix}$$

$$\geq \operatorname{rank}\begin{pmatrix} B^* & C & D \end{pmatrix} - \operatorname{rank}\begin{pmatrix} C & D \end{pmatrix},$$

and thus by using Lemma 5.1.5 we obtain

$$i_+(M) \geq p, \quad i_-(M) \geq q.$$

Next, the upper bounds in (5.1.10) and (5.1.11) follow directly from interlacing. For (5.1.13), one only needs to notice that

$$i_+(M) + i_-(M) = \operatorname{rank} M \leq n_1 + n_3 + r.$$

It remains to show the necessity of (5.1.12). By Lemma 5.1.5,

$$2i_+(M) \geq \pi + \gamma + \pi' + \gamma'.$$

As $i_+(M) = n_1 + n_2 + n_3 - i_-(M) - i_0(M)$, we obtain by subtracting this equality from the above inequality, and after rearranging, that

$$i_+(M) - i_-(M) \geq \pi + \gamma + \pi' + \gamma' + i_0(M) - n_1 - n_2 - n_3.$$

As $\pi + \delta + \nu = n_1 + n_2$ and $\pi' + \delta' + \nu' = n_2 + n_3$, we obtain

$$i_+(M) - i_-(M) \geq -\nu - \nu' - \delta - \delta' + \gamma + \gamma' + i_0(M) + n_2. \tag{5.1.17}$$

Observe that

$$\operatorname{rank} M \leq \operatorname{rank}\begin{pmatrix} A & B & X \\ B^* & C & D \end{pmatrix} + \operatorname{rank}\begin{pmatrix} B^* & C & D \\ X^* & D^* & E \end{pmatrix} - \operatorname{rank}\begin{pmatrix} B^* & C & D \end{pmatrix}$$

$$= \gamma + \pi + \nu + \gamma' + \pi' + \nu' - r,$$

and thus

$$i_0(M) = n_1 + n_2 + n_3 - \operatorname{rank} M \geq \delta - \gamma + \delta' - \gamma' - n_2 + r. \tag{5.1.18}$$

Combining (5.1.17) and (5.1.18) gives

$$i_+(M) - i_-(M) \geq -\nu - \nu' + r.$$

The other inequality in (5.1.12) follows similarly.

For the converse statement, first write

$$C = \begin{pmatrix} C_1 & 0 \\ 0 & 0 \end{pmatrix} : \begin{matrix} \operatorname{Ran} C \\ \oplus \\ \operatorname{Ker} C \end{matrix} \rightarrow \begin{matrix} \operatorname{Ran} C \\ \oplus \\ \operatorname{Ker} C \end{matrix},$$

and decompose $B = \begin{pmatrix} B_1 & B_2 \end{pmatrix}$ and $D = \begin{pmatrix} D_1 \\ D_2 \end{pmatrix}$ accordingly. Then, by using Schur complements, one sees that

$$i(M(X)) = (i_+(C), 0, i_-(C)) + i(H),$$

where

$$H = \begin{pmatrix} F & B_2 & Y \\ B_2^* & 0 & D_2 \\ Y^* & D_2^* & G \end{pmatrix}$$

and

$$F = A - B_1 C_1^{-1} B_1^*, \quad G = E - D_1^* C_1^{-1} D_1, \quad Y = X - B_1 C_1^{-1} D_1.$$

Next, apply Lemma 5.1.8 to obtain invertible matrices S, T_1, and T_2 such that

$$S \begin{pmatrix} B_2^* & D_2 \end{pmatrix} \begin{pmatrix} T_1 & 0 \\ 0 & T_2 \end{pmatrix} = \begin{pmatrix} I_a & 0 & 0 & 0 & 0 & 0 \\ 0 & I_b & 0 & 0 & I_b & 0 \\ 0 & 0 & 0 & 0 & 0 & I_c \\ 0 & 0 & 0 & 0 & 0 & 0 \end{pmatrix},$$

where

$$\begin{aligned} a &= \operatorname{rank} \begin{pmatrix} B_2^* & D_2 \end{pmatrix} - \operatorname{rank} (D_2), \\ c &= \operatorname{rank} \begin{pmatrix} B_2^* & D_2 \end{pmatrix} - \operatorname{rank} (B_2^*), \\ b &= \operatorname{rank} (B_2^*) + \operatorname{rank} (D_2) - \operatorname{rank} \begin{pmatrix} B_2^* & D_2 \end{pmatrix}. \end{aligned}$$

If we decompose $T_1^* F T_1, T_2^* G T_2$, and $T_1^* Y T_2$ accordingly, we obtain that H is congruent to

$$\widehat{H} = \begin{pmatrix} F_{11} & F_{12} & F_{13} & I_a & 0 & 0 & 0 & Y_{11} & Y_{12} & Y_{13} \\ F_{21} & F_{22} & F_{23} & 0 & I_b & 0 & 0 & Y_{21} & Y_{22} & Y_{23} \\ F_{31} & F_{32} & F_{33} & 0 & 0 & 0 & 0 & Y_{31} & Y_{32} & Y_{33} \\ I_a & 0 & 0 & 0 & 0 & 0 & 0 & 0 & 0 & 0 \\ 0 & I_b & 0 & 0 & 0 & 0 & 0 & 0 & I_b & 0 \\ 0 & 0 & 0 & 0 & 0 & 0 & 0 & 0 & 0 & I_c \\ 0 & 0 & 0 & 0 & 0 & 0 & 0 & 0 & 0 & 0 \\ Y_{11}^* & Y_{21}^* & Y_{31}^* & 0 & 0 & 0 & 0 & G_{11} & G_{12} & G_{13} \\ Y_{12}^* & Y_{22}^* & Y_{32}^* & 0 & I_b & 0 & 0 & G_{21} & G_{22} & G_{23} \\ Y_{13}^* & Y_{23}^* & Y_{33}^* & 0 & 0 & I_c & 0 & G_{31} & G_{32} & G_{33} \end{pmatrix}.$$

Using a congruence we can see that $i(\widehat{H}) = i(H')$, where

$$H' = \begin{pmatrix} 0 & 0 & 0 & I_a & 0 & 0 & 0 & 0 & 0 & 0 \\ 0 & F_{22} & F_{23} & 0 & I_b & 0 & 0 & Y_{21} & Y_{22} & 0 \\ 0 & F_{32} & F_{33} & 0 & 0 & 0 & 0 & Y_{31} & Y_{32} & 0 \\ I_a & 0 & 0 & 0 & 0 & 0 & 0 & 0 & 0 & 0 \\ 0 & I_b & 0 & 0 & 0 & 0 & 0 & 0 & I_b & 0 \\ 0 & 0 & 0 & 0 & 0 & 0 & 0 & 0 & 0 & I_c \\ 0 & 0 & 0 & 0 & 0 & 0 & 0 & 0 & 0 & 0 \\ 0 & Y_{21}^* & Y_{31}^* & 0 & 0 & 0 & 0 & G_{11} & G_{12} & 0 \\ 0 & Y_{22}^* & Y_{32}^* & 0 & I_b & 0 & 0 & G_{21} & G_{22} & 0 \\ 0 & 0 & 0 & 0 & 0 & I_c & 0 & 0 & 0 & 0 \end{pmatrix}.$$

The principal submatrix in H' supported in rows and columns $1,2,4,5,6,10$ has inertia $(a+b+c,0,a+b+c)$, and computing the Schur complement supported in rows and columns 3, 7, 8, 9 we obtain

$$H^{''} = \begin{pmatrix} F_{33} & 0 & Y_{31} & Y_{32}-F_{23}^* \\ 0 & 0 & 0 & 0 \\ Y_{31}^* & 0 & G_{11} & G_{12}-Y_{21}^* \\ Y_{32}^*-F_{23} & 0 & G_{12}^*-Y_{21} & G_{22}-Y_{22}-Y_{22}^*+F_{22} \end{pmatrix}. \qquad (5.1.19)$$

Notice that as all Y_{ij} are to be chosen freely, the completion problem (5.1.19) is of the type

$$\begin{pmatrix} F_{33} & 0 & ? & ? \\ 0 & 0 & 0 & 0 \\ ? & 0 & G_{11} & ? \\ ? & 0 & ? & ? \end{pmatrix} \cong \begin{pmatrix} F_{33} & ? & ? & 0 \\ ? & G_{11} & ? & 0 \\ ? & ? & ? & 0 \\ 0 & 0 & 0 & 0 \end{pmatrix},$$

where the diagonal zero is of size dim Ker $\begin{pmatrix} B \\ D_2^* \end{pmatrix}$. Denoting

$$H^{'''} = \begin{pmatrix} F_{33} & ? & ? \\ ? & G_{11} & ? \\ ? & ? & ? \end{pmatrix},$$

which is of size $(p_1+p_2+p_3)\times(p_1+p_2+p_3)$, say, it follows from Proposition 5.1.4 that the possible inertia of $H^{'''}$ is described via

$$\max\{i_+(F_{33}), i_+(G_{11})\} \le i_+(H^{'''}) \le p_3 + \min\{i_+(F_{33})+p_2, i_+(G_{11})+p_1\},$$

$$\max\{i_-(F_{33}), i_-(G_{11})\} \le i_-(H^{'''}) \le p_3 + \min\{i_-(F_{33})+p_2, i_-(G_{11})+p_1\},$$

$$-p_3 - i_-(F_{33}) - i_-(G_{11}) \le i_+(H^{'''}) - i_-(H^{'''}) \le p_3 + i_+(F_{33}) + i_+(G_{11}),$$

$$i_+(H^{'''}) + i_-(H^{'''}) = \text{rank} H^{'''} \le p_1+p_2+p_3.$$

Now use that

$$i(M(X)) = (i_+(C), 0, i_-(C)) + i(H)$$

and

$$\begin{aligned} i(H) = i(\widehat{H}) = i(H') &= (a+b+c, 0, a+b+c) + i(H^{''}) \\ &= \left(a+b+c, \dim \text{Ker} \begin{pmatrix} B \\ D_2^* \end{pmatrix}, a+b+c\right) + i(H^{'''}). \end{aligned}$$

We also note that

$$i\begin{pmatrix} A & B \\ B^* & C \end{pmatrix} = i(C_1) + i(F_{33}) + \left(a+b, c+\dim \text{Ker} \begin{pmatrix} B \\ D_2^* \end{pmatrix}, a+b\right),$$

$$i\begin{pmatrix} C & D \\ D^* & E \end{pmatrix} = i(C_1) + i(G_{11}) + \left(b+c, a+\dim \text{Ker} \begin{pmatrix} B \\ D_2^* \end{pmatrix}, b+c\right),$$

$$a+b+c = r - \text{rank} C, \quad p_1 = n_1 - a - b, \quad p_2 = n_3 - b - c, \quad p_3 = b.$$

Combining these observations now yields the theorem. □

If C is invertible in (5.1.9) and one chooses $X = BC^{-1}D$, one obtains a Hermitian completion whose inverse has a 0 in the $(1,3)$ (and $(3,1)$) position (see Theorem 2.2.3 for the positive definite case). This particular completion with a banded inverse satisfies

$$i\begin{pmatrix} A & B & BC^{-1}D \\ B^* & C & D \\ D^*C^{-1}B^* & D^* & E \end{pmatrix} = i\begin{pmatrix} A & B \\ B^* & C \end{pmatrix} + i\begin{pmatrix} C & D \\ D^* & E \end{pmatrix} - i\,(C). \tag{5.1.20}$$

In Theorem 5.1.12 we will see that this result holds for the more general case where the inverse has a chordal zero pattern. Before stating the result we need to introduce some additional notions regarding chordal graphs.

Theorem 5.1.9 *A graph $G = (V, E)$ is chordal if and only if there exists a tree $T = (\mathcal{K}, \mathcal{E}(T))$, the vertex set of which is the collection $\mathcal{K}$ of all maximal cliques of G, and for each vertex $v \in V$, the set $\mathcal{K}_v = \{K \in \mathcal{K} : v \in K\}$ is connected (and hence a subtree of T).*

Proof. We prove the necessity by induction on n, the number of vertices of G. For $n = 1$ the statement is trivial. We assume it is true for graphs for which the number of vertices is $\leq n-1$. Let now $G = (V, E)$ be a chordal graph with n vertices. By Theorem 1.2.4, there exists a perfect scheme $\sigma = [v_1, \dots, v_n]$ of G. Let $T' = (\mathcal{K}', \mathcal{E}(T'))$ be a tree with the desired properties for the graph $G' = G|\{v_2, \dots, v_n\}$. We have two distinct situations.

(i) $\mathrm{Adj}(v_1)$ is a maximal clique in G'. Then a tree T for G is obtained by changing in T' the vertex $\mathrm{Adj}(v_1)$ to $\{v_1\} \cup \mathrm{Adj}(v_1)$.

(ii) $\mathrm{Adj}(v_1)$ is not a maximal clique in G'. Consider then a new vertex corresponding to $\{v_1\} \cup \mathrm{Adj}(v_1)$ and join it with any vertex containing $\mathrm{Adj}(v_1)$. Since v_1 is simplicial, the tree T has the desired properties.

The sufficiency part of the theorem is straightforward and it is left as an exercise (see Exercise 5.12.3). □

Any tree given by Theorem 5.1.9 is referred to as a *clique tree* for $G = (V, E)$. We remark that in most cases the clique tree is not uniquely determined by $G = (V, E)$. A graph $G = (V, E)$ is called an *intersection graph* if there exist a set S and subsets S_g, $g \in G$, of S such that $(g_1, g_2) \in E$ if and only if $S_{g_1} \cap S_{g_2} \neq \emptyset$.

Corollary 5.1.10 *A graph is chordal if and only if it is the intersection graph of subtrees of a tree.*

Proof. Let $G = (V, E)$ be a chordal graph and let $T = (\mathcal{K}, \mathcal{E}(T))$ be a clique tree of it. If $u, v \in V$, then $(u, v) \in E$ if and only if $\mathcal{K}_u \cap \mathcal{K}_v \neq \emptyset$. □

The following corollary immediately follows from the construction in the proof of Theorem 5.1.9.

Corollary 5.1.11 *Let $G = (V, E)$ be a chordal graph and let $T = (\mathcal{K}, \mathcal{E}(T))$ be a clique tree of G. Then the set $\mathcal{S}$ of all minimal vertex separators of G is*

$$\mathcal{S} = \{\alpha \cap \beta \ : \ (\alpha, \beta) \in \mathcal{E}(T)\}.$$

The *multiplicity of a separator* is the number of times it appears as a separator of different vertices, namely, the number of times it appears as intersection of neighboring cliques in any clique tree. Exercise 5.12.4 requests an example of maximal cliques and minimal vertex separators for a certain chordal graph. Recall from Section 1.2 that for a matrix $M = (m_{ij})_{i,j=1}^{n}$ and a set $\alpha \subseteq \{1, \ldots, n\}$, we let $M|\alpha$ denote the principal submatrix $M|\alpha = (m_{ij})_{i,j\in\alpha}$.

Theorem 5.1.12 *Let G be a chordal graph on vertices $\{1, \ldots, n\}$ and let M be a nonsingular $n \times n$ Hermitian matrix such that $M^{-1} \in \mathcal{H}_G$. Let $\mathcal{K}$ denote the collection of all maximal cliques of G and $\mathcal{S}$ denote the collection of all minimal vertex separators of G. Then the inertia $i(M)$ satisfies the identity*

$$i(M) = \sum_{\alpha\in\mathcal{K}} i(M|\alpha) - \sum_{\beta\in\mathcal{S}} m(\beta) i(M|\beta), \tag{5.1.21}$$

where $m(\beta)$ is the multiplicity of a separator β.

The simplest nontrivial case covered in the above theorem concerns the case of two maximal cliques; this case yields equality (5.1.20).

The following proposition will be applied to M^{-1} in the proof of Theorem 5.1.12. For a set α we let α^c denote its complement; that is, if $\alpha \subseteq \{1, \ldots, n\}$ then $\alpha^c = \{1, \ldots, n\} \setminus \alpha$.

Proposition 5.1.13 *Let G be a chordal graph on vertices $\{1, \ldots, n\}$ and let $N \in \mathcal{H}_G$. Let $\mathcal{K}$ denote the collection of all maximal cliques of G and $\mathcal{S}$ denote the collection of all minimal vertex separators of G. Then the following identity holds:*

$$\sum_{\alpha\in\mathcal{K}} i(N|\alpha^c) = \sum_{\beta\in\mathcal{S}} m(\beta) i(N|\beta^c), \tag{5.1.22}$$

where $m(\beta)$ is the multiplicity of a separator β.

Proof. If the graph is complete, there is nothing to prove. When there are two maximal cliques, we have that

$$N = \begin{pmatrix} P & Q & 0 \\ Q^* & R & S \\ 0 & S^* & T \end{pmatrix},$$

and (5.1.22) states that

$$i(P) + i(T) = i \begin{pmatrix} P & 0 \\ 0 & T \end{pmatrix},$$

which is obviously true.

The general case follows by induction, using the procedure by which we constructed the clique tree of a chordal graph. Let $G = (V, E)$ be a chordal graph and let $\sigma = [v_1, \ldots, v_n]$ be a perfect scheme for G. If $\{v_1, \ldots, v_k\}$ are all simplicial, but v_{k+1} is not, then a clique tree of G can be obtained by joining to the clique tree $T' = (\mathcal{K}', \mathcal{E}(T'))$ of $\{v_{k+1}, \ldots, v_n\}$, the maximal clique γ of G containing $\{v_1, \ldots, v_k\}$. Let

$$N = \begin{pmatrix} N_{11} & N_{12} & 0 \\ N_{12}^* & N_{22} & N_{23} \\ 0 & N_{31} & N_{33} \end{pmatrix}$$

be the decomposition of N with respect to

$$\sigma = \{v_1, \ldots, v_k\} \cup (\gamma \setminus \{v_1, \ldots, v_k\}) \cup (\sigma \setminus \gamma).$$

The following holds:

$$\sum_{\alpha \in \mathcal{K}'} i(N_{33}|\alpha^c) = \sum_{\beta \in \mathcal{S}'} i(N_{33}|\beta^c), \tag{5.1.23}$$

$\mathcal{S}'$ being the set of all minimal vertex separators of $G' = G|\{v_{k+1}, \ldots, v_n\}$. The following differences exist between the graphs G and G'.

(i) The set $\delta = \gamma \setminus \{v_1, \ldots, v_k\}$ is a minimal vertex separator in G.

(ii) The set $\sigma \setminus \gamma$ is the complement of a maximal clique in G.

(iii) The set $\{v_1, \ldots, v_k\}$ is a component of the complement of a maximal clique in G that does not lie in G'.

Adding

$$i(N|\sigma \setminus \gamma) + i(N|\{v_1, \ldots, v_k\}) = i(N|\delta^c)$$

to (5.1.23), one obtains (5.1.22) for the graph G. □

We need the following lemma, which is known as the *nullity lemma.*

Lemma 5.1.14 *Consider*

$$\begin{pmatrix} A & B \\ C & D \end{pmatrix}^{-1} = \begin{pmatrix} P & Q \\ R & S \end{pmatrix},$$

where A is of size $p \times q$, and P is of size $q \times p$. Then $\dim \operatorname{Ker} C = \dim \operatorname{Ker} R$.

Proof. Since $CP = -DR$, $P[\operatorname{Ker} R] \subseteq \operatorname{Ker} C$. Likewise, since $RA = -SC$, we get $A[\operatorname{Ker} C] \subseteq \operatorname{Ker} R$. Consequently,

$$AP[\operatorname{Ker} R] \subseteq A[\operatorname{Ker} C] \subseteq \operatorname{Ker} R.$$

Since $AP + BR = I$, $AP[\operatorname{Ker} R] = \operatorname{Ker} R$, thus

$$A[\operatorname{Ker} C] = \operatorname{Ker} R.$$

This yields $\dim \operatorname{Ker} C \geq \dim \operatorname{Ker} R$. By reversing the roles of C and R one obtains also that $\dim \operatorname{Ker} R \geq \dim \operatorname{Ker} C$. This gives $\dim \operatorname{Ker} R = \dim \operatorname{Ker} C$, yielding the lemma. □

Lemma 5.1.15 *Let*

$$A = \begin{pmatrix} A_{11} & A_{12} \\ A_{21} & A_{22} \end{pmatrix}$$

be invertible and Hermitian, and

$$A^{-1} = \begin{pmatrix} B_{11} & B_{12} \\ B_{21} & B_{22} \end{pmatrix}.$$

Then

$$i_+(B_{22}) = i_+(A) - i_{\geq 0}(A_{11}), \quad i_-(B_{22}) = i_-(A) - i_{\leq 0}(A_{11}), \tag{5.1.24}$$

and

$$i_0(B_{22}) = i_0(A_{11}). \tag{5.1.25}$$

Proof. Equation (5.1.25) follows directly from Lemma 5.1.14.

Decomposing A_{11} with respect to the decomposition $\mathrm{Ran}A_{11} \oplus \mathrm{Ker}A_{11}$, we may write A and A^{-1} as

$$A = \begin{pmatrix} C_{11} & 0 & C_{13} \\ 0 & 0 & C_{23} \\ C_{31} & C_{32} & A_{22} \end{pmatrix}, \quad A^{-1} = \begin{pmatrix} D_{11} & D_{12} & D_{13} \\ D_{21} & D_{22} & D_{23} \\ D_{31} & D_{32} & B_{22} \end{pmatrix},$$

with C_{11} invertible. So, for instance, $A_{21} = \begin{pmatrix} C_{31} & C_{32} \end{pmatrix}$, and $i(C_{11}) = (i_+(A_{11}), 0, i_-(A_{11}))$. By Lemma 5.1.1 we have that

$$i \begin{pmatrix} D_{22} & D_{23} \\ D_{32} & B_{22} \end{pmatrix} = i(A) - i(C_{11}).$$

In particular, $\left(\begin{smallmatrix} D_{22} & D_{23} \\ D_{32} & B_{22} \end{smallmatrix}\right)$ is invertible. Next, as $i_0(B_{22}) = i_0(A_{11}) =$ size of D_{22}, we obtain from Lemma 5.1.2 applied to $\left(\begin{smallmatrix} D_{22} & D_{23} \\ D_{32} & B_{22} \end{smallmatrix}\right)$ that

$$i \begin{pmatrix} D_{22} & D_{23} \\ D_{32} & B_{22} \end{pmatrix} = (i_{\geq 0}(B_{22}), 0, i_{\leq 0}(B_{22})).$$

The equalities in (5.1.24) now follow. □

Proof of Theorem 5.1.12. The proof is obtained by combining Proposition 5.1.13 and Lemma 5.1.15 as follows. First notice that

$$\begin{aligned} &\sum_{\alpha \in \mathcal{K}} i_0(M|\alpha) - \sum_{\beta \in \mathcal{S}} m(\beta) i_0(M|\beta) \\ &= \sum_{\alpha \in \mathcal{K}} i_0(M^{-1}|\alpha^c) - \sum_{\beta \in \mathcal{S}} m(\beta) i_0(M^{-1}|\beta^c) = 0 = i_0(M), \end{aligned}$$

where in the first equality we use (5.1.25) from Lemma 5.1.15 and in the second equality we use Proposition 5.1.13.

Next,

$$\sum_{\alpha\in\mathcal{K}} i_+(M|\alpha) - \sum_{\beta\in\mathcal{S}} m(\beta) i_+(M|\beta)$$

$$= \sum_{\alpha\in\mathcal{K}} (i_+(M) - i_{\geq 0}(M^{-1}|\alpha^c)) - \sum_{\beta\in\mathcal{S}} m(\beta)(i_+(M) - i_{\geq 0}(M^{-1}|\beta^c))$$

$$= \left(\text{card}\mathcal{K} - \sum_{\beta\in\mathcal{S}} m(\beta) \right) i_+(M) - (\sum_{\alpha\in\mathcal{K}} i_{\geq 0}(M^{-1}|\alpha^c)$$

$$- \sum_{\beta\in\mathcal{S}} m(\beta) i_{\geq 0}(M^{-1}|\beta^c)) = i_+(M).$$

The first equality is due to (5.1.24), while the second is due to Proposition 5.1.13 and the fact that $\text{card}\mathcal{K} - \sum_{\beta\in\mathcal{S}} m(\beta) = 1$. The latter is the true because $\text{card}\mathcal{K}$ is the number of vertices and $\sum_{\beta\in\mathcal{S}} m(\beta)$ the number of edges of a clique tree of G. For i_- the same arguments apply. □

5.2 RANKS OF COMPLETIONS

As we have seen in Section 5.1, as we explore inertia of Hermitian completions the ranks of certain prescribed submatrices play a role. In order to derive results regarding inertia of banded partial matrices beyond the 3×3 block case, we first need to develop some results regarding ranks of completions. We do that in this section. The results in this section concern partial matrices without any Hermitian requirement.

For $n \in \mathbb{N}$ we let $\underline{\underline{n}}$ denote the set $\{1, \dots, n\}$. Let us recall the following definitions. Let $n, m, \nu_1, \dots, \nu_n, \mu_1, \dots, \mu_m$ be nonnegative integers. The *pattern* of specified entries in a partial block matrix will be described by a set $J \subset \underline{\underline{n}} \times \underline{\underline{m}}$. A pattern K that is a subset of J will be called a *subpattern* of J. Let now A_{ij}, $(i, j) \in J$, be given matrices of size $\nu_i \times \mu_j$. We will allow ν_i and μ_j to equal 0. The collection of matrices $\mathcal{A} = \{A_{ij} : (i, j) \in J\}$ is called a *partial block matrix with the pattern J*. When all the blocks are of size 1×1 (i.e., $\nu_i = \mu_j = 1$ for all i and j), we will simply talk about a *partial matrix*. Clearly, any block matrix as above may be viewed as a partial matrix of size $N \times M$ as well, where $N = \nu_1 + \cdots + \nu_n$, $M = \mu_1 + \cdots + \mu_m$.

Let a partial matrix $\mathcal{A} = \{A_{ij} : (i, j) \in J\}$ with A_{ij} of size $\nu_i \times \mu_j$, be given. In this section we are interested in the possible ranks of a completion of such a partial matrix. Therefore we introduce the minimal and maximal rank of a partial matrix. The *minimal rank* of $\mathcal{A}$ (notation: min rank($\mathcal{A}$)) is defined by

$$\text{min rank}(\mathcal{A}) = \min\{\text{rank}\, B : B \,\text{is a completion of}\, \mathcal{A}\}.$$

Similarly, the *maximal rank* of $\mathcal{A}$ (notation: max rank($\mathcal{A}$)) is defined by

$$\text{max rank}(\mathcal{A}) = \max\{\text{rank}\, B : B \,\text{is a completion of}\, \mathcal{A}\}.$$

Note that any rank between the minimal and the maximal rank can be achieved. Indeed if $A_{\min}$ and $A_{\max}$ are completions of $\mathcal{A}$ of minimal and maximal possible rank, respectively, then by changing the entries from $A_{\min}$ one at the time to ultimately become $A_{\max}$, one must along the way have come across completions of $\mathcal{A}$ with any rank in between min rank($\mathcal{A}$) and max rank($\mathcal{A}$).

We start with addressing the minimal rank completion problem. For triangular patterns there is a simple formula for the minimal rank.

Theorem 5.2.1 *The partial matrix $\mathcal{A} = \{A_{ij} : 1 \leq j \leq i \leq n\}$, with A_{ij} of size $\nu_i \times \mu_j$, has minimal rank*

$$\min \operatorname{rank} \mathcal{A} = \sum_{i=1}^{n} \operatorname{rank} \begin{pmatrix} A_{i1} & \dots & A_{ii} \\ \vdots & & \vdots \\ A_{n1} & \dots & A_{ni} \end{pmatrix} - \sum_{i=1}^{n-1} \operatorname{rank} \begin{pmatrix} A_{i+1,1} & \dots & A_{i+1,i} \\ \vdots & & \vdots \\ A_{n1} & \dots & A_{ni} \end{pmatrix}. \tag{5.2.1}$$

In addition, if $\nu_1 > 0$ and $\mu_n > 0$, the minimal rank completion is unique if and only if

$$\operatorname{rank} \begin{pmatrix} A_{i1} & \dots & A_{ii} \\ \vdots & & \vdots \\ A_{n1} & \dots & A_{ni} \end{pmatrix} = \operatorname{rank} A_{n1}, \ i = 1, \dots, n.$$

Proof. We let $\operatorname{col}_i(M)$ denote the ith scalar column of the matrix M. For $p = 1, \dots, n$ we let $J_p \subseteq \{1, \dots, \mu_p\}$ be a smallest possible set such that the columns

$$\operatorname{col}_i \begin{pmatrix} A_{pp} \\ \vdots \\ A_{np} \end{pmatrix}, \ i \in J_p, \tag{5.2.2}$$

satisfy

$$\operatorname{Span} \left\{ \operatorname{col}_i \begin{pmatrix} A_{pp} \\ \vdots \\ A_{np} \end{pmatrix} : i \in J_p \right\} + \operatorname{Ran} \begin{pmatrix} A_{p1} & \cdots & A_{p,p-1} \\ \vdots & & \vdots \\ A_{n1} & \cdots & A_{n,p-1} \end{pmatrix} = \operatorname{Ran} \begin{pmatrix} A_{p1} & \cdots & A_{pp} \\ \vdots & & \vdots \\ A_{n1} & \cdots & A_{np} \end{pmatrix}.$$

Note that the number of elements in J_p equals

$$\operatorname{rank} \begin{pmatrix} A_{p1} & \cdots & A_{pp} \\ \vdots & & \vdots \\ A_{n1} & \cdots & A_{np} \end{pmatrix} - \operatorname{rank} \begin{pmatrix} A_{p1} & \cdots & A_{p,p-1} \\ \vdots & & \vdots \\ A_{n1} & \cdots & A_{n,p-1} \end{pmatrix}.$$

Thus $\sum_{p=1}^{n} \operatorname{card} J_p$ equals the right-hand side of (5.2.1). It is clear that regardless of the choice for A_{ij}, $i < j$, the collection of columns

$$\operatorname{col}_i \begin{pmatrix} A_{1p} \\ \vdots \\ A_{np} \end{pmatrix}, \; i \in J_p, \; p = 1, \ldots, n, \tag{5.2.3}$$

will be linearly independent. This gives that the minimal rank is greater than or equal to the right-hand side of (5.2.1). On the other hand, when one has identified the columns (5.2.2) one can freely choose entries above these columns. Once such a choice is made, every other column of the matrix can be written as a linear combination of the columns (5.2.3), and thus a so constructed completion has rank equal to the right-hand side of (5.2.1). This yields (5.2.1).

Now assume that $\nu_1 > 0$ and $\mu_n > 0$. The above procedure shows that if any of the numbers

$$\operatorname{rank} \begin{pmatrix} A_{p1} & \cdots & A_{pp} \\ \vdots & & \vdots \\ A_{n1} & \cdots & A_{np} \end{pmatrix} - \operatorname{rank} \begin{pmatrix} A_{p1} & \cdots & A_{p,p-1} \\ \vdots & & \vdots \\ A_{n1} & \cdots & A_{n,p-1} \end{pmatrix}, \; p = 2, \ldots, n, \tag{5.2.4}$$

is positive, we do not have a unique completion (as there are some elements in the $\nu_1 \times (\mu_2 + \cdots + \mu_n)$ matrix $\begin{pmatrix} A_{12} & \cdots & A_{1n} \end{pmatrix}$ that are to be chosen freely). By a similar argument but now using rows instead of columns, we get that if any of the numbers

$$\operatorname{rank} \begin{pmatrix} A_{p1} & \cdots & A_{pp} \\ \vdots & & \vdots \\ A_{n1} & \cdots & A_{np} \end{pmatrix} - \operatorname{rank} \begin{pmatrix} A_{p+1,1} & \cdots & A_{p+1,p} \\ \vdots & & \vdots \\ A_{n1} & \cdots & A_{n,p} \end{pmatrix}, p = 1, \ldots, n-1, \tag{5.2.5}$$

is greater than zero, then we also do not have a unique minimal rank completion (as there are some elements in the $(\nu_1 + \cdots + \nu_{n-1}) \times \mu_n$ matrix $(A_{in})_{i=1}^{n-1}$ that are to be chosen freely). Thus for the minimal rank completion to be unique, we need that the numbers in (5.2.4) and (5.2.5) are all zero. But then, by (5.2.1) we obtain that a minimal rank completion $(A_{ij})_{i,j=1}^{n}$ of $\mathcal{A}$ has rank equal to

$$\operatorname{rank} \begin{pmatrix} A_{11} \\ \vdots \\ A_{n1} \end{pmatrix} = \operatorname{rank} \begin{pmatrix} A_{n1} & \cdots & A_{nn} \end{pmatrix}.$$

Letting

$$B_{11} = \begin{pmatrix} A_{11} \\ \vdots \\ A_{n-1,1} \end{pmatrix}, \; B_{12} = (A_{ij})_{i=1,j=2}^{n-1\;n}, \; B_{22} = \begin{pmatrix} A_{n2} & \cdots & A_{nn} \end{pmatrix},$$

we get that

$$\operatorname{rank} \begin{pmatrix} B_{11} & B_{12} \\ A_{n1} & B_{22} \end{pmatrix} = \operatorname{rank} \begin{pmatrix} A_{n1} & B_{22} \end{pmatrix} = \operatorname{rank} \begin{pmatrix} B_{11} \\ A_{n1} \end{pmatrix}.$$

In particular,

$$\min \operatorname{rank} \begin{pmatrix} B_{11} & ? \\ A_{n1} & B_{22} \end{pmatrix} \leq \operatorname{rank} \begin{pmatrix} A_{n1} & B_{22} \end{pmatrix} = \operatorname{rank} \begin{pmatrix} B_{11} \\ A_{n1} \end{pmatrix}.$$

As, by the 2×2 case of (5.2.1), the left-hand side equals

$$\operatorname{rank} \begin{pmatrix} A_{n1} & B_{22} \end{pmatrix} + \operatorname{rank} \begin{pmatrix} B_{11} \\ A_{n1} \end{pmatrix} - \operatorname{rank} A_{n1},$$

we obtain that $\operatorname{rank} \begin{pmatrix} B_{11} \\ A_{n1} \end{pmatrix} - \operatorname{rank} A_{n1} \leq 0$, and thus

$$\operatorname{rank} \begin{pmatrix} A_{n1} & B_{22} \end{pmatrix} = \operatorname{rank} A_{n1} = \operatorname{rank} \begin{pmatrix} B_{11} \\ A_{n1} \end{pmatrix}.$$

But then it is easy to see that

$$\min \operatorname{rank} \begin{pmatrix} B_{11} & ? \\ A_{n1} & B_{22} \end{pmatrix} \tag{5.2.6}$$

is achieved by a unique minimal rank completion, and it equals

$$\operatorname{rank} \begin{pmatrix} B_{11} & B_{12} \\ A_{n1} & B_{22} \end{pmatrix}.$$

But then this implies that $\mathcal{A}$ has a unique minimal rank completion as well, as two different minimal rank completions of $\mathcal{A}$ would imply two different rank completions of (5.2.6). □

One can give a description for the set of all minimal rank completions of $\mathcal{A}$; this description is given in Exercise 5.12.10. As we will see there, some entries are completely free to be chosen, and the other entries depend polynomially on these freely chosen entries. For nontriangular patterns, rational functions come into play. Indeed, for the 2×2 partial matrix with 1s on the diagonal and the off-diagonal entries to be chosen, the set of minimal rank completions is given by

$$\begin{pmatrix} 1 & x \\ \frac{1}{x} & 1 \end{pmatrix}, \quad x \neq 0.$$

For general patterns finding the minimal rank is more involved. Consider, for instance, the partial matrices

$$\mathcal{A}_1 = \begin{pmatrix} 1 & 1 & ? \\ ? & 1 & 1 \\ 1 & ? & 1 \end{pmatrix}, \quad \mathcal{A}_2 = \begin{pmatrix} 1 & 1 & ? \\ ? & 1 & 1 \\ 2 & ? & 1 \end{pmatrix}.$$

Then one easily checks that $\min \operatorname{rank} \mathcal{A}_1 = 1$ (take a completion with all 1s) and $\min \operatorname{rank} \mathcal{A}_2 = 2$. However, all the prescribed submatrices of both $\mathcal{A}_1$ and $\mathcal{A}_2$ have rank 1. Thus for this particular pattern no general formula for the minimal rank exists that is in terms of ranks of prescribed submatrices. The same holds for the $n \times n$ bidiagonal circulant pattern

$$\begin{pmatrix} * & * & & & & \\ & * & * & & ? & \\ & & \ddots & & \ddots & \\ & ? & & & * & * \\ * & & & & & * \end{pmatrix}, \tag{5.2.7}$$

where again one can choose $\mathcal{A}_1$ to have all given entries equal to 1, and $\mathcal{A}_2$ to have all but one given entries equal to 1 and the remaining entry equal to 2. It is conjectured that these patterns (5.2.7) form the only obstacle in obtaining a formula for the minimal rank in terms of ranks of prescribed submatrices. We will see in Exercise 5.12.2 that this relates to the notion of bipartite chordality.

The next question is what general formula for the minimal rank one may expect when subpatterns of the form (5.2.7) do not appear. In Theorem 5.2.2 we present this general formula. It is based on the simple observation that when one replaces some of the known entries by unknown entries one makes the minimal rank smaller. Now, if one removes known entries in such a way that the resulting pattern is (permutation equivalent) to a triangular pattern as in Theorem 5.2.1, one obtains by applying this result a lower bound for the minimal rank of the original matrix. For instance,

$$\min \operatorname{rank} \begin{pmatrix} 1 & 1 & 1 \\ 1 & 2 & ? \\ 1 & ? & 1 \end{pmatrix}$$

$$\geq \max \left\{ \min \operatorname{rank} \begin{pmatrix} 1 & 1 & 1 \\ 1 & 2 & ? \\ 1 & ? & ? \end{pmatrix}, \min \operatorname{rank} \begin{pmatrix} 1 & 1 & 1 \\ 1 & ? & ? \\ 1 & ? & 1 \end{pmatrix} \right\} = 2.$$

As we shall see, for some patterns this lower bound is actually achieved. We will state this now more precisely.

Let $T \subseteq \{1, \dots, n\} \times \{1, \dots, m\}$ be a pattern. We say that T is *triangular* if whenever $(i, j) \in T$ and $(k, l) \in T$, then $(i, l) \in T$ or $(k, j) \in T$. In other words, a subpattern of the form

$$\begin{pmatrix} * & ? \\ ? & * \end{pmatrix}$$

is not allowed. Denote $\operatorname{ind}_j(T) = \{i : (i, j) \in T\}$. For a pattern $T \subset \{1, \dots, n\} \times \{1, \dots, m\}$, one sees (as in the proof of Theorem 4.1.4) that T is triangular if and only if for all $1 \leq j_1, j_2 \leq m$ we have $\operatorname{ind}_{j_1}(T) \subseteq \operatorname{ind}_{j_2}(T)$ or $\operatorname{ind}_{j_2}(T) \subseteq \operatorname{ind}_{j_1}(T)$. In other words, the pattern T is triangular if and only if the sets $\operatorname{ind}_j(T)$, $1 \leq j \leq m$, may be linearly ordered by inclusion. This is equivalent to the fact that after suitable permutations of its rows and columns, the pattern can be transformed into a block lower triangular one, with the blocks having appropriate sizes. Clearly, an analogous result holds for the "rows" of a triangular pattern T.

Next we say that T is a *triangular subpattern* of J if (1) $T \subseteq J$ and (2) T is triangular. The pattern T is called a *maximal triangular subpattern* of J if $T \subseteq S \subseteq J$ with S a triangular subpattern of J implies that $T = S$. We now have the following result. For $T \subseteq J$ and $\mathcal{A} = \{A_{ij} : (i,j) \in J\}$ a partial matrix, let $\mathcal{A} \mid T$ denote the partial matrix $\mathcal{A} \mid T = \{A_{ij} : (i,j) \in T\}$.

Theorem 5.2.2 *Let* $J \subseteq \{1,\ldots,n\} \times \{1,\ldots,m\}$ *be a pattern, and let* $\mathcal{A} = \{A_{ij} : (i,j) \in J\}$ *be a partial block matrix. Then*

$$\min \operatorname{rank}\mathcal{A} \geq \max\{\min \operatorname{rank}(\mathcal{A} \mid T) : T \subseteq J \text{ maximal triangular}\}. \quad (5.2.8)$$

When $J = J_n = \{(1,i),(i,i),(i,1) : 1 \leq i \leq n\}$ *then equality holds in* (5.2.8).

There are other patterns for which equality holds in (5.2.8). For instance, in Exercise 5.12.9 we will see that this is true for banded patterns.

Let us illustrate this result with an example.

Example 5.2.3 Given the partial matrix

$$\mathcal{M} = \begin{pmatrix} P & Q & R \\ T & ? & S \\ U & V & ? \end{pmatrix},$$

we have

$$\min \operatorname{rank}\mathcal{M} = \max\left\{\min \operatorname{rank}\begin{pmatrix} P & Q & R \\ T & ? & S \\ U & ? & ? \end{pmatrix}, \min \operatorname{rank}\begin{pmatrix} P & Q & R \\ T & ? & ? \\ U & V & ? \end{pmatrix}\right\}$$

$$= \operatorname{rank}\begin{pmatrix} P & Q & R \end{pmatrix} + \begin{pmatrix} P \\ T \\ U \end{pmatrix} + \max\left\{\operatorname{rank}\begin{pmatrix} P & R \\ T & S \end{pmatrix} - \operatorname{rank}\begin{pmatrix} P & R \end{pmatrix}\right.$$

$$\left. - \operatorname{rank}\begin{pmatrix} P \\ T \end{pmatrix}, \operatorname{rank}\begin{pmatrix} P & Q \\ U & V \end{pmatrix} - \operatorname{rank}\begin{pmatrix} P & Q \end{pmatrix} - \operatorname{rank}\begin{pmatrix} P \\ U \end{pmatrix}\right\}.$$

Note that to obtain this result one permutes columns 2 and 3 in the partial matrix $\mathcal{M}$ and applies Theorem 5.2.2 in the case $J = J_3$. We will come across this pattern in Theorem 5.4.4.

Proof. The inequality in (5.2.8) is obvious as one always has that if $T \subseteq J$ then $\min \operatorname{rank}\mathcal{A} \geq \min \operatorname{rank}(\mathcal{A} \mid T)$.

Next, we prove that equality holds in (5.2.8) for the patterns J_n. The cases $n = 1, 2$ are trivial. To check for $n = 3$, let $\mathcal{A}$ have pattern J_3. Using allowable elementary (i.e., such that they do not introduce unspecified entries into the specified spots of $\mathcal{A}$) row and column operations we can bring $\mathcal{A}$ to the form

$$\begin{pmatrix} I_\alpha & 0 & 0 & 0 & 0 & 0 & 0 & 0 \\ 0 & 0 & 0 & 0 & I_\beta & 0 & 0 & 0 \\ 0 & 0 & 0 & 0 & 0 & 0 & I_\gamma & 0 \\ 0 & 0 & 0 & 0 & 0 & 0 & 0 & 0 \\ 0 & I_\epsilon & 0 & 0 & 0 & 0 & ? & ? \\ 0 & 0 & 0 & 0 & 0 & X & ? & ? \\ 0 & 0 & I_\eta & 0 & ? & ? & 0 & 0 \\ 0 & 0 & 0 & 0 & ? & ? & 0 & Y \end{pmatrix}$$

of size $(\alpha+\beta+\gamma+\delta+\varepsilon+\zeta+\eta+\upsilon)\times(\alpha+\varepsilon+\eta+\iota+\beta+\kappa+\gamma+\lambda)$, say. Now J_3 has two maximal triangles $K = J_3 \setminus \{(3,3)\}$ and $L = J_3 \setminus \{(2,2)\}$. It is easy to compute min rank$(\mathcal{A}, J_3) = \alpha+\beta+\gamma+\varepsilon+\eta+\max\{\text{rank}X, \text{rank}Y\} = \max\{\alpha+\beta+\gamma+\varepsilon+\eta+\text{rank}X, \alpha+\beta+\gamma+\varepsilon+\eta+\text{rank}Y\} = \max\{\min \text{rank}(\mathcal{A}|K), \min \text{rank}(\mathcal{A}|L)\}$ and hence equality holds in (5.2.8).

We now proceed by induction on $n \geq 4$. Note that

$$J_n = J_{n-1} \cup \{(n,1),(1,n),(n,n)\}.$$

By the induction hypothesis equality holds in (5.2.8) for the pattern J_{n-1}. The maximal triangular subpatterns of J_n are $M_i, i = 2,\ldots,n$, where $M_i = \{(1,j),(j,1) : j = 1,\ldots,n\} \cup \{(i,i)\}$. We will first show that equality holds in (5.2.8) for the pattern $J_{n-1} \cup \{(n,1)\}$. Let $\mathcal{A} = \{A_{ij} : (i,j) \in J_{n-1} \cup \{(n,1)\}\}$ be a partial block matrix. Note that if T is maximal triangular in $J_{n-1} \cup \{(n,1)\}$, then $T \setminus \{(n,1)\}$ is maximal triangular in J_{n-1}, and

$$\min \text{rank}(\mathcal{A}|T) = \min \text{rank}(\mathcal{A}|T \setminus \{(n,1)\}) + \text{rank}\begin{pmatrix} A_{11} \\ \vdots \\ A_{n1} \end{pmatrix} - \text{rank}\begin{pmatrix} A_{11} \\ \vdots \\ A_{n-1,1} \end{pmatrix}. \tag{5.2.9}$$

Let now A_{ij}, $1 \leq i,j \leq n-1$ be chosen such that $(A_{ij})_{i,j=1}^{n-1}$ is a minimal rank completion of $\{A_{ij} : (i,j) \in J_{n-1}\}$. By the induction assumption we get that

$$\text{rank}(A_{ij})_{i,j=1}^{n-1} = \max\{\min \text{rank}(\mathcal{A}|S), S \text{ maximal triangular in } J_{n-1}\}. \tag{5.2.10}$$

Consider the partial matrix

$$\mathcal{B} = \{A_{ij} : (i,j) \in (\{1,\ldots,n-1\} \times \{1,\ldots,n-1\}) \cup \{(n,1)\}\}.$$

As $\mathcal{B}$ has a triangular pattern we get that

$$\min \text{rank}\mathcal{B} = \text{rank}(A_{ij})_{i,j=1}^{n-1} + \text{rank}\begin{pmatrix} A_{11} \\ \vdots \\ A_{n1} \end{pmatrix} - \text{rank}\begin{pmatrix} A_{11} \\ \vdots \\ A_{n-1,1} \end{pmatrix}. \tag{5.2.11}$$

Thus $\mathcal{B}$ has a completion with rank equal to the right-hand side of (5.2.11). But then this same completion is a completion of $\mathcal{A}$ with rank equal to the right-hand side of (5.2.11), which by (5.2.9) and (5.2.10) equals

$$\max\{\min \text{rank}(\mathcal{A}|T) : T \text{ maximal triangular in } J_{n-1} \cup \{(n,1)\}\}.$$

This proves equality in (5.2.8) for $J_{n-1} \cup \{(n,1)\}$.

A similar argument shows that that equality holds in (5.2.8) for the pattern $J_{n-1} \cup \{(n,1),(1,n)\} = J_n \setminus \{(n,n)\}$. Now let $\mathcal{A} = \{A_{ij} : (i,j) \in J_n\}$ be given. Find a minimal rank completion $(B_{ij})_{i,j=1}^n$ of $\mathcal{A}|(J_n \setminus \{(n,n)\})$. As equality in (5.2.8) holds for $J_n \setminus \{(n,n)\}$, we have that

$$\text{rank}(B_{ij})_{i,j=1}^n = \max_{i=2,\ldots,n-1} \min \text{rank}(\mathcal{A}|M_i). \tag{5.2.12}$$

Put $B_{nn} = A_{nn}$ and consider the partial matrix $\mathcal{B} = \{B_{ij} : (i,j) \in J \cup (\{1,\dots,n-1\} \times \{1,\dots,n-1\})\}$. The partial matrix $\mathcal{B}$ has a block pattern J_3, and thus by the first part of the proof we have the equality

$$\min \text{rank}\mathcal{B} = \max\{\min \text{rank}(\mathcal{B}|T), \min \text{rank}(\mathcal{B}|M_n)\}, \qquad (5.2.13)$$

where $T = (\{1,\dots,n-1\} \times \{1,\dots,n-1\}) \cup \{(n,1),(1,n)\}$. Note that M_n is one of the maximal triangular subpatterns of J_n identified before. Using (5.2.12) we now get that $\mathcal{B}$ has a completion of rank at most

$$\max\{\min \text{rank}(\mathcal{B}|T), \min \text{rank}(\mathcal{B}|M_n)\} = \max_{i=2,\dots,n} \min \text{rank}(\mathcal{A}|M_i).$$

As a completion of $\mathcal{B}$ is also a completion of $\mathcal{A}$, thus we obtain equality in (5.2.8). $\square$

For the maximal rank the situation is easier as in this case we can derive a formula for any pattern. We say that a pattern $J \subset \underline{\underline{n}} \times \underline{\underline{m}}$ is *rectangular* if it is of the form $J_1 \times J_2$ with $J_1 \subset \underline{\underline{n}}$ and $J_2 \subset \underline{\underline{m}}$. First observe that for the $n \times n$ partial matrix

$$\mathcal{A} = \begin{pmatrix} A & ? \\ ? & ? \end{pmatrix}$$

with A a $p \times q$ specified matrix we have that

$$\max \text{rank}\mathcal{A} = \min\{n, m, (n-p) + (m-q) + \text{rank}A\}$$

Indeed, one can easily convince oneself of this by considering without loss of generality the case when $A = \left(\begin{smallmatrix} I_a & 0 \\ 0 & 0 \end{smallmatrix}\right)$ where $a = \text{rank}A$. We are thus led to associate with every partial matrix $\mathcal{A}$ with a rectangular pattern J as above the number $\rho(\mathcal{A}) = (n-p) + (m-q) + \text{rank}A$. This definition extends to trivial rectangular patterns ($p = 0$ or $q = 0$).

The maximal rank of a partial matrix is now given via the following formula.

Theorem 5.2.4 *Let $\mathcal{A}$ be a $p \times q$ partial matrix with pattern $J \subset \{1,\dots,n\} \times \{1,\dots,m\}$. Then*

$$\max \text{rank}\mathcal{A} = \min \rho(\mathcal{A}|K),$$

where K runs over all the maximal rectangular subpatterns of J, including the trivial ones of size $0 \times m$ and $n \times 0$.

Proof. Because of basic known inequalities for the rank of a conventional matrix $\max \text{rank}(\mathcal{A}) \leq \min \rho(\mathcal{A}|K)$; thus, we have to show that

$$\max \text{rank}(A, J) \geq \min \rho(\mathcal{A}|K).$$

Let K be a maximal rectangular subpattern of J for which the above minimum is obtained. Up to row and column permutations we have

$$\mathcal{A} = \begin{pmatrix} U & V & * \\ W & ? & * \\ * & * & * \end{pmatrix}, \quad \mathcal{A}|K = \begin{pmatrix} U & V & ? \\ ? & ? & ? \\ ? & ? & ? \end{pmatrix},$$

of size $(\alpha + 1 + \beta) \times (\gamma + \delta + \varepsilon)$, say. Here $*$ represents irrelevant partially defined blocks. The basic recurring step is to show that one can always find X such that

$$\operatorname{rank} \begin{pmatrix} U & V \\ W & X \end{pmatrix} = \operatorname{rank} \begin{pmatrix} U & V \end{pmatrix} + 1. \tag{5.2.14}$$

For if (5.2.14) is true we can continue by induction on the number of rows and columns in $\mathcal{A}$. To prove (5.2.14) we examine three possible cases.

Case 1: $\operatorname{rank} \left(\begin{smallmatrix} U \\ W \end{smallmatrix} \right) > \operatorname{rank} U$. Here (5.2.14) holds for all X, due to Theorem 5.2.1.

Case 2: $\operatorname{rank} \left(\begin{smallmatrix} U \\ W \end{smallmatrix} \right) = \operatorname{rank} U > \operatorname{rank} \begin{pmatrix} U & V \end{pmatrix} - \delta$, where δ is the number of columns of V. This means that up to column permutation $V = \begin{pmatrix} V' & V" \end{pmatrix}$, where $\operatorname{rank} \begin{pmatrix} U & V' \end{pmatrix} = \operatorname{rank} \begin{pmatrix} U & V \end{pmatrix}$ and V' is of size $\alpha \times \eta$, with $\eta = \operatorname{rank} \begin{pmatrix} U & V \end{pmatrix} - \operatorname{rank} U < \delta$. Under the same partition, $X = \begin{pmatrix} X' & X" \end{pmatrix}$, where we choose X' arbitrary. We have

$$\operatorname{rank} \begin{pmatrix} U & V' \\ W & X' \end{pmatrix} = \operatorname{rank} \begin{pmatrix} U & V' \end{pmatrix} = \operatorname{rank} \begin{pmatrix} U & V \end{pmatrix}.$$

But then, Theorem 5.2.1 yields that $\left(\begin{smallmatrix} U & V' & V" \\ W & X' & ? \end{smallmatrix} \right)$ has a unique minimal rank completion. Choose any $X"$ different than the minimal rank 1, and (5.2.14) is obtained.

Case 3: $\operatorname{rank} \left(\begin{smallmatrix} U \\ W \end{smallmatrix} \right) = \operatorname{rank} U = \operatorname{rank} \begin{pmatrix} U & V \end{pmatrix} - \delta$. This case can never occur, as it leads to the following contradiction. Let L be the rectangular subpattern of J yielding the partial matrix

$$\mathcal{A}|L = \begin{pmatrix} U & ? & ? \\ W & ? & ? \\ ? & ? & ? \end{pmatrix}.$$

Then

$$\rho(\mathcal{A}|L) = \operatorname{rank} U + \beta + \delta + \epsilon = \rho(\mathcal{A}|K) - 1,$$

contradicting the choice of K. □

We end this section with a result that connects the minimal ranks of triangular partial matrices that come from a matrix and its inverse. One can see the result as a generalization of the statement that the inverse of an upper triangular matrix is upper triangular. As we shall see in Exercise 5.12.23, one of the special cases gives that the inverse of a upper Hessenberg matrix has a lower triangular part that has a minimal rank completion of rank 1.

Theorem 5.2.5 *Let $T = (T_{ij})_{i,j=1}^n$ be an invertible block matrix with T_{ij} of size $\nu_i \times \mu_j$, where $\nu_i \geq 0$, $\mu_j \geq 0$ and $N = \nu_i + \cdots + \nu_n = \mu_i + \cdots + \mu_n$. Put $T^{-1} = (S_{ij})_{i,j=1}^n$ where S_{ij} is of size $\mu_i \times \nu_j$. Then*

$$\text{min rank} \begin{pmatrix} T_{11} & ? & \cdots & ? \\ T_{21} & T_{22} & \cdots & ? \\ \vdots & & \ddots & \vdots \\ T_{n1} & T_{n2} & \cdots & T_{nn} \end{pmatrix}$$

$$+ \min \operatorname{rank} \begin{pmatrix} ? & ? & \cdots & ? \\ S_{21} & ? & \cdots & ? \\ \vdots & \ddots & \ddots & \vdots \\ S_{n1} & \cdots & S_{n,n-1} & ? \end{pmatrix} = N.$$

Proof of Theorem 5.2.5. The nullity lemma (Lemma 5.1.14) covers the $n = 2$ case. Indeed, if

$$T^{-1} = \begin{pmatrix} T_{11} & T_{12} \\ T_{21} & T_{22} \end{pmatrix}^{-1} = \begin{pmatrix} S_{11} & S_{12} \\ S_{21} & S_{22} \end{pmatrix},$$

we get from Lemma 5.1.14 that

$$\nu_2 - \operatorname{rank} T_{21} = \mu_2 - \operatorname{rank} S_{21}. \tag{5.2.15}$$

As T is invertible we have that $\begin{pmatrix} T_{11} \\ T_{21} \end{pmatrix}$ and $\begin{pmatrix} T_{21} & T_{22} \end{pmatrix}$ are full rank, so (5.2.15) gives

$$\operatorname{rank} \begin{pmatrix} T_{11} \\ T_{21} \end{pmatrix} + \operatorname{rank} \begin{pmatrix} T_{21} & T_{22} \end{pmatrix} - \operatorname{rank} T_{21} + \operatorname{rank} S_{21} = \mu_1 + \nu_2 + \mu_2 - \nu_2 = N,$$

which is exactly the statement of the theorem when $n = 2$.

The general case follows by observing that Lemma 5.1.14 yields that for $1 \leq p, q \leq n$

$$\operatorname{rank} \begin{pmatrix} T_{p1} & \cdots & T_{pq} \\ \vdots & & \vdots \\ T_{n1} & \cdots & T_{nq} \end{pmatrix} + \sum_{j=q+1}^{n} \mu_j \tag{5.2.16}$$

$$= \operatorname{rank} \begin{pmatrix} S_{q+1,1} & \cdots & S_{q+1,p-1} \\ \vdots & & \vdots \\ S_{n1} & \cdots & S_{n,p-1} \end{pmatrix} + \sum_{j=p}^{n} \nu_j.$$

Now apply Theorem 5.2.1 to the lower triangular partial matrices $\{T_{ij} : 1 \leq j \leq i \leq n\}$ and $\{S_{ij} : 1 \leq j < i \leq n\}$ and use (5.2.16) to get the desired formula. □

5.3 MINIMAL NEGATIVE AND POSITIVE SIGNATURE

In this section we will focus on the question of the minimal negative and positive inertia among all Hermitian completions of a partial Hermitian matrix.

We have the following theorem. For $I \subseteq \mathbb{N}$ we will denote $I \times I$ by I^2. A pattern $J \subseteq \underline{\underline{n}}^2$ is called *symmetric* if $(i, j) \in J$ implies $(j, i) \in J$.

Theorem 5.3.1 *Let J be a symmetric pattern in $\underline{\underline{n}}^2$ with $(i, i) \in J, i = 1, \ldots, n$, and whose underlying graph is chordal. Let $\mathcal{A}$ be a partial Hermitian matrix. Then* *(i)* $\min\{i_{\geq 0}(A) : A \textit{ is a Hermitian completion of } \mathcal{A}\} = \max_{K^2 \subseteq J} i_{\geq 0}(\mathcal{A}|K^2)$;

(ii) if in addition $A|K^2$ *is nonsingular for every maximal clique* K^2 *in* J, *then*

$$\min\{i_+(A) : A \textit{ is a Hermitian completion of } \mathcal{A}\} = \max_{K^2 \subseteq J} i_+(\mathcal{A}|K^2).$$

Similar results hold for $i_{\leq 0}$ *and* i_-. *If* $\mathcal{A}$ *is a banded Hermitian block Toeplitz partial matrix, then the completion with the desired inertia may be chosen to be block Toeplitz as well. Finally, if in the above statements* $\mathcal{A}$ *is a real symmetric partial matrix, then the above completions may be chosen to be real symmetric as well.*

Proof. By Theorem 5.1.6 the result is true for the 3×3 block case. Moreover, under the condition that $\begin{pmatrix} A & B \\ B^* & C \end{pmatrix}$ and $\begin{pmatrix} C & D \\ D^* & E \end{pmatrix}$ are invertible and after fixing $l \in \{-, +\}$, one can make M invertible and satisfying

$$i_l(M) = \max\left\{ i_l \begin{pmatrix} A & B \\ B^* & C \end{pmatrix}, i_l \begin{pmatrix} C & D \\ D^* & E \end{pmatrix} \right\}.$$

Using a perfect elimination scheme one can prove the theorem by induction and repeatedly using Theorem 5.1.6 and the above observation.

In the case of a block Toeplitz matrix one carries out the above algorithm in such a way that one fills up the matrix one diagonal at a time, choosing the same block entry along each diagonal. This works as when one fills up a diagonal it is the same 3×3 block problem at each step, so the unknown entry can be chosen to be the same. □

As can be seen from Theorem 5.1.6 it is crucial in Theorem 5.3.1(ii) to assume that $A|K^2$ is nonsingular for every maximal clique $K^2 \subseteq J$, as otherwise in applying Lemma 5.1.5 one may have $\delta > 0$. We will now develop a version of Theorem 5.3.1(ii) that does not require this invertibility condition. We will however restrict ourselves to banded patterns. Before we can state the result we need to develop results for banded patterns.

We call $J \subseteq \underline{\underline{n}}^2$ *banded* if $(i, j) \in J$, $(k, l) \in J$, $i \leq k$ and $j \geq l$ imply

$$\{i, \ldots, k\} \times \{l, \ldots, j\} \subseteq J.$$

Note that patterns associated with p-banded partial matrices, as defined in Section 2.1, are banded patterns, but that not every banded pattern is p-banded for some p. We will refer to the pattern

$$D_n = \{(i, i) : i \in \underline{\underline{n}}\}$$

as the *diagonal*. Given several patterns $J_p \subseteq \underline{\underline{n}}_p \times \underline{\underline{m}}_p$, $p = 1, \ldots, s$, we define the *direct sum*

$$J_1 \oplus \cdots \oplus J_s \subseteq \underline{N} \times \underline{M}$$

with $N = \sum_{p=1}^s n_p$ and $M = \sum_{p=1}^s m_p$, as the pattern

$$J_1 \oplus \cdots \oplus J_s = \bigcup_{p=1,\dots,s} \left\{ \left(i + \sum_{k=1}^{s-1} n_k,\ j + \sum_{k=1}^{s-1} m_k \right) : (i,j) \in J_s \right\}. \quad (5.3.1)$$

Conversely, if a pattern J can be written in the form (5.3.1) for some choice of s, $n_1, \dots, n_s$, $m_1, \dots, m_s$, and $J_k \subseteq \underline{\underline{n}}_k \times \underline{\underline{m}}_k (k = 1, \dots, s)$, we say that J *decomposes* as the direct sum $J_1 \oplus \cdots \oplus J_s$, and refer to the patterns $J_1, \dots, J_s$ as the *summands.*

Lemma 5.3.2 *Each symmetric banded pattern $J \subseteq \underline{\underline{n}}^2$ can be decomposed as a direct sum $J_1 \oplus \cdots \oplus J_s$, where each summand J_k, $k = 1, \dots, s$, is a symmetric banded pattern in $\underline{\underline{n}}_k^2$ which is either the empty set or contains the diagonal D_{n_k}.*

This result is easy to verify, and is left to the reader. Based on this lemma we may often assume that a symmetric banded pattern contains the diagonal.

As before, a pattern $R \subseteq \underline{\underline{n}} \times \underline{\underline{m}}$ is called *rectangular* if it is of the form $R = K \times L$ with $K \subseteq \underline{\underline{n}}$ and $L \subseteq \underline{\underline{m}}$; equivalently, R is rectangular if $(i,j) \in R$ and $(k,l) \in R$ imply that $(i,l) \in R$ (and $(k,j) \in R$). The sets K and L will be referred to as the *rows* and *columns covered by* R, respectively. Note that R is *symmetric rectangular* (that is, symmetric and rectangular) if and only if it is of the form $K \times K$. A (symmetric) rectangular pattern will also be referred to as a *(symmetric) rectangle.* We say that a (symmetric) rectangular pattern $R \subseteq J$ is *maximal (symmetric) rectangular* in J if $R \subseteq S \subseteq J$ and S is (symmetric) rectangular, then $R = S$.

We have the following proposition.

Proposition 5.3.3 *Each symmetric triangular pattern T (i.e., T is symmetric and triangular) has a unique maximal symmetric rectangle.*

The proof of this proposition follows immediately from the following lemma.

Lemma 5.3.4 *Let T be a symmetric triangular pattern. If $P \times P \subseteq T$ and $Q \times Q \subseteq T$, then $(P \cup Q) \times (P \cup Q) \subseteq T$.*

Proof. Let $p, q \in P \cup Q$. Then (p,p) and (q,q) belong to $(P \times P) \cup (Q \times Q) \subseteq T$. Since T is triangular, either $(p,q) \in T$ or $(q,p) \in T$. Since T is symmetric, we get in fact that both (p,q) and (q,p) belong to T. □

The unique maximal symmetric rectangle of a symmetric triangular pattern T will be denoted by $R_{\text{symm}}(T)$. The rows/columns covered by $R_{\text{symm}}(T)$ are denoted by $rs(T)$. Thus

$$R_{\text{symm}}(T) = rs(T) \times rs(T) = rs(T)^2.$$

Lemma 5.3.5 *If T is a symmetric triangular pattern, then for all $(r,s) \in T$ we have $r \in rs(T)$ or $s \in rs(T)$ (i.e., $T \subseteq (rs(T) \times \underline{\underline{n}}) \cup (\underline{\underline{n}} \times rs(T))$).*

Proof. Let $(r, s) \in T$. Then also $(s, r) \in T$, and consequently either $(r, r) \in T$ or $(s, s) \in T$. Let us say that $(r, r) \in T$. By Lemma 5.3.4 (with $P = \{r\}$ and $Q = rs(T)$) we get that $(\{r\} \times rs(T))^2 \subseteq T$. Since $rs(T)^2$ is maximal symmetric rectangular in T, it follows that $r \in rs(T)$. □

It follows from Lemma 5.3.5 that if $T \neq \emptyset$ then $rs(T) \neq \emptyset$ as well.

Lemma 5.3.6 *If T is a symmetric triangular pattern and $rs(T) \subseteq P$, then $T \cup P^2$ is a symmetric triangular pattern as well.*

Proof. Since both T and P^2 are triangular and symmetric, it suffices to show that $(r, s) \in T$ and $(p, q) \in P^2$ imply that $(r, q) \in T \cup P^2$ or $(p, s) \in T \cup P^2$. So suppose that $(r, s) \in T$ and $(p, q) \in P^2$. By Lemma 5.3.5 we have that $r \in rs(T)$ or $s \in rs(T)$. But then either $(r, q) \in rs(T) \times P \subseteq P^2 \subseteq T \cup P^2$ or $(p, s) \in T \cup P^2$. □

For a symmetric banded pattern J we let $\mathcal{T}(J)$ denote the set of maximal symmetric triangular patterns in J.

Corollary 5.3.7 *Let J be a symmetric banded pattern and $T \in \mathcal{T}(J)$. Then $rs(T)^2$ is a maximal symmetric rectangle in J.*

Proof. Choose a maximal symmetric rectangle R in J containing $R_{\text{symm}}(T)$. By Lemma 5.3.6 the pattern $T \cup R$ is also a symmetric triangular pattern in J. By maximality of T in J, we must have that $T \cup R = T$, thus $R \subseteq T$. But then it follows from Corollary 5.3.3 that $R = R_{\text{symm}}(T)$. □

For a symmetric triangular pattern $T \subseteq \underline{\underline{n}} \times \underline{\underline{n}}$ we let $R(T)$ denote the set of maximal rectangles in $T \cap (\underline{\underline{n}} \times rs(T))$ together with $rs(T)^2$ (if not already present). Note that $R(T)$ does *not* contain all maximal rectangles in T. The following lemma shows that $rs(T)^2$ is always present among the maximal rectangles in $T \cap (\underline{\underline{n}} \times rs(T))$ when T is a maximal triangular pattern in a symmetric banded one.

Lemma 5.3.8 *For any nonempty symmetric banded pattern $J \subseteq \underline{\underline{n}} \times \underline{\underline{n}}$ and any $T \in \mathcal{T}(J)$ we have that $R_{\text{symm}}(T)$ is a maximal rectangle in $T \cap (\underline{\underline{n}} \times rs(T))$.*

Proof. Suppose

$$R_{\text{symm}}(T) \subseteq R \subseteq T \cap (\underline{\underline{n}} \times rs(T)),$$

with R rectangular. Then $R = P \times rs(T)$ with $rs(T) \subseteq P$. Suppose there exists a $p \in P \setminus rs(T)$. Then $(\{p\} \cup rs(T)) \times rs(T) \subseteq T$, and since T is symmetric also $rs(T) \times (\{p\} \cup rs(T)) \subseteq T$.

We claim that $\{(p, p)\} \cup T$ is still triangular. Indeed, if $(r, s) \in T$, we get by Lemma 5.3.5 that $r \in rs(T)$ or $s \in rs(T)$. But then $(r, p) \in rs(T) \times (\{p\} \cup rs(T)) \subseteq T \cup \{(p, p)\}$ or $(p, s) \in T \cup \{(p, p)\}$, proving that $T \cup \{(p, p)\}$ is triangular. Since $p \notin rs(T) \neq \emptyset$, there exists an $r \in rs(T)$ with $r < p$ or $r > p$. Without loss of generality we assume that $r < p$. Then $(r, p) \in T \subseteq J$ and $(p, r) \in T \subseteq J$, thus, since J is banded,

$$\{r, \ldots, p\} \times \{r, \ldots, p\} \subseteq J.$$

In particular, $(p,p) \in J$. But then $T \cup \{(p,p)\}$ is a symmetric triangular pattern in J, and thus by maximality of T it follows that $(p,p) \in T$. Lemma 5.3.5 now implies that $p \in rs(T)$, giving a contradiction. □

Not for every symmetric triangular pattern T we have that $rs(T)^2$ is a maximal rectangle in $T \cap (\underline{n} \times rs(T))$. For example, $T = \{(1,1),(1,2),(2,1)\} \subseteq \underline{2}^2$.

The following proposition supplies us with the main tools to state our results.

Proposition 5.3.9 *Let T be a symmetric triangular pattern. Then the members of $R(T)$ can be ordered uniquely as $R_1, \ldots, R_m$ (with $m = \mathrm{card}R(T)$) such that the number of rows that R_i covers is strictly smaller than the number of rows that R_{i+1} covers, $i = 1, \ldots, m-1$. Moreover, $R_1 = R_{\mathrm{symm}}(T)$.*

Proof. Since $T \cap (\underline{n} \times rs(T))$ is a triangular pattern, we can order the members of $rs(T)$ as $\{j_1, \ldots, j_s\}$ such that

$$\mathrm{ind}_{j_1}(T) \subseteq \mathrm{ind}_{j_2}(T) \subseteq \cdots \subseteq \mathrm{ind}_{j_s}(T) \tag{5.3.2}$$

(see the proof of Theorem 4.1.4). Note that from Lemma 5.3.8 it follows that $rs(T) \subseteq \mathrm{ind}_{j_1}(T)$. If $rs(T) = \mathrm{ind}_{j_1}(T)$, put $\widetilde{m} = \mathrm{card}R(T)$. Otherwise put $\widetilde{m} = \mathrm{card}R(T) - 1$. Write now

$$rs(T) = \{j_1, \ldots, j_{n_1}\} \cup \{j_{n_1+1}, \ldots, j_{n_2}\} \cup \cdots \cup \{j_{n_{\widetilde{m}-1}+1}, \ldots, j_{n_{\widetilde{m}}}\}$$

$$=: P_1 \cup \cdots \cup P_{\widetilde{m}},$$

where $n_1, \ldots, n_{\widetilde{m}} = s$ are such that

$$\mathrm{ind}_{j_{n_i}}(T) \neq \mathrm{ind}_{j_{n_i+1}}(T).$$

Choose now $R_1 = R_{\mathrm{symm}}$, and

$$R_{i+\epsilon} = \mathrm{ind}_{j_{n_i}} \times (P_1 \cup \cdots \cup P_i),\ i = 1, \ldots, \widetilde{m},$$

where $\epsilon = \mathrm{card}R(T) - \widetilde{m}$. It is easy to verify that this choice of R_i has the required properties. □

An example is useful at this point.

Example 5.3.10 Let $J = \{(i,j) \in \underline{5} \times \underline{5} : |i-j| \leq 2\}$ and $T = J \setminus (\underline{2} \times \underline{2}) \in \mathcal{T}(J)$. This may be depicted as

$$\begin{array}{ccccc} \cdot & \cdot & \times & & \\ \cdot & \cdot & \times & \times & \\ \times & \times & \times & \times & \times \\ & \times & \times & \times & \times \\ & & \times & \times & \times \end{array}$$

Then

$$R(T) = \left\{\{3,4,5\}^2, \{2,3,4,5\} \times \{3,4\}, \underline{5} \times \{3\}\right\}.$$

We need to develop some further properties regarding banded patterns and their maximal symmetric triangular subpatterns, as they will be needed to state and prove the main result of this section, Theorem 5.3.19.

Lemma 5.3.11 *Let J be a symmetric banded pattern in $\underline{\underline{n}}^2$, and suppose that for $i_1, i_2 \in \underline{\underline{n}}$ we have that*

$$i_2 \in \{j : (i_1, j) \in J\} \subseteq \{j : (i_2, j) \in J\}.$$

Then $(i_1, j) \in T \in \mathcal{T}(J)$ implies $(i_2, j) \in T$. In other words, for every $T \in \mathcal{T}(J)$ we have that

$$\{j : (i_1, j) \in T\} \subseteq \{j : (i_2, j) \in T\}.$$

Proof. Let J, i_1, i_2, and T be as above, and denote $K = \{j : (i_1, j) \in J\}$. Then $i_2 \in K$. We will show that

$$T \cup (\{i_2\} \times K) \cup (K \times \{i_2\}) \tag{5.3.3}$$

is still triangular. Since $(\{i_2\} \times K) \cup (K \times \{i_2\})$ is triangular (since $i_2 \in K$), it suffices to show that $(p, q) \in T$ and $k \in K$ imply that (p, k) or (i_2, q) belongs to (5.3.3). To this end let p, q, and k be as above. Since $(i_1, k) \in T$ either $(i_1, q) \in T$ or $(p, k) \in T$. In the latter case we are done, so assume $(i_1, q) \in T (\subseteq J)$. But then $q \in K$, and thus (i_2, q) belongs to (5.3.3). □

Lemma 5.3.12 *Let Δ be a triangular pattern in $\underline{\underline{n}}^2$ and let $i_1, i_2 \in \underline{\underline{n}}$ be such that*

$$\{j : (i_1, j) \in \Delta\} \subseteq \{j : (i_2, j) \in \Delta\}. \tag{5.3.4}$$

If (i_1, j) belongs to a maximal rectangle R in Δ, then (i_2, j) also belongs to R. In other words, for every maximal rectangle R of Δ we have that

$$\{j : (i_1, j) \in R\} \subseteq \{j : (i_2, j) \in R\}. \tag{5.3.5}$$

Proof. Let Δ, i_1, i_2, and $R = K \times L$ be as above. If the set on the left-hand side of (5.3.5) is empty, then there is nothing to prove. So assume that $\{j : (i_1, j) \in R\}$ is nonempty, which then implies that this set equals L. Thus L is a subset of the left-hand side of (5.3.4), and thus also of the right-hand side of (5.3.4). Consequently,

$$(\{i_2\} \cup K) \times L \subseteq \Delta.$$

Since R is a maximal rectangle in Δ, we must have that $i_2 \in K$. Thus $\{i_2\} \times L \subseteq R$, proving the claim. □

Note that in Lemma 5.3.12 we did not require Δ to be symmetric. We shall apply the result to $\Delta = T \cap (\underline{\underline{n}} \times rs(T))$, where T is a symmetric triangular pattern.

The last results culminate in the following auxiliary proposition, which is easy to comprehend when explained with pictures. The version that follows is in words and symbols. For a symmetric pattern J we denote

$$\gamma_J(i) = \min\{k : (i, k) \in J\},$$
$$\zeta_J(i) = \max\{k : (i, k) \in J\}.$$

Proposition 5.3.13 *Let J be a symmetric banded pattern in $\underline{\underline{n}}$ containing the diagonal D_n, and let $T \in \mathcal{T}(J)$ be such that T contains an element (i, j) with $j \geq \gamma_J(n)$ and $i > \zeta_J(\gamma_J(n) - 1)$. Denote*

$$Q = \{\zeta_J(\gamma_J(n) - 1) + 1, \ldots, n\} \times \{j\}.$$

Then $Q \subseteq T$, and for any maximal rectangle R in $T \cap (\underline{\underline{n}} \times rs(T))$ we have that

$$Q \cap R \neq \emptyset \text{ implies } \{j + 1, \ldots, \zeta_J(\gamma_J(n) - 1)\} \times \{j\} \subseteq R.$$

Consequently, if $R_1, R_2 \in R(T)$ then either $R_1 \cap R_2 \cap Q = \emptyset$ or $\{j + 1, \ldots, \zeta_J(\gamma_J(n) - 1)\} \times \{j\} \subseteq R_1 \cap R_2$.

Proof. First note that $\{\gamma_J(n), \ldots, n\}^2 \subseteq J$ (since $(\gamma_J(n), n)$ and $(n, \gamma_J(n))$ belong to J and J is banded). Thus for $k \geq \gamma_J(n)$ we have

$$\{p : (k, p) \in J\} \supseteq \{\gamma_J(n), \ldots, n\}.$$

Furthermore, since $i > \zeta_J(\gamma_J(n) - 1)$,

$$\{p : (i, p) \in J\} = \{\gamma_J(n), \ldots, n\}.$$

But then it follows from Lemma 5.3.11 (with $i_1 = i$ and $i_2 = k$) that $\{p : (i, p) \in T\} \subseteq \{p : (k, p) \in T\}$. Now $(i, j) \in T$ yields that $(k, j) \in T$ for all $k \geq \gamma(n)$. In particular, $Q \subseteq T$. In addition, it follows that

$$\{p : (i, p) \in T \cap (\underline{\underline{n}} \times rs(T))\} \subseteq \{p : (k, p) \in T \cap (\underline{\underline{n}} \times rs(T))\}.$$

Lemma 5.3.12 now yields that $(i, p) \in R \in R(T)$ implies that $\{\gamma_J(n), \ldots, n\} \times \{p\} \in R$. In fact, this argument extends in exactly the same way to the case when i is replaced by another integer i_1, with $\zeta_J(\gamma_J(n) - 1) < i_1 \leq n$. But then the second part of the proposition also follows. □

Lemma 5.3.14 *Let J be a symmetric banded pattern in $\underline{\underline{n}}^2$ and $T \in \mathcal{T}(J)$. Then $rs(T)^2$ is a maximal rectangle in J.*

Proof. Suppose $rs(T)^2 \subseteq rs(T) \times K \subseteq J$. Then also $K \times rs(T) \subseteq J$. Thus

$$(\min K, \ \max rs(T)) \in J,$$
$$(\max rs(T), \ \min K) \in J.$$

Therefore

$$rs(T)^2 \subseteq \{\min K, \ldots, rs(T)\}^2 \subseteq J.$$

By Corollary 5.3.7 it follows that

$$rs(T) = \{\min K, \ldots, \ \max rs(T)\}.$$

So $\min K \in rs(T)$. Similarly, $\max K \in rs(T)$. Thus

$$K \subseteq \{\min rs(T), \ldots, \max rs(T)\} = rs(T).$$

The last equality follows from

$$(\min rs(T), \max rs(T)) \in J$$

and J being symmetric and banded. □

Lemma 5.3.15 *Let J be a symmetric banded pattern, $T \in \mathcal{T}(J)$, and R a maximal rectangle in $T \cap (\underline{\underline{n}} \times rs(T))$. Then R is a maximal rectangle in T.*

Proof. Suppose $R = K \times L \subseteq \widetilde{K} \times \widetilde{L} \subseteq T$. If $\widetilde{L} \subseteq rs(T)$ we are done, since $R = K \times L \subseteq \widetilde{K} \times \widetilde{L} \subseteq T \cap (\underline{\underline{n}} \times rs(T))$ implies that $R = \widetilde{K} \times \widetilde{L}$. Suppose $\widetilde{L} \not\subseteq rs(T)$. By Lemma 5.3.5 we must have that $\widetilde{K} \subseteq rs(T)$. Thus also $K \subseteq rs(T)$. But then $K = L = rs(T)$, since $K \times L \subseteq rs(T)^2 \subseteq T \cap (\underline{\underline{n}} \times rs(T))$ and $K \times L$ is a maximal rectangle in $T \cap (\underline{\underline{n}} \times rs(T))$. But then

$$R = rs(T)^2 \subseteq \widetilde{K} \times \widetilde{L} \subseteq T \subseteq J.$$

And so by Lemma 5.3.14 we get that $R = \widetilde{K} \times \widetilde{L}$. □

Let $\mathcal{A} = \{a_{ij} : (i,j) \in J\}$. When $R = P \times Q$ is a rectangle in J we may view $\mathcal{A}|R$ as the usual matrix

$$(a_{ij})_{(i,j)\in R} = (a_{ij})_{i\in P,\ j\in Q}.$$

As such we may define

$$\text{rank } (\mathcal{A}|R) = \text{ rank } [(a_{ij_{(i,j)\in R}}].$$

When R is in addition symmetric, and $\mathcal{A}$ is a Hermitian partial matrix, we may introduce

$$i_\pm(\mathcal{A}|R) = i_\pm[(a_{ij})_{(i,j)\in R}].$$

We are now ready to introduce the main inertial notions for partial matrices. Let $\mathcal{A} = \{a_{ij} : (i,j) \in J\}$ be a partial Hermitian matrix. For a symmetric triangular subpattern T of J we define the *triangular negative/positive inertia* of $\mathcal{A}|T$ by

$$ti_\pm(\mathcal{A}|T) = i_\pm(\mathcal{A}|R_1) + \sum_{j=1}^{m-1} (\text{rank } (\mathcal{A}|R_{j+1}) - \text{ rank } (\mathcal{A}|R_{j+1} \cap R_j)),$$

and

$$R(T) = \{R_1, \ldots, R_m\}$$

is ordered increasingly according to the number of rows covered (see Proposition 5.3.9).

Example 5.3.16 For J and T as in Example 5.3.10, we have for $\mathcal{A} = \{a_{ij} : (i,j) \in J\}$ that

$$ti_-(\mathcal{A}|T) = i_- \begin{pmatrix} a_{33} & a_{34} & a_{35} \\ a_{43} & a_{44} & a_{45} \\ a_{53} & a_{54} & a_{55} \end{pmatrix} + \text{rank} \begin{pmatrix} a_{23} & a_{24} \\ a_{33} & a_{34} \\ a_{43} & a_{44} \\ a_{53} & a_{54} \end{pmatrix}$$

$$- \text{ rank} \begin{pmatrix} a_{33} & a_{34} \\ a_{43} & a_{44} \\ a_{53} & a_{54} \end{pmatrix} + \text{ rank} \begin{pmatrix} a_{13} \\ a_{23} \\ a_{33} \\ a_{43} \\ a_{53} \end{pmatrix} - \text{ rank} \begin{pmatrix} a_{23} \\ a_{33} \\ a_{43} \\ a_{53} \end{pmatrix}.$$

Proposition 5.3.17 *Let J be a symmetric pattern, $\mathcal{A} = \{a_{ij} : (i,j) \in J\}$ a Hermitian partial matrix, and T a symmetric triangular subpattern. Then for any Hermitian completion A of $\mathcal{A}$ we have*

$$i_\pm(A) \geq ti_\pm(\mathcal{A}|T). \tag{5.3.6}$$

Proof. Let $R(T) = \{R_1, \dots, R_m\}$ be ordered as in Proposition 5.3.9. Perform, if necessary, a permutation similarity such that

$$R_i = \{1, \dots, n_i\} \times \{1, \dots, m_i\},\ i = 1, \dots, m,$$

with $n_1 < n_2 < \cdots < n_m$ and $(n_1 =) m_1 > m_2 > \cdots > m_m$. Decompose A as

$$\begin{pmatrix} A_{11} & A_{12} & A_{13} \\ A_{12}^* & A_{22} & A_{23} \\ A_{13}^* & A_{23}^* & A_{33} \end{pmatrix}$$

with

$$A_{11} = (a_{ij})_{i,j=1}^{m_2}, A_{22} = (a_{ij})_{i,j=m_2+1}^{m_1}, A_{33} = (a_{ij})_{i,j=m_1+1}^{n},$$

and A_{12}, A_{13}, and A_{23} defined accordingly. By Lemma 5.1.5 it follows that

$$\begin{aligned} i_-(A) &\geq i_-\begin{pmatrix} A_{11} & A_{12} \\ A_{12}^* & A_{22} \end{pmatrix} + \text{ rank} \begin{pmatrix} A_{11} \\ A_{12}^* \\ A_{13}^* \end{pmatrix} - \text{rank } \begin{pmatrix} A_{11} \\ A_{12}^* \end{pmatrix} \\ &= i_-\left(\mathcal{A}|R_{\text{symm}}(T)\right) + \text{ rank} \begin{pmatrix} A_{11} \\ A_{12}^* \\ A_{13}^* \end{pmatrix} - \text{rank } (\mathcal{A}|R_1 \cap R_2) \end{aligned} \tag{5.3.7}$$

By Theorem 5.2.1 it further follows that

$$\text{rank} \begin{pmatrix} A_{11} \\ A_{12}^* \\ A_{13}^* \end{pmatrix} \geq \sum_{j=2}^{m} \text{rank}(\mathcal{A}|R_j) - \sum_{j=3}^{m} \text{rank}(\mathcal{A}|R_{j-1} \cap R_j). \tag{5.3.8}$$

Combining (5.3.7) and (5.3.8) gives (5.3.6). □

Let J be a symmetric pattern and $\mathcal{A} = \{a_{ij} : (i,j) \in J\}$ a partial Hermitian matrix. Define the *negative/positive triangular inertia* of $\mathcal{A}$ by

$$ti_-(\mathcal{A}) = \max_{T \in \mathcal{T}(J)} ti_-(\mathcal{A}|T),\ ti_+(\mathcal{A}) = \max_{T \in \mathcal{T}(J)} ti_+(\mathcal{A}|T).$$

Corollary 5.3.18 *Let J be a symmetric pattern and $\mathcal{A}$ a partial Hermitian matrix with pattern J. Then for any Hermitian completion A of $\mathcal{A}$ we have that*

$$i_-(A) \geq ti_-(\mathcal{A}),\ i_+(A) \geq ti_+(\mathcal{A}).$$

We now get to the main result of this section.

Theorem 5.3.19 *Let J be a symmetric banded pattern in $\underline{\underline{n}}^2$, and $\mathcal{A} = \{a_{ij} : (i,j) \in J\}$ a partial Hermitian matrix. Then*

$$\min\{i_\pm(A) : A \text{ is a Hermitian completion of } \mathcal{A}\} = ti_\pm(\mathcal{A}).$$

In fact, $\mathcal{A}$ has a Hermitian completion $A = (a_{ij})_{i,j=1}^n$ such that for each $k \in \{1,\dots,n\}$ the matrix $(a_{ij})_{i,j=k}^n$ is a completion of $\mathcal{A}|(J \cap \{k,\dots,n\}^2)$ with minimal possible negative/positive inertia. If $\mathcal{A}$ is a real symmetric partial matrix, then the above completion may be chosen to be real symmetric as well.

We will make use of the following observations.

Lemma 5.3.20 *Let A_{ij}, $1 \leq j \leq i \leq n$, be complex matrices of size $1 \times p_j$ (with possibly $p_j = 0$). If*

$$\begin{aligned}
&\min \operatorname{rank} \begin{pmatrix} A_{11} & & & ? \\ A_{21} & A_{22} & & \\ \vdots & \vdots & \ddots & \\ A_{n1} & A_{n2} & \cdots & A_{nn} \end{pmatrix} \\
&= \min \operatorname{rank} \begin{pmatrix} A_{21} & A_{22} & & ? \\ \vdots & \vdots & \ddots & \\ A_{n1} & A_{n2} & \cdots & A_{nn} \end{pmatrix} \\
&= \min \operatorname{rank} \begin{pmatrix} A_{21} & A_{22} & & ? \\ \vdots & \vdots & \ddots & \\ A_{n-1,1} & A_{n-1,2} & \cdots & A_{n-1,n-1} \end{pmatrix} + 1 \\
&= \min \operatorname{rank} \begin{pmatrix} A_{11} & & & ? \\ A_{21} & A_{22} & & \\ \vdots & \vdots & \ddots & \\ A_{n-1,1} & A_{n-1,2} & \cdots & A_{n-1,n-1} \end{pmatrix}.
\end{aligned} \tag{5.3.9}$$

Then $A_{11}, A_{n1} \notin$ rowspace $\begin{pmatrix} A_{21} \\ \vdots \\ A_{n-1,1} \end{pmatrix}$, and there exist $s_1, \dots, s_n \in \mathbb{C}$ with $s_1 \neq 0$ and $s_n \neq 0$ such that

$$\begin{pmatrix} s_1 & \cdots & s_n \end{pmatrix} \begin{pmatrix} A_{11} \\ \vdots \\ A_{n1} \end{pmatrix} = 0.$$

If A_{ij}, $1 \leq j \leq i \leq n$, are real matrices, then $s_1, \dots, s_n$ can be chosen to be real as well.

Proof. The first equality in (5.3.9) shows that

$$A_{11} \in \text{rowspace} \begin{pmatrix} A_{21} \\ \vdots \\ A_{n,1} \end{pmatrix}$$

(use the minimal rank formula). By the same token, the third equality in (5.3.9) shows that

$$A_{11} \notin \text{ rowspace } \begin{pmatrix} A_{21} \\ \vdots \\ A_{n-1,\,1} \end{pmatrix}.$$

These two observations imply that

$$A_{11} = s_2 A_{21} + \cdots + s_n A_{n1}$$

for some $s_2, \ldots, s_n \in \mathbb{C}$ with $s_n \neq 0$. In addition, we get that

$$A_{n1} \notin \text{ rowspace } \begin{pmatrix} A_{21} \\ \vdots \\ A_{n-1,1} \end{pmatrix};$$

otherwise

$$\text{rowspace } \begin{pmatrix} A_{21} \\ \vdots \\ A_{n1} \end{pmatrix} = \text{ rowspace } \begin{pmatrix} A_{21} \\ \vdots \\ A_{n-1,1} \end{pmatrix}.$$

Clearly, $s_2, \ldots s_n$ can be chosen to be real, when $A_{11}, \ldots, A_{n1}$ are real. □

Lemma 5.3.21 *Let $A = A^* \in \mathbb{C}^{(n-1)\times(n-1)}$, $b \in \mathbb{C}^{n-1}$, and $c \in \mathbb{R}$. Then*

$$i_- \begin{pmatrix} A & b \\ b^* & c \end{pmatrix} = i_-(A) \tag{5.3.10}$$

if and only if there exist $\mu \geq 0$ and $s \in \mathbb{C}^{n-1}$ such that

$$s^* \begin{pmatrix} A & b \end{pmatrix} = \begin{pmatrix} b^* & c - \mu \end{pmatrix}. \tag{5.3.11}$$

Proof. If (5.3.11) holds then

$$\begin{pmatrix} A & b \\ b^* & c \end{pmatrix} = \begin{pmatrix} I & 0 \\ s^* & 1 \end{pmatrix} \begin{pmatrix} A & 0 \\ 0 & \mu \end{pmatrix} \begin{pmatrix} I & s \\ 0 & 1 \end{pmatrix}.$$

Thus

$$i_- \begin{pmatrix} A & b \\ b^* & c \end{pmatrix} = i_-(A) + i_-(\mu) = i_-(A).$$

Conversely, if (5.3.10) holds, decompose A as

$$A = \begin{pmatrix} A_{11} & 0 \\ 0 & 0 \end{pmatrix} : \begin{matrix} \text{Im}A \\ \bigoplus \\ \text{Ker}A \end{matrix} \rightarrow \begin{matrix} \text{Im}A \\ \bigoplus \\ \text{Ker}A \end{matrix}$$

and

$$b^* = \begin{pmatrix} b_1^* & b_2^* \end{pmatrix} : \begin{matrix} \text{Im}A \\ \bigoplus \\ \text{Ker}A \end{matrix} \rightarrow \mathbb{C}$$

accordingly. Then, using a Schur complement,

$$i_-\begin{pmatrix} A_{11} & 0 & b_1 \\ 0 & 0 & b_2 \\ b_1^* & b_2^* & c \end{pmatrix} = i_-\begin{pmatrix} A_{11} & 0 & 0 \\ 0 & 0 & b_2 \\ 0 & b_2^* & c - b_1^*A_{11}^{-1}b_1 \end{pmatrix}$$

$$= i_-(A_{11}) + i_-\begin{pmatrix} 0 & b_2 \\ b_2^* & c - b_1^*A_{11}^{-1}b_1 \end{pmatrix}$$

$$= i_-(A) + i_-\begin{pmatrix} 0 & b_2 \\ b_2^* & c - b_1^*A_{11}^{-1}b_1 \end{pmatrix}.$$

Since (5.3.10) holds we must have that

$$i_-\begin{pmatrix} 0 & b_2 \\ b_2^* & c - b_1^*A_{11}^{-1}b_1 \end{pmatrix} = 0,$$

so $b_2 = 0$ and $c - b_1^*A_{11}^{-1}b_1 \geq 0$. Put now $\mu = c - b_1^*A_{11}^{-1}b_1$ and $s = A_{11}^{-1}b_1$. Then (5.3.11) holds. □

Lemma 5.3.22 *Let $J \subseteq \underline{\underline{n}} \times \underline{\underline{m}}$ be such that for some $p \leq q$ we have $\underline{\underline{n}} \times \{p, \ldots, q\} \subseteq J$. Then any maximal triangular pattern in J contains $\underline{\underline{n}} \times \{p, \ldots, q\}$.*

Proof. Let $T \subseteq J$ be triangular. We claim that $T \cup (\underline{\underline{n}} \times \{p, \ldots, q\})$ is also triangular. Indeed, suppose that (k, l) and (r, s) belong to $T \cup (\underline{\underline{n}} \times \{p, \ldots, q\})$. If both l and s are not in $\{p, \ldots, q\}$, then (k, l) and (r, s) belong to T, and thus either $(k, s) \in T$ or $(r, l) \in T$. Otherwise, if l or s belongs to $\{p, \ldots, q\}$, then (k, s) or (r, l) belongs to $\underline{\underline{n}} \times \{p, \ldots, q\}$. □

Recall that

$$\gamma_J(j) = \min\{i : (i, j) \in J\},$$
$$\zeta_J(j) = \max\{i : (i, j) \in J\}.$$

Proof of Theorem 5.3.19 Only in this proof we will use the notation

$$\mathcal{A} = \{((i, j), a_{ij}) \ : \ (i, j) \in J\}$$

for a partial matrix, emphasizing the location as well as the value of the prescribed entries. The advantage of this notation is that when we change the value of an entry, we can easily convey this as follows. For instance,

$$\{((i, j), a_{ij}) \ : \ (i, j) \in J \setminus \{(i_0, j_0)\}\} \cup \{((i_0, j_0), \alpha)\}$$

denotes the partial matrix obtained from $\mathcal{A}$ by replacing the value in the (i_0, j_0)th entry by α.

Notice that by Lemma 5.3.2 we may assume that $D_n \subseteq J$. We prove the result by induction on the size of n. When $n = 1$ the theorem is trivially true. So suppose the theorem holds for partial Hermitian banded matrices of size $(n-1) \times (n-1)$ (including the "In fact" part).

Let now $\mathcal{A} = \{((i, j), a_{ij}) : (i, j) \in J\}$ be an $n \times n$ partial Hermitian banded matrix, and let $k = ti_-(\mathcal{A})$ and

$$\widetilde{J} = J \cap (\underline{\underline{n-1}})^2.$$

Denote $\widetilde{\mathcal{A}} = \left\{((i,j), a_{ij}) : (i,j) \in \widetilde{J}\right\}$. By the induction hypothesis there exists a Hermitian completion $\widetilde{A} = (a_{ij})_{i,\ j=1}^{n-1}$ of $\widetilde{\mathcal{A}}$ such that $i_-\left(\widetilde{A}\right) = ti_-\left(\widetilde{\mathcal{A}}\right)$, and for each $p \in \{1, \ldots, n-1\}$ the matrix

$$(a_{ij})_{i,\ j=p}^{n-1}$$

is a Hermitian completion of

$$\left\{((i,j), a_{ij}) : (i,j) \in \widetilde{J} \cap \{p, \ldots, n-1\} \times \{p, \ldots, n-1\}\right\}$$

with minimal possible negative eigenvalues.

We distinguish 3 cases.

Case 1. $i_-\,(a_{ij})_{i,\ j=\gamma(n)}^{n} = i_-\,(a_{ij})_{i,\ j=\gamma(n)}^{n-1}$. By Lemma 5.3.21 there exists a $\mu \geq 0$ and $s_{\gamma_J(n)}, \ldots, s_{n-1} \in \mathbb{C}$ such that

$$\begin{pmatrix} a_{n\gamma_J(n)} & \cdots & a_{n,n-1} & a_{nn} - \mu \end{pmatrix}$$

$$= \begin{pmatrix} s_{\gamma(n)} & \cdots & s_{n-1} \end{pmatrix} \begin{pmatrix} a_{\gamma_J(n),\gamma_J(n)} & \cdots & a_{\gamma_J(n),n} \\ \vdots & & \vdots \\ a_{n-1,\gamma_J(n)} & \cdots & a_{n-1,n} \end{pmatrix}.$$

Put now

$$\begin{pmatrix} a_{n1} & \cdots & a_{n,\gamma_J(n)-1} \end{pmatrix}$$

$$= \begin{pmatrix} s_{\gamma_J(n)} & \cdots & s_{n-1} \end{pmatrix} \cdot \begin{pmatrix} a_{\gamma_J(n),1} & \cdots & a_{\gamma_J(n),\gamma_J(n)-1} \\ \vdots & & \vdots \\ a_{n-1,1} & \cdots & a_{n-1,\gamma_J(n)-1} \end{pmatrix}.$$

Then by Lemma 5.3.21 we conclude that for $m = 1, \ldots, \gamma_J(n)$

$$i_- \begin{pmatrix} a_{mm} & \cdots & a_{mn} \\ \vdots & & \vdots \\ a_{nm} & \cdots & a_{nn} \end{pmatrix} = i_- \begin{pmatrix} a_{mm} & \cdots & a_{m,n-1} \\ \vdots & & \vdots \\ a_{n-1,m} & \cdots & a_{n-1,n-1} \end{pmatrix},$$

and thus the completion has the required properties.

For the remaining cases we will need the following triangular subpattern of J:

$$T_1 := J \cap \left(\underline{\underline{n}} \times \{\gamma_J(n), \ldots, n\}\right).$$

Case 2. $i_-\,(a_{ij})_{i,j=\gamma_J(n)}^{n} = i_-\,(a_{ij})_{i,j=\gamma_J(n)}^{n-1} + 1$ and there is a $\mu \geq 0$ such that

$$\min \operatorname{rank}\left(\{((i,j), a_{ij}) : (i,j) \in T_1 \setminus \{(n,n)\}\} \cup \{((n,n), a_{nn} - \mu)\}\right)$$

$$= \min \operatorname{rank}\left\{((i,j), a_{ij}) : (i,j) \in T_1 \cap \left(\underline{\underline{n-1}} \times \underline{\underline{n}}\right)\right\}.$$

Let m be as large as possible such that there exists a $\mu \geq 0$ such that the minimal rank of the partial matrix

$$\left\{((i,j), a_{ij}) : (i,j) \in \left(T_1 \cap \{m, \ldots, n\} \times \underline{\underline{n}}\right) \setminus \{(n,n)\}\right\} \cup \{((n,n), a_{nn} - \mu)\}$$

equals the minimal rank of

$$\{((i,j),a_{ij}) : (i,j) \in T_1 \cap \{m,\dots,n-1\} \times \underline{n}\},$$

and choose $\mu \geq 0$ accordingly. Note that by the assumptions in this case, m is well defined and

$$\gamma_J(\gamma_J(n)) \leq m \leq \gamma_J(n) - 1.$$

Because of the choice of m, by Lemma 5.3.20 there exist $s_m, \dots, s_n$ with $s_m \neq 0$ and $s_n \neq 0$ such that

$$\begin{pmatrix} s_m & \cdots & s_n \end{pmatrix} \begin{pmatrix} a_{m,\gamma_J(n)} & \cdots & a_{m,\zeta_J(m)} \\ \vdots & & \vdots \\ a_{n,\gamma_J(n)} & \cdots & a_{n,\zeta_J(m)} \end{pmatrix} = \begin{pmatrix} 0 & \cdots & 0 \end{pmatrix}.$$

Let now

$$\begin{pmatrix} a_{n,m+1} & \cdots & a_{n,\gamma_J(n)-1} \end{pmatrix}$$

$$= \begin{pmatrix} -\frac{s_m}{s_n} & \cdots & -\frac{s_{n-1}}{s_n} \end{pmatrix} \cdot \begin{pmatrix} a_{m,m+1} & \cdots & a_{m,\gamma_J(n)-1} \\ \vdots & & \vdots \\ a_{n-1,m+1} & \cdots & a_{n-1,\gamma_J(n)-1} \end{pmatrix}.$$

Consider the 3×3 block matrix

$$\begin{pmatrix} \alpha & \beta & \chi \\ \beta^* & \gamma & \delta \\ \chi^* & \delta^* & \epsilon \end{pmatrix},$$

where

$$\begin{aligned} \alpha &= a_{m,m}, & \beta &= (a_{m,m+1} \cdots a_{m,\zeta_J(m)}), \\ \gamma &= (a_{ij})_{i,j=m+1}^{\zeta_J(m)}, & \delta &= (a_{ij})_{i=m+1,j=\zeta_J(m)+1}^{\zeta_J(m) \quad n}, \\ \epsilon &= (a_{ij})_{i,j=\zeta_J(m)+1}^{n}, \end{aligned}$$

and χ is considered as unspecified. We claim that

$$\min_{\chi} i_- \begin{pmatrix} \alpha & \beta & \chi \\ \beta^* & \gamma & \delta \\ \chi^* & \delta^* & \epsilon \end{pmatrix} = i_{<0}(a_{ij})_{i,\ j=m+1}^{n-1} + 1.$$

Indeed, by Theorem 5.1.6,

$$\min i_- \begin{pmatrix} \alpha & \beta & \chi \\ \beta^* & \gamma & \delta \\ \chi^* & \delta^* & \epsilon \end{pmatrix}$$

$$= \max \left\{ i_- \begin{pmatrix} \alpha & \beta \\ \beta^* & \gamma \end{pmatrix} + \operatorname{rank} \begin{pmatrix} \beta \\ \gamma \\ \delta^* \end{pmatrix} - \operatorname{rank} \begin{pmatrix} \beta \\ \gamma \end{pmatrix}, \right. \tag{5.3.12}$$

$$\left. i_- \begin{pmatrix} \gamma & \delta \\ \delta^* & \epsilon \end{pmatrix} + \operatorname{rank} \begin{pmatrix} \beta \\ \gamma \\ \delta^* \end{pmatrix} - \operatorname{rank} \begin{pmatrix} \gamma \\ \delta^* \end{pmatrix} \right\}.$$

Since $\beta \in \text{rowspace}\begin{pmatrix}\gamma\\ \delta^*\end{pmatrix}$, we have

$$\text{rank}\begin{pmatrix}\beta\\ \gamma\\ \delta^*\end{pmatrix} = \text{rank}\begin{pmatrix}\gamma\\ \delta^*\end{pmatrix}.$$

Since $\beta \notin \text{rowspace}\,\gamma$ (use Lemma 5.3.20 and that $m < \gamma(n)$), we have

$$\text{rank}\begin{pmatrix}\beta\\ \gamma\end{pmatrix} = \text{rank}\,\gamma + 1.$$

and

$$i_{\leq 0}\begin{pmatrix}\alpha & \beta\\ \beta^* & \gamma\end{pmatrix} = i_-(\gamma) + 1.$$

Since $\beta \notin \text{rowspace}\,(a_{ij})_{i=m+1,\ j=m+1}^{n-1\ \zeta_J(m)}$ (again we use Lemma 5.3.20 and that $m < \gamma_J(n)$), and $\beta \in \text{rowspace}\begin{pmatrix}\gamma\\ \delta^*\end{pmatrix}$ we get that

$$\text{rank}\begin{pmatrix}\beta\\ \gamma\\ \delta^*\end{pmatrix} = \text{rank}\begin{pmatrix}a_{m+1,m+1} & \cdots & a_{m+1,\zeta_J(m)}\\ \vdots & & \vdots\\ a_{n-1,m+1} & \cdots & a_{n-1,\zeta_J(m)}\end{pmatrix} + 1.$$

So the right-hand side of (5.3.12) equals

$$\max\Big\{i_-(\gamma) + 1 + \text{ rank}\,(a_{ij})_{i=m+1,\ j=m+1}^{n-1\quad \zeta_J(m)} - \text{ rank}\,\gamma,$$

$$i_-\begin{pmatrix}\gamma & \delta\\ \delta^* & \epsilon\end{pmatrix}\Big\} = i_-(a_{ij})_{i,j=m+1}^{n-1} + 1.$$

Here we used that

$$\begin{pmatrix} a_{n,m+1} & \cdots & a_{n,n-1}\end{pmatrix} \notin \text{ rowspace}\,(a_{ij})_{i,j=m+1}^{n-1},$$

such that

$$i_-\begin{pmatrix}\gamma & \delta\\ \delta^* & \epsilon\end{pmatrix} = i_-(a_{ij})_{i,j=m+1}^{n-1} + 1.$$

Choose a completion $(b_{ij})_{i,j=m}^{n}$ of

$$\begin{pmatrix}\alpha & \beta & ?\\ \beta^* & \gamma & \delta\\ ? & \delta^* & \epsilon\end{pmatrix}$$

with minimal negative inertia. Note that $b_{ij} = a_{ij}$ for $i, j \geq m$ with

$$(i,j) \notin (\{m\} \times \{\zeta_J(m)+1,\ldots,n\} \cup \{\zeta_J(m)+1,\ldots,n\} \times \{m\}).$$

Now for $p = m, \ldots, \gamma_J(n)$ we have that $(b_{ij})_{i,j=p}^{n}$ is a completion of

$$\{((i,j), b_{ij}) : (i,j) \in J \cap \{p,\ldots,n\}^2\}$$

with minimal negative inertia. Indeed, for $p = m+1, \ldots, \gamma_J(n)$ this follows from the induction assumption and the fact that

$$\begin{pmatrix}b_{n,\gamma_J(n)}\\ \vdots\\ b_{n,\zeta_J(m)}\end{pmatrix}^T \notin \text{rowspace}\begin{pmatrix}b_{m+1,\gamma_J(n)} & \cdots & b_{m+1,\zeta_J(m)}\\ \vdots & & \vdots\\ b_{n-1,\gamma_J(n)} & \cdots & b_{n-1,\zeta_J(m)}\end{pmatrix}.$$

For $p = m$ this follows from the observations that

$$\begin{aligned}
&i_-(b_{ij})_{i,j=m}^{n} = i_-(b_{ij})_{i,j=m+1}^{n-1} + 1 \\
&= ti_- \left\{((i,j), b_{ij}) : J \cap \{m+1, \dots, n-1\} \times \{m+1, \dots, n-1\}\right\} + 1 \\
&= ti_- \left\{((i,j), b_{ij}) : J \cap \{m+1, \dots, n\}^2\right\}.
\end{aligned}$$

For $i < m$ or $j < m$ let $b_{ij} = a_{ij}$. Furthermore,

$$\begin{aligned}
&ti_- \left\{((i,j), b_{ij}) : (i,j) \in J \cup \{m, \dots, n\}^2\right\} \\
&= ti_- \{((i,j), b_{ij}) : (i,j) \in J\} \\
&(= ti_-\{((i,j), a_{ij}) : (i,j) \in J\}).
\end{aligned} \tag{5.3.13}$$

In order to prove (5.3.13) we will show that

$$\begin{aligned}
&ti_-\{((i,j), a_{ij}) : (i,j) \in J\} \\
&\geq ti_- \left\{((i,j), a_{ij}) : (i,j) \in \left(J \cap (\underline{\underline{n-1}})^2\right) \cup \{m, \dots, n-1\}^2\right\} \\
&= ti_- \left\{((i,j), b_{ij}) : (i,j) \in \left(J \cap (\underline{\underline{n-1}})^2\right) \cup \{m, \dots, n-1\}^2\right\} \\
&= ti_- \left\{((i,j), b_{ij}) : (i,j) \in J \cup \{m, \dots, n\}^2\right\} \\
&\geq ti_- \{((i,j), a_{ij}) : (i,j) \in J\}.
\end{aligned} \tag{5.3.14}$$

The inequalities in (5.3.14) are trivial. In order to prove the first equality in (5.3.14), first note that

$$\begin{pmatrix} a_{m,m+1} \\ \vdots \\ a_{m,\zeta_J(m)} \end{pmatrix}^T \notin \text{rowspace} \begin{pmatrix} a_{m+1,\, m+1} & \cdots & a_{m+1,\zeta_J(m)} \\ \vdots & & \vdots \\ a_{n-1,\, m+1} & \cdots & a_{n-1,\zeta_J(m)} \end{pmatrix}.$$

Thus for any $Y \in \mathbb{C}^{1 \times q}$ there exists an $S \in \mathbb{C}^{(\zeta_J(m)-m) \times q}$ such that

$$\begin{aligned}
(a_{m,m+1} \cdots a_{m,\zeta(m)})S &= Y, \\
(a_{ij})_{i=m+1, j=m+1}^{n-1 \quad \zeta(m)} S &= 0.
\end{aligned}$$

Let $Y = (b_{m,\zeta_J(m)+1} \cdots b_{m,n-1}) - (a_{m,\zeta_J(m)+1} \cdots a_{m,n-1})$, and choose $S = (S_{ij})_{i=m+1, j=\zeta_J(m)+1}^{\zeta_J(m) \quad n-1}$ accordingly. Then for all $s \leq m+1$, $r \geq \zeta_J(m)+1$ and $m+1 \leq l \leq \zeta_J(m)$, we have that

$$\begin{aligned}
&\left[(a_{ij})_{i=m+1,\; j=s}^{l \quad\; r}\right] \begin{pmatrix} I_{m+1-s} & 0 & 0 \\ 0 & I_{\zeta_J(m)-m} & \widehat{S} \\ 0 & 0 & I_{r-\zeta_J(m)} \end{pmatrix} \\
&= \left[(b_{ij})_{i=m+1,\; j=s}^{l \quad\; r}\right],
\end{aligned}$$

where $\widehat{S} = (S_{ij})_{i=m+1,\; j=\zeta_J(m)+1}^{\zeta_J(m) \quad r}$. Thus

$$\operatorname{rank} (a_{ij})_{i=m+1,\; j=s}^{l \quad\; r} = \operatorname{rank} (b_{ij})_{i=m+1,\; j=s}^{l \quad\; r}. \tag{5.3.15}$$

Furthermore, the only rectangle appearing in $R(T)$ for some $T \in \mathcal{T}((J \cap (\underline{\underline{n-1}})^2) \cup \{m, \dots, n-1\}^2)$ which contains both elements from $\{m\} \times$

$\{\zeta_J(m)+1,\ldots,n-1\}$ and $\{\zeta_J(m)+1,\ldots,n-1\}\times\{m\}$ is $\{m,\ldots,n-1\}^2$, and here we have

$$i_-(a_{ij})_{i,j=m}^{n-1} = i_-(b_{ij})_{i,j=m}^{n-1}. \tag{5.3.16}$$

By Proposition 5.3.13 (applied to $(J\cap(\underline{n-1})^2)\cup\{m,\ldots,n-1\}^2$) and the above observation, the only possible difference in the terms of the expressions for

$$ti_-\left\{((i,j),a_{ij}) : (i,j)\in\left(J\cap(\underline{\underline{n-1}})^2\right)\cup\{m,\ldots,n-1\}^2\right\} \tag{5.3.17}$$

and

$$ti_-\left\{((i,j),b_{ij}) : (i,j)\in\left(J\cap(\underline{\underline{n-1}})^2\right)\cup\{m,\ldots,n-1\}^2\right\} \tag{5.3.18}$$

might be the difference between left-hand sides and right-hand sides of (5.3.15) and (5.3.16). Consequently, it follows that (5.3.17) and (5.3.18) are equal.

To see that the second equality in (5.3.14) holds, note that if for $T\in\mathcal{T}(J)$ a member of $R(T)$ contains a part of the nth row, it will also contain the corresponding part (i.e., the part within the same columns) of rows $m,\ldots,n-1$. Since

$$i_-(b_{ij})_{i,j=m}^{n} = i_-(b_{ij})_{i,j=m}^{n-1}$$

and for $K\subseteq\{m,\ldots,n-1\}$,

$$\text{rank }(b_{ij})_{i=m,\ j\in K}^{n} = \text{rank }(b_{ij})_{i=m,\ j\in K}^{n-1},$$

the second equality in (5.3.14) follows.

We may thus conclude that (5.3.13) is true. Now we may apply Case 1 to the partial matrix

$$\left\{((i,j),b_{ij}) : (i,j)\in J\cup\{m,\ldots,n\}^2\right\}$$

and obtain the desired completion. This finishes Case 2.

Case 3. For all $\mu\geq 0$ we have that

$$\begin{aligned}&\min\text{ rank }\left(\{((i,j),a_{ij}) : (i,j)\in T_1\setminus\{(n,n)\}\}\cup\{((n,n),a_{nn}-\mu)\}\right)\\ &\quad>\min\text{ rank }\left\{((i,j),a_{ij}) : (i,j)\in T_1\cap(\underline{\underline{n-1}}\times\underline{\underline{n}})\right\}.\end{aligned}$$

Let $\widetilde{q}$ be as large as possible such that

$$\begin{aligned}&\min\text{ rank }\left(\{((i,j),a_{ij}) : (i,j)\in T_1\setminus\{(n,\widetilde{q}+1),\ldots,(n,n)\}\right) =\\ &\quad=\min\text{ rank }\left\{((i,j),a_{ij}) : (i,j)\in T_1\cap\{(\underline{\underline{n-1}}\times\underline{\underline{n}})\right\}.\end{aligned}$$

Denote $\widetilde{p}=\widetilde{q}+1$. Note that $\gamma_J(n)\leq\widetilde{p}\leq n$. Furthermore, the choice of $\widetilde{p}$ implies that

$$\text{rank }(a_{ij})_{i=\gamma_J(\widetilde{p}),j=\gamma_J(n)}^{n-1\quad\widetilde{p}} = \text{rank }(a_{ij})_{i=\gamma_J(\widetilde{p}),\ j=\gamma_J(n)}^{n\quad\widetilde{p}} - 1. \tag{5.3.19}$$

Consider the pattern

$$J_1=\{\gamma_J(\widetilde{p}),\ldots,n\}^2\cap J.$$

By Lemma 5.3.22 any maximal triangular pattern in J_1 contains $\{\gamma_J(\widetilde{p}),\ldots,n\} \times \{\gamma_J(n),\ldots,\widetilde{p}\}$. But then it follows from (5.3.19) that

$$ti_-\left(\{((i,j),a_{ij}):(i,j)\in J_1\}\right)$$
$$= ti_-\left(\{((i,j),a_{ij}):(i,j)\in \{\gamma(\widetilde{p}),\ldots,n-1\}^2\cap J_1\}\right)+1. \quad (5.3.20)$$

Recall from the induction assumption that $\widetilde{A}=(a_{ij})_{i,j=1}^{n-1}$ is a Hermitian completion of $\widetilde{\mathcal{A}}$ such that for each $p\in\{1,\ldots,n-1\}$ the matrix $(a_{ij})_{i,j=p}^{n-1}$ is a Hermitian completion of

$$\left\{((i,j),a_{ij}):(i,j)\in \widetilde{J}\cap\{p,\ldots,n-1\}^2\right\}$$

with minimal possible negative eigenvalues. Moreover,

$$i_-(\widetilde{A}) = ti_-(\widetilde{\mathcal{A}}) =: l.$$

Let

$$j_s=\max\{\nu : i_-(a_{ij})_{i,j=\nu}^{n-1}=l-s+1\}, \qquad s=1,\ldots,l.$$

In case $j_1\geq\gamma_J(\widetilde{p})$, we obtain that

$$ti_-\left(\{((i,j),a_{ij}):(i,j)\in\{\gamma_J(\widetilde{p}),\ldots,n-1\}^2\cap J_1\}\right)=l,$$

and thus by (5.3.20) that

$$ti_-(\mathcal{A})\geq l+1.$$

On the other hand,

$$ti_-(\mathcal{A})\leq ti_-(\widetilde{\mathcal{A}})+1=l+1.$$

Thus

$$l+1=ti_-(\mathcal{A})=ti_-(\widetilde{\mathcal{A}})+1.$$

But then any choice for $(a_{n1}\cdots a_{n,\gamma(n)-1})$ will yield a desired completion for $\mathcal{A}$. In case $j_1<\gamma_J(\widetilde{p})$, choose q as large as possible such that

(i) $j_q<\gamma_J(\widetilde{p})$;

(ii) $\operatorname{rank}\,(a_{ij})_{i=j_q,\ j=\gamma_J(n)}^{n\quad \zeta_J(j_q)} = \operatorname{rank}\,(a_{ij})_{i=j_q,\ j=\gamma_J(n)}^{n-1\quad \zeta_J(j_q)}$.

Because of the choice of $\widetilde{p}$ such a q must exist.

Change now $(a_{j_q,\zeta_J(j_q)+1}\ \cdots\ a_{j_q,n-1})$ to $(b_{j_q,\zeta_J(j_q)+1}\ \cdots\ b_{j_q,n-1})$ so that

$$\operatorname{rank}\,(b_{ij})_{i=j_q,\ j=\gamma_J(n)}^{n\quad n-1}=\operatorname{rank}\,(b_{ij})_{i=j_q,\ j=\gamma_J(n)}^{n-1\quad n-1},$$

where for $(i,j)\notin\{j_q\}\times\{\zeta_J(j_q)+1,\ldots,n-1\}\cup\{\zeta_J(j_q)+1,\ldots,n-1\}\times\{j_q\}$ we let $b_{ij}=a_{ij}$. Note that $b_{ij}=a_{ij}$ for $(i,j)\in J$. This is possible because of (ii). Let $J_2=J\cup\{j_q,\ldots,n-1\}^2$. Now, with $\mathcal{B}=\{((i,j),b_{ij}):(i,j)\in J_2\}$ (which has $ti_-(\mathcal{B})=ti_-(\mathcal{A})$), we can apply Case 2 to get the desired result. Or to reduce it to Case 1, choose $(b_{n,j_q}\cdots b_{n,\gamma_J(n-1)})$ such that one obtains a Hermitian completion with minimal negative inertia of

$$\{((i,j),b_{ij}):(i,j)\in\{j_q,\ldots,n-1\}^2\cup\{\gamma_J(n),\ldots,n\}^2\}.$$

Then one can apply Case 1 to

$$\{((i,j),b_{ij}):(i,j)\in J\cup\{j_q,\ldots,n\}^2\}.$$

This finishes the proof for the complex case. For the real case observe that all the arguments remain the same when the underlying field is the real line.
□

5.4 INERTIA OF HERMITIAN MATRIX EXPRESSIONS

In this section we will study the inertia of various Hermitian expressions. In Chapter 2 we have already briefly encountered the classical (positive definite) versions of the Stein and Lyapunov equations (see (2.3.15), Exercise 2.9.44). We start this section with indefinite versions of the Stein equation and use it to prove Krein's theorem concerning roots of zeros of orthogonal polynomials with respect to an indefinite inner product. In Exercise 5.12.14 we give the indefinite statement of Lyapunov's theorem and point out its proof.

Proposition 5.4.1 *Let A, X be $m \times m$ matrices such that $A = A^*$ is invertible and*

$$H := A - X^*AX > 0. \tag{5.4.1}$$

Then X has no eigenvalues on the unit circle $\mathbb{T}$ and the number of eigenvalues of X outside the unit circle equals the number of negative eigenvalues of A.

Proof. Let us first show that $A > 0$ if and only if the spectrum $\sigma(X)$ of X lies in $\mathbb{D}$. Assuming that $A > 0$ we get that

$$0 < A^{-\frac{1}{2}}HA^{-\frac{1}{2}} = I - (A^{-\frac{1}{2}}X^*A^{\frac{1}{2}})(A^{\frac{1}{2}}XA^{-\frac{1}{2}}),$$

and thus $\|A^{\frac{1}{2}}XA^{-\frac{1}{2}}\| < 1$. Consequently, $\sigma(X) = \sigma(A^{\frac{1}{2}}XA^{-\frac{1}{2}}) \subset \mathbb{D}$. Conversely, suppose that $\sigma(X) \subset \mathbb{D}$. Then $X^n \to 0$ as $n \to \infty$. Rewriting the Stein equation and reusing it over and over again we get that

$$\begin{aligned} A &= H + X^*AX = H + X^*HX + X^{*2}AX^2 \\ &= \cdots = \sum_{k=0}^{n-1} X^{*k}HX^k + X^{*n}AX^n \to \sum_{k=0}^{\infty} X^{*k}HX^k \geq H > 0, \end{aligned} \tag{5.4.2}$$

showing that $A > 0$. Analogously, we get that $A < 0$ if and only if the spectrum of X lies outside the unit circle.

Now, suppose that (5.4.1) holds with A indefinite and invertible. Suppose that x is an eigenvector of X with eigenvalue λ. Then (5.4.1) yields that

$$0 < x^*Hx = (1 - |\lambda|^2)x^*Ax,$$

and thus $|\lambda| \neq 1$. Let us next put X is Jordan canonical form; that is, let S be an invertible matrix such that

$$S^{-1}XS = \begin{pmatrix} J_1 & 0 \\ 0 & J_2 \end{pmatrix},$$

where $J_1 \in \mathbb{C}^{p \times p}$ and $J_2 \in \mathbb{C}^{(m-p)\times(m-p)}$ and the eigenvalues of J_1 and J_2 lie inside and outside the unit circle, respectively. Decomposing $S^*HS = (H_{ij})_{i,j=1}^2$ and $S^{*-1}AS^{-1} = (A_{ij})_{i,j=1}^2$ accordingly, we get that

$$0 < H_{11} = A_{11} - J_1^*A_{11}J_1 \ , \ 0 < H_{22} = A_{22} - J_2^*A_{22}J_2.$$

Using the first paragraph of the proof, we get that $A_{11} > 0$ and $A_{22} < 0$. But then it follows by the interlacing inequalities that A has p positive eigenvalues and $m - p$ negative eigenvalues. □

We can now state and prove Krein's theorem.

Theorem 5.4.2 *Let $C_j \in \mathbb{C}^{m\times m}$, $j = -n, \ldots, n$ be given, and assume that the Toeplitz block matrix*

$$T_n := \begin{pmatrix} C_0 & \cdots & C_{-n} \\ \vdots & \ddots & \vdots \\ C_n & \cdots & C_0 \end{pmatrix}$$

is invertible. Let

$$\begin{pmatrix} X_0 \\ X_1 \\ \vdots \\ X_n \end{pmatrix} = T_n^{-1} \begin{pmatrix} I \\ 0 \\ \vdots \\ 0 \end{pmatrix}, \quad \begin{pmatrix} Y_{-n} \\ \vdots \\ Y_{-1} \\ Y_0 \end{pmatrix} = T_n^{-1} \begin{pmatrix} 0 \\ \vdots \\ 0 \\ I \end{pmatrix}, \tag{5.4.3}$$

and put $X(z) = \sum_{i=0}^{n} X_i z^i$ and $Y(z) = \sum_{i=-n}^{0} Y_i z^i$. Suppose moreover that $X_0 > 0$ and $Y_0 > 0$. Then $X(z)$ and $Y(z)$ are invertible for $z \in \mathbb{T}$. Moreover, $\det X(z)$ *and* $\det Y(z)$ *have exactly as many roots (counting multiplicities) outside the unit circle as T_n has negative eigenvalues.*

Proof. As in the proof of Theorem 2.3.1, we let K be the companion type matrix belonging to $z^n I + \sum_{j=0}^{n-1} U_{n-j} := z^n X(\frac{1}{z}) X_0^{-1}$, that is,

$$K = \begin{pmatrix} -U_1 & I & \cdots & 0 \\ \vdots & \vdots & \ddots & \vdots \\ -U_{n-1} & 0 & \cdots & I \\ -U_n & 0 & \cdots & 0 \end{pmatrix}.$$

As in the proof of Theorem 2.3.1 we get that

$$\begin{pmatrix} T_{n-1} & K^{*n} T_{n-1} \\ T_{n-1} K^n & T_{n-1} \end{pmatrix} = T_{2n-1} = (C_{i-j})_{i,j=0}^{2n}, \tag{5.4.4}$$

where

$$C_k = \begin{pmatrix} C_{k-1} & \cdots & C_{k-n} \end{pmatrix} T_{n-1}^{-1} \begin{pmatrix} C_1 \\ \vdots \\ C_n \end{pmatrix}, \quad k > n. \tag{5.4.5}$$

By the choice of C_k, $k > n$, we have by applying observation (5.1.20) k times, that $i(T_{n+k}) = k(i(T_n) - i(T_{n-1})) + i(T_n)$. Thus $i(T_{2n-1}) = n(i(T_n)) - (n - 1)i(T_{n-1})$. But then by taking a Schur complement in (5.4.4) we get that $i(T_{n-1} - K^{*n} T_{n-1} T_{n-1}^{-1} T_{n-1} K^n) = n(i(T_n) - i(T_{n-1}))$. As $X_0 > 0$, we have that $i(T_n) - i(T_{n-1}) = (m, 0, 0)$. But then we get that

$$T_{n-1} - K^{*n} T_{n-1} K^n > 0. \tag{5.4.6}$$

But now the theorem follows from Proposition 5.4.1 and Lemma 2.3.3. □

Several Hermitian matrix expressions can be viewed as Schur complements in a block matrix. This observation allows one to obtain results for matrix (in)equalities from the statements in Section 5.1. We shall illustrate this phenomena to Riccati inequalities. We encountered the Riccati equation in

Section 2.7 where we were interested in Hermitian solutions which in turn led us to find a spectral factorization of a nonnegative definite matrix polynomial on the real line. If we drop the Hermitian requirement of the solution we can employ the techniques in this chapter to find a solution as follows.

Proposition 5.4.3 *Let $A = A^* \in \mathbb{C}^{m\times m}$, $B \in \mathbb{C}^{m\times n}$ and $C = C^* \in \mathbb{C}^{n\times n}$. Consider the Riccati expression $P(Z) = A + BZ^* + ZB^* + ZCZ^*$ where $Z \in \mathbb{C}^{m\times n}$. Put*

$$H = \begin{pmatrix} A & B \\ B^* & C \end{pmatrix}, \quad r = \operatorname{rank} \begin{pmatrix} B \\ C \end{pmatrix}.$$

Then the possible inertia $i(P(Z)) = (\pi, \nu, \delta)$ of $P(Z)$ are described by the following inequalities:

$$\max\{0, i_+(H) - r\} \leq \pi \leq \min\{m, i_+(H)\},$$

$$\max\{0, i_-(H) - r\} \leq \nu \leq \min\{m, i_-(H)\},$$

$$-i_-(H) \leq \pi - \nu \leq i_+(H), \pi + \nu \leq m.$$

In particular, there exists a Z such that $A + BZ^* + ZB^* + ZCZ^* = 0$ if and only if $\max\{i_+(H), i_-(H)\} \leq r$. When A, B, and C are real matrices, Z may be chosen to be real as well to obtain the desired inertia.

Proof. Note that $P(Z)$ appears as the Schur complement supported in the first row and column of

$$M = \begin{pmatrix} A & B & Z \\ B^* & C & -I_n \\ Z^* & -I_n & 0 \end{pmatrix}. \tag{5.4.7}$$

But then one can apply Theorem 5.1.6 to M and obtain the possible inertia for $P(Z)$ by observing that $i(M) = i(P(Z)) + (n, n, 0)$. □

Clearly, when one takes $B = 0$ in the statement above one obtains a Stein type result, and when $C = 0$ one obtains a Lyapunov type result.

Let us end this section with a result that is the mirror image of Proposition 5.1.4, namely, the situation of a 3×3 Hermitian block matrix of which two diagonal entries are unknown. Subsequently we will apply the result to address the Hermitian expression $A - BXB^* - CYC^*$, where X and Y are the unknowns.

Theorem 5.4.4 *Consider the $(n_1+n_2+n_3)\times(n_1+n_2+n_3)$ partial Hermitian matrix*

$$M = \begin{pmatrix} P & Q & R \\ Q^* & ? & S \\ R^* & S^* & ? \end{pmatrix}.$$

Then all possible inertia $i(M)$ are described by the following inequalities

$$i_+(M) \geq i_+(P) + \operatorname{rank} \begin{pmatrix} P & Q & R \end{pmatrix} - \operatorname{rank} P = \operatorname{rank} \begin{pmatrix} P & Q & R \end{pmatrix} - i_-(P), \tag{5.4.8}$$

$$i_-(M) \geq i_-(P) + \text{rank}\begin{pmatrix} P & Q & R \end{pmatrix} - \text{rank}P = \text{rank}\begin{pmatrix} P & Q & R \end{pmatrix} - i_+(P), \tag{5.4.9}$$

$$i_0(M) \geq n_1 - \text{rank}\begin{pmatrix} P & Q & R \end{pmatrix}, \tag{5.4.10}$$

$$i_+(M) + i_-(M) \geq 2 \ \text{rank}\begin{pmatrix} P & Q & R \end{pmatrix} + \text{rank}\begin{pmatrix} P & R \\ Q^* & S \end{pmatrix}$$
$$- \text{rank}\begin{pmatrix} P & R \end{pmatrix} - \text{rank}\begin{pmatrix} P \\ Q^* \end{pmatrix}. \tag{5.4.11}$$

Moreover, if the prescribed matrices are real then a completion with the desired inertia can be chosen to be real as well.

Proof. The necessity of conditions (5.4.8)–(5.4.11) follows easily from earlier results. Indeed, (5.4.8) and (5.4.9) follow from Lemma 5.1.5 with the choice $A = P$ and $X = \begin{pmatrix} Q & R \end{pmatrix}$. Next, clearly

$$v \in \text{Ker}\begin{pmatrix} P \\ Q^* \\ R^* \end{pmatrix} \text{ implies } \begin{pmatrix} v \\ 0 \\ 0 \end{pmatrix} \in \text{Ker}M,$$

so

$$i_0(M) \geq \dim \text{Ker}\begin{pmatrix} P \\ Q^* \\ \mathcal{R}^* \end{pmatrix} = n_1 - \text{rank}\begin{pmatrix} P & Q & R \end{pmatrix},$$

yielding (5.4.10). Finally (5.4.11) follows from the minimal rank result Theorem 5.2.2 where we use that

$$\text{rank}M \geq \min \text{rank}\begin{pmatrix} P & Q & R \\ Q^* & ? & S \\ R^* & ? & ? \end{pmatrix}.$$

To prove that (5.4.8)–(5.4.11) are the only restrictions on the inertia of M we proceed similarly as in the proof of Theorem 5.1.6. First, decompose P as

$$P = \begin{pmatrix} P_1 & 0 \\ 0 & 0 \end{pmatrix} : \begin{matrix} \text{Ran}P \\ \oplus \\ \text{Ker}P \end{matrix} \to \begin{matrix} \text{Ran}P \\ \oplus \\ \text{Ker}P \end{matrix},$$

and write

$$Q = \begin{pmatrix} Q_1 \\ Q_2 \end{pmatrix}, \ R = \begin{pmatrix} R_1 \\ R_2 \end{pmatrix}$$

accordingly. Next, apply Lemma 5.1.8 to find invertible matrices A, B_1, B_2 such that

$$A\begin{pmatrix} Q_2 & R_2 \end{pmatrix}\begin{pmatrix} B_1 & 0 \\ 0 & B_2 \end{pmatrix} = \begin{pmatrix} I_a & 0 & 0 & 0 & 0 & 0 \\ 0 & I_b & 0 & 0 & I_b & 0 \\ 0 & 0 & 0 & 0 & 0 & 0 \\ 0 & 0 & 0 & 0 & 0 & I_c \end{pmatrix}.$$

Notice that $a+b+c = \operatorname{rank}\begin{pmatrix} Q_2 & R_2 \end{pmatrix} = \operatorname{rank}\begin{pmatrix} P & Q & R \end{pmatrix} - \operatorname{rank} P$, and that the zero block row has $\gamma := n_1 - \operatorname{rank}\begin{pmatrix} P & Q & R \end{pmatrix}$ scalar rows. Now, by using a congruence one can see that it suffices to determine the possible inertia of

$$\widetilde{M} = \begin{pmatrix} P_1 & 0 & 0 & 0 & 0 & 0 & 0 & 0 & 0 & 0 & 0 \\ 0 & 0 & 0 & 0 & 0 & I_a & 0 & 0 & 0 & 0 & 0 \\ 0 & 0 & 0 & 0 & 0 & 0 & I_b & 0 & 0 & I_b & 0 \\ 0 & 0 & 0 & 0 & 0 & 0 & 0 & 0 & 0 & 0 & 0 \\ 0 & 0 & 0 & 0 & 0 & 0 & 0 & 0 & 0 & 0 & I_c \\ 0 & I_a & 0 & 0 & 0 & X_{11} & X_{12} & X_{13} & S_{11} & S_{12} & S_{13} \\ 0 & 0 & I_b & 0 & 0 & X_{21} & X_{22} & X_{23} & S_{21} & S_{22} & S_{23} \\ 0 & 0 & 0 & 0 & 0 & X_{31} & X_{32} & X_{33} & S_{31} & S_{32} & S_{33} \\ 0 & 0 & 0 & 0 & 0 & S_{11}^* & S_{21}^* & S_{31}^* & Y_{11} & Y_{12} & Y_{13} \\ 0 & 0 & I_b & 0 & 0 & S_{12}^* & S_{22}^* & S_{32}^* & Y_{21} & Y_{22} & Y_{23} \\ 0 & 0 & 0 & 0 & I_c & S_{13}^* & S_{23}^* & S_{33}^* & Y_{31} & Y_{32} & Y_{33} \end{pmatrix},$$

where $X_{ij}, Y_{ij}, 1 \geq i,j \geq 3$, are the unknowns. Next, again by congruence, one can make all of S equal to zero, except for S_{31}; indeed, use I_a to annihilate S_{11}, S_{12}, S_{13}; I_b to annihilate S_{21}, S_{22}, S_{32}; and I_c to annihilate S_{23}, S_{33}. Next, we can write (after an equivalence)

$$S_{31} = \begin{pmatrix} I_d & 0 \\ 0 & 0 \end{pmatrix},$$

where

$$d = \operatorname{rank}\begin{pmatrix} P & R \\ Q^* & S \end{pmatrix} - \operatorname{rank}\begin{pmatrix} P & R \end{pmatrix} - \operatorname{rank}\begin{pmatrix} P \\ Q^* \end{pmatrix} + \operatorname{rank} P.$$

This results in the following matrix $\widetilde{\widetilde{M}}$, which has the same inertia as $\widetilde{M}$ (and M). For notational convenience we denote the unknowns again as X_{ij} and Y_{ij}, although they are different from before. Thus we get that $\widetilde{\widetilde{M}}$ equals

$$\begin{pmatrix} P_1 & 0 & 0 & 0 & 0 & 0 & 0 & 0 & 0 & 0 & 0 & 0 & 0 \\ 0 & 0 & 0 & 0 & 0 & I_a & 0 & 0 & 0 & 0 & 0 & 0 & 0 \\ 0 & 0 & 0 & 0 & 0 & 0 & I_b & 0 & 0 & 0 & 0 & I_b & 0 \\ 0 & 0 & 0 & 0 & 0 & 0 & 0 & 0 & 0 & 0 & 0 & 0 & 0 \\ 0 & 0 & 0 & 0 & 0 & 0 & 0 & 0 & 0 & 0 & 0 & 0 & I_c \\ 0 & I_a & 0 & 0 & 0 & X_{11} & X_{12} & X_{13} & X_{14} & 0 & 0 & 0 & 0 \\ 0 & 0 & I_b & 0 & 0 & X_{21} & X_{22} & X_{23} & X_{24} & 0 & 0 & 0 & 0 \\ 0 & 0 & 0 & 0 & 0 & X_{31} & X_{32} & X_{33} & X_{34} & I_d & 0 & 0 & 0 \\ 0 & 0 & 0 & 0 & 0 & X_{41} & X_{42} & X_{43} & X_{44} & 0 & 0 & 0 & 0 \\ 0 & 0 & 0 & 0 & 0 & 0 & 0 & I_d & 0 & Y_{11} & Y_{12} & Y_{13} & Y_{14} \\ 0 & 0 & 0 & 0 & 0 & 0 & 0 & 0 & 0 & Y_{21} & Y_{22} & Y_{23} & Y_{24} \\ 0 & 0 & I_b & 0 & 0 & 0 & 0 & 0 & 0 & Y_{31} & Y_{32} & Y_{33} & Y_{34} \\ 0 & 0 & & 0 & I_c & 0 & 0 & 0 & 0 & Y_{41} & Y_{42} & Y_{43} & Y_{44} \end{pmatrix}.$$

Choosing the off-diagonal blocks of X and Y equal to zero, and computing the Schur complement supported in rows and columns $1, 4, 8, 9, 10, 11, 12$, we obtain

$$N := P_1 \oplus 0_\gamma \oplus \begin{pmatrix} X_{33} & I_d \\ I_d & Y_{11} \end{pmatrix} \oplus X_{44} \oplus Y_{22} \oplus (Y_{33} - X_{22}). \tag{5.4.12}$$

Clearly,

$$\begin{aligned} i_+(N) \geq i_+(P_1) &= i_+(P), \\ i_-(N) \geq i_-(P_1) &= i_-(P), \\ i_0(N) \geq \gamma = n_1 - \operatorname{rank}\begin{pmatrix} P & Q & R \end{pmatrix}, \\ i_+(N) + i_-(N) &= \operatorname{rank} N \geq d, \end{aligned}$$

are the only restrictions on the inertia of N even when the unknowns are restricted to be real. As, by the Schur complement rule,

$$i(M) = i(\widetilde{\widetilde{M}}) = (a + b + c, a + b + c, 0) + i(N),$$

we obtain that (5.4.8)–(5.4.11) are the only restrictions on $i(M)$. □

We can now apply this result to address the possible inertia of the Hermitian expression $A - BXB^* - CYC^*$.

Corollary 5.4.5 *Let $A = A^* \in \mathbb{C}^{n\times n}, B \in \mathbb{C}^{n\times m}$, and $C \in \mathbb{C}^{n\times p}$ be given. For $X = X^* \in \mathbb{C}^{m\times m}$ and $Y = Y^* \in \mathbb{C}^{p\times p}$, consider the expression*

$$N(X, Y) = A - BXB^* - CYC^*.$$

The possible inertia $i(N(X, Y)) = (\pi, \nu, \delta)$ for $N(X, Y)$ is given by the inequalities

$$\pi \geq \operatorname{rank}\begin{pmatrix} A & B & C \end{pmatrix} - i_-\begin{pmatrix} A & B & C \\ B^* & 0 & 0 \\ C^* & 0 & 0 \end{pmatrix}, \tag{5.4.13}$$

$$\nu \geq \operatorname{rank}\begin{pmatrix} A & B & C \end{pmatrix} - i_+\begin{pmatrix} A & B & C \\ B^* & 0 & 0 \\ C^* & 0 & 0 \end{pmatrix}, \tag{5.4.14}$$

$$\delta \geq n - \operatorname{rank}\begin{pmatrix} A & B & C \end{pmatrix}, \tag{5.4.15}$$

$$\begin{aligned} \pi + \nu \geq{} & 2 \operatorname{rank}\begin{pmatrix} A & B & C \end{pmatrix} + \operatorname{rank}\begin{pmatrix} A & C \\ B^* & 0 \end{pmatrix} \\ & - \operatorname{rank}\begin{pmatrix} A & B & C \\ B^* & 0 & 0 \end{pmatrix} - \operatorname{rank}\begin{pmatrix} A & B & C \\ C^* & 0 & 0 \end{pmatrix}. \end{aligned} \tag{5.4.16}$$

In particular, there exist Hermitian X and Y such that $N(X, Y) = 0$ if and only if

$$\begin{aligned} & \operatorname{rank}\begin{pmatrix} A & B & C \\ B^* & 0 & 0 \end{pmatrix} + \operatorname{rank}\begin{pmatrix} A & B & C \\ C^* & 0 & 0 \end{pmatrix} \\ & \geq 2\operatorname{rank}\begin{pmatrix} A & B & C \end{pmatrix} + \operatorname{rank}\begin{pmatrix} A & C \\ B^* & 0 \end{pmatrix}, \end{aligned} \tag{5.4.17}$$

and

$$\min\left\{ i_+ \begin{pmatrix} A & B & C \\ B^* & 0 & 0 \\ C^* & 0 & 0 \end{pmatrix}, i_- \begin{pmatrix} A & B & C \\ B^* & 0 & 0 \\ C^* & 0 & 0 \end{pmatrix} \right\} \tag{5.4.18}$$

$$\geq \operatorname{rank} \begin{pmatrix} A & B & C \end{pmatrix}.$$

When A, B, and C are real matrices, X and Y may be chosen to be real as well to obtain the desired inertia.

Proof. The expression $N = N(X, Y)$ is equal to the Schur complement of

$$M = \begin{pmatrix} A & B & C & 0 & 0 \\ B^* & 0 & 0 & I_m & 0 \\ C^* & 0 & 0 & 0 & I_p \\ 0 & I_m & 0 & X & 0 \\ 0 & 0 & I_p & 0 & Y \end{pmatrix}$$

supported in the first row and column. Thus $i(M) = i(N) + (m+p, m+p, 0)$. The result now follows directly by applying Theorem 5.4.4. $\square$

5.5 BOUNDS FOR EIGENVALUES OF HERMITIAN COMPLETIONS

In the previous section we looked at the inertia of Hermitian completions. In this section we will look at single eigenvalues. For instance, we will answer what the least upper bound and the greatest lower bound of the possible values of the pth largest eigenvalue of a Hermitian completion are. If C is a $n \times n$ Hermitian matrix, we let

$$\lambda_1(C) \geq \cdots \geq \lambda_n(C)$$

denote the eigenvalues of C in nonincreasing order, and let

$$\mu_1(C) \leq \cdots \leq \mu_n(C)$$

denote the eigenvalues of C in nondecreasing order. This notation seems somewhat superfluous at first sight but simplifies the statement of the results in this section.

In this section we will not require that diagonal entries are specified as we did in Chapters 1 and 2. Therefore we will use a slightly more general notion of chordality than before. We say that a graph $G = (V, E)$ is called *chordal* if every circuit $[v_1, \ldots, v_p]$ of length $p \geq 4$ such that $\{v_i, v_i\} \in E$ for $i = 1, \ldots, p$ has a *chord*, that is, there exists an edge between v_i and v_j for some $i < j - 1$ and $\{i, j\} \neq \{1, p\}$. This definition of chordal graphs is different from that in Chapters 1 and 2, as we require chordality only in the subgraph formed by the subset of vertices v such that $\{v, v\}$ is an edge. So, for instance, the partial matrix

$$\begin{pmatrix} * & * & ? & ? & * & * \\ * & * & * & * & ? & * \\ ? & * & * & * & * & ? \\ ? & * & * & ? & * & * \\ * & ? & * & * & ? & * \\ * & * & ? & * & * & ? \end{pmatrix}$$

has a graph that is chordal (in the sense of this section). Indeed, we only need to look at the principal submatrix where all the diagonal entries are specified. In this case this is the principal submatrix corresponding to rows and columns 1, 2, and 3. There the pattern is banded, and thus chordal.

Theorem 5.5.1 *Let A be an $n \times n$ partial Hermitian matrix, and $G = (V, E)$ denote its associated graph. Suppose that G is chordal. Then for $p = 1, \ldots, n$,*

$$\inf\{\lambda_p(B) : B = B^* \text{ is a completion of } A\} \tag{5.5.1}$$

$$= \max\{\lambda_p(A|K) : K \text{ is a maximal clique with } \text{card}K \geq p\},$$

and

$$\sup\{\mu_p(B) : B = B^* \text{ is a completion of } A\} \tag{5.5.2}$$

$$= \min\{\mu_p(A|K) : K \text{ is a maximal clique with } \text{card}K \geq p\}.$$

Here we use the convention that the minimum (resp. maximum) over the empty set is ∞ (resp. $-\infty$).

Observe that Theorem 5.5.1 is valid (provided the associated graph is chordal) also for block partial Hermitian matrices $A = [A_{ij}]_{i,j=1}^{p}$, that is, such that each entry A_{ij} is either a fully specified $n_i \times n_j$ matrix, or a free variable $n_i \times n_j$ matrix. The Hermitian property here means that $A_{ij} = A_{ji}^*$ whenever the entry A_{ij} is fully specified. The block case can be easily reduced to the situation of Theorem 5.5.1, by the following simple observation. The associated graph of a block partial Hermitian matrix $A = [A_{ij}]_{i,j=1}^{n}$ is chordal if and only if the associated graph of $\widehat{A}$ is chordal, where $\widehat{A}$ is the $(n_1 + \cdots + n_p) \times (n_1 + \cdots + n_p)$ partial matrix obtained from A by regarding the entries in each block A_{ij} as entries in A itself. This fact is easy to check since one replaces vertices by cliques and connects vertices in two different cliques if and only if in the original graph there is an edge between the corresponding vertices.

Before proving Theorem 5.5.1 it will be convenient to have the following lemmas. The following is due to Ostrowski.

Lemma 5.5.2 *Let $A = A^*, S \in \mathbb{C}^{n \times n}$. Assume that S is invertible. Then*

$$\lambda_p(S^*AS) \leq \|S\|^2 \lambda_p(A).$$

Proof. Using Sylvester's inertia law the pth eigenvalue of $S^*(A - \lambda_p(A)I)S$ is zero. Using Weyl's inequality we now get that

$$0 = \lambda_p(S^*AS - \lambda_p(A)S^*S) \geq \lambda_p(S^*AS) + \lambda_n(-\lambda_p(A)S^*S)$$
$$\geq \lambda_p(S^*AS) - \lambda_p(A)\|S\|^2.$$

□

Lemma 5.5.3 *Let $A = A^*, C = C^*$, and B be matrices of sizes $r \times r, q \times q$, and $r \times q$, respectively. Then for $p \leq q$, we have*

$$\lim_{\beta \to -\infty} \lambda_p \begin{pmatrix} A + \beta I & B \\ B^* & C \end{pmatrix} = \lambda_p(C).$$

Proof. Let $\beta < -\lambda_1(A)$, and consider the equality

$$\begin{pmatrix} A + \beta I & B \\ B^* & C \end{pmatrix} = \begin{pmatrix} I & 0 \\ B^*(A + \beta I)^{-1} & I \end{pmatrix} \begin{pmatrix} A + \beta I & 0 \\ 0 & C - B^*(A + \beta I)^{-1}B \end{pmatrix}$$
$$\times \begin{pmatrix} I & (A + \beta I)^{-1}B \\ 0 & I \end{pmatrix}. \tag{5.5.3}$$

Since $-(A + \beta I)$ is positive definite, for negative β with $|\beta|$ sufficiently large

$$\lambda_p \begin{pmatrix} A + \beta I & 0 \\ 0 & C - B^*(A + \beta I)^{-1}B \end{pmatrix} = \lambda_p(C - B^*(A + \beta I)^{-1}B)$$
$$\leq \lambda_p(C) + \|B^*(A + \beta I)^{-1}B\|,$$

where in the last step we use Weyl's inequality. On the other hand, applying Lemma 5.5.2 to equality (5.5.3) we obtain

$$\lambda_p \begin{pmatrix} A + \beta I & B \\ B^* & C \end{pmatrix}$$
$$\leq \left\| \begin{pmatrix} I & (A + \beta I)^{-1}B \\ 0 & I \end{pmatrix} \right\|^2 \lambda_p \begin{pmatrix} A + \beta I & 0 \\ 0 & C - B^*(A + \beta I)^{-1}B \end{pmatrix}.$$

Combining this with the preceding inequality, it follows that

$$\lim_{\beta \to -\infty} \sup \lambda_p \begin{pmatrix} A + \beta I & B \\ B^* & C \end{pmatrix} \leq \lambda_p(C).$$

The opposite inequality

$$\lim_{\beta \to -\infty} \inf \lambda_p \begin{pmatrix} A + \beta I & B \\ B^* & C \end{pmatrix} \geq \lambda_p(C)$$

is easily seen by the interlacing inequalities for eigenvalues of Hermitian matrices (recall that these inequalities state that for a $k \times k$ principal submatrix X of an $n \times n$ Hermitian matrix Y the inequalities

$$\lambda_{p+n-k}(Y) \leq \lambda_p(X) \leq \lambda_p(Y), \quad p = 1, \ldots, k$$

hold). □

Proof of Theorem 5.5.1. The proof of (5.5.2) is easily reduced to (5.5.1), by replacing A with $-A$. So we will prove (5.5.1) only. The inequality

$$\inf\{\lambda_p(B) : B = B^* \text{ is a completion of } A\} \geq \max\{\lambda_p(A|K) \ : \ K \text{ is a maximal clique with } \mathrm{card}K \geq p\}$$

is a simple consequence of the interlacing inequalities for the eigenvalues of B and the eigenvalues of its principal submatrix $A|K$. It remains to prove the opposite inequality. Let V' be the subset of V consisting of all $v \in V$ with $(v, v) \in E$. If $p > \mathrm{card}V'$, there are no cliques of size p and the inf should be $-\infty$. Furthermore, if we specify the diagonal entries $(v, v), v \in V \setminus V'$, with β, and let β tend to $-\infty$ we conclude that the pth eigenvalue of the corresponding completions tends to $-\infty$ as well. Thus (5.5.1) holds in the case $p > \mathrm{card}V$. Let us now assume that $p \leq \mathrm{card}V'$ (in particular, $V' \neq \emptyset$). We claim

$$\inf\{\lambda_p(B) : B = B^* \text{ is a completion of } A\} \tag{5.5.4}$$
$$= \inf\{\lambda_p(B') : B' = B'^* \text{ is a completion of } A|V'\}.$$

Indeed, by the interlacing inequalities $\geq$ holds in (5.5.4). For the inequality $\leq$, let $B = B^*$ is a completion of A be such that the (v, v) entry of B is equal to $\beta, v \in V \setminus V'$. Letting $\beta \to -\infty$ and using Lemma 5.5.3, the inequality $\leq$ in (5.5.4) follows.

We have reduced the proof of (5.5.1) to the proof of the following inequalities:

$$\inf\{\lambda_p(B') : B' = B'^* \text{ is a completion of } A|V'\} \tag{5.5.5}$$
$$\leq \max\{\lambda_p(A|K) : K \text{ is a maximal clique, } \mathrm{card}K \geq p\},$$

$p = 1, \ldots, \mathrm{card}V'$. Pick $\epsilon > 0$. First assume that the right-hand side of (5.5.5) equals $N < \infty$ (p is fixed). Let $A = A - (N + \epsilon)I$. Then for any clique $K \subseteq V'$ the matrix $A|K$ has fewer than p nonnegative eigenvalues. By Theorem 5.3.1 there is a Hermitian completion $\widehat{B}|V'$ of $\widehat{A}|V'$ such that $\widehat{B}|V'$ has fewer than p nonnegative eigenvalues. Thus $\lambda_p(\widehat{B}|V') < 0$ and therefore

$$\lambda_p(\widehat{B}|V') + (N + \epsilon)I) < N + \epsilon.$$

As $\widehat{B}|V' + (N + \epsilon)I$ is a completion of $A|V'$, we have

$$\inf\{\lambda_p(B') : B' = B'^* \text{ is a completion of } A|V'\} < N + \epsilon,$$

and (5.5.5) is proved in the case the right-hand side is finite. Next, suppose that K has no maximal cliques with $\mathrm{card}K \geq p$ in V', that is, the right-hand side of (5.5.5) equals $-\infty$. Let $N \in \mathbb{R}$ and $\widehat{A} = A - NI$. Note now that the maximal number of nonnegative eigenvalues of $\widehat{A}|K$, where K is a clique in V', is $< p$. Therefore by Theorem 5.3.1 there is a Hermitian completion $\widehat{B}|V'$ of $\widehat{A}|V'$ such that $\widehat{B}|V'$ has fewer than p nonnegative eigenvalues. Thus

$$\lambda_p(\widehat{B}|V' + NI) < N.$$

As $\widehat{B}|V' + NI$ is a completion of $A|V$ we get

$$\inf\{\lambda_p(B') : B' = B'^* \text{ is a completion of } A|V'\} < N.$$

As $N \in \mathbb{R}$ is arbitrary, we obtain

$$\inf\{\lambda_p(B') : B' = B'^* \text{ is a completion of } A|V'\} = -\infty,$$

proving (5.5.5) in the case that the right-hand side equals $-\infty$. This finishes the proof. □

Easy examples show that "inf" and "sup" in Theorem 5.5.1 are not attainable in general, that is, cannot be replaced by "min" and "max," respectively. For example, consider

$$A = \begin{pmatrix} ? & 1 \\ 1 & 0 \end{pmatrix}.$$

Then

$$\inf\left\{\lambda_1\begin{pmatrix} x & 1 \\ 1 & 0 \end{pmatrix} : x \text{ is real}\right\} = 0,$$

as expected by (5.5.1), but this infimum is not attainable. We point out some cases when "inf" and "sup" are attainable. The positive semidefinite completion problem in Chapter 1 gives such a result.

Proposition 5.5.4 *Let A and G be as in Theorem 5.5.1, and assume that the diagonal in A is prescribed. Then*

$$\inf\{\lambda_1(B) : B = B^* \textit{ is a completion of } A\}$$

and

$$\sup\{\lambda_n(B) : B = B^* \textit{ is a completion of } A\}$$

are attainable.

We remark that the associated graph being chordal is in a sense a necessary condition for this result. Indeed, as it follows from Theorem 1.2.10, for any nonchordal (undirected) graph G on the vertices $\{1, \ldots, n\}$ there exists a partial Hermitian matrix A whose associated graph is G and such that

$$\max\{\lambda_1(A|K) : K \subseteq V \text{ is a clique}\} < 0$$

but

$$\inf\{\lambda_1(B) : B = B^* \text{ is a completion of } A\} \geq 0.$$

Analogously, there exists a partial Hermitian matrix A with the associated graph G such that

$$\min\{\mu_1(A|K) : K \subseteq V \text{ is clique}\} > 0$$

but

$$\sup\{\lambda_n(B) : B = B^* \text{ is a completion of } A\} \leq 0.$$

The following example shows that a strict inequality in (5.5.1) can hold also for other eigenvalues (not only the first and last eigenvalues), if the graph is not chordal.

Example 5.5.5 Let

$$A = \begin{pmatrix} 1 & 1 & ? & -1 \\ 1 & 1 & 1 & ? \\ ? & 1 & 1 & 1 \\ -1 & ? & 1 & 1 \end{pmatrix}.$$

The associated graph here is a four cycle which is nonchordal. We obviously have

$$\max\{\lambda_2(A|K) : K \subseteq V \text{ clique with } \operatorname{card} K \geq 2\} = 0.$$

On the other hand, for every Hermitian completion B of A the inequality

$$\lambda_2(B) \geq \max\left\{\lambda_2\begin{pmatrix} 1 & 1 & x \\ 1 & 1 & 1 \\ \overline{x} & 1 & 1 \end{pmatrix}, \lambda_2\begin{pmatrix} 1 & x & -1 \\ \overline{x} & 1 & 1 \\ -1 & 1 & 1 \end{pmatrix} : x \in \mathbb{C}\right\}, \tag{5.5.6}$$

holds. Now

$$\det\begin{pmatrix} 0 & 1 & x \\ 1 & 0 & 1 \\ \overline{x} & 1 & 0 \end{pmatrix} = 2\mathrm{Re}\, x, \ \det\begin{pmatrix} 0 & x & -1 \\ \overline{x} & 0 & 1 \\ -1 & 1 & 0 \end{pmatrix} = -2\mathrm{Re}\, x,$$

and therefore the matrix

$$Q_1 = \begin{pmatrix} 0 & 1 & x \\ 1 & 0 & 1 \\ \overline{x} & 1 & 0 \end{pmatrix}$$

has 1 negative eigenvalue if $\mathrm{Re}\, x < 0$ and 2 negative eigenvalues if $\mathrm{Re}\, x > 0$; also,

$$Q_2 = \begin{pmatrix} 0 & x & -1 \\ \overline{x} & 0 & 1 \\ -1 & 1 & 0 \end{pmatrix}$$

has 2 negative eigenvalues if $\mathrm{Re}\, x < 0$ and 1 negative eigenvalue if $\mathrm{Re}\, x > 0$. It follows that

$$\max\{\lambda_2(Q_1), \lambda_2(Q_2)\} > 0$$

for $\mathrm{Re}\, x \neq 0$, and therefore

$$\max\{\lambda_2(Q_1), \lambda_2(Q_2)\} \geq 0.$$

Hence the right-hand side of (5.5.6) is not smaller than 1, and we have

$$\inf\{\lambda_2(B) : B = B^* \text{ is a completion of } A\} \geq 1.$$

Next, we look at the Toeplitz case.

Theorem 5.5.6 *Let $a_{-(p-1)}, \ldots, a_0, \ldots, a_{p-1}, p \leq n$, be given numbers in $\mathbb{C}$ such that $\overline{a}_{-j} = a_j$ let A be the corresponding $n \times n$ partial Hermitian Toeplitz matrix, and let $A_p = (a_{j-i})_{i,j=0}^{p-1}$ be the $p \times p$ maximal specified submatrix of A. For $k \leq p$ we have*

$$\inf\{\lambda_k(B) : B = B^* \text{ is a Toeplitz completion of } A\} = \lambda_k(A_p) \tag{5.5.7}$$

and

$$\sup\{\mu_k(B) : B = B^* \text{ is a Toeplitz completion of } A\} = \mu_k(A_p). \quad (5.5.8)$$

When $p < k \leq n$ *then*

$$\inf\{\lambda_k(B) : B = B^* \text{ is a Toeplitz completion of } A\} = -\infty \quad (5.5.9)$$

and

$$\sup\{\mu_k(B) : B = B^* \text{ is a Toeplitz completion of } A\} = \infty. \quad (5.5.10)$$

Proof. Let $k \leq p$. We prove (5.5.7). The equality (5.5.8) can be obtained from (5.5.7) by replacing A by $-A$. First note that $\geq$ in (5.5.7) follows from the interlacing inequalities. To prove the inequality $\leq$, let $N = \lambda_k(A_p)$, and $0 < \epsilon_1 < \epsilon$. Put $\widehat{A} = A-(N+\epsilon_1)I_n$, where $a_p = \overline{a}_{-p}, \dots, a_{n-1} = \overline{a}_{-(n-1)}$ are considered to be free variables in $\mathbb{C}$. Since $A_p - NI_p$ has fewer than k positive eigenvalues, Theorem 5.3.1 yields that for all ϵ_1 with the possible exception of a finite number of values (those that would make the prescribed principal matrix singular), we can find numbers $a_p = \overline{a}_{-p}, \dots, a_{n-1} = \overline{a}_{-(n-1)}$ such that with this choice $\widehat{A}$ has fewer than k positive eigenvalues. Thus $\lambda_k(\widehat{A}) \leq 0$, and therefore

$$\lambda_k(\widehat{A} + (N + \epsilon_1)I_n) \leq N + \epsilon_1.$$

As $\widehat{A} + (N+\epsilon_1)I_n$ is a Hermitian Toeplitz completion of A, the inequality $\leq$ in (5.5.7) follows. This proves (5.5.7). Let $p < k \leq n$. To show that (5.5.9) and (5.5.10) holds, one argues in the same way as in the last paragraph of the proof of Theorem 5.5.1. □

5.6 BOUNDS FOR SINGULAR VALUES OF COMPLETIONS OF PARTIAL TRIANGULAR MATRICES

For a given $n \times m$ matrix X we denote by

$$s_1(X) \geq s_2(X) \geq \cdots$$

the singular values of X ordered in nonincreasing order, where by convention $s_k(X) = 0$ if $k > \min(n, m)$.

Theorem 5.6.1 *Let*

$$A = \begin{pmatrix} A_{11} & & ? \\ \vdots & \ddots & \\ A_{n1} & \cdots & A_{nn} \end{pmatrix} \quad (5.6.1)$$

be a partial block triangular matrix (with specified blocks A_{ij}, $i \geq j$*). Then for* $k = 1, 2, \dots$ *we have*

$$\inf\{s_k(B) \mid B \text{ is a completion of } A\} = \quad (5.6.2)$$

$$\max_{i=1,\dots,n} \left\{ s_k \begin{pmatrix} A_{i1} & \cdots & A_{ii} \\ \vdots & & \vdots \\ A_{n1} & \cdots & A_{ni} \end{pmatrix} \right\}.$$

For $k = 1$ the statement was covered in Theorem 4.1.1.

Proof. As $s_k(X) \geq s_k(Y)$ for any submatrix Y of a matrix X, the inequality $\geq$ in (5.6.2) is evident. For the proof of the opposite inequality, consider the partial Hermitian matrix

$$\widehat{A} = \begin{pmatrix} 0 & A \\ A^* & 0 \end{pmatrix}.$$

Clearly, $\widehat{A}$ is a partial block banded matrix and thus the associated graph of $\widehat{A}$ is chordal. We may therefore apply Theorem 5.5.1, and conclude that for every $\epsilon > 0$ and given r there exists a Hermitian completion

$$\widehat{B} = \begin{pmatrix} 0 & B \\ B^* & 0 \end{pmatrix}$$

of $\widehat{A}$ such that

$$\lambda_r(\widehat{B}) < \epsilon + \max\{\lambda_r(\widehat{A}|K) \ : \ K \text{ is a maximal clique with } \mathrm{card}K \geq r\}$$

(here we assume r is such that the graph G associated with $\widehat{A}$ indeed has a clique of cardinality r). Since for a $k \times k$ matrix X the k largest eigenvalues of $\left(\begin{smallmatrix} 0 & X \\ X^* & 0 \end{smallmatrix}\right)$ coincide with the k singular values of X, the expression above translates into $s_r(B) < \epsilon + \alpha$, where α is the right-hand side of (5.6.2). This proves the theorem. □

It is instructive to compare formula (5.6.2) with the minimal rank formula from Theorem 5.2.1:

$$\min\{\mathrm{rank}B \mid B \text{ is a completion of } A\}$$

$$= \sum_{i=1}^{n} \mathrm{rank} \begin{pmatrix} A_{i1} & \cdots & A_{ii} \\ \vdots & & \vdots \\ A_{n1} & \cdots & A_{ni} \end{pmatrix} - \sum_{i=1}^{n-1} \mathrm{rank} \begin{pmatrix} A_{i+1,1} & \cdots & A_{i+1,i} \\ \vdots & & \vdots \\ A_{n1} & \cdots & A_{ni} \end{pmatrix}.$$

Let

$$k_0 = \min\{\mathrm{rank}B : B \text{ is a completion of } A\}.$$

Clearly,

$$\inf\{s_k(B) : B \text{ is a completion of } A\} = 0$$

for every $k > k_0$. However, it can happen that

$$\inf\{s_k(B) \ : \ B \text{ is a completion of } A\} = 0$$

also for $k \leq k_0$ (of course, in these cases the infimum is not achievable). For instance, consider

$$A = \begin{pmatrix} 1 & ? & ? & \cdots & ? \\ 0 & 1 & ? & \cdots & ? \\ 0 & 0 & 1 & \cdots & ? \\ \vdots & \vdots & \vdots & & \vdots \\ 0 & 0 & 0 & \cdots & 1 \end{pmatrix}. \tag{5.6.3}$$

Every completion of A has norm at least 1 and rank n. However, by Theorem 5.6.1 we have

$$\inf\{s_2(B) : B \text{ is a completion of } A\} = 0. \tag{5.6.4}$$

In Exercise 5.12.36 we will outline how a completion B of A with $s_2(B) < \epsilon$ may be constructed. As this example shows, the infimum is (5.6.2) is generally not achievable. We note that it is achievable when $k = 1$ (for the simple reason that the set of matrices X with $s_1(X)$ not exceeding a given constant is a compact set) and when

$$k > \sum_{i=1}^{n} \operatorname{rank} \begin{pmatrix} A_{i1} & \cdots & A_{ii} \\ \vdots & & \vdots \\ A_{n1} & \cdots & A_{ni} \end{pmatrix} - \sum_{i=1}^{n-1} \operatorname{rank} \begin{pmatrix} A_{i+1,1} & \cdots & A_{i+1,i} \\ \vdots & & \vdots \\ A_{n1} & \cdots & A_{ni} \end{pmatrix}.$$

Indeed, if k satisfies this inequality, by Theorem 5.2.1 there exists a completion B of A of rank smaller than k; then $s_k(B) = 0$, and the left-hand side of (5.6.2) is obviously zero.

We now consider the description of all patterns of specified entries for which the equalities (5.6.2) hold. It was proved in Theorem 4.1.4 that for a given pattern P the equality

$$\inf\{s_1(B) : B \text{ completion of } A\} \tag{5.6.5}$$

$$= \max\{s_1(A_0) : A_0 \text{ fully specified rectangular submatrix of } A\}$$

holds for every partial matrix A with pattern P if and only if P has the following form, possibly bordered by rows and columns consisting of ?s, and up to permutations of rows and columns:

$$P = \begin{pmatrix} P_1 & ? & \cdots & ? \\ ? & P_2 & \cdots & ? \\ \vdots & \vdots & \ddots & \vdots \\ ? & ? & \cdots & P_r \end{pmatrix}, \tag{5.6.6}$$

where each P_i is a block triangular pattern. We get a result analogous to (5.6.5) for all singular values.

Theorem 5.6.2 *Let P be a pattern given by (5.6.6), possibly bordered by rows and columns consisting entirely of ?s and up to permutations of rows and columns. Then for all partial matrices A with pattern P, and for $k = 1, 2, \ldots$ the equality*

$$\inf\{s_k(B) : B \text{ is a completion of } A\} \tag{5.6.7}$$

$$= \max\{s_k(A_0) : A_0 \text{ a fully specified rectangular submatrix of } A\}$$

holds

For the proof we need a lemma.

Lemma 5.6.3 *Let $X_1, \ldots, X_r$ be matrices of sizes $l_1 \times m_1, \ldots, l_r \times m_r$, respectively, and let*

$$s = \max\{s_k(X_1), s_k(X_2), \ldots, s_k(X_r)\}$$

(the integer $k \geq 1$ is fixed). Then there exist matrices $Y_{ij}, i \neq j, 1 \leq i, j \leq r$, where the size of Y_{ij} is $l_i \times m_j$, such that

$$s_k \begin{pmatrix} X_1 & Y_{12} & Y_{13} & \cdots & Y_{1r} \\ Y_{21} & X_2 & Y_{23} & \cdots & Y_{2r} \\ Y_{31} & Y_{32} & X_3 & \cdots & Y_{3r} \\ \vdots & \vdots & \vdots & & \vdots \\ Y_{r1} & Y_{r2} & Y_{r2} & \cdots & X_r \end{pmatrix} = s.$$

Proof. We can assume $k > 1$ (if $k = 1$, take $Y_{ij} = 0$). Let Z_i be a matrix of rank $\leq k - 1$ such that

$$\|X_i - Z_i\| = s_k(X_i), \ i = 1, \ldots, r.$$

Write $Z_i = V_i W_i$, where the matrix V_i has $k-1$ columns and the matrix W_i has $k - 1$ rows, and put

$$Y_{ij} = V_i W_j, \ i \neq j.$$

Then

$$\left\| \begin{pmatrix} X_1 & Y_{12} & \cdots & Y_{1r} \\ Y_{21} & X_2 & \cdots & Y_{2r} \\ \vdots & \vdots & & \vdots \\ Y_{r1} & Y_{r2} & \cdots & X_r \end{pmatrix} - \begin{pmatrix} V_1 \\ V_2 \\ \vdots \\ V_r \end{pmatrix} \begin{pmatrix} W_1 & W_2 & \cdots & W_r \end{pmatrix} \right\|$$

$$= \left\| \begin{pmatrix} X_1 - Z_1 & 0 & \cdots & 0 \\ 0 & X_2 - Z_2 & \cdots & 0 \\ \vdots & \vdots & & \vdots \\ 0 & 0 & \cdots & X_r - Z_r \end{pmatrix} \right\| = s,$$

and therefore

$$s_k \begin{pmatrix} X_1 & Y_{12} & \cdots & Y_{1r} \\ Y_{21} & X_2 & \cdots & Y_{2r} \\ \vdots & \vdots & & \vdots \\ Y_{r1} & Y_{r2} & \cdots & X_r \end{pmatrix} \leq s.$$

Since the opposite inequality is obvious, we are done. □

Proof of Theorem 5.6.2. Observe that the hypotheses and the conclusion of the theorem are invariant under permutations of rows and columns in P, so we can assume P is given by (5.6.6). We can also assume that there are no rows and columns consisting entirely of ?s (indeed, such rows and columns are filled with zeros when constructing the desired completions). Let A be a partial matrix with pattern P, and fix k. Note that every fully specified rectangular submatrix A_0 of A must satisfy

$$\inf\{s_k(B) : B \text{ completion of } A\} \geq s_k(A_0).$$

Next, note that

$$A = \begin{pmatrix} A_1 & ? & \cdots & ? \\ ? & A_2 & \cdots & ? \\ \vdots & \vdots & \ddots & \vdots \\ ? & ? & \cdots & A_r \end{pmatrix},$$

where A_i has pattern P_i, $i = 1, \ldots, r$. Note that every fully specified rectangular submatrix A_0 of A must be a submatrix of one of $A_1, A_2, \ldots, A_r$. For $\epsilon > 0$, use Theorem 5.6.1 to find a completion B_i of A_i such that $s_k(B_i) < \epsilon + \max\{s_k(A_0) \mid A_0$ fully specified rectangular submatrix of $A\}$. By Lemma 5.6.3 (with $X_i = B_i, i = 1, \ldots, r$) find a completion B of A such that

$$s_k(B) = \max\{s_k(B_1), \ldots, s_k(B_r)\}.$$

It remains to compare with the previous inequality and use the arbitrariness of ϵ. □

We end this section with the following result regarding the approximation of a Hankel operator with a low rank Hankel operator. For an operator T we denote

$$s_m(T) = \inf\{\|T - R\| \ : \ \text{rank} R < m\}, \ m \in \mathbb{N}.$$

Theorem 5.6.4 *Let $\Gamma : H^2 \to H^2_-$ be a compact Hankel operator, and let $m \in \mathbb{N}$. Then there exists a Hankel operator Γ_m of rank less than m such that $\|\Gamma - \Gamma_m\| = s_m(\Gamma)$.*

Proof. Let m be such that $s_1(\Gamma) = \cdots = s_{m-1}(\Gamma) > s = s_m(\Gamma)$. It suffices to prove the result for this m, as one can proceed by induction. If $s = 0$ we can simply take $\Gamma_m = \Gamma$; thus assume that $s > 0$. Let $\mu \in \mathbb{N}$ be such that $s_m(\Gamma) = s_{m+1}(\Gamma) = \cdots = s_{m+\mu-1}(\Gamma) > s_{m+\mu}(\Gamma)$. Let now $0 \neq v \in H^2$, $w \in H^2_-$ be such that $\Gamma v = sw$ and $\Gamma^* w = sv$. As in the proof of Corollary 4.3.9 one can show that $\frac{w}{v}$ is independent of the choice of v, w above, and that $|\frac{w}{v}| \equiv 1$. In other words, for all nonzero x in $E = \{x \in H^2 \ : \ \Gamma^*\Gamma x = s^2 x\}$ we have that $\frac{1}{s}\frac{\Gamma x}{x} = \frac{w}{v}$. Notice that $\dim E = \mu$. We now define $\Gamma_m = -H_{s\frac{w}{v}} + \Gamma$, where H_ϕ stands for the Hankel operator with symbol ϕ (see Section 4.3). Then, by definition, Γ_m is Hankel and $\|\Gamma - \Gamma_m\| = s$. It remains to show that $\text{rank}\Gamma_m \leq m - 1$. Notice that $H_{s\frac{w}{v}} x = \Gamma x$ for all $x \in E$. By the Hankel structure it follows that $\Gamma_m z^n x = 0$ for all $n \in \mathbb{N}_0$ and $x \in E$. Consider now the shift invariant space spanned by $\{z^n x \ : \ n \in \mathbb{N}_0, x \in E\}$, which by the Beurling-Lax-Halmos theorem (Theorem 4.11.1) is of the form θH^2 for some inner θ. Thus we have that θH^2 lies in the kernel of Γ_m. Thus $\text{rank}\Gamma_m \leq \dim(H_2 \ominus \theta H_2)$. It remains to show that $d := \dim(H_2 \ominus \theta H_2) < m$. Suppose instead that $d \geq m$. Then there exist inner $\theta_1, \ldots, \theta_m = \theta$ such that $\theta_{j+1}\theta_j^{-1} \in H^\infty$, $\theta_{j+1}\theta_j^{-1} \neq$ const, and $\theta_1 \neq$ const (e.g., if $\theta = \prod_{k=1}^d \frac{z - z_k}{1 - z\overline{z_k}}$ and $m = d$, then put $\theta_j = \prod_{k=1}^j \frac{z - z_k}{1 - z\overline{z_k}}$).

Let Θ be the multiplication operator with symbol θ. Then it is not hard to see that for every $x \in \text{Span}\{E, \theta_1^{-1}E, \dots, \theta_m^{-1}E\}$ satisfies $(\Gamma\Theta)^*\Gamma\Theta x = s^2 x$. As this span has at least dimension $\mu + m$, we have that $\Gamma\Theta$ has at least $\mu + m$ singular values equal to s. But as, $s_{m+\mu}(\Gamma\Theta) \leq s_{m+\mu}(\Gamma) < s$, we have a contradiction. Thus we obtain that $d < m$, finishing the proof. □

Note that for a Hankel matrix the standard rank k approximation using singular vectors does not need to give a Hankel matrix. For instance, for

$$\Gamma = \begin{pmatrix} 0 & 1 & 0 \\ 1 & 0 & 0 \\ 0 & 0 & 0 \end{pmatrix},$$

a rank 1 approximation is

$$\begin{pmatrix} 0 & 0 & 0 \\ 1 & 0 & 0 \\ 0 & 0 & 0 \end{pmatrix},$$

which is not Hankel. This does not contradict Theorem 5.6.4 as by taking $\Gamma_2 = 0$ we do get a Hankel matrix satisfying $\text{rank}\Gamma_2 < 2$ and $\|\Gamma - \Gamma_2\| = s_2(\Gamma)(= 1)$.

5.7 MOMENT PROBLEMS FOR REAL MEASURES ON THE UNIT CIRCLE

In this section we consider an indefinite version of the trigonometric moment problem (see Theorem 1.3.6 for the classical (definite) version). Given are $c_j \in \mathbb{C}$, $j = -n, \dots, n$, with $c_j = \overline{c}_{-j}$, $j = 0, \dots, n$. We are looking for a real measure $\mu = \mu_+ - \mu_-$ on the unit circle $\mathbb{T}$, where $\mu_\pm$ are positive measures on $\mathbb{T}$, such that the jth moment of μ coincides with c_j, that is,

$$c_j = \int_{\mathbb{T}} z^j d\mu(z), \; j = -n, \dots, n. \tag{5.7.1}$$

In addition, we are interested in measures μ where $\mu_\pm$ are supported in the least number of points. We therefore introduce the following terminology. We call $\mu = \mu_+ - \mu_-$ an *(m_+, m_-) extension of $(c_j)_{j=-n}^n$* if (5.7.1) holds and $\mu_\pm$ are positive measures supported on exactly $m_\pm$ points on the unit circle. From Theorem 1.3.6 we get the following reformulation of the problem.

Proposition 5.7.1 *Let $c_j \in \mathbb{C}$, $j = -n, \dots, n$, with $c_j = \overline{c}_{-j}$, $j = 0, \dots, n$, be given. Let T_n be the Toeplitz matrix $T_n = (c_{i-j})_{i,j=1}^n$. Then the following are equivalent.*

(i) There exists an (m_+, m_-) extension of $(c_j)_{j=-n}^n$.

(ii) There exist positive semidefinite Toeplitz matrices $T_+, T_- \in \mathcal{H}_{n+1}$ with $\text{rank}T_\pm = m_\pm$ such that $T_n = T_+ - T_-$.

(iii) *The Toeplitz matrix T_n can be factored as $T_n = RDR^*$, where R is a Vandermonde matrix,*

$$R = \begin{pmatrix} 1 & 1 & \dots & 1 \\ \alpha_1 & \alpha_2 & \dots & \alpha_r \\ \vdots & \vdots & & \vdots \\ \alpha_1^n & \alpha_2^n & \dots & \alpha_r^n \end{pmatrix},$$

with $\alpha_j \in \mathbb{T}$ for $j = 1, \dots, r := m_+ + m_-$, and $\alpha_j \neq \alpha_p$ for $j \neq p$, and D is a $r \times r$ diagonal matrix with m_+ positive entries and m_- negative entries.

(iv) *There exist $\alpha_j \in \mathbb{T}$, $j = 1, \dots, m_+ + m_-$, with $\alpha_j \neq \alpha_p$ for $j \neq p$, and $\rho_j > 0$, $j = 1, \dots, m_+ + m_-$, such that*

$$c_l = \sum_{j=1}^{m_+} \rho_j \alpha_j^l - \sum_{s=m_++1}^{m_++m_-} \rho_s \alpha_s^l, \quad |l| \leq n. \tag{5.7.2}$$

Proof. Assume (i) holds. Then there exist $\alpha_j \in \mathbb{T}$ and $\rho_j > 0$, $j = 1, \dots, m_+ + m_-$ such that

$$\mu = \mu_+ - \mu_-, \mu_+ = \sum_{j=1}^{m_+} \rho_j \delta_{\alpha_j}, \mu_- = \sum_{s=m_++1}^{m_++m_-} \rho_j \delta_{\alpha_j},$$

and (5.7.1) is satisfied. But then (iv) holds.

(iv) $\Rightarrow$ (iii): Letting D have diagonal entries

$$\rho_1, \dots, \rho_{m_+}, -\rho_{m_++1}, \dots, -\rho_{m_++m_-},$$

this implication is direct.

(iii) $\Rightarrow$ (ii): Write

$$R = \begin{pmatrix} R_1 & R_2 \end{pmatrix}, \ D = \begin{pmatrix} D_1 & 0 \\ 0 & D_2 \end{pmatrix},$$

with R_1 having m_+ columns, and D_1 of size $m_+ \times m_+$. Putting, $T_+ = R_1 D_1 R_1^*$ and $T_- = -R_2 D_2 R_2^*$, we get the desired $T_\pm$.

(ii) $\Rightarrow$ (i): By applying Theorem 1.3.6 to both T_+ and T_- we obtain positive Borel measures μ_+ and μ_- supported in m_+ and m_- points, respectively, with moments corresponding to the entries in the Toeplitz matrices $T_\pm$. But then $\mu = \mu_+ - \mu_-$ gives the desired (m_+, m_-) extension. □

From this result we can draw the following conclusion.

Corollary 5.7.2 *Let $c_j \in \mathbb{C}$, $j = -n, \dots, n$, with $c_j = \overline{c}_{-j}$, $j = 0, \dots, n$, be given. Let T_n be the Toeplitz matrix $T_n = (c_{i-j})_{i,j=1}^n$. If there exists an (m_+, m_-) extension of $(c_j)_{j=-n}^n$, then $i_+(T_n) \leq m_+$ and $i_-(T_n) \leq m_-$.*

Proof. Use Theorem 5.7.1 to write $T_n = T_+ - T_-$ with T_+, T_- positive semidefinite Toeplitz matrices with rank$T_\pm = m_\pm$. Let $j > m_+$. Using the

Courant-Fischer theorem we get that the jth largest eigenvalue λ_j of T_n satisfies

$$\lambda_j = \min_{w_1,w_2,\ldots,w_{n+1-j}\in\mathbb{C}^{n+1}} \max_{\substack{x\neq 0,\ x\in\mathbb{C}^{n+1}\\ x\perp w_1,w_2,\ldots,w_{n+1-j}}} \frac{x^*T_nx}{x^*x}$$

$$\leq \min_{w_1,w_2,\ldots,w_{n+1-j}\in\mathbb{C}^{n+1}} \max_{\substack{x\neq 0,\ x\in\mathbb{C}^{n+1}\\ x\perp w_1,w_2,\ldots,w_{n+1-j}}} \frac{x^*T_+x}{x^*x} = 0,$$

where in the last step we use that rank $T_+ = m_+$. This gives $i_+(T_n) \leq m_+$. The other inequality is derived in a similar way. □

The following example shows, however, that one can not always find a (m_+, m_-) extension such that $i_\pm(T_n) = m_\pm$.

Example 5.7.3 Consider the given data $c_0 = c_1 = c_{-1} = 1, c_2 = c_{-2} = 3$. Then the Toeplitz matrix

$$T_2 = (c_{i-j})_{i,j=0}^2 = \begin{pmatrix} 1 & 1 & 3 \\ 1 & 1 & 1 \\ 3 & 1 & 1 \end{pmatrix}$$

has inertia $i(T_2) = (2, 0, 1)$. We can write

$$T_2 = \frac{3}{2}\begin{pmatrix} 1 & 1 & 1 \\ 1 & 1 & 1 \\ 1 & 1 & 1 \end{pmatrix} - \frac{1}{2}\begin{pmatrix} 1 & i & -1 \\ -i & 1 & i \\ -1 & -i & 1 \end{pmatrix}$$

$$+\frac{1}{2}\begin{pmatrix} 1 & -1 & 1 \\ -1 & 1 & -1 \\ 1 & -1 & 1 \end{pmatrix} - \frac{1}{2}\begin{pmatrix} 1 & -i & -1 \\ i & 1 & -i \\ -1 & i & 1 \end{pmatrix},$$

yielding that

$$\mu = \frac{3}{2}\delta_1 - \frac{1}{2}\delta_{-i} + \frac{1}{2}\delta_{-1} - \frac{1}{2}\delta_i$$

is a $(2, 2)$ extension for $(c_j)_{j=-2}^2$. Using the following result we will see why there does not exist a $(2, 1)$ extension even though $i_-(T_2) = 1$.

Theorem 5.7.4 *A sequence $(c_j)_{j=-n}^n$, $c_{-j} = \bar{c}_j$ such that $1 \leq i_+(T_n) = m_+$, $1 \leq i_-(T_n) = m_-$, and $m_+ + m_- \leq n$ admits an (m_+, m_-) extension if and only if* rank T_n = rank T_{n-1} *and all roots of the polynomial*

$$P(z) = \det \begin{pmatrix} c_0 & \bar{c}_1 & \cdots & \bar{c}_k \\ c_1 & c_0 & \cdots & \bar{c}_{k-1} \\ \vdots & \vdots & \vdots & \vdots \\ c_{k-1} & c_{k-2} & \cdots & \bar{c}_1 \\ 1 & z & \cdots & z^k \end{pmatrix} \tag{5.7.3}$$

are different and lie on the unit circle, where k is the largest number $0 \leq k \leq n$ such that T_{k-1} is invertible. In this case $k = m_+ + m_-$.

Before we prove this result, let us continue the above example.

Example 5.7.3 continued. Suppose that $(c_j)_{j=-2}^2$ has a $(2,1)$ extension. Then there exists a $c_3 = \overline{c}_{-3}$ such that $T_3 = (c_{i-j})_{i,j=0}^3$ satisfies $i_+(T_3) = 2$, $i_-(T_3) = 1$. Thus $i_0(T_3)$ must equal 1, and thus $\det T_3 = 0$. This easily gives that $c_3 = \overline{c}_{-3} = 7 + ix$ for some $x \in \mathbb{R}$. Then, with $P(z)$ defined in Theorem 5.7.4 (which now depends on x), we get

$$P_x(z) = z^3 - \left(2 - \frac{ix}{2}\right) z^2 - \left(2 + \frac{ix}{2}\right) z + 1.$$

If its roots $z_1 = e^{i\alpha}, z_2 = e^{i\beta}, z_3 = e^{i\gamma}$ are on the unit circle, then we get that $e^{i\alpha}e^{i\beta}e^{i\gamma} = -1$, and thus $\gamma = \pi - (\alpha + \beta) (\operatorname{mod} 2\pi)$. Using now that $z_1 + z_2 + z_3 = 2 - \frac{ix}{2}$, we get that

$$\cos\alpha + \cos\beta + \cos\gamma = \cos\alpha + \cos\beta - \cos(\alpha + \beta) = 2,$$

which is easily seen to have no solutions.

Proof of Theorem 5.7.4. If $(c_j)_{j=-n}^n$ admits an (m_+, m_-) extension, then (5.7.2) holds. But then one easily checks that $\alpha_1, \ldots, \alpha_{m_+ + m_-}$ are the roots of $P(z)$, where $k = m_+ + m_-$. The "only if" part of the theorem now easily follows.

To prove the "if" part, let us write $P(z) = v_0 + \cdots + v_k z^k$, and suppose that its roots $\alpha_l, l = 1, \ldots, k$, are all different and lie on the unit circle. We first note that $\operatorname{rank} T_{k-1} = \operatorname{rank} T_k = \cdots = \operatorname{rank} T_n$ implies that

$$T_n \begin{pmatrix} v_0 \\ \vdots \\ v_k \\ 0_{n-k} \end{pmatrix} = 0 = T_n \begin{pmatrix} 0 \\ v_0 \\ \vdots \\ v_k \\ 0_{n-k-1} \end{pmatrix} = 0 = \cdots = T_n \begin{pmatrix} 0_{n-k} \\ v_0 \\ \vdots \\ v_k \end{pmatrix} = 0. \tag{5.7.4}$$

Indeed, $v = (v_i)_{i=0}^k$ spans $\operatorname{Ker} T_k$ (as this kernel is one-dimensional and v lies in it). Moreover, since the first $k+1$ columns of T_n are linearly dependent, and since T_k is a submatrix of this, v must give the coefficients yielding this linear dependence. This gives the first equality in (5.7.4). The same argument for columns $2, \ldots, k+2$ of T_n, yields the second equality, and so on.

For a fixed $l \in \underline{\underline{k}}$ we let $p^{(l)}(z) = p_0^{(l)} + \cdots + p_{k-1}^{(l)} z^{k-1}$ be the unique polynomial such that $p_l(\alpha_l) = 1$ and $p_l(\alpha_i) = 0, i \neq l$. Put

$$\rho_l = \begin{pmatrix} \overline{p}_0^{(l)} & \cdots & \overline{p}_{k-1}^{(l)} \end{pmatrix} \begin{pmatrix} c_0 & \cdots & c_{-k+1} \\ \vdots & \ddots & \vdots \\ c_{k-1} & \cdots & c_0 \end{pmatrix} \begin{pmatrix} p_0^{(l)} \\ \vdots \\ p_{k-1}^{(l)} \end{pmatrix} \in \mathbb{R}. \tag{5.7.5}$$

We claim that with these choices of ρ_l and α_l we have that

$$c_j = \sum_{l=1}^k \rho_l \alpha_l^j, \ |j| \leq n. \tag{5.7.6}$$

We first observe that

$$\rho_l = \begin{pmatrix} \overline{p}_0^{(l)} & \dots & \overline{p}_{k-1}^{(l)} \end{pmatrix} \begin{pmatrix} c_0 & \dots & c_{-k+1} \\ \vdots & \ddots & \vdots \\ c_{k-1} & \dots & c_0 \end{pmatrix} \begin{pmatrix} 1 \\ 0 \\ \vdots \\ 0 \end{pmatrix}. \tag{5.7.7}$$

Indeed $p^{(l)}(z) - 1 = (z - \alpha_l)r(z)$, where $r(z) = r_0 + \cdots + r_{k-2}z^{k-2}$ is some polynomial of degree at most $k - 2$. But then

$$\begin{pmatrix} p_0^{(l)} \\ \vdots \\ p_{k-1}^{(l)} \\ 0 \end{pmatrix} - \begin{pmatrix} 1 \\ 0 \\ \vdots \\ 0 \end{pmatrix} = -\alpha_l \begin{pmatrix} r_0 \\ \vdots \\ r_{k-2} \\ 0 \\ 0 \end{pmatrix} + \begin{pmatrix} 0 \\ r_0 \\ \vdots \\ r_{k-2} \\ 0 \end{pmatrix}.$$

This in turn yields

$$\begin{aligned}
&\begin{pmatrix} \overline{p}_0^{(l)} & \dots & \overline{p}_{k-1}^{(l)} & 0 \end{pmatrix} \begin{pmatrix} c_0 & \dots & c_{-k-1} \\ \vdots & \ddots & \vdots \\ c_{k+1} & \dots & c_0 \end{pmatrix} \left[\begin{pmatrix} p_0^{(l)} \\ \vdots \\ p_k^{(l)} \\ 0 \end{pmatrix} - \begin{pmatrix} 1 \\ 0 \\ \vdots \\ 0 \end{pmatrix} \right] \\
&= -\alpha_l \begin{pmatrix} \overline{p}_0^{(l)} & \dots & \overline{p}_{k-1}^{(l)} & 0 \end{pmatrix} \begin{pmatrix} c_0 & \dots & c_{-k} \\ \vdots & \ddots & \vdots \\ c_k & \dots & c_0 \end{pmatrix} \begin{pmatrix} r_0 \\ \vdots \\ r_{k-2} \\ 0 \\ 0 \end{pmatrix} \\
&\quad + \begin{pmatrix} \overline{p}_0^{(l)} & \dots & \overline{p}_{k-1}^{(l)} & 0 \end{pmatrix} \begin{pmatrix} c_0 & \dots & c_{-k} \\ \vdots & \ddots & \vdots \\ c_k & \dots & c_0 \end{pmatrix} \begin{pmatrix} 0 \\ r_0 \\ \vdots \\ r_{k-2} \\ 0 \end{pmatrix} \\
&= \left[\begin{pmatrix} \overline{p}_0^{(l)} & \dots & \overline{p}_{k-1}^{(l)} & 0 \end{pmatrix} - \alpha_l \begin{pmatrix} 0 & \overline{p}_0^{(l)} & \dots & \overline{p}_{k-1}^{(l)} \end{pmatrix} \right] \\
&\qquad \times \begin{pmatrix} c_0 & \dots & c_{-k} \\ \vdots & \ddots & \vdots \\ c_k & \dots & c_0 \end{pmatrix} \begin{pmatrix} 0 \\ r_0 \\ \vdots \\ r_{k-2} \\ 0 \end{pmatrix}.
\end{aligned} \tag{5.7.8}$$

As $p^{(l)}(z) - \overline{\alpha}_l z p^{(l)}(z)$ equals 0 for $z \in \{\alpha_1, \dots, \alpha_k\}$ we must have that $p^{(l)}(z) - \overline{\alpha}_l z p^{(l)}(z) = \alpha P(z)$ for some α. But then (5.7.8) equals

$$\overline{\alpha} \begin{pmatrix} \overline{v}_0 & \dots & \overline{v}_k \end{pmatrix} \begin{pmatrix} c_0 & \dots & c_{-k} \\ \vdots & \ddots & \vdots \\ c_k & \dots & c_0 \end{pmatrix} \begin{pmatrix} 0 \\ r_0 \\ \vdots \\ r_{k-2} \\ 0 \end{pmatrix} = 0.$$

Thus the right-hand sides of (5.7.5) and (5.7.7) are the same. This proves (5.7.7).

Next, we show that for $q(z) = \sum_{i=0}^{n} q_i z^i$ we have

$$\begin{pmatrix} 1 & 0 & \dots & 0 \end{pmatrix} T_n \begin{pmatrix} q_0 \\ \vdots \\ q_n \end{pmatrix} = \sum_{i=1}^{k} \rho_l q(\alpha_l). \tag{5.7.9}$$

Indeed, write $q(z) = h(z)P(z) + w(z)$ with $\deg w \leq k-1$. Then $q(\alpha_l) = w(\alpha_l)$. Moreover,

$$w(z) = \sum_{l=1}^{k} w(\alpha_l) p^{(l)}(z),$$

as the two sides coincide for $z = \alpha_1, \dots, \alpha_k$. Using (5.7.4) we now get that

$$\begin{pmatrix} 1 & 0 & \dots & 0 \end{pmatrix} T_n \begin{pmatrix} q_0 \\ \vdots \\ q_n \end{pmatrix}$$

$$= \sum_{l=1}^{k} w(\alpha_l) \begin{pmatrix} 1 & 0 & \dots & 0 \end{pmatrix} T_k \begin{pmatrix} p_0^{(l)} \\ \vdots \\ p_k^{(l)} \end{pmatrix} = \sum_{l=1}^{k} \varrho_l q(\alpha_l).$$

Finally, taking $q(z) = z^p$, $p = 0, \dots, n$, we get from (5.7.9) that

$$c_p = \sum_{i=1}^{k} \rho_l \alpha_l^p, \quad p = 0, \dots, n.$$

Using that $\rho_l \in \mathbb{R}$, $\alpha_l \in \mathbb{T}$, and $c_l = \overline{c}_{-l}$, we obtain the remaining equalities in (5.7.6) as well.

It remains to show that $k = m_- + m_+$ and that m_+ of the ρs are positive, and the remaining ones are negative. First we note that using (5.7.6) we get that $T_{k-1} = RDR^*$, where

$$R = \begin{pmatrix} 1 & 1 & \dots & 1 \\ \alpha_1 & \alpha_2 & \dots & \alpha_k \\ \vdots & \vdots & & \vdots \\ \alpha_1^{k-1} & \alpha_2^{k-1} & \dots & \alpha_k^{k-1} \end{pmatrix}, \quad D = \operatorname{diag}(\rho_l)_{l=1}^{k}.$$

As the αs are all different, we have that R is invertible (which also follows from T_{k-1} being invertible). But then the inertia of T_{k-1} and D are the same. As $i_\pm(T_n) = m_\pm$ and rankT_{k-1} = rankT_n, we must have that $i_\pm(T_{k-1}) = m_\pm$. Finally, as T_{k-1} is invertible, we get that $m_+ + m_- = k$. □

It should be noted that in the above result we only deal with the case when T_n is a singular matrix. When T_n is invertible, we can proceed as follows. Construct a singular and Hermitian one step extension T_{n+1} of T_n. It is not hard to see that the entry $c_{n+1}(x)$ depends on the parameter x as described below, a and b being some constants depending on the given data:

$$c_{n+1}(x) = \begin{cases} a + bx, & x \in \mathbb{R}, \qquad \det T_{n-1} = 0, \\ a + be^{ix}, & x \in [0, 2\pi], \quad \det T_{n-1} \neq 0. \end{cases}$$

The question now becomes whether there exists a parameter x, such that the polynomial (depending on x),

$$P_x(z) = \det \begin{pmatrix} c_0 & \overline{c}_1 & \dots & \overline{c}_n & \overline{c}_{n+1}(x) \\ c_1 & c_0 & \dots & \overline{c}_{n-1} & \overline{c}_n \\ \vdots & \vdots & \ddots & \vdots & \vdots \\ c_n & c_{n-1} & \dots & t_0 & \overline{c}_1 \\ 1 & z & \dots & z^n & z^{n+1} \end{pmatrix}$$

has all its roots on $\mathbb{T}$. By using Theorem 1.3.7 we can reformulate this as solving a matrix inequality $\sigma(x) \geq 0$, where $\sigma(x)$ is constructed from the data. We illustrate this on the above example.

Example 5.7.3 continued. With the given data

$$T_2 = \begin{pmatrix} 1 & 1 & 3 \\ 1 & 1 & 1 \\ 3 & 1 & 1 \end{pmatrix}$$

we have seen that

$$P_x(z) = z^3 - \left(2 - \frac{ix}{2}\right) z^2 - \left(2 + \frac{ix}{2}\right) z + 1.$$

If we let z_1, z_2, z_3 be the roots of $P_x(z)$ and put $\sigma_j(x) = \sum_{k=1}^{3} z_k^j$, we get that

$$\sigma_0(x) = 3,$$

$$\sigma_1(x) = z_1 + z_2 + z_3 = 2 - \frac{xi}{2},$$

$$\sigma_2(x) = z_1^2 + z_2^2 + z_3^2 = (z_1 + z_2 + z_3)^2 - 2(z_1 z_2 + z_2 z_3 + z_1 z_3)$$
$$= 8 - xi - \frac{x^2}{4},$$

and

$$\sigma_3(x) = z_1^3 + z_2^3 + z_3^3$$
$$= (z_1 + z_2 + z_3)^3 - 3(z_1 z_2 + z_2 z_3 + z_1 z_3)(z_1 + z_2 + z_3) + 3 z_1 z_2 z_3$$
$$= \left(2 - \frac{xi}{2}\right)^3 + 3\left(2 - \frac{xi}{2}\right)\left(2 + \frac{xi}{2}\right) - 3.$$

Theorem 1.3.7 now says that z_1, z_2, z_3 are different points on $\mathbb{T}$ if

$$\sigma(x) = \begin{pmatrix} 3 & 2+\frac{xi}{2} & 8+xi-\frac{x^2}{4} & \sigma_3(x) \\ 2-\frac{xi}{2} & 3 & 2+\frac{xi}{2} & 8+xi-\frac{x^2}{4} \\ 8-xi-\frac{x^2}{4} & 2-\frac{xi}{2} & 3 & 2+\frac{xi}{2} \\ \overline{\sigma_3(x)} & 8-xi-\frac{x^2}{4} & 2-\frac{xi}{2} & 3 \end{pmatrix}$$

is positive semidefinite and of rank 3. To obtain that $\sigma(x) \geq 0$ we certainly need

$$0 \leq \det \begin{pmatrix} 3 & 8+xi-\frac{x^2}{4} \\ 8-xi-\frac{x^2}{4} & 3 \end{pmatrix} = 9 - \left(8 - \frac{x^2}{4}\right)^2 - x^2,$$

which is easily seen not to have real solutions. Thus the given data does not have a (2,1) extension, which we had observed earlier in another way.

When $m_- = 1$, in order to find out whether some given data admits an $(m, 1)$ extension, we can also try to determine directly whether there exist $\rho > 0$ and $\alpha \in \mathbb{T}$ such that the matrix

$$T_n + \rho \begin{pmatrix} 1 & \overline{\alpha} & \dots & \overline{\alpha}^n \\ \alpha & 1 & \dots & \overline{\alpha}^{n-1} \\ \vdots & \vdots & \ddots & \vdots \\ \alpha^n & \alpha^{n-1} & \dots & 1 \end{pmatrix} \geq 0$$

is a singular positive semidefinite matrix.

Example 5.7.5 Consider the data defining the Toeplitz matrix

$$T_2 = \begin{pmatrix} 1 & 1 & -1 \\ 1 & 1 & 1 \\ -1 & 1 & 1 \end{pmatrix}.$$

We are looking for $\rho > 0$ and $\alpha \in \mathbb{T}$ such that

$$\begin{pmatrix} 1 & 1 & -1 \\ 1 & 1 & 1 \\ -1 & 1 & 1 \end{pmatrix} + \rho \begin{pmatrix} 1 & \overline{\alpha} & \overline{\alpha}^2 \\ \alpha & 1 & \overline{\alpha} \\ \alpha^2 & \alpha & 1 \end{pmatrix}$$

$$= \begin{pmatrix} 1+\rho & 1+\rho\overline{\alpha} & -1+\rho\overline{\alpha}^2 \\ 1+\rho\alpha & 1+\rho & 1+\rho\overline{\alpha} \\ -1+\rho\alpha^2 & 1+\rho\alpha & 1+\rho \end{pmatrix} \tag{5.7.10}$$

is a singular positive semidefinite matrix. In particular, its determinant is 0, yielding that

$$\rho(\alpha^2 + \overline{\alpha}^2) - 2\rho(\alpha + \overline{\alpha}) - 2 = 0,$$

where we used that $\alpha \in \mathbb{T}$. Writing $\alpha = \cos\theta + i\sin\theta$, we obtain the equation

$$\rho(2\cos^2\theta - 2\cos\theta - 1) = 1,$$

which has a solution for every θ such that $2\cos^2\theta - 2\cos\theta - 1 > 0$. For instance, considering $\theta = \pi$, we have that $\rho = \frac{1}{3}$. For this particular choice, (5.7.10) becomes

$$\begin{pmatrix} \frac{4}{3} & \frac{2}{3} & -\frac{2}{3} \\ \frac{2}{3} & \frac{4}{3} & \frac{2}{3} \\ -\frac{2}{3} & \frac{2}{3} & \frac{4}{3} \end{pmatrix} \geq 0.$$

For the latter matrix we now get $\alpha_1 = e^{i\frac{\pi}{3}}$, $\alpha_2 = e^{-i\frac{\pi}{3}}$, and $\rho_1 = \rho_2 = \frac{2}{3}$. Consequently,

$$\mu = \frac{2}{3}\delta_{e^{i\frac{\pi}{3}}} + \frac{2}{3}\delta_{e^{-i\frac{\pi}{3}}} - \frac{1}{3}\delta_{-1}$$

is a $(2,1)$ extension of the given data.

5.8 EUCLIDEAN DISTANCE MATRIX COMPLETIONS

In this section we consider a completion problem for distance matrices. A matrix $D = (d_{ij})_{i,j=1}^n$ is called a (*Euclidean*) *distance matrix* if there exist $P_1, \dots, P_n \in \mathbb{R}^k$ such that $d_{ij} = \|P_i - P_j\|^2$, where $\|\cdot\|$ denotes the Euclidean distance. The following result gives a characterization of distance matrices.

Theorem 5.8.1 *A real symmetric matrix $D = (d_{ij})_{i,j=1}^n$, with $d_{ii} = 0$, $i = 1, \dots, n$, is a distance matrix if and only if D is negative semidefinite on the orthogonal complement of the vector $e = \begin{pmatrix} 1 & \cdots & 1 \end{pmatrix}^T$. This is equivalent to the statement that the bordered matrix*

$$\begin{pmatrix} 0 & e^T \\ e & D \end{pmatrix} \tag{5.8.1}$$

has only one positive eigenvalue, or to the fact that the Schur complement of (5.8.1) supported in the last $n-1$ rows and columns is negative semidefinite. Furthermore, the rank of this Schur complement is the minimum dimension k of an affine space in which the points $P_1, \dots, P_n$ may lie. In this case we say that D is a distance matrix in $\mathbb{R}^k$.

Proof. Assume without loss of generality that $P_1 = 0$ and consider the distance matrix

$$D = \begin{pmatrix} 0 & \|P_2\|^2 & \cdots & \|P_n\|^2 \\ \|P_2\|^2 & 0 & \cdots & \|P_2 - P_n\|^2 \\ \|P_3\|^2 & \|P_3 - P_2\|^2 & \cdots & \|P_3 - P_n\|^2 \\ \vdots & \vdots & & \vdots \\ \|P_n\|^2 & \|P_n - P_2\|^2 & \cdots & 0 \end{pmatrix}.$$

Taking the Schur complement supported in the last $n-1$ rows and columns of $\left(\begin{smallmatrix} 0 & e^T \\ e & D \end{smallmatrix}\right)$, one obtains

$$\begin{pmatrix} 0 & \|P_2 - P_3\|^2 & \cdots & \|P_2 - P_n\|^2 \\ \|P_3 - P_2\|^2 & 0 & \cdots & \|P_3 - P_n\|^2 \\ \vdots & \vdots & \ddots & \vdots \\ \|P_n - P_2\|^2 & \|P_n - P_3\|^2 & \cdots & 0 \end{pmatrix}$$

$$-\begin{pmatrix} 1 & \|P_2\|^2 \\ 1 & \|P_3\|^2 \\ \vdots & \vdots \\ 1 & \|P_n\|^2 \end{pmatrix} \begin{pmatrix} 0 & 1 \\ 1 & 0 \end{pmatrix} \begin{pmatrix} 1 & 1 & \cdots & 1 \\ \|P_2\|^2 & \|P_3\|^2 & \cdots & \|P_n\|^2 \end{pmatrix},$$

which equals the matrix

$$\begin{aligned}\left(\|P_i - P_j\|^2 - \|P_i\|^2 - \|P_j\|^2\right)_{i,j=2}^n &= -2\left(P_i^T P_j\right)_{i,j=2}^n \\ &= -2 \begin{pmatrix} P_2^T \\ P_3^T \\ \vdots \\ P_n^T \end{pmatrix} \begin{pmatrix} P_2 & P_3 & \cdots & P_n \end{pmatrix},\end{aligned}$$

which is negative semidefinite of rank k, where k is the dimension of Span$\{P_i : i = 1, \ldots, n\}$. The converse is similar. □

Note that in Theorem 5.8.1 one can also take the Schur complement supported in rows and columns $2, \ldots, j-1, j+1, \ldots, n+1$ for some $j \in \{2, \ldots, n+1\}$, and arrive at the same result.

We call an $n \times n$ real partial matrix $A = (a_{ij})_{i,j=1}^n$ a *partial distance matrix* in $\mathbb{R}^k$ if

(i) a_{ii} is specified as 0, $i = 1, \ldots, n$, and a_{ji} is specified (and equal to a_{ij}) if and only a_{ij} is; and

(ii) every fully specified principal submatrix of A is itself a distance matrix in $\mathbb{R}^k$.

The *distance matrix completion problem* then asks which partial distance matrices have distance matrix completions. It is clear that assumption (ii) above is necessary for this. This "inheritance property" is similar to the previously studied positive semidefinite completion problem.

The main result of this section is that if the undirected graph of the specified entries of a partial distance matrix is chordal, then it necessarily has a distance matrix completion. This is based upon an analysis of the case of partial distance matrices with one pair of symmetrically placed unspecified entries, the "one variable problem," together with a one-step-at-a-time technology previously used in the proof of Theorem 2.2.4. For any non-chordal graph there exist partial distance matrices without distance matrix completions.

In the last part of this section we are concerned with connections between positive semidefinite completions and distance matrix completions.

For any $m \times n$ matrix A, let $A^{(-1)}$ denote the *Moore-Penrose inverse* of A, which is the unique matrix such that $AA^{(-1)}$ and $A^{(-1)}A$ are Hermitian, $AA^{(-1)}A = A$, and $A^{(-1)}AA^{(-1)} = A^{(-1)}$ (see Exercise 5.12.24).

We start with the following lemma.

Lemma 5.8.2 *Let*

$$R = \begin{pmatrix} a & B & ? \\ B^T & C & D \\ ? & D^T & f \end{pmatrix}$$

be a real partial positive semidefinite matrix, with rank $\left(\begin{smallmatrix} a & B \\ B^T & C \end{smallmatrix}\right) = p$ *and* rank $\left(\begin{smallmatrix} C & D \\ D^T & f \end{smallmatrix}\right) = q$*, with (necessarily)* $|p-q| \le 1$*. Then there is a real positive semidefinite completion* F *of* R *such that* rank$F = \max\{p,q\}$*. Moreover, this completion is unique if and only if* rank$C = p$ *or* rank$C = q$*.*

Proof. Let U be an orthogonal matrix that diagonalizes C, namely $U^TCU = Y$, in which Y is a positive semidefinite diagonal matrix. Let $\widehat{U} = 1 \oplus U \oplus 1$ and

$$\widehat{R} = \widehat{U}^T R \widehat{U} = \begin{pmatrix} a & \widehat{B} & x \\ \widehat{B}^T & Y & \widehat{D} \\ x & \widehat{D}^T & f \end{pmatrix},$$

in which $\widehat{B} = BU$, $\widehat{D} = U^TD$, and with x denoting the unknown entry. Since $\widehat{U}$ is orthogonal, the set of numbers that make R positive semidefinite coincides with that making $\widehat{R}$ positive semidefinite and rank$R$ = rank$\widehat{R}$. Since $\left(\begin{smallmatrix} a & \widehat{B} \\ \widehat{B}^T & Y \end{smallmatrix}\right)$ and $\left(\begin{smallmatrix} Y & \widehat{D} \\ \widehat{D}^T & f \end{smallmatrix}\right)$ are positive semidefinite, all entries (in the rows and columns corresponding to diagonal entries of Y that are 0) equal zero. We may eliminate those rows and columns and then assume that Y is invertible. Then $\widehat{R}$ is positive semidefinite if and only if the Schur complement

$$S = \begin{pmatrix} a - \widehat{B}Y^{-1}\widehat{B}^T & x - \widehat{B}Y^{-1}\widehat{D} \\ x - \widehat{D}^TY^{-1}\widehat{B}^T & f - \widehat{D}^TY^{-1}\widehat{D} \end{pmatrix}$$

is positive semidefinite. Note that rank$\widehat{R}$ = rank$C$ + rank$S$. If rank$C = p$ or rank$C = q$ then $a - \widehat{B}Y^{-1}\widehat{B}^T = 0$ and $f - \widehat{D}^TY^{-1}\widehat{D}^T = 0$, respectively, and thus we have to choose $x = \widehat{B}Y^{-1}\widehat{D}$. If rank$C < \max\{p,q\}$, the problem has two solutions given via

$$|x - \widehat{B}Y^{-1}\widehat{D}|^2 = (a - \widehat{B}Y^{-1}\widehat{B}^T)(f - \widehat{D}^TY^{-1}\widehat{D}),$$

which realize the completion of S to a rank 1 positive semidefinite matrix. □

The basic step in the study of the Euclidean distance matrix completion problem is the analysis of the situation with only one unspecified entry. This is described by the next proposition.

Proposition 5.8.3 *The partial distance matrix*

$$R = \begin{pmatrix} 0 & D_{12} & ? \\ D_{12}^T & D_{22} & D_{23} \\ ? & D_{23}^T & 0 \end{pmatrix}$$

admits at least one completion to a distance matrix F*. Moreover, if* $\left(\begin{smallmatrix} 0 & D_{12} \\ D_{12}^T & D_{22} \end{smallmatrix}\right)$ *and* $\left(\begin{smallmatrix} D_{22} & D_{23} \\ D_{23}^T & 0 \end{smallmatrix}\right)$ *are distance matrices in* $\mathbb{R}^p$ *(resp.* $\mathbb{R}^q$*), then a completion* F *can be chosen to be a distance matrix in* $\mathbb{R}^s$*,* $s = \max\{p,q\}$*.*

Proof. Let x denote the unknown entry. Without loss of generality, we may assume that R is at least 3×3, since, otherwise, we may complete with any positive number. Thus, R has at least one fully specified row and column. Interchange the first two rows and the first two columns of R, and then we have to complete the partial distance matrix

$$\widetilde{R} = \begin{pmatrix} 0 & d_{12} & \widetilde{D}_{13} & d_{14} \\ d_{12} & 0 & \widetilde{D}_{23} & x \\ \widetilde{D}_{13}^T & \widetilde{D}_{23}^T & \widetilde{D}_{33} & \widetilde{D}_{34} \\ d_{14} & x & \widetilde{D}_{34}^T & 0 \end{pmatrix}$$

to a distance matrix in $\mathbb{R}^s$. By the remark at the beginning of this section, this latter problem is equivalent to finding completions of the partial matrix

$$\widetilde{D} = \begin{pmatrix} 0 & 1 & 1 & e^T & 1 \\ 1 & 0 & d_{12} & \widetilde{D}_{13} & d_{14} \\ 1 & d_{12} & 0 & \widetilde{D}_{23} & x \\ e & \widetilde{D}_{13}^T & \widetilde{D}_{23}^T & \widetilde{D}_{33} & \widetilde{D}_{34} \\ 1 & d_{14} & x & \widetilde{D}_{34}^T & 0 \end{pmatrix}$$

to a matrix in which the Schur complement supported in block rows and columns 3,4,5, is negative semidefinite and has rank s. This Schur complement is of the form

$$S = \begin{pmatrix} a & B & x - d_{12} - d_{14} \\ B^T & C & D \\ x - d_{12} - d_{14} & D^T & f \end{pmatrix},$$

in which $\left(\begin{smallmatrix} a & B \\ B^T & C \end{smallmatrix}\right)$ and $\left(\begin{smallmatrix} C & D \\ D^T & f \end{smallmatrix}\right)$ are negative semidefinite and have ranks less than or equal to s. Then, any negative semidefinite completion of S of rank s given by Lemma 5.8.2 provides a solution to our distance completion problem. □

Corollary 5.8.4 *The partial distance matrix in* $\mathbb{R}^k$,

$$R = \begin{pmatrix} 0 & D_{12} & ? \\ D_{12}^T & D_{22} & D_{23} \\ ? & D_{23}^T & 0 \end{pmatrix},$$

admits a unique completion to a distance matrix in $\mathbb{R}^k$ *if and only if* $\operatorname{rank}\left(\begin{smallmatrix} 0 & e^T \\ e & D_{22} \end{smallmatrix}\right) = k + 2$.

Proof. Follows from the proof of Proposition 5.8.3 and the uniqueness part of Lemma 5.8.2. □

The main result of this section is the following:

Theorem 5.8.5 *Every partial distance matrix in* $\mathbb{R}^k$*, the graph of whose specified entries is chordal, admits a completion to a distance matrix in* $\mathbb{R}^k$.

Proof. Based on Proposition 5.8.3, the proof can be done by the same one-entry-at-a-time completion procedure as the proof of Theorem 2.2.4. □

Proposition 5.8.6 *Given any nonchordal graph $G = (V, E)$, $V = \{1, \ldots, n\}$, there exists a partial distance matrix $R = (r_{ij})_{i,j=1}^n$ such that R has no completion to a distance matrix.*

Proof. Assume that the vertices $1, 2, \ldots, k \geq 4$ form a chordless cycle in G. Define the partial distance matrix R by:

$$r_{ij} = \begin{cases} 0 \text{ if } (i,j) \in E \text{ and } k+1 \leq i,j \leq n, \\ 0 \text{ if } |i-j| = 1,\ 1 \leq i \leq j \leq k, \\ 1 \text{ for any other } (i,j) \in E. \end{cases}$$

Then any fully specified principal submatrix of R is either

$$0, \begin{pmatrix} 0 & e^T \\ e & 0 \end{pmatrix}, \begin{pmatrix} 0 & 0 & e^T \\ 0 & 0 & e^T \\ e & e & 0 \end{pmatrix}, \text{ or } \begin{pmatrix} 0 & 1 & e^T \\ 1 & 0 & e^T \\ e & e & 0 \end{pmatrix},$$

each of them being a distance matrix. Thus R is a partial distance matrix, but R does not admit a completion to a distance matrix. Indeed, the upper left $k \times k$ principal submatrix cannot be completed to a distance matrix since otherwise there would exist points $P_1, \ldots, P_k$ such that $\|P_i - P_{i+1}\| = 0$ for $i = 1, \ldots, k-1$ and $\|P_1 - P_k\| = 1$, a contradiction. □

Theorem 5.8.7 *Let R be a partial distance matrix in $\mathbb{R}^k$, the graph $G = (V, E)$ of whose specified entries is chordal, and let $\mathcal{S}$ be the collection of all minimal vertex separators of G. Then R admits a unique completion to a distance matrix in $\mathbb{R}^k$ if and only if*

$$\begin{pmatrix} 0 & e^T \\ e & R|S \end{pmatrix} \tag{5.8.2}$$

has rank equal to $k+2$ for any $S \in \mathcal{S}$.

Proof. We prove the result by induction on the cardinality m of the complement of the edge set of G. If $m = 1$, the result follows from Corollary 5.8.4. Assume the result is true for any chordal graph whose complement has cardinality less than m. Let $G = (V, E)$ be a chordal graph such that the cardinality of the complement of E is m and let R be a partial distance matrix satisfying (5.8.2). Let S be an arbitrary minimal vertex separator of G. By Corollary 5.1.11, there exist vertices u and v belonging to different connected components of G_{V-S} with the property that $S \subset \operatorname{Adj}(u)$ and $S \subset \operatorname{Adj}(v)$.

We first prove that the graph $G' = (V, E \cup \{(u,v)\})$ is also chordal. Assuming the contrary there exists a chordless cycle $[u, x_1, \ldots, x_k, v]$, $k \geq 2$, in G'. By the definition of a minimal vertex separator, at least one $x_l \in S$, $1 \leq l \leq k$. This implies that $(x_l, u), (x_l, v) \in E$, a contradiction, showing that G' is chordal.

Then $S \cup \{u, v\}$ is the unique maximal clique in G' that is not a clique in G. Consider the principal submatrix $R|S \cup \{u,v\}$ having only one pair of symmetrically placed unspecified entries. Complete $R|S \cup \{u,v\}$ to a distance matrix in $\mathbb{R}^k$ to obtain a partial distance matrix $\widetilde{R}$ having G' as the graph of its specified entries.

If rank $\left(\begin{smallmatrix} 0 & e^T \\ e & R|S \end{smallmatrix}\right) < k+2$, by Proposition 5.8.3, $R|S \cup \{u, v\}$ has more than one completion to a distance matrix in $\mathbb{R}^k$, and so R admits more than one completion.

If R satisfies (5.8.2), then $\widetilde{R}$ constructed above is uniquely determined. Since any minimal vertex separator of G' contains a minimal vertex separator of G, $\widetilde{R}$ also satisfies condition (5.8.2). By the assumption made for $m-1$, $\widetilde{R}$ admits a unique completion to a distance matrix in $\mathbb{R}^k$. This implies that R also admits a unique completion to a distance matrix in $\mathbb{R}^k$. □

Let $0 < m < n$ be given integers. Since the graph $G = (V, E)$ with $E = \{(i,j) \ : \ 0 < |i-j| \leq m\}$ is chordal, Theorem 5.8.5 and Proposition 5.8.4 have the following consequence in the "band" case.

Corollary 5.8.8 *Any partial distance matrix $R = (r_{ij})_{i,j=1}^n$ in $\mathbb{R}^k$, with r_{ij} specified if and only if $|i-j| \leq m$, admits a completion to a distance matrix in $\mathbb{R}^k$. Moreover, the completion is unique if and only if all the matrices*

$$\begin{pmatrix} 0 & e^T \\ e & R|\{l, \ldots, l+m-1\} \end{pmatrix}$$

have rank equal to $k+2$ for any $l = 1, \ldots, n-m+1$.

5.8.1 Circumhyperspheres

Using the above results, one may easily derive the following results concerning the existence of a circumhypersphere for a set of points (i.e. a hypersphere containing all points of the set).

Theorem 5.8.9 *Let D be a distance matrix corresponding to distinct points $P_1, \ldots, P_{n+1}$ in $\mathbb{R}^n$. Then there is a circumhypersphere for $P_1, \ldots, P_{n+1}$ if and only if $e^T D^{(-1)} e > 0$. If the points are in general position, then the radius is uniquely given by $r_0 = (2e^T D^{(-1)} e)^{-\frac{1}{2}}$. Otherwise, the radius can be chosen to equal any value $\geq r_0$.*

Proof. By using the results in Section 5.1, one easily sees that the number $i_+(M)$ of positive eigenvalues of a partitioned matrix $M = \left(\begin{smallmatrix} A & B \\ B^* & C \end{smallmatrix}\right)$ satisfies

$$i_+(M) \geq i_+(C) + i_+(A - BC^{(-1)}B^*). \tag{5.8.3}$$

Let D be a distance matrix corresponding to the points $P_1, \ldots, P_{n+1}$ in $\mathbb{R}^n$. As the points are different, D contains a principal submatrix of the form $\left(\begin{smallmatrix} 0 & a \\ a & 0 \end{smallmatrix}\right)$ with $a > 0$. This gives that $i_+(D) \geq 1$. On the other hand, by Theorem 5.8.1, the bordered matrix $\left(\begin{smallmatrix} 0 & e^T \\ e & D \end{smallmatrix}\right)$ has exactly one positive eigenvalue. Thus we have that $i_+(D) = 1$.

In order to find a circumhypersphere for $P_1, \ldots, P_{n+1}$, we need to identify a point $O \in \mathbb{R}^n$ and a radius $r > 0$ such that $\|O - P_i\| = r$ for $i = 1, \ldots, n+1$. By Theorem 5.8.1 this is equivalent to the condition that the bordered matrix

$$\widehat{D} = \begin{pmatrix} 0 & 1 & e^T \\ 1 & 0 & r^2 e^T \\ e & r^2 e & D \end{pmatrix}$$

has exactly one positive eigenvalue and the Schur complement of $\widehat{D}$ supported in the last $n+1$ rows and columns has rank $\leq n$. Let $a = -e^T D^{(-1)} e$. By using (5.8.3) we have that

$$i_+(\widehat{D}) \geq i_+(D) + i_+ \begin{pmatrix} a & 1 + r^2 a \\ 1 + r^2 a & r^4 a \end{pmatrix}.$$

As $i_+(D) = 1$, we must have that $\begin{pmatrix} a & 1+r^2 a \\ 1+r^2 a & r^4 a \end{pmatrix}$ is negative semidefinite. This yields that $a < 0$ and that $r \geq \frac{1}{\sqrt{-2a}} =: r_0$. When we choose $r = r_0$ it is easy to check that $\widehat{D}$ and $\begin{pmatrix} 0 & e^T \\ e & D \end{pmatrix}$ have the same rank, and thus we find the desired existence of the point O. In case the Schur complement of $\begin{pmatrix} 0 & e^T \\ e & D \end{pmatrix}$ supported in the last n rows and columns has rank strictly smaller than n, one has the flexibility to allow an additional negative eigenvalue in $\widehat{D}$. Thus in this case one may choose $r \geq r_0$. □

5.8.2 Connections with Positive Semidefinite Completions

Lemma 5.8.10 *Let $A = (a_{ij})_{i,j=1}^n$ be a real symmetric matrix such that $a_{ii} = 1$ for $i = 1, \ldots, n$. Then A is positive semidefinite if and only if there are n points $P_1, \ldots, P_n$ on a hypersphere of radius $\frac{\sqrt{2}}{2}$ in $\mathbb{R}^k$, $k = \text{rank} A$, such that $\|P_i - P_j\| = \sqrt{1 - a_{ij}}$ for any $i, j = 1, \ldots, n$.*

Proof. By Theorem 5.8.1, the existence of the points $P_1, \ldots, P_n$ and O in $\mathbb{R}^k$ such that $\|P_i - P_j\| = \sqrt{1 - a_{ij}}$ and $\|O - P_i\| = \frac{\sqrt{2}}{2}$ is equivalent to the condition that the Schur complement supported in the last n rows and columns of the matrix

$$\begin{pmatrix} 0 & 1 & 1 & 1 & \cdots & 1 \\ 1 & 0 & \frac{1}{2} & \frac{1}{2} & \cdots & \frac{1}{2} \\ 1 & \frac{1}{2} & 0 & 1 - a_{12} & \cdots & 1 - a_{1n} \\ 1 & \frac{1}{2} & 1 - a_{12} & 0 & \cdots & 1 - a_{2n} \\ \vdots & \vdots & \vdots & \vdots & \vdots & \vdots \\ 1 & \frac{1}{2} & 1 - a_{1n} & 1 - a_{2n} & \cdots & 0 \end{pmatrix}$$

is negative semidefinite and has rank k. A straightforward computation shows that this latter Schur complement equals $-A$. This completes the proof. □

Using Lemma 5.8.10, Theorem 2.2.4 yields the following result.

Theorem 5.8.11 *Let $G = (V, E)$ be a chordal graph and $\{K_j\}_{j=1}^m$ the maximal cliques of G. Consider points $P_1, \ldots, P_n$ in $\mathbb{R}^{n-1}$ satisfying the conditions:*

(i) the distances $\|P_i - P_j\|$ are specified if and only if $(i, j) \in E$;

(ii) each of the subsets $\{P_l\}_{l \in K_j}$, $j = 1, \ldots, m$, lies on a hypersphere of radius $R > 0$.

Then points $\widetilde{P_1}, \ldots, \widetilde{P_n}$ can be chosen to lie on a hypersphere of radius R, such that

$$\|\widetilde{P_i} - \widetilde{P_j}\| = \|P_i - P_j\|, (i, j) \in E. \tag{5.8.4}$$

Proof. We may assume that $R = \frac{\sqrt{2}}{2}$; otherwise we multiply all P_i by $\frac{\sqrt{2}}{2R}$. Put $a_{ii} = 1$, $i = 1, \ldots, n$, and $a_{ij} = \|P_i - P_j\|^2, (i, j) \in E$. Condition (ii), in conjunction with Lemma 5.8.10, gives that $A = (a_{ij})_{i,j=1}^n$ is a partially positive semidefinite matrix with a chordal graph. Thus by Theorem 2.2.4 A has a positive semidefinite completion $\widetilde{A} = (\tilde{a}_{ij})_{i,j=1}^n$, say. By Lemma 5.8.10 there now exist points $\widetilde{P_1}, \ldots, \widetilde{P_n}$ on a hypersphere of radius $R = \frac{\sqrt{2}}{2}$, such that $\|\widetilde{P_i} - \widetilde{P_j}\| = \sqrt{1 - \tilde{a}_{ij}}$, $i, j = 1, \ldots, n$. But then (5.8.4) follows. □

The conclusion of Theorem 5.8.11 is not valid when the graph G is not chordal. Consider, for example, G to be the simple cycle of length 4 and the points A, B, C, D such that $\|A - B\| = \|B - C\| = \|C - D\| = 2$ and $\|A - D\| = 0$. Then each of the pairs $\{A, B\}$, $\{B, C\}$, $\{C, D\}$ and $\{A, D\}$ lies on a sphere of radius 1, but the smallest radius of a sphere on which $A = D$, B and C may lie is $\frac{2\sqrt{3}}{3} > 1$.

5.9 NORMAL COMPLETIONS

The classes of Hermitian matrices and normal matrices have some important similarities but also have some significant differences. In this section we will explore the normal completion problems, which, as we will see, is vastly different from the Hermitian completion problem. In Section 5.11 the normal completion problem will be related to the separability problem that appears in quantum computation.

One can consider several variations of the normal completion problem. We will focus on the *minimal normal completion problem.* Given $A \in \mathbb{C}^{n \times n}$, find a smallest possible normal matrix with A as a principal submatrix. In other words, find a normal completion of

$$\begin{pmatrix} A & ? \\ ? & ? \end{pmatrix} : \begin{matrix} \mathbb{C}^n \\ \oplus \\ \mathbb{C}^k \end{matrix} \to \begin{matrix} \mathbb{C}^n \\ \oplus \\ \mathbb{C}^k \end{matrix}$$

of smallest possible size (thus smallest possible k). We shall call this smallest number k the *normal defect* of A, and denote it by $\mathrm{nd}(A)$. Clearly $\mathrm{nd}(A) \geq 0$, and $\mathrm{nd}(A) = 0$ if and only if A is normal. In addition, for $A \in \mathbb{C}^{n\times n}$ we have that $\mathrm{nd}(A) \leq n$ as the matrix

$$\begin{pmatrix} A & A^* \\ A^* & A \end{pmatrix}$$

is normal. We can actually give stricter bounds than these, which the following theorem shows. For matrices C and D, their *commutator* is denoted by $[C, D] = CD - DC$.

Theorem 5.9.1 *Given $A \in \mathbb{C}^{n\times n}$. Then*

$$\max\{i_+([A, A^*]), i_-([A, A^*])\} \leq \mathrm{nd}(A) \leq \mathrm{rank}\left(||A||^2 I_n - A^*A\right).$$

Proof. Let $A_{ext} := \left(\begin{smallmatrix} A & A_{12} \\ A_{21} & A_{22} \end{smallmatrix}\right)$ be a normal completion of size $(n+q) \times (n+q)$. From the normality of A_{ext} we get

$$AA^* - A^*A = A_{21}^* A_{21} - A_{12} A_{12}^*. \tag{5.9.1}$$

Let us denote the eigenvalues of the Hermitian matrices $A_{21}^*A_{21} - A_{12}A_{12}^*$ and $A_{21}^*A_{21}$ by $\lambda_1 \leq \cdots \leq \lambda_n$ and $\mu_1 \leq \cdots \leq \mu_n$, respectively. By the Courant-Fischer theorem we get for $1 \leq j \leq n$

$$\begin{aligned} \lambda_j &= \min_{w_1, w_2, \ldots, w_{n-j} \in \mathbb{C}^n} \max_{\substack{x \neq 0,\ x \in \mathbb{C}^n \\ x \perp w_1, w_2, \ldots, w_{n-j}}} \frac{x^*(A_{21}^*A_{21} - A_{12}A_{12}^*)x}{x^*x} \\ &\leq \min_{w_1, w_2, \ldots, w_{n-j} \in \mathbb{C}^n} \max_{\substack{x \neq 0,\ x \in \mathbb{C}^n \\ x \perp w_1, w_2, \ldots, w_{n-j}}} \frac{x^*A_{21}^*A_{21}x}{x^*x} = \mu_j. \end{aligned}$$

Since $\mu_{n-q} = 0$, we get $\lambda_{n-q} \leq 0$. Thus $i_+([A, A^*]) \leq q$. A similar argument can be carried out by looking at the eigenvalues of $A_{21}^*A_{21} - A_{12}A_{12}^*$ and $-A_{12}A_{12}^*$, which will give $i_-([A, A^*]) \leq q$. This proves the lower bound.

For the upper bound, write the singular value decomposition of A as

$$A = \begin{pmatrix} U_1 & U_2 \end{pmatrix} \begin{pmatrix} ||A||I & 0 \\ 0 & \Sigma \end{pmatrix} \begin{pmatrix} V_1^* \\ V_2^* \end{pmatrix},$$

where $U = (U_1\ U_2)$ and $V = (V_1\ V_2)$ are unitary and Σ is diagonal with $||\Sigma|| < ||A||$. Note that Σ has size $d = \mathrm{rank}(\|A\|^2 I_n - A^*A)$. Then

$$A_{ext} = \begin{pmatrix} A & U_2\left(||A||^2 I_d - \Sigma^2\right)^{1/2} \\ \left(||A||^2 I_d - \Sigma^2\right)^{1/2} V_2^* & -\Sigma \end{pmatrix} \tag{5.9.2}$$

is $||A||$ times a unitary matrix. Thus (5.9.2) is normal. This shows that $\mathrm{nd}(A) \leq \mathrm{rank}(\|A\|^2 I_n - A^*A)$. □

Neither bound is strict in general. Let us start with an example for the lower bound.

Example 5.9.2 Let

$$A = \begin{pmatrix} 0 & 0 & 0 & \sqrt{2} \\ 1 & 0 & 0 & 0 \\ 0 & 1 & 0 & 0 \\ 0 & 0 & 1 & 0 \end{pmatrix}.$$

Then $\max\{i_+([A, A^*]), i_-([A, A^*])\} = 1$, but $\mathrm{nd}(A) > 1$. Indeed, suppose that

$$N = \begin{pmatrix} 0 & 0 & 0 & \sqrt{2} & y_1 \\ 1 & 0 & 0 & 0 & y_2 \\ 0 & 1 & 0 & 0 & y_3 \\ 0 & 0 & 1 & 0 & y_4 \\ \overline{x_1} & \overline{x_2} & \overline{x_3} & \overline{x_4} & t \end{pmatrix}$$

is normal. Then using the (1,1), (2,1), and (2,2) entries of the equality $NN^* = N^*N$, we get that

$$|x_1|^2 = 1 + |y_1|^2, \ \overline{x_2}x_1 = y_2\overline{y_1}, \ |x_2|^2 = |y_2|^2.$$

This can only happen when $x_2 = y_2 = 0$. Similarly, using the (3,3), (3,4), and (4,4) entries entries of the equality $NN^* = N^*N$, we get that $x_3 = y_3 = 0$. But now we get that the (2,5) entry of the equality $NN^* = N^*N$ yields that $x_1 = 0$, which is in contradiction with $|x_1|^2 = 1 + |y_1|^2$.

For the upper bound it is even simpler to find a case where the bound is not strict. For instance, one can take a normal A with eigenvalues of different magnitude, for example, $A = \left(\begin{smallmatrix} 1 & 0 \\ 0 & 0 \end{smallmatrix}\right)$. Then $\mathrm{nd}(A) = 0 < \mathrm{rank}(\|A\|^2 I_n - A^*A)$. This somewhat trivial example may be circumvented as follows. When A is *unitarily reducible*,, that is, when

$$A = U^* \begin{pmatrix} A_1 & 0 \\ 0 & A_2 \end{pmatrix} U$$

with U unitary and A_1 and A_2 of nontrivial size, we have that $\mathrm{nd}(A) \leq \mathrm{nd}(A_1) + \mathrm{nd}(A_2)$. Thus, in this case we can apply the upper bound in Theorem 5.9.1 to each of A_1 and A_2 and get a better upper bound. Even with this improvement, the upper bound is not sharp in general as the following example shows.

Example 5.9.3 Let

$$A = \begin{pmatrix} 0 & 1 & 0 \\ 1 & 0 & 1 \\ 0 & 1 & \frac{3i}{2} \end{pmatrix}.$$

Then $\mathrm{nd}(A) = 1$ as

$$A_{\mathrm{ext}} = \begin{pmatrix} 0 & 1 & 0 & 0 \\ 1 & 0 & 1 & 1 \\ 0 & 1 & \frac{3i}{2} & -\frac{3i}{2} \\ 0 & 1 & -\frac{3i}{2} & \frac{3i}{2} \end{pmatrix}$$

is normal. Next, it is easy to see that $\mathrm{rank}(\|A\|^2 I_n - A^*A) = 2$ as the largest singular value of A is greater than the next one. Finally, A is not unitarily reducible, as A does not have an eigenvector that is orthogonal to the other two.

Example 5.9.3 is a nice illustration of the following result, which gives necessary and sufficient conditions when $\mathrm{nd}(A) = 1$.

Theorem 5.9.4 *Let A be a $n \times n$ matrix and let P denote the orthogonal projection onto $\mathrm{Ker}[A, A^*]$. Then the following are equivalent:*

(i) $\mathrm{nd}(A) = 1$;

(ii) there exist linearly independent $x, y \in \mathbb{C}^n$ such that $[A, A^*] = xx^* - yy^*$ and x, y, Ax, A^*y are linearly dependent;

(iii) $\mathrm{rank}[A, A^*] = 2$ *and for all $x, y \in \mathbb{C}^n$ such that $[A, A^*] = xx^* - yy^*$ we have that for some $t \in \mathbb{D}$ the vectors $x, y, Ax + tAy, A^*y + \bar{t}A^*x$ are linearly dependent;*

(iv) $\mathrm{rank}[A, A^*] = 2$ *and for all $x, y \in \mathbb{C}^n$ such that $[A, A^*] = xx^* - yy^*$ there exist $a, b \in \mathbb{C}$, with $|a|^2 - |b|^2 = 1$ such that*

$$aPAx + bPAy = \bar{a}PA^*y + \bar{b}PA^*x; \tag{5.9.3}$$

(v) there exist linearly independent $x, y \in \mathbb{C}^n$ such that $[A, A^*] = xx^* - yy^*$ and $PAx = PA^*y$.

In case (ii) holds we can choose $z \in \mathbb{C}$ and $\gamma \in \mathbb{T}$ such that

$$\begin{pmatrix} A & \gamma y \\ x^* & z \end{pmatrix} \tag{5.9.4}$$

is normal. Finally, if A is a real $n \times n$ matrix and $\mathrm{nd}(A) = 1$ *then a minimal normal completion can be chosen to be real as well.*

We first prove a lemma.

Lemma 5.9.5 *Suppose that $A \in \mathbb{C}^{n \times n}$, $x, y \in \mathbb{C}^n$ are such that $[A, A^*] = xx^* - yy^*$. Then*

$$x^*x = y^*y, x^*Ax = y^*Ay, x^*A^2x = y^*A^2y, x^*AA^*x = y^*A^*Ay. \tag{5.9.5}$$

Proof. Since $[A, A^*]$ has trace equal to 0, the vectors x and y satisfy $x^*x = y^*y$. Next, let $A(t) = A + tA^*$, $t \in \mathbb{R}$. One easily sees that for $t \neq 1, -1$,

$$\frac{1}{1 - t^2}[A(t), A(t)^*] = xx^* - yy^*.$$

But then it follows that

$$\begin{aligned} x^*A(t)^n x - y^*A(t)^n y &= \mathrm{tr}(A(t)^n(xx^* - yy^*)) \\ &= \mathrm{tr}\left(\frac{1}{1 - t^2}A(t)^n[A(t)A(t)^* - A(t)^*A(t)]\right) = 0. \end{aligned}$$

Checking both sides the coefficients of t^j, $0 \leq j \leq n$, for the cases when $n = 1, 2$, gives the equalities

$$x^*Ax = y^*Ay, x^*A^2x = y^*A^2y, x^*(AA^* + A^*A)x = y^*(AA^* + A^*A)y, \tag{5.9.6}$$

and their adjoints. Next, observe that

$$x^*(AA^* - A^*A)x = x^*(xx^* - yy^*)x = (x^*x)^2 - |x^*y|^2$$
$$= y^*yy^*y - y^*xx^*y = y^*(A^*A - AA^*)y.$$

Together with $x^*(AA^* + A^*A)x = y^*(AA^* + A^*A)y$, this gives $x^*AA^*x = y^*A^*Ay$. Thus we have established (5.9.5). □

One can prove in fact the more general formula

$$x^*w(A, A^*)x = y^*\overleftarrow{w}(A, A^*)y, \tag{5.9.7}$$

where $w(A, A^*)$ is a word in A and A^*, and $\overleftarrow{w}(A, A^*)$ is the reverse of $w(A, A^*)$. We will come back to this in the exercises.

Proof of Theorem 5.9.4. (i) ⇒ (v): Since $\mathrm{nd}(A) = 1$ the matrix A has a minimal normal completion $N = \begin{pmatrix} A & y \\ x^* & z \end{pmatrix}$. Then $NN^* = N^*N$ yields that $AA^* + yy^* = A^*A + xx^*$ and $Ax + y\overline{z} = A^*y + xz$. Thus $[A, A^*] = xx^* - yy^*$. As A is not normal, we must have that $[A, A^*] \neq 0$. Since $[A, A^*]$ is Hermitian and has trace 0, the rank of $[A, A^*]$ is at least 2, implying that x, y are linearly independent. But then it follows that $\mathrm{Ran}[A, A^*] = \mathrm{Span}\{x, y\}$. Next, we have that $P(Ax + y\overline{z}) = P(A^*y + xz)$. As x, y are in the orthogonal complement of the range of P we thus get $PAx = PA^*y$.

(v) ⇒ (iv): Let x, y be as in (v). Let $[A, A^*] = uu^* - vv^*$ for some $u, v \in \mathbb{C}^n$. Clearly $\mathrm{Ran}[A, A^*] \subseteq \mathrm{Span}\{x, y\} \cap \mathrm{Span}\{u, v\}$, and as $\mathrm{rank}[A, A^*] = 2$, we get that $\mathrm{Span}\{x, y\} = \mathrm{Span}\{u, v\}$. Thus $\begin{pmatrix} x & y \end{pmatrix} = \begin{pmatrix} u & v \end{pmatrix} B$ for some $B \in \mathbb{C}^{2\times 2}$. As $xx^* - yy^* = uu^* - vv^*$, we get that

$$\begin{pmatrix} u & v \end{pmatrix} J \begin{pmatrix} u^* \\ v^* \end{pmatrix} = \begin{pmatrix} u & v \end{pmatrix} BJB^* \begin{pmatrix} u^* \\ v^* \end{pmatrix},$$

where $J = \begin{pmatrix} 1 & 0 \\ 0 & -1 \end{pmatrix}$. Since $\begin{pmatrix} u & v \end{pmatrix}$ has a left inverse, we obtain $BJB^* = J$; that is, B is *J-unitary*. One easily checks that B must now be of the form

$$B = \begin{pmatrix} \alpha & \overline{\beta} \\ e^{i\theta}\beta & e^{i\theta}\overline{\alpha} \end{pmatrix},$$

with $|\alpha|^2 - |\beta|^2 = 1$. This yields that

$$x = \alpha u + e^{i\theta}\beta v, y = \overline{\beta}u + e^{i\theta}\overline{\alpha}v.$$

As $PAx = PA^*y$ we thus get that

$$\alpha PAu + e^{i\theta}\beta PAv = \overline{\beta}PA^*u + e^{i\theta}\overline{\alpha}PA^*v.$$

Dividing by $e^{\frac{1}{2}i\theta}$ and putting $a = \alpha e^{-\frac{1}{2}i\theta}$ and $b = \beta e^{\frac{1}{2}i\theta}$ yields the desired result.

(iv) ⇒ (iii): Let x, y be such that $[A, A^*] = xx^* - yy^*$. By (iv) there exist a, b with $|a|^2 - |b|^2 = 1$ such that (5.9.3) holds. Clearly $a \neq 0$. Let $t = \frac{b}{a} \in \mathbb{D}$. Then

$$PAx + tPAy = \frac{\overline{a}}{a}(PA^*y + \overline{t}PA^*x).$$

Thus $PAx + tPAy$ and $PA^*y + \bar{t}PA^*x$ are linearly dependent. As P is an orthogonal projection with kernel equal to $\mathrm{Ran}[A, A^*] = \mathrm{Span}\{x, y\}$, we get that the vectors $x, y, Ax + tAy, A^*y + \bar{t}A^*x$ span a space of dimension at most 3, yielding (iii).

(iii) $\Rightarrow$ (ii): As $\mathrm{rank}[A, A^*] = 2$ and $[A, A^*]$ is Hermitian with trace 0, we have that $[A, A^*] = xx^* - yy^*$ for some linearly independent x, y (use for instance the spectral decomposition of $[A, A^*]$). By (iii) there exists a $t \in \mathbb{D}$ such that $x, y, Ax + tAy, A^*y + \bar{t}A^*x$ are linearly dependent. We may represent t as $t = \frac{\sqrt{r^2-1}}{r}e^{i\theta}$ for some $r \geq 1$. Put now

$$u = rx + \sqrt{r^2 - 1}e^{i\theta}y, \; v = ry + \sqrt{r^2 - 1}e^{-i\theta}x.$$

Then u, v, Au, A^*v are linearly dependent, and $[A, A^*] = xx^* - yy^* = uu^* - vv^*$.

(ii) $\Rightarrow$ (i): Let $[A, A^*] = xx^* - yy^*$ such that x, y, Ax, A^*y are linearly dependent. Let us find the appropriate γ and z such that (5.9.4) is normal. As the vectors x, y, Ax, A^*y are linearly dependent, there must exist complex numbers a, b, c, d, not all equal to zero, such that

$$aA^*y + by = cAx + dx. \tag{5.9.8}$$

Notice that a and c cannot both be zero, as x and y are linearly independent. We now claim that

$$\bar{a}Ax + \bar{b}x = \bar{c}A^*y + \bar{d}y \tag{5.9.9}$$

holds as well. In order to show (5.9.9), it suffices to show that

$$v^*(\bar{a}Ax + \bar{b}x) = v^*(\bar{c}A^*y + \bar{d}y) \tag{5.9.10}$$

for all $v \in \{x, y, Ax, A^*y\}$. But this follows from combining (5.9.8) and (5.9.5) in Lemma 5.9.5. For instance, when we take $v = Ax$, we get

$$\bar{a}x^*A^*Ax + \bar{b}x^*A^*x = (\bar{a}y^*A + \bar{b}y^*)A^*y = (\bar{c}x^*A^* + x^*\bar{d})A^*y,$$

establishing (5.9.10) for $v = Ax$. For other v the argument is analogous. This gives (5.9.9).

Let now $\alpha \in \mathbb{T}$ be such that $\bar{c}$ and αa have the same argument. Then $\alpha a + \bar{c} \neq 0$, as not both a and c are zero. Multiplying (5.9.8) by α and adding it to (5.9.9) we get

$$(\alpha a + \bar{c})A^*y + (\alpha b + \bar{d})y = (\alpha c + \bar{a})Ax + (\alpha d + \bar{b})x. \tag{5.9.11}$$

Letting now $\gamma = \frac{\alpha a + \bar{c}}{\alpha c + \bar{a}} \in \mathbb{T}$ and $z = -\frac{\alpha d + \bar{b}}{\alpha c + \bar{a}} \in \mathbb{C}$, one can easily check that (5.9.4) is normal.

Note that if A, x, y are real then the construction in the proof of (ii)$\Rightarrow$(i) yields a real normal matrix (5.9.4).

Finally, we show that if A is real and $\mathrm{nd}(A) = 1$, then a minimal normal completion (5.9.4) can be chosen to be real as well. Clearly, if A is real and $\mathrm{rank}[A, A^*] = 2$, then we can always find real vectors x and y such that $[A, A^*] = xx^* - yy^*$ (use for instance the spectral decomposition of

$[A, A^*]$). As $\mathrm{nd}(A) = 1$ we obtain by (iv) that there exist complex a, b with $|a|^2 - |b|^2 = 1$ such that (5.9.3) holds. But then

$$(\mathrm{Re}\ a)PAx + (\mathrm{Re}\ b)PAy = (\mathrm{Re}\ a)PA^*y + (\mathrm{Re}\ b)PA^*x, \tag{5.9.12}$$

$$(\mathrm{Im}\ a)PAx + (\mathrm{Im}\ b)PAy = -(\mathrm{Im}\ a)PA^*y - (\mathrm{Im}\ b)PA^*x. \tag{5.9.13}$$

If $|\mathrm{Re}\ a| > |\mathrm{Re}\ b|$ then one may multiply (5.9.12) by a real scalar to obtain real α and β with $|\alpha|^2 - |\beta|^2 = 1$ such that

$$\alpha PAx + \beta PAy = \alpha PA^*y + \beta PA^*x. \tag{5.9.14}$$

Otherwise, as $|a|^2 - |b|^2 = 1$, we will have that $|\mathrm{Im}\ a| > |\mathrm{Im}\ b|$. Now by multiplying (5.9.13) by a scalar we obtain real α and β with $|\alpha|^2 - |\beta|^2 = 1$ such that

$$\alpha PAx + \beta PA(-y) = \alpha PA^*(-y) + \beta PA^*x. \tag{5.9.15}$$

In case of (5.9.14) introduce $u = \alpha x + \beta y, v = \alpha y + \beta x$, and in case of (5.9.15) introduce $u = \alpha x - \beta y, v = -\alpha y + \beta x$. Then u, v are real vectors such that $[A, A^*] = uu^* - vv^*$ and u, v, Au, A^*v are linearly dependent. Then, by the construction in the proof of (ii)$\Rightarrow$(i) one may construct a real minimal normal completion. □

While in the case $\mathrm{nd}(A) = 1$ one may always find a real minimal normal completion when A is real, it is not clear that this also holds when $\mathrm{nd}(A) > 1$.

Clearly, an indefinite Hermitian matrix $[A, A^*]$ of rank 2 can always be written as $xx^* - yy^*$ for some vectors x and y. It can happen, though, that for some choices of x and y the vectors x, y, Ax, A^*y are linearly independent, while for other choices they are not. The following example illustrates this.

Example 5.9.6 Let

$$A = \begin{pmatrix} 0 & 1 & 0 & 0 \\ 0 & 0 & 1 & 0 \\ 0 & 0 & 0 & 1 \\ 0 & 0 & 0 & 0 \end{pmatrix}.$$

Then $[A, A^*] = xx^* - yy^* = zz^* - ww^*$, where

$$x = \begin{pmatrix} 1 \\ 0 \\ 0 \\ 0 \end{pmatrix}, \ y = \begin{pmatrix} 0 \\ 0 \\ 0 \\ 1 \end{pmatrix}, \ z = \begin{pmatrix} \frac{5}{3} \\ 0 \\ 0 \\ \frac{4}{3} \end{pmatrix}, \ w = \begin{pmatrix} \frac{4}{3} \\ 0 \\ 0 \\ \frac{5}{3} \end{pmatrix}.$$

Also, x, y, Ax, A^*y are linearly dependent, while z, w, Az, A^*w are not.

Using Theorem 5.9.4 we can now establish a procedure to determine whether $\mathrm{nd}(A) = 1$ or not. Indeed, after checking that $\mathrm{rank}[A, A^*] = 2$ one can choose x, y such that $[A, A^*] = xx^* - yy^*$. By (iv) in Theorem 5.9.4 it remains to check whether there exist complex a, b with $|a|^2 - |b|^2 = 1$ such that (5.9.3) holds. Introduce

$$PAx =: u = u_R + iu_I, \ PAy =: v = v_R + iv_I, \tag{5.9.16}$$

$$PA^*y =: q = q_R + iq_I, \; PA^*x =: w = w_R + iw_I,$$

where $u_R, u_I, v_R, v_I, q_R, q_I, w_R, w_I \in \mathbb{R}^n$. Also write $a = a_I + ia_I, b = b_R + ib_I$, with $a_R, a_I, b_R, b_I \in \mathbb{R}$. Equation (5.9.3) now becomes

$$(a_R + ia_I)(u_R + iu_I) + (b_R + ib_I)(v_R + iv_I) \tag{5.9.17}$$

$$= (a_R - ia_I)(q_R + iq_I) + (b_R - ib_I)(w_R + iw_I).$$

If we introduce

$$Q = \begin{pmatrix} u_R - q_R & -u_I - q_I & v_R - w_R & -v_I - w_I \\ u_I - q_I & u_R + q_R & v_I - w_I & v_R + w_R \end{pmatrix}, \tag{5.9.18}$$

then (5.9.17) is equivalent to

$$Q \begin{pmatrix} a_R \\ a_I \\ b_R \\ b_I \end{pmatrix} = 0. \tag{5.9.19}$$

Let $m = \text{rank}Q$. If $m = 0$ then clearly (5.9.19) has a solution with $a_R^2 + a_I^2 - b_R^2 - b_I^2 = 1$, and thus $\text{nd}(A) = 1$. If $m = 4$, then (5.9.19) has no nontrivial solutions and thus $\text{nd}(A) > 1$. In case $1 \le m \le 3$, let $F \in \mathbb{R}^{4\times(4-m)}$ such that the columns of F span the kernel of Q. Write $F = \left(\begin{smallmatrix} F_1 \\ F_2 \end{smallmatrix}\right)$ with $F_i \in \mathbb{R}^{2\times(4-m)}$, $i = 1, 2$. Then there exist a, b with $|a|^2 - |b|^2 = 1$ such that (5.9.3) holds if and only if there exists an $h \in \mathbb{R}^{4-m}$ such that

$$\gamma := h^T(F_1^T F_1 - F_2^T F_2)h > 0. \tag{5.9.20}$$

Indeed, in that case one may let

$$\begin{pmatrix} a_R \\ a_I \\ b_R \\ b_I \end{pmatrix} = \frac{1}{\sqrt{\gamma}} \begin{pmatrix} F_1 \\ F_2 \end{pmatrix} h,$$

to obtain $a = a_I + ia_I, b = b_R + ib_I$ that satisfy (iv) in Theorem 5.9.4. Clearly, h exists such that (5.9.20) if and only if $F_1^T F_1 - F_2^T F_2$ has a positive eigenvalue. Finally observe that if $m = 1$, then we can take $0 \ne h \in \text{Ker}F_2$ to arrive at (5.9.20). Thus we arrive at the following procedure.

Procedure for checking whether $\text{nd}(A) = 1$**:** Let $A \in \mathbb{C}^{n\times n}$ be given.

1. If $\text{rank}[A, A^*] \ne 2$ then $\text{nd}(A) \ne 1$. If $\text{rank}[A, A^*] = 2$ then continue with Step 2.

2. Write $[A, A^*] = xx^* - yy^*$, and introduce u, v, q, w as in (5.9.16). Build the matrix Q as in (5.9.18), and put $m = \text{rank}Q$. If $m \le 1$ then $\text{nd}(A) = 1$. If $m = 4$ then $\text{nd}(A) > 1$. If $2 \le m \le 3$ then continue with Step 3.

3. Let $F \in \mathbb{R}^{4\times(4-m)}$ be such that the columns of F span the kernel of Q, and write $F = \begin{pmatrix} F_1 \\ F_2 \end{pmatrix}$ with $F_i \in \mathbb{R}^{2\times(4-m)}$, $i = 1, 2$. If $F_1^T F_1 - F_2^T F_2 \leq 0$ then $\text{nd}(A) > 1$. Otherwise, $\text{nd}(A) = 1$.

For 3×3 matrices one can easily determine the normal defect.

Corollary 5.9.7 *Let A be a 3×3 matrix. Then*

$$\text{nd}(A) = \max\{i_+([A, A^*]), i_-([A, A^*])\} = \lceil \text{rank}[A, A^*]/2 \rceil.$$

Proof. When $[A, A^*] = 0$, clearly $\text{nd}(A) = 0$. Notice that when $[A, A^*] \neq 0$ it must have at least one negative and one positive eigenvalue as $\text{tr}[A, A^*] = 0$. When $\text{rank}[A, A^*] = 3$, then both the lower bound and the upper bound in Theorem 5.9.1 are equal to 2, yielding $\text{nd}(A) = 2$. Finally, when $\text{rank}[A, A^*] = 2$, we get from Theorem 5.9.4 that $\text{nd}(A) = 1$, as in a 3-dimensional space the four vectors x, y, Ax, A^*y are automatically linearly dependent. □

As the above shows, it is hard in general to determine the normal defect of a matrix. Even the following seemingly simple problem is still open: does it hold in general that $\text{nd}(\text{diag}(A_i)_{i=1}^m) = \sum_{i=1}^m \text{nd}(A_i)$?

We end this section with a Matlab procedure that first generates a 3×3 matrix with normal defect 2, and subsequently finds a minimal normal extension for it.

```
A=rand(3,3); B=A*A'-A'*A; [U,D]=eig(B);
if D(2,2) < 0, y=U(:,1)*sqrt(-D(1,1));yy=U(:,1); z=U(:,2)*sqrt(-D(2,2));
   zz=U(:,2); x=U(:,3)*sqrt(D(3,3)),
else A=A'; x=U(:,1)*sqrt(-D(1,1)); y=U(:,2)*sqrt(D(2,2));yy=U(:,2);
   z=U(:,3)*sqrt(D(3,3));zz=U(:,3); end
norm(x*x'-y*y'-z*z'-(A*A'-A'*A))
K= [1 -sqrt(2)*i sqrt(2)*i;-sqrt(2)*i 0 1; -sqrt(2)*i -1 2]; M=[y z x]*K';
y=M(:,1); z=M(:,2); x=M(:,3);
norm(x*x'-y*y'-z*z'-(A*A'-A'*A))
SS=[x y z]; abc=inv(SS)*A*x; def=inv(SS)*A'*y;
ghi=inv(SS)*A'*z; F2=[abc(1)-conj(def(2)) -conj(ghi(2)); ...
   -conj(def(3)) abc(1)-conj(ghi(3))];
E2=[def(1)-conj(abc(2));ghi(1)-conj(abc(3))];
vv=inv(F2)*E2; Tconj=[def(2)-vv(1)*abc(2) ghi(2)-vv(2)*abc(2); ...
   def(3)-vv(1)*abc(3) ghi(3)-vv(2)*abc(3)];
NN=[A y z; conj(vv)*x' Tconj']; norm(NN*NN'-NN'*NN)
M=[x'*x 0 0; 0 -y'*y -y'*z;0 -z'*y -z'*z];
J=-eye(3);J(1,1)=1; Q=[abc def ghi]; MM=[M Q';Q J];
D=[sqrt(x'*x) 0 0 ; 0 sqrt(y'*y) 0 ; 0 0 sqrt(z'*z)];
DD=[D zeros(3);zeros(3) eye(3)]; inv(DD)*MM*inv(DD)
```

5.10 APPLICATION TO MINIMAL REPRESENTATION OF DISCRETE SYSTEMS

Consider the discrete time system

$$\begin{cases} x_{k+1} = A_k x_k + B_k u_k, & x_0 = 0, \quad k = 0, \dots, n-1, \\ y_{k+1} = C_k x_{k+1}, \end{cases}$$

with $\{u_k\}_{k=0}^{n-1}$ the sequence of vector inputs in $\mathbb{R}^n$, $\{x_k\}_{k=0}^{n}$ the sequence of vector states in the *state space* $\mathbb{R}^p$ and $\{y_k\}_{k=1}^{n}$ the sequence of vector outputs $\mathbb{R}^m$. The dimension p of the state space is called the *order of the system.* The input-output map is given by $\mathrm{col}(y_i)_{i=1}^{n} = T\mathrm{col}(u_i)_{i=0}^{n-1}$, where

$$T = \begin{pmatrix} C_0B_0 & 0 & \cdots & 0 \\ C_1A_1B_0 & C_1B_1 & \ddots & 0 \\ C_2A_2A_1B_0 & C_2A_2B_1 & \ddots & \vdots \\ \vdots & \vdots & & 0 \\ C_{n-1}A_{n-1}\cdots A_1B_0 & C_{n-1}A_{n-1}\cdots A_2B_1 & \cdots & C_{n-1}B_{n-1} \end{pmatrix}.$$

With any completion X of the lower triangular partial matrix

$$\mathcal{A} = \begin{pmatrix} C_0B_0 & & & ? \\ C_1A_1B_0 & C_1B_1 & & \\ \vdots & \vdots & \ddots & \\ C_{n-1}A_{n-1}\cdots A_1B_0 & C_{n-1}A_{n-1}\cdots A_2B_1 & \cdots & C_{n-1}B_{n-1} \end{pmatrix} \tag{5.10.1}$$

we can associate a system with the same input-output behavior as the original system, as follows. Write $q = \mathrm{rank}\ X$, and make a rank decomposition $X = \mathrm{col}(F_i)_{i=0}^{n-1}\mathrm{row}(G_i)_{i=0}^{n-1}$ of X, where F_i and G_i are of sizes $m \times q$ and $q \times n$, respectively. Then the system

$$\begin{cases} x_{k+1} = x_k + G_k u_k, & x_0 = 0, \quad k = 0, \dots, n-1, \\ y_{k+1} = F_k x_{k+1} \end{cases}$$

has the same input-output behavior as the original system. This leads to a problem of choosing a completion X of (5.10.1) of minimal possible rank. Thus, Theorem 5.2.1 may be applied. This leads to the following result.

Theorem 5.10.1 *Given the discrete time system*

$$\theta_1 = \begin{cases} x_{k+1} = A_k x_k + B_k u_k, & x_0 = 0, \quad k = 0, \dots, n-1, \\ y_{k+1} = C_k x_{k+1}. \end{cases}$$

Let $\mathcal{A}$ be as in (5.10.1) *and put $\alpha = \min\ \mathrm{rank}\mathcal{A}$. Then there exists a discrete time system*

$$\theta_2 = \begin{cases} x'_{k+1} = A'_k x'_k + B'_k u_k, & x'_0 = 0, \quad k = 0, \dots, n-1, \\ y'_{k+1} = C'_k x'_{k+1}, \end{cases}$$

of order α such that θ_1 and θ_2 have the same input-output behavior. Moreover, α is the smallest number with this property.

5.11 THE SEPARABILITY PROBLEM IN QUANTUM INFORMATION

In this section we introduce the so-called "separability problem" that appears in quantum information and connect it to normal completions. We will use the notation $\otimes$ for the Kronecker product; that is, $(a_{ij})_{i,j=1}^n \otimes B := (a_{ij}B)_{i,j=1}^n$. Recall that if $A \geq 0, B \geq 0$, then $A \otimes B \geq 0$. Similarly, $A > 0, B > 0$ implies that $A \otimes B > 0$. One way to see this is by realizing that the eigenvalues of $A \otimes B$ are exactly all possible products of an eigenvalue of A with an eigenvalue of B.

Given positive integers $N_1, \ldots, N_d$, the $N_1 \times \cdots \times N_d$ separability problem asks whether a complex positive semidefinite matrix of size $(\prod_{i=1}^d N_i) \times (\prod_{i=1}^d N_i)$ lies in the cone generated by matrices

$$Q_1 \otimes \cdots \otimes Q_d,$$

where Q_i is a $N_i \times N_i$ positive semidefinite matrix, $i = 1, \ldots, d$. We will focus on the case when $d = 2$, and write $N_1 = N$, $N_2 = M$. Thus we are concerned with matrices

$$\begin{pmatrix} A_{11} & \cdots & A_{1N} \\ \vdots & & \vdots \\ A_{N1} & \cdots & A_{NN} \end{pmatrix} \geq 0, \tag{5.11.1}$$

where A_{ij} are $M \times M$ matrices. The $N \times M$ separability problem asks when we can write (5.11.1) as

$$\sum_{i=1}^{K} Q_{1i} \otimes Q_{2i}, \tag{5.11.2}$$

where K is some positive integer and Q_{1i} and Q_{2i} are positive semidefinite matrices of size $N \times N$ and $M \times M$, respectively. If this is possible, we say that the matrix (5.11.1) is $N \times M$ *separable*. In case of separability one may always choose Q_{1i} and Q_{2i} to be of rank 1, and we refer to K as the *number of states* in the separable representation (5.11.2).

It is easy to give an example of a positive semidefinite matrix that is not separable. For instance,

$$A = \begin{pmatrix} 1 & 0 & 0 & 1 \\ 0 & 0 & 0 & 0 \\ 0 & 0 & 0 & 0 \\ 1 & 0 & 0 & 1 \end{pmatrix}$$

is not 2×2 separable. Indeed, if $A = \sum_{i=1}^k Q_{1i} \otimes Q_{2i}$ with Q_{1i}, Q_{2i} 2×2 positive semidefinite matrices, then

$$A^\Gamma := \sum_{i=1}^{k} Q_{1i} \otimes Q_{2i}^T = \begin{pmatrix} 1 & 0 & 0 & 0 \\ 0 & 0 & 1 & 0 \\ 0 & 1 & 0 & 0 \\ 0 & 0 & 0 & 1 \end{pmatrix}$$

should also be positive semidefinite (as $Q_{2i}^T \geq 0$), giving a contradiction. By the same reasoning, if A in (5.11.1) is $N \times M$ separable then

$$A^\Gamma := \begin{pmatrix} A_{11}^T & \dots & A_{1N}^T \\ \vdots & & \vdots \\ A_{N1}^T & \dots & A_{NN}^T \end{pmatrix} \geq 0. \tag{5.11.3}$$

The matrix A^Γ is known as the *partial transpose* of A, and checking that $A^\Gamma \geq 0$ is known as the *Peres test.*

Let us show that the $2 \times M$ separability problem reduces to a normal completion problem as follows. In this case we are concerned with matrices

$$P = \begin{pmatrix} A & B \\ B^* & C \end{pmatrix} \geq 0. \tag{5.11.4}$$

The Peres test concerns checking that $P^\Gamma \geq 0$, which is equivalent to checking whether

$$\overline{P^\Gamma} = \begin{pmatrix} A & B^* \\ B & C \end{pmatrix} \geq 0. \tag{5.11.5}$$

We first reduce the problem to the case when $A = I$. The notation $A^{(-\frac{1}{2})}$ is used for the Moore-Penrose inverse of $A^{\frac{1}{2}}$.

Proposition 5.11.1 *Let A, B, and C be $M \times M$ matrices such that* (5.11.4) *and* (5.11.5) *hold. Let $\mathcal{H} = \mathrm{Ker} A \subseteq \mathbb{C}^M$, S be the Schur complement of A supported on $\mathcal{H}$, and $P : \mathbb{C}^M \to \mathcal{H}^\perp$ the orthogonal projection onto $\mathcal{H}^\perp = \mathrm{Ran} A$. Then*

$$\begin{pmatrix} A & B \\ B^* & C \end{pmatrix} \geq 0 \tag{5.11.6}$$

is $2 \times M$ separable if and only if

$$\begin{pmatrix} I_{\mathcal{H}^\perp} & PA^{(-\frac{1}{2})}BA^{(-\frac{1}{2})}P^* \\ PA^{(-\frac{1}{2})}B^*A^{(-\frac{1}{2})}P^* & PA^{(-\frac{1}{2})}SA^{(-\frac{1}{2})}P^* \end{pmatrix} \geq 0 \tag{5.11.7}$$

is $2 \times \dim \mathcal{H}^\perp$ separable. Here $I_{\mathcal{H}^\perp}$ denotes the identity operator on $\mathcal{H}^\perp$.

Proof. Let $\mathcal{H} = \mathrm{Ker} A$ and decompose $A = (A_{ij})_{i,j=1}^2$, $B = (B_{ij})_{i,j=1}^2$, and $C = (C_{ij})_{i,j=1}^2$ with respect to the decomposition $\mathcal{H} \oplus \mathcal{H}^\perp$. Then by (5.11.4) and (5.11.5) we have that $A_{11}, A_{12}, A_{21}, B_{11}, B_{12}$ and B_{21} are all zero. Thus

$$\begin{pmatrix} A & B \\ B^* & C \end{pmatrix} = \begin{pmatrix} 0 & 0 & 0 & 0 \\ 0 & A_{22} & 0 & B_{22} \\ 0 & 0 & C_{11} & C_{12} \\ 0 & B_{22}^* & C_{21} & C_{22} \end{pmatrix}. \tag{5.11.8}$$

Let now

$$Q = \begin{pmatrix} I & 0 \\ -C_{21}C_{11}^{(-1)} & I \end{pmatrix}.$$

Then

$$\begin{pmatrix} QAQ^* & QBQ^* \\ QB^*Q^* & QCQ^* \end{pmatrix} = \begin{pmatrix} 0 & 0 & 0 & 0 \\ 0 & A_{22} & 0 & B_{22} \\ 0 & 0 & C_{11} & 0 \\ 0 & B_{22}^* & 0 & C_{22} - C_{21}C_{11}^{(-1)}C_{12} \end{pmatrix}. \quad (5.11.9)$$

It is an elementary fact that, due to the invertibility of Q, (5.11.8) is separable if and only if (5.11.9) is. In addition, we claim that (5.11.9) is separable if and only if

$$\begin{pmatrix} A_{22} & B_{22} \\ B_{22}^* & C_{22} - C_{21}C_{11}^{(-1)}C_{12} \end{pmatrix} \quad (5.11.10)$$

is. Indeed, if (5.11.9) is written as $\sum_{i=1}^K Q_{1i} \otimes Q_{2i}$, with $Q_{1i} \geq 0$ and $Q_{2i} \geq 0$, then (5.11.10) equals $\sum_{i=1}^K Q_{1i} \otimes PQ_{2i}P^*$. Conversely, if (5.11.10) is written as $\sum_{i=1}^K Q_{1i} \otimes Q_{2i}$, then (5.11.9) equals

$$\sum_{i=1}^K Q_{1i} \otimes P^*Q_{2i}P + \begin{pmatrix} 0 & 0 \\ 0 & 1 \end{pmatrix} \otimes \begin{pmatrix} C_{11} & 0 \\ 0 & 0 \end{pmatrix}.$$

Finally, as $A_{22}^{(-\frac{1}{2})}$ is invertible on $\mathcal{H}^\perp$, we get that (5.11.10) is separable if and only if

$$\begin{pmatrix} I_{H^\perp} & A_{22}^{(-\frac{1}{2})}B_{22}A_{22}^{(-\frac{1}{2})} \\ A_{22}^{(-\frac{1}{2})}B_{22}^*A_{22}^{(-\frac{1}{2})} & A_{22}^{(-\frac{1}{2})}(C_{22} - C_{21}C_{11}^{(-1)}C_{12})A_{22}^{(-\frac{1}{2})} \end{pmatrix}$$

is separable. Since this latter matrix equals (5.11.7) we are done. □

Theorem 5.11.2 *Let $Q, \Lambda \in \mathbb{C}^{M\times M}$ be such that*

$$\begin{pmatrix} I & Q \\ Q^* & \Lambda \end{pmatrix} \geq 0 \,, \begin{pmatrix} I & Q^* \\ Q & \Lambda \end{pmatrix} \geq 0. \quad (5.11.11)$$

Then

$$\begin{pmatrix} I & Q \\ Q^* & \Lambda \end{pmatrix} \quad (5.11.12)$$

is separable if and only if there exists a normal matrix

$$R = \begin{pmatrix} Q & S \\ * & * \end{pmatrix}$$

such that $SS^* \leq \Lambda - QQ^*$. In that case, the minimal number of states in a separable representation of (5.11.12) is at most

$$\text{size}\, R + \text{rank}(\Lambda - QQ^* - SS^*). \quad (5.11.13)$$

Proof. Suppose that (5.11.12) is $2 \times M$ separable. Let $x_i \in \mathbb{C}^2$ and $y_i \in \mathbb{C}^M$ be such that such that

$$\begin{pmatrix} I & Q \\ Q^* & \Lambda \end{pmatrix} = \sum_{i=1}^K x_i x_i^* \otimes y_i y_i^*.$$

Since for $\mu \in \mathbb{C} \setminus \{0\}$ we have that $x_i x_i^* \otimes y_i y_i^* = (\frac{1}{\mu} x_i)(\frac{1}{\mu} x_i^*) \otimes (\mu y_i)(\mu y_i)^*$, we may always normalize x_i such that its first nonzero entry is 1. Thus we may assume that $x_i = \begin{pmatrix} 1 & x_{2i} \end{pmatrix}^T$ or $x_i = \begin{pmatrix} 0 & * \end{pmatrix}^T$. In other words, we may write (with possibly other x_i, y_i)

$$\begin{pmatrix} I & Q \\ Q^* & \Lambda \end{pmatrix} = \sum_{i=1}^{K_1} x_i x_i^* \otimes y_i y_i^* + \sum_{i=1}^{K_2} w_i w_i^* \otimes z_i z_i^*,$$

where $x_i = \begin{pmatrix} 1 & x_{2i} \end{pmatrix}^T$, $i = 1, \ldots, K_1$, and $w_i = \begin{pmatrix} 0 & 1 \end{pmatrix}^T$, $i = 1, \ldots, K_2$. Let $Y = \begin{pmatrix} y_1 & \cdots & y_{K_1} \end{pmatrix}$ and $N_2 = \operatorname{diag}(x_{2i})_{i=1}^{K_1}$. Putting $N_1 = I$, we have that

$$\begin{pmatrix} I & Q \\ Q^* & \Lambda \end{pmatrix} = (Y N_i N_j^* Y^*)_{i,j=1}^2 + \begin{pmatrix} 0 & 0 \\ 0 & \sum_{i=1}^{K_2} z_i z_i^* \end{pmatrix}. \tag{5.11.14}$$

Since Y is a co-isometry (by the upper left corner of (5.11.14)), we may choose V such that $\left(\begin{smallmatrix} Y \\ V \end{smallmatrix}\right)$ is unitary (e.g., $V = (I - Y^* Y)^{\frac{1}{2}}$). Put

$$R = \begin{pmatrix} Y \\ V \end{pmatrix} N_2^* \begin{pmatrix} Y^* & V^* \end{pmatrix} = \begin{pmatrix} Q & S \\ * & * \end{pmatrix},$$

where $S = Y N_2^* V^*$. Then R is normal. Moreover,

$$RR^* = \begin{pmatrix} Y \\ V \end{pmatrix} N_2^* N_2 \begin{pmatrix} Y^* & V^* \end{pmatrix} = \begin{pmatrix} QQ^* + SS^* & * \\ * & * \end{pmatrix}.$$

Thus by using (5.11.14),

$$\Lambda - QQ^* - SS^* = Y N_2 N_2^* Y^* + \sum_{i=1}^{K_2} z_i z_i^* - Y N_2 N_2^* Y^* \geq 0.$$

This the "only if" part.

Conversely, suppose that

$$R = \begin{pmatrix} Q & S \\ * & * \end{pmatrix}$$

is normal with $SS^* \leq \Lambda - QQ^*$. We may write

$$R = \begin{pmatrix} Y \\ V \end{pmatrix} N_2 \begin{pmatrix} Y^* & V^* \end{pmatrix}$$

with Y of size $M \times K$, $\left(\begin{smallmatrix} Y \\ V \end{smallmatrix}\right)$ unitary, and $N_2 = \operatorname{diag}(x_{2j})_{i,j=1}^K$ diagonal. Put $N_1 = I_K = R_1$ and $S_1 = 0$. Note that

$$\begin{pmatrix} Y \\ V \end{pmatrix} N_2 N_2^* \begin{pmatrix} Y^* & V^* \end{pmatrix} = RR^* = \begin{pmatrix} QQ^* + SS^* & * \\ * & * \end{pmatrix}, i, j = 1, \ldots, N,$$

and thus $QQ^* + SS^* = Y N_2 N_2^* Y^*$. Let now $x_i = \begin{pmatrix} 1 & x_{2i} \end{pmatrix}^T$, $i = 1, \ldots, K$, and write $Y = \begin{pmatrix} y_1 & \cdots & y_K \end{pmatrix}$. Then

$$\begin{pmatrix} I & Q \\ Q^* & QQ^* + SS^* \end{pmatrix} = \sum_{j=1}^{K} x_j x_j^* \otimes y_j y_j^*$$

is $2 \times M$ separable. Since $\Lambda - QQ^* - SS^* \geq 0$, we may write

$$\Lambda - QQ^* - SS^* = \sum_{j=1}^{L} z_i z_i^*,$$

where $z_i \in \mathbb{C}^M$ and L is as small as possible. Put $w_i = \binom{0}{1} \in \mathbb{C}^2$, $i = 1, \ldots, L$. Then

$$\begin{pmatrix} I & Q \\ Q^* & \Lambda \end{pmatrix} = \sum_{j=1}^{K} x_j x_j^* \otimes y_j y_j^* + \sum_{j=1}^{L} w_i w_i^* \otimes z_i z_i^*$$

is $2 \times M$ separable. Moreover, the number of states in this representation equals (5.11.13). This proves the result. □

In the special case that $Q\Lambda = \Lambda Q$, a normal completion of

$$\begin{pmatrix} Q & (\Lambda - QQ^*)^{\frac{1}{2}} \\ ? & ? \end{pmatrix}$$

can easily be given.

Proposition 5.11.3 *Let Q and Λ be $M \times M$ matrices such that $Q\Lambda = \Lambda Q^*$ and*

$$\begin{pmatrix} I & Q \\ Q^* & \Lambda \end{pmatrix} \geq 0 \, , \; \begin{pmatrix} I & Q^* \\ Q & \Lambda \end{pmatrix} \geq 0. \tag{5.11.15}$$

Then

$$\begin{pmatrix} Q & (\Lambda - QQ^*)^{\frac{1}{2}} \\ (\Lambda - Q^*Q)^{\frac{1}{2}} & -Q^* \end{pmatrix} \tag{5.11.16}$$

is normal.

Proof. Since $Q\Lambda = \Lambda Q$ and Λ is Hermitian, we get by taking adjoints that $Q^*\Lambda = \Lambda Q^*$. Moreover, since $Q(\Lambda - Q^*Q) = (\Lambda - QQ^*)Q$, we get by functional calculus that

$$Q(\Lambda - Q^*Q)^{\frac{1}{2}} = (\Lambda - QQ^*)^{\frac{1}{2}} Q.$$

Similarly (or by taking adjoints),

$$Q^*(\Lambda - QQ^*)^{\frac{1}{2}} = (\Lambda - Q^*Q)^{\frac{1}{2}} Q^*.$$

It is now straightforward to check that (5.11.16) is normal. □

Corollary 5.11.4 *Let Q and Λ be $M \times M$ matrices such that $Q\Lambda = \Lambda Q$ and* (5.11.15) *holds. Then*

$$\begin{pmatrix} I & Q \\ Q^* & \Lambda \end{pmatrix}$$

is separable. In this case a separable representation can be chosen with at most $2M$ states.

Proof. Combine Theorem 5.11.2 and Proposition 5.11.3, where we choose $S = (\Lambda - QQ^*)^{\frac{1}{2}}$. □

Corollary 5.11.5 *Let B, $C \in \mathbb{C}^{n\times n}$ be such that*

$$M = \begin{pmatrix} I_n & B^* \\ B & C \end{pmatrix} \geq 0, \quad \widetilde{M} = \begin{pmatrix} I_n & B \\ B^* & C \end{pmatrix} \geq 0,$$

and suppose that $\text{rank} M = \text{rank}\widetilde{M} = n + 1$. *Write*

$$C - BB^* = xx^*, \quad C - B^*B = yy^*$$

for some vectors $x, y \in \mathbb{C}^n$. Then M is $2 \times n$ separable if and only if x, y, B^*x, By are linearly dependent. In this case, the minimal number of states in a separable representation of M is at most $n + 1$.

Proof. First notice that $B^*B - BB^* = xx^* - yy^*$.

Suppose that x, y, B^*x, By are linearly dependent. Then by Theorem 5.9.4 there exists a normal matrix

$$N = \begin{pmatrix} B & \nu x \\ y^* & z \end{pmatrix},$$

where $|\nu| = 1$. But as $(\nu x)(\nu x)^* = C - BB^*$ it follows from Theorem 5.11.2 that M is $2 \times n$ separable, and that the minimal number of states in a separable representation of M is $n + 1$.

Conversely, suppose that M is $2 \times n$ separable. By Theorem 5.11.2 there exists a normal matrix $N = (\begin{smallmatrix} B & S \\ T & P \end{smallmatrix})$ such that $BB^* + SS^* \leq C$. But then $SS^* \leq xx^*$ and thus $S = xv^*$ with $\|v\| \leq 1$. Also $B^*B + T^*T = BB^* + SS^* \leq C$, and thus $T^*T \leq yy^*$, yielding $T = yw^*$ with $\|w\| \leq 1$. In addition, $BT^* + SP^* = B^*S + T^*P$. In particular, $\text{Ran}(BT^* + SP^*) = \text{Ran}(B^*S + T^*P)$. Note that $\text{Ran}(BT^* + SP^*) \subseteq \text{Span}\{By, x\}$ and $\text{Ran}(B^*S + T^*P) \subseteq \text{Span}\{B^*x, y\}$. But then it follows easily that x, y, B^*x, By are linearly dependent. Indeed, if $BT^* + SP^* = B^*S + T^*P \neq 0$, then $\text{Span}\{By, x\}$ and $\text{Span}\{B^*x, y\}$ must have a nontrivial intersection, and if $BT^* + SP^* = B^*S + T^*P = 0$, then $\text{Span}\{By, x\}$ and $\text{Span}\{B^*x, y\}$ are both at most one-dimensional. □

In low dimensional cases passing the Peres test is equivalent to being separable.

Theorem 5.11.6 *Let A, B, C be $n \times n$ matrices with $n \leq 3$, such that*

$$M = \begin{pmatrix} A & B^* \\ B & C \end{pmatrix} \geq 0, \quad \widetilde{M} = \begin{pmatrix} A & B \\ B^* & C \end{pmatrix} \geq 0. \tag{5.11.17}$$

Then M is $2 \times n$ separable.

We will use a result on cones. A *face* F of a convex cone C is a subset of C such that if $sc_1 + (1 - s)c_2 \in F$ for some $c_1, c_2 \in C$ and $s \in (0, 1)$, we have that $c_1, c_2 \in F$.

Lemma 5.11.7 *Let K be a convex cone in a real vector space $\mathcal{H}$ of finite dimension N, and let $\mathcal{L} \subseteq \mathcal{H}$ be a subspace of dimension n. Let $K' = K \cap \mathcal{L}$ and x the generator of an extreme ray in K'. Then the minimal face in K containing x has dimension at most $N - n + 1$.*

Proof. Let F be the minimal face in K containing x. We first show that x is in the relative interior of F. Indeed, let $\mathcal{M}$ be the smallest linear space containing F. Note that any cone $\widetilde{F} \subset F$, $\widetilde{F} \neq F$, satisfying $[\widetilde{x}+\widetilde{y} \in \widetilde{F}, \widetilde{x}, \widetilde{y} \in K] \Rightarrow \widetilde{x}, \widetilde{y} \in \widetilde{F}$, does not contain x. Suppose that x is in the boundary of F as a subset of $\mathcal{M}$. Consider a hyperplane H through x with F on one side. Then $\widetilde{F} := F \cap H$ is a smaller cone with the above property containing x, which is a contradiction.

Now let $F' = F \cap \mathcal{L}$. We claim that dim $F' = 1$. Indeed, x is in the relative interior of $F' \subset K'$, and thus F' is a face of K' containing x. If dim $F' > 1$, then x cannot generate an extreme ray of K'.

To finish the proof, let d be the dimension of F. Then $F \cap \mathcal{L}$ has dimension $1 = d' \geq d + n - N$, and thus $d \leq N - n + 1$ follows. □

Notice that if we consider the cone PSD_n of $n \times n$ complex positive semidefinite matrices, then the minimal face containing $M \geq 0$ is the cone $F = \{GCG^* \colon C \in \text{PSD}_k\}$, where $M = GG^*$ with $\text{Ker}G = \{0\}$ and $k = \text{rank}M$. In particular, the real dimension of this minimal face is $(\text{rank}M)^2$.

Proof of Theorem 5.11.6. Since the case $n < 3$ can be embedded in the case $n = 3$, we will focus on the latter. As the $2 \times n$ separable matrices form a convex cone, it suffices to prove the result for pairs $(M, \widetilde{M})$ that generate extreme rays in the cone of pairs of matrices as in (5.11.17). If we apply Lemma 5.11.7 with the choices of $K = \text{PSD}_6 \times \text{PSD}_6$ and $\mathcal{L}$ the subspace

$$\left\{ \left(\begin{pmatrix} A & B^* \\ B & C \end{pmatrix}, \begin{pmatrix} A & B \\ B^* & C \end{pmatrix} \right) \right\}$$

in the (real) vector space of pairs of Hermitian matrices of size 6×6, then K' is the cone of pairs of matrices as in (5.11.17). By Lemma 5.11.7 the minimal faces in K containing extreme rays of K' cannot have dimension greater than $72 - 36 + 1 = 37$. However, the minimal face in K containing $(M, \widetilde{M})$ (which generates an extreme ray in K') has dimension $(\text{rank}M)^2 + (\text{rank}\widetilde{M})^2$, and hence the vector $(\text{rank}M, \text{rank}\widetilde{M}) \in \mathbb{R}^2$ lies in the closed disk of radius $\sqrt{37}$ centered at the origin. This now gives that either $\min\{\text{rank}M, \text{rank}\widetilde{M}\} \leq 3$ or $\max\{\text{rank}M, \text{rank}\widetilde{M}\} = 4$. Next, as above we can assume that $A = I$. If now $\min\{\text{rank}M, \text{rank}\widetilde{M}\} \leq 3$ we have that $C = BB^* = B^*B$, and thus B is normal, which yields by Theorem 5.11.2 that M is $2 \times n$ separable. On the other hand, if $\max\{\text{rank}M, \text{rank}\widetilde{M}\} = 4$ we can conclude by Corollary 5.11.5 that M is $2 \times n$ separable (as 4 vectors in $\mathbb{C}^n$ are always linearly dependent when $n \leq 3$). □

It should be noted that the dual form of the above statement is the following: if $\Phi : \mathbb{C}^{2\times 2} \to \mathbb{C}^{n\times n}$ is a positive linear map (thus $\Phi(\text{PSD}_2) \subseteq \text{PSD}_n$)

and $n \leq 3$, then Φ must be decomposable. That is, Φ must be of the form $\Phi(M) = \sum_{i=1}^{k} R_i M R_i^* + \sum_{i=1}^{l} S_i M^T S_i^*$.

The statement in Theorem 5.11.6 is no longer true for $n \geq 4$, as the following example shows.

Example 5.11.8 Let

$$Q = \begin{pmatrix} 0 & 1 & 0 & 0 \\ 0 & 0 & 1 & 0 \\ 0 & 0 & 0 & 1 \\ 0 & 0 & 0 & 0 \end{pmatrix}, \ \Lambda = \begin{pmatrix} \frac{25}{9} & 0 & 0 & \frac{20}{9} \\ 0 & 1 & 0 & 0 \\ 0 & 0 & 1 & 0 \\ \frac{20}{9} & 0 & 0 & \frac{25}{9} \end{pmatrix}.$$

Note that

$$\begin{pmatrix} I & Q \\ Q^* & \Lambda \end{pmatrix} \geq 0, \ \begin{pmatrix} I & Q^* \\ Q & \Lambda \end{pmatrix} \geq 0.$$

Next,

$$\Lambda - QQ^* = \begin{pmatrix} \frac{16}{9} & 0 & 0 & \frac{20}{9} \\ 0 & 0 & 0 & 0 \\ 0 & 0 & 0 & 0 \\ \frac{20}{9} & 0 & 0 & \frac{25}{9} \end{pmatrix} = \begin{pmatrix} \frac{4}{3} \\ 0 \\ 0 \\ \frac{5}{3} \end{pmatrix} \begin{pmatrix} \frac{4}{3} & 0 & 0 & \frac{5}{3} \end{pmatrix}.$$

Let $u = \begin{pmatrix} \frac{4}{3} & 0 & 0 & \frac{5}{3} \end{pmatrix}^T$. Thus $SS^* \leq \Lambda - QQ^*$ implies that $S = uv^*$ for some vector v with $\|v\| \leq 1$. Suppose that

$$R = \begin{pmatrix} Q & S \\ T & W \end{pmatrix}$$

is normal. Then

$$T^*T = QQ^* - Q^*Q + SS^* = \begin{pmatrix} 1 + \frac{16}{9}\|v\|^2 & 0 & 0 & \frac{20}{9}\|v\|^2 \\ 0 & 0 & 0 & 0 \\ 0 & 0 & 0 & 0 \\ \frac{20}{9}\|v\|^2 & 0 & 0 & \frac{25}{9}\|v\|^2 - 1 \end{pmatrix}.$$

This matrix is positive semidefinite only if $\|v\|^2 \geq 1$. As we also had that $\|v\|^2 \leq 1$, we must have $\|v\|^2 = 1$. But then $T^*T = yy^*$ where $y = \begin{pmatrix} \frac{5}{3} & 0 & 0 & \frac{4}{3} \end{pmatrix}^T$, and thus $T = xy^*$ for some vector x with $\|x\| = 1$. As R is normal we also have the equality $QT^* + SW^* = Q^*S + T^*W$. Looking at the third row of this equality, we get $\frac{4}{3}x^* = 0$ which is in contradiction with $\|x\| = 1$. Thus, by Theorem 5.11.2, $\begin{pmatrix} I & Q \\ Q^* & \Lambda \end{pmatrix}$ is not 2×4 separable.

5.12 EXERCISES

1 Let A be an invertible $n \times n$ matrix. For $\alpha, \beta \subset \{1, \ldots, n\}$, let $A(\alpha, \beta)$ denote the submatrix of A whose entries lie in rows indexed by α and columns indexed by β. Prove that

$$\dim \mathrm{Ker} A(\alpha, \beta) = \dim \mathrm{Ker} A^{-1}(\beta^c, \alpha^c).$$

2 A *bipartite graph* is an undirected graph $G = (V, E)$ in which the vertex set can be partitioned as $V = X \cup Y$ such that each edge in E has one endpoint in X and one in Y. We denote bipartite graphs by $G = (X, Y, E)$. With the specified/unspecified patterns of a partial matrix $(A_{ij})_{i,j=1}^{m,n}$ we associate the bipartite graph $G = (X, Y, E)$, where $X = \{1, \ldots, m\}$, $Y = \{1, \ldots, n\}$, and $(u, v) \in E$ if and only if $u \in X$, $v \in Y$, and $A_{u,v}$ is specified.

(a) Show that cycles in a bipartite graph always have even length.

(b) Show that a minimal cycle of length $2n$ corresponds exactly to a partially defined submatrix as in (5.2.7).

(c) We say that a bipartite graph is *bipartite chordal* if and only if the graph does not contain chordless cycles of length 6 or larger. Show that a partial matrix has a bipartite graph that is bipartite chordal if and only if there is no submatrix which after permutation has the form (5.2.7).

3 Prove that every graph that is the intersection graph of a family of subtrees of a tree is chordal. (Note: (i, j) is an edge in the graph if and only if subtree i and subtree j intersect.)

4 Construct a clique tree for the graph $G = (V, E)$, with $V = \{1, 2, 3, 4, 5, 6, 7\}$ and $E = \{(1, 2), (2, 3), (3, 4), (3, 6), (4, 5), (6, 7)\}$.

5 Let A be an invertible $n \times n$ matrix. If $G = (V, E)$ is a graph with $V = \{1, \ldots, n\}$, the notation $A^{-1} \in M_G$ means that $(A^{-1})_{ij} = (A^{-1})_{ji} = 0$ for every $i \neq j$, $(i, j) \notin E$. Prove that if $G = (V, E)$ is chordal and $T = (\mathcal{K}, \mathcal{E}(T))$ is a clique tree for G, then $A^{-1} \in M_G$ implies that

$$\det A = \frac{\prod_{\alpha \in \mathcal{K}} \det(A|\alpha)}{\prod_{(\alpha,\beta) \in \mathcal{E}(T)} \det(A|\alpha \cap \beta)},$$

provided that the denominator is nonzero. (Hint: use Theorem 5.1.9 and induction.)

6 Show that

$$\min \operatorname{rank} \begin{pmatrix} A & B & ? \\ ? & C & D \end{pmatrix} = \max \left\{ \min \operatorname{rank} \begin{pmatrix} A & B \\ ? & C \end{pmatrix}, \min \operatorname{rank} \begin{pmatrix} B & ? \\ C & D \end{pmatrix} \right\}.$$

7 Let A and B be square matrices, with A invertible. Consider the completion problem

$$\begin{pmatrix} A & ? \\ ? & B \end{pmatrix}^{-1} = \begin{pmatrix} ? & C \\ D & ? \end{pmatrix},$$

where C and D are of appropriate size. Show that there is a solution if and only if there exists an invertible Z such that $ZBZ - Z = DAC$. In that case, provide a completion. In addition, show that there is a one-to-one correspondence between all completions and all invertible Z satisfying $ZBZ - Z = DAC$.

8 Consider the completion problem

$$\begin{pmatrix} A & B \\ B^T & ? \end{pmatrix}^{-1} = \begin{pmatrix} ? & ? \\ ? & C \end{pmatrix},$$

where $A = A^T, B, C = C^T$ are real matrices of sizes $n \times n$, $n \times m$, and $m \times m$, respectively. Show that there is a real symmetric solution to this completion problem if and only if

$$\operatorname{rank} \begin{pmatrix} A & B \end{pmatrix} = n, \; n - \operatorname{rank} A = m - \operatorname{rank} C, \; \operatorname{Ran} BC \subseteq \operatorname{Ran} A.$$

Moreover, show that the solution is unique if and only if A and C are invertible, and find the solution in that case.

9 Use the ideas in the proof of Theorem 5.3.19 to prove the following. Let $\mathcal{A}$ be a $n \times m$ partial matrix with banded pattern J. Show that $\mathcal{A}$ has a completion $A = (a_{ij})_{i=1,j=1}^{n \;\; m}$, such that equality in (5.2.8) holds, and such that for each $k \in \{1, \dots, n\}$ the matrix $A_k = (a_{ij})_{i=k,j=1}^{n \;\; m}$ is a minimal rank completion of

$$\{a_{ij} \; : \; (i,j) \in J \cap (\{k, \dots, n\} \times \{1, \dots, m\})\}.$$

10

(a) Show that all minimal rank completions of

$$\begin{pmatrix} ? & ? & ? \\ 1 & 0 & ? \\ 0 & 1 & 1 \end{pmatrix}$$

are

$$\begin{pmatrix} x_1 & x_2 & x_1x_3 + x_2 \\ 1 & 0 & x_3 \\ 0 & 1 & 1 \end{pmatrix}.$$

(b) Show for $\mathcal{A}$ as in Theorem 5.2.1 there exist matrix polynomials $A_{ij} = A_{ij}(x_1, \dots, x_k)$, $1 \le i < j \le n$, with the degree of each x_i either one or zero, such that the set of minimal rank completions of $\mathcal{A}$ is given by

$(A_{ij})_{i,j=1}^n$ where $x_1, \ldots, x_k$ can be chosen freely. The number k is equal to

$$k = \sum_{j=1}^{n-1} \sum_{i=1}^{j} \nu_i (r_{j+1,j+1} - r_{j+1,j}) + \sum_{j=1}^{n-1} \sum_{i=j+1}^{n} \mu_i (r_{jj} - r_{j+1,j})$$

$$- \sum_{j=1}^{n-1} \sum_{i=j}^{n-1} (r_{jj} - r_{j+1,j})(r_{i+1,i+1} - r_{i+1,i}),$$

where

$$r_{pq} = \operatorname{rank} \begin{pmatrix} A_{p1} & \cdots & A_{pq} \\ \vdots & & \vdots \\ A_{n1} & \cdots & A_{nq} \end{pmatrix}.$$

11 Consider the partial matrix

$$A = \begin{pmatrix} 1 & ? & ? \\ ? & 1 & ? \\ -1 & ? & 1 \end{pmatrix}.$$

Show that there exists a completion of A that is a Toeplitz matrix of rank 1, but that such a completion cannot be chosen to be real.

12 Consider the partial matrix

$$\mathcal{A} = \begin{pmatrix} A_{11} & A_{12} & ? \\ A_{21} & ? & A_{23} \\ ? & A_{32} & A_{33} \end{pmatrix}$$

with A_{ij} invertible $n \times n$ matrices.

(a) Show that $\min \operatorname{rank} \mathcal{A} = \lceil \frac{r}{2} \rceil$, where

$$r = \operatorname{rank} \begin{pmatrix} A_{11} & A_{12} & 0 \\ A_{21} & 0 & -A_{23} \\ 0 & -A_{32} & -A_{33} \end{pmatrix}.$$

(b) Show that $\min \operatorname{rank} \mathcal{A} = n$ if and only if $A_{21} A_{11}^{-1} A_{12} = A_{23} A_{33}^{-1} A_{32}$. In that case, show that the minimal rank completion of $\mathcal{A}$ is unique and determine the unknown entries of this unique minimal rank completion.

13 Here we consider a completion problem that involves a condition on the product of completions.

(a) Let $\mathcal{A}, \mathcal{B}$ and $\mathcal{C}$ be partial matrices of sizes $n \times p$, $p \times m$ and $n \times m$, respectively. Show that there exist completions A, B and C of $\mathcal{A}, \mathcal{B}$ and $\mathcal{C}$, respectively, such that $AB = C$ if and only if there exists a completion of $\begin{pmatrix} I_p & \mathcal{B} \\ \mathcal{A} & \mathcal{C} \end{pmatrix}$ with rank equal to p.

(b) Use part (a) and Exercise 5.12.9 to give necessary and sufficient conditions for the existence of completions T and S of

$$\begin{pmatrix} T_{11} & ? \\ T_{21} & T_{22} \end{pmatrix} \text{ and } \begin{pmatrix} ? & S_{12} \\ ? & ? \end{pmatrix}$$

such that $ST = I$.

14 Let A have no eigenvalues on the unit circle, and let $C = -(A^* + I)(A^* - I)^{-1}$. Why is C well defined? Show that A satisfies the Stein equation $H - A^*HA = V$, with V positive definite, if and only if C satisfies a Lyapunov equation $CH + HC^* = G$ with G positive definite. Use this observation and Proposition 5.4.1 to prove the following Lyapunov equation result. *Suppose $CH + HC^* = G$ with $H = H^*$ and G positive definite. Then C has no purely imaginary eigenvalues, H is invertible, and the number of eigenvalues of C in the right half-plane coincides with the number of positive eigenvalues of H.*

15 Given $A, B \in \mathbb{C}^{n\times n}$ we consider the problem of finding A_{ext}, B_{ext} of smallest possible size, with

$$A_{ext} = \begin{pmatrix} A & A_{12} \\ A_{21} & A_{22} \end{pmatrix}, \; B_{ext} = \begin{pmatrix} B & B_{12} \\ B_{21} & B_{22} \end{pmatrix}, \tag{5.12.1}$$

such that $[A_{ext}, B_{ext}] = 0$. We shall call the smallest possible number q the *commuting defect* of A and B, and denote it $\mathrm{cd}(A, B)$.

(a) Show that $\mathrm{cd}(A, B) \geq \frac{1}{2}\mathrm{rank}([A, B])$.

A related problem is the following. Given $A, B \in \mathbb{C}^{n\times n}$, how do we find $A_{ext}, B_{ext} \in \mathbb{C}^{(n+q)\times(n+q)}$ as in (5.12.1) of smallest possible size such that $A_{ext}B_{ext} = \alpha I_{n+q}$ for some $\alpha \neq 0$. We shall call this smallest number q the *inverse defect* and denote it by $\mathrm{id}(A, B)$.

(b) Show that $\mathrm{cd}(A, B) \leq \mathrm{id}(A, B)$, and give an example where the inequality is strict.

(c) Show that $\mathrm{id}(A, B) = \mathrm{rank}(\alpha I_n - AB)$, where α is a nonzero eigenvalue of AB with highest possible geometric multiplicity.

16 Let $A_{11} = A_{11}^* \in \mathbb{C}^{p\times p}$, $A_{12}^* \in \mathbb{C}^{p\times q}$, $A_{22} = A_{22}^* \in \mathbb{C}^{q\times q}$, $B_{11}^* \in \mathbb{C}^{p\times k}$, $B_{21}^* \in \mathbb{C}^{q\times k}$, $B_{22}^* \in \mathbb{C}^{q\times l}$, $C_{11} = C_{11}^* \in \mathbb{C}^{k\times k}$, $C_{12}^* \in \mathbb{C}^{k\times l}$, and $C_{22}^* \in \mathbb{C}^{l\times l}$ be given. Use Theorem 5.3.19 to show that there exist $B_{12}^* \in \mathbb{C}^{p\times l}$, and $Z \in \mathbb{C}^{(p+q)\times(k+l)}$ such that

$$A + BZ^* + ZB^* + ZCZ^* \geq 0$$

if and only if

$$i_-\begin{pmatrix} A_{11} & A_{12} & B_{11} \\ A_{12}^* & A_{22} & B_{21} \\ B_{11}^* & B_{21}^* & C_{11} \end{pmatrix} + \mathrm{rank}\begin{pmatrix} A_{12} & B_{11} \\ A_{22} & B_{21} \\ B_{21}^* & C_{11} \\ B_{22}^* & C_{12}^* \end{pmatrix}$$

$$\leq l \, + \, \mathrm{rank} \begin{pmatrix} A_{12} & B_{11} \\ A_{22} & B_{21} \\ B_{21}^* & C_{11} \end{pmatrix} + \, \mathrm{rank} \begin{pmatrix} B_{11} \\ B_{21} \\ C_{11} \\ C_{12}^* \end{pmatrix},$$

$$i_- \begin{pmatrix} A_{22} & B_{21} & B_{22} \\ B_{21}^* & C_{11} & C_{12} \\ B_{22}^* & C_{12}^* & C_{22} \end{pmatrix} + \, \mathrm{rank} \begin{pmatrix} A_{12} & B_{11} \\ A_{22} & B_{21} \\ B_{21}^* & C_{11} \\ B_{22}^* & C_{12}^* \end{pmatrix}$$

$$\leq l \, + \, \mathrm{rank} \begin{pmatrix} A_{22} & B_{21} \\ B_{21}^* & C_{11} \\ B_{22}^* & C_{12}^* \end{pmatrix} + \, \mathrm{rank} \begin{pmatrix} B_{11} \\ B_{21} \\ C_{11} \\ C_{12}^* \end{pmatrix},$$

and

$$i_- \begin{pmatrix} A_{22} & B_{21} & B_{22} \\ B_{21}^* & C_{11} & C_{12} \\ B_{22}^* & C_{12}^* & C_{22} \end{pmatrix} \leq \, \mathrm{rank} \begin{pmatrix} B_{21} & B_{22} \\ C_{11} & C_{12} \\ C_{12}^* & C_{22} \end{pmatrix}.$$

Hint: Use the partial matrix

$$\begin{pmatrix} A_{11} & A_{12} & B_{11} & ? & ? & ? \\ A_{12}^* & A_{22} & B_{21} & B_{22} & ? & ? \\ B_{11}^* & B_{21}^* & C_{11} & C_{12} & -I_k & 0 \\ ? & B_{22}^* & C_{12}^* & C_{22} & 0 & -I_l \\ ? & ? & -I_k & 0 & 0 & 0 \\ ? & ? & 0 & -I_l & 0 & 0 \end{pmatrix}, \tag{5.12.2}$$

and observe that in this 6×6 (block) pattern the 4 maximal symmetric triangular patterns are

$$T_1 = (\underline{3} \times \underline{3}) \cup \{(2,4),(4,2)\} \cup (\{3\} \times \{4,5,6\}) \cup (\{4,5,6\} \times \{3\}),$$
$$T_2 = (\{2,3\} \times \underline{4}) \cup (\underline{4} \times \{2,3\}) \cup (\{3\} \times \underline{6}) \cup (\underline{6} \times \{3\}) \cup \{(4,4)\},$$
$$T_3 = (\{3,4\} \times \{2,3,4,5,6\}) \cup (\{2,3,4,5,6\} \times \{3,4\}) \cup \{(1,3),(2,2),(3,1)\},$$
$$T_4 = \{3,4,5,6\}^2 \cup \{(1,3),(2,3),(2,4),(3,1),(3,2),(4,2)\}.$$

17 Consider the matrix inequality

$$A - BD^* - CC^* + CF^*D^* - DB^* + DFC^* + DFF^*D^* - DED^* \geq 0. \tag{5.12.3}$$

(a) Show that the left-hand side of (5.12.3) is the Schur complement supported in first block row and column of

$$\begin{pmatrix} A & B & C & D \\ B^* & E & F & I \\ C^* & F^* & I & 0 \\ D^* & I & 0 & 0 \end{pmatrix}.$$

(b) Prove the following. Given are $n \times n$ complex matrices A, B, C, D, E of which A and E are Hermitian. There exists an $n \times n$ matrix F such that (5.12.3) holds if and only if

$$A - BD^* - DB^* + DED^* \geq 0,$$

$$\text{rank} \begin{pmatrix} A - BD^* & C & D \end{pmatrix} = \text{rank} \begin{pmatrix} A - BD^* & D \end{pmatrix},$$

and

$$i_- \begin{pmatrix} A - CC^* & D \\ D^* & 0 \end{pmatrix} + \text{rank} \begin{pmatrix} A - BD^* & C & D \end{pmatrix}$$

$$\leq \text{rank} \begin{pmatrix} A & C & D \\ D^* & 0 & 0 \end{pmatrix}.$$

Hint: Use the partial matrix

$$\begin{pmatrix} E & B^* & I_n & ? \\ B & A & D & C \\ I_n & D^* & 0 & 0 \\ ? & C^* & 0 & I_n \end{pmatrix}. \tag{5.12.4}$$

Note that the two maximal symmetric triangular in this pattern are

$$T_1 = \underline{3} \times \underline{3} \cup \{(4,2), (4,3), (3,4), (2,4)\}$$

and

$$T_2 = \{2,3,4\}^2 \cup \{(1,2), (1,3), (3,1), (2,1)\}.$$

18 In this exercise we study the skew-adjoint/skew-symmetric expression

$$N(X, Y) = A - BXB^* - CYC^*,$$

where $A = -A^*$ and B, C are real or complex matrices of appropriate sizes. We put

$$k_{\min} = 2\,\text{rank} \begin{pmatrix} A & B & C \end{pmatrix} + \text{rank} \begin{pmatrix} A & B \\ C^* & 0 \end{pmatrix}$$

$$-\text{rank} \begin{pmatrix} A & B & C \\ B^* & 0 & 0 \end{pmatrix} - \text{rank} \begin{pmatrix} A & B & C \\ C^* & 0 & 0 \end{pmatrix}.$$

(a) Let

$$A = \begin{pmatrix} 0 & 0 \\ 0 & 0 \end{pmatrix}, \; B = \begin{pmatrix} 1 & 0 \\ 0 & 0 \end{pmatrix}, \; C = \begin{pmatrix} 0 & 0 \\ 0 & 1 \end{pmatrix}.$$

Show that for $X = -X^T, Y = -Y^T \in \mathbb{R}^{2 \times 2}$ we always have that $\text{rank} N(X, Y) = 0$, while if we let $X = -X^*, Y = -Y^* \in \mathbb{C}^{2 \times 2}$, we have that the possible ranks of $N(X, Y)$ are 0, 1 and 2.

(b) Show that

$$\{\operatorname{rank} N(X,Y) \ : \ X=-X^* \text{ and } \ Y=-Y^* \text{ complex}\}$$
$$=\left\{s \mid s \text{ is an integer }, k_{\min} \leq s \leq \operatorname{rank}\begin{pmatrix} A & B & C \end{pmatrix}\right\}.$$

(c) Show that when

$$\operatorname{rank}\begin{pmatrix} A & B & C \\ B^T & 0 & 0 \end{pmatrix} + \operatorname{rank}\begin{pmatrix} A & B & C \\ C^T & 0 & 0 \end{pmatrix}$$

$$= \operatorname{rank}\begin{pmatrix} A & B & C \end{pmatrix} + \operatorname{rank}\begin{pmatrix} A & B & C \\ B^T & 0 & 0 \\ C^T & 0 & 0 \end{pmatrix},$$

$$\operatorname{rank}\begin{pmatrix} A & B & C \end{pmatrix} = \operatorname{rank}\begin{pmatrix} A & C \\ B^T & 0 \end{pmatrix},$$

and

$$\operatorname{rank}\begin{pmatrix} A & B & C \\ B^T & 0 & 0 \end{pmatrix} - \operatorname{rank}\begin{pmatrix} A & C \\ B^T & 0 \end{pmatrix},$$

$$\operatorname{rank}\begin{pmatrix} A & B & C \\ C^T & 0 & 0 \end{pmatrix} - \operatorname{rank}\begin{pmatrix} A & C \\ B^T & 0 \end{pmatrix}$$

are both odd, then

$$\{\operatorname{rank} N(X,Y) \ : \ X=-X^T \text{ and } \ Y=-Y^T \text{ real}\}$$
$$=\left\{s : s \text{ is even integer }, k_{\min} \leq s \leq \operatorname{rank}\begin{pmatrix} A & B & C \end{pmatrix} - 2\right\},$$

and otherwise,

$$\{\operatorname{rank} N(X,Y) \ : \ X=-X^T \text{ and } \ Y=-Y^T \text{ real}\}$$
$$=\left\{s : s \text{ is even integer }, k_{\min} \leq s \leq \operatorname{rank}\begin{pmatrix} A & B & C \end{pmatrix}\right\}.$$

(d) Derive similar results for the expression $\widetilde{N}(X) = A - BXC^* + CX^*B^*$ with $A=-A^*, B, C$ given.

19 Consider the partial matrix

$$\mathcal{A} := \begin{pmatrix} 6 & 3 & ? & 1 \\ 3 & 1 & 1 & ? \\ ? & 1 & 2 & 3 \\ 1 & ? & 1 & 1 \end{pmatrix}.$$

(a) Show that all fully specified submatrices have rank 2.

(b) Show that the min $\operatorname{rank}\mathcal{A} = 3$.

(c) Can one find a partial matrix of minimal rank 2 with all nonzero prescribed entries and such that all the fully specified submatrices have rank 1?

20 Consider the lower triangular partial scalar matrix $\mathcal{A} = \{a_{ij} : 1 \leq j \leq i \leq n\}$, and assume that there does not exist an $1 < r \leq n$ such that $a_{kl} = 0$, $r \leq k \leq n$, $1 \leq l \leq r-1$.

(a) Show that there exists a completion of $\mathcal{A}$ with eigenvalues $\lambda_1, \ldots, \lambda_n \in \mathbb{C}$ if and only if $\sum_{i=1}^n \lambda_i = \sum_{i=1}^n a_{ii}$.

(b) Show that the completion in (a) may be chosen to be nonderogatory (i.e., each eigenvalue only has one Jordan block).

(c) Formulate and prove a more general version of (a) where blocks of zeros are allowed in the lower triangular part.

21 Let $\Lambda_k(M)$ denote the sum of the largest k eigenvalues of the Hermitian matrix M. For $A \in \mathbb{C}^{n\times n}$, put

$$v_k(A) = \min \left\{ \Lambda_k \begin{pmatrix} Z & A \\ A^* & -Z \end{pmatrix} : Z = Z^* \right\}.$$

Show that $v_k(A)$, $1 \leq k \leq n$, defines a norm on $n \times n$ matrices that is invariant under unitary similarity.

22 Let $\mathcal{T} = \{t_{ij} : 1 \leq j \leq i \leq n\}$ be a scalar-valued lower triangular partial matrix. Show that $\min \operatorname{rank}(\mathcal{T}) = n$ if and only if $t_{ii} \neq 0, i = 1, \ldots, n$, and $t_{ij} = 0$ for $i > j$.

23 Let $p \geq 0$ and $A = (a_{ij})_{i,j=1}^N$ be an $N \times N$ scalar matrix with inverse $B = (b_{ij})_{i,j=1}^N$. Show that $a_{ij} = 0$ for all i and j with $j > i + p$, and $a_{ij} \neq 0$, $j = i + p$, if and only if there exist an $N \times p$ matrix F and a $p \times N$ matrix G such that $b_{ij} = (FG)_{ij}$, $i < j + p$. In particular, if $p = 1$ (so A is lower Hessenberg) then $b_{ij} = F_i G_j$, $1 \leq i \leq j \leq N$, where $F_1, \ldots, F_N, G_1, \ldots, G_N$ are scalars. (In this case, one says that B is *lower semiseparable.*)

24 Prove that for every $m \times n$ matrix A, there exists a unique $n \times m$ matrix $A^{(-1)}$ which simultaneously verifies the following conditions:

(i) $AA^{(-1)}A = A$,

(ii) $A^{(-1)}AA^{(-1)} = A^{(-1)}$,

(iii) $(AA^{(-1)})^* = AA^{(-1)}$, and

(iv) $(A^{(-1)}A)^* = A^{(-1)}A$.

(v) Show that if $A = U\Sigma V^*$ is a singular value decomposition with nonzero singular values $s_1(A) \geq \cdots \geq s_r(A) > 0$, then $A^{(-1)} = V\Sigma^{(-1)}U^*$, where $\Sigma^{(-1)}$ is the $n \times m$ matrix with (i,i)th entry equal to $\frac{1}{s_i(A)}$, $i = 1, \ldots, r$, and zero otherwise.

$A^{(-1)}$ is called the *Moore-Penrose inverse of* A.

25 Given $A \in \mathcal{H}_n$ and $B \in \mathbb{C}^{n\times m}$, show that

$$\min_{X\in\mathbb{C}^{m\times n}} \operatorname{rank}(A - BX - X^*B^*) = \operatorname{rank}\begin{pmatrix} A & B \\ B^* & 0 \end{pmatrix} - 2\operatorname{rank}B. \tag{5.12.5}$$

In addition, show that any matrix of the form

$$X = B^{(-1)}A - \frac{1}{2}B^{(-1)}ABB^{(-1)} + UB^* + (I - B^{(-1)}B)V,$$

with $U = -U^* \in \mathbb{C}^{m\times m}$ and $V \in \mathbb{C}^{m\times n}$ arbitrary, gives the minimal possible rank in (5.12.5).

26 Consider the partial matrix

$$\begin{pmatrix} a & ? \\ b & c \end{pmatrix}. \tag{5.12.6}$$

Show that all normal completions of (5.12.6) are obtained by choosing

$$? = \begin{cases} \frac{a-c}{\overline{a}-\overline{c}}\overline{b} & \text{when } a \neq c, \\ |b|e^{i\varphi} & \text{when } a = c, \end{cases}$$

where $\varphi \in [0, 2\pi]$ is to be chosen freely.

27 Consider the 2×2 partial block diagonal matrix

$$\begin{pmatrix} A & ? \\ ? & B \end{pmatrix}. \tag{5.12.7}$$

with A and B normal with disjoint spectra $\sigma(A) = \{\alpha_1, \dots, \alpha_n\}$ and $\sigma(B) = \{\beta_1, \dots, \beta_m\}$. Show that the set of all normal completions of (5.12.7) are in one-to-one correspondence with the set

$$\Bigg\{Y \in \mathbb{C}^{m\times n} : \sum_{k=1}^{m} \left(1 - \sigma_{ik}^2/\sigma_{jk}^2\right) \overline{y}_{ki}y_{kj} = 0, \tag{5.12.8}$$

$$\sum_{k=1}^{n} \left(1 - \sigma_{kj}^2/\sigma_{ki}^2\right) y_{ik}\overline{y}_{jk} = 0, i = 1, \dots, m,\ j = 1, \dots, n\Bigg\},$$

where $\sigma_{ij} = \operatorname{sign}(\alpha_i - \beta_j)$. Here the sign function is given via

$$\operatorname{sign}(\epsilon) = \begin{cases} \frac{\epsilon}{|\epsilon|} & \text{when } \epsilon \neq 0, \\ 1 & \text{when } \epsilon = 0. \end{cases}$$

28 In this exercise we relate the normal completion problem to a structured Hermitian completion problem.

(a) Let A be an $n \times n$ matrix. Show that A is normal if and only if

$$i_-\begin{pmatrix} -I_n & A & 0 \\ A^* & 0 & A \\ 0 & A^* & I_n \end{pmatrix} = n.$$

(b) Let $\mathcal{A}$ be a $n \times n$ partial matrix. Show that $\mathcal{A}$ has a normal completion if and only if the partial matrix

$$\begin{pmatrix} -I_n & \mathcal{A} & 0 \\ \mathcal{A}^* & 0 & \mathcal{A} \\ 0 & \mathcal{A}^* & I_n \end{pmatrix} = n. \tag{5.12.9}$$

has a Hermitian completion with block entries $(1,2)$ and $(2,3)$ equal to one another that has negative inertia (smaller than or) equal to n.

29 Let $A \in \mathbb{C}^{n\times n}$ be of the form $A := \begin{pmatrix} 0 & a_1 & & 0 \\ & \ddots & \ddots & \\ & & \ddots & a_{n-1} \\ 0 & & & 0 \end{pmatrix}$ with either $|a_1| = \cdots = |a_l| > \cdots > |a_{n-1}| > 0$ or $0 < |a_1| < \cdots < |a_{n-l}| = \cdots = |a_{n-1}|$, where $1 \le l \le n-1$. Determine $\text{nd}(A)$.

30 Given $A \in \mathbb{C}^{n\times n}$. Let A be strictly upper triangular with $a_{i,i+1} \neq 0$ for $1 \le i \le n-1$. Show that A is not unitarily reducible (i.e., A is not unitarily similar to the direct sum of two (or more) nontrivial matrices).

31

(a) Given are complex numbers $\alpha_1, \ldots, \alpha_n$ and $\beta_1, \ldots, \beta_{n-1}$ and arrange them such that $\alpha_1, \ldots, \alpha_q$ are all distinct from $\beta_1, \ldots, \beta_{q-1}$, and such that $\alpha_j = \beta_{j-1}, j = q+1, \ldots, n$. Let $B = \text{diag}(\beta_i)_{i=1}^{n-1}$. Show that there exists a $n \times n$ normal matrix of the form $\left(\begin{smallmatrix} B & ? \\ ? & ? \end{smallmatrix}\right)$ with eigenvalues $\alpha_1, \ldots, \alpha_n$ if and only if the $2q-1$ points $\alpha_1, \ldots, \alpha_q, \beta_1, \ldots, \beta_{q-1}$ are distinct and lie on a line, and that every segment on this line limited by two adjacent α_i's $(1 \le i \le q)$ contains exactly one β_i $(1 \le i \le q-1)$.

(b) Show that there exists a normal completion N of

$$\begin{pmatrix} \frac{5+8i}{10} & 0 & ? & ? \\ 0 & \frac{5+2i}{10} & ? & ? \\ ? & ? & ? & ? \\ ? & ? & ? & ? \end{pmatrix}$$

with eigenvalues $0, 1, i, 1+i$, but that any such N cannot have a 3×3 upper left submatrix that is normal as well. (Hint: use a unitary similarity with a unitary whose first two columns are equal to

$$\frac{1}{\sqrt{10}} \begin{pmatrix} 1 & 1 & 2 & 2 \end{pmatrix}^T, \frac{1}{\sqrt{10}} \begin{pmatrix} 2 & -2 & -1 & 1 \end{pmatrix}^T.)$$

32 Consider the partial matrix $W = \left(\begin{smallmatrix} I & B \\ B^* & ? \end{smallmatrix}\right)$. Show that among all positive definite completions the smallest condition for W is given by $(\|B\| +$

$\sqrt{1+\|B\|^2})^2$. Provide an X yielding this condition number. (Recall that the *condition number* of a matrix is given by the quotient of the largest singular value and the smallest singular value, $\frac{s_{\max}}{s_{\min}}$.)

33 Consider the $(n+(N-n)) \times (m+(N-m))$ partial matrix $W(X) = \left(\begin{smallmatrix} A & B \\ C & X \end{smallmatrix}\right)$, where X is the $N-n) \times (N-m)$ unknown block. Put

$$\alpha = \min \left\{ s_n \begin{pmatrix} A & B \end{pmatrix}, s_m \begin{pmatrix} A \\ C \end{pmatrix} \right\},$$

and assume that $\alpha > 0$. Here s_j denote the singular values with s_1 being the largest. Show that

$$\inf_X \|W(X)^{-1}\| = \frac{1}{\alpha},$$

and that the infimum is attained in the case that α is not a singular value of A.

34 Let $a_j = \overline{a}_{-j} \in \mathbb{C}$, $j = -k, \ldots, k$, be given complex numbers such that $A := (a_{i-j})_{i,j=0}^k$ is invertible. Let $i_-, i_+ \in \mathbb{N}_0$ be such that $i_- + i_+ = n$, $i_- \geq i_-(A)$, and $i_+ \geq i_+(A)$. Show that there exist $a_j = \overline{a}_{-j} \in \mathbb{C}$, $k < |j| < n$, such that $M := (a_{i-j})_{i,j=0}^{n-1}$ satisfies $i_-(M) = i_-$ and $i_+(M) = i_+$. Moreover, if a_j, $|j| \leq k$, are real then a_j, $k < |j| < n$, can be chosen to be real as well.

35 Show that for A in Example 5.9.2 all ways of writing $[A, A^*] = xx^* - yy^*$ yields vectors x and y such that $\{x, y, Ax, A^*y\}$ is linearly independent.

36 Let x be a real number, and consider the completion of A in (5.6.3), given by

$$B(x) = \begin{pmatrix} 1 & b_{12}(x) & b_{13}(x) & \cdots & b_{1n}(x) \\ 0 & 1 & b_{23}(x) & \cdots & b_{2}n(x) \\ 0 & 0 & 1 & \cdots & b_{3n}(x) \\ \vdots & \vdots & \vdots & & \vdots \\ 0 & 0 & 0 & \cdots & 1 \end{pmatrix},$$

where

$$b_{ij}(x) = x^{2^{n-i} - 2^{n-j}}.$$

Define

$$C(x) = [c_j(x) c_i(x)]_{i,j=1}^n,$$

where

$$c_j(x) = \left(\sum_{k=1}^{j} (x^{2^{n-k+1} - 2^{n-j+1}}) \right)^{\frac{1}{2}}.$$

Clearly, $C(x)$ is a rank 1 matrix for all x. Show that

$$\lim_{x\to\infty}[C(x) - B(x)^T B(x)] = 0. \tag{5.12.10}$$

Use this to show that (5.6.4) holds.

37 Give necessary and sufficient conditions for a partial matrix

$$\begin{pmatrix} A & B \\ C & ? \end{pmatrix}$$

to have a completion M with $s_2(M) \leq 1$.

38 Let A be the partial matrix (5.6.1) and let β_k be the right-hand side of (5.6.2). Show that for all $\epsilon > 0$ there exist B_{ij}, $1 \leq j \leq i \leq n$, such that the partial matrix

$$B = \begin{pmatrix} B_{11} & & ? \\ \vdots & \ddots & \\ B_{n1} & \cdots & B_{nn} \end{pmatrix}$$

has a completion of rank less than k, and the partial matrix $A - B$ has a completion of norm less than $\beta_k + \epsilon$.

39 Let A be the partial matrix (5.6.1) and put

$$\alpha = \max_{i=1,\ldots,n} \operatorname{rank} \begin{pmatrix} A_{i1} & \cdots & A_{ii} \\ \vdots & & \vdots \\ A_{n1} & \cdots & A_{ni} \end{pmatrix}.$$

Show that for all $\epsilon > 0$ there exist B_{ij}, $1 \leq j \leq i \leq n$, such that the partial matrix

$$B = \begin{pmatrix} B_{11} & & ? \\ \vdots & \ddots & \\ B_{n1} & \cdots & B_{nn} \end{pmatrix}$$

has minimal rank α and the partial matrix $A - B$ has a completion of norm less than ϵ.

40 Let $\lambda_1 \leq 0 \leq \lambda_2 \leq \cdots \leq \lambda_n$ be given such that $\sum_{i=1}^n \lambda_i = 0$. For $2 \leq n \leq 6$, show that there exists a distance matrix with eigenvalues $\lambda_1, \ldots, \lambda_n$.

41

(a) Show that the positive semidefinite Hankel

$$\begin{pmatrix} S_0 & S_1 & \cdots & S_n \\ S_1 & S_2 & \cdots & S_{n+1} \\ \vdots & \vdots & \iddots & \vdots \\ S_n & S_{n+1} & \cdots & S_{2n} \end{pmatrix},$$

with $m \times m$ blocks, is $n \times m$ separable. (Hint: use Theorem 2.7.9.)

(b) Use the maps Φ and Ψ from Exercise 1.6.38 and the above result for Hankels, to prove the same for Toeplitz matrices.

42 Let $c_0 = c_1 = c_{-1} = 1$ and $c_2 = c_{-2} = 4$. Does $(c_j)_{j=-2}^{2}$ have a $(2,2)$ extension. How about a $(2,1)$ extension?

43 In this exercise we outline an alternative proof for the "if" part of Theorem 5.7.4. Let T_n and $P(z) = v_0 + \cdots + v_k z^k$ be as in Theorem 5.7.4.

(a) Show that $v_k \neq 0$.

(b) Introduce the companion matrix

$$C = \begin{pmatrix} 0 & 1 & \cdots & & 0 \\ & & \ddots & \ddots & \\ & & & 0 & 1 \\ -\frac{v_0}{v_k} & \cdots & & \cdots & -\frac{v_{k-1}}{v_k} \end{pmatrix}.$$

Show that eigenvalues of C are the roots of P.

(c) Show that

$$\mathrm{col}(c_i)_{i=1}^{k} = C\mathrm{col}(c_i)_{i=0}^{k-1}. \tag{5.12.11}$$

(d) Denoting the roots of P by $\alpha_1, \ldots, \alpha_k \in \mathbb{T}$, show that $C = VDV^{-1}$, where

$$D = \mathrm{diag}(\alpha_i)_{i=1}^{k}, V = \begin{pmatrix} 1 & \cdots & 1 \\ \alpha_1 & \cdots & \alpha_k \\ \vdots & & \vdots \\ \alpha_1^{k-1} & \cdots & \alpha_k^{k-1} \end{pmatrix}.$$

(e) Complete the proof of the "if" part of Theorem 5.7.4.

44 Find the representation (5.7.2) for

$$\begin{pmatrix} 1 & -i & -\frac{1}{2i} \\ i & 0 & -i \\ \frac{1}{2i} & i & 0 \end{pmatrix}.$$

45 For the following matrices A, determine whether $\mathrm{nd}(A) = 1$ or not. In the case when $\mathrm{nd}(A) = 1$, determine all minimal normal completions.

(a)

$$A = \begin{pmatrix} 0 & 0 & 1 & -i \\ 2 & 0 & 0 & 0 \\ 0 & 1 & \frac{1}{\sqrt{2}} & \frac{i}{\sqrt{2}} \\ 0 & -i & \frac{i}{\sqrt{2}} & -\frac{1}{\sqrt{2}} \end{pmatrix}.$$

(b)

$$A = \begin{pmatrix} 0 & 0 & \frac{1}{\sqrt{2}} & \frac{i}{\sqrt{2}} \\ 0 & 0 & 1 & i \\ 1 & \frac{1}{\sqrt{2}} & \frac{\sqrt{3}}{2} & -\frac{\sqrt{3}}{2}i \\ i & \frac{i}{\sqrt{2}} & -\frac{\sqrt{3}}{2}i & -\frac{\sqrt{3}}{2} \end{pmatrix}.$$

(c)

$$A = \begin{pmatrix} 1 & 0 & 0 \\ 0 & 1 & 1 \\ 1 & 0 & 1 \end{pmatrix}.$$

For this matrix, give also all real minimal normal completions. (Answer: all minimal normal completions of A have the form

$$B = \begin{pmatrix} 1 & 0 & 0 & \mu x_1 \\ 0 & 1 & 1 & \mu x_2 \\ 1 & 0 & 1 & 0 \\ \overline{\mu} x_2 & \overline{\mu} x_1 & 0 & 1 \end{pmatrix},$$

with arbitrary $\mu \in \mathbb{T}$, and $x_1 \in \mathbb{C}$, $x_2 \in \mathbb{R}$ satisfying $|x_1|^2 - x_2^2 = 1$. Note that some real minimal normal completions are found by using complex μ, x_1, and x_2 only.)

(d)

$$A = \begin{pmatrix} 0 & 1 & 0 \\ 1 & 0 & 1 \\ 0 & 1 & \frac{3}{2}i \end{pmatrix}.$$

(Answer: all minimal normal completions of A have the form

$$B = \begin{pmatrix} 0 & 1 & 0 & 0 \\ 1 & 0 & 1 & \mu\frac{h_1+h_2}{\sqrt{2}} \\ 0 & 1 & \frac{3}{2}i & \mu\frac{2h_3+i(h_2-h_1)}{\sqrt{2}} \\ 0 & \overline{\mu}\frac{h_1+h_2}{\sqrt{2}} & \overline{\mu}\frac{2h_3+i(h_2-h_1)}{\sqrt{2}} & \frac{h_1h_3+5h_2h_3+i(3-2h_1h_2+2h_2^2)}{3} \end{pmatrix},$$

with arbitrary $\mu \in \mathbb{T}$, and $h_1, h_2, h_3 \in \mathbb{R}$: $h_1^2 - h_2^2 = 3$.)

46 Determine all completions of the 3×3 (front cover) partial matrix

$$\begin{pmatrix} 16 & 17 & 18 \\ 16 & ? & 18 \\ 16 & 17 & 18 \end{pmatrix}$$

for which the normal defect is minimal.

47 Consider the $n \times n$ weighted shift matrix

$$A = \begin{pmatrix} 0 & a_1 & 0 & \dots & 0 \\ \vdots & \ddots & \ddots & \ddots & \vdots \\ \vdots & & \ddots & \ddots & 0 \\ \vdots & & & \ddots & a_{n-1} \\ 0 & \dots & \dots & \dots & 0 \end{pmatrix}$$

with weights $a_j \in \mathbb{T}$.

(a) Show that for $n \geq 4$ all its minimal normal completions B are also minimal unitary completions and have the form

$$B = \begin{pmatrix} 0 & a_1 & 0 & \dots & 0 & 0 \\ \vdots & \ddots & \ddots & \ddots & \vdots & \vdots \\ \vdots & & \ddots & \ddots & 0 & \vdots \\ \vdots & & & \ddots & a_{n-1} & 0 \\ 0 & \dots & \dots & \dots & 0 & \zeta \\ \rho & 0\dots & \dots & \dots & \dots & 0 \end{pmatrix}$$

with $\zeta, \rho \in \mathbb{T}$.

(b) Show that for $n = 2$ all minimal normal completions are given by

$$B = \begin{pmatrix} 0 & a_1 & \mu x_2 \\ 0 & 0 & \mu x_1 \\ \overline{\mu} x_1 & \overline{\mu} x_2 & \overline{a}_1 x_2^2 + a_1 x_1 \overline{x}_2 \end{pmatrix},$$

where $\mu \in \mathbb{T}$ and $x_1, x_2 \in \mathbb{C}$: $|x_1|^2 - |x_2|^2 = 1$ are arbitrary.

(c) Show that for $n = 3$ all minimal normal completions are given by

$$B = \begin{pmatrix} 0 & a_1 & 0 & \mu x_2 \\ 0 & 0 & a_2 & 0 \\ 0 & 0 & 0 & \mu x_1 \\ \overline{\mu} x_1 & 0 & \overline{\mu} x_2 & 0 \end{pmatrix},$$

where $\mu \in \mathbb{T}$ and $x_1, x_2 \in \mathbb{C}$: $|x_1|^2 - |x|_2^2 = 1$, $\overline{a}_1 x_2 = a_2 \overline{x}_2$ are arbitrary.

48 Let (A_1, A_2) be a pair of Hermitian matrices of size $n \times n$. We define the *commuting Hermitian defect of* A_1 *and* A_2, denoted $\operatorname{chd}(A_1, A_2)$, as the smallest p such that there exist commuting Hermitian matrices

$$B_1 = \begin{pmatrix} A_1 & * \\ * & * \end{pmatrix} \text{ and } B_2 = \begin{pmatrix} A_2 & * \\ * & * \end{pmatrix}$$

of size $(n+p)\times(n+p)$. We call such a pair (B_1, B_2) of size $(n+\operatorname{chd}(A_1, A_2))\times(n + \operatorname{chd}(A_1, A_2))$ a *minimal commuting Hermitian completion of* (A_1, A_2).

(a) Show that (B_1, B_2) is a commuting Hermitian completion of a pair (A_1, A_2) of Hermitian matrices if and only if $B = B_1 + iB_2$ is a normal completion of $A = A_1 + iA_2$.

(b) Show that $\mathrm{chd}(A_1, A_2) = \mathrm{nd}(A_1 + iA_2)$.

(c) Show that $\mathrm{chd}(A_1, A_2) \geq \max\{i_+(i[A_1, A_2]), i_-(i[A_1, A_2])\}$.

(d) Show that $\mathrm{chd}(A_1, A_2) = 1$ if and only if $\mathrm{rank}(A_1A_2 - A_2A_1) = 2$ and the equation

$$PA_1(t_1u_1 - \bar{t}_1u_2) = iPA_2(t_2u_1 + \bar{t}_2u_2) \tag{5.12.12}$$

has a solution pair $t_1, t_2 \in \mathbb{C}$ satisfying

$$\mathrm{Re}(\bar{t}_1t_2) = d. \tag{5.12.13}$$

Here $u_1, u_2 \in \mathbb{C}^n$ are the unit eigenvectors of the matrix $2i(A_1A_2 - A_2A_1)$ corresponding to its nonzero eigenvalues $\lambda_1 = d(> 0)$ and $\lambda_2 = -d$, and $P = I_n - u_1u_1^* - u_2u_2^*$ is the orthogonal projection of $\mathbb{C}^n$ onto $\mathrm{Ker}(A_1A_2 - A_2A_1)$.

(e) Show that if $\mathrm{chd}(A_1, A_2) = 1$ then the pair (B_1, B_2) of matrices

$$B_1 = \begin{pmatrix} A_1 & \frac{\mu}{2}(t_2u_1 + \bar{t}_2u_2) \\ \frac{\bar{\mu}}{2}(\bar{t}_2u_1^* + t_2u_2^*) & z_1 \end{pmatrix}, \tag{5.12.14}$$

$$B_2 = \begin{pmatrix} A_2 & \frac{\mu}{2i}(t_1u_1 - \bar{t}_1u_2) \\ -\frac{\bar{\mu}}{2i}(\bar{t}_1u_1^* - t_1u_2^*) & z_2 \end{pmatrix} \tag{5.12.15}$$

is a minimal commuting Hermitian completion of (A_1, A_2). Here t_1 and t_2 satisfy (5.12.12) and (5.12.13), $\mu \in \mathbb{T}$ is arbitrary,

$$z_1 = u_1^*A_1u_1 - \frac{1}{d}\left(\mathrm{Im}(t_2^2u_2^*A_2u_1) + \mathrm{Re}(t_1t_2u_2^*A_1u_1)\right), \tag{5.12.16}$$

and

$$z_2 = u_1^*A_2u_1 - \frac{1}{d}\left(\mathrm{Im}(t_1^2u_2^*A_1u_1) - \mathrm{Re}(t_1t_2u_2^*A_2u_1)\right). \tag{5.12.17}$$

(f) Show that all minimal commuting Hermitian completions of (A_1, A_2) arise in the above way.

49 The *Gauss-Lucas theorem* states that the roots of the derivative of a polynomial lie in the convex hull of the roots of the polynomial. In this exercise we shall use normal matrices to prove this. Let $p(z)$ be a polynomial of degree n and let $\alpha_1, \ldots, \alpha_n$ be its roots. Let $N = U^*AU$ be the normal matrix where $A = \mathrm{diag}(\alpha_i)_{i=1}^n$ and U is a unitary matrix with first column equal to $\frac{1}{\sqrt{n}}\begin{pmatrix}1 & \cdots & 1\end{pmatrix}^T$.

(a) Let M be the matrix obtained from N by deleting the first row and column. Show (using Kramer's rule, for instance) that

$$\frac{n\det(zI-M)}{\det(zI-N)}=\sum_{i=1}^{n}\frac{1}{z-\alpha_i}.$$

(b) Use (a) to show that the eigenvalues of M are the roots of p'.

(c) Recall that the numerical range of a matrix B is given by

$$W(B)=\{x^*Bx\ :\ \|x\|=1\}.$$

Use now the (obvious) inclusion $W(M)\subseteq W(N)$ to prove the Gauss-Lucas theorem.

50 Let

$$\rho_\alpha=\frac{1}{7}\begin{pmatrix} \frac{2}{3} & 0 & 0 & 0 & \frac{2}{3} & 0 & 0 & 0 & \frac{2}{3}\\ 0 & \frac{\alpha}{3} & 0 & 0 & 0 & 0 & 0 & 0 & 0\\ 0 & 0 & \frac{5-\alpha}{3} & 0 & 0 & 0 & 0 & 0 & 0\\ 0 & 0 & 0 & \frac{5-\alpha}{3} & 0 & 0 & 0 & 0 & 0\\ \frac{2}{3} & 0 & 0 & 0 & \frac{2}{3} & 0 & 0 & 0 & \frac{2}{3}\\ 0 & 0 & 0 & 0 & 0 & \frac{\alpha}{3} & 0 & 0 & 0\\ 0 & 0 & 0 & 0 & 0 & 0 & \frac{\alpha}{3} & 0 & 0\\ 0 & 0 & 0 & 0 & 0 & 0 & 0 & \frac{5-\alpha}{3} & 0\\ \frac{2}{3} & 0 & 0 & 0 & \frac{2}{3} & 0 & 0 & 0 & \frac{2}{3} \end{pmatrix},$$

where $0\le\alpha\le 5$. We want to investigate when ρ_α is 3×3 separable.

(a) Show that ρ_α passes the Peres test if and only if $1\le\alpha\le 4$.

(b) Let

$$Z=\begin{pmatrix} 1 & 0 & 0 & 0 & -1 & 0 & 0 & 0 & -1\\ 0 & 0 & 0 & 0 & 0 & 0 & 0 & 0 & 0\\ 0 & 0 & 2 & 0 & 0 & 0 & 0 & 0 & 0\\ 0 & 0 & 0 & 2 & 0 & 0 & 0 & 0 & 0\\ -1 & 0 & 0 & 0 & 1 & 0 & 0 & 0 & -1\\ 0 & 0 & 0 & 0 & 0 & 0 & 0 & 0 & 0\\ 0 & 0 & 0 & 0 & 0 & 0 & 0 & 0 & 0\\ 0 & 0 & 0 & 0 & 0 & 0 & 0 & 2 & 0\\ -1 & 0 & 0 & 0 & -1 & 0 & 0 & 0 & 1 \end{pmatrix}.$$

Show that for $x,y\in\mathbb{C}^3$ we have that $(x\otimes y)^*Z(x\otimes y)\ge 0$.

(c) Show that $\operatorname{tr}(\rho_\alpha Z)=\frac{1}{7}(3-\alpha)$, and conclude that ρ_α is not 3×3 separable for $3<\alpha\le 5$.

(d) Show that ρ_α is not 3×3 separable for $0\le\alpha<2$.

(e) Show that ρ_α is 3×3 separable for $2\le\alpha\le 3$.

(f) Show that the map $\Lambda : \mathbb{C}^{3\times 3} \to \mathbb{C}^{3\times 3}$ defined by

$$\Lambda((a_{ij})_{i,j=1}^3) := \begin{pmatrix} a_{11} + 2a_{22} & -a_{12} & -a_{13} \\ -a_{12} & a_{22} + 2a_{33} & -a_{23} \\ -a_{31} & -a_{32} & a_{33} + 2a_{11} \end{pmatrix}$$

is positive (i.e., $\Lambda(\mathrm{PSD}_3) \subseteq \mathrm{PSD}_3$) but is not completely positive; see Exercise 5.12.51 for a more general statement.

(g) Write $\rho_\alpha = (\rho_{ij})_{i,j=1}^3$ with $\rho_{ij} \in \mathbb{C}^{3\times 3}$. Show that $(\Lambda(\rho_{ij}))_{i,j=1}^3$ is not positive semidefinite for $\alpha > 3$, providing an alternative proof for the fact that ρ_α is not 3×3 separable for $\alpha > 3$.

51 A linear map $\Phi : \mathbb{C}^{n\times n} \to \mathbb{C}^{m\times m}$ is called *decomposable* if there exist matrices S_i and T_i such that for all M we have

$$\Phi(M) = \sum_{i=1}^{s} S_i^* M S_i + \sum_{j=1}^{t} T_j^* M^T T_j.$$

Clearly, every decomposable map is positive.

For $a, b, c \geq 0$, consider the map

$\Phi_{a,b,c}((x_{ij})_{i,j=1}^3)$

$$= \begin{pmatrix} ax_{11} + bx_{22} + cx_{33} & -x_{12} & -x_{13} \\ -x_{21} & ax_{22} + bx_{33} + cx_{11} & -x_{23} \\ -x_{31} & -x_{32} & ax_{33} + bx_{11} + cx_{22} \end{pmatrix}.$$

(a) Show that $\Phi_{a,b,c}$ is positive if and only if (i) $a \geq 0$; (ii) $a + b + c \geq 2$; and (iii) $bc \geq (1 - a)^2$ if $0 \leq a \leq 1$.

(b) Show that $\Phi_{a,b,c}$ is decomposable if and only if (i) $a \geq 0$; and (ii) $\sqrt{bc} \geq \frac{2-a}{2}$ if $0 \leq a \leq 2$.

52 Let

$$M = \begin{pmatrix} 0 & 0 & \frac{1}{2\sqrt{2}} & \frac{1}{2\sqrt{2}} \\ 0 & 0 & \frac{1}{2\sqrt{2}} & \frac{-i}{2\sqrt{2}} \\ \frac{1}{2\sqrt{2}} & \frac{1}{2\sqrt{2}} & 0 & \frac{1-i}{4} \\ \frac{1}{2\sqrt{2}} & \frac{i}{2\sqrt{2}} & \frac{1+i}{4} & 0 \end{pmatrix}, \quad N = \frac{1}{6}\begin{pmatrix} 0 & 1 & 1 & -1 \\ 1 & 0 & -1 & 1 \\ 1 & -1 & 0 & 1 \\ -1 & 1 & 1 & 0 \end{pmatrix},$$

and define the linear map

$$\Phi(X) = I \odot X + M \odot X + N \odot X^T.$$

Show that $\Phi(X)$ is a positive map, but not completely positive (in fact, Φ is not decomposable).

53 For $0 \le b \le 1$, let

$$\rho_b = \frac{1}{7b+1}\begin{pmatrix} b & 0 & 0 & 0 & 0 & b & 0 & 0 \\ 0 & b & 0 & 0 & 0 & 0 & b & 0 \\ 0 & 0 & b & 0 & 0 & 0 & 0 & b \\ 0 & 0 & 0 & b & 0 & 0 & 0 & 0 \\ 0 & 0 & 0 & 0 & \frac{1+b}{2} & 0 & 0 & \frac{\sqrt{1-b^2}}{2} \\ b & 0 & 0 & 0 & 0 & b & 0 & 0 \\ 0 & b & 0 & 0 & 0 & 0 & b & 0 \\ 0 & 0 & b & 0 & \frac{\sqrt{1-b^2}}{2} & 0 & 0 & \frac{1+b}{2} \end{pmatrix}.$$

Determine for which b the matrix ρ_b is 2×4 separable.

54 A matrix is $2\times2\times2$ separable if it lies in the cone generated by matrices of the form $A \otimes B \otimes C$ with $A, B, C \in \mathrm{PSD}_2$. Put

$$R = I - x_1x_1^* - x_2x_2^* - x_3x_3^* - x_4x_4^*,$$

where

$$x_1 = \begin{pmatrix} 1 \\ 0 \end{pmatrix} \otimes \begin{pmatrix} 0 \\ 1 \end{pmatrix} \otimes \begin{pmatrix} \frac{1}{2}\sqrt{2} \\ \frac{1}{2}\sqrt{2} \end{pmatrix}, \quad x_2 = \begin{pmatrix} 0 \\ 1 \end{pmatrix} \otimes \begin{pmatrix} \frac{1}{2}\sqrt{2} \\ \frac{1}{2}\sqrt{2} \end{pmatrix} \otimes \begin{pmatrix} 1 \\ 0 \end{pmatrix},$$

$$x_3 = \begin{pmatrix} \frac{1}{2}\sqrt{2} \\ \frac{1}{2}\sqrt{2} \end{pmatrix} \otimes \begin{pmatrix} 1 \\ 0 \end{pmatrix} \otimes \begin{pmatrix} 0 \\ 1 \end{pmatrix}, \quad x_4 = \begin{pmatrix} \frac{1}{2}\sqrt{2} \\ -\frac{1}{2}\sqrt{2} \end{pmatrix} \otimes \begin{pmatrix} \frac{1}{2}\sqrt{2} \\ -\frac{1}{2}\sqrt{2} \end{pmatrix} \otimes \begin{pmatrix} \frac{1}{2}\sqrt{2} \\ -\frac{1}{2}\sqrt{2} \end{pmatrix}.$$

Show that R is not $2 \times 2 \times 2$ separable.

Hint: Let

$$Z = \begin{pmatrix} 1 & -1 & -1 & 1 & -1 & 1 & 1 & -1 \\ -1 & 4 & 1 & 0 & 1 & 3 & -1 & 1 \\ -1 & 1 & 4 & 3 & 1 & -1 & 0 & 1 \\ 1 & 0 & 3 & 4 & -1 & 1 & 1 & -1 \\ -1 & 1 & 1 & -1 & 4 & 0 & 3 & 1 \\ 1 & 3 & -1 & 1 & 0 & 4 & 1 & -1 \\ 1 & -1 & 0 & 1 & 3 & 1 & 4 & -1 \\ -1 & 1 & 1 & -1 & 1 & -1 & -1 & 1 \end{pmatrix},$$

and show that $\mathrm{trace}(RZ) = -\frac{3}{8}$ but

$$(v \otimes w \otimes z)^* Z (v \otimes w \otimes z) \ge 0,$$

for all $v, w, z \in \mathbb{C}^{2\times2}$.

5.13 NOTES

Section 5.1

The first part of Section 5.1 follows the paper [134], in which Theorem 5.1.6 is a key result. It is an open problem how Theorem 5.1.6 generalizes for $n \times n$ banded patterns, $n \geq 4$. Earlier results regarding the inertia of a 3×3 partial block matrix appear in [253] and [144]. Some of the early results on inertia, such as formulas (5.1.3) and (5.1.4) and Lemmas 5.1.1 and 5.1.2, go back to [305]. Lemma 5.1.8 is due to [445]. Theorem 5.1.12 is due to [335], and generalizes a result in [204] for the banded case. Theorem 5.1.9 was first obtained in [243] (see also [276]). Exercise 5.12.1 is a well-known result (it is in fact Lemma 5.1.14, which goes back to [289] and [221]; see also [532] and [552]), while Exercise 5.12.5 is a theorem in [82].

Section 5.2

Section 5.2 is based on [344], [562], [564], and [135]. The minimal rank completion problem was introduced in [344], and was motivated by the problem of minding minimal realizations for input-output operators; see Section 5.10 for the discrete case. Aside from Theorems 5.2.1 and 5.2.5 (see also [575]), the papers [344] and [562] also address the minimal rank completion problem for integral operators. The rank completion problem was then further pursued in [135], where a formula for the maximal possible rank was obtained for all patterns, and where the minimal rank problem was addressed for different patterns. The conjecture in [135], that equality holds in (5.2.8) only for all partial matrices with a pattern whose graph is bipartite chordal, is still open. Exercise 5.12.23 is inspired by [45]; see also [501]. Exercise 5.12.10 is based on [564].

Some of the results on minimal rank completions that we have not included concern the following. In [566] it was shown that equality in (5.2.8) holds for banded matrices; see Exercise 5.12.9. In [568] the minimal rank completion was studied in the class of Toeplitz matrices, and it was shown that for triangular and banded Toeplitz partial matrices, the minimal rank completion can always to be chosen to be Toeplitz as well. The result has the following consequence for the partial realization problem. Consider the following variation of the classical partial realization problem. Let $M_1, M_2, \ldots, M_{p-1}, M_p$ be a given finite sequence of $r \times s$ matrices. A system $\Sigma = (A, B, C)$, where A, B, and C are matrices of sizes $q \times q$, $q \times s$, and $r \times q$, respectively, is called a *t-realization* if $CA^{i-1+t}B = M_i, i = 1, \ldots, p$. The classical case is when $t = 0$; see [351] (see also, e.g., [104] and [264]). The number q is called the *dimension* of the t-realization. The problem is to find for a given sequence of matrices $M_1, \ldots, M_p$ a realization with lowest possible dimension. This lowest possible dimension we shall denote by $\delta_t(M_1, \ldots, M_p)$, and refer to it as the *degree* of $M_1, \ldots, M_p$.

The following relates this partial realization problem to the Toeplitz minimal rank completion problem. *Let $M_1, M_2, \ldots, M_p$ be a sequence of $r \times s$*

matrices, and let t be a nonnegative integer. Then

$$\delta_t(M_1, \ldots, M_p) = \min \operatorname{rank}\mathcal{A},$$

where $\mathcal{A}$ is the $(t+p) \times (t+p)$ Toeplitz partial block matrix

$$\mathcal{A} = \begin{pmatrix} M_p & ? & ? & \cdots & ? & \cdots & ? \\ M_{p-1} & M_p & ? & \cdots & ? & \cdots & ? \\ \vdots & \vdots & \ddots & \ddots & & & \vdots \\ M_1 & M_2 & \cdots & M_p & ? & & ? \\ ? & M_1 & M_2 & \cdots & M_p & \cdots & ? \\ \vdots & \vdots & \ddots & \ddots & & \ddots & \vdots \\ ? & ? & \cdots & M_1 & M_2 & \cdots & M_p \end{pmatrix}. \tag{5.13.1}$$

See [214] for a parametrization of the set of all solutions in the case that $t = 0$.

Exercise 5.12.11 comes from an example in [568].

The theory of minimal rank completions has recently become an area of high interest, first of all because of its relation with semiseparable and quasiseparable matrices. In many computational problems such matrices appear, and this structure can be taken advantage of. Useful references to access this area of research are [349], [198], [553]. In addition, recent algorithms have been developed to find minimal rank completions via minimizing the nuclear norm. The latter minimization makes it a convex problem and thus convex optimization techniques may be used. This recent interest is due to applications in predicting consumer preferences (e.g, the so-called Netflix problem), camera surveillance, and many others. Useful references for this area include [117], [118], [480], [116].

Section 5.3

Section 5.3 is based on [569].

Section 5.4

Theorem 5.4.2 appears in [273] and [24]. The scalar version goes back to [375]. The proof presented here relies on the Stein equation with a positive definite right-hand side (see (5.4.6)); for results on the Stein equation with an indefinite right-hand side see, for instance, [172], [400], and [522]. The remainder of the section is based on part of [569] and [132]. Exercise 5.12.18 is based on [132], where also all possible inertia of these expressions are derived as well as for other expressions. The earlier paper [541] contained conjectures regarding these results, which were correct for the complex case. The case when three or more unknown matrices appear, such as an expression like $A - BXB^* - CYC^* - DZD^*$, seems substantially more difficult. Using a Schur complement trick converts this problem to a minimal rank completion problem where now three block diagonal entries are unknown. As the underlying graph is not bipartite chordal, the problem is significantly more difficult (see the remarks above for Section 5.2). Exercise 5.12.25 is based on [541, Theorem 3.1].

Section 5.5

Section 5.5 is based on [274].

Section 5.6

This section is also mostly based on [274]. Theorem 5.6.4 is due to [5], and is also valid when Γ is not compact. The proof given here is based on the proof given in [457], where the noncompact case was proven as well. An open problem is how to generalize Theorem 5.6.4 to a multivariable setting. In [290], [585], and [278] different algorithms are described to find the singular value decomposition of a finite rank Hankel operator. The answer to Exercise 5.12.37 is in [439].

One interesting application of the problem of minimizing the kth singular value of a partial triangular matrix appears in the recovery of a 3D rigid structure from motion under occlusion. In this problem we have a set of images on which several feature points of a rigid 3D structure appear. As images capture the scene at time intervals and the 3D structure is moving, some feature points will not be visible on several of the images due to occlusion. When there are N feature points and K images, we obtain a $2K \times N$ observation matrix; namely, for each image the x and y coordinates of each feature point are recorded. In a noise-free setting this leads to a rank 4 matrix, due to the rigidity of the object. The occlusion has the effect that after permutation the observation matrix contains a triangular pattern of unknown entries. Thus we arrive at the problem of trying to fill in the unknown entries with the object of finding a rank 4 matrix. As the data will contain noise, we really are looking for a completion of the matrix that minimizes the 5th singular value. The papers [14], [15], [13] have more information on this application.

Section 5.7

Section 5.7 is based on [57]. Starting with [332], the problem of extending a given sequence $(c_j)_{j=-n}^{n}$, with $c_{-j} = \overline{c}_j$ to a sequence $(c_j)_{j\in\mathbb{Z}}$ such that all Toeplitz matrices $T_q = (c_{i-j})_{i,j=0}^{q}$ for $q \geq 0$ are Hermitian and have less than or equal to a given number k of negative eigenvalues, has extensively been studied. In [332], such extensions which are also bounded were characterized for the case of $G = \mathbb{Z}$, while in [510] the general case of locally compact Abelian groups was addressed. These results are not related so far to moment problems for real measures on the unit circle, but even so, in [545] the results of [119] were extended to the case of Hermitian Toeplitz block matrices, and an announcement was made toward applying such results to moment problems in a future work. Part of Theorem 5.7.1 was inspired by [162]. Other related papers are [203] (which includes the answer to Exercise 5.12.44) and [133]. We mention that in [145], [146] Krein space techniques were used to study indefinite extensions of Hermitian block matrices. There is a strong indication that such techniques might be used to solve matrix-valued indefinite moment problems as well. Finally, in [245] a Hankel variation is discussed where the extension is required to have a minimal number of negative eigenvalues.

Section 5.8

The presentation of Section 5.8 follows closely that of [55], where all results except Lemma 5.8.2 and Theorem 5.8.9 were initially proved. Lemma 5.8.2 originates in [202] and Theorem 5.8.9 in [277]. Our proof of Theorem 5.8.9 is also taken from [55]. The subject of distance matrices is of great importance in applied mathematics. There are many studies dedicated to them, from the classical reference [96] to applications in molecular conformation [155], [255], [256], protein structures [425], and statistical shape analysis [399]. The Euclidean matrix completion problem is motivated by the molecular conformation problem. The characterization of distance matrices at the beginning of Section 5.8 is taken from [303]. For more connections between Euclidean distance matrix completions and positive semidefinite completions see [392], [393], [394], and [328]. Further developments regarding Euclidean distance matrix completions can be found in [334], [392], [393], [394], [20], [549], [19], [328], and [402]. The answer to Exercise 5.12.40 may be found in [304]. It is still unknown whether the same result holds for $n = 7$.

Section 5.9

The minimal normal completion problem was introduced in [569], inspired by the paper [298]. The upper bound in Theorem 5.9.1 was obtained there. The lower bound in Theorem 5.9.1 was later obtained in [359]. The remainder of the section, which concerns the case when a matrix can be made normal by adding one row and column to it, was taken from [350]. It is still an open problem how to recognize matrices with a normal defect equal to k, $k \geq 2$. For instance, is the normal defect of

$$\begin{pmatrix} 0 & 1 & 0 & 0 \\ 0 & 0 & 3 & 0 \\ 0 & 0 & 0 & 2 \\ 0 & 0 & 0 & 0 \end{pmatrix}$$

2 or 3? Another open problem is whether the equality $\mathrm{nd}\left(\begin{smallmatrix} A & 0 \\ 0 & B \end{smallmatrix}\right) = \mathrm{nd}(A) + \mathrm{nd}(B)$ holds in general; the inequality $\leq$ is obvious. Exercises 5.12.45, 5.12.47, 5.12.48 are inspired by examples and results in [350]. The result in exercise 5.12.47 originated in [359]. Exercise 5.12.26 shows that any 2×2 triangular partial matrix has a normal completion. In [236] it was shown that any scalar $n \times n$ triangular partial matrix has a normal completion. Exercise 5.12.15 is based on [359]. See also [173] for earlier work on the commuting defect and its application to cubature formulas. Exercise 5.12.49 uses ideas from [461] and [413]. Exercise 5.12.31 uses the results of [212]. See [234], [235] for the question of finding a normal completion when some columns are prescribed, and [329, 476] for recent results on normal matrices with a normal principal block.

Section 5.10

Section 5.10 discusses the original motivation for the minimal rank completion problem. The references [262], [263], [344] give a more complete background on these system-theoretic applications of the minimal rank completion problem. If the original system is allowed to be slightly perturbed, then it is exactly the problem solved in [493].

Section 5.11

Section 5.11 is based on [572] and [350, Section 5]. The separability problem appears in quantum computation. The book [436] gives an excellent exposition on quantum computation. Theorem 5.11.6 is in its dual form due to [580], who proved: if $\Phi : \mathbb{C}^{2\times 2} \to \mathbb{C}^{n\times n}$ is a positive linear map (thus $\Phi(\mathrm{PSD}_2) \subseteq \mathrm{PSD}_n$) and $n \leq 3$, then Φ must be decomposable; that is, Φ must be of the form $\Phi(M) = \sum_{i=1}^{k} R_i M R_i^* + \sum_{i=1}^{l} S_i M^T S_i^*$. The test in Theorem 5.11.6 is referred to as the Peres test; see [462] and [326]. Example 5.11.8 is also due to [580]. Lemma 5.11.7 was taken from [318, Lemma 2.6]. See also [508] for the connection of the separability problem and commuting normal matrices. For a formulation of the separability problem in terms of continuous positive semidefinite functions on products of compact groups, please see [370]. The papers [148], [147], and [149] give characterizations of concepts regarding quantum states (pure states, entropy, Peres test, Kraus representation) using the Schur parameters as discussed in Section 2.5.

Exercises

Exercise 5.12.21 was taken from [405]. Exercise 5.12.32 is based on an example in [205], and Exercise 5.12.33 is based on [206, Theorem 4].

Exercise 5.12.34 is based on [337, Theorem 9].

Exercise 5.12.20 is based on [73] (parts (a) and (c)) and [487] (part (b)). The question what possible Jordan canonical forms completion of triangular partial matrices may have was also considered in [487], and a conjecture was stated there that was later solved in [378]. Other papers that address this conjecture are [521], [341], and [342]. Many other results exist where a partial matrix is given and where a completion is to be found with specified eigenvalues; a recent overview article is [154].

Exercises 5.12.38 and 5.12.39 are based, respectively, on Theorems 2.1 and 4.1 in [493].

Exercise 5.12.41 is based on [287]; see also [288]. Exercise 5.12.12 is based on results in [540] (part (a)) and [222] (part (b)). The problem in Exercise 5.12.13 is treated in much greater generality in [193].

Exercise 5.12.50 is based on an example in [325, Section 4.6], which was further worked out in [180]. The linear map Λ was shown to be positive in [129]. The solution to Exercise 5.12.51 may be found in [127] and the solution to Exercise 5.12.52 may be found in [406, Section 4].

Exercise 5.12.7 is Proposition 3.1 in [86]. Exercise 5.12.8 is based on [163], where there is also a description of the set of all solutions. Exercise 5.12.54 is based on an example from [181].

Bibliography

[1] Yuri I. Abramovich, Ben A. Johnson, and Nicholas K. Spencer. Two-dimensional multivariate parametric models for radar applications. I. Maximum-entropy extensions for Toeplitz-block matrices. *IEEE Trans. Signal Process.*, 56(11):5509–5526, 2008. (Cited on p. 251.)

[2] Yuri I. Abramovich, Ben A. Johnson, and Nicholas K. Spencer. Two-dimensional multivariate parametric models for radar applications. II. Maximum-entropy extensions for Hermitian-block matrices. *IEEE Trans. Signal Process.*, 56(11):5527–5539, 2008. (Cited on p. 251.)

[3] V. M. Adamjan, D. Z. Arov, and M. G. Kreĭn. Infinite Hankel matrices and generalized Carathéodory-Fejér and I. Schur problems. *Funkcional. Anal. i Priložen.*, 2(4):1–17, 1968. (Cited on pp. 353, 355.)

[4] V. M. Adamjan, D. Z. Arov, and M. G. Kreĭn. Infinite Hankel matrices and generalized problems of Carathéodory-Fejér and F. Riesz. *Funkcional. Anal. i Priložen.*, 2(1):1–19, 1968. (Cited on p. 353.)

[5] V. M. Adamjan, D. Z. Arov, and M. G. Kreĭn. Analytic properties of the Schmidt pairs of a Hankel operator and the generalized Schur-Takagi problem. *Mat. Sb. (N.S.)*, 86(128):34–75, 1971. (Cited on p. 472.)

[6] V. M. Adamjan, D. Z. Arov, and M. G. Kreĭn. Infinite Hankel block matrices and related problems of extension. *Izv. Akad. Nauk Armjan. SSR Ser. Mat.*, 6(2-3):87–112, 1971. (Cited on p. 353.)

[7] Jim Agler. Some interpolation theorems of Nevanlinna-Pick type. *Unpublished manuscript*, 1988. (Cited on p. 355.)

[8] Jim Agler. On the representation of certain holomorphic functions defined on a polydisc. In *Topics in operator theory: Ernst D. Hellinger memorial volume*, volume 48 of *Oper. Theory Adv. Appl.*, pages 47–66. Birkhäuser, Basel, 1990. (Cited on pp. 177, 250, 355.)

[9] Jim Agler, J. William Helton, Scott McCullough, and Leiba Rodman. Positive semidefinite matrices with a given sparsity pattern. *Linear Algebra Appl.*, 107:101–149, 1988. (Cited on pp. 65, 67.)

[10] Jim Agler and John E. McCarthy. Nevanlinna-Pick interpolation on the bidisk. *J. Reine Angew. Math.*, 506:191–204, 1999. (Cited on p. 356.)

[11] Jim Agler and John E. McCarthy. *Pick interpolation and Hilbert function spaces*, volume 44 of *Graduate Studies in Mathematics*. American Mathematical Society, Providence, RI, 2002. (Cited on p. 356.)

[12] Jim Agler and John E. McCarthy. What Hilbert spaces can tell us about bounded functions in the bidisk. arXiv:0901.0907. (Cited on p. 356.)

[13] Pedro M. Q. Aguiar, Marko Stošić, and João Xavier. Globally optimal solution to exploit rigidity when recovering structure from motion under occlusion. In *IEEE International Conference on Image Processing – ICIP'08*, 2008. (Cited on p. 472.)

[14] Pedro M. Q. Aguiar, Marko Stošić, and João Xavier. On singular values of partially prescribed matrices. *Linear Algebra Appl.*, 429(8-9):2136–2145, 2008. (Cited on p. 472.)

[15] Pedro M. Q. Aguiar, Marko Stošić, and João Xavier. Spectrally optimal factorization of incomplete matrices. In *IEEE Computer Society Conference on Computer Vision and Patter Recoginition - CVPR*, 2008. (Cited on p. 472.)

[16] N. I. Aheizer and M. Krein. *Some questions in the theory of moments.* translated by W. Fleming and D. Prill. Translations of Mathematical Monographs, Vol. 2. American Mathematical Society, Providence, R.I., 1962. (Cited on p. 173.)

[17] N. I. Akhiezer. *The classical moment problem and some related questions in analysis.* Translated by N. Kemmer. Hafner, New York, 1965. (Cited on pp. 65, 173, 173.)

[18] Akram Aldroubi, Anatoly Baskakov, and Ilya Krishtal. Slanted matrices, Banach frames, and sampling. *J. Funct. Anal.*, 255(7):1667–1691, 2008. (Cited on p. 355.)

[19] Abdo Y. Alfakih. On the uniqueness of Euclidean distance matrix completions. *Linear Algebra Appl.*, 370:1–14, 2003. (Cited on p. 473.)

[20] Abdo Y. Alfakih, Amir Khandani, and Henry Wolkowicz. Solving Euclidean distance matrix completion problems via semidefinite programming. *Comput. Optim. Appl.*, 12(1-3):13–30, 1999. Computational optimization—a tribute to Olvi Mangasarian, Part I. (Cited on p. 473.)

[21] Charalambos D. Aliprantis and Rabee Tourky. *Cones and duality*, volume 84 of *Graduate Studies in Mathematics*. American Mathematical Society, Providence, RI, 2007. (Cited on p. 65.)

[22] D. Alpay and C. Dubi. Carathéodory Fejér interpolation in the ball with mixed derivatives. *Linear Algebra Appl.*, 382:117–133, 2004. (Cited on p. 354.)

[23] Daniel Alpay, Vladimir Bolotnikov, and Philippe Loubaton. On a new positive extension problem for block Toeplitz matrices. *Linear Algebra Appl.*, 268:247–287, 1998. (Cited on p. 171.)

[24] Daniel Alpay and Israel Gohberg. On orthogonal matrix polynomials. In *Orthogonal matrix-valued polynomials and applications (Tel Aviv, 1987–88)*, volume 34 of *Oper. Theory Adv. Appl.*, pages 25–46. Birkhäuser, Basel, 1988. (Cited on p. 471.)

[25] Daniel Alpay and Philippe Loubaton. The partial trigonometric moment problem on an interval: the matrix case. *Linear Algebra Appl.*, 225:141–161, 1995. (Cited on p. 67.)

[26] C.-G. Ambrozie. Finding positive matrices subject to linear restrictions. *Linear Algebra Appl.*, 426(2-3):716–728, 2007. (Cited on p. 66.)

[27] Brian D. O. Anderson and John B. Moore. *Linear optimal control.* Prentice-Hall, Englewood Cliffs, N.J., 1971. (Cited on p. 68.)

[28] W. N. Anderson, Jr. and R. J. Duffin. Series and parallel addition of matrices. *J. Math. Anal. Appl.*, 26:576–594, 1969. (Cited on p. 172.)

[29] W. N. Anderson, Jr. and G. E. Trapp. Shorted operators. II. *SIAM J. Appl. Math.*, 28:60–71, 1975. (Cited on p. 172.)

[30] William N. Anderson, Jr. and George E. Trapp. The extreme points of a set of positive semidefinite operators. *Linear Algebra Appl.*, 106:209–217, 1988. (Cited on p. 172.)

[31] T. Andô. On a pair of commutative contractions. *Acta Sci. Math. (Szeged)*, 24:88–90, 1963. (Cited on p. 356.)

[32] T. Andô. Truncated moment problems for operators. *Acta Sci. Math. (Szeged)*, 31:319–334, 1970. (Cited on p. 173.)

[33] T. Ando. Structure of operators with numerical radius one. *Acta Sci. Math. (Szeged)*, 34:11–15, 1973. (Cited on p. 172.)

[34] T. Ando. Generalized Schur complements. *Linear Algebra Appl.*, 27:173–186, 1979. (Cited on p. 171.)

[35] T. Ando. Extreme points of a positive operator ball. In *Operator theory and related topics, Vol. II (Odessa, 1997)*, volume 118 of *Oper. Theory Adv. Appl.*, pages 53–66. Birkhäuser, Basel, 2000. (Cited on p. 172.)

[36] Tsuyoshi Ando. Extreme points of an intersection of operator intervals. *Sūrikaisekikenkyūsho Kōkyūroku*, (939):54–63, 1996. Research on nonlinear analysis and convex analysis (Japanese) (Kyoto, 1995). (Cited on p. 172.)

[37] James R. Angelos, Carl C. Cowen, and Sivaram K. Narayan. Triangular truncation and finding the norm of a Hadamard multiplier. *Linear Algebra Appl.*, 170:117–135, 1992. (Cited on p. 358.)

[38] M. Anoussis. Interpolating operators in nest algebras. *Proc. Amer. Math. Soc.*, 114(3):707–710, 1992. (Cited on p. 357.)

[39] M. Anoussis, E. G. Katsoulis, R. L. Moore, and T. T. Trent. Interpolation problems for ideals in nest algebras. *Math. Proc. Cambridge Philos. Soc.*, 111(1):151–160, 1992. (Cited on p. 357.)

[40] Gr. Arsene, Zoia Ceauşescu, and T. Constantinescu. Schur analysis of some completion problems. *Linear Algebra Appl.*, 109:1–35, 1988. (Cited on p. 172.)

[41] Gr. Arsene and A. Gheondea. Completing matrix contractions. *J. Operator Theory*, 7(1):179–189, 1982. (Cited on p. 353.)

[42] A.P. Artjomenko. *Hermitian positive functions and positive functionals.* Dissertation. Odessa State University, 1941. (Cited on p. 252.)

[43] William Arveson. Interpolation problems in nest algebras. *J. Functional Analysis*, 20(3):208–233, 1975. (Cited on pp. 352, 356, 357, 357, 357, 357.)

[44] William B. Arveson. Subalgebras of C^*-algebras. *Acta Math.*, 123:141–224, 1969. (Cited on p. 224.)

[45] Edgar Asplund. Inverses of matrices $\{a_{ij}\}$ which satisfy $a_{ij} = 0$ for $j > i+p$. *Math. Scand.*, 7:57–60, 1959. (Cited on p. 470.)

[46] Robert Azencott and Didier Dacunha-Castelle. *Series of irregular observations.* Applied Probability. A Series of the Applied Probability Trust. Springer, New York, 1986. Forecasting and model building. (Cited on p. 255.)

[47] Mihály Bakonyi. Completion of operator partial matrices associated with chordal graphs. *Integral Equations Operator Theory*, 15(2):173–185, 1992. (Cited on p. 170.)

[48] Mihály Bakonyi. *Completion of partial operator matrices.* Ph.D. thesis, Mathematics, College of William and Mary. 1992. (Cited on p. 352.)

[49] Mihály Bakonyi. A remark on Nehari's problem. *Integral Equations Operator Theory*, 22(1):123–125, 1995. (Cited on p. 357.)

[50] Mihály Bakonyi. The extension of positive definite operator-valued functions defined on a symmetric interval of an ordered group. *Proc. Amer. Math. Soc.*, 130(5):1401–1406 (electronic), 2002. (Cited on p. 252.)

[51] Mihály Bakonyi. Nehari and Carathéodory-Fejér type extension results for operator-valued functions on groups. *Proc. Amer. Math. Soc.*, 131(11):3517–3525 (electronic), 2003. (Cited on p. 355.)

[52] Mihály Bakonyi and Tiberiu Constantinescu. Inheritance principles for chordal graphs. *Linear Algebra Appl.*, 148:125–143, 1991. (Cited on pp. 170, 173.)

[53] Mihály Bakonyi and Tiberiu Constantinescu. *Schur's algorithm and several applications*, volume 261 of *Pitman Research Notes in Mathematics Series.* Longman Scientific & Technical, Harlow, 1992. (Cited on pp. 65, 171, 172, 172, 353.)

[54] Mihály Bakonyi and Tiberiu Constantinescu. Research problem: the completion number of a graph. *Linear Multilinear Algebra*, 53(3):189–192, 2005. (Cited on p. 170.)

[55] Mihály Bakonyi and Charles R. Johnson. The Euclidean distance matrix completion problem. *SIAM J. Matrix Anal. Appl.*, 16(2):646–654, 1995. (Cited on p. 473.)

[56] Mihály Bakonyi, Victor G. Kaftal, Gary Weiss, and Hugo J. Woerdeman. Bounds for operator/Hilbert-Schmidt norm minimization using entropy. *Indiana Univ. Math. J.*, 46(2):405–425, 1997. (Cited on p. 357.)

[57] Mihály Bakonyi and Ekaterina V. Lopushanskaya. Moment problems for real measures on the unit circle. In *Recent advances in operator theory in Hilbert and Krein spaces*, volume 198 of *Oper. Theory Adv. Appl.*, pages 49–60. Birkhäuser, Basel, 2010. (Cited on p. 472.)

[58] Mihály Bakonyi and Geir Naevdal. On the matrix completion method for multidimensional moment problems. *Acta Sci. Math. (Szeged)*, 64(3-4):547–558, 1998. (Cited on pp. 66, 67, 251.)

[59] Mihály Bakonyi and Geir Nævdal. The finite subsets of $\mathbf{Z}^2$ having the extension property. *J. London Math. Soc. (2)*, 62(3):904–916, 2000. (Cited on p. 66.)

[60] Mihály Bakonyi, Leiba Rodman, Ilya M. Spitkovsky, and Hugo J. Woerdeman. Positive matrix functions on the bitorus with prescribed Fourier coefficients in a band. *J. Fourier Anal. Appl.*, 5(1):21–44, 1999. (Cited on p. 252.)

[61] Mihály Bakonyi and Dan Timotin. On a conjecture of Cotlar and Sadosky on multidimensional Hankel operators. *C. R. Acad. Sci. Paris Sér. I Math.*, 325(10):1071–1075, 1997. (Cited on p. 354.)

[62] Mihály Bakonyi and Dan Timotin. On an extension problem for polynomials. *Bull. London Math. Soc.*, 33(5):599–605, 2001. (Cited on p. 358.)

[63] Mihály Bakonyi and Dan Timotin. The central completion of a positive block operator matrix. In *Operator theory, structured matrices, and dilations*, volume 7 of *Theta Ser. Adv. Math.*, pages 69–83. Theta, Bucharest, 2007. (Cited on pp. 172, 173.)

[64] Mihály Bakonyi and Dan Timotin. Extensions of positive definite functions on free groups. *J. Funct. Anal.*, 246(1):31–49, 2007. (Cited on p. 253.)

[65] Mihály Bakonyi and Dan Timotin. Extensions of positive definite functions on amenable groups. *Canad. Math. Bull.*, 54:3–11, 2011. (Cited on p. 252.)

[66] Mihály Bakonyi and Hugo J. Woerdeman. The central method for positive semi-definite, contractive and strong Parrott type completion problems. In *Operator theory and complex analysis (Sapporo, 1991)*, volume 59 of *Oper. Theory Adv. Appl.*, pages 78–95. Birkhäuser, Basel, 1992. (Cited on pp. 172, 173, 353, 353, 353.)

[67] Mihály Bakonyi and Hugo J. Woerdeman. Linearly constrained interpolation in nest algebras. *Indiana Univ. Math. J.*, 42(4):1297–1304, 1993. (Cited on p. 356.)

[68] Mihály Bakonyi and Hugo J. Woerdeman. On the strong Parrott completion problem. *Proc. Amer. Math. Soc.*, 117(2):429–433, 1993. (Cited on p. 353.)

[69] Mihály Bakonyi and Hugo J. Woerdeman. Maximum entropy elements in the intersection of an affine space and the cone of positive definite matrices. *SIAM J. Matrix Anal. Appl.*, 16(2):369–376, 1995. (Cited on pp. 66, 67.)

[70] Mihály Bakonyi and Hugo J. Woerdeman. Extensions of positive definite Fredholm operators with partially zero inverse. *Math. Nachr.*, 193:5–18, 1998. (Cited on p. 66.)

[71] Radu Balan, Peter G. Casazza, Christopher Heil, and Zeph Landau. Density, overcompleteness, and localization of frames. I. Theory. *J. Fourier Anal. Appl.*, 12(2):105–143, 2006. (Cited on p. 355.)

[72] J. A. Ball. Linear systems, operator model theory and scattering: multivariable generalizations. In *Operator theory and its applications (Winnipeg, MB, 1998)*, volume 25 of *Fields Inst. Commun.*, pages 151–178. American Mathematical Society, Providence, RI, 2000. (Cited on p. 250.)

[73] J. A. Ball, I. Gohberg, L. Rodman, and T. Shalom. On the eigenvalues of matrices with given upper triangular part. *Integral Equations Operator Theory*, 13(4):488–497, 1990. (Cited on p. 474.)

[74] J. A. Ball, W. S. Li, D. Timotin, and T. T. Trent. A commutant lifting theorem on the polydisc. *Indiana Univ. Math. J.*, 48(2):653–675, 1999. (Cited on p. 250.)

[75] Joseph A. Ball. Multidimensional circuit synthesis and multivariable dilation theory. *Multidimens. Syst. Signal Process.*, 22(1–3):27–44, 2011. (Cited on p. 356.)

[76] Joseph A. Ball, Israel Gohberg, and Leiba Rodman. *Interpolation of rational matrix functions*, volume 45 of *Operator Theory: Advances and Applications*. Birkhäuser, Basel, 1990. (Cited on p. 356.)

[77] Joseph A. Ball and J. William Helton. A Beurling-Lax theorem for the Lie group $\mathrm{U}(m, n)$ which contains most classical interpolation theory. *J. Operator Theory*, 9(1):107–142, 1983. (Cited on p. 359.)

[78] Joseph A. Ball, Cora Sadosky, and Victor Vinnikov. Scattering systems with several evolutions and multidimensional input/state/output systems. *Integral Equations Operator Theory*, 52(3):323–393, 2005. (Cited on pp. 251, 356.)

[79] Joseph A. Ball and Tavan T. Trent. Unitary colligations, reproducing kernel Hilbert spaces, and Nevanlinna-Pick interpolation in several variables. *J. Funct. Anal.*, 157(1):1–61, 1998. (Cited on pp. 250, 356, 356.)

[80] Wayne W. Barrett. A theorem on inverses of tridiagonal matrices. *Linear Algebra Appl.*, 27:211–217, 1979. (Cited on p. 65.)

[81] Wayne W. Barrett and Philip J. Feinsilver. Inverses of banded matrices. *Linear Algebra Appl.*, 41:111–130, 1981. (Cited on p. 65.)

[82] Wayne W. Barrett and Charles R. Johnson. Determinantal formulae for matrices with sparse inverses. *Linear Algebra Appl.*, 56:73–88, 1984. (Cited on p. 470.)

[83] Wayne W. Barrett, Charles R. Johnson, and Raphael Loewy. The real positive definite completion problem: cycle completability. *Mem. Amer. Math. Soc.*, 122(584):viii+69, 1996. (Cited on p. 170.)

[84] Wayne W. Barrett, Charles R. Johnson, and Raphael Loewy. Critical graphs for the positive definite completion problem. *SIAM J. Matrix Anal. Appl.*, 20(1):117–130 (electronic), 1999. (Cited on p. 170.)

[85] Wayne W. Barrett, Charles R. Johnson, and Michael Lundquist. Determinantal formulae for matrix completions associated with chordal graphs. *Linear Algebra Appl.*, 121:265–289, 1989. (Cited on pp. 170, 173.)

[86] Wayne W. Barrett, Charles R. Johnson, Michael E. Lundquist, and Hugo J. Woerdeman. Completing a block diagonal matrix with a partially prescribed inverse. *Linear Algebra Appl.*, 223/224:73–87, 1995. Special issue honoring Miroslav Fiedler and Vlastimil Pták. (Cited on p. 474.)

[87] Wayne W. Barrett, Charles R. Johnson, and Pablo Tarazaga. The real positive definite completion problem for a simple cycle. *Linear Algebra Appl.*, 192:3–31, 1993. Computational linear algebra in algebraic and related problems (Essen, 1992). (Cited on p. 170.)

[88] A. Ben-Artzi, R. L. Ellis, I. Gohberg, and D. C. Lay. The maximum distance problem and band sequences. *Linear Algebra Appl.*, 87:93–112, 1987. (Cited on p. 170.)

[89] H. Bercovici, C. Foias, and A. Tannenbaum. Spectral variants of the Nevanlinna-Pick interpolation problem. In *Signal processing, scattering and*

operator theory, and numerical methods (Amsterdam, 1989), volume 5 of *Progr. Systems Control Theory*, pages 23–45. Birkhäuser Boston, Boston, 1990. (Cited on p. 356.)

[90] Christian Berg, Jens Peter Reus Christensen, and Paul Ressel. *Harmonic analysis on semigroups*, volume 100 of *Graduate Texts in Mathematics*. Springer, New York, 1984. Theory of positive definite and related functions. (Cited on pp. 66, 237.)

[91] A. Berman. *Cones, matrices and mathematical programming*. Volume 79 of Lecture Notes in Economics and Mathematical Systems. Springer, Berlin, 1973. (Cited on p. 65.)

[92] S. Bernstein. Sur une classe de polynomes orthogonaux. *Commun. Kharkow*, 4, 1930. (Cited on p. 250.)

[93] Arne Beurling. On two problems concerning linear transformations in Hilbert space. *Acta Math.*, 81:17, 1948. (Cited on p. 359.)

[94] David P. Blecher and Vern I. Paulsen. Explicit construction of universal operator algebras and applications to polynomial factorization. *Proc. Amer. Math. Soc.*, 112(3):839–850, 1991. (Cited on p. 360.)

[95] Grigoriy Blekherman. Nonnegative polynomials and sums of squares. arXiv:1010.3465. (Cited on p. 67.)

[96] Leonard M. Blumenthal. *Theory and applications of distance geometry*. Chelsea, New York, 1970. (Cited on p. 473.)

[97] S. Bochner. *Vorlesungen über Fourierische Integral*. Akademische Verlagsgesellschaft, Leipzig, 1932. (Cited on p. 255.)

[98] S. Bochner. Monotone Funktionen, Stieltjesschee Integrale und harmonische Analyse. *Math. Anal.*, 108:378–410, 1933. (Cited on p. 255.)

[99] S. Bochner and R. S. Phillips. Absolutely convergent Fourier expansions for non-commutative normed rings. *Ann. of Math. (2)*, 43:409–418, 1942. (Cited on p. 171.)

[100] V. Bolotnikov. Degenerate Stieltjes moment problem and associated J-inner polynomials. *Z. Anal. Anwendungen*, 14(3):441–468, 1995. (Cited on p. 174.)

[101] Vladimir Bolotnikov. On degenerate Hamburger moment problem and extensions of nonnegative Hankel block matrices. *Integral Equations Operator Theory*, 25(3):253–276, 1996. (Cited on p. 173.)

[102] Vladimir Bolotnikov. Interpolation for multipliers on reproducing kernel Hilbert spaces. *Proc. Amer. Math. Soc.*, 131(5):1373–1383 (electronic), 2003. (Cited on p. 354.)

[103] N. K. Bose. *Applied multidimensional systems theory*. Van Nostrand Reinhold Electrical/Computer Science and Engineering Series. Van Nostrand Reinhold, New York, 1982. (Cited on p. 66.)

[104] O. H. Bosgra. On parametrizations for the minimal partial realization problem. *Systems Control Lett.*, 3(4):181–187, 1983. (Cited on p. 470.)

[105] Albrecht Bőttcher and Bernd Silbermann. *Analysis of Toeplitz operators*. Springer Monographs in Mathematics. Springer, Berlin, second edition, 2006. Prepared jointly with Alexei Karlovich. (Cited on p. 171.)

[106] Stephen Boyd and Laurent El Ghaoui. Method of centers for minimizing generalized eigenvalues. *Linear Algebra Appl.*, 188/189:63–111, 1993. (Cited on p. 66.)

[107] Stephen Boyd, Laurent El Ghaoui, Eric Feron, and Venkataramanan Balakrishnan. *Linear matrix inequalities in system and control theory*, volume 15 of *SIAM Studies in Applied Mathematics*. Society for Industrial and Applied Mathematics (SIAM), Philadelphia, 1994. (Cited on pp. 66, 68.)

[108] Stephen Boyd and Lieven Vandenberghe. *Convex optimization*. Cambridge University Press, Cambridge, 2004. (Cited on pp. 65, 66.)

[109] Marek Bożejko. Positive-definite kernels, length functions on groups and a noncommutative von Neumann inequality. *Studia Math.*, 95(2):107–118, 1989. (Cited on p. 253.)

[110] Marek Bożejko and Roland Speicher. Completely positive maps on Coxeter groups, deformed commutation relations, and operator spaces. *Math. Ann.*, 300(1):97–120, 1994. (Cited on p. 253.)

[111] M. S. Brodskiĭ. Unitary operator colligations and their characteristic functions. *Uspekhi Mat. Nauk*, 33(4(202)):141–168, 256, 1978. (Cited on p. 356.)

[112] Ramón Bruzual and Marisela Domínguez. Extensions of operator valued positive definite functions on an interval of $\mathbf{Z}^2$ with the lexicographic order. *Acta Sci. Math. (Szeged)*, 66(3-4):623–631, 2000. (Cited on p. 252.)

[113] Adhemar Bultheel and Marc Van Barel. *Linear algebra, rational approximation and orthogonal polynomials*, volume 6 of *Studies in Computational Mathematics*. North-Holland, Amsterdam, 1997. (Cited on p. 171.)

[114] J.P. Burg. *Maximum Entropy Spectral Analysis*. Ph.D. Thesis. Geophysics, Stanford University, 1975. (Cited on p. 65.)

[115] A. Calderón and R. Pepinsky. On the phases of fourier coefficients for positive real periodic functions. In *Computing methods and the phase problem in X-ray crystal analysis*, pages 339–348. X-Ray Crystal Analysis Laboratory, Department of Physics, Pennsylvania State College, 1952. (Cited on pp. 65, 66.)

[116] Emmanuel J. Candès, Xiaodong Li, Yi Ma, and John Wright. Robust principal component analysis? arXiv:0912.3599. (Cited on p. 471.)

[117] Emmanuel J. Candès and Benjamin Recht. Exact matrix completion via convex optimization. *Foundations of Computational Mathematics*, 9:717–772, 2009. (Cited on p. 471.)

[118] Emmanuel J. Candès and Terrence Tao. The power of convex relaxation: Near-optimal matrix completion. *IEEE Trans. Inform. Theory*, 56(5):2053–2080, 2010. (Cited on p. 471.)

[119] C. Carathéodory. Über den Variabilitätsbereich der Fourierischen Konstanten von positiven harmonischen Funktionen. *Rend. Circ. Mat. Palermo*, 32:193–217, 1911. (Cited on pp. 65, 472.)

[120] C. Carathéodory and L. Fejér. Über den Zusammenhang der Extremen von harmonischen Funktionen mit ihren Koeffizienten und über den Picard-Landauschen Satz. *Rend. Circ. Mat. Palermo*, 32:218–239, 1911. (Cited on pp. 65, 353.)

[121] Lennart Carleson. Interpolations by bounded analytic functions and the corona problem. *Ann. of Math. (2)*, 76:547–559, 1962. (Cited on p. 357.)

[122] Glaysar Castro. *Coefficient de réflexion généralisés, extension de covariance multidimensionelles et autres applications.* PhD thesis, Université de Paris-Sud Centre d'Orsay, 1997. (Cited on p. 251.)

[123] Glaysar Castro and Valérie Girardin. Maximum of entropy and extension of covariance matrices for periodically correlated and multivariate processes. *Statist. Probab. Lett.*, 59(1):37–52, 2002. (Cited on p. 251.)

[124] Zoia Ceauşescu and Ciprian Foiaş. On intertwining dilations. V. *Acta Sci. Math. (Szeged)*, 40(1-2):9–32, 1978. (Cited on p. 172.)

[125] Li Chai and Li Qiu. Multirate periodic systems and constrained analytic function interpolation problems. *SIAM J. Control Optim.*, 43(6):1972–1986 (electronic), 2005. (Cited on p. 356.)

[126] Sun-Yung A. Chang and Robert Fefferman. A continuous version of duality of H^1 with BMO on the bidisc. *Ann. of Math. (2)*, 112(1):179–201, 1980. (Cited on p. 354.)

[127] Sung Je Cho, Seung-Hyeok Kye, and Sa Ge Lee. Generalized Choi maps in three-dimensional matrix algebra. *Linear Algebra Appl.*, 171:213–224, 1992. (Cited on p. 474.)

[128] Man Duen Choi. Completely positive linear maps on complex matrices. *Linear Algebra Appl.*, 10:285–290, 1975. (Cited on p. 253.)

[129] Man Duen Choi. Positive semidefinite biquadratic forms. *Linear Algebra Appl.*, 12(2):95–100, 1975. (Cited on p. 474.)

[130] M.D. Choi, M. Giesinger, J.A. Holbrook, and D.W Kribs. Geometry of higher-rank numerical ranges. *Linear Multilinear Algebra*, 56(1-2):53–64, 2008. (Cited on p. 174.)

[131] J. P. R. Christensen and J. Vesterstrøm. A note on extreme positive definite matrices. *Math. Ann.*, 244(1):65–68, 1979. (Cited on p. 66.)

[132] Delin Chu, Y. S. Hung, and Hugo J. Woerdeman. Inertia and rank characterizations of some matrix expressions. *SIAM J. Matrix Anal. Appl.*, 31(3):1187–1226, 2009. (Cited on p. 471.)

[133] S. Ciccariello and A. Cervellino. Generalization of a theorem of Carathéodory. *J. Phys. A*, 39(48):14911–14928, 2006. (Cited on pp. 65, 472.)

[134] Nir Cohen and Jerome Dancis. Inertias of block band matrix completions. *SIAM J. Matrix Anal. Appl.*, 19(3):583–612 (electronic), 1998. (Cited on p. 470.)

[135] Nir Cohen, Charles R. Johnson, Leiba Rodman, and Hugo J. Woerdeman. Ranks of completions of partial matrices. In *The Gohberg anniversary collection, Vol. I (Calgary, AB, 1988)*, volume 40 of *Oper. Theory Adv. Appl.*, pages 165–185. Birkhäuser, Basel, 1989. (Cited on p. 470.)

[136] R. R. Coifman, R. Rochberg, and Guido Weiss. Factorization theorems for Hardy spaces in several variables. *Ann. of Math. (2)*, 103(3):611–635, 1976. (Cited on p. 354.)

[137] Brian J. Cole and John Wermer. Pick interpolation, von Neumann inequalities, and hyperconvex sets. In *Complex potential theory (Montreal, PQ, 1993)*, volume 439 of *NATO Adv. Sci. Inst. Ser. C Math. Phys. Sci.*, pages 89–129. Kluwer Academic, Dordrecht, 1994. (Cited on p. 356.)

[138] Brian J. Cole and John Wermer. Interpolation in the bidisk. *J. Funct. Anal.*, 140(1):194–217, 1996. (Cited on p. 356.)

[139] Brian J. Cole and John Wermer. Ando's theorem and sums of squares. *Indiana Univ. Math. J.*, 48(3):767–791, 1999. (Cited on pp. 251, 356.)

[140] T. Constantinescu. On the structure of positive Toeplitz forms. In *Dilation theory, Toeplitz operators, and other topics (Timişoara/Herculane, 1982)*, volume 11 of *Oper. Theory Adv. Appl.*, pages 127–149. Birkhäuser, Basel, 1983. (Cited on p. 172.)

[141] T. Constantinescu. Schur analysis of positive block-matrices. In *I. Schur methods in operator theory and signal processing*, volume 18 of *Oper. Theory Adv. Appl.*, pages 191–206. Birkhäuser, Basel, 1986. (Cited on p. 172.)

[142] T. Constantinescu. Operator Schur algorithm and associated functions. *Math. Balkanica (N.S.)*, 2(2-3):244–252, 1988. (Cited on p. 171.)

[143] T. Constantinescu. *Schur parameters, factorization and dilation problems*, volume 82 of *Operator Theory: Advances and Applications*. Birkhäuser, Basel, 1996. (Cited on p. 172.)

[144] T. Constantinescu and A. Gheondea. The negative signature of some Hermitian matrices. *Linear Algebra Appl.*, 178:17–42, 1993. (Cited on p. 470.)

[145] T. Constantinescu and A. Gheondea. On the indefinite trigonometric moment problem of I. S. Iohvidov and M. G. Kreĭn. *Math. Nachr.*, 171:79–94, 1995. (Cited on p. 472.)

[146] T. Constantinescu and A. Gheondea. Representations of Hermitian kernels by means of Krein spaces. II. Invariant kernels. *Comm. Math. Phys.*, 216(2):409–430, 2001. (Cited on p. 472.)

[147] T. Constantinescu and V. Ramakrishna. Parametrizing quantum states and channels. *Quantum Inf. Process.*, 2(3):221–248, 2003. (Cited on p. 474.)

[148] Tiberiu Constantinescu and Viswanath Ramakrishna. On a parameterization of purifications of a qubit. *Quantum Inf. Process.*, 1(5):409–424 (2003), 2002. (Cited on p. 474.)

[149] Tiberiu Constantinescu and Viswanath Ramakrishna. Some parametrizations of mixed states motivated by quantum information. In *Lagrangian and Hamiltonian methods for nonlinear control 2003*, pages 301–304. IFAC, Laxenburg, 2003. (Cited on p. 474.)

[150] John B. Conway. *Functions of one complex variable*, volume 11 of *Graduate Texts in Mathematics*. Springer, New York, second edition, 1978. (Cited on pp. 331, 359.)

[151] Mischa Cotlar and Cora Sadosky. Nehari and Nevanlinna-Pick problems and holomorphic extensions in the polydisk in terms of restricted BMO. *J. Funct. Anal.*, 124(1):205–210, 1994. (Cited on p. 356.)

[152] Mischa Cotlar and Cora Sadosky. Two distinguished subspaces of product BMO and Nehari-AAK theory for Hankel operators on the torus. *Integral Equations Operator Theory*, 26(3):273–304, 1996. (Cited on p. 354.)

[153] M. J. Crabb and A. M. Davie. Von Neumann's inequality for Hilbert space operators. *Bull. London Math. Soc.*, 7:49–50, 1975. (Cited on p. 356.)

[154] Glória Cravo. Matrix completion problems. *Linear Algebra Appl.*, 430(8-9):2511–2540, 2009. (Cited on p. 474.)

[155] G. M. Crippen and T. F. Havel. *Distance geometry and molecular conformation*, volume 15 of *Chemometrics Series*. Research Studies Press, Chichester, 1988. (Cited on p. 473.)

[156] Raúl Curto, Carlos Hernández, and Elena de Oteyza. Contractive completions of Hankel partial contractions. *J. Math. Anal. Appl.*, 203(2):303–332, 1996. (Cited on pp. 352, 353.)

[157] Raúl E. Curto and Lawrence A. Fialkow. Recursiveness, positivity, and truncated moment problems. *Houston J. Math.*, 17(4):603–635, 1991. (Cited on p. 173.)

[158] Raúl E. Curto and Lawrence A. Fialkow. Solution of the truncated complex moment problem for flat data. *Mem. Amer. Math. Soc.*, 119(568):x+52, 1996. (Cited on p. 254.)

[159] Raúl E. Curto and Lawrence A. Fialkow. Flat extensions of positive moment matrices: recursively generated relations. *Mem. Amer. Math. Soc.*, 136(648):x+56, 1998. (Cited on p. 254.)

[160] Raúl E. Curto and Lawrence A. Fialkow. The quadratic moment problem for the unit circle and unit disk. *Integral Equations Operator Theory*, 38(4):377–409, 2000. (Cited on p. 254.)

[161] Raúl E. Curto and Lawrence A. Fialkow. Truncated K-moment problems in several variables. *J. Operator Theory*, 54(1):189–226, 2005. (Cited on p. 254.)

[162] George Cybenko. Moment problems and low rank Toeplitz approximations. *Circuits Systems Signal Process.*, 1(3-4):345–366, 1982. (Cited on p. 472.)

[163] Hua Dai. Completing a symmetric 2×2 block matrix and its inverse. *Linear Algebra Appl.*, 235:235–245, 1996. (Cited on p. 474.)

[164] Ingrid Daubechies. *Ten lectures on wavelets*, volume 61 of *CBMS-NSF Regional Conference Series in Applied Mathematics*. Society for Industrial and Applied Mathematics (SIAM), Philadelphia, 1992. (Cited on p. 251.)

[165] Kenneth R. Davidson. *Nest algebras*, volume 191 of *Pitman Research Notes in Mathematics Series*. Longman Scientific & Technical, Harlow, 1988. Triangular forms for operator algebras on Hilbert space. (Cited on p. 352.)

[166] Kenneth R. Davidson and Marc S. Ordower. Some exact distance constants. *Linear Algebra Appl.*, 208/209:37–55, 1994. (Cited on p. 352.)

[167] Kenneth R. Davidson and Stephen C. Power. Failure of the distance formula. *J. London Math. Soc. (2)*, 32(1):157–165, 1985. (Cited on p. 352.)

[168] Chandler Davis. An extremal problem for extensions of a sesquilinear form. *Linear Algebra Appl.*, 13(1-2):91–102, 1976. Collection of articles dedicated to Olga Taussky Todd. (Cited on pp. 358, 359.)

[169] Chandler Davis, W. M. Kahan, and H. F. Weinberger. Norm-preserving dilations and their applications to optimal error bounds. *SIAM J. Numer. Anal.*, 19(3):445–469, 1982. (Cited on p. 353.)

[170] Etienne de Klerk. *Aspects of semidefinite programming*, volume 65 of *Applied Optimization*. Kluwer Academic, Dordrecht, 2002. Interior point algorithms and selected applications. (Cited on p. 66.)

[171] Leonede De-Michele and Alessandro Figà-Talamanca. Positive definite functions on free groups. *Amer. J. Math.*, 102(3):503–509, 1980. (Cited on p. 253.)

[172] Luz M. DeAlba and Charles R. Johnson. Possible inertia combinations in the Stein and Lyapunov equations. *Linear Algebra Appl.*, 222:227–240, 1995. (Cited on p. 471.)

[173] Ilan Degani, Jeremy Schiff, and David J. Tannor. Commuting extensions and cubature formulae. *Numer. Math.*, 101(3):479–500, 2005. (Cited on p. 473.)

[174] Philippe Delsarte, Yves V. Genin, and Yves G. Kamp. Orthogonal polynomial matrices on the unit circle. *IEEE Trans. Circuits and Systems*, CAS-25(3):149–160, 1978. (Cited on p. 171.)

[175] Philippe Delsarte, Yves V. Genin, and Yves G. Kamp. Planar least squares inverse polynomials. I. Algebraic properties. *IEEE Trans. Circuits and Systems*, 26(1):59–66, 1979. (Cited on p. 250.)

[176] Philippe Delsarte, Yves V. Genin, and Yves G. Kamp. Half-plane Toeplitz systems. *IEEE Trans. Inform. Theory*, 26(4):465–474, 1980. (Cited on p. 250.)

[177] Patrick Dewilde, Augusto C. Vieira, and Thomas Kailath. On a generalized Szegő-Levinson realization algorithm for optimal linear predictors based on a network synthesis approach. *IEEE Trans. Circuits and Systems*, 25(9):663–675, 1978. Special issue on the mathematical foundations of system theory. (Cited on p. 171.)

[178] Jacques Dixmier. *Les C^*-algèbres et leurs représentations.* Volume 29 of Cahiers Scientifiques. Gauthier-Villars, Paris. (Cited on pp. 354, 357, 357.)

[179] Jacques Dixmier. *C^*-algebras.* Translated from the French by Francis Jellett, volume 15 of North-Holland Mathematical Library. North-Holland, Amsterdam, 1977. (Cited on p. 252.)

[180] Andrew C. Doherty, Pablo A. Parrilo, and Frederico M. Spedalieri. Distinguishing separable and entangled states. *Phys. Rev. Lett.*, 88(18):187904, 4, 2002. (Cited on p. 474.)

[181] Andrew C. Doherty, Pablo A. Parrilo, and Frederico M. Spedalieri. Detecting multipartite entanglement. *Phys. Rev. A*, 71(032333), 2005. (Cited on p. 474.)

[182] Marisela Dominguez. Interpolation and prediction problems for connected compact abelian groups. *Integral Equations Operator Theory*, 40(2):212–230, 2001. (Cited on p. 355.)

[183] R. G. Douglas. On majorization, factorization, and range inclusion of operators on Hilbert space. *Proc. Amer. Math. Soc.*, 17:413–415, 1966. (Cited on p. 171.)

[184] Ronald G. Douglas. *Banach algebra techniques in operator theory.* volume 49 of Pure and Applied Mathematics. Academic Press, New York, 1972. (Cited on p. 354.)

[185] John C. Doyle, Keith Glover, Pramod P. Khargonekar, and Bruce A. Francis. State-space solutions to standard H_2 and H_∞ control problems. *IEEE Trans. Automat. Control*, 34(8):831–847, 1989. (Cited on p. 354.)

[186] Michael A. Dritschel. On factorization of trigonometric polynomials. *Integral Equations Operator Theory*, 49(1):11–42, 2004. (Cited on pp. 172, 251, 252, 255, 255, 255.)

[187] Michael A. Dritschel and Hugo J. Woerdeman. Outer factorizations in one and several variables. *Trans. Amer. Math. Soc.*, 357(11):4661–4679 (electronic), 2005. (Cited on pp. 171, 172, 172, 172.)

[188] Bogdan Dumitrescu. *Positive trigonometric polynomials and signal processing applications.* Signals and Communication Technology. Springer, Dordrecht, 2007. (Cited on p. 66.)

[189] Peter L. Duren. *Theory of H^p spaces.* volume 38 of Pure and Applied Mathematics. Academic Press, New York, 1970. (Cited on pp. 171, 353, 359.)

[190] Harry Dym. *J contractive matrix functions, reproducing kernel Hilbert spaces and interpolation*, volume 71 of *CBMS Regional Conference Series in Mathematics.* Published for the Conference Board of the Mathematical Sciences, Washington, DC, 1989. (Cited on pp. 353, 356.)

[191] Harry Dym. On Hermitian block Hankel matrices, matrix polynomials, the Hamburger moment problem, interpolation and maximum entropy. *Integral Equations Operator Theory*, 12(6):757–812, 1989. (Cited on p. 173.)

[192] Harry Dym and Israel Gohberg. Extensions of band matrices with band inverses. *Linear Algebra Appl.*, 36:1–24, 1981. (Cited on p. 65.)

[193] Harry Dym and Israel Gohberg. Extension of matrix valued functions and block matrices. *Indiana Univ. Math. J.*, 31(5):733–765, 1982. (Cited on p. 474.)

[194] Harry Dym and Israel Gohberg. Extensions of kernels of Fredholm operators. *J. Analyse Math.*, 42:51–97, 1982/83. (Cited on p. 170.)

[195] Harry Dym and Israel Gohberg. A maximum entropy principle for contractive interpolants. *J. Funct. Anal.*, 65(1):83–125, 1986. (Cited on p. 357.)

[196] Yuriy M. Dyukarev, Bernd Fritzsche, Bernd Kirstein, Conrad Mädler, and Helge C. Thiele. On distinguished solutions of truncated matricial Hamburger moment problems. *Complex Anal. Oper. Theory*, 3(4):759–834, 2009. (Cited on p. 173.)

[197] Torsten Ehrhardt and Cornelis V. M. van der Mee. Canonical factorization of continuous functions on the d-torus. *Proc. Amer. Math. Soc.*, 131(3):801–813 (electronic), 2003. (Cited on p. 252.)

[198] Y. Eidelman and I. Gohberg. On a new class of structured matrices. *Integral Equations Operator Theory*, 34(3):293–324, 1999. (Cited on p. 471.)

[199] R. L. Ellis, I. Gohberg, and D. C. Lay. Extensions with positive real part. A new version of the abstract band method with applications. *Integral Equations Operator Theory*, 16(3):360–384, 1993. (Cited on p. 170.)

[200] Robert L. Ellis and Israel Gohberg. *Orthogonal systems and convolution operators*, volume 140 of *Operator Theory: Advances and Applications.* Birkhäuser, Basel, 2003. (Cited on p. 171.)

[201] Robert L. Ellis, Israel Gohberg, and David Lay. Band extensions, maximum entropy and the permanence principle. In *Maximum entropy and Bayesian methods in applied statistics (Calgary, Alta., 1984)*, pages 131–155. Cambridge University, Cambridge, 1986. (Cited on p. 173.)

[202] Robert L. Ellis and David C. Lay. Rank-preserving extensions of band matrices. *Linear and Multilinear Algebra*, 26(3):147–179, 1990. (Cited on p. 473.)

[203] Robert L. Ellis and David C. Lay. Factorization of finite rank Hankel and Toeplitz matrices. *Linear and Multilinear Algebra*, 173(1):19–38, 1992. (Cited on p. 472.)

[204] Robert L. Ellis, David C. Lay, and Israel Gohberg. On negative eigenvalues of selfadjoint extensions of band matrices. *Linear and Multilinear Algebra*, 24(1):15–25, 1988. (Cited on p. 470.)

[205] L. Elsner, C. He, and V. Mehrmann. Minimizing the condition number of a positive definite matrix by completion. *Numer. Math.*, 69(1):17–23, 1994. (Cited on p. 474.)

[206] Ludwig Elsner, Chun Yang He, and Volker Mehrmann. Minimization of the norm, the norm of the inverse and the condition number of a matrix by completion. *Numer. Linear Algebra Appl.*, 2(2):155–171, 1995. (Cited on p. 474.)

[207] Jacob C. Engwerda, André C. M. Ran, and Arie L. Rijkeboer. Necessary and sufficient conditions for the existence of a positive definite solution of the matrix equation $X + A^*X^{-1}A = Q$. *Linear Algebra Appl.*, 186:255–275, 1993. (Cited on p. 172.)

[208] Jőrg Eschmeier, Linda Patton, and Mihai Putinar. Carathéodory-Fejér interpolation on polydisks. *Math. Res. Lett.*, 7(1):25–34, 2000. (Cited on p. 250.)

[209] A. Eshelman and H. J. Woerdeman. A note on the truncated matrix Hamburger problem. *Preprint.* (Cited on p. 173.)

[210] Ruy Exel. Hankel matrices over right ordered amenable groups. *Canad. Math. Bull.*, 33(4):404–415, 1990. (Cited on p. 252.)

[211] Pierre Eymard. L'algèbre de Fourier d'un groupe localement compact. *Bull. Soc. Math. France*, 92:181–236, 1964. (Cited on p. 253.)

[212] Ky Fan and Gordon Pall. Imbedding conditions for Hermitian and normal matrices. *Canad. J. Math.*, 9:298–304, 1957. (Cited on p. 473.)

[213] L. Fejér. Über trigonometrische polynome. *J. Reine Angew. Math.*, 146:53–82, 1916. (Cited on p. 65.)

[214] Sven Feldmann and Georg Heinig. Parametrization of minimal rank block Hankel matrix extensions and minimal partial realizations. *Integral Equations Operator Theory*, 33(2):153–171, 1999. (Cited on p. 471.)

[215] Sarah H. Ferguson. The Nehari problem for the Hardy space on the torus. *J. Operator Theory*, 40(2):309–321, 1998. (Cited on p. 354.)

[216] Sarah H. Ferguson and Michael T. Lacey. A characterization of product BMO by commutators. *Acta Math.*, 189(2):143–160, 2002. (Cited on p. 354.)

[217] Sarah H. Ferguson and Cora Sadosky. Characterizations of bounded mean oscillation on the polydisk in terms of Hankel operators and Carleson measures. *J. Anal. Math.*, 81:239–267, 2000. (Cited on p. 354.)

[218] Lawrence Fialkow and Jiawang Nie. Positivity of Riesz functionals and solutions of quadratic and quartic moment problems. *J. Funct. Anal.*, 258(1):328–356, 2010. (Cited on p. 254.)

[219] Miroslav Fiedler. A remark on positive definite matrices. *Časopis Pěst. Mat.*, 85:75–77, 1960. (Cited on pp. 66, 170.)

[220] Miroslav Fiedler. Matrix inequalities. *Numer. Math.*, 9:109–119, 1966. (Cited on p. 170.)

[221] Miroslav Fiedler and Thomas L. Markham. Completing a matrix when certain entries of its inverse are specified. *Linear Algebra Appl.*, 74:225–237, 1986. (Cited on p. 470.)

[222] Miroslav Fiedler and Thomas L. Markham. Rank-preserving diagonal completions of a matrix. *Linear Algebra Appl.*, 85:49–56, 1987. (Cited on p. 474.)

[223] Alessandro Figà-Talamanca and Massimo A. Picardello. *Harmonic analysis on free groups*, volume 87 of *Lecture Notes in Pure and Applied Mathematics*. Marcel Dekker, New York, 1983. (Cited on p. 253.)

[224] E. Fischer. Über das Carathéodorysche Problem, Potenzreihen mit positivem reelen Teil betreffend. *Rend. Circ. Mat. Palermo*, 32:240–256, 1911. (Cited on p. 173.)

[225] C. Foias and A. E. Frazho. *The commutant lifting approach to interpolation problems*, volume 44 of *Oper. Theory Adv. Appl.* Birkhäuser, Basel, 1990. (Cited on pp. 172, 253, 353, 353.)

[226] C. Foias and A. E. Frazho. Commutant lifting and simultaneous H^∞ and L^2 suboptimization. *SIAM J. Math. Anal.*, 23(4):984–994, 1992. (Cited on p. 357.)

[227] C. Foias, A. E. Frazho, I. Gohberg, and M. A. Kaashoek. *Metric constrained interpolation, commutant lifting and systems*, volume 100 of *Oper. Theory Adv. Appl.* Birkhäuser, Basel, 1998. (Cited on p. 356.)

[228] C. Foias, A. E. Frazho, and W. S. Li. The exact H^2 estimate for the central H^∞ interpolant. In *New aspects in interpolation and completion theories*, volume 64 of *Oper. Theory Adv. Appl.*, pages 119–156. Birkhäuser, Basel, 1993. (Cited on p. 357.)

[229] C. Foias and A. Tannenbaum. A strong Parrott theorem. *Proc. Amer. Math. Soc.*, 106(3):777–784, 1989. (Cited on p. 353.)

[230] Yung Kuan Foo and Ian Postlethwaite. An H^∞-minimax approach to the design of robust control systems. II. All solutions, all-pass form solutions and the "best" solution. *Systems Control Lett.*, 7(4):261–268, 1986. (Cited on p. 359.)

[231] Bruce A. Francis. *A course in H_∞ control theory*, volume 88 of *Lecture Notes in Control and Information Sciences*. Springer, Berlin, 1987. (Cited on p. 359.)

[232] A. E. Frazho, K. M. Grigoriadis, and S. M. Kherat. Alternating projection methods for mixed H^2 and H^∞ Nehari problems. *IEEE Trans. Automat. Control*, 40(12):2127–2131, 1995. (Cited on p. 357.)

[233] Arthur E. Frazho and Mario A. Rotea. A remark on mixed L^2/L^∞ bounds. *Integral Equations Operator Theory*, 15(2):343–348, 1992. (Cited on p. 357.)

[234] Stephen H. Friedberg and Arnold J. Insel. Hyponormal 2×2 matrices are subnormal. *Linear Algebra Appl.*, 175:31–38, 1992. (Cited on p. 473.)

[235] Stephen H. Friedberg and Arnold J. Insel. Characterizations of subnormal matrices. *Linear Algebra Appl.*, 231:1–13, 1995. (Cited on p. 473.)

[236] Shmuel Friedland. Normal matrices and the completion problem. *SIAM J. Matrix Anal. Appl.*, 23(3):896–902 (electronic), 2001/02. (Cited on p. 473.)

[237] Bernd Fritzsche and Bernd Kirstein. Representations of central matrix-valued Carathéodory functions in both nondegenerate and degenerate cases. *Integral Equations Operator Theory*, 50(3):333–361, 2004. (Cited on p. 172.)

[238] Bernd Fritzsche, Bernd Kirstein, and Lutz Klotz. Completion of non-negative Block operators in Banach spaces. *Positivity*, 3(4):389–397, 1999. (Cited on p. 172.)

[239] Paul A. Fuhrmann. On the Corona theorem and its application to spectral problems in Hilbert space. *Trans. Amer. Math. Soc.*, 132:55–66, 1968. (Cited on p. 357.)

[240] Jean-Pierre Gabardo. Trigonometric moment problems for arbitrary finite subsets of $\mathbf{Z}^n$. *Trans. Amer. Math. Soc.*, 350(11):4473–4498, 1998. (Cited on p. 65.)

[241] Radu Gadidov. The central intertwining lifting and strict contractions. *Proc. Amer. Math. Soc.*, 124(12):3813–3817, 1996. (Cited on p. 357.)

[242] W. A. Gardner. *Introduction to random processes with applications to signals and systems.* Macmillan, New York, NY, 1986. (Cited on p. 174.)

[243] Fănică Gavril. The intersection graphs of subtrees in trees are exactly the chordal graphs. *J. Combinatorial Theory Ser. B*, 16:47–56, 1974. (Cited on p. 470.)

[244] Y. Genin, Y. Hachez, Yu. Nesterov, and P. Van Dooren. Convex optimization over positive polynomials and filter design. In *Proceedings UKACC Int. Conf. Control 2000,* Paper SS41, 2000. (Cited on p. 68.)

[245] Yves V. Genin. Hankel matrices, positive functions and related questions. *Linear Algebra Appl.*, 314(1-3):137–164, 2000. (Cited on p. 472.)

[246] Yves V. Genin and Yves G. Kamp. Two-dimensional stability and orthogonal polynomials on the hypercircle. *Proc. IEEE*, 65(6):873–881, 1977. (Cited on p. 250.)

[247] Jeffrey S. Geronimo and Ming-Jun Lai. Factorization of multivariate positive Laurent polynomials. *J. Approx. Theory*, 139(1-2):327–345, 2006. (Cited on pp. 252, 255.)

[248] Jeffrey S. Geronimo and Hugo J. Woerdeman. Positive extensions, Fejér-Riesz factorization and autoregressive filters in two variables. *Ann. of Math. (2)*, 160(3):839–906, 2004. (Cited on pp. 250, 251, 251, 251, 251.)

[249] Jeffrey S. Geronimo and Hugo J. Woerdeman. The operator valued autoregressive filter problem and the suboptimal Nehari problem in two variables. *Integral Equations Operator Theory*, 53(3):343–361, 2005. (Cited on pp. 250, 354.)

[250] Jeffrey S. Geronimo and Hugo J. Woerdeman. Two-variable polynomials: intersecting zeros and stability. *IEEE Trans. Circuits Syst. I Regul. Pap.*, 53(5):1130–1139, 2006. (Cited on p. 251.)

[251] Jeffrey S. Geronimo and Hugo J. Woerdeman. Two variable orthogonal polynomials on the bicircle and structured matrices. *SIAM J. Matrix Anal. Appl.*, 29(3):796–825 (electronic), 2007. (Cited on pp. 172, 251, 251.)

[252] L. Ya. Geronimus. *Orthogonal polynomials: Estimates, asymptotic formulas, and series of polynomials orthogonal on the unit circle and on an interval.* Authorized translation from the Russian. Consultants Bureau, New York, 1961. (Cited on p. 253.)

[253] Aurelian Gheondea. One-step completions of Hermitian partial matrices with minimal negative signature. *Linear Algebra Appl.*, 173:99–114, 1992. (Cited on p. 470.)

[254] Valérie Girardin and Abdellatif Seghier. Constructions d'extensions de suites définies positives. Cas multidimensionnel. *C. R. Acad. Sci. Paris Sér. I Math.*, 313(2):71–74, 1991. (Cited on p. 251.)

[255] W. Glunt, T. L. Hayden, and W.-M. Liu. The embedding problem for predistance matrices. *Bull. Math. Biology*, 53(5):769–796, 1991. (Cited on p. 473.)

[256] W. Glunt, T. L. Hayden, and M. Raydan. Molecular conformations from distance matrices. *J. Comp. Chem.*, 14(1):114–120, 1993. (Cited on p. 473.)

[257] Roger Godement. Les fonctions de type positif et la théorie des groupes. *Trans. Amer. Math. Soc.*, 63:1–84, 1948. (Cited on p. 253.)

[258] Michel X. Goemans. Semidefinite programming in combinatorial optimization. *Math. Programming*, 79(1-3, Ser. B):143–161, 1997. Lectures on mathematical programming (ismp97) (Lausanne, 1997). (Cited on p. 66.)

[259] I. Gohberg. The factorization problem for operator functions. *Izv. Akad. Nauk SSSR Ser. Mat.*, 28:1055–1082, 1964. (Cited on p. 171.)

[260] I. Gohberg, S. Goldberg, and M. A. Kaashoek. *Classes of linear operators. Vol. II*, volume 63 of *Oper. Theory Adv. Appl.* Birkhäuser, Basel, 1993. (Cited on p. 353.)

[261] I. Gohberg and G. Heinig. Inversion of finite Toeplitz matrices consisting of elements of a noncommutative algebra. *Rev. Roumaine Math. Pures Appl.*, 19:623–663, 1974. (Cited on p. 171.)

[262] I. Gohberg and M. A. Kaashoek. Time varying linear systems with boundary conditions and integral operators. I. The transfer operator and its properties. *Integral Equations Operator Theory*, 7(3):325–391, 1984. (Cited on p. 474.)

[263] I. Gohberg and M. A. Kaashoek. Minimal representations of semiseparable kernels and systems with separable boundary conditions. *J. Math. Anal. Appl.*, 124(2):436–458, 1987. (Cited on p. 474.)

[264] I. Gohberg, M. A. Kaashoek, and L. Lerer. Minimal rank completion problems and partial realization. In *Recent advances in mathematical theory of systems, control, networks and signal processing, I (Kobe, 1991)*, pages 65–70. Mita, Tokyo, 1992. (Cited on p. 470.)

[265] I. Gohberg, M. A. Kaashoek, and Hugo J. Woerdeman. The band method for positive and contractive extension problems. *J. Operator Theory*, 22(1):109–155, 1989. (Cited on p. 170.)

[266] I. Gohberg, M. A. Kaashoek, and Hugo J. Woerdeman. The band method for positive and strictly contractive extension problems: an alternative version and new applications. *Integral Equations Operator Theory*, 12(3):343–382, 1989. (Cited on pp. 170, 353, 354.)

[267] I. Gohberg, M. A. Kaashoek, and Hugo J. Woerdeman. The band method for several positive extension problems of nonband type. *J. Operator Theory*, 26(1):191–218, 1991. (Cited on p. 170.)

[268] I. Gohberg, M. A. Kaashoek, and Hugo J. Woerdeman. A maximum entropy principle in the general framework of the band method. *J. Funct. Anal.*, 95(2):231–254, 1991. (Cited on pp. 170, 171, 357.)

[269] I. Gohberg, P. Lancaster, and L. Rodman. *Matrix polynomials*. Academic Press [Harcourt Brace], New York, 1982. Computer Science and Applied Mathematics. (Cited on p. 174.)

[270] I. Gohberg, P. Lancaster, and L. Rodman. *Indefinite linear algebra and applications*. Birkhäuser, Basel, 2005. (Cited on p. 174.)

[271] I. Gohberg and H. J. Landau. Prediction and the inverse of Toeplitz matrices. In *Approximation and computation (West Lafayette, IN, 1993)*, volume 119 of *Internat. Ser. Numer. Math.*, pages 219–229. Birkhäuser Boston, Boston, 1994. (Cited on p. 174.)

[272] I. Gohberg and H. J. Landau. Prediction for two processes and the Nehari problem. *J. Fourier Anal. Appl.*, 3(1):43–62, 1997. (Cited on p. 353.)

[273] I. Gohberg and L. Lerer. Matrix generalizations of M. G. Kreĭn theorems on orthogonal polynomials. In *Orthogonal matrix-valued polynomials and applications (Tel Aviv, 1987–88)*, volume 34 of *Oper. Theory Adv. Appl.*, pages 137–202. Birkhäuser, Basel, 1988. (Cited on p. 471.)

[274] I. Gohberg, L. Rodman, T. Shalom, and H. J. Woerdeman. Bounds for eigenvalues and singular values of matrix completions. *Linear and Multilinear Algebra*, 33(3-4):233–249, 1993. (Cited on pp. 359, 472, 472.)

[275] I. Gohberg and A. Semencul. The inversion of finite Toeplitz matrices and their continual analogues. *Mat. Issled.*, 7(2(24)):201–223, 290, 1972. (Cited on p. 171.)

[276] Martin Charles Golumbic. *Algorithmic graph theory and perfect graphs*. Academic Press [Harcourt Brace], New York, 1980. With a foreword by Claude Berge, Computer Science and Applied Mathematics. (Cited on pp. 65, 470.)

[277] J. C. Gower. Properties of Euclidean and non-Euclidean distance matrices. *Linear Algebra Appl.*, 67:81–97, 1985. (Cited on p. 473.)

[278] W. B. Gragg and L. Reichel. On singular values of Hankel operators of finite rank. *Linear Algebra Appl.*, 121:53–70, 1989. (Cited on p. 472.)

[279] William L. Green and T. D. Morley. The extreme points of order intervals of positive operators. *Adv. in Appl. Math.*, 15(3):360–370, 1994. (Cited on p. 172.)

[280] Frederick P. Greenleaf. *Invariant means on topological groups and their applications.* Van Nostrand Mathematical Studies, No. 16. Van Nostrand Reinhold Co., New York, 1969. (Cited on p. 252.)

[281] Ulf Grenander and Gábor Szegő. *Toeplitz forms and their applications.* Chelsea, New York, second edition, 1984. (Cited on p. 65.)

[282] S. A. Grigoryan. On a theorem of Sarason. *Izv. Nats. Akad. Nauk Armenii Mat.*, 28(4):34–41 (1995), 1993. (Cited on p. 354.)

[283] Anatolii Grinshpan, Dmitry S. Kaliuzhnyi-Verbovetskyi, Victor Vinnikov, and Hugo J. Woerdeman. Classes of tuples of commuting contractions satisfying the multivariable von Neumann inequality. *J. Funct. Anal.*, 256(9):3035–3054, 2009. (Cited on p. 356.)

[284] Karlheinz Gröchenig. Localization of frames, Banach frames, and the invertibility of the frame operator. *J. Fourier Anal. Appl.*, 10(2):105–132, 2004. (Cited on p. 355.)

[285] Robert Grone, Charles R. Johnson, Eduardo M. de Sá, and Henry Wolkowicz. Positive definite completions of partial Hermitian matrices. *Linear Algebra Appl.*, 58:109–124, 1984. (Cited on pp. 65, 66, 170, 170, 170.)

[286] Robert Grone, Stephen Pierce, and William Watkins. Extremal correlation matrices. *Linear Algebra Appl.*, 134:63–70, 1990. (Cited on p. 66.)

[287] L. Gurvits. Factorization of positive matrix polynomials and the separability of composite quantum systems. *Preprint.* (Cited on pp. 173, 474.)

[288] Leonid Gurvits and Howard Barnum. Separable balls around the maximally mixed multipartite quantum states. *Phys. Rev. A*, 68:042312, 2003. (Cited on pp. 173, 474.)

[289] William H. Gustafson. A note on matrix inversion. *Linear Algebra Appl.*, 57:71–73, 1984. (Cited on p. 470.)

[290] Martin H. Gutknecht. On complex rational approximation. II. The Carathéodory-Fejér method. In *Computational aspects of complex analysis (Braunlage, 1982)*, volume 102 of *NATO Adv. Sci. Inst. Ser. C: Math. Phys. Sci.*, pages 103–132. Reidel, Dordrecht, 1983. (Cited on p. 472.)

[291] Uffe Haagerup. An example of a nonnuclear C^*-algebra, which has the metric approximation property. *Invent. Math.*, 50(3):279–293, 1978/79. (Cited on p. 253.)

[292] Yvan Hachez and Hugo J. Woerdeman. Approximating sums of squares with a single square. *Linear Algebra Appl.*, 399:187–201, 2005. (Cited on p. 251.)

[293] Yvan Hachez and Hugo J. Woerdeman. The Fischer-Frobenius transformation and outer factorization. In *Operator theory, structured matrices, and dilations: Tiberiu Constantinescu Memorial Volume, Eds: M. Bakonyi, A. Gheondea, M. Putinar and J. Rovnyak*, pages 181–203. Theta Foundation, Bucharest, 2007. (Cited on p. 174.)

[294] D. Hadwin, D. R. Larson, and D. Timotin. Approximation theory and matrix completions. *Linear Algebra Appl.*, 377:165–179, 2004. (Cited on p. 352.)

[295] G. D. Halikias, D. J. N. Limebeer, and K. Glover. A state-space algorithm for the superoptimal Hankel-norm approximation problem. *SIAM J. Control Optim.*, 31(4):960–982, 1993. (Cited on p. 359.)

[296] P. R. Halmos. Shifts on Hilbert spaces. *J. Reine Angew. Math.*, 208:102–112, 1961. (Cited on p. 359.)

[297] P. R. Halmos. Quasitriangular operators. *Acta Sci. Math. (Szeged)*, 29:283–293, 1968. (Cited on p. 357.)

[298] P. R. Halmos. Subnormal suboperators and the subdiscrete topology. In *Anniversary volume on approximation theory and functional analysis (Oberwolfach, 1983)*, volume 65 of *Internat. Schriftenreihe Numer. Math.*, pages 49–65. Birkhäuser, Basel, 1984. (Cited on p. 473.)

[299] H. Hamburger. Uber eine Erweiterung des Stieltjesschen Momentproblems i, ii, iii. *Math. Ann.*, 81:235–319, 1920. (Cited on p. 173.)

[300] H. Hankel. *Über eine besondere Classe der symmetrische Determinanten.* Dissertation, Göttingen, 1861. Leipziger. (Cited on p. 353.)

[301] Philip Hartman. On completely continuous Hankel matrices. *Proc. Amer. Math. Soc.*, 9:862–866, 1958. (Cited on p. 354.)

[302] E. K. Haviland. On the momentum problem for distribution functions in more than one dimension. *Amer. J. Math.*, 57(3):562–568, 1935. (Cited on p. 254.)

[303] T. L. Hayden and Jim Wells. Approximation by matrices positive semidefinite on a subspace. *Linear Algebra Appl.*, 109:115–130, 1988. (Cited on pp. 352, 353, 473.)

[304] Thomas L. Hayden, Robert Reams, and James Wells. Methods for constructing distance matrices and the inverse eigenvalue problem. *Linear Algebra Appl.*, 295(1-3):97–112, 1999. (Cited on p. 473.)

[305] Emilie V. Haynsworth. Determination of the inertia of a partitioned Hermitian matrix. *Linear Algebra Appl.*, 1(1):73–81, 1968. (Cited on p. 470.)

[306] Ming He and Michael K. Ng. Toeplitz and positive semidefinite completion problem for cycle graph. *Numer. Math. J. Chinese Univ. (English Ser.)*, 14(1):67–78, 2005. (Cited on p. 170.)

[307] Henry Helson. *Lectures on invariant subspaces.* Academic Press, New York, 1964. (Cited on pp. 171, 359.)

[308] Henry Helson and David Lowdenslager. Prediction theory and Fourier series in several variables I. *Acta Math.*, 99:165–202, 1958. (Cited on pp. 252, 255.)

[309] J. W. Helton, S. Pierce, and L. Rodman. The ranks of extremal positive semidefinite matrices with given sparsity pattern. *SIAM J. Matrix Anal. Appl.*, 10(3):407–423, 1989. (Cited on pp. 65, 66, 66, 67.)

[310] J. William Helton. "Positive" noncommutative polynomials are sums of squares. *Ann. of Math. (2)*, 156(2):675–694, 2002. (Cited on p. 254.)

[311] J. William Helton, Daniel Lam, and Hugo J. Woerdeman. Sparsity patterns with high rank extremal positive semidefinite matrices. *SIAM J. Matrix Anal. Appl.*, 15(1):299–312, 1994. (Cited on p. 65.)

[312] J. William Helton and Mihai Putinar. Positive polynomials in scalar and matrix variables, the spectral theorem, and optimization. In *Operator theory, structured matrices, and dilations*, volume 7 of *Theta Ser. Adv. Math.*, pages 229–306. Theta, Bucharest, 2007. (Cited on p. 254.)

[313] J. William Helton and Hugo J. Woerdeman. Symmetric Hankel operators: minimal norm extensions and eigenstructures. *Linear Algebra Appl.*, 185:1–19, 1993. (Cited on p. 353.)

[314] G. Herglotz. Über Potenzreihen mit positivem, reellen Teil im Einhetkreis. *Leipziger Berichte, Math.-Pys.*, Kl. 63:501–511, 1911. (Cited on pp. 17, 65.)

[315] Daniel Hershkowitz. Positive semidefinite pattern decompositions. *SIAM J. Matrix Anal. Appl.*, 11(4):612–619, 1990. (Cited on p. 65.)

[316] Nicholas J. Higham. Computing the nearest correlation matrix—a problem from finance. *IMA J. Numer. Anal.*, 22(3):329–343, 2002. (Cited on p. 67.)

[317] D. Hilbert. Über die Darstellung definiter Formen als Summe von Formenquadraten. *Math. Ann.*, 32:342–350, 1888. (Cited on p. 66.)

[318] Roland Hildebrand. Semidefinite descriptions of low-dimensional separable matrix cones. *Linear Algebra Appl.*, 429(4):901–932, 2008. (Cited on p. 474.)

[319] Richard D. Hill and Steven R. Waters. On the cone of positive semidefinite matrices. *Linear Algebra Appl.*, 90:81–88, 1987. (Cited on p. 65.)

[320] Kenneth Hoffman. *Banach spaces of analytic functions.* Dover, New York, 1988. Reprint of the 1962 original. (Cited on pp. 171, 353, 359.)

[321] John A. Holbrook. Inequalities of von Neumann type for small matrices. In *Function spaces (Edwardsville, IL, 1990)*, volume 136 of *Lecture Notes in Pure and Applied Mathematics*, pages 189–193. Dekker, New York, 1992. (Cited on p. 356.)

[322] John A. Holbrook. Schur norms and the multivariate von Neumann inequality. In *Recent advances in operator theory and related topics (Szeged, 1999)*, volume 127 of *Oper. Theory Adv. Appl.*, pages 375–386. Birkhäuser, Basel, 2001. (Cited on pp. 356, 359.)

[323] Roger A. Horn and Charles R. Johnson. *Matrix analysis.* Cambridge University Press, Cambridge, 1985. (Cited on p. 66.)

[324] Roger A. Horn and Charles R. Johnson. *Topics in matrix analysis.* Cambridge University Press, Cambridge, 1991. (Cited on p. 174.)

[325] M. Horodecki, P. Horodecki, and R. Horodecki. Mixed-state entanglement and quantum communication. In *Quantum information: an introduction to basic theoretical concepts and experiments*, volume 173 of *Springer Tracts in Modern Physics*, pages 151–195. Springer, New York, 2001. (Cited on p. 474.)

[326] Michał Horodecki, Paweł Horodecki, and Ryszard Horodecki. Separability of mixed states: necessary and sufficient conditions. *Phys. Lett. A*, 223(1-2):1–8, 1996. (Cited on p. 474.)

[327] Jin Chuan Hou. On the spectra of the positive completions for operator matrices. *J. Operator Theory*, 33(2):299–315, 1995. (Cited on p. 170.)

[328] Hong-Xuan Huang, Zhi-An Liang, and Panos M. Pardalos. Some properties for the Euclidean distance matrix and positive semidefinite matrix completion problems. *J. Global Optim.*, 25(1):3–21, 2003. Dedicated to Professor J. B. Rosen on his 80th birthday. (Cited on p. 473.)

[329] Arnold J. Insel. Levels of subnormality. *Linear Algebra Appl.*, 262:27–53, 1997. (Cited on p. 473.)

[330] I. S. Iohvidov. The Fischer-Frobenius transformation. *Teor. Funciĭ Funcional. Anal. i Priložen.*, (15):203–212, 1972. (Cited on p. 67.)

[331] I. S. Iohvidov. *Hankel and Toeplitz matrices and forms.* Birkhäuser Boston, 1982. Algebraic theory. Translated from the Russian by G. Philip A. Thijsse, with an introduction by I. Gohberg. (Cited on p. 173.)

[332] I. S. Iohvidov and M. G. Kreĭn. Spectral theory of operators in space with indefinite metric. I. *Trudy Moskov. Mat. Obšč.*, 5:367–432, 1956. (Cited on p. 472.)

[333] Vlad Ionescu, Cristian Oară, and Martin Weiss. *Generalized Riccati theory and robust control.* John Wiley, Chichester, 1999. A Popov function approach. (Cited on p. 68.)

[334] Charles R. Johnson, Catherine A. Jones, and Brenda K. Kroschel. The Euclidean distance completion problem: cycle completability. *Linear and Multilinear Algebra*, 39(1-2):195–207, 1995. (Cited on p. 473.)

[335] Charles R. Johnson and Michael Lundquist. An inertia formula for Hermitian matrices with sparse inverses. *Linear Algebra Appl.*, 162/164:541–556, 1992. Directions in matrix theory (Auburn, AL, 1990). (Cited on p. 470.)

[336] Charles R. Johnson, Michael Lundquist, and Geir Nævdal. Positive definite Toeplitz completions. *J. London Math. Soc. (2)*, 59(2):507–520, 1999. (Cited on pp. 31, 66, 67.)

[337] Charles R. Johnson and Leiba Rodman. Inertia possibilities for completions of partial Hermitian matrices. *Linear and Multilinear Algebra*, 16(1-4):179–195, 1984. (Cited on p. 474.)

[338] Charles R. Johnson and Leiba Rodman. Completion of partial matrices to contractions. *J. Funct. Anal.*, 69(2):260–267, 1986. (Cited on p. 352.)

[339] Charles R. Johnson and Leiba Rodman. Chordal inheritance principles and positive definite completions of partial matrices over function rings. In *Contributions to operator theory and its applications (Mesa, AZ, 1987)*, volume 35 of *Oper. Theory Adv. Appl.*, pages 107–127. Birkhäuser, Basel, 1988. (Cited on p. 173.)

[340] Charles R. Johnson and Leiba Rodman. Completion of Toeplitz partial contractions. *SIAM J. Matrix Anal. Appl.*, 9(2):159–167, 1988. (Cited on pp. 352, 353.)

[341] C. Jordán, J. R. Torregrosa, and A. Urbano. On the Jordan form of completions of partial upper triangular matrices. *Linear Algebra Appl.*, 254:241–250, 1997. (Cited on p. 474.)

[342] Cristina Jordán, Juan R. Torregrosa, and Ana M. Urbano. On the Rodman-Shalom conjecture regarding the Jordan form of completions of partial upper triangular matrices. *Electron. J. Linear Algebra*, 3:103–118 (electronic), 1998. (Cited on p. 474.)

[343] Il Bong Jung, Eungil Ko, Chunji Li, and Sang Soo Park. Embry truncated complex moment problem. *Linear Algebra Appl.*, 375:95–114, 2003. (Cited on p. 254.)

[344] M. A. Kaashoek and Hugo J. Woerdeman. Unique minimal rank extensions of triangular operators. *J. Math. Anal. Appl.*, 131(2):501–516, 1988. (Cited on pp. 470, 474.)

[345] V. Kaftal, D. Larson, and G. Weiss. Quasitriangular subalgebras of semifinite von Neumann algebras are closed. *J. Funct. Anal.*, 107(2):387–401, 1992. (Cited on p. 357.)

[346] T. Kailath, A. Vieira, and M. Morf. Inverses of Toeplitz operators, innovations, and orthogonal polynomials. *SIAM Rev.*, 20(1):106–119, 1978. (Cited on p. 171.)

[347] Thomas Kailath. *Linear systems*. Prentice-Hall, Englewood Cliffs, NJ, 1980. Prentice-Hall Information and System Sciences Series. (Cited on p. 68.)

[348] Thomas Kailath. A theorem of I. Schur and its impact on modern signal processing. In *I. Schur methods in operator theory and signal processing*, volume 18 of *Oper. Theory Adv. Appl.*, pages 9–30. Birkhäuser, Basel, 1986. (Cited on p. 255.)

[349] Thomas Kailath and Vadim Olshevsky. Displacement-structure approach to polynomial Vandermonde and related matrices. *Linear Algebra Appl.*, 261:49–90, 1997. (Cited on p. 471.)

[350] Dmitry S. Kaliuzhnyi-Verbovetskyi, Ilya M. Spitkovsky, and Hugo J. Woerdeman. Matrices with normal defect one. *Oper. Matrices*, 3(3):401–438, 2009. (Cited on pp. 473, 474.)

[351] R. E. Kalman, P. L. Falb, and M. A. Arbib. *Topics in mathematical system theory*. McGraw-Hill, New York, 1969. (Cited on p. 470.)

[352] D. S. Kalyuzhnyĭ. The von Neumann inequality for linear matrix functions of several variables. *Mat. Zametki*, 64(2):218–223, 1998. (Cited on p. 250.)

[353] Dmitry S. Kalyuzhnyĭ-Verbovetzkiĭ. Carathéodory interpolation on the non-commutative polydisk. *J. Funct. Anal.*, 229(2):241–276, 2005. (Cited on pp. 255, 256.)

[354] Tosio Kato. *Perturbation theory for linear operators*. Springer, Berlin, second edition, 1976. volume 132 of Grundlehren der Mathematischen Wissenschaften. (Cited on p. 171.)

[355] E. G. Katsoulis, R. L. Moore, and T. T. Trent. Interpolation in nest algebras and applications to operator corona theorems. *J. Operator Theory*, 29(1):115–123, 1993. (Cited on p. 357.)

[356] Ahmet H. Kayran. Two-dimensional orthogonal lattice structures for autoregressive modeling of random fields. *IEEE Trans. Signal Process.*, 44(4):963–978, 1996. (Cited on p. 255.)

[357] V. L. Kharitonov and J. A. Torres Muñoz. Recent results on the robust stability of multivariate polynomials. *IEEE Trans. Circuits Systems I Fund. Theory Appl.*, 49(6):715–724, 2002. Special issue on multidimensional signals and systems. (Cited on p. 250.)

[358] Toru Kidera and Shizuo Miyajima. Contractive completions of operator matrices. *SUT J. Math.*, 27(1):49–71, 1991. (Cited on p. 353.)

[359] David P. Kimsey and Hugo J. Woerdeman. Minimal normal and commuting completions. *Int. J. Inf. Syst. Sci.*, 4(1):50–59, 2008. (Cited on p. 473.)

[360] David P. Kimsey and Hugo J. Woerdeman. The multivariable matrix valued K-moment problem on $\mathbb{R}^d$, $\mathbb{C}^d$, and $\mathbb{T}^d$. preprint. (Cited on p. 254.)

[361] Greg Knese. Rational inner functions in the Schur-Agler class of the polydisk. *Publicacions Matemàtiques*, page to appear. (Cited on p. 359.)

[362] Greg Knese. Bernstein-Szegő measures on the two dimensional torus. *Indiana Univ. Math. J.*, 57(3):1353–1376, 2008. (Cited on p. 251.)

[363] Greg Knese. Kernel decompositions for Schur functions on the polydisk. arXiv:0909.1828. (Cited on p. 251.)

[364] Greg Knese. Polynomials on the two dimensional torus. preprint. (Cited on p. 251.)

[365] Greg Knese. Stable symmetric polynomials and the Schur-Agler class. preprint. (Cited on p. 359.)

[366] Tosiro Koga. Synthesis of finite passive n-ports with prescribed real matrices of several variables. *IEEE Trans. Circuit Theory*, CT-15:2–23, 1968. (Cited on p. 250.)

[367] Chiaki Kojima, Kiyotsugu Takaba, Osamu Kaneko, and Paolo Rapisarda. A characterization of solutions of the discrete-time algebraic Riccati equation based on quadratic difference forms. *Linear Algebra Appl.*, 416(2-3):1060–1082, 2006. (Cited on p. 174.)

[368] Paul Koosis. *Introduction to H_p spaces*, volume 115 of *Cambridge Tracts in Mathematics*. Cambridge University Press, Cambridge, second edition, 1998. With two appendices by V. P. Havin [Viktor Petrovich Khavin]. (Cited on pp. 171, 353, 359.)

[369] A. Korányi and L. Pukánszky. Holomorphic functions with positive real part on polycylinders. *Trans. Amer. Math. Soc.*, 108:449–456, 1963. (Cited on p. 250.)

[370] J. K. Korbicz, J. Wehr, and M. Lewenstein. Entanglement of positive definite functions on compact groups. *Comm. Math. Phys.*, 281(3):753–774, 2008. (Cited on p. 474.)

[371] Selcuk Koyuncu and Hugo J. Woerdeman. The inverse of a two-level positive definite Toeplitz operator matrix. In *I. Gohberg memorial volume*, Eds: H. Dym, M. A. Kaashoek, P. Lancaster, H. Langer and L. Lerer, Oper. Theory Adv. Appl., to appear. (Cited on p. 250.)

[372] Jon Kraus and David R. Larson. Reflexivity and distance formulae. *Proc. London Math. Soc. (3)*, 53(2):340–356, 1986. (Cited on p. 352.)

[373] M. Krein. Sur le problème du prolongement des fonctions hermitiennes positives et continues. *C. R. (Doklady) Acad. Sci. URSS (N.S.)*, 26:17–22, 1940. (Cited on p. 252.)

[374] M. Krein. The theory of self-adjoint extensions of semi-bounded Hermitian transformations and its applications. I. *Rec. Math. [Mat. Sbornik] N.S.*, 20(62):431–495, 1947. (Cited on p. 171.)

[375] M. G. Kreĭn. Distribution of roots of polynomials orthogonal on the unit circle with respect to a sign-alternating weight. *Teor. Funkciĭ Funkcional. Anal. i Priložen. Vyp.*, 2:131–137, 1966. (Cited on p. 471.)

[376] M. G. Krein and A. A. Nudelman. *The Markov moment problem and extremal problems.* volume 50 of Translations of Mathematical Monographs. American Mathematical Society, Providence, RI, 1977. Ideas and problems of P. L.

Čebyšev and A. A. Markov and their further development, Translated from the Russian by D. Louvish. (Cited on pp. 67, 173, 358.)

[377] L. Kronecker. Zur Theorie der Elimination einer Variabilen aus zwei algebraische Gleichungen. *Kőnig Preuss. Akad. Wiss. (Berlin)*, pages 535–600, 1881. (Cited on p. 354.)

[378] M. Krupnik and A. Leibman. Jordan structures of strictly lower triangular completions of nilpotent matrices. *Integral Equations Operator Theory*, 23(4):459–471, 1995. (Cited on p. 474.)

[379] Anton Kummert. Synthesis of 3-D lossless first-order one ports with lumped elements. *IEEE Trans. Circuits and Systems*, 36(11):1445–1449, 1989. (Cited on p. 359.)

[380] Anton Kummert. Synthesis of two-dimensional lossless m-ports with prescribed scattering matrix. *Circuits Systems Signal Process.*, 8(1):97–119, 1989. (Cited on p. 355.)

[381] Anton Kummert. 2-D stable polynomials with parameter-dependent coefficients: generalizations and new results. *IEEE Trans. Circuits Systems I Fund. Theory Appl.*, 49(6):725–731, 2002. Special issue on multidimensional signals and systems. (Cited on p. 255.)

[382] Huibert Kwakernaak. A polynomial approach to minimax frequency domain optimization of multivariable feedback systems. *Internat. J. Control*, 44(1):117–156, 1986. (Cited on p. 359.)

[383] Michael T. Lacey. Lectures on Nehari's theorem on the polydisk. In *Topics in harmonic analysis and ergodic theory*, volume 444 of *Contemp. Math.*, pages 185–213. American Mathematical Society, Providence, RI, 2007. (Cited on p. 354.)

[384] Peter Lancaster and Leiba Rodman. *Algebraic Riccati equations*. Oxford Science Publications. Clarendon Press, New York, 1995. (Cited on p. 174.)

[385] Peter Lancaster and Miron Tismenetsky. *The theory of matrices*. Computer Science and Applied Mathematics. Academic Press, Orlando, FL, second edition, 1985. (Cited on p. 174.)

[386] E. C. Lance. Some properties of nest algebras. *Proc. London Math. Soc. (3)*, 19:45–68, 1969. (Cited on p. 357.)

[387] Heinz Langer. Factorization of operator pencils. *Acta Sci. Math. (Szeged)*, 38(1-2):83–96, 1976. (Cited on p. 174.)

[388] David R. Larson. Nest algebras and similarity transformations. *Ann. of Math. (2)*, 121(3):409–427, 1985. (Cited on p. 250.)

[389] Jean B. Lasserre. A sum of squares approximation of nonnegative polynomials. *SIAM Rev.*, 49(4):651–669, 2007. (Cited on p. 252.)

[390] Sneh Lata, Meghna Mittal, and Vern I. Paulsen. An operator algebraic proof of Agler's factorization theorem. *Proc. Amer. Math. Soc.*, 137(11):3741–3748, 2009. (Cited on p. 360.)

[391] Alan J. Laub and W. F. Arnold. Generalized eigenproblem algorithms and software for algebraic Riccati equations. *Proc. IEEE*, 72:1746–1754, 1984. (Cited on p. 174.)

[392] Monique Laurent. A connection between positive semidefinite and Euclidean distance matrix completion problems. *Linear Algebra Appl.*, 273:9–22, 1998. (Cited on p. 473.)

[393] Monique Laurent. A tour d'horizon on positive semidefinite and Euclidean distance matrix completion problems. In *Topics in semidefinite and interior-point methods (Toronto, ON, 1996)*, volume 18 of *Fields Inst. Commun.*, pages 51–76. American Mathematical Society, Providence, RI, 1998. (Cited on p. 473.)

[394] Monique Laurent. Polynomial instances of the positive semidefinite and Euclidean distance matrix completion problems. *SIAM J. Matrix Anal. Appl.*, 22(3):874–894 (electronic), 2000. (Cited on p. 473.)

[395] Monique Laurent. On the sparsity order of a graph and its deficiency in chordality. *Combinatorica*, 21(4):543–570, 2001. (Cited on p. 67.)

[396] Monique Laurent. Sums of squares, moment matrices and optimization over polynomials. In *Emerging applications of algebraic geometry*, volume 149 of *IMA Vol. Math. Appl.*, pages 157–270. Springer, New York, 2009. (Cited on p. 252.)

[397] Peter D. Lax. Translation invariant spaces. *Acta Math.*, 101:163–178, 1959. (Cited on p. 359.)

[398] David B. Leep and Colin L. Starr. Polynomials in $\mathbb{R}[x, y]$ that are sums of squares in $\mathbb{R}(x, y)$. *Proc. Amer. Math. Soc.*, 129(11):3133–3141 (electronic), 2001. (Cited on p. 66.)

[399] S. Lele and J.T. Richtsmeier. *An invariant approach to statistical analysis of shapes.* Interdisciplinary Statistics series. Chapman and Hall/CRC Pres, London, 2001. (Cited on p. 473.)

[400] Leonid Lerer and André C. M. Ran. A new inertia theorem for Stein equations, inertia of invertible Hermitian block Toeplitz matrices and matrix orthogonal polynomials. *Integral Equations Operator Theory*, 47(3):339–360, 2003. (Cited on p. 471.)

[401] Hanoch Lev-Ari, Sydney R. Parker, and Thomas Kailath. Multidimensional maximum-entropy covariance extension. *IEEE Trans. Inform. Theory*, 35(3):497–508, 1989. (Cited on pp. 250, 255.)

[402] Chi-Kwong Li and Tom Milligan. Uniqueness of the solutions of some completion problems. *Linear Algebra Appl.*, 392:91–102, 2004. (Cited on p. 473.)

[403] Chi-Kwong Li and Nung-Sing Sze. Canonical forms, higher rank numerical ranges, totally isotropic subspaces, and matrix equations. *Proc. Amer. Math. Soc.*, 136(9):3013–3023, 2008. (Cited on p. 174.)

[404] Chi-Kwong Li and Bit Shun Tam. A note on extreme correlation matrices. *SIAM J. Matrix Anal. Appl.*, 15(3):903–908, 1994. (Cited on p. 66.)

[405] Chi-Kwong Li and Hugo J. Woerdeman. A research problem: a generalized numerical radius. *Linear and Multilinear Algebra*, 41(1):41–47, 1996. (Cited on p. 474.)

[406] Chi-Kwong Li and Hugo J. Woerdeman. Special classes of positive and completely positive maps. *Linear Algebra Appl.*, 255:247–258, 1997. (Cited on p. 474.)

[407] X. Li and E. B. Saff. On Nevai's characterization of measures with almost everywhere positive derivative. *J. Approx. Theory*, 63(2):191–197, 1990. (Cited on p. 172.)

[408] D. J. N. Limebeer, G. D. Halikias, and K. Glover. State-space algorithm for the computation of superoptimal matrix interpolating functions. *Internat. J. Control*, 50(6):2431–2466, 1989. (Cited on p. 359.)

[409] R. Loewy. Extreme points of a convex subset of the cone of positive semidefinite matrices. *Math. Ann.*, 253(3):227–232, 1980. (Cited on p. 66.)

[410] B. A. Lotto and T. Steger. Von Neumann's inequality for commuting, diagonalizable contractions. II. *Proc. Amer. Math. Soc.*, 120(3):897–901, 1994. (Cited on p. 356.)

[411] Philippe Loubaton. A regularity criterion for lexicographical prediction of multivariate wide-sense stationary processes on $\mathbf{Z}^2$ with non-full-rank spectral densities. *J. Funct. Anal.*, 104(1):198–228, 1992. (Cited on p. 255.)

[412] Michael E. Lundquist and Charles R. Johnson. Linearly constrained positive definite completions. *Linear Algebra Appl.*, 150:195–207, 1991. (Cited on p. 67.)

[413] S. M. Malamud. Inverse spectral problem for normal matrices and the Gauss-Lucas theorem. *Trans. Amer. Math. Soc.*, 357(10):4043–4064 (electronic), 2005. (Cited on p. 473.)

[414] J. G. Marcano and M. D. Morán. The Arov-Grossman model and the Burg multivariate entropy. *J. Fourier Anal. Appl.*, 9(6):623–647, 2003. (Cited on p. 171.)

[415] Marvin Marcus and Henryk Minc. *A survey of matrix theory and matrix inequalities*. Dover, New York, 1992. Reprint of the 1969 edition. (Cited on p. 66.)

[416] A. S. Markus. *Introduction to the spectral theory of polynomial operator pencils*, volume 71 of *Translations of Mathematical Monographs*. American Mathematical Society, Providence, RI, 1988. Translated from the Russian by H. H. McFaden, Translation edited by Ben Silver, With an appendix by M. V. Keldysh. (Cited on p. 174.)

[417] Murray Marshall. *Positive polynomials and sums of squares*, volume 146 of *Mathematical Surveys and Monographs*. American Mathematical Society, Providence, RI, 2008. (Cited on p. 252.)

[418] Thomas L. Marzetta. Two-dimensional linear prediction: autocorrelation arrays, minimum-phase prediction error filters, and reflection coefficient arrays. *IEEE Trans. Acoust. Speech Signal Process.*, 28(6):725–733, 1980. (Cited on p. 255.)

[419] P. Masani. Wiener's contributions to generalized harmonic analysis, prediction theory and filter theory. *Bull. Amer. Math. Soc.*, 72(1, pt. 2):73–125, 1966. (Cited on p. 255.)

[420] Attila Máté, Paul Nevai, and Vilmos Totik. Strong and weak convergence of orthogonal polynomials. *Amer. J. Math.*, 109(2):239–281, 1987. (Cited on p. 172.)

[421] Scott McCullough. Factorization of operator-valued polynomials in several non-commuting variables. *Linear Algebra Appl.*, 326(1-3):193–203, 2001. (Cited on pp. 253, 254, 254.)

[422] Scott McCullough. Extremal properties of outer polynomial factors. *Proc. Amer. Math. Soc.*, 132(3):815–825 (electronic), 2004. (Cited on p. 174.)

[423] Scott McCullough and Mihai Putinar. Noncommutative sums of squares. *Pacific J. Math.*, 218(1):167–171, 2005. (Cited on p. 254.)

[424] R. L. Moore and T. T. Trent. Factorization along commutative subspace lattices. *Integral Equations Operator Theory*, 25(2):224–234, 1996. (Cited on p. 250.)

[425] Jorge J. Moré and Zhijun Wu. Distance geometry optimization for protein structures. *J. Global Optim.*, 15(3):219–234, 1999. (Cited on p. 473.)

[426] T. S. Motzkin. The arithmetic-geometric inequality. In *Inequalities (Proc. Sympos. Wright-Patterson Air Force Base, Ohio, 1965)*, pages 205–224. Academic Press, New York, 1967. (Cited on p. 66.)

[427] Geir Nævdal. On the completion of partially given triangular Toeplitz matrices to contractions. *SIAM J. Matrix Anal. Appl.*, 14(2):545–552, 1993. (Cited on p. 352.)

[428] Geir Nævdal and Hugo J. Woerdeman. Partial matrix contractions and intersections of matrix balls. *Linear Algebra Appl.*, 175:225–238, 1992. (Cited on p. 353.)

[429] Geir Nævdal and Hugo J. Woerdeman. Cone inclusion numbers. *SIAM J. Matrix Anal. Appl.*, 19(3):613–639 (electronic), 1998. (Cited on p. 352.)

[430] Zeev Nehari. On bounded bilinear forms. *Ann. of Math. (2)*, 65:153–162, 1957. (Cited on p. 353.)

[431] H. Nelis, E. Deprettere, and P. Dewilde. Approximate inversion of partially specified positive definite matrices. In *Numerical linear algebra, digital signal processing and parallel algorithms (Leuven, 1988)*, volume 70 of *NATO Adv. Sci. Inst. Ser. F Comput. Systems Sci.*, pages 559–567. Springer, Berlin, 1991. (Cited on p. 250.)

[432] Yurii Nesterov. Squared functional systems and optimization problems. In *High performance optimization*, volume 33 of *Appl. Optim.*, pages 405–440. Kluwer Academic, Dordrecht, 2000. (Cited on p. 68.)

[433] Yurii Nesterov and Arkadii Nemirovskii. *Interior-point polynomial algorithms in convex programming*, volume 13 of *SIAM Studies in Applied Mathematics*. Society for Industrial and Applied Mathematics (SIAM), Philadelphia, 1994. (Cited on p. 66.)

[434] Paul Nevai. Characterization of measures associated with orthogonal polynomials on the unit circle. *Rocky Mountain J. Math.*, 19(1):293–302, 1989. Constructive Function Theory—86 Conference (Edmonton, AB, 1986). (Cited on p. 172.)

[435] R. Nevalinna. Über beschräkte Funktionen die im gegebenen Punkten vorbereibte Werte ahnnemen. *Ann. Acad. Sci. Fenn. Ser. A*, 13(1):141–160, 1919. (Cited on p. 355.)

[436] Michael A. Nielsen and Isaac L. Chuang. *Quantum computation and quantum information.* Cambridge University Press, Cambridge, 2000. (Cited on p. 474.)

[437] L. N. Nikol'skaya and Yu. B. Farforovskaya. Toeplitz and Hankel matrices as Hadamard-Schur multipliers. *Algebra i Analiz*, 15(6):141–160, 2003. (Cited on p. 358.)

[438] N. K. Nikol'skiĭ. *Treatise on the shift operator*, volume 273 of *Grundlehren der Mathematischen Wissenschaften.* Springer, Berlin, 1986. Spectral function theory, with an appendix by S. V. Hruščev [S. V. Khrushchëv] and V. V. Peller, Translated from the Russian by Jaak Peetre. (Cited on pp. 353, 354, 359.)

[439] David Ogle and Nicholas Young. The Parrott problem for singular values. In *Recent advances in operator theory (Groningen, 1998)*, volume 124 of *Oper. Theory Adv. Appl.*, pages 481–503. Birkhäuser, Basel, 2001. (Cited on p. 472.)

[440] Vadim Olshevsky, Ivan Oseledets, and Eugene Tyrtyshnikov. Tensor properties of multilevel Toeplitz and related matrices. *Linear Algebra Appl.*, 412(1):1–21, 2006. (Cited on p. 250.)

[441] Vadim Olshevsky, Ivan Oseledets, and Eugene Tyrtyshnikov. Superfast inversion of two-level Toeplitz matrices using Newton iteration and tensor-displacement structure. In *Recent advances in matrix and operator theory*, volume 179 of *Oper. Theory Adv. Appl.*, pages 229–240. Birkhäuser, Basel, 2008. (Cited on p. 250.)

[442] David Opěla. A generalization of Andô's theorem and Parrott's example. *Proc. Amer. Math. Soc.*, 134(9):2703–2710 (electronic), 2006. (Cited on p. 356.)

[443] Lavon B. Page. Bounded and compact vectorial Hankel operators. *Trans. Amer. Math. Soc.*, 150:529–539, 1970. (Cited on p. 354.)

[444] Lavon B. Page. Applications of the Sz.-Nagy and Foiaş lifting theorem. *Indiana Univ. Math. J.*, 20:135–145, 1970/1971. (Cited on p. 353.)

[445] C. C. Paige and M. A. Saunders. Towards a generalized singular value decomposition. *SIAM J. Numer. Anal.*, 18(3):398–405, 1981. (Cited on p. 470.)

[446] A. Papoulis. *Signal Analysis.* McGraw-Hill, New York, 1977. (Cited on p. 68.)

[447] Pablo A. Parrilo. *Structured semidefinite programs and semialgebraic geometry methods in robustness and optimization.* California Institute of Technology, 2000. (Cited on p. 66.)

[448] Pablo A. Parrilo. Semidefinite programming relaxations for semialgebraic problems. *Math. Program.*, 96(2, Ser. B):293–320, 2003. Algebraic and geometric methods in discrete optimization. (Cited on p. 252.)

[449] Stephen Parrott. On a quotient norm and the Sz.-Nagy - Foiaş lifting theorem. *J. Funct. Anal.*, 30(3):311–328, 1978. (Cited on pp. 352, 353.)

[450] Vern I. Paulsen. Representations of function algebras, abstract operator spaces, and Banach space geometry. *J. Funct. Anal.*, 109(1):113–129, 1992. (Cited on p. 250.)

[451] Vern I. Paulsen. *Completely bounded maps and operator algebras*, volume 78 of *Cambridge Studies in Advanced Mathematics*. Cambridge University Press, Cambridge, 2002. (Cited on pp. 253, 355, 360.)

[452] Vern I. Paulsen, Stephen C. Power, and Roger R. Smith. Schur products and matrix completions. *J. Funct. Anal.*, 85(1):151–178, 1989. (Cited on pp. 65, 170, 255, 255.)

[453] Vern I. Paulsen and Leiba Rodman. Positive completions of matrices over C^*-algebras. *J. Operator Theory*, 25(2):237–253, 1991. (Cited on p. 170.)

[454] V. V. Peller. Hankel operators of class $\mathcal{S}_p$ and their applications (rational approximation, Gaussian processes, the problem of majorization of operators). *Mat. Sb. (N.S.)*, 113(155)(4(12)):538–581, 637, 1980. (Cited on pp. 353, 358.)

[455] V. V. Peller. Smooth Hankel operators and their applications (ideals $\mathfrak{S}_p$, Besov classes, random processes). *Dokl. Akad. Nauk SSSR*, 252(1):43–48, 1980. (Cited on p. 353.)

[456] V. V. Peller. Vectorial Hankel operators, commutators and related operators of the Schatten-von Neumann class γ_p. *Integral Equations Operator Theory*, 5(2):244–272, 1982. (Cited on p. 353.)

[457] V. V. Peller. *Hankel operators and their applications*. Springer Monographs in Mathematics. Springer, New York, 2003. (Cited on pp. 353, 354, 354, 359, 472.)

[458] V. V. Peller and S. V. Khrushchëv. Hankel operators, best approximations and stationary Gaussian processes. *Uspekhi Mat. Nauk*, 37(1(223)):53–124, 176, 1982. (Cited on pp. 331, 359.)

[459] V. V. Peller and S. R. Treil. Superoptimal singular values and indices of infinite matrix functions. *Indiana Univ. Math. J.*, 44(1):243–255, 1995. (Cited on p. 359.)

[460] V. V. Peller and N. J. Young. Superoptimal analytic approximations of matrix functions. *J. Funct. Anal.*, 120(2):300–343, 1994. (Cited on pp. 354, 358, 359, 359, 359.)

[461] Rajesh Pereira. Differentiators and the geometry of polynomials. *J. Math. Anal. Appl.*, 285(1):336–348, 2003. (Cited on p. 473.)

[462] Asher Peres. Separability criterion for density matrices. *Phys. Rev. Lett.*, 77(8):1413–1415, 1996. (Cited on p. 474.)

[463] Georg Pick. Über die Beschränkungen analytischer Funktionen, welche durch vorgegebene Funktionswerte bewirkt werden. *Math. Ann.*, 77(1):7–23, 1915. (Cited on p. 355.)

[464] Gilles Pisier. *Similarity problems and completely bounded maps*, volume 1618 of *Lecture Notes in Mathematics*. Springer, Berlin, 1996. (Cited on p. 356.)

[465] Gelu Popescu. Positive-definite functions on free semigroups. *Canad. J. Math.*, 48(4):887–896, 1996. (Cited on p. 253.)

[466] Gelu Popescu. Structure and entropy for Toeplitz kernels. *C. R. Acad. Sci. Paris Sér. I Math.*, 329(2):129–134, 1999. (Cited on p. 253.)

[467] Gelu Popescu. Structure and entropy for positive-definite Toeplitz kernels on free semigroups. *J. Math. Anal. Appl.*, 254(1):191–218, 2001. (Cited on p. 253.)

[468] Gelu Popescu. Multivariable Nehari problem and interpolation. *J. Funct. Anal.*, 200(2):536–581, 2003. (Cited on p. 354.)

[469] Vasile-Mihai Popov. *Hyperstability of control systems.* volume 204 of Die Grundlehren der mathematischen Wissenschaften. Editura Academiei, Bucharest, 1973. Translated from the Romanian by Radu Georgescu. (Cited on p. 68.)

[470] Stanislav Popovych. Trace-positive complex polynomials in three unitaries. *Proc. Amer. Math. Soc.*, 138(10), 2010. (Cited on p. 255.)

[471] I. Postlethwaite, M.-C. Tsai, and D.-W. Gu. State-space approach to discrete-time super-optimal H^∞ control problems. *Internat. J. Control*, 49(1):247–268, 1989. (Cited on p. 359.)

[472] S. C. Power. *Hankel operators on Hilbert space*, volume 64 of *Research Notes in Mathematics.* Pitman (Advanced Publishing Program), Boston, 1982. (Cited on p. 353.)

[473] Mihai Putinar. A two-dimensional moment problem. *J. Funct. Anal.*, 80(1):1–8, 1988. (Cited on p. 173.)

[474] Mihai Putinar. Positive polynomials on compact semi-algebraic sets. *Indiana Univ. Math. J.*, 42(3):969–984, 1993. (Cited on p. 252.)

[475] Mihai Putinar and Florian-Horia Vasilescu. Solving moment problems by dimensional extension. *Ann. of Math. (2)*, 149(3):1087–1107, 1999. (Cited on pp. 252, 255.)

[476] João Filipe Queiró and António Leal Duarte. Imbedding conditions for normal matrices. *Linear Algebra Appl.*, 430(7):1806–1811, 2009. (Cited on p. 473.)

[477] E. A. Rahmanov. The asymptotic behavior of the ratio of orthogonal polynomials. *Mat. Sb. (N.S.)*, 103(145)(2):237–252, 319, 1977. (Cited on p. 172.)

[478] E. A. Rakhmanov. The asymptotic behavior of the ratio of orthogonal polynomials. II. *Mat. Sb. (N.S.)*, 118(160)(1):104–117, 143, 1982. (Cited on p. 172.)

[479] A. C. M. Ran and L. Rodman. Factorization of matrix polynomials with symmetries. *SIAM J. Matrix Anal. Appl.*, 15(3):845–864, 1994. (Cited on p. 174.)

[480] Benjamin Recht, Maryam Fazel, and Pablo A. Parrilo. Guaranteed minimum rank solutions to linear matrix equations via nuclear norm minimization. *SIAM Rev.*, 52(3):471–501 (electronic), 2010. (Cited on p. 471.)

[481] Michael Reed and Barry Simon. *Methods of modern mathematical physics. II. Fourier analysis, self-adjointness.* Academic Press [Harcourt Brace], New York, 1975. (Cited on pp. 139, 173, 173, 173, 173.)

[482] Bruce Reznick. Uniform denominators in Hilbert's seventeenth problem. *Math. Z.*, 220(1):75–97, 1995. (Cited on p. 252.)

[483] Bruce Reznick. Some concrete aspects of Hilbert's 17th Problem. In *Real algebraic geometry and ordered structures (Baton Rouge, LA, 1996)*, volume 253 of *Contemp. Math.*, pages 251–272. American Mathematical Society, Providence, RI, 2000. (Cited on pp. 66, 67.)

[484] Raphael M. Robinson. Some definite polynomials which are not sums of squares of real polynomials. In *Selected questions of algebra and logic (collection dedicated to the memory of A. I. Mal'cev) (Russian)*, pages 264–282. Izdat. "Nauka" Sibirsk. Otdel., Novosibirsk, 1973. (Cited on p. 67.)

[485] R. Tyrrell Rockafellar. *Convex analysis*. Princeton Landmarks in Mathematics. Princeton University Press, Princeton, NJ, 1997. Reprint of the 1970 original, Princeton Paperbacks. (Cited on p. 65.)

[486] Leiba Rodman. *An introduction to operator polynomials*, volume 38 of *Oper. Theory Adv. Appl.* Birkhäuser, Basel, 1989. (Cited on p. 174.)

[487] Leiba Rodman and Tamir Shalom. Jordan forms of completions of partial upper triangular matrices. *Linear Algebra Appl.*, 168:221–249, 1992. (Cited on p. 474.)

[488] Leiba Rodman and Ilya M. Spitkovsky. Almost periodic factorization and corona theorem. *Indiana Univ. Math. J.*, 47(4):1243–1256, 1998. (Cited on p. 357.)

[489] Leiba Rodman, Ilya M. Spitkovsky, and Hugo J. Woerdeman. Carathéodory-Toeplitz and Nehari problems for matrix valued almost periodic functions. *Trans. Amer. Math. Soc.*, 350(6):2185–2227, 1998. (Cited on pp. 252, 355.)

[490] Leiba Rodman, Ilya M. Spitkovsky, and Hugo J. Woerdeman. Multiblock problems for almost periodic matrix functions of several variables. *New York J. Math.*, 7:117–148 (electronic), 2001. (Cited on p. 252.)

[491] Leiba Rodman, Ilya M. Spitkovsky, and Hugo J. Woerdeman. Abstract band method via factorization, positive and band extensions of multivariable almost periodic matrix functions, and spectral estimation. *Mem. Amer. Math. Soc.*, 160(762):viii+71, 2002. (Cited on p. 174.)

[492] Leiba Rodman, Ilya M. Spitkovsky, and Hugo J. Woerdeman. Contractive extension problems for matrix valued almost periodic functions of several variables. *J. Operator Theory*, 47(1):3–35, 2002. (Cited on pp. 354, 355, 355, 355.)

[493] Leiba Rodman and Hugo J. Woerdeman. Perturbations, singular values, and ranks of partial triangular matrices. *SIAM J. Matrix Anal. Appl.*, 16(1):278–288, 1995. (Cited on p. 474.)

[494] Murray Rosenblatt. A multi-dimensional prediction problem. *Ark. Mat.*, 3:407–424, 1958. (Cited on p. 171.)

[495] Marvin Rosenblum. Vectorial Toeplitz operators and the Fejér-Riesz theorem. *J. Math. Anal. Appl.*, 23:139–147, 1968. (Cited on p. 171.)

[496] Marvin Rosenblum. A corona theorem for countably many functions. *Integral Equations Operator Theory*, 3(1):125–137, 1980. (Cited on p. 357.)

[497] Marvin Rosenblum and James Rovnyak. *Hardy classes and operator theory*. Oxford Mathematical Monographs. Clarendon Press, New York, 1985. Oxford Science Publications. (Cited on pp. 171, 174, 353.)

[498] Mario A. Rotea and Pramod P. Khargonekar. H^2-optimal control with an H^∞-constraint: the state feedback case. *Automatica J. IFAC*, 27(2):307–316, 1991. (Cited on p. 357.)

[499] James Rovnyak. Ideals of square summable power series. *Proc. Amer. Math. Soc.*, 13:360–365, 1962. (Cited on p. 359.)

[500] James Rovnyak and Michael A. Dritschel. The operator Fejér-Riesz theorem. In *A glimpse at Hilbert space operators. Paul R. Halmos in memoriam*, volume 207 of *Oper. Theory Adv. Appl.*, pages 223–254. Springer, Basel, 2010. (Cited on p. 172.)

[501] Pál Rózsa, Roberto Bevilacqua, Francesco Romani, and Paola Favati. On band matrices and their inverses. *Linear Algebra Appl.*, 150:287–295, 1991. (Cited on p. 470.)

[502] Walter Rudin. *Fourier analysis on groups.* Interscience Tracts in Pure and Applied Mathematics, No. 12. Interscience Publishers (a division of John Wiley and Sons), New York-London, 1962. (Cited on pp. 252, 355.)

[503] Walter Rudin. The extension problem for positive-definite functions. *Illinois J. Math.*, 7:532–539, 1963. (Cited on pp. 65, 66, 254.)

[504] Walter Rudin. *Function theory in polydiscs.* W. A. Benjamin, New York, 1969. (Cited on p. 356.)

[505] Walter Rudin. Spaces of type $H^\infty + C$. *Ann. Inst. Fourier (Grenoble)*, 25(1):vi, 99–125, 1975. (Cited on p. 354.)

[506] L. A. Sahnovič. Effective construction of noncontinuable Hermite-positive functions of several variables. *Funktsional. Anal. i Prilozhen.*, 14(4):55–60, 96, 1980. (Cited on p. 66.)

[507] L. A. Sakhnovich. *Interpolation theory and its applications*, volume 428 of *Mathematics and its Applications.* Kluwer Academic, Dordrecht, 1997. (Cited on pp. 66, 356.)

[508] Jan Samsonowicz, Marek Kuś, and Maciej Lewenstein. Separability, entanglement, and full families of commuting normal matrices. *Phys. Rev. A*, 76(2):022314, Aug 2007. (Cited on p. 474.)

[509] Donald Sarason. Generalized interpolation in H^∞. *Trans. Amer. Math. Soc.*, 127:179–203, 1967. (Cited on pp. 353, 354, 354, 355, 358.)

[510] Zoltán Sasvári. *Positive definite and definitizable functions*, volume 2 of *Mathematical Topics.* Akademie Verlag, Berlin, 1994. (Cited on pp. 65, 66, 66, 252, 253, 255, 255, 355, 472.)

[511] Thomas Sauer. Polynomial interpolation of minimal degree. *Numer. Math.*, 78(1):59–85, 1997. (Cited on p. 254.)

[512] Konrad Schmüdgen. The K-moment problem for compact semi-algebraic sets. *Math. Ann.*, 289(2):203–206, 1991. (Cited on p. 252.)

[513] I. Schur. Über Potenzreinen die im innern des Einhesskreises beschränkt sind, I. *J. Reine Angew. Math.*, 147:205–232, 1917. (Cited on p. 172.)

[514] I. Schur. Über Potenzreinen die im innern des Einhesskreises beschränkt sind, II. *J. Reine Angew. Math.*, 148:151–163, 1918. (Cited on p. 172.)

[515] Markus Schweighofer. Optimization of polynomials on compact semialgebraic sets. *SIAM J. Optim.*, 15(3):805–825 (electronic), 2005. (Cited on p. 252.)

[516] A. Seghier. Extension de fonctions de type positif et entropie associée. Cas multidimensionnel. *Ann. Inst. H. Poincaré Anal. Non Linéaire*, 8(6):651–675, 1991. (Cited on p. 251.)

[517] Ioana Serban and Mohamed Najim. Multidimensional systems: BIBO stability test based on functional Schur coefficients. *IEEE Trans. Signal Process.*, 55(11):5277–5285, 2007. (Cited on p. 250.)

[518] Naomi Shaked-Monderer. Extremal positive semidefinite matrices with given sparsity pattern. *Linear and Multilinear Algebra*, 36(4):287–292, 1994. (Cited on pp. 65, 66.)

[519] Yu. L. Shmul'yan and R. N. Yanovskaya. Blocks of a contractive operator matrix. *Izv. Vyssh. Uchebn. Zaved. Mat.*, (7):72–75, 1981. (Cited on p. 353.)

[520] J. A. Shohat and J. D. Tamarkin. *The problem of moments.* volume 2 of American Mathematical Society Mathematical surveys. American Mathematical Society, New York, 1943. (Cited on p. 173.)

[521] Fernando C. Silva. On a conjecture about the Jordan form of completions of partial upper triangular matrices. *Linear Algebra Appl.*, 260:319–321, 1997. (Cited on p. 474.)

[522] Fernando C. Silva and Rita Simões. On the Lyapunov and Stein equations. II. *Linear Algebra Appl.*, 426(2-3):305–311, 2007. (Cited on p. 471.)

[523] Barry Simon. *Orthogonal polynomials on the unit circle*, volume 54 of *American Mathematical Society Colloquium Publications.* American Mathematical Society, Providence, RI, 2005. (Cited on pp. 171, 172, 253.)

[524] P. G. Spain. Tracking poles, representing Hankel operators, and the Nehari problem. *Linear Algebra Appl.*, 223/224:637–694, 1995. Special issue honoring Miroslav Fiedler and Vlastimil Pták. (Cited on p. 353.)

[525] T. P. Speed and H. T. Kiiveri. Gaussian Markov distributions over finite graphs. *Ann. Statist.*, 14(1):138–150, 1986. (Cited on p. 66.)

[526] Ilya M. Spitkovsky and Hugo J. Woerdeman. The Carathéodory-Toeplitz problem for almost periodic functions. *J. Funct. Anal.*, 115(2):281–293, 1993. (Cited on p. 252.)

[527] Jovan Stefanovski. Polynomial J-spectral factorization in minimal state space. *Automatica J. IFAC*, 39(11):1893–1901, 2003. (Cited on p. 174.)

[528] T.-J. Stieltjes. Recherches sur les fractions continues. *Ann. Fac. Sci. Toulouse Sci. Math. Sci. Phys.*, 8(4):J1–J122, 1894. (Cited on p. 173.)

[529] Jan Stochel. Solving the truncated moment problem solves the full moment problem. *Glasg. Math. J.*, 43(3):335–341, 2001. (Cited on p. 173.)

[530] Jan Stochel and Franciszek Hugon Szafraniec. The complex moment problem and subnormality: a polar decomposition approach. *J. Funct. Anal.*, 159(2):432–491, 1998. (Cited on pp. 252, 255.)

[531] Gilbert Strang. The discrete cosine transform. *SIAM Rev.*, 41(1):135–147 (electronic), 1999. (Cited on p. 67.)

[532] Gilbert Strang and Tri Nguyen. The interplay of ranks of submatrices. *SIAM Rev.*, 46(4):637–646 (electronic), 2004. (Cited on p. 470.)

[533] T. J. Suffridge and T. L. Hayden. Approximation by a Hermitian positive semidefinite Toeplitz matrix. *SIAM J. Matrix Anal. Appl.*, 14(3):721–734, 1993. (Cited on p. 67.)

[534] Béla Sz.-Nagy and Ciprian Foiaş. *Harmonic analysis of operators on Hilbert space*. Translated from the French and revised. North-Holland, Amsterdam, 1970. (Cited on pp. 171, 329, 359, 359.)

[535] Béla Sz.-Nagy and Ciprian Foiaş. On contractions similar to isometries and Toeplitz operators. *Ann. Acad. Sci. Fenn. Ser. A I Math.*, 2:553–564, 1976. (Cited on p. 357.)

[536] G. Szegő. Über Orthogonalsysteme von Polynomen. *Math. Z.*, 4(1-2):139–151, 1919. (Cited on p. 250.)

[537] Gabor Szegő. *Orthogonal polynomials*. American Mathematical Society, New York, 1939. volume 23 of American Mathematical Society Colloquium Publications. (Cited on p. 171.)

[538] Bit Shun Tam. On the structure of the cone of positive operators. *Linear Algebra Appl.*, 167:65–85, 1992. Sixth Haifa Conference on Matrix Theory (Haifa, 1990). (Cited on p. 65.)

[539] Joseph L. Taylor. A general framework for a multi-operator functional calculus. *Advances in Math.*, 9:183–252, 1972. (Cited on p. 356.)

[540] Yongge Tian. The minimum rank of a 3×3 partial block matrix. *Linear Multilinear Algebra*, 50(2):125–131, 2002. (Cited on p. 474.)

[541] Yongge Tian and Yonghui Liu. Extremal ranks of some symmetric matrix expressions with applications. *SIAM J. Matrix Anal. Appl.*, 28(3):890–905 (electronic), 2006. (Cited on p. 471.)

[542] Dan Timotin. Completions of matrices and the commutant lifting theorem. *J. Funct. Anal.*, 104(2):291–298, 1992. (Cited on pp. 252, 253.)

[543] Dan Timotin. A note on Parrott's strong theorem. *J. Math. Anal. Appl.*, 171(1):288–293, 1992. (Cited on p. 353.)

[544] Miron Tismenetsky. Factorizations of Hermitian block Hankel matrices. *Linear Algebra Appl.*, 166:45–63, 1992. (Cited on p. 255.)

[545] Miron Tismenetsky. Matrix generalizations of a moment problem theorem. I. The Hermitian case. *SIAM J. Matrix Anal. Appl.*, 14(1):92–112, 1993. (Cited on p. 472.)

[546] Serguei Treil. On superoptimal approximation by analytic and meromorphic matrix-valued functions. *J. Funct. Anal.*, 131(2):386–414, 1995. (Cited on p. 359.)

[547] William F. Trench. An algorithm for the inversion of finite Toeplitz matrices. *J. Soc. Indust. Appl. Math.*, 12:515–522, 1964. (Cited on p. 171.)

[548] Harry L. Trentelman and Paolo Rapisarda. New algorithms for polynomial J-spectral factorization. *Math. Control Signals Systems*, 12(1):24–61, 1999. (Cited on p. 174.)

[549] Michael W. Trosset. Distance matrix completion by numerical optimization. *Comput. Optim. Appl.*, 17(1):11–22, 2000. (Cited on p. 473.)

[550] Walter Van Assche. Rakhmanov's theorem for orthogonal matrix polynomials on the unit circle. *J. Approx. Theory*, 146(2):227–242, 2007. (Cited on p. 172.)

[551] Cornelis V. M. van der Mee, Sebastiano Seatzu, and Giuseppe Rodriguez. Spectral factorization of bi-infinite multi-index block Toeplitz matrices. *Linear Algebra Appl.*, 343/344:355–380, 2002. Special issue on structured and infinite systems of linear equations. (Cited on p. 250.)

[552] Raf Vandebril and Marc Van Barel. A note on the nullity theorem. *J. Comput. Appl. Math.*, 189(1-2):179–190, 2006. (Cited on p. 470.)

[553] Raf Vandebril, Marc Van Barel, and Nicola Mastronardi. *Matrix computations and semiseparable matrices. Vol. 1.* Johns Hopkins University Press, Baltimore, 2008. Linear systems. (Cited on p. 471.)

[554] Lieven Vandenberghe and Stephen Boyd. Semidefinite programming. *SIAM Rev.*, 38(1):49–95, 1996. (Cited on p. 66.)

[555] N. Th. Varopoulos. On an inequality of von Neumann and an application of the metric theory of tensor products to operators theory. *J. Functional Analysis*, 16:83–100, 1974. (Cited on pp. 356, 359.)

[556] F.-H. Vasilescu. Hamburger and Stieltjes moment problems in several variables. *Trans. Amer. Math. Soc.*, 354(3):1265–1278 (electronic), 2002. (Cited on p. 252.)

[557] Alexander Volberg. Factorization of polynomials with estimates of norms. In *Current trends in operator theory and its applications*, volume 149 of *Oper. Theory Adv. Appl.*, pages 569–585. Birkhäuser, Basel, 2004. (Cited on p. 358.)

[558] Gilbert Walker. On periodicity in series of related terms. *Proc. Royal Soc. London. Series A*, 131(818), 1931. (Cited on p. 170.)

[559] P. Whittle. On stationary processes in the plane. *Biometrika*, 41:434–449, 1954. (Cited on p. 255.)

[560] N. Wiener and P. Masani. The prediction theory of multivariate stochastic processes. I. The regularity condition. *Acta Math.*, 98:111–150, 1957. (Cited on p. 255.)

[561] Jan C. Willems. Least squares stationary optimal control and the algebraic Riccati equation. *IEEE Trans. Automatic Control*, AC-16:621–634, 1971. (Cited on p. 68.)

[562] Hugo J. Woerdeman. The lower order of lower triangular operators and minimal rank extensions. *Integral Equations Operator Theory*, 10(6):859–879, 1987. (Cited on p. 470.)

[563] Hugo J. Woerdeman. *Matrix and operator extensions*, volume 68 of *CWI Tract*. Stichting Mathematisch Centrum Centrum voor Wiskunde en Informatica, Amsterdam, 1989. (Cited on p. 170.)

[564] Hugo J. Woerdeman. Minimal rank completions for block matrices. *Linear Algebra Appl.*, 121:105–122, 1989. (Cited on p. 470.)

[565] Hugo J. Woerdeman. Strictly contractive and positive completions for block matrices. *Linear Algebra Appl.*, 136:63–105, 1990. (Cited on pp. 352, 353.)

[566] Hugo J. Woerdeman. Minimal rank completions of partial banded matrices. *Linear and Multilinear Algebra*, 36(1):59–68, 1993. (Cited on p. 470.)

[567] Hugo J. Woerdeman. Superoptimal completions of triangular matrices. *Integral Equations Operator Theory*, 20(4):491–501, 1994. (Cited on p. 358.)

[568] Hugo J. Woerdeman. Toeplitz minimal rank completions. *Linear Algebra Appl.*, 202:267–278, 1994. (Cited on pp. 470, 471.)

[569] Hugo J. Woerdeman. Hermitian and normal completions. *Linear and Multilinear Algebra*, 42(3):239–280, 1997. (Cited on pp. 471, 473.)

[570] Hugo J. Woerdeman. The Carathéodory-Toeplitz problem with partial data. *Linear Algebra Appl.*, 342:149–161, 2002. (Cited on p. 171.)

[571] Hugo J. Woerdeman. Positive Carathéodory interpolation on the polydisc. *Integral Equations Operator Theory*, 42(2):229–242, 2002. (Cited on p. 250.)

[572] Hugo J. Woerdeman. The separability problem and normal completions. *Linear Algebra Appl.*, 376:85–95, 2004. (Cited on p. 474.)

[573] Hugo J. Woerdeman. Estimates of inverses of multivariable Toeplitz matrices. *Oper. Matrices*, 2(4):507–515, 2008. (Cited on p. 250.)

[574] Hugo J. Woerdeman. The higher rank numerical range is convex. *Linear Multilinear Algebra*, 56(1-2):65–67, 2008. (Cited on p. 174.)

[575] Hugo J. Woerdeman. A matrix and its inverse: revisiting minimal rank completions. In *Recent advances in matrix and operator theory*, volume 179 of *Oper. Theory Adv. Appl.*, pages 329–338. Birkhäuser, Basel, 2008. (Cited on p. 470.)

[576] Hugo J. Woerdeman. A general Christoffel-Darboux type formula. *Integral Equations Operator Theory*, 67(2):203–213, 2010. (Cited on p. 250.)

[577] Hugo J. Woerdeman, Jeffrey S. Geronimo, and Glaysar Castro. A numerical algorithm for stable 2d autoregressive filter design. *Signal Processing*, 83:1299–1308, 2003. (Cited on p. 251.)

[578] Herman Wold. *A study in the analysis of stationary time series.* Almqvist and Wiksell, Stockholm, second edition, 1954. With an appendix by Peter Whittle. (Cited on p. 359.)

[579] John W. Woods. Two-dimensional Markov spectral estimation. *IEEE Trans. Information Theory*, IT-22(5):552–559, 1976. (Cited on p. 251.)

[580] S. L. Woronowicz. Positive maps of low dimensional matrix algebras. *Rep. Math. Phys.*, 10(2):165–183, 1976. (Cited on p. 474.)

[581] Shao-Po Wu, Stephen Boyd, and Lieven Vandenberghe. FIR filter design via spectral factorization and convex optimization. In *Applied and computational control, signals, and circuits, Vol. 1*, volume 1 of *Appl. Comput. Control Signals Circuits*, pages 215–245. Birkhäuser Boston, Boston, 1999. (Cited on p. 68.)

[582] Mihalis Yannakakis. Computing the minimum fill-in is NP-complete. *SIAM J. Algebraic Discrete Methods*, 2(1):77–79, 1981. (Cited on p. 67.)

[583] David A. Yopp and Richard D. Hill. Extremals and exposed faces of the cone of positive maps. *Linear Multilinear Algebra*, 53(3):167–174, 2005. (Cited on p. 67.)

[584] D. C. Youla. On the factorization of rational matrices. *IRE Trans.*, IT-7:172–189, 1961. (Cited on p. 68.)

[585] N. J. Young. The singular-value decomposition of an infinite Hankel matrix. *Linear Algebra Appl.*, 50:639–656, 1983. (Cited on p. 472.)

[586] N. J. Young. The Nevanlinna-Pick problem for matrix-valued functions. *J. Operator Theory*, 15(2):239–265, 1986. (Cited on pp. 358, 359.)

[587] G. Udny Yule. On a method of investigating periodicities in disturbed series, with special reference to wolfer's sunspot numbers. *Phil. Trans. R. Soc. A.*, 226:267–298, 1927. (Cited on p. 170.)

[588] Fuzhen Zhang, editor. *The Schur complement and its applications*, volume 4 of *Numerical Methods and Algorithms.* Springer, New York, 2005. (Cited on pp. 171, 172.)

[589] K. Zhou, D.C. Doyle, and K Glover. *Robust and optimal control.* Prentice-Hall, Upper Saddle River, NJ, 1996. (Cited on p. 68.)

[590] Juan Carlos Zúñiga and Didier Henrion. A Toeplitz algorithm for polynomial J-spectral factorization. *Automatica J. IFAC*, 42(7):1085–1093, 2006. (Cited on p. 174.)

[591] A. Zygmund. *Trigonometric series. Vols. I, II.* Cambridge University Press, New York, second edition, 1959. (Cited on p. 358.)

Subject Index

Notation Index

Princeton Series in Applied Mathematics

The Princeton Series in Applied Mathematics publishes high quality advanced texts and monographs in all areas of applied mathematics. Books include those of a theoretical and general nature as well as those dealing with the mathematics of specific applications areas and real-world situations.

Chaotic Transitions in Deterministic and Stochastic Dynamical Systems: Applications of Melnikov Processes in Engineering, Physics, and Neuroscience, Emil Simiu

Selfsimilar Processes, Paul Embrechts and Makoto Maejima

Self-Regularity: A New Paradigm for Primal-Dual Interior Point Algorithms, Jiming Peng, Cornelis Roos, and Tamas Terlaky

Analytic Theory of Global Bifurcation: An Introduction, Boris Buffoni and John Toland

Entropy, Andreas Greven, Gerhard Keller, and Gerald Warnecke, editors

Auxiliary Signal Design for Failure Detection, Stephen L. Campbell and Ramine Nikoukhah

Thermodynamics: A Dynamical Systems Approach, Wassim M. Haddad, VijaySekhar Chellaboina, and Sergey G. Nersesov

Optimization: Insights and Applications, Jan Brinkhuis and Vladimir Tikhomirov

Max Plus at Work, Modeling and Analysis of Synchronized Systems: A Course on Max-Plus Algebra and Its Applications, Bernd Heidergott, Geert Jan Olsder, and Jacob van der Woude

Impulsive and Hybrid Dynamical Systems: Stability, Dissipativity, and Control, Wassim M. Haddad, VijaySekhar Chellaboina, and Sergey G. Nersesov

The Traveling Salesman Problem: A Computational Study, David L. Applegate, Robert E. Bixby, Vasek Chvatal, and William J. Cook

Positive Definite Matrices, Rajendra Bhatia

Genomic Signal Processing, Ilya Shmulevich and Edward Dougherty

Wave Scattering by Time-Dependent Perturbations: An Introduction, G. F. Roach

Algebraic Curves over a Finite Field, J.W.P. Hirschfeld, G. Korchmáros, and F. Torres

Distributed Control of Robotic Networks: A Mathematical Approach to Motion Coordination Algorithms, Francesco Bullo, Jorge Cortés, and Sonia Martínez

Robust Optimization, Aharon Ben-Tal, Laurent El Ghaoui, and Arkadi Nemirovski

Control Theoretic Splines: Optimal Control, Statistics, and Path Planning, Magnus Egerstedt and Clyde Martin

Matrices, Moments, and Quadrature with Applications, Gene Golub and Gérard Meurant

Totally Nonnegative Matrices, Shaun M. Fallat and Charles R. Johnson

Matrix Completions, Moments, and Sums of Hermitian Squares, Mihály Bakonyi and Hugo J. Woerdeman

Modern Anti-windup Synthesis: Control Augmentation for Actuator Saturation, Luca Zaccarian and Andrew W. Teel

www.ingramcontent.com/pod-product-compliance
Ingram Content Group UK Ltd.
Pitfield, Milton Keynes, MK11 3LW, UK
UKHW031839161224
452263UK00003B/116